Second International Seminar on Earth Sciences and Technology (ISEST-2023)

IOP Conference Series: Earth and Environmental Science
Volume 1245

Bandung, Indonesia
7 - 8 March 2023

Editors:

Riantini Virtriana
Brian Bramanto
Muhammad Rais Abdillah
Reynaldy Fifariz
Idham Andri Kurniawan

ISBN: 978-1-7138-7980-0
ISSN: 1755-1307

Printed from e-media with permission by:

Curran Associates, Inc.
57 Morehouse Lane
Red Hook, NY 12571

Some format issues inherent in the e-media version may also appear in this print version.

Institute of Physics
Dirac House, Temple Back
Bristol BS1 6BE UK

Phone: 44 1 17 929 7481
Fax: 44 1 17 920 0979

techtracking@iop.org

Additional copies of this publication are available from:

Curran Associates, Inc.
57 Morehouse Lane
Red Hook, NY 12571 USA
Phone: 845-758-0400
Fax: 845-758-2633
Email: curran@proceedings.com
Web: www.proceedings.com

TABLE OF CONTENTS

PREFACE

The 2023 2nd International Seminar on Earth Science and Technology (ISEST 2023) was hosted by the Faculty of Earth Sciences and Technology, Institut Teknologi Bandung (ITB), Indonesia. The seminar organized by the Executive Committees from Bandung, Indonesia. ISEST 2023 was successfully held in the form of a hybrid conference on March 7th – 8th, 2023. In the modern era, the interest of earth science and technology discipline has shifted following the urge in achieving a more sustainable future. The world is encouraging the future generation to take the present decision into consideration in the hope of maintaining the existence of natural resources, while combating the climate issue. ISEST 2023 served as an international forum, which continued disseminating innovative ideas, discussions, and academic exchanges among researchers. It aimed to integrate sustainability into the field of earth science and technology. The generous support for ISEST 2023 is provided by the Faculty of Earth Sciences and Technology, Institut Teknologi Bandung.

ISEST has been a trademark of the Faculty of Earth Sciences and Technology, Institut Teknologi Bandung, since 2021. Due to the COVID-19 pandemic, we could only hold our first scientific conference fully online back in December 2021. So, it is such an honor to take part in bringing the academic and scientific atmosphere back to our community in the form of a hybrid conference this year.

The theme of this year's conference is "Advancing Earth sciences, embracing technology". We, as the only faculty in Indonesia that integrates multi-layered and interdisciplinary Earth sciences, would like to encourage our community to share their most recent scientific discovery and technological advancements. Our topics, ranging from atmospheric sciences, oceanic sciences, solid Earth sciences, and interdisciplinary geosciences, are chosen based on the wide discipline we would like to cover from each of our scientific community. We hope that our conference could be the source of wider knowledge sharing and networking between universities, research agencies, and governmental institutions.

Within this conference, we cater more than 16 parallel sessions, with keynote and invited speakers from Indonesia, Japan, Australia, the United Kingdom, and South Korea. Most of the papers and posters presented in the conference would be submitted to the IOP Conference Series: Earth and Environmental Sciences. We hope that the publication will truly advance Earth sciences in each of our disciplines while embracing the continuously developing technology.

Gabriella Alodia, Ph.D.
Chairperson of the 2nd ISEST

IOP Conf. Series: Earth and Environmental Science 1245 (2023) 011001 doi:10.1088/1755-1315/1245/1/011001

2nd International Seminar on Earth Science and Technology (ISEST)

Executive Committee

Chair
Gabriella Alodia, Ph.D.

Secretariat
Dr. Alfita Puspa Handayani

Treasurer
Ratri Widyastuti, S.T., M.T.

Logistic and General Affair
Dr. Astyka Pamumpuni, S.T., M.T.
Alfend Rudyawan, S.T., M.Sc., Ph.D.

Event, content, and programs
Dr.rer.nat. Rima Rachmayani, S.Si., M.Si.
Mika Rizki Puspaningrum, Ph.D.

Members

Dr.techn. Dhota Pradipta, S.T., M.T.
Dr.rer.nat Suliskania Nurfitri, S.Si., M.Si.
Sella Lestari Nurmaulia, S.T., M.T.
Augusman Marthin Leonard, S.T., MBA.
R. Tedi Kusmawan, S.T.
Tri Sugiyarti Irianingsih, S.Sos.
Endang Ratna Kusumawati, A.Md.
Alqinthara Nuragnia, S.T
Tania Septi Anggraini, M.T
Azhari Barbarosa, S.T
Indra Andra Dinata, M.T
Firman Sauqi Nur Sabila, S.T
Alifah Hanif, S.T
Ben William Rogers, S.T
Abdillah Baraas
Aditya Kurnia Utama, S.T
Hananda Dwi Mahetran, S.T
Arash Anggita Sari
Fajar Ariansyah Muharam
Hanna Fauziya Salama
Khanifah Afifi
Ainun Azhari
Rahayu Lestari
Gandhi Napitupulu
Lisa Ayu Cahyaningtyas, S.T
Lutfiyah Jannatunnisa
Suci Farissa Nabilah, S.T
Nadila Novandaru
Kalingga Titon, M.T

IOP Conf. Series: Earth and Environmental Science 1245 (2023) 011001 doi:10.1088/1755-1315/1245/1/011001

International Reviewer

Dr. Alvin C. G. Varquez
Tokyo Institute of Technology, Japan

Dr. Arnadi Murtiyoso
ETH Zürich, Switzerland

Lino Garda Denaro, Ph.D.
National Taiwan University, Taiwan

Dr. techn. Zohreh Adavi, M.Sc.
Department of Geodesy and Geoinformation, TU Wien, Austria

Nur Fajar Trihantoro, M.Geosp.Sc.
Royal Melbourne Institute of Technology University, Australia

National Reviewer

Prof. Dr. Irwan Meilano, S.T., M.Sc
Prof. Agus M. Ramdhan, S.T, M.T, Ph.D
Dr. rer.nat. Mutiara R. Putri, S.Si, M.Si
Prof. Ir. Benyamin Sapiie, Ph.D.
Prof. Ir. Ketut Wikantika, M.Eng, Ph.D
Prof. Dr. Ir. Eddy Ariyono Subroto
Prof. Dr. Aswan, S.T., M.T.
Prof. Dr. Ir. Deny Juanda Puradimaja, DEA.
Prof. Dr. Ir. Yan Rizal R, Dipl. Geol.
Prof. Ir. Hasanuddin Zainal Abidin, M.Sc., Ph.D.
Prof. Dr.Eng. Nining Sari Ningsih, MS.
Prof. Dr. Ir. Rubiyanto Kapid
Prof. Ir. Lambok Hutasoit, M.Sc., Ph.D
Dr. Ir. Irwan Gumilar
Mika Rizki Puspaningrum, Ph.D
Dr. Astyka Pamumpuni
Reynaldy Fifariz, S.T., M.T., Ph.D.
Dr. Idham Andri Kurniawan, S.T., M.T.
Dr. Riantini Virtriana
Dr. techn. Nabila S.E. Putri
Dr. Suliskania Nurfitri
Brian Bramanto, Ph.D
Dr. Techn. Dhota Pradipta
Dr. Muhammad Rais Abdillah`
Dr. Rusmawan Suwarman
Dr. Endra Gunawan
Dr. Karina Aprilia Sujatmiko
Gabriella Alodia, Ph.D
Dr. Ayi Tarya
Dr. Faruq Khadami
Dr. rer.nat Rima Rachmayani
Miga Magenika Julian, M.T
Alfend Rudyawan, Ph.D
Dr. Deni Suwardhi

IOP Conf. Series: Earth and Environmental Science 1245 (2023) 011001 doi:10.1088/1755-1315/1245/1/011001

IOP Conf. Series: Earth and Environmental Science 1245 (2023) 011001 doi:10.1088/1755-1315/1245/1/011001

Steering Committee

IOP Conf. Series: Earth and Environmental Science 1245 (2023) 011001 doi:10.1088/1755-1315/1245/1/011001

Keynote Speakers and Invited Speakers

Keynote Speakers

Dr. Alvin C.G. Varquez
Tokyo Institute of Technology

Dr. Agus Santoso
University of New South Wales, Australia

Prof. Robert Hall
Royal Holloway University of London, United Kingdom

Prof. Phil Cummins
Australian National University, Australia

Invited Speakers

Prof. Wonkook Kim, Ph.D.
Pusan National University

Prof. Nining Sari Ningsih
Oceanography Research Group, Institut Teknologi Bandung

Prof. Aswan
Palaeontology and Quaternary Geology Research Group, Institut Teknologi Bandung

Dr. Irwan Gumilar
Geodesy Research Group, Institut Teknologi Bandung

Dr. Agus Setiawan
National Research and Innovation Agency (BRIN)

Dr. Akhmad Riqqi
Remote Sensing and GIS Research Group, Institut Teknologi Bandung

Dr. Wiwin Windupranata
Hydrography Research Group, Insititut Teknologi Bandung

Dr. Irwan Meilano
Surveying and Cadastral Research Group, Institut Teknologi Bandung

Dr. Supartoyo
Center of Volcanology and Geological Hazard Mitigation (PVMGB)

Dr. Dimas Salomo
Indonesia Agency for Meteorology Climatology and Geophysics (BMKG)

Supari, Ph.D.
Indonesia Agency for Meteorology Climatology and Geophysics (BMKG)

IOP Conf. Series: Earth and Environmental Science 1245 (2023) 011002 doi:10.1088/1755-1315/1245/1/011002

Peer Review Statement

All papers published in this volume have been reviewed through processes administered by the Editors. Reviews were conducted by expert referees to the professional and scientific standards expected of a proceedings journal published by IOP Publishing.

- **Type of peer review:** Single Anonymous
- **Conference submission management system:** Morressier
- **Number of submissions received:** 59
- **Number of submissions sent for review:** 57
- **Number of submissions accepted:** 52
- **Acceptance Rate (Submissions Accepted / Submissions Received × 100):** 88.1
- **Average number of reviews per paper:** 2.15
- **Total number of reviewers involved:** 29
- **Contact person for queries:**
  Name: Brian Bramanto, Ph.D.
  Email: brian.bramanto@itb.ac.id
  Affiliation: Institut Teknologi Bandung

The marine fossils and paleoecological significance of the Southern edge of South Sumatra Basin in Linggapura Lampung, Indonesia

D G Harbowo[1,2,3,a], D Pratama[2], B Priadi[2,4], T Julian[2], D J P Sihombing[2], E S Sitinjak[3,5]

[1] Sedimentology, Stratigraphy, and Geodynamic Research Group, Department of Geology, Institut Teknologi Sumatera, 35352, Wayhui, Lampung, Indonesia.
[2] Department of Geology, Institut Teknologi Sumatera, 35352, Wayhui, Lampung, Indonesia.
[3] Doctoral Program in Department of Geological Engineering, Faculty of Earth Science and Technology, Institut Teknologi Bandung, Jl Ganesha, 10, Bandung, 40312, Jawa Barat, Indonesia.
[4] Department of Geological Engineering, Faculty of Earth Science dan Technology, Institut Teknologi Bandung, Jl. Ganesha 10, Bandung, 40312, Jawa Barat, Indonesia.
[5] Reservoir Technical Services Division, Baker Hughes, East Ahmadi, Block 8 Plot 42, 47582, Kuwait
[a] Correspondent author: danni.gathot@gl.itera.ac.id

Abstract. The South Sumatra Basin is identified as a Paleogene-Neogene sedimentary basin with Mesozoic basements. It is located in the eastern Barisan Mountains, covering Jambi, South Sumatra, and Linggapura (Lampung), Indonesia. Linggapura represents the southernmost part of the South Sumatra Basin. The basin consists of two well-known rock formations, namely the Talangakar and Baturaja Formations. In this study, we investigated these formations through fossil records and paleoecological interpretations. Our investigation was based on measured stratigraphy and fossil observations along the Way Penandingan River in Linggapura, Lampung. We discovered that the bottom of this area consists of Cretaceous granitoid, which forms a nonconformity with the sedimentary rocks in this area. The Talangakar Formation in this region comprises quartzy-grained sandstone and conglomerates containing fossilized remains of terrestrial plants/vegetation. These plant fossils have undergone coalification, with some being silicified. The Batujara Formation in Linggapura is dominated by bioclasts, primarily consisting of Miocene shoreface-lagoonal larger benthic foraminifera, along with other marine and ichnofossils. Based on our findings, we suggest that the paleoecological condition of this area is associated with shoreline transgression during the Oligo-Miocene period. The hotter Miocene climate contributed to a rise in sea levels, resulting in shoreline transgression. Gradually, the area transformed from a terrestrial environment to a highly bioproductive marine environment, at least from the Late Oligocene to the Early Miocene. This study represents an important milestone in better understanding the paleoenvironmental conditions at the edge of the South Sumatra Basin throughout its geological history.

Keyword: fossil, paleoecology, paleoenvironment, south sumatra basin, carbonate, lampung.

1. Introduction

The South Sumatra Basin, located in the Sumatra region, is a back-arc basin known for its significant reserves of strategic fossil resources, including coal, hydrocarbons, and natural gas [1,2]. The formation of this basin can be attributed to a series of tectonic regime changes that have been active since the late Jurassic [3,4]. The period of maximum extension and sediment accumulation in the basin occurred in the lower to middle Miocene, followed by subsequent tectonic compressional inversion [1,5]. The presence of hydrocarbons and coal in the basin indicates the deposition and accumulation of organic carbon influenced by paleoecological conditions and ecosystem productivity during that time [6,7]. These organic materials eventually underwent fossilization along with subsequent geological processes.

Referring to the evolutionary phases of the South Sumatra Basin, the early-to mid-Miocene period is a critical phase related to the optimum development of basin dimensions [2,8]. The size of the basin dimensions is one of the factors determining the potential volume capacity of sediment accumulation [9]. In this study, we focused on assessing the period at the southernmost edge of the South Sumatra Basin by interpreting sedimentation environments and ecological conditions based on litostratigraphical records and fossil findings. The primary objective of this study was to trace the fossil record and interpret the paleoecological conditions at the southernmost margin of the South Sumatra Basin (**Figure 1**), particularly during the peak period of basin formation. The results of this study will contribute to further research aimed at understanding paleoenvironmental changes, including paleobathymetry and paleogeography, that occurred during the evolution of the South Sumatra Basin.

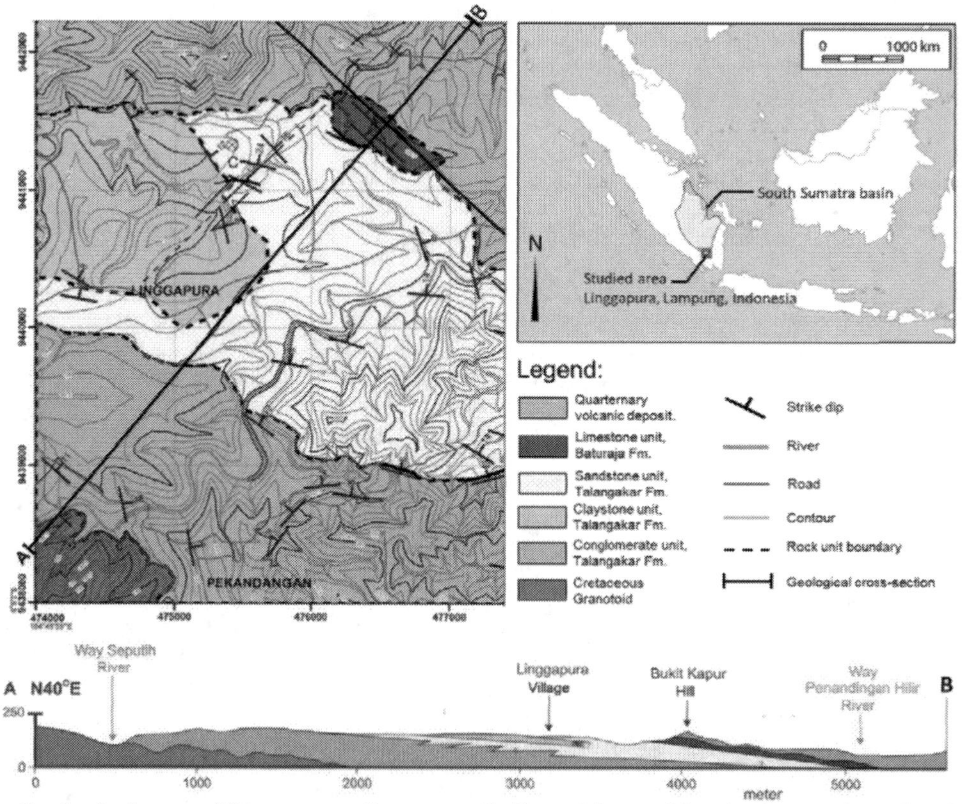

Figure 1. Geological map of Linggapura, Lampung (left), and its position in the southern South Sumatra Basin (right). The geological cross-section and the position of the Limestone of Baturaja Formation in the studied area (below), after [10].

Linggapura, Lampung, is the southernmost boundary of the South Sumatra Basin [11,12]. The region comprises several geological formations, including Cretaceous granitoids, Oligo-Miocene Talangakar Formation, Miocene Baturaja Formation, Plio-Pleistocene Kasai Formation, and Quaternary volcanic deposits. The Talangakar Formation is classified as a syn-rift deposit within the South Sumatra Basin, while the Baturaja Formation is considered a post-rift deposit [13,14]. Both formations have been reported to contain a variety of fossils [15]. In contrast, in the Talangakar Formation, the fossil composition tended to consist of terrestrial organisms, whereas in the Baturaja Formation, it tended to consist of marine organisms. Unfortunately, changes in the fossil composition and paleoecological conditions during formation deposition and basin evolution have rarely been studied.

2. Method

The study was conducted by stratigraphic measurements along the Penandingan River, Linggapura, Lampung (**Figure 1**). These measurements were presented in lithostratigraphic profiles that included information on thickness, texture, sedimentary structure, fossils, and other relevant features (**Figure 2**).

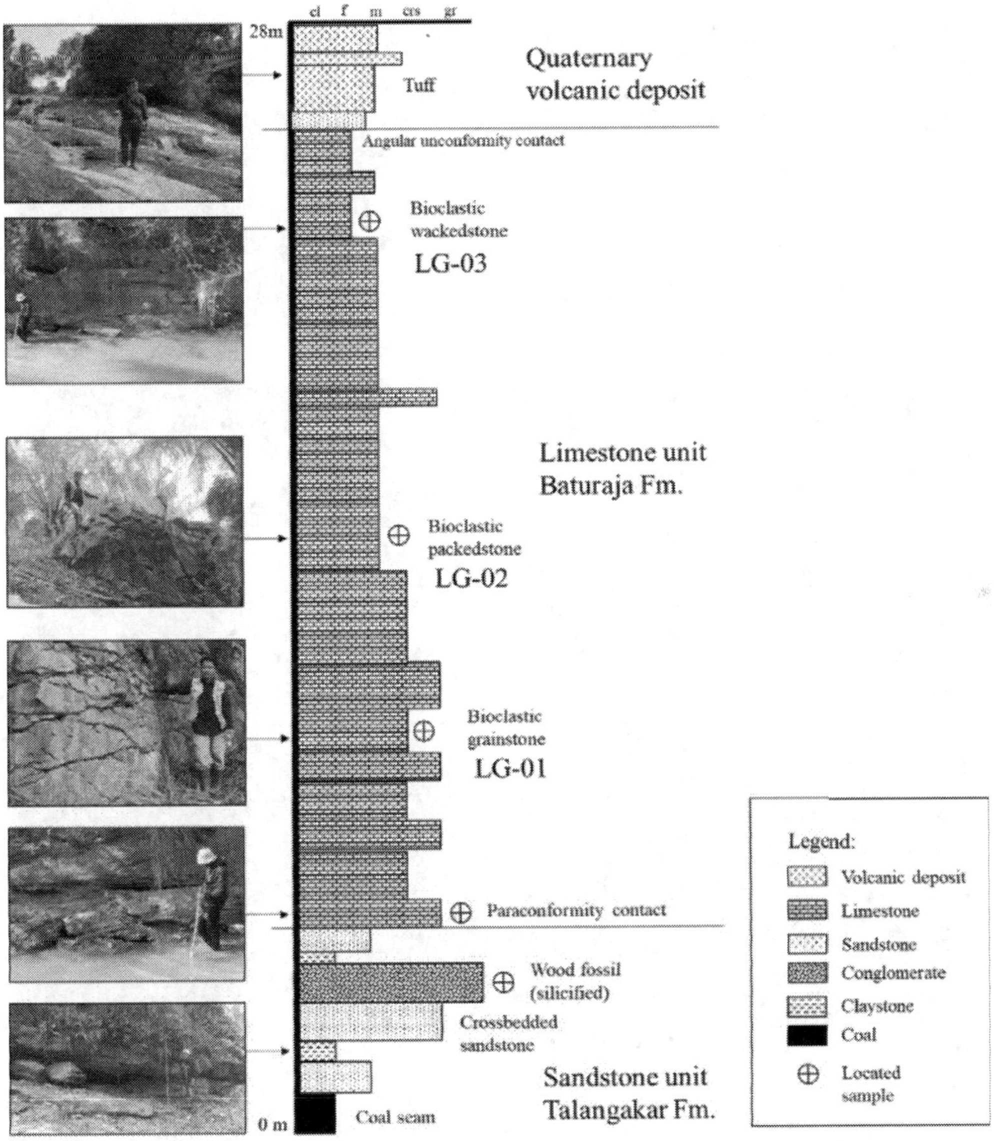

Figure 2. Litostratigraphy profile in the studied area along Way Penanding River, Linggapura (captured by authors)

IOP Conf. Series: Earth and Environmental Science 1245 (2023) 012001 doi:10.1088/1755-1315/1245/1/012001

Representative samples were collected for thin-section preparation, and these samples were subsequently observed under a light microscope (**Figure 3-5**). The observations were carefully documented to interpret the depositional environment and paleoecological conditions based on fossil findings and rock texture.

3. Result

Based on the results of measured stratigraphy in the Penandingan River, Linggapura, three major rock units were identified in the study area: the sandstone units of the Talangakar Formation, the Limestone unit of the Baturaja Formation, and the Quaternary volcanic deposits (**Figure 2**). The limestone unit is situated between the Talangakar Formation and the Quaternary volcanic deposits. It has an approximate thickness of 20 meters and a strike dip of N330°E/25°NE.

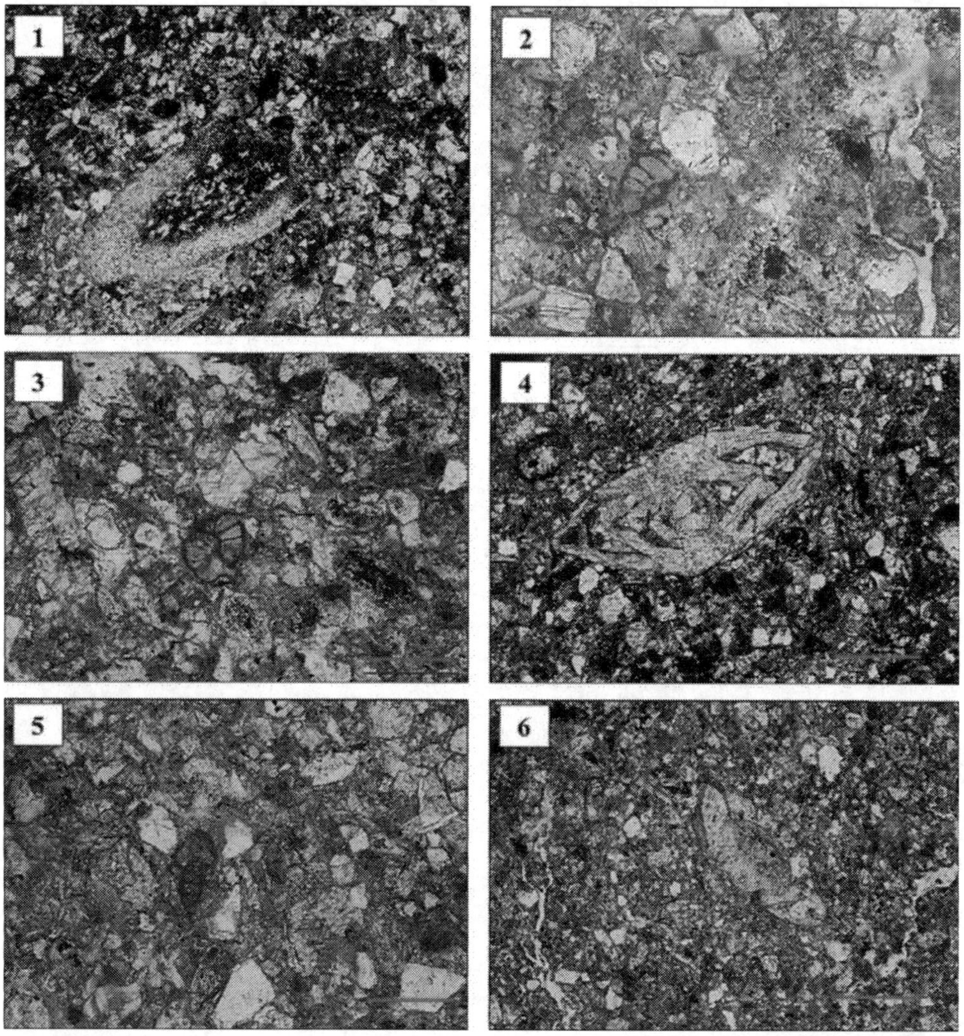

Figure 3. Thin sections of bioclastic grainstone with siliciclastics (LG-01) showed skeletal fragmentary spiky gastropod shells (1), axial sections of uni-biserial foraminifera (2), median sections of trochospiral foraminifera (3), axial sections of larger benthic foraminifera (order: Rotaliida) (4), frontal sections of planispiral foraminifera (5), and an oblique section of Milioida foraminifera. Horizontal red line: scale 1 mm.

The presence of limestone layers on top of the Talangakar Formation indicates a shoreline transgression that occurred after the deposition of the terrestrial Talangakar Formation. Additionally, some silicified woods (*Terminalioxylon* origin) were discovered between the Talangakar and Baturaja Formations. The presence of these wood fossils suggests an association with a terrestrial environment prior to the deposition of the Baturaja Formation.

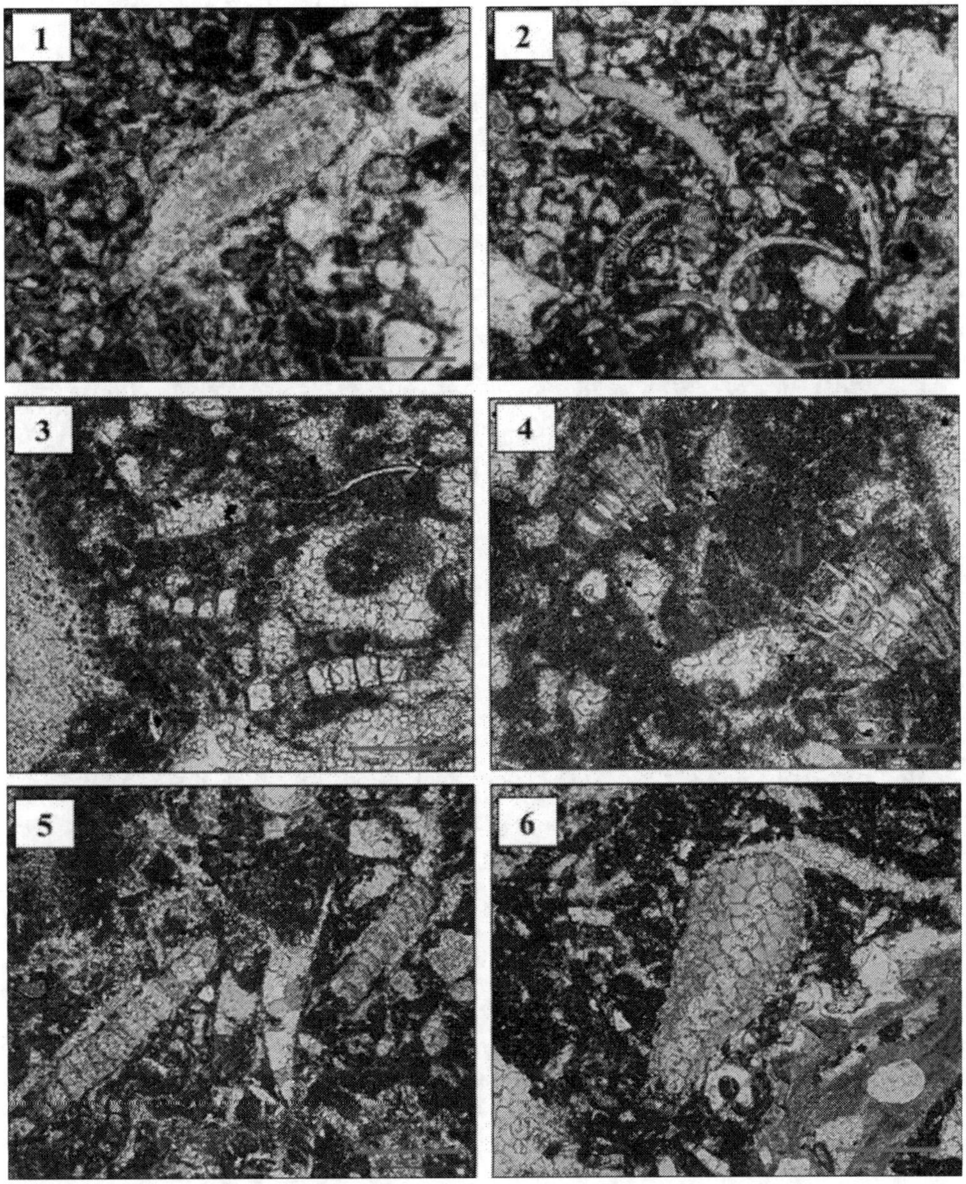

Figure 4. Thin sections of bioclastic packstone filled with micrite (LG-02) showed diverse skeletons around sclerosponges, as calcite crystalized in larger benthic foraminifera (1.a), various bivalve shell sections (2.b), axial section of tubulae-septa of corals (3.c dan 5), axial section of corallite spicula (4.d), and biserial benthic foraminifera (family: Bolivinitidae) (6).

Lithostratigraphically, the limestone of the Baturaja Formation in Linggapura can be categorized into three facies. Based on the observations, all facies were found to be associated with marine fossils as bioclasts. The sequence, from oldest to youngest, consists of bioclast grainstone (**Figure 3**), bioclast packstone (**Figure 4**), and bioclast wackestone (**Figure 5**).

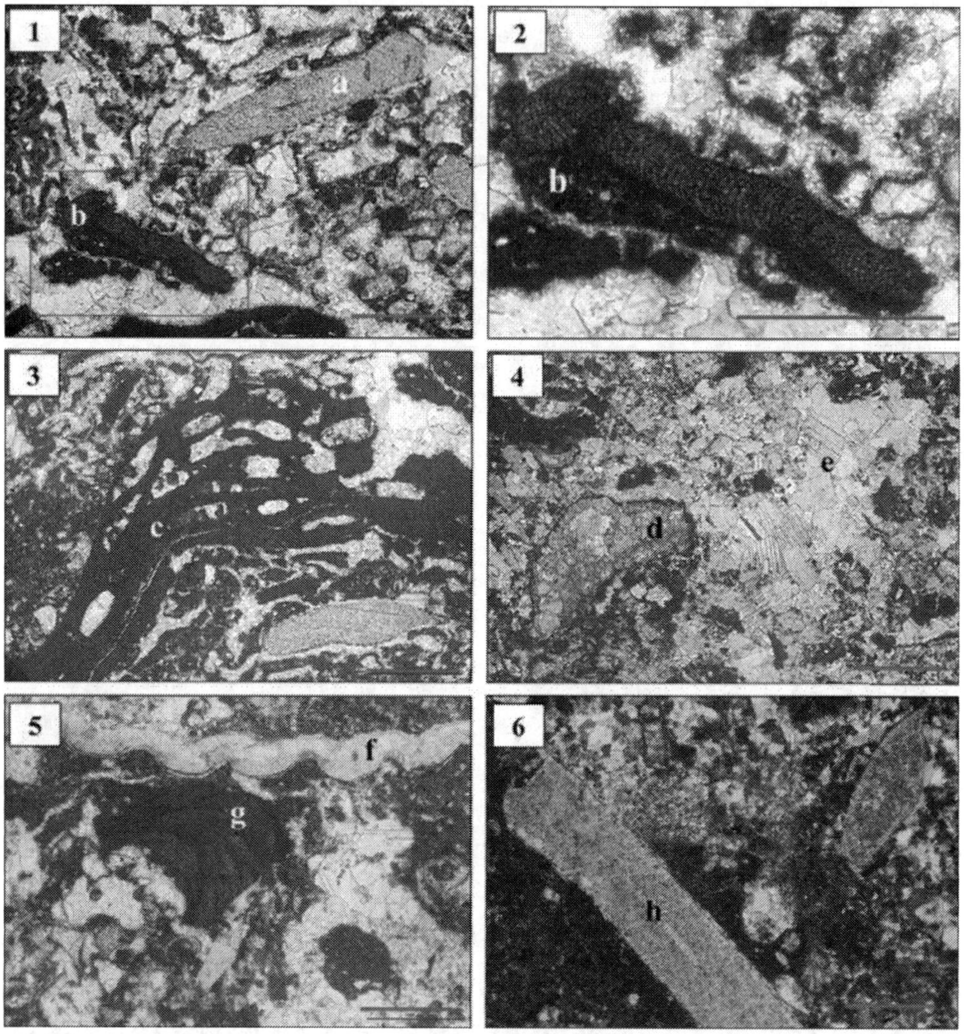

Figure 5. The thin sections of bioclastic wackestone (LG-03) showed axial sections of larger benthic foraminifera (1.a), and rhizocorallium-ichnofossil molluscs (1.b/2.b (zoomed), planolite–ichnofossils (3.c), intergranular quartz (4.d), and calcite (4.e) as grains and fragments and repichnia-ichnofossils (5. f), and fossilized encrusted coralline red algae (5.g), and crustacean tubular-skeletal remains (6.h).

In thin-section observations, the sample (LG-01) from the bioclast grainstone facies was found to be composed of calcareous-siliciclastic angular grains and skeletal remains of marine organisms, including gastropod spiky shells and larger benthic foraminifera (**Figure 3**). These observations suggest that the sedimentation environment of this facies is associated with a marine environment, influenced by dominant terrestrial sediment sources and high-to mid-current depositional environments. Ecologically, gastropods with spiky shells have adapted to cope with tidal current dynamics, particularly during the transition from high to low tides. The spiky shell features help these gastropods resist drifting away easily. Such environments include rocky shorelines, coral reefs, and seagrass beds. Moreover, the presence of larger benthic foraminifera (such as Rotaliida and Millioida) indicates marine conditions characterized by rapid changes in salinity, often associated with arid lagoonal or shoreface environments.

The presence of angular grains in LG-01, along with the clastic composition and ecological characteristics of the fossils, suggests that the depositional environment of this facies was likely near the shoreline. In contrast, the LG-02 sample indicates an environment that was further away (more seaward) from the shoreline. LG-02 is classified as a bioclastic packstone, characterized by a large number of marine fossils surrounded by micrites (**Figure 4**). Marine fossils found in LG-02 include corals, larger benthic foraminifera, and various bivalves. The diverse range of fossils in these rock layers indicates that the depositional environment was an ecosystem with a high ecological productivity.

The presence of micrites intermixed with clastic grains suggests the occurrence of low-current sedimentation environments at certain times. Such environments are commonly found in deep bathymetry. The presence of coral and coralite in this sample indicates that the sedimentation took place within a reef ecosystem. This particular environment is characterized by relatively high nutrient levels, stable salinity, and sufficient sunlight, which are essential for the growth of reef ecosystems. The mild sea currents in this environment allowed for the deposition of finer suspended carbonate grains (micrite).

Compared to the shallower and more landward depositional environment represented by LG-01, the deeper and more seaward environment of LG-02 indicates that shoreline transgression and relative sea-level rise occurred during deposition. This continuous indication persists in the deposition of LG-03. Our findings show that samples from LG-03 primarily consist of bioclastic wackestone, composed of various ichnofossils and an increased presence of micrite (**Figure 5**). This suggests that the depositional environment of LG-03 is deeper, but still influenced by the supply of terrestrial sediment sources, particularly in a littoral environment.

Ichnofossils such as rhizocorallium, planoliter, and rephicnia indicate backreef-lagoonal environments. In such environments, the substrate tends to be finer and rich in suspended nutrients. Organisms such as mollusks, arthropods/crustaceans, and other epifaunal species often graze on the substrate, feeding on attached algae and fragments of broken corals. On thicker substrates, some infaunal organisms burrow as a life strategy for sheltering, grazing, and hunting. Crustaceans, in particular, employ this strategy to camouflage themselves before capturing prey. The presence of tubular crustacean skeletal remains on the rocks also indicates the presence of crustaceans. These organisms often feed on planktonic algae, foraminifera, or other smaller nektonic organisms, such as small fish. We interpret that this paleoenvironment is also associated with the presence of coraline algae, as indicated by traces of encrusted coralline algae preserved in the samples. Coralline algae thrive in submerged conditions and require exposure to sunlight. Algae play a significant ecological role as autotrophic producers in the reef environment, contributing to the food pyramid and material cycle.

4. Discussion

The Talangakar dan Baturaja Formation in Linggapura Lampung was previously traced by the Geology of the Kota Agung Sheet, published by the Indonesian Geology Survey Agency (PSG) [11]. Linggapura was the most southern area that exposed this formation to the earth's surface. This formation is known to have coal-bearing mudstones, which consist of conglomerate, sandstone, and claystone facies, formed during the Oligo-Miocene. Coal is often present in seams. Based on maceral analysis [12], the depositional environment of this facies occurred in a fluvial-paralytic environment with a wet swampy forest ecosystem that gradually changed into a sublittoral-littoral environment. Next, to resume the paleoecological conditions or their changes, we illustrated a schematic transition during the deposition of the formations (**Figure 6**).

In this study, conglomerate layers containing silicified fossilized wood fragments were discovered in the upper part of the Talangakar Formation in Linggapura. The process of silicification in this condition is associated with the presence of hydrological flows that carry dissolved silica with a more alkaline pH [16]. The dissolved silica can originate from volcanic material or the leaching process of certain silicate minerals in the surrounding area [17,18]. These silicate minerals may have originated from the weathering of granitoids or older basement rocks that have been present since the Cretaceous period and are widely exposed in the southern part of Sumatra [4,19,20]. The granitoid body began to weather physically, chemically, and biologically, especially when it exposed in warmer climate occur during the Oligo-Miocene [21].

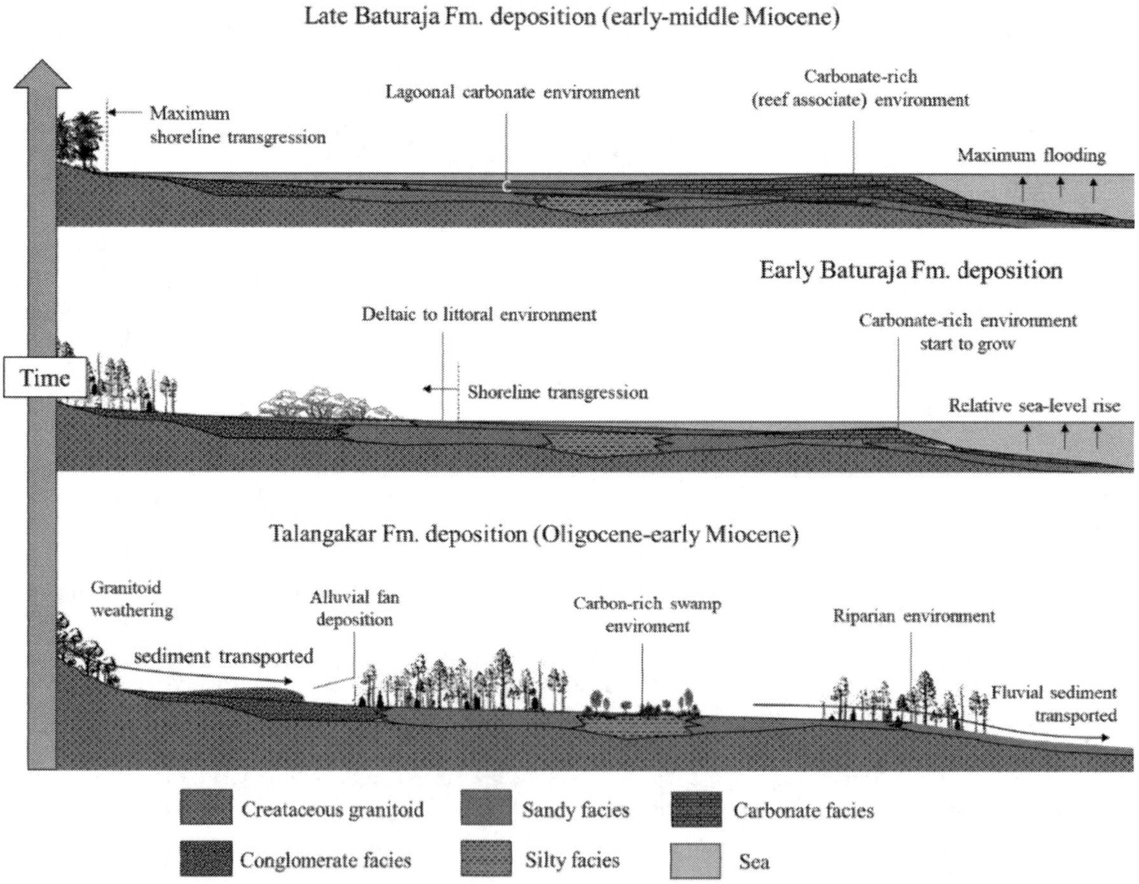

Figure 6. Proposed schematic representation of the geological process and paleoecological changes during the deposition of the Talangakar to Baturaja Formation in the Linggapura region (not scaled).

Starting in the Miocene, the Earth experienced relatively high temperatures, resulting in the melting of ice at the poles [22,23]. This ice melt led to a significant global rise in sea levels known as the Miocene Climate Optimum (MCO). The rising sea levels caused transgression in coastal areas worldwide, resulting in changes to ecosystems and depositional environments. In the study area, there was a simultaneous shift from terrestrial to marine environments during this period. This transition gradually led to the formation of a carbonate environment in the seaward region. As the shoreline reached this area, a mixture of terrestrial and marine sediments occurred. Siliciclastic materials, primarily transported by fluvial activity, carried granitoid-siliciclastic and other terrestrial debris from the landward side. The angular shape of these siliciclastic materials indicates a relatively short transportation distance from their source outcrop. The marine currents in this environment were also relatively strong. In the shallow environment characterized by tides and the warm Miocene climate, the deposition (bed) environment experienced high evaporation [22,24]. This resulted in increased salinity in the depositional environment, similar to a deltaic to lagoonal-littoral environment. Such ecological conditions were ideal for the growth of foraminifera, specifically Miliolida or Rotalida/Numulites [25]. Large foraminifera are typically abundant in shallow, warm, highly saline, and well-oxygenated marine environments. They thrived in the Miocene tropical and subtropical seas, particularly in areas with high organic matter and nutrient inputs. These conditions are represent in the bioclastic grainstone LG-01, that found in Linggapura.

During the Middle Miocene, there was a significant rise in sea levels, leading to the flooding of shores and landward expansion. The warmer Miocene climate, known as the Miocene Climate Optimum (MCO), played a role in the development and vertical growth of reef ecosystems in response to the rising sea levels. As the bathymetry of certain depositional environments deepened due to the sea level rise, the water column conditions became calmer, resulting in the easier sedimentation of finer carbonate particles. The presence of nutrient-rich carbonate suspensions in the water column attracted substrate-grazing organisms, particularly mollusks. These shelled mollusks adapted to this environment by filtering the suspended particles from the water column. They often inhabited areas around coral formations. Additionally, some benthic foraminifera were present in this environment. These ecological characteristics were observed in the bioclastic packstone LG-02 of the Baturaja Formation in Linggapura.

Compared to the other samples, LG-03 exhibits shallower conditions and a relatively lower abundance of clay-sized material (calcite micro-sparite) compared to LG-02. Although not as prominent as in LG-01, interclastic siliciclastic material can be observed in LG-03. In terms of ichnofossils, organisms with substrate grazing behavior were commonly found in LG-03. Some of these organisms were encrusted with red algae and skeletal remains of crustaceans. This group of organisms thrives in reef-associated environments, although with calmer currents and a more seaward location compared to LG-01. Based on these findings, we propose that LG-03 represents a back-reef ecological environment, specifically a muddy sand apron or foreshore.

Referring to previous research, the depositional environment of the Baturaja Formation is generally interpreted as being associated with back-reef, reef-mound, and/or fore-reef environments [26–28]. In the study area of Linggapura, which is located at the edge of the South Sumatra Basin, a back-reef environment is likely to be present. Additionally, other observations from thin-section analysis of cores drilled in several wells in the South Sumatra Basin suggest that this facies exhibits varied porosity. The larger porosity could be attributed to the intergranular spaces between clastic grains. However, the variations in porosity, which typically manifest as intracrystalline, intergranular, or vuggy porosity, are primarily caused by the interaction between meteoric water and marine water. This interaction leads to the recrystallization of carbonate minerals (dolomitization), resulting in smaller porosities. Other findings indicate that the most appropriate carbonate facies model for the Baturaja Formation is an isolated platform following the Rimmed Shelf Accretionary model proposed by Tucker and Wright [29]. When carbonate rocks are exposed to the atmosphere, the intensified interaction between meteoric water and isolated seawater triggers extensive recrystallization, thereby transforming primary porosity into secondary porosity. The reef depositional environment has contributed to the Baturaja Formation becoming a hydrocarbon reservoir. In contrast, if the depositional environment is a lagoon, it is typically dominated by mudstone lithology. Mudstone, being very fine-grained, exhibits minimal porosity, making it an unfavorable environment for hydrocarbon retention. The depositional model proposed by Susilowati and Suyoto [30] illustrates that a lagoon positioned far from seawater interaction and in a calm environment promotes the proper deposition of very fine-grained mudstone.

Based on a previous review, it is likely that the carbonate facies forming the Baturaja Formation in Linggapura represent a rimmed carbonate environment dominated by bioclastics. This is in contrast to the limestone facies of the Baturaja Formation in South Sumatra, located 50 km north of Linggapura, which predominantly consists of boundstone [31]. The presence of boundstone suggests the existence of a massive barrier reef environment. The higher clastic content observed in the limestone at Linggapura is attributed to the stronger water currents and its relatively closer proximity to the shoreline. Terrestrial clastic materials also make a significant contribution to the sediment supply during this period. However, this environment is supported by favorable paleoecological conditions that facilitate high bioproductivity within the ecosystem. Various marine fossils were discovered in the samples, indicating the presence of a relatively complex food chain ecosystem. This ecological condition also contributes to the accumulation of organic carbon and carbonates in this type of environment.

5. Conclusion

The limestone found in Linggapura primarily comprises marine fossils associated with shore-to-shallow carbonate environments. These fossils include larger benthic foraminifera, massive corals, intertidal-shallow marine mollusks, crustaceans, encrusted remains of red algae, and various shore-related ichnofossils. The carbonate rocks exhibit clastic textures, mainly consisting of bioclastic grainstone, packstone, and wackestone, which indicate the formation of rimmed carbonate environments. The deposition of the Baturaja Formation in Linggapura was influenced by a hotter climate and sea-level rise during the Miocene. These factors played a role in the transgressive nature of the depositional environment. To gain a more comprehensive understanding of the paleobathymetric changes that have occurred, we recommend conducting a detailed geochemical study of the stratigraphy in this area, supported by a thorough paleontological review.

6. References

[1] Duo L and Wen Z 2021 Classification and exploration potential of sedimentary basins based on the superposition and evolution process of prototype basins *Pet. Explor. Dev.* **48** 1271–88. DOI: 10.1016/s1876-3804(21)60286-0.

[2] Pulunggono A 1984 Sumatran Microplates, their characteristics and their role in the evolution of the Central and South Sumatra Basins *Proc. Indon Petrol. Assoc., 13th Ann. Conv.* (Jakarta: Indonesian Petroleum Association (IPA)) pp 121–43. DOI: 10.29118/ipa.2126.121.143.

[3] Hu G, Kong X, Yang L, Zhang J, Zhu H, Ma Y, Li M, Hong G and Bai Z 2020 Prediction on Weathered Basement Reservoirs: A Case Study from South Sumatra Basin pp 2018–29. DOI: 10.1007/978-981-15-2485-1_182.

[4] Sapiie B 2017 A New Approach in Exploring a Basement-Fractured Reservoir in the Sumatra Back-Arc Basin *Proc. Indon. Petrol. Assoc., 41st Ann. Conv., 2017* (Jakarta: Indonesian Petroleum Association (IPA)). DOI: 10.29118/ipa.50.17.260.g.

[5] Idarwati -, Purwanto H S, Sutriyono E, Prasetyadi C and Jati S N 2021 History Woyla Arc of the Garba Complex: Implications for Tectonic Evolution of the South Sumatra Region, Indonesia *J. Geosci. Environ. Prot.* **09** 118–32. DOI: 10.4236/gep.2021.912008.

[6] Raghumani Singh Y, Atamajit Singh K, Reshma Devi N, Elliott Arnold T and Abbott M B 2021 Hydrocarbon source rock potential and paleoecological studies of the Surma sediments of Manipur region, India: Insights from palynology and organic geochemistry *J. Asian Earth Sci. X* **6** 100072. DOI: 10.1016/j.jaesx.2021.100072.

[7] Mansour A, Geršlová E, Sýkorová I and Vöröš D 2020 Hydrocarbon potential and depositional paleoenvironment of a Middle Jurassic succession in the Falak-21 well, Shushan Basin, Egypt:

Integrated palynological, geochemical and organic petrographic approach *Int. J. Coal Geol.* **219** 103374. DOI: 10.1016/j.coal.2019.103374.

[8] Metcalfe I 2011 Tectonic framework and Phanerozoic evolution of Sundaland *Gondwana Res.* **19** 3–21. DOI: 10.1016/j.gr.2010.02.016.

[9] Balázs A, Matenco L, Magyar I, Horváth F and Cloetingh S 2016 The link between tectonics and sedimentation in back-arc basins: New genetic constraints from the analysis of the Pannonian Basin *Tectonics* **35** 1526–59. DOI: 10.1002/2015tc004109.

[10] Pratama D 2021 *Geologi, Studi Fasies, Lingkungan Pengendapan, dan Karakteristik Batugamping serta Batuan Sedimen Penyerta pada Jalur Lintasan Way Penandingan, Linggapura, Lampung Tengah, Lampung* (Department of Geology, Institut Teknologi Sumatra)

[11] Amin T C, Sidarto S and Santosa & Gunawan W 1994 Geologi lembar Kotaagung, Sumatera *Bandung Pus. Penelit. dan Pengemb. Geol.*

[12] Kusnama K and Pangabean H 2009 Karakteristik Batubara dan Batuan Sedimen Pembawanya, Formasi Talangakar, di daerah Lampung Tengah *Indones. J. Geosci.* **4.** DOI: 10.17014/ijog.vol4no2.20095.

[13] Ginger D and Fielding K The petroleum systems and future potential of the South Sumatra basin *Proc. Indon Petrol. Assoc., 30th Ann. Conv.* (Jakarta: Indonesian Petroleum Association (IPA)). DOI: 10.29118/ipa.2226.05.g.039.

[14] Kasim S and Armstrong J 2015 Oil-oil correlation of the South Sumatra Basin reservoirs *J. Pet. Gas Eng.* **6** 54–61. DOI: 10.5897/jpge2013.0162.

[15] Amijaya H, Schwarzbauer J and Littke R 2006 Organic geochemistry of the Lower Suban coal seam, South Sumatra Basin, Indonesia: Palaeoecological and thermal metamorphism implications *Org. Geochem.* **37** 261–79. DOI: 10.1016/j.orggeochem.2005.10.012.

[16] Mustoe G 2017 Wood Petrifaction: A New View of Permineralization and Replacement *Geosciences* **7** 119. DOI: 10.3390/geosciences7040119.

[17] Harbowo D G, Priadi B, Julian T, Amelia R N, Sihombing D J P and Kencana F S 2021 A preliminary study on the element abundance in the Hulusimpang Formation, Way Kalianda, Pesawaran, Lampung, Indonesia *IOP Conf. Ser. Earth Environ. Sci.* **882** 012078. DOI: 10.1088/1755-1315/882/1/012078.

[18] Harbowo D G, Nahar R N F A, Sari D, Julian T, Kuswara T A, Abimayu R A, Lajona R A and Huzaifah S 2022 The Significances of Cretaceous Petrified Wood Fossils from Padangcermin, Lampung in Paleoenvironmental Perspectives *IOP Conf. Ser. Earth Environ. Sci.* **1047** 012016. DOI: 10.1088/1755-1315/1047/1/012016.

[19] Natalia H C, Harbowo D G and Ikhram R 2021 Potensi Geodiversity di Sekitar Kawasan Anak Krakatau-Way Kambas, Provinsi Lampung, Sebagai Kandidat Geopark Indonesia *J. Sci. Appl. Technol.* **5** 47–57. DOI: 10.35472/jsat.v5i1.318.

[20] Jansen D, Sihombing P, Harbowo D G, Priadi B and Ardhianto L 2021 Chemostratigraphy of Paleozoic Carbonate in Natar , South Lampung , Indonesia *Proc. Jt. Conv. Bandung* 1–4.

[21] O'Brien C L, Huber M, Thomas E, Pagani M, Super J R, Elder L E and Hull P M 2020 The enigma of Oligocene climate and global surface temperature evolution *Proc. Natl. Acad. Sci.* **117** 25302–9. DOI: 10.1073/pnas.2003914117.

[22] You Y, Huber M, Müller R D, Poulsen C J and Ribbe J 2009 Simulation of the Middle Miocene Climate Optimum *Geophys. Res. Lett.* **36** L04702. DOI: 10.5194/egusphere-egu23-4102.

[23] Warny S, Askin R A, Hannah M J, Mohr B A R, Raine J I, Harwood D M and Florindo F 2009 Palynomorphs from a sediment core reveal a sudden remarkably warm Antarctica during the middle Miocene *Geology* **37** 955–8. DOI: 10.1130/g30139a.1.

[24] Herold N, Huber M, Müller R D and Seton M 2012 Modeling the Miocene climatic optimum: Ocean circulation *Paleoceanography* **27**. DOI: 10.1029/2010pa002041.

[25] Hallock P and Glenn E C 1986 Larger Foraminifera: A Tool for Paleoenvironmental Analysis of Cenozoic Carbonate Depositional Facies *Palaios* **1** 55. DOI: 10.2307/3514459.

[26] Aswan, Abdurrachman M, Sapta Fitriana B, Fery Mustofa M, Dwijo Santoso W, Rudyawan A,

Rahayu W D, Hamdani A, Rohiman Y and Zin Oo T 2017 Paleoenvironmental Study of Miocene Sediments from JTB-1 and NRM-1 wells, in West Ogan Komering Block, Meraksa Area, South Sumatra Basin *IOP Conf. Ser. Earth Environ. Sci.* **71** 012033. DOI: 10.1088/1755-1315/71/1/012033.

[27] Maryanto S 2007 Petrografi dan Proses Diagenesis Batugamping Formasi Baturaja di Lintasan Air Saka, OKU Selatan, Sumatera Selatan *J. Sumber Daya Geol.* **17** 13–31

[28] Irzon R, Maryanto S, Syafri I, Kurnia K, Jazuli H H and Hartanto P 2022 Wackestones and Grainstones Geochemistry from Baturaja Formation, South Sumatra Province, Indonesia: Origin and Depositional Environment *Sains Malaysiana* **51** 1005–15. DOI: 10.17576/jsm-2022-5104-05.

[29] Tucker M E and Wright V P 2009 *Carbonate sedimentology* (John Wiley & Sons)

[30] Susilowati T and Suyoto S 2014 Model Fasies Karbonat Formasi Baturaja, Lapangan Danendra, Cekungan Sumatra Selatan *J. Ilm. Kebumian MTG* **2**

[31] Amir V 2011 Facies architecture and depositional relationship of Batu Raja carbonates in Letang, Rawa, and Tengah field, Corridor Block, South Sumatra *Proc. Indon Petrol. Assoc., 35th Ann. Conv.* (Jakarta: Indonesian Petroleum Association (IPA)). DOI: doi.org/10.29118/ipa.168.11.g.090

Acknowledgments

We would like to express our gratitude to all colleagues who participated in the implementation of this research and contributed valuable discussions.

ANALYSIS OF SOIL DEFORMATION DUE TO OIL AND GAS EXPLORATION AND HYDROCARBONS MICRO SEEPAGE IN NGASEM DISTRICT, BOJONEGORO REGENCY

H Sadewa Wiguna[1*], N Hayati[1]

[1] Department of Geomatics Engginering, Institut Teknologi Sepuluh November, Surabaya, Indonesia, 60111

*Corresponding author: sadewawigunahusein@gmail.com

Abstract. *Land subsidence is the result of the extraction of liquid in the pores of the rock from the compression of the sediment. The decrease occurred due to the exploitation of oil and gas wells in the area which took and caused reduced sediment compression. Aside from the result of excessive extraction of hydrocarbons, hydrocarbon micro seepage can also trigger an anomaly in the ground level. This anomaly is due to the abundance of hydrocarbons in the reservoir and the dynamic nature of oil and gas accumulation, which results in the absence of a perfect rock seal. This study aims to determine the average rate of subsidence in Ngasem District and to analyze the relationship between soil deformation and oil and gas exploration activities and hydrocarbon micro seepage. The method used in this research is InSAR Time Series with LiCSAR Sentinel 1 SLC and Generic data. Atmospheric Correction Online Service (GACOS) which was used from 2014 to. Meanwhile, maps of the distribution of hydrobarone micro seeps were obtained from Exxon Mobil and Exploration Wells for the East Java Basin were obtained from the Ministry of Energy and Mineral Resources. The results showed that the deformation of the land surface in the Ngasem sub-district was very diverse due to the presence of geological structures, exploration wells and potential seepage of micro-hydrocarbons. From the results of deformation processing in the Ngasem District area, it has the highest rising speed in the study area at 77mm/year while for the greatest decrease at -85 mm/year. Exxon Mobill's exploration wells experienced a subsidence of -30mm/year, while the potential for micro-hydrocarbon seepage increased by 60mm/year. From the results of the analysis of the decline that occurred in exploration wells due to oil and gas exploitation.*

***Keywords**:* Hydrocarbon, Micro Seepage, Radar Remote Sensing,*SBAS, InSAR*

Introduction

Land subsidence is the result of the extraction of liquid in the pores of the rock from the compression of the sediment. This phenomenon can be found in various parts of the world including in California, Arizona, Nevada, Italy, Mexico, Bangladesh, China and the coasts of Texas. Several studies [1];[2];[3] argue that the extraction of hydrocarbons has a very large role for the Harris-Galveston area. The research has facts from the Goose Creek Field which showed a 1 meter drop from 1975 to 1925.

Apart from the result of excessive extraction of hydrocarbons, according to [4] that hydrocarbon micro seepage can also trigger an anomaly in the ground level. This anomaly is caused by the abundance of hydrocarbons in the reservoir and the dynamic nature of oil and gas accumulation, which results in the absence of a perfect rock seal. The phenomenon of hydrocarbon micro seepage is one of the phenomena of mass migration of hydrocarbons that occurs due to a large pressure differential from the reservoir [5];[6]. This phenomenon refers to the process of mass migration of hydrocarbons which occurs very slowly, invisible, but alkane gas accumulates [7]. Hydrocarbon micro seeps predominantly consist of methane (CH_4), ethane (C_2H_6), propane (C_2H_8), butane (C_2H_{10}) and pentane (C_2H_5) gas content [8]. These hydrocarbons can interact with the stratigraphic column and produce a variety of physical, chemical, mineralogical, geobotanical and microbiological variations on the Earth's surface [9].

The use of the Global navigation satellite system (GNSS) in determining deformation is considered inadequate due to the large area of a Basin which can be included in several districts causing difficulties in determining subsidence caused by oil and gas exploration and anomalies caused by hydrocarbons. Utilization of technological developments can provide a separate solution for observing soil deformation using Earth Observation (EO) from the European Commission, opening up new opportunities in monitoring land subsidence that occurs on the surface of the earth. Newer satellite constellations, such as Sentinel from the Copernicus program, provide free data services that allow observations of the entire Earth's surface at unprecedented frequencies and by using a variety of sensors, both optical and radar [10]. Radar technology on the Sentinel 1 Synthetic Aperture Radar (SAR) satellite works in all weather and data acquisition during the day and night, where the sensor can penetrate clouds to allow detection of subsidence. Therefore, in this study, spatial and temporal analysis of land subsidence will be carried out by utilizing unwrapped interferogram data from SAR Sentinel 1 satellite imagery that has been processed. Where later the area of Ngasem Subdistrict can be distinguished where there is an anomaly due to seepage of hydrocarbons and a decrease that occurs due to oil and gas exploration. It is hoped that this research will better understand the phenomena that occur, and be able to analyze the differences in land subsidence that occur in the study area.

Study Area

The research area is included in the Cepu Block area which is an oil field on the border of Central Java and East Java shown on Figure 1. The Cepu Block is an oil and gas contract area covering the Bojonegoro Regency - East Java, Blora Regency - Central Java, and Tuban Regency - East Java. Prior to the recent discovery of sizable oil reserves in the Cepu region and its surroundings, namely in the Regencies of Bojonegoro and Tuban. One of them is the Banyu Urip project which is under the auspices of Exxon Mobil Cepu Limited in Bojonegoro Regency.

IOP Conf. Series: Earth and Environmental Science 1245 (2023) 012002 doi:10.1088/1755-1315/1245/1/012002

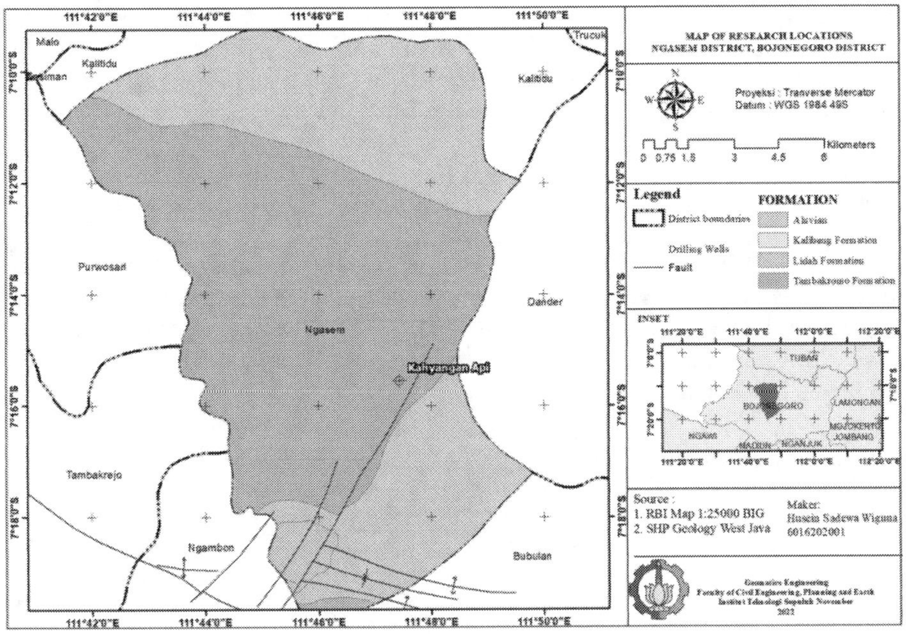

Figure 1. Research Location Map

Materials and Methods

This study uses the SBAS Time Series InSAR method to be able to find out the deformation that occurs in the Ngasem District area. The data used is LiCSAR Sentinel 1 SLC data with an acquisition date of 1 January 2016 – 9 August 2022. After obtaining LiCSAR image data from the COMETLiCS web portal (https://comet.nerc.ac.uk/comet-lics-portal/) then Pre-processing is carried out which will pass through several stages shown on Figure 2. The first stage is Convert Geotiff and Downsample. This stage converts to single precision floating point format without using additional information to carry out further processing. For the downsampling process it can be used for LiCSAR products (~100m) so that it will reduce the amount of data and is very useful for testing process parameters and making data adjustments that will be used in high resolution or higher. After changing the next step is Mask and Clip Interferogram to cut areas that are only used in research

Furthermore, the data processing is divided into 2. The first processing uses the Small Baseline Subset (SBAS) method which will use the LiCSBAS software. Furthermore, the data from the SBAS results were processed using GIS Software (ArcMap) to overlay and clip in certain areas so that only cross-sections could be made in the study area. LiCSBAS will perform a quality check on each LiCSAR image to reduce errors that occur during processing. Next, iterations are carried out to detect bad unwrapped interferogram data. Unwrapped data will have errors that will significantly affect the time series processing and must be removed before processing. By using a comparison of 3 images and 3 interferograms, you will see a comparison of the smallest RMS and the closest pixel to be selected and not selected as error data. After that, calculation/estimation of speed is performed using pixel data from the surface to generate the required data transfer of the stack of unwrapped interferograms. This step is used to cover data shortages resulting from the acquisition of the Sentinel 1 image which is less than perfect or does not exist at a certain time so that later it is hoped that using this method can at least approach the data that was acquired directly. From the iterative data, there is still some residual noise such as tropospheric noise residue, ionosphere noise and orbital error so that it will be reduced by using a 2-dimensional Gaussian kernel with space and time variables. Using this method is expected to facilitate the application of filters for short time series data and will be able to distinguish the original nonlinear displacement.

The results of the deformation velocity generated from LiCSBAS were then evaluated with maps of exploration wells and maps of hydrocarbon micro seepage. exploration data per year from 2014 – 2021 which will be compared with land subsidence data around exploration wells. From these results, statistical processing is carried out to see the link/correlation between the 2 variables whether they are related or not. This will be a benchmark for the surrounding area if large-scale extraction of hydrocarbons results in very significant reductions. Actions must be taken to prevent extreme declines for the area.

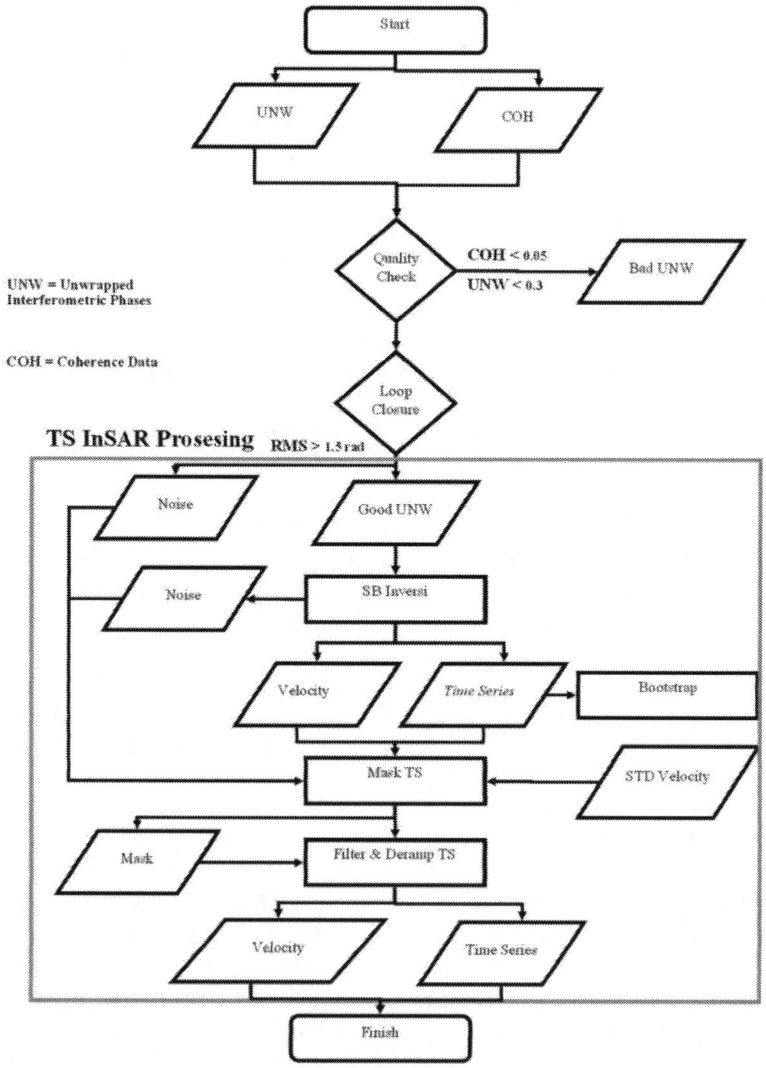

Figure 2. Research methods

Result and Discussion

From the data processed by LiCSBAS software, there are outputs in the form of speed, average coherence, cumulative displacement, standard deviation, number of interferograms that do not experience repetition and maximum length of interferogram relationships. Processing speed data using GACOS atmospheric correction is very influential on the results obtained. Even though the interferogram error has been controlled using a loop closure quality check, GACOS is still needed because it can correct signal turbulence caused by tropospheric total delays. The outcome data is very useful for interpreting in that area the results obtained are very good or vice versa [11].

LiCSAR data using Sentinel 1 satellite data has a temporal resolution of 12 days and a spatial resolution of 5 meters. From this spatial resolution, Sentinel 1 data is used to compare micro seepage for detailed analysis which is very difficult due to the difference in resolution of the 2 results. In the dissertation belonging to [4] active remote sensing can only analyze large land surface rise due to the large reservoir area and not too small to be interpreted by the Sentinel 1 satellite. With a temporal resolution of 12 days and exploration well hydrocarbons are taken every day Land subsidence will occur every day and can be monitored continuously by satellite.

If we compare it with plotting on the standard deviation data shown on Figure 3 before and after GACOS, the output from LiCSBAS shows some changes, especially in the unwrapped interferogram. In the scatter plot comparison of Figure 3, there are significant changes in some of the data, which initially had a standard deviation of 20, dropping to 15 or 10. This shows that the higher the standard deviation, the farther the range of values is from the average value. This makes the data will at one point tend to experience large deviations. Making the standard value closer to 0 will make the value more accurate due to the smaller range of values.

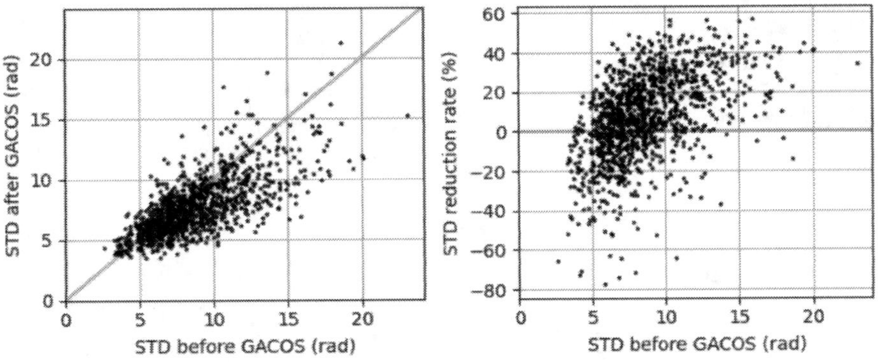

Figure 3 Scatter Plot of Standard Deviation Data Before and After GACOS

These results are in line with the conclusion [11] that GACOS is highly recommended to improve accuracy in time series. However, in several studies, the results of the GACOS correction actually resulted in a negative effect on the image, so that the relationship between corrected and uncorrected GACOS is very important for deformation speed [12].

The results of the coherence value show how far each pixel is similar between the master and slave images with a value between 0 to 1. Loss or blank data at some pixels (zero coherence) is caused by decorellation so that the coherence value decreases. Decorellation has various types such as spatial decorellation. This correlation is due to the better height of the target due to the increasing length of the baseline. In addition, there is Doppler centroid decorellation, this decorelation occurs due to the speed of the vehicle changing, causing variations in speed when it returns to sensing the place [13]. According to [14], decorelation can be caused by the volume factor due to the uneven surface which causes the backscatter to spread large, this usually occurs in vegetated surface areas. In addition to volume decorellation, satellite systems can produce decorelations due to the existing system. However, this decorellation will only have an impact on coherent interferometrics when the return signal is very weak. Figure 4 shows some areas that do not have pixel information due to low coherence values due to decorrelation. One example of low coherence (zero) is shown in several areas of the Ngasem District. Ngasem Subdistrict, where the majority of land use is built-up land and rice fields. From rice fields and production forests, vegetation growth or leaf movement can cause different phases of the two data collections. However, differences in conditions during planting, growing and harvesting also show phase variations throughout the time series data and can reduce coherence values, this condition occurs around Ngasem District which is located in a paddy field planted with rice.

From the data processed by LiCSBAS software, there are outputs in the form of speed, average coherence, standard deviation, number of interferograms that do not experience repetition and maximum length of interferogram relationships. The outcome data is very useful for interpreting in that area the results obtained are very good or vice versa. The results of the coherence value show how far each pixel is similar between the master and slave images with a value between 0 to 1. Loss or blank data at some pixels (zero coherence) is caused by temporal decorrelation so that the coherence value decreases. Figure 4 shows some areas that do not have pixel information due to low coherence values due to temporal decorrelation. One example of low coherence (zero) is shown in several areas of the Ngasem District.

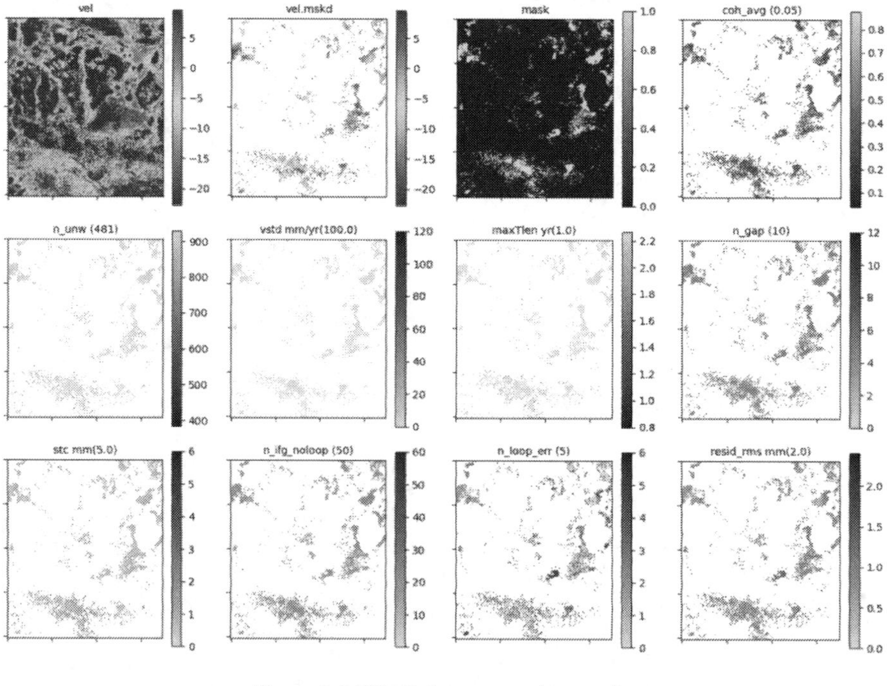

Figure 4. LiCSAR data processing results

Referring to Figure 4 with the subtitle maxTLen which shows the length of the interferogram connections in the research area based on data from 2016 to 2022 processed by LiCSBAS using a predetermined default threshold, the results are 2.2 years. By following the default threshold for coherence of 0.05 and unwrapped data of 0.3, there are several interferogram pairs that do not fit into the predetermined criteria so that from the image the interferogram pairs shown in Figure 5 have 2 different colors indicating acceptance and rejection of the threshold. . The red color in the image shows that the interferogram pair on that date is not used so that it will reduce the time series and make the start of processing from 2016 – 2022. Figure 5 Meanwhile the blue color can be continued for processing because it meets predetermined threshold standards. However, the black straight line upwards is a gap in the interferogram network due to noise or no data on that date. The existence of a gap in the interferogram pair date 20151114_20151208 20151208_20160406 20160711_20160921 20160921_20161108 20161108_20161202 20161202_20161226 20161226_20170119 20170119_20170224 20170425_20170507.

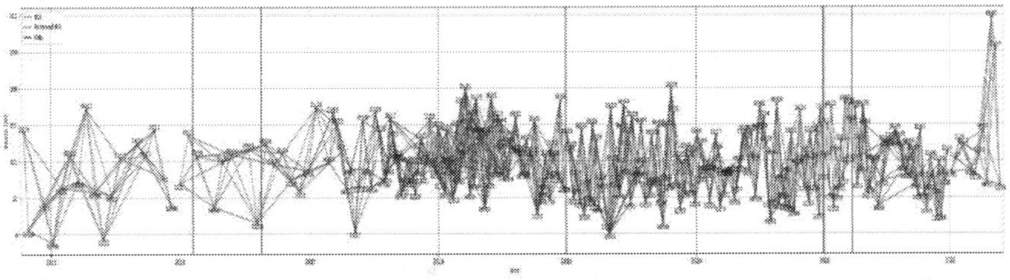

Figure 5. LicSAR Data Interferogram Pair 2014 – 2022

From the deformation results it was found that the highest deformation speed in the study area was at 77mm/year which is depicted in red. Meanwhile, the biggest decline is at a speed of -85 mm/year which is depicted in green. In the Ngasem sub-district itself, there is the Kahyangan Api and there is also an exploration well owned by Exxon Mobil Bojonegoro. From the results of processing using the SBAS method on LiCSAR data, the results show that the subsidence in the area around the exploration wells varies greatly. In the cross-section of the Southeast - Northwest direction that passes through the existence of exploration wells, the decrease in height varies and there tends to be an increase from -10 mm/year to -40 mm/year. The exploration well area itself has decreased by -30 mm/year.

An incision that is in the direction of Southwest - Northeast which passes through several geological structures in the form of anticlines and synclines and passes through faults and ends in the Kahyangan Api. From the results of the incision, it shows an increase of 45 mm/year in the anticline area while the decrease in the heavenly area is -40 mm/year. It is known that the structure of the anticline and syncline greatly influences the deformation and can be seen from the results of the incisions of the two anticlines that they both have an increase of 45 mm/year while the syncline itself has an increase of only 20 mm/year. Meanwhile for faults, the difference between the foot wall and the hanging wall is very visible in the incision. For the hanging wall of the fault there is absolutely no change, while for the foot wall it has decreased by around -10 to -20 mm/year so that it can be concluded that the fault is a normal fault. In the Kahyangan Api area, it experienced a very significant decrease with a decrease of -40mm/year, which was the largest decrease in the incision. This is due to the continuous reduction of the gas contained in the formation thereby reducing the existing pressure and compressing the soil.

From the results of Exxon Mobil data regarding micro-hydrocarbon seepage, the bottom yield in the southern area of Ngasem District has a very high seepage potential. To test the potential results with the phenomenon of deformation, cross sections are made in the direction of

IOP Conf. Series: Earth and Environmental Science 1245 (2023) 012002 doi:10.1088/1755-1315/1245/1/012002

Northwest - Southeast. From the results of the cross section that there is an increase of 60mm/year. In line with research [4] that in areas experiencing micro-hydrocarbon seepage will experience anomaly in ground level. Where the area in the middle will experience the highest increase and will decrease exponentially to areas that do not experience seepage. From the results it shows that the highest increase is at 4.7 kilometers from Kahyangan Api and shows the highest increase in that area of 60mm/year.

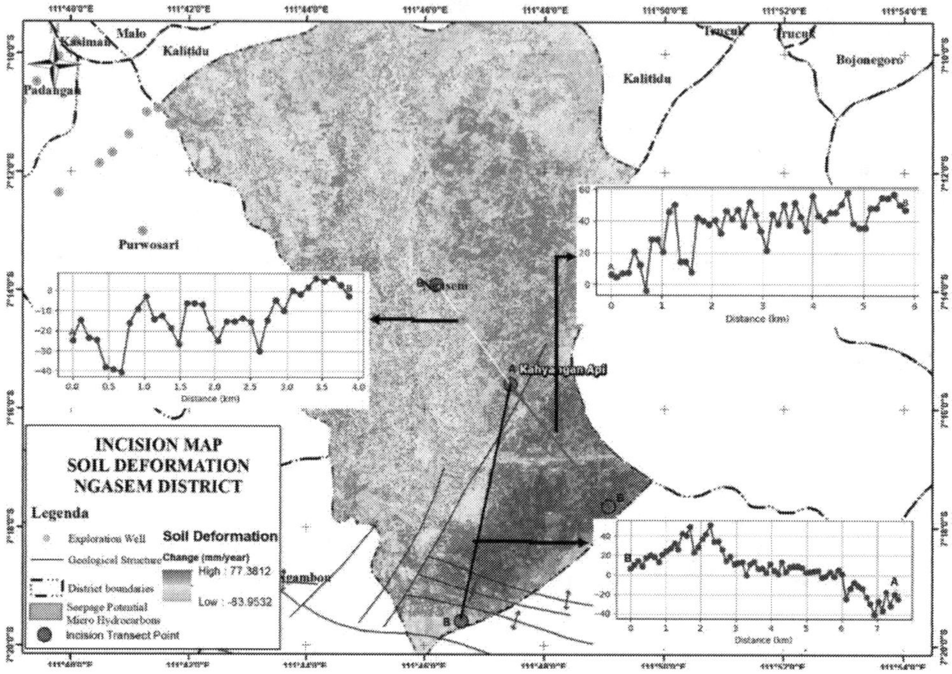

Figure 6. Land Deformation Cross Section Map

From the processed data, there are several places that have decreased or vice versa experienced an increase in the soil surface. According to the results obtained, the resulting soil deformation in Ngasem District comes from gas and hydrocarbons that are in formations that are pressing upwards. From the speed time series results, the highest potential for hydrocarbon seepage shows uplift results. This is in line with research from [15] that the migration of hydrocarbons is dynamic and there is no perfect seal cap, causing oil and gas traps in formations to be expressed directly by symptoms of anomalies on the earth's surface. Due to the accumulation of hydrocarbons in the reservoir. Causes a large pressure difference from the reservoir source and causes an anomaly on the ground surface due to the large upward pressure force from the reservoir [16].

The highest potential of the seepage zone is in the south of Ngasem District where the area has 2 geological structures that can become hydrocarbon traps. The first structure is anticline and syncline geological structure. The anticline structure in this area becomes a hydrocarbon trap which stores gas and hydrocarbons in it causing great pressure and suppresses the seal rock so that the uplift phenomenon is faster than the syncline structure which has no pressure. This is very much in line with what was stated by [4] that oil and gas will tend to press against the formation so that the hydrocarbons seep onto the statigraphic surface.

In line with the existence of the Kahyangan Api from the Ngasem District which continues to emit gas to the surface. As a result of the gas pressure being released continuously so that the formation experiences a lack of pressure and tends to solidify. As a result of this, the area around Kahyangan Api continues to experience very fast land subsidence -24.6 mm/year. The decline was also accelerated by Exxon Mobill's exploratory wells which produced 2 million barrels per year. With very large production, even though it has been replaced with the same specific gravity, over time it will lose the same pressure as hydrocarbons and cause a decrease that tends to be significant [17].

The amount of oil and gas production from exploration wells greatly affects land subsidence in the area. According to [18] that the extraction of oil that is in the formation will cause a decrease even though it has been replaced with drilling mud so that the pressure in the formation will tend to be stable. However, the pore pressure that is created due to oil and natural gas will not be the same as the drilling mud so that it will experience a loss of density and experience compression so that it will experience a significant decrease.

Table 1. Oil and Gas Production and Land Subsidence in Exxon Mobill Exploration Wells

Years	Crude Oil (Barrel)	Line Of Sight Displacement (mm)
2016	2.052.234	-22.47
2017	2.842.770	-31.39
2018	2.505.854	-55.36
2019	2.812.254	-81.09
2020	3.304.153	-83.03
2021	3.117.978	-88.8

When compared to the results of oil and gas production at Exxon Mobil's wells taken from 2016 - 2021 sourced from Oil and Gas in the annual figures released by the Ministry of Energy and Mineral Resources as shown in Table 4.1. Oil and gas production always increases from year to year. Oil and gas production tends to increase resulting in a decrease in the area around exploration wells as shown in Table 1.

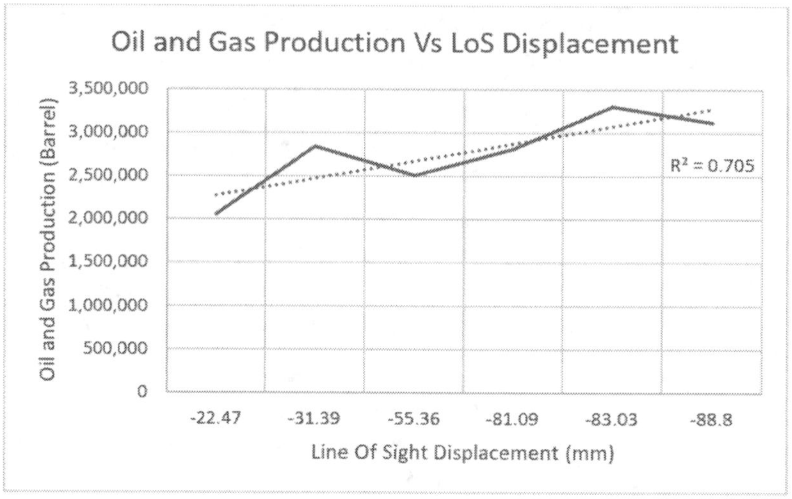

Figure 7 Graph of Oil and Gas Production Against Soil Deformation

From the results of oil and gas production, to see how much the level of relationship between the 2 variables is, the Pearson correlation method is used. From the correlation results, the value is -0.774. This value can be interpreted as a relationship between 2 variables where an increase in one variable causes a decrease in the value of the other variable. Vice versa, the smaller the value of a variable, the greater the value of other variables so that the relationship between the two variables between vertical deformation and oil and gas production is reversed [19]. This can be seen from Table 1. From 2016 to 2021, Exxon Mobill's production will increase every year, accompanied by faster land subsidence. In 2019, when well production rose by 300,000 thousand barrels, it caused a significant decrease of -81.09. The decrease is 30 mm different from the previous year. According to [20] the enhanced oil recovery in this formation did not take hydrocarbons from the formation causing a sudden extreme decline in 2019.

Conclusion and Recommendation

In the results of this study, the deformation of the land surface in Ngasem District is very diverse due to the geological structure, exploration wells and potential seepage of micro-hydrocarbons. From the results of deformation processing in the Ngasem District area, the rate of increase is the highest in the research area at 77mm/year while the largest settlement is at -85 mm/year. In Exxon Mobil's exploration wells, there is a subsidence of -30mm/year while for potential seepage of micro-hydrocarbons has increased by 60mm/year.

Reference

[1] Sharp, J.M.Jr. and Hill, D.W., 1995. Land Subsidence Along the Northeastern Texas Gulf Coast: Effects of Deep Hydrocarbon Production. Environmental Geology, V. 25: 181-191.

[2] Kreitler, C.W., 1976. *Faulting and Land Subsidence From Ground-Water and Hydrocarbon Production, Houston-Galveston, Texas*. Land Subsidence Symposium, Proceedings of the Anaheim Symposium. IAHS publication No. 121, pp. 435-446.

[3] Holzer, T.L. and Bluntzer, R.L., 1984. *Land Subsidence Near Oil and Gas Fields, Houston, Texas*. Ground Water, V.22(No. 4): 450-459.

[4] Hong, Y. (1999): *Imaging Spectrometry for Hydrocarbon Microseepage*. Delft, Thesis TU Delf University.

[5] Yang, Q., Zhao, W., Dixon, T. H., Amelung, F., Han, W. S., & Li, P. (2015). InSAR monitoring of ground deformation due to CO2 injection at an enhanced oil recovery site, West Texas. International Journal of Greenhouse Gas Control, 41, 20–28. https://doi.org/10.1016/j.ijggc.2015.06.016

[6] Chen, S., Ying, Z., Liang, Z., Yanli, L., dan Chao, Z. (2017): *Hydrocarbon Micro-Seepage Detection by Altered Minerals Mapping from Airborne Hyper Spectral Data in Xifeng Oilfield, China*. Journal of Earth Science, 28, 656- 665.

[7] Asadzadeh, S. dan Carlos, R . (2017): *Spectral Remote Sensing for Onshore Seepage Characterization: A Critical Overview*. Earth Science Reviews, 168,48-72.

[8] Lammoglia., T., Carlos, R d. S. F., dan Filho. (2008): *Characterization of Hydrocarbon Microseepages in The Tucano Basin, (Brazil) Through Hyperspectral Classification and Neural Network Analysis of Advanced Spaceborne Thermal Emission and Reflection Radiometer (ASTER) Data*. The International Archives of the Photogrammetry, Remote Sensing and Spatial Information Sciences, XXXVII, 1195-1200.

[9] Salati, S., Frank, v. R., Freek, v. d. M., dan Babak N. (2014): *Detection of Alteration Induced by Onshore Gas Seeps from ASTER and WorldView-2 Data.* Remote Sensing, 6, 3188-3209.

[10] Fajrin, A. M. (2019, Oktober 10). Pemanfaatan Citra Sentinel-1 SAR Untuk Deteksi Banjir Studi Kasus Pangkalan Koto Baru Sumatera Barat. Seminar Nasional SPI-4,, 10-14.

[11] Morishita, Y.; Lazecky, M.; Wright, T.J.; Weiss, J.R.; Elliott, J.R.; Hooper, (2020) "A. LiCSBAS: An Open-Source InSAR Time Series Analysis Package Integrated with the LiCSAR Automated Sentinel-1 InSAR Processor". *Remote Sens*, vol. *12*, hal. 424.

[12] Wang, X.; Aoki, Y.; Chen, J. (2019), "Surface deformation of Asama volcano, Japan, detected by time series InSAR combining persistent and distributed scatterers, 2014-2018". Earth Planets Space, vol. 71, hal. 121.

[13]Aboutorabi, S. (2016). "Interferometric Synthetic Aperture Radar Temporal De-correlation". August.

[14] Woodhouse. L, (2006), Introduction to Microwave Remote Sensing. Taylors & Francis.

[15] Susantoro, T. M., Ketut, W., Asep, S., Agus, H. H. (2017), "Pemetaan Mineral Lempung untuk Identifikasi Anomali Potensi Rembesan Mikro di Lapangan Migas, Cekungan Jawa Barat Utara Menggunakan Data Landsat-8". Seminar Nasional Penginderaan Jauh.

[16] Chen, S., Ying, Z., Liang, Z., Yanli, L., dan Chao, Z. (2017), "Hydrocarbon Micro-Seepage Detection by Altered Minerals Mapping from Airborne Hyper Spectral Data in Xifeng Oilfield, China". Journal of Earth Science, Vol. 28, hal. 656- 665.

[17] De Luca, C., Bonano, M., Casu, F., Manunta, M., Manzo, M., Onorato, G., Zinno, I., & Lanari, R. (2018). "The parallel SBAS-DINSAR processing chain for the generation of national scale sentinel-1 deformation time-series". Procedia Computer Science, vol. 138, hal. 326–331.

[18] Kreitler, C.W., (1976), "Faulting and Land Subsidence From Ground-Water and Hydrocarbon Production, Houston-Galveston, Texas". Land Subsidence Symposium, Proceedings of the Anaheim Symposium. IAHS publication No. 121, pp. 435-446.

[19] Hyne,N.J. (2015), Petroleum Geology Exploration, Drilling & Production 3rd edition, Penwell Corporation., Oklahoma

[20] Chang, L., Ku, O., & Hanssen, R. F. (2019). "Identification of deformation pattern changes caused by enhanced oil recovery (EOR) using InSAR. International Journal of Remote Sensing, hal. 95–1505.

Identification and Characterization of Gas Hydrate using Marine Seismic and Well Data: South Makassar Strait case study

A W Muntashir[1*], T B Nainggolan[2] and E Supriyana[1]

[1]Department of Geophysics, Padjadjaran University, Jl. Raya Bandung Sumedang Km. 21, Jatinangor, 45363, Indonesia
[2]Research Center for Geological Resources, National Research and Innovation Agency, Jl. Sangkuriang 21, Bandung, 40135, Indonesia

[*]rvrclm@gmail.com

Abstract. Gas hydrate has the potential as an alternative unconventional resource to current oil and gas. Identification of gas hydrate is made possible by applying the seismic method with the main indicator of the bottom simulating reflector (BSR). BSR is characterized on the seismic section by the presence of high amplitudes, reverse polarity reflections, and cutting through the stratigraphy. In this study, the presence and thickness of the gas hydrate zone have been identified using five 2D seismic tracks, namely CM02-57, SM93-06 SM93-08, SM93-09 and SM93-11, and the SULTAN-1 well. The gas hydrate zone was identified at a depth of about 2300 m. The existence of the gas hydrate zone is followed by the presence of a gas chimney indicator as a free gas migration path. The gas hydrate zones are estimated to be 15 m, 28 m, 33 m, 23 m, and 30 m thick from the respective seismic tracks CM02-57, SM93-06 SM93-08, SM93-09 and SM93-11.

Keywords: gas hydrate, bottom simulating reflector, post-stack time migration, South Makassar Strait

1. Introduction

Gas hydrate holds a significant potential energy source. It is due to the energy density (volume of methane at standard conditions per rock volume) of gas hydrate and 2 to 5 times greater energy density than conventional natural gas [1]. Gas hydrate also has fewer carbon emissions compared to other conventional fuels. The development of gas hydrate will assist in the development of Indonesian Renewable Energy by the Ministry of Energy and Mineral Resources. Gas hydrates have the potential to form in areas that have low temperatures and high pressures such as deep sea areas in tropical climate countries [2]. Makassar Strait stretches for 600 km from north-south with a width of between 100-200 kilometers with a water column of more than 2000 m [3]. Adjacent to the western part of the Makassar Strait, there is the Kutai Basin, which is one of the areas that have the most hydrocarbon-proven resources in Indonesia [4]. Therefore, the South Makassar Strait Basin meets the parameters to be an environment for the formation of gas hydrate. It is necessary to identify gas hydrate based on geological and geophysical aspects to utilize the potential of gas hydrate as an alternative energy source to replace conventional energy sources in the future. This research focuses on the Tanjung Formation which consists of sandstone at Lower Tanjung, shale and limestone at Middle Tanjung and Upper Tanjung. The carbonate reservoir is in the Berai Formation. The Berai Formation is from the Early Oligocene to the Early Miocene. This formation is mostly composed of carbonate rocks that have accumulated from Lower Berai to Upper Berai. In the Martaban-1 and Berlian-1 wells, carbonate rocks of the Lower Berai Formation were found deposited. The research area is situated in the South Makassar Strait between 118^0E – 119^0E and 3^0S – 4^0S with the objective of this research is to estimate the thickness and position of the gas hydrate zone in the South Makassar Strait basin (**Figure 1**).

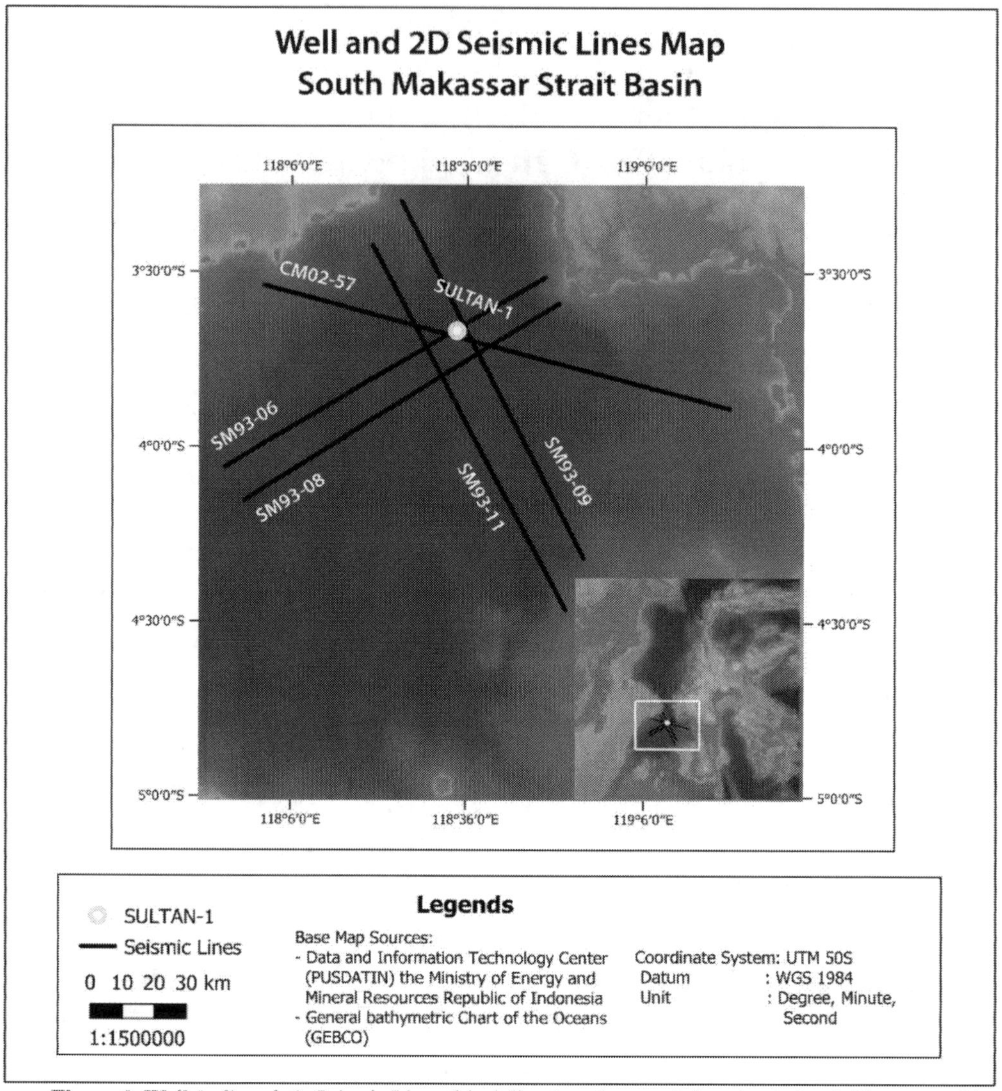

Figure 1. Well (yellow dot), Seismic Lines (black lines), and research location index (white rectangle)

2. Data and Methodology

2.1. Well Data
Well-log data and information such as marker formation, and corrected check-shot of one well, namely SULTAN-1 are acquired in this study. Gamma ray and sonic logs are needed for lithology and well seismic ties.

2.2. Seismic Data
Five 2D post-stack seismic lines are acquired in this study, namely CM02-57, SM93-06, SM93-08, SM93-09 and SM93-11 which are nearest to SULTAN-1 well. These seismic sections are used to identify and determine bottom simulating reflection layers of the South Makassar Strait Basin.

2.3. Gas Hydrate
Gas hydrate is a solid material resulting from a combination of water molecules and one or more hydrocarbon gases or non-hydrocarbon gases. In gas hydrate, gas molecules are trapped in a crystal structure of water molecules [5]. In a sea sedimentary environment, gas hydrate is stable at temperatures between 10 and 15 degrees Celsius, at sea depths of more than 500 m and it can be stable to depths of several hundred meters below the seabed [6]. However, the physical nature of in-situ gas hydrate is still very diverse. Gas hydrate was observed to fill the pore spaces between quartz rock grains, as solids filling fractures, in the form of nodules dissemination in rock grains, and the form of massive units formed from gas hydrate with little sediment. Gas hydrates usually occur due to free gas migrating to the gas hydrate stability zone, so free gas is usually found below this zone [7].

2.4. Bottom Simulating Reflector
Bottom Simulating Reflector (BSR) is a representation of reduced acoustic impedance at the bottom of the gas hydrate stability zone where free gas accumulates beneath relatively impermeable hydrate sediments [8]. In acoustic waves, the BSR is shown as a bright area (peak) on the impedance of the recorded wave, distinguishing the surface of the sediment where there is a gas hydrate and free gas [9]. The physical characteristics of BSR on the seismic section are high amplitude, polarity reversal, cutting through the stratigraphy, and looks like a seismic multiple [10].

3. Results and Discussion
Seismic interpretation is the thoughtful procedure to convey the geologic meaning of seismic data by extracting subsurface information from it, such as structural, formation boundaries, faults, and migration patterns [11]. Based on geological seismic interpretation and Direct Hydrocarbon Indicator (DHI) through bright spot identification, out of 2D post-stack seismic lines that have been interpreted, three of them indicate contrast seismic amplitude appearance of strong bottom simulating reflections that average depth of 2300 m. BSR is located in the Early Pliocene rock formation (Tanjung Formation) which was dominated by shale and intercalated with sandstone and limestone. The result is correlated with SULTAN-1 well data that states the theoretical base of the gas hydrate stability zone. The Geothermal gradient is 8°C/km higher than the prediction (80°C/km). All seismic lines show the appearance of BSRs characterized by cuts through the stratigraphy as we can see that BSRs will not show discontinuity even when there are fractures. It corresponds with the migration of free gas as indicated by the presence of a gas chimney originating from the top carbonate horizon which acts as a gas reservoir. The appearance of a gas chimney and bright spot is important for interpreting the presence of BSRs because gas hydrate is formed due to free gas migrating to the gas hydrate stability zone and accumulating until the methane concentration exceeds the concentration at which methane will dissolve in water. So an indicator of the presence of gas and the direction of migration of the gas. the presence of a gas chimney is also consistent with the presence of fractures which are migration routes of free gas. Seismic section line SM93-06 shows a strong bottom simulating reflection at a depth of 2300 m. This gas hydrate layer came from forceful disruption of low-permeability interface from deeper disrupted sediment continuity while permeating to the seafloor via various faults, fractures and chimney structures (**Figure 2**).

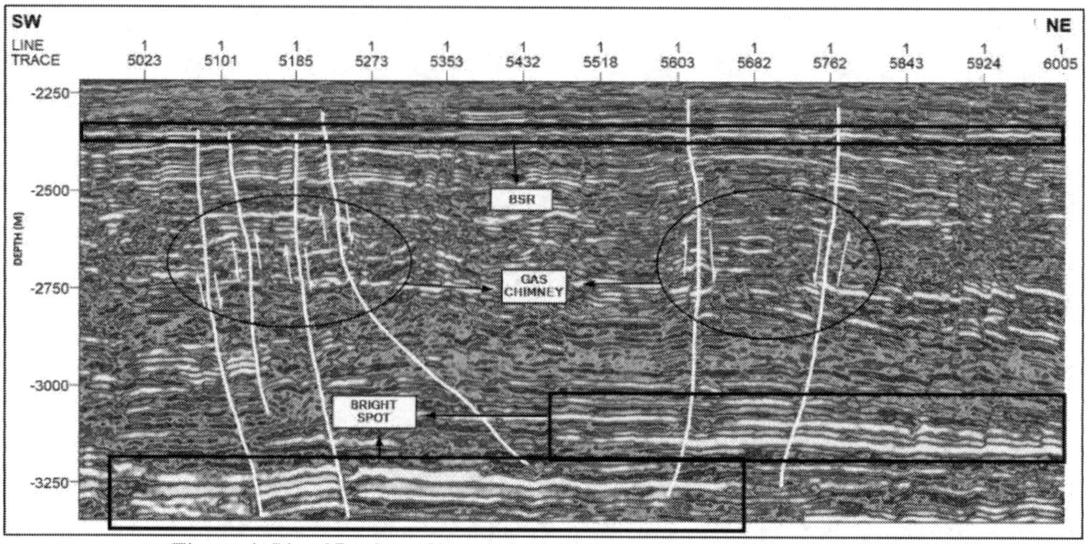

Figure 2. Identification of bottom simulating reflector, gas chimney, bright spot, fracture (white lines) and fracture direction (white arrows) of line SM93-06

Figure 3 shows the occurrence of gas hydrate probably originated from natural gas in much deeper sediment which was migrated throughout faults and stabilized at a depth of 2300 m. Positive anomalies of seismic amplitude (orange color) indicate the presence of gas hydrates, whereas the negative anomalies (blue colour) indicate the presence of free gas on line SM93-08

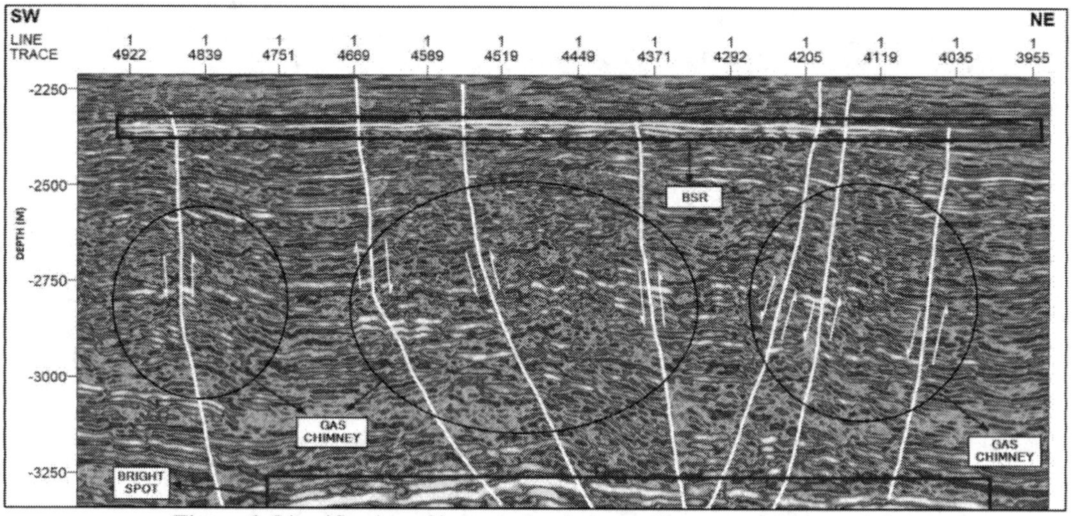

Figure 3. Identification of bottom simulating reflector, gas chimney, bright spot, fracture (white lines) and fracture direction (white arrows) of line SM93-08

Bottom simulating reflection appearance on lines SM93-09 and SM93-11 show successful gas migration throughout faults (**Figures 4 & 5**). This causes evidence of a gas chimney on the seismic section. Negative anomalies show an indication of free gas and positive anomalies show an indication of gas hydrate. Stabilized gas might cause gas hydrate, mostly methane to migrate up from deeper sediment where it was thermally derived from deeply organic matter.

IOP Conf. Series: Earth and Environmental Science 1245 (2023) 012003 doi:10.1088/1755-1315/1245/1/012003

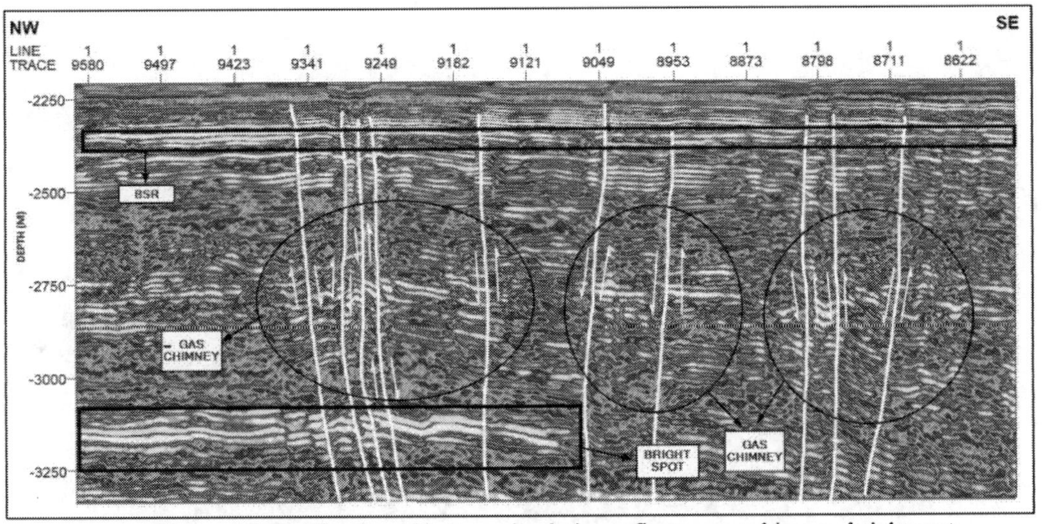

Figure 4. Identification of bottom simulating reflector, gas chimney, bright spot, fracture (white lines) and fracture direction (white arrows) of line SM93-09

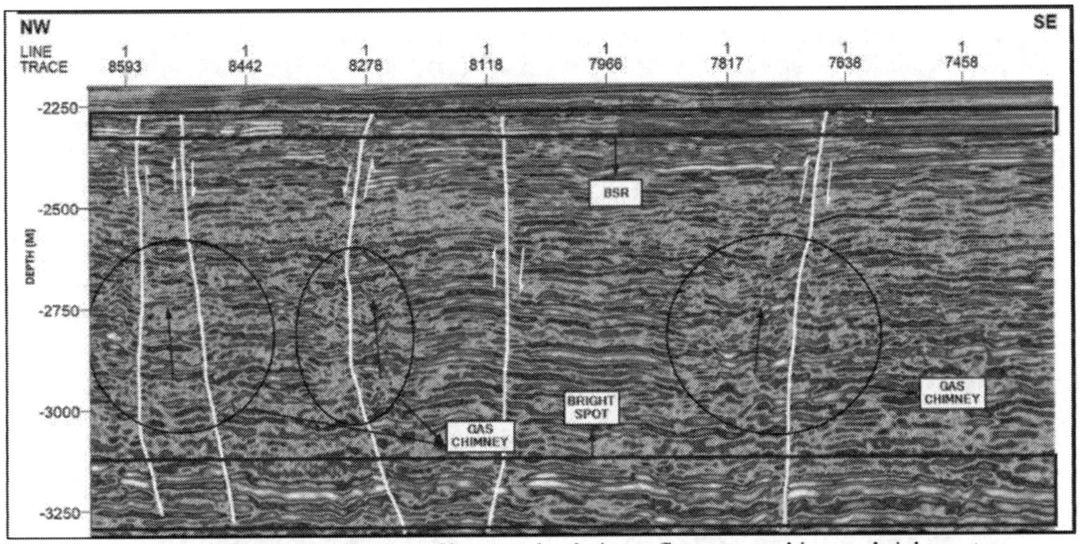

Figure 5. Identification of bottom simulating reflector, gas chimney, bright spot, fracture (white lines) and fracture direction (white arrows) of line SM93-11

Seismic section line CM02-57 shows the bottom simulating reflection at a depth of 2290 m. This gas hydrate layer was accumulated from natural gas of deeper sediment throughout faults (**Figure 6**). Gas hydrate accumulation reaches the equilibrium phase and migrated gas is discontinued. This causes an inactive gas chimney on the seismic section.

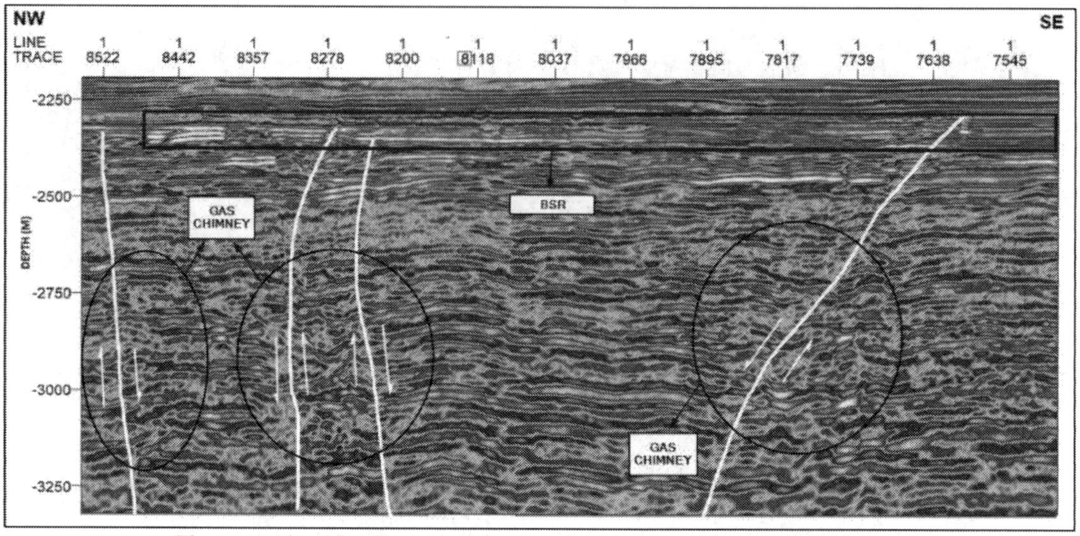

Figure 6. Identification of bottom simulating reflector, gas chimney, bright spot, fracture (white lines) and fracture direction (white arrows) of line CM02-57

The BSR thickness is approximated by performing a well-seismic tie to convert the TWT depth to measured depth. The seismic line that is used for a well-seismic tie is CM02-57 because it is the closest line to the SULTAN-1 (**Figure 7**). A well-seismic tie is a process through which seismic data wavelets are extracted and compared with sonic log and checkshot data from the adjacent well [12]. The result of a well-seismic tie obtained a correlation of 0.689 which means it is good to be used as a reference for depth in seismic data. It was found that BSR was in the Early Pliocene rock formation which was dominated by shale rock, with sandstone and limestone intercalation. The presence of fractures is needed for free gas migration because shale has a low permeability. It is suspected that this shale layer is the barrier so that the free gas does not migrate out and is retained to form a gas hydrate, BSR thickness approximation is carried out from one seismic wave (peak and trough).

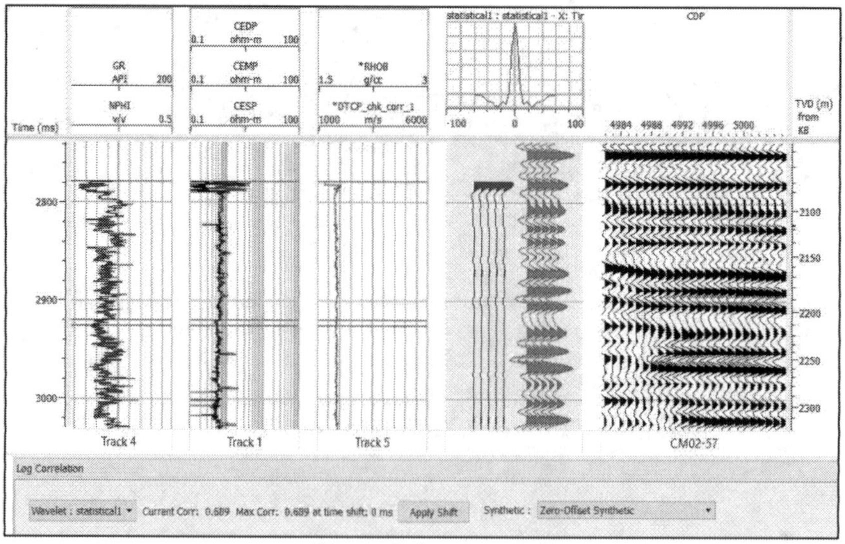

Figure 7. Log correlation

Table 1. BSR location and thickness

No	Seismic Line	CDP Range	Depth (m)	Gas Hydrate Zone Thickness (m)
1	CM02-57	4797 – 5914	2290	15
2	SM93-06	3461 - 7103	2300	28
3	SM93-08	1411 - 8078	2300	33
4	SM93-09	6525 - 9963	2320	23
5	SM93-11	5664 - 8732	2280	30

The gridding method is applied in the gas hydrate zone depth structure map. This map indicates the delineation of BSR geological structures and approximates the distribution of the gas hydrate zone. Based on qualitative analysis, BSR distributed on the theoretical base of the gas hydrate stability zone, which is around 2300 m, spreads almost throughout South Makassar Strait, especially in the fold belt in the northeastern part of the basin. The process of forming methane hydrate came from sediment that fill the fold belt. This fold belt consists of several thrust sheets that form a long anticlinal structure and initiate the formation of a mini-basin. The deposition process consists of a series of spill and fill turbidity processes. It has two significant effects on the accumulation of gas hydrate in the West Sulawesi fold belt. Firstly, turbidite flows carrying coarse-grained materials provide an effective reservoir for hydrate formation, primarily acting as a transport medium for gas migration as well as a steady supply of water. Secondly, turbidity flows bring plant material (terrigenous) from the mainland as a source of organic material for methane-producing bacteria in the ocean. BSR is located in the Early Pliocene rock formation (Tanjung Formation) which was dominated by shale and intercalated with sandstone and limestone. It can be estimated that at a depth of 2300 m, sufficient temperature and pressure are obtained for the formation of gas hydrate in the South Makassar Strait Basin.

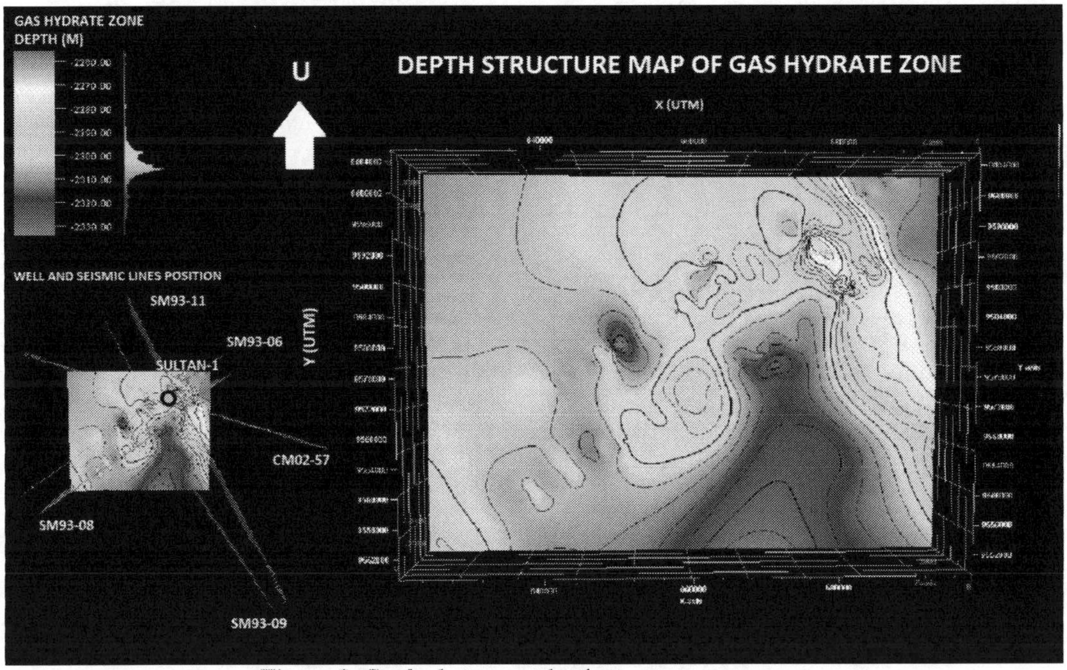

Figure 8. Gas hydrate zone depth structure map

4. Conclusion

Based on the results of the interpretation and depth structure map, BSR in the South Makassar Strait Basin is shown to be at a depth of around 2300 m and is associated with the presence of a gas chimney. BSR is located in the Early Pliocene rock formations which are dominated by shale and interspersed with sandstones and limestones. The presence of fractures helps the gas to freely migrate into the gas hydrate stability zone due to the low permeability of the shale. The presence of gas hydrate is estimated in five seismic lines. The CM02-57 track is identified with BSR in the CDP range 4797-5914 at a depth of 2290 m with an estimated thickness of 15 m. Track SM93-06 is identified with BSR in the range CDP 3461-7103 at a depth of 2300 m with an estimated thickness of 28 m. Track SM93-08 is estimated to have gas hydrate in the range CDP 1411-8078 at a depth of 2300 m with an estimated thickness of 33 m. Track SM93-09 is estimated to have gas hydrate in the range CDP 6525-9963 at a depth of 2320 m with an estimated thickness of 23 m. And track SM93-11 is estimated to have BSR in the range CDP 5664-8732 at a depth of 2280 m with an estimated thickness of 30 m. The five trajectories show a good potential distribution of gas hydrate in the South Makassar Strait Basin.

Acknowledgment

Sincere appreciation and gratitude to the honorable Head of the Marine Geological Institute for supervising and supporting the authors. Special appreciation to Pusdatin - Ministry of Energy and Mineral Resources for providing us seismic data and well log.

References

[1] Kvenvolden K A 1993 Gas hydrates-geological perspective and global change *Reviews of Geophysics* vol 31 pp 173-187

[2] Wang X and Pan D 2017 Application of AVO attribute inversion technology to gas hydrate identification in the Shenhu area, South China Sea *Marine and Petroleum Geology* vol 80 pp 23-31

[3] Wijaya P H and Kusnida D 2009 Tinjauan geoteknik Selat Makassar Utara, implikasinya terhadap potensi hidrokarbon laut dalam Cekungan Kutai Kalimantan Timur *Jurnal Geologi Kelautan* vol 7 pp 109-121

[4] Hall R 2009 Hydrocarbon basins in SE Asia: Understanding why they are there *Petroleum Geoscience* vol 15 pp 131-146

[5] Suess E, Torres M E, Bohrmann G, Collier R W, Greinert J, Linke P, Rehder G, Trehu A, Wallmann K, Winckler G and Zuleger E 1999 Gas hydrate destabilization: enhanced dewatering, benthic material turnover and large methane plumes at the Cascadia convergent margin *Earth and Planetary Science Letters* vol 170 pp 1-15

[6] Triarso E and Troa R A 2016 Indikasi keberadaan gas hidrat pada cekungan busur muka Simeulue dan potensinya sebagai sumber energi masa depan Jurnal Kelautan Nasional vol 11 pp 127-140

[7] Sloan E D and Koh C A 2007 *Clathrate hydrate of natural gases* Boca Raton CRC Press 752 p

[8] Tucholke B E, Bryan G M and Ewing J I 1977 Gas-hydrate horizons detected in seismic-profiler data from the Western North Atlantic *American Association of Petroleum Geologists Bulletin* vol 61 pp 698–707

[9] Rogers R 2015 *Deep ocean sediment–hydrate relationships* In: *Offshore gas hydrates: Origins, development and production* Gulf Professional Publishing 381 p

[10] Ashi J, Tokuyama H and Taira A 2002 Distribution of methane hydrate BSRs and its implication for the prism growth in the Nankai Trough *Marine Geology* vol 187 pp 177- 91

[11] Ustiawan A B, Nainggolan T B and Setyawan R 2019 Interpretasi struktur geologi di Perairan Aru Selatan, Maluku berdasarkan data seismik 2D multi-channel *Jurnal Geosains dan Teknologi* vol 2 pp 53-60

[12] Sidiq A P, Manik H M and Nainggolan T B 2019 Studi komparasi metode migrasi seismik dalam mengkarakterisasi reservoir migas di Blok Kangean, Laut Bali menggunakan inversi impedansi akustik berbasis model *Jurnal Ilmu dan Teknologi Kelautan Tropis* vol 11 pp 205-219

Carbon Emissions Prediction from Land Change Model in Bandung Regency, Indonesia

I Indrarini and D N Handiani
[1]Geodetic Engineering, Faculty of Civil Engineering and Planning, Institut Teknologi Nasional Bandung

indrariniira@gmail.com

Abstract. Carbon emissions in Bandung Regency is potentially increasing and can harm future environment in the area. Our study aims to predict carbon emissions in Bandung Regency based on land changes model of the area. We utilized Cellular Automata Markov method which predicted land use in the years of 2027 and 2036. This land use prediction used overlay intersect method and were analysed using two scenarios: backward and forward looking. The backward looking scenario analysis combined the land used map of 2019 and two land use maps prediction of 2027 and 2036. Meanwhile, the forward looking scenario analysis combined the land used map of 2019 and the urban planning plan of Bandung Regency in the years of 2016 to 2036. Carbon emission were calculate using stock difference method for both scenarios. Afterwards, the carbon emission results were classified using quartile method. The result in the backward scenario showed that the carbon emission is dominated compare to sequestration, with total emission of 811,330.81 tonnes of CO_2^{eq} for the years of 2019 to 2027 and 998,288.22 tonnes of CO_2^{eq} for the years of 2019 to 2036. In the forward looking scenario, carbon sequestration is dominated compare to emission, with highest sequestration around -804,125.22 tonnes of CO_2^{eq}, while highest emission is around 245,506.51 tonnes CO_2^{eq}. Overall, both simulation showed that the highest emission occurs in Cilengkrang-Cimenyan development area, and the carbon sequestration occurs in Baleendah, Cileunyi-Rancaekek, and Margahayu-Margaasih development areas.

Keywords: Carbon Emissions, Land Change, Spatial Modeling, Bandung Regency

1. Introduction

Global warming has been correlated with the occurrence of natural disasters in several areas in Indonesia, such as floods, landslides, droughts, and several ecological disasters that harm human livelihoods [1]. It has been assumed, that anthropogenic activities are one of the contributor for the increase of global warming through high release of greenhouse gases into the atmosphere and reduced CO_2 absorption by green areas. CO_2 gas is the largest contributor based on data released by the Intergovernmental Panel on Climate Change (IPCC) and accounts for about 76 percent of total greenhouse gas emissions [2-3]. Balanced amount of CO_2 in the earth's atmosphere is needed to provide a warm effect on earth. However, if the amount is excessive, it will result in an imbalance between the energy received and released. This will affect the carbon cycle and has the potential to cause temperature

changes [4]. Human activities that cause land changes can resulted in the reduction of green areas that serve as CO_2 absorber which can then further increase the temperature in the area [5].
The effects of global warming can be felt in local areas, especially in urban areas with high population density and activity. One of them is in Bandung Regency. This Regency observes annual population increase. For example, in 2016 the population of Bandung Regency was 3,596,623 people [6]; and increased to 3,666,156 people in 2021 with an area of 1,762.40 km^2. So that in 5 years there was an increase of 39.45 people/km^2 [7]. The high population growth in Bandung Regency has resulted in an increasing need for housing which further trigger land changes. Green areas such as agriculture and plantations are often converted into residential areas. Continuity of this land change will have the potential for further increase of carbon emission in Bandung Regency and might cause negative impact on the environment, so proper mitigation to reduce carbon emissions is crucial. One way to identify proper mitigation approaches for reducing carbon emission in Bandung Regency is by predicting carbon emissions based on land change.
This study aims to predict carbon emissions based on a spatial model of land change in Bandung Regency using backward and forward looking approaches. The model is dynamic in time so it can simulate or predict future conditions [8]. Using this approach, we will be able to see the future trend of increasing emission values and identify areas that needs to be prioritized in carbon emission reduction actions.

2. Material and Method

2.1. Study Area

The location of this research is in Bandung Regency which is located at coordinates 107^{0}22'–108^{0}50' East Longitude and 60^{0}41'–70^{0}19' South Latitude. The total area of Bandung Regency is 1,762.4 km^2 consisting of 31 districts.

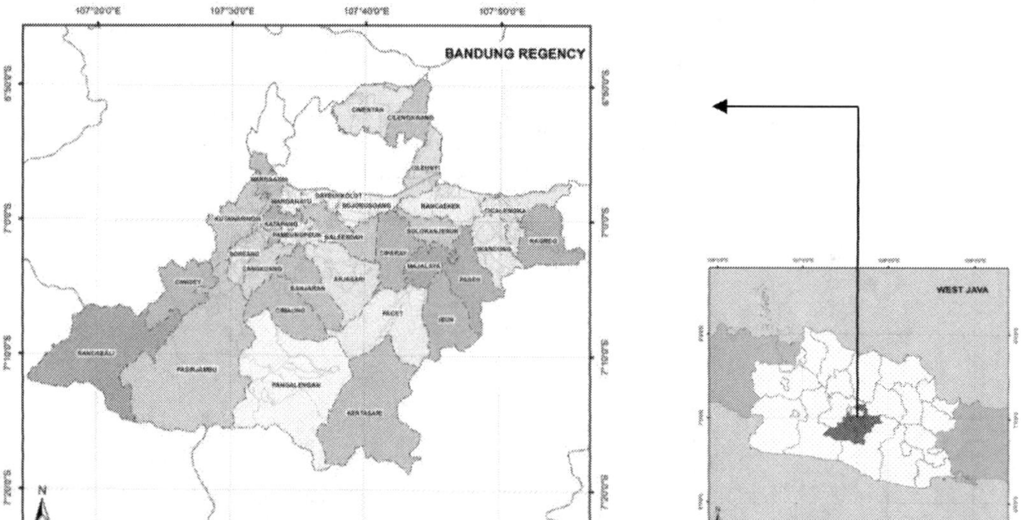

Figure 1. Study area in Bandung Regency

2.2. Data

The data used in this study are 1) three-time series of land use in Bandung Regency in 2004, 2011, and 2019; 2) the urban planning plan of Bandung Regency in the years of 2016 to 2036. Both data are in 1:50.000 scale shapefile format obtained from the Public Works and Spatial Planning Office.

2.3. Methodology

We performed two scenarios in this study: backward looking scenario and forward looking scenario, to see possible future carbon estimates.

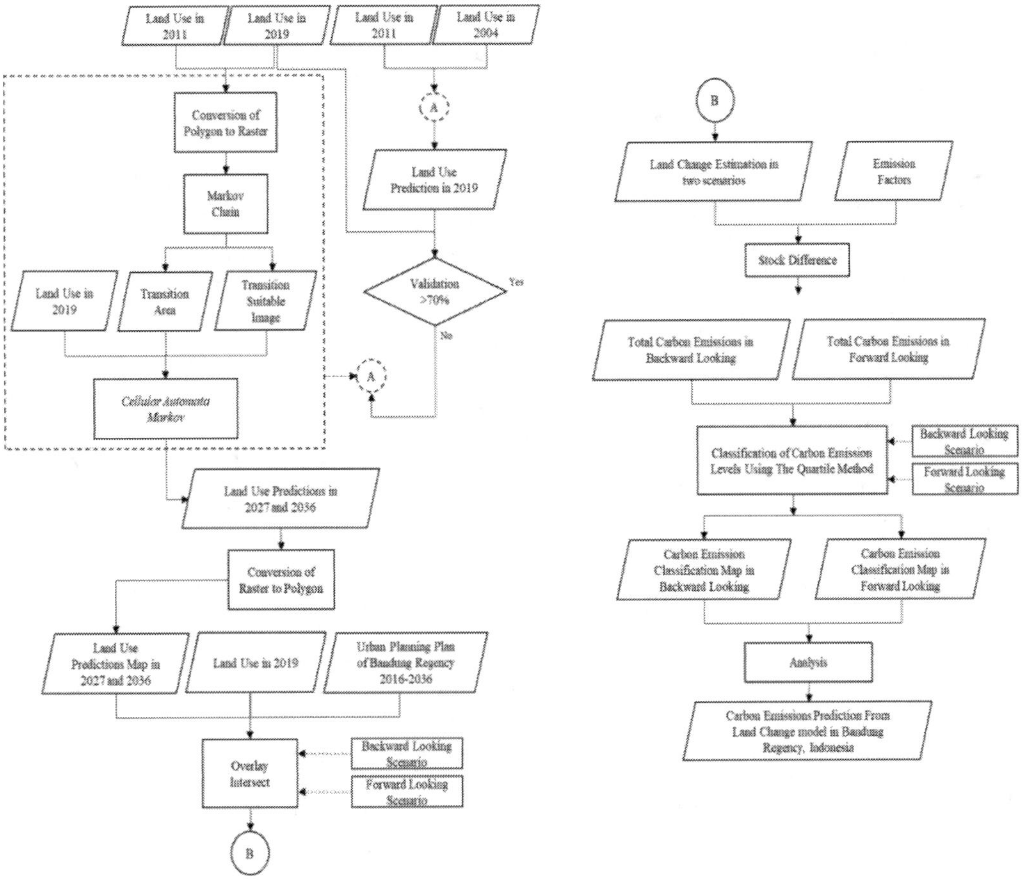

Figure 2. Methodology

The steps are: land use prediction, land change estimation, carbon emission calculation, and analysis of carbon emission estimation based on land change.

2.3.1. Land use prediction. Prediction of land use is carried out using data from the year 2019; and the time lapse data between the prediction year and the base years: 2004 and 2011. The data processing used Idrisi Selva software with the Markov Chain method, Cellular Automata Markov, and validation tests. Markov Chain analysis a pair of land maps in two-time series and produces a transition probability matrix, a transition area matrix, and a transition suitable image. The transition probability matrix is a matrix that records the probability that each land class will change to every other class. The transition area matrix is a matrix that records the number of pixels that are expected to change from each class during a specified time [9]. The next process is to run the Cellular Automata Markov method to get land use predictions in 2027 and 2036 with input data for land use based on the year and two data from the previous Markov Chain. After obtaining the model, a validation test was carried out between the existing land use in 2019 and the results of the 2019 prediction model to find out how accurate the predictions were.

2.3.2. Land change estimation. Land changes were processed by the intersect overlay method using ArcGIS software. Data processing is carried out in two scenarios by overlapping two land use map data.

The backward looking scenario combines the land used map of 2019 land use with the predicted years 2027 and 2036; while the forward looking scenario combines the land used map of 2019 land use with the urban planning plan of Bandung Regency in the years of 2016 to 2036. After overlapping the two scenarios, the attribute data for land change is generated.

2.3.3. Carbon emission calculation. There are several factors in the calculation of carbon emissions, some of the previous studies carried out: calculating carbon emissions from Municipal Solid Waste (MSW) using the method recommended by IPCC (2006) and divided into four parts according to the four treatment modes (sanitation landfill, simple landfill) [10]; Calculation of carbon emissions using the Logarithmic Mean Division Index (LMDI) model is used to describe the main factors which include the proportion of direct energy, the unit value of energy consumption, value creation effects, indirect carbon intensity, and output scale effects [11]; and identification of traces of carbon emissions based on household domestic use results using primary and secondary carbon emission calculations each year and the calculation of the Specific Emission Factor (SEF) for each household [12].
This study, we calculated carbon emissions from land use change based on the Technical Guidelines Calculation of Baseline Emissions and Removals of Greenhouse Gases in the Land-Based Sector: Book I Scientific Basis [13], emissions are calculated following the IPCC Guidelines for National Greenhouse Gas Inventories 2006. Land that has changed over time is calculated based on the difference in carbon stocks between the initial year and the following years. Two approaches can be used: (1) Stock difference and (2) Gain and loss [14].

2.4. Analysis of carbon emission estimation based on land change. To calculate the carbon emissions, we used the stock difference method for both scenarios. The stock difference is a calculation method by estimating the difference in carbon stocks at certain time intervals. Land use that does not change within a certain period of time consider to have no emissions or zero emissions, while land that has changed will emit a certain amount of carbon. The application of the calculation of carbon emissions are shown in Figure 3 where we multiply activity data in the form of area of land change and emission factors in the form of carbon stock values [14]. The emission factor for each class uses national [13], sub-national emission factors [15], previous research [16], and assumptions that match the characteristics of other classes. The result of the calculation of carbon emissions based on land change is the result of the reduction between the initial land change carbon emissions and the next year's land change carbon emissions.

Figure 3. Calculation of carbon emissions
Source: IPCC (2006)

The analysis is based on how much carbon emissions are released for each land change. So that carbon emissions on the initial land use are reduced by carbon emissions from next year's land use. So, it can be concluded that if the reduction results in a negative emission number (-), sequestration will occur; positive (+) then emission occurs; and (0) which means non emission. Then to get the classification of carbon emission levels from the results of previous calculations using the quartile method. Classification using the quartile method uses the equation as shown in Equation (1). Classification of carbon emission levels will produce five levels: very low, low, medium, high, and very high.

$$Q_i = \frac{i(n+1)}{4} \quad (1)$$

Description:
Qi : the value of i quartile; i : quartile type; and n : amount of data

3. Result and Discussion

The predictions of land use in 2027 and 2036 were based on simulations based on land changes occurred in 2011 and 2019. The prediction results for 2027 and 2036 can be used if the validation results meet the requirements. Validation was carried out between the existing year 2019 and the predicted year 2019 using the Cellular Automata-Markov method which were applied for predictions in 2027 and 2036. Prediction 2019 (Figure 4) is a simulation of land change in 2004 and 2011. The validation process in Idrisi Selva software obtained a Kappa value (Kstandard) of 0.87 or 87% which met the requirements greater than 70% [17]. This value is included in the standard agreement percentage range of >75% which means the model is classified to be "very good" [18]. Thus, the results of land use predictions in 2027 (Figure 5) and 2036 (Figure 6) can be used for the next process of land changes that occurred.

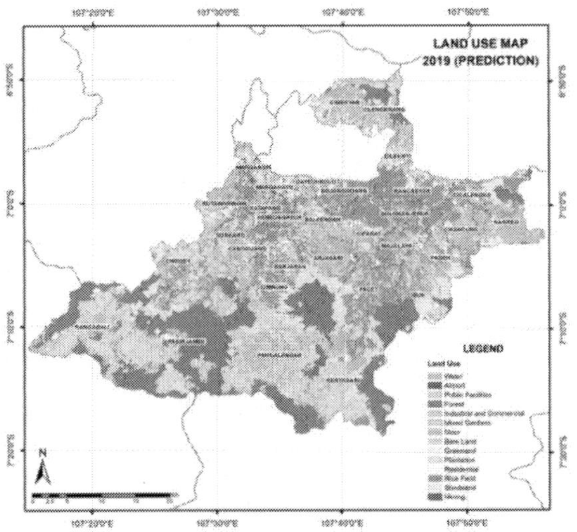

Figure 4. Bandung Regency Land Use Prediction Map in 2019

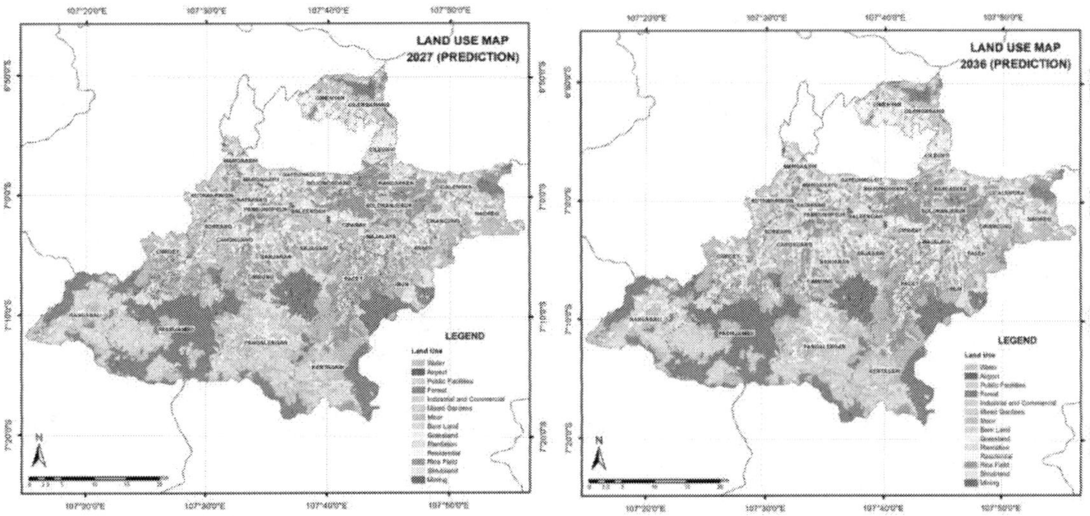

Figure 5. Bandung Regency land use prediction map in 2027 & 2036

3.1. *Total carbon emissions in backward looking scenario*

For the backward scenario, land change is calculated from 2019 to 2027 and 2019 to 2036. Bandung Regency experienced changes in land use, either increase of land area, or land area decrease between 2019 to the predicted years of 2027 and 2036. Land use classification is divided into 14 classes, namely water (BA), airport (BDR), public facilities (FS), forest (HT), industrial and commercial (IP), mixed gardens (KC), moor (LD), bare land (LT), grassland (PR), plantation (KB), residential (PM), rice field (SW), shrubland (SB), and mining (TB).

Land change from 2019 to 2027 (Table 1) emitted 811,330.81 tonnes of CO_2^{eq} (Table 2). The largest carbon emissions (marked in yellow in Tables 1 and 2) occurred due to the conversion of forest land to moor area of 4,157.01 ha with emissions of 312,815.00 tonnes of CO_2^{eq}. Meanwhile, the largest sequestration (marked with green highlights in Tables 1 and 2) was caused by the change of rice fields into residential covering an area of 7,862.82 ha with a sequestration value of -15,725.63 tonnes of CO_2^{eq}.

Table 1. Area of land change in Bandung Regency of 2019-2027 (ha)

		2027														
		BA	BDR	FS	HT	IP	KC	LD	LT	PR	KB	PM	SW	SB	TB	Area of 2019
	BA	1.086,30														1.086,30
	BDR		77,19	1,33			0,09		0,27			0,25	0,06			79,19
	FS			4.135,19												4.135,19
	HT	1,64		4,20	30.783,05	0,37	1.862,68	**4.157,01**	92,86	0,02	57,20	17,04	258,55	73,14	0,01	37.307,77
	IP	3,12		8,22	0,36	1.952,64	2,01	3,06	2,21		4,09	6,94	3,66	1,49	4,62	1.992,42
	KC	66,92	0,10	8,95	0,22	44,36	6.499,47	84,00	44,37	0,01	4,49	2.252,86	124,09	27,70	33,77	9.191,30
2019	LD	21,66		16,95	1,29	37,25	76,01	19.344,03	545,04	0,45	131,42	4.039,43	177,33	3,62	2,30	24.396,78
	LT	9,21		3,47	2,95	3,91	3,40	2,87	714,27		32,12	147,87	14,31	1,36	75,77	1.011,51
	PR	1,18	5,18	0,26			1,68	1,66	4,55	35,76		27,87	0,03	1,09	0,01	79,26
	KB	107,66		33,95	52,35	3,43	797,86	308,22	2,84		18.611,42	1.887,00	53,11	395,25	2,72	22.255,81
	PM	137,11	0,48	592,38	9,21	1.578,84	53,37	69,50	14,15	0,01	35,40	22.937,94	237,89	19,24	4,19	25.689,70
	SW	117,00		32,50	4,71	696,64	87,04	143,22	307,27		31,36	**7.862,82**	25.167,61	57,55	2,65	34.510,35
	SB	28,30		14,95	68,94	5,67	68,60	5.178,36	15,55	0,09	30,34	587,94	198,00	8.179,65	0,10	14.376,48
	TB	9,48		0,04			1,36	5,20				1,32	0,45		107,80	125,65
	Area of 2027	1.589,58	82,95	4.852,38	30.923,08	4.323,11	9.453,57	29.297,13	1.743,37	36,34	18.937,84	39.769,28	26.235,09	8.760,07	233,93	176.237,70

Table 2. Total carbon emissions in 2027 (ton CO_2^{eq})

		2027														
		BA	BDR	FS	HT	IP	KC	LD	LT	PR	KB	PM	SW	SB	TB	Total
	BA	0														0
	BDR		0	-5,30			-2,55		-0,68			-1,02	-0,11			-9,67
	FS			0				-0,00			-0,00					0
	HT	140,04		340,96	0	30,23	102.913,25	**312.815,00**	7.683,72	1,74	1.272,76	1.384,33	21.524,01	4.041,00	0,95	452.148,00
	IP	12,50		0	-29,43	0	-52,31	-18,38	3,31		-241,14	0	7,32	-38,62	18,47	-338,29
	KC	2.007,51	3,03	232,68	-12,00	1.153,39	0	1.680,05	1.220,17	0,26	-148,10	58.574,35	3.474,47	0	1.013,00	69.198,81
2019	LD	216,64		101,69	-97,27	223,51	-1.520,14	0	4.087,82	2,68	-6.965,37	24.236,60	1.418,64	-72,37	22,95	21.655,38
	LT	23,03		-5,21	-243,89	-5,87	-93,35	-21,56	0		-1.943,51	-221,81	7,15	-37,37	189,43	-2.352,94
	PR	4,72	20,71	0			-43,54	-9,95	6,82	0		0	0,06	-28,32	0,05	-49,45
	KB	6.782,70		2.003,22	-1.164,83	202,13	26.329,48	16.335,44	171,72		0	111.333,20	3.239,97	13.043,19	171,12	178.447,32
	PM	548,42	1,91	0	-748,23	0	-1.387,63	-416,98	21,23	0	-2.088,45	0	475,78	-500,18	16,77	-4.077,35
	SW	234,00		-64,99	-391,65	-1.393,28	-2.437,23	-1.145,75	-153,63		-1.912,88	**-15.725,63**	0	-1.611,30	5,31	-24.597,04
	SB	848,84		388,64	-3.809,10	147,54	0	103.567,18	427,55	2,37	-1.001,16	15.286,35	5.544,11	0	2,99	121.405,32
	TB	0		-0,15			-40,90	-52,04				-5,29	-0,90		0	-99,28
	Total	10.818,40	25,65	2.991,54	-6.496,39	357,65	123.665,09	432.733,01	13.468,03	7,06	-13.027,85	194.861,08	35.690,50	14.796,02	1.441,03	811.330,81

Land change from 2019 to 2036 (Table 3) emitted 998,288.22 tonnes of CO_2^{eq} (Table 4). The largest carbon emissions (marked in yellow in Tables 3 and 4) occurred due to the conversion of forest land to moor area of 4,618.67 ha with emissions of 347,555.19 tonnes of CO_2^{eq}. Meanwhile, the largest sequestration (marked with green highlights in Tables 3 and 4) was caused by the change of rice fields into residential areas covering 11,146.81 ha with a sequestration value of -22,293.61 tonnes of CO_2^{eq}.

Table 3. Area of land change in Bandung Regency of 2019-2036 (ha)

		2036														
		BA	BDR	FS	HT	IP	KC	LD	LT	PR	KB	PM	SW	SB	TB	Area of 2019
	BA	1.086,30														1.086,30
	BDR		75,02	1,90			0,08	0,01	1,41			0,52	0,26			79,19
	FS			4.135,19								-				4.135,19
	HT	1,72		4,55	30.112,46	1,44	2.007,17	**4.618,67**	121,65		57,45	24,66	300,98	57,01	0,01	37.307,77
	IP	3,38		7,43	0,45	1.950,94	1,04	3,28	1,44		1,48	11,72	4,15	0,53	6,60	1.992,42
	KC	92,48	0,06	10,53	0,43	86,26	5.358,60	72,81	33,03		6,01	3.364,16	109,68	9,90	47,36	9.191,30
2019	LD	90,55	0,01	18,83	5,81	162,62	56,74	17.191,50	434,87	0,28	134,66	6.139,25	127,09	1,64	32,95	24.396,78
	LT	11,90		4,33	1,97	5,01	3,30	6,73	566,98		95,50	238,86	6,61	0,01	70,32	1.011,51
	PR	3,25	7,01	0,20			2,16	1,15	0,06	24,49		40,40		0,52	0,01	79,26
	KB	137,61		100,40	56,60	62,46	828,12	265,23	1,46		17.946,75	2.500,36	45,52	309,58	1,73	22.255,81
	PM	105,53	0,43	546,43	9,22	1.693,49	25,96	45,64	9,09	0,14	64,69	23.024,96	156,57	3,30	4,26	25.689,70
	SW	204,18		172,56	2,19	1.027,86	63,69	130,60	301,93		30,30	**11.146,81**	21.398,61	19,22	12,42	34.510,35
	SB	66,42		16,21	61,91	20,68	70,96	7.187,02	88,25	0,06	38,83	1.344,09	220,97	5.260,95	0,12	14.376,48
	TB	3,09		0,04		0,03	1,22	0,11				1,14	0,15		119,87	125,65
	Area of 2036	1.806,39	82,52	5.018,59	30.251,03	5.010,79	8.419,03	29.522,73	1.560,17	24,97	18.375,68	47.836,92	22.370,59	5.662,65	295,65	176.237,70

Table 4. Total carbon emissions in 2036 (ton of CO_2^{eq})

		2036														
		BA	BDR	FS	HT	IP	KC	LD	LT	PR	KB	PM	SW	SB	TB	Total
	BA	0		0												0
	BDR		0	-7,61			-2,49	-0,04	-3,52			-2,07	-0,51			-16,24
	FS			0				-0,00			0	0				0
	HT	146,35		369,26	0	116,95	110.896,05	**347.555,19**	10.066,40		1.278,34	2.003,48	25.056,81	3.149,84	0,95	500.639,64
	IP	13,50		0	-36,19	0	-27,08	-19,66	2,16		-87,12	0	8,31	-13,74	26,40	-133,42
	KC	2.774,40	1,86	273,78	-23,56	2.242,79	0	1.456,25	908,37		-198,24	87.468,11	3.070,89	0	1.420,73	99.395,39
2019	LD	905,47	0,07	112,96	-437,20	975,72	-1.134,73	0	3.261,51	1,70	-7.136,97	36.835,52	1.016,70	-32,70	329,51	34.697,56
	LT	29,75		-6,49	-162,58	-7,51	-90,65	-50,44	0		-5.777,93	-358,29	3,31	-0,30	175,80	-6.245,34
	PR	13,00	28,03	0			-56,21	-6,90	0,09	0		0		-13,62	0,05	-35,58
	KB	8.669,53		5.923,63	-1.259,33	3.685,30	27.327,95	14.056,97	88,24		0	147.521,28	2.776,56	10.216,03	108,70	219.114,86
	PM	422,10	1,71	0	-748,72	-	-674,91	-273,82	13,64	0	-3.816,76	0	313,14	-85,76	17,04	-4.832,34
	SW	408,35		-345,12	-182,44	-2.055,71	-1.783,29	-1.044,81	-150,96		-1.848,41	**-22.293,61**	0	-538,01	24,83	-29.809,20
	SB	1.992,63		421,48	-3.420,66	537,68	0	143.740,33	2.426,97	1,47	-1.281,48	34.946,28	6.187,21	0	3,68	185.555,59
	TB	0		-0,15		-0,14	-36,47	-1,06				-4,55	-0,31		0	-42,68
	Total	15.375,08	31,66	6.741,73	-6.270,69	5.495,08	134.418,16	505.412,01	16.612,90	3,17	-18.868,57	286.116,15	38.432,11	12.681,73	2.107,70	998.288,22

3.2. Total Carbon Emissions in Forward Looking Scenario

Land changes calculations comparing 2019 prediction map to the urban planning plan of Bandung Regency in the years of 2016 to 2036 (Table 5) tend to result in sequestration compared to emissions, as much as -804,125.22 tonnes of CO_2^{eq} (Table 6). The largest sequestration (marked by highlights in Table 5 and Table 6) occurred due to land conversion from shrubland to protected forest covering an area of 7,927.10 ha with a sequestration of -334,048.16 tonnes of CO_2^{eq}. The shrub is in a protected zone of the urban planning plan in Bandung regency. In spatial planning, land conversion in protected zones is limited. The shrub land in the protected zones probably originated from forest. Their fuction had changed due to illegal logging activities. The absence of reforestation efforts has resulted overgrown shrubs in the former logging area, and it takes a long time within natural process to return into forest. Based on the spatial policy, shrubs that are inside the protected zone will be reallocated to forest [19]. Based on Government Regulation of The Republic of Indonesia Number 24 (2010), it is about the use of forest area. It explains that the reforestation is an effort to plant type of forest trees species in the areas where forests are ruin. Hence, it restores their function again. These areas are bare land, reeds, or shrubs [20].

Meanwhile, the largest emission (marked with yellow highlights in Table 5 and Table 6) occurred due to the conversion of forest land to protected forest covering an area of 18,726.66 ha with emissions of 245,506.51 tonnes of CO_2^{eq}. We use a regulation from the Ministry of Environment which states that the emission factor for forest land cover types on a regional scale (Java Island) has higher secondary forest carbon stocks than primary forest carbon stocks [15].

Table 5. Area of land change in Bandung Regency of 2019 - urban planning plan (ha)

		Urban Planning Plan of Bandung Regency																			
		CA	HL	HP	HPT	HR	FS	K	PM	P-1	P-2	HK	KPI	IK	P-4	RTH	KPR	TB	THR	TWA	Area of 2019
2019	BA																1.086,30				1.086,30
	BDR							0,14	0,25			78,79									79,18
	FS						4.135,19														4.135,19
	HT	13.287,31	**18.726,66**	162,03	23,40	3.007,56		1,04	25,37	82,48	240,35	143,76	0			1,24	1,35	986,47	191,81	426,95	37.307,77
	IP	4,23	23,26	0,26		51,98	20,71	90,33	185,19	92,60	10,03	2,22	1.468,79	1,17		39,49	0,88			1,28	1.992,42
	KC	35,87	765,46	1.527,58		2.687,46	31,60	136,48	1.644,56	1.302,23	773,25	16,30	81,06	3,89	4,58	165,04	11,51	0,01	1,12	3,30	9.191,30
	LD	453,38	5.075,73	1.611,62		5.810,97	111,84	370,71	1.699,80	3.938,42	5.084,16	34,42	26,50	17,22	38,27	100,64	5,27	6,13	8,81	2,88	24.396,78
	LT	7,24	50,11	54,03		229,91	0,71	50,60	411,28	79,03	42,10	5,70	50,24	0,49	1,88	21,39	0,37			6,46	1.011,51
	PR								5,60	0,04		13,67				59,95					79,26
	KB	115,95	878,29	234,49	30,19	17.749,18		87,21	374,72	1.680,33	1.009,42	25,38	1,94	2,15	18,41	10,40	11,61		21,82	4,34	22.255,81
	PM	13,83	74,74	128,93	2,45	1.223,42	116,58	1.255,85	17.987,95	2.703,56	366,76	155,20	1.110,41	52,12	11,87	458,33	15,77	0,16	2,43	9,35	25.689,70
	SW	17,37	509,25	153,41		1.336,61	140,67	965,27	9.261,81	19.520,51	283,38	36,69	1.471,33	398,60		376,54	32,62	6,22	0,09		34.510,35
	SB	1.166,57	7.927,10	372,67	0	2.764,73	6,42	32,24	455,53	1.065,40	384,83	0,12	25,15	6,54		80,82	6,51	42,71	1,32	37,82	14.376,48
	TB			40,70		8,45		2,32	40,47	9,07	22,70		1,93			0,01					125,65
	Area of Plan	15.101,75	34.030,60	4.285,72	56,03	34.870,27	4.563,72	2.992,19	32.092,53	30.473,67	8.216,97	512,25	4.237,34	482,16	75,01	1.313,84	1.172,18	1.041,70	227,40	492,37	176.237,70

Table 6. Total carbon emissions in urban planning plan (ton of CO_2^{eq})

		Urban Planning Plan of Bandung Regency																			
		CA	HL	HP	HPT	HR	FS	K	PM	P-1	P-2	HK	KPI	IK	P-4	RTH	KPR	TB	THR	TWA	Total
2019	BA																0				0
	BDR							- 0,57	- 1,01			- 315,16									- 316,75
	FS						0														0
	HT	174.196,69	**245.506,51**	1.629,97	235,43	66.918,24		84,38	2.061,02	6.866,38	18.086,51	11.680,14	0			102,46	114,94	2.959,40	575,44	1.280,86	532.298,35
	IP	- 288,29	- 1.585,21	- 18,80		- 3.066,60	0	0	0	185,21	- 60,17	0	0	4,67		59,23	3,53			- 99,75	- 4.866,17
	KC	- 1.511,66	- 32.256,58	- 69.031,54		- 88.686,25	821,59	3.548,48	42.758,65	36.462,34	15.465,07	423,80	2.107,45	116,55	119,20	4.538,52	345,32	- 0,35	- 58,44	- 172,21	- 85.010,06
	LD	- 28.172,82	-315.405,90	-105.061,62		-307.981,30	671,05	2.224,28	10.198,81	31.507,38	0	206,54	159,01	172,21	229,64	754,76	52,65	- 443,17	-636,81	- 207,77	-711.733,07
	LT	- 504,17	- 3.489,65	- 3.927,11		- 13.909,54	- 1,06	- 75,90	- 616,93	39,52	- 315,71	- 8,56	- 75,35	1,21	- 2,82	0	0,91			- 514,76	- 23.399,91
	PR								0	0,08		0				89,92					90,00
	KB	- 1.059,75	- 8.027,55	- 2.858,46	-367,95	0		5.145,54	22.108,57	102.499,81	53.499,17	1.497,41	114,31	135,72	1.085,91	628,87	731,29		-419,98	- 83,58	174.629,33
	PM	- 942,38	- 5.092,50	- 9.178,53	-174,03	- 72.181,77	0	0	0	5.407,11	- 2.200,55	0	0	208,48	0	687,50	63,08	- 12,42	-190,28	- 731,72	- 84.338,00
	SW	- 1.218,43	- 35.718,75	- 11.227,99		- 81.532,91	- 281,34	-1.930,54	-18.523,62	0	- 2.267,02	- 73,37	-2.942,66	797,19		- 188,27	65,23	- 499,15	- 7,27		-155.548,89
	SB	- 49.159,18	-334.048,16	- 16.840,98	0	- 91.236,22	167,03	838,28	11.843,66	29.831,11	7.696,54	3,19	653,85	196,10		2.222,54	195,26	-2.231,75	- 68,81	-1.976,25	-441.913,80
	TB			- 3.059,92		- 532,30		- 9,28	- 161,89	- 18,15	- 226,98		- 7,71			- 0,03					- 4.016,25
	Total	91.340,01	-490.117,78	-219.574,99	-306,56	-592.208,64	1.377,28	9.824,69	69.667,25	212.780,78	89.676,85	13.413,99	8,90	1.632,13	1.431,92	8.895,50	1.572,22	- 227,44	-806,15	-2.505,18	-804.125,22

The classification of carbon emission levels is detailed in Table 7.

Table 7. Carbon emission level classification

Classification	Carbon Emission Value
Very Low	0.01 – 17.04
Low	17.05 – 175.80
Medium	175.81 – 848.84
High	848.85 – 5,923.63
Very High	5,923.64 – 347,555.19

The carbon emission classification map can be seen in Figure 7, Figure 8, and Figure 9. The largest total carbon emissions and very high levels in the backward looking scenario occurs in forest land that was converted to moorlands because forests have a higher carbon stock value than moors (85.25 t/ha and 10 t/ha, respectively). This indicates carbon stock decrease which will lead to further emissions.

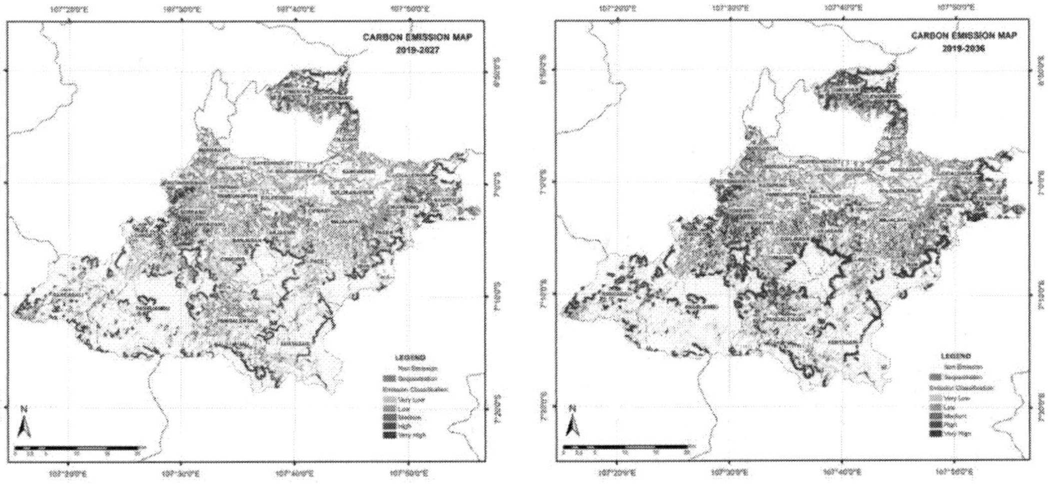

Figure 6. Carbon emission map 2019-2027 & 2019-2036

The forward looking scenario tends to result in sequestration due to land changes from low carbon stocks to high carbon stocks, with the largest sequestration occurs on shrubland with a carbon stock value of 30 t/ha to protected forest with a carbon stock value of 72.14 t/ha.

The Cilengkrang-Cimenyan area is the largest emitter in both backward and forward-looking scenarios. This is because that forest land, moors, and mixed gardens have high carbon stocks have been turned into residential areas. Meanwhile, the Baleendah, Cileunyi-Rancaekek, and Margahayu-Margaasih areas produced sequestration compared to emissions in both backward and forward-looking scenarios. This is because domination of changes rice fields to residential areas. The residential carbon stocks are higher compare to rice fields. The residential carbon stock value is 4 t/ha, and the rice fields value is 2 t/ha [13]. The higher carbon stock value could relate to the property developer's obligation to provide green areas of 30 percent of the total development area. It ensures the ecosystem balance, and the carbon cycle in the residential environment must be good [21]. Meanwhile, rice fields not only have a low carbon stock but are also considered as one of the main sources of methane emissions. However, regarding the carbon stock of residential and rice fields, further studies need to be done.

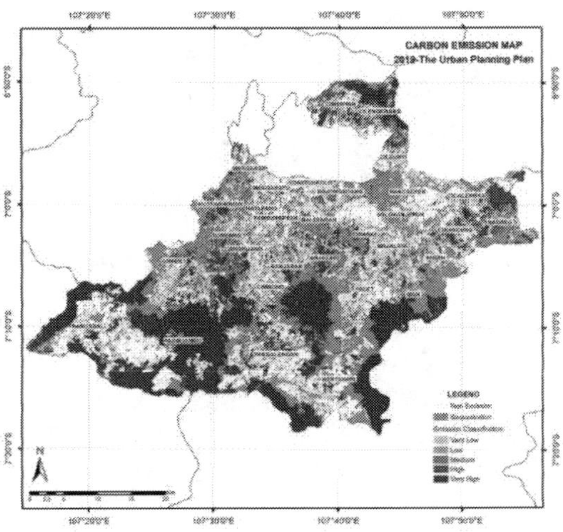

Figure 7. Carbon emission map 2019 - the urban planning plan

Overall, we found that the forward looking scenario is a good approach to predict carbon emissions resulting from land changes that occurred in the backward looking scenario. We can also conclude that the spatial pattern plan that has been prepared in the urban planning plan of Bandung Regency in the years of 2016 to 2036 is a spatial planning policy that promotes environmental conservation through low emission areas.
The results of this study can also be used as an input for stakeholders in formulating an action plan to reduce carbon emissions in the strategic areas, such as in the Cilengkrang-Cimenyan area as the largest emission contributor area. The Bandung Regency Government is also advised to monitor and act decisively on irregularities in land change and protect agricultural land.

4. Conclusion

We found that up to the backward looking scenario showed emission of 811,330.81 tonnes of CO_2^{eq} from land change in 2019 to 2027, and emissions of 998288.22 tonnes of CO_2^{eq} from land change in 2019 to 2036. Forward looking scenarios tend to result in sequestration compared to emissions of -804,125.22 tonnes of CO_2^{eq} with the largest emission 245,506.51 tonnes of $CO2^{eq}$. The Cilengkrang-Cimenyan area is the largest contributor to emissions because forest land, moor, and mixed gardens that have high carbon stocks are converted into residential, while the Baleendah, Cileunyi-Rancaekek, and Margahayu-Margaasih areas are the sequestration contributors because they are dominated by conversion of rice fields into residential areas. Forward looking scenarios can anticipate predictions of carbon emissions due to land changes that occurred in backward looking scenarios.

5. References

[1] V. Thomas and R. Lopez, "Global Increase in Climate-Related Disasters," 2015.
[2] G. Yoga, "Dampak Emisi Karbon," Aug. 17, 2021. https://www.icdx.co.id/news-detail/publication/dampak-emisi-karbon (accessed Apr. 18, 2022).
[3] C2ES, "Global Emissions," *Center for Climate and Energy Solutions*. https://www.c2es.org/content/international-emissions/ (accessed Sep. 20, 2022).
[4] P. Wacana, "Bencana Ekologi Sebagai Dampak Perubahan Iklim Global dan Upaya Peredaman Risiko Bencana | Badan Penanggulangan Bencana Daerah," *BPBD*, 2014. https://bpbd.bulelengkab.go.id/informasi/detail/artikel/bencana-ekologi-sebagai-dampak-perubahan-iklim-global-dan-upaya-peredaman-risiko-bencana-1 (accessed May 12, 2022).
[5] U. F. Kurniawati, "Dampak Perubahan Penggunaan Lahan Terhadap Besaran Stok Karbon di Kota Surabaya," *J. Penataan Ruang*, vol. 16, no. 1, pp. 54–58, 2021, doi: 10.12962/j2716179x.v16i1.8951.
[6] BPS, "Kabupaten Bandung Dalam Angka," 2017.
[7] BPS, "Kabupaten Bandung Dalam Angka," 2022.
[8] K. Munibah, S. R. . Sitorus, E. Rustiadi, H. Hardjomidjojo, and K. Gandasasmita, "Model Spasial Perubahan Penggunaan Lahan dan Arahan Penggunaan Lahan Berwawasan Lingkungan (Studi Kasus DAS Cidanau, Provinsi Banten)," *repository.ipb.ac.id*, 2008. https://repository.ipb.ac.id/handle/123456789/46847 (accessed Apr. 21, 2022).
[9] T. Takada, A. Miyamoto, and S. F. Hasegawa, "Derivation of a Yearly Transition Probability Matrix for Land-Use Dynamics and its Applications," *Landsc. Ecol.*, vol. 25, pp. 561–572, 2010.
[10] Z. Wang and L. Geng, "Carbon Emissions Calculation from Municipal Solid Wasste and Influensing Factors Analysis in China," *J. Clean. Prod.*, 2015.
[11] Q. Du, X. Lu, Y. Li, M. Wu, L. Bai, and M. Yu, "Carbon Emissions in China's Construction Industry: Calculations, Factors, and Regions," *Int. J. Environ. Res. Public Health*, 2018.
[12] O. Nursetiyani, "Identifikasi Jejak Emisi Gas Karbon Dioksida (CO2) Berdasarkan Hasil Penggunaan Domestik Rumah Tangga Pada Kawasan Permukiman di Kecamatan Cikampek Kabupaten Karawang," 2021.
[13] F. Agus *et al.*, *Pedoman Teknis Penghitungan Baseline Emisi dan Serapan Gas Rumah Kaca*

Sektor Berbasis Lahan: Buku I Landasan Ilmiah. 2013.
[14] IPCC, "2006 IPCC Guidelines for National Greenhouse Gas Inventories," S. Eggleston, L. Buendia, K. Miwa, T. Ngara, and T. Tanabe, Eds. Hayama: Institute for Global Environmental Strategies (IGES), 2006.
[15] A. Tosiani, *Buku kegiatan serapan dan emisi karbon*. 2015.
[16] M. Fahreza, "Identifikasi Alokasi Pemanfaatan Ruang yang Rendah Emisi Karbon Berbasis Lahan di Kota Palembang," 2018.
[17] K. S. Kurasaki, "Intercoder Reliability for Validating Conclusions Drawn from Open-Ended Interview Data," *Field methods*, vol. 12(3), pp. 179–194, 2000.
[18] J. L. Fleiss, B. Levin, and M. C. Paik, *Statistical Methods for Rates and Proportions*. New Jersey, USA: John Wiley & Sons, Inc., 2003.
[19] Danu, Syaifullah, Ruwedah, Saktiawan, Arham, and Naufal, *Analisis Bentang Alam dan Bentang Kehidupan Industri Veneer Berbasis Masyarakat di Hutan Rakyat Kabupaten Luwu Timur*. Sulawesi Community Foundation, 2017.
[20] *Peraturan Pemerintah Republik Indonesia Nomor 24 Tahun 2010 Tentang Penggunaan Kawasan Hutan*. 2010.
[21] Republik Indonesia, *Undang-Undang No. 26 tahun 2007 tentang Penataan Ruang*. 2007.

IOP Conf. Series: Earth and Environmental Science 1245 (2023) 012005 doi:10.1088/1755-1315/1245/1/012005

Identification of Landslides Controlling and Triggering Factors using the Analytical Hierarchy Process (AHP) in Dermasuci Village, Pangkah District, Tegal Regency, Central Java Province - Indonesia

B T Widodo[1], H Setiawan[2*] and I W Warmada[2]**

[1] Master Program in Geological Engineering, Gadjah Mada University, Indonesia
[2] Department of Geological Engineering, Gadjah Mada University, Indonesia

brenanto.tsa1999@mail.ugm.ac.id[1], hendy.setiawan@ugm.ac.id[2*]and warmada@ugm.ac.id[2**]

Abstract. The National Disaster Management Agency (BNPB) stated that in 2022 around 634 landslides occurred in Indonesia. One of the regions in Indonesia that experienced landslide disasters is Dermasuci Village, Pangkah District, Tegal Regency, Central Java Province. The first landslide disaster occurred on February 5, 2022, and the second landslide occurred on February 10, 2022. The landslide disasters caused damage to 61 residents' houses and four road sections. The objective of this study is to identify the controlling and triggering factors that mainly contributed to the landslide event. These two factors are determined based on the order of their static and dynamic properties using the analytical hierarchy process (AHP) method. Controlling and triggering factors of the landslide in the study area are determined based on geological and engineering geological conditions which were obtained from surface mapping results. The controlling factors obtained are the rock bedding inclination, geological structure in the form of faults, slope inclination, engineering properties of soil and rock, land use, and groundwater table. The triggering factors that are known and commonly occur are rainfall in the form of antecedents and intensity, and earthquakes that can induce landslides.

1. Introduction

Indonesia is one of the countries that have a high level of vulnerability to natural disasters in the world. Indonesia ranks 38th out of 181 countries with the highest natural disaster risk index[21]. As of December 31, 2022, there were 3,544 natural disaster events occurred in Indonesia; where the landslide disaster which is the main topic of this study, ranks third with a total of 634 incidents[11]. One of the areas in Indonesia that experienced landslides throughout 2022 is Dermasuci Village, Pangkah District, Tegal Regency, Central Java Province. The first landslide disaster in Dermasuci Village occurred on February 5, 2022, and the second landslide occurred on February 10, 2022. The landslide disaster caused damage to 61 residents' houses and four road sections[17]. The administrative map for the Dermasuci Village and the location points where the landslide occurred is shown in Figure 1.

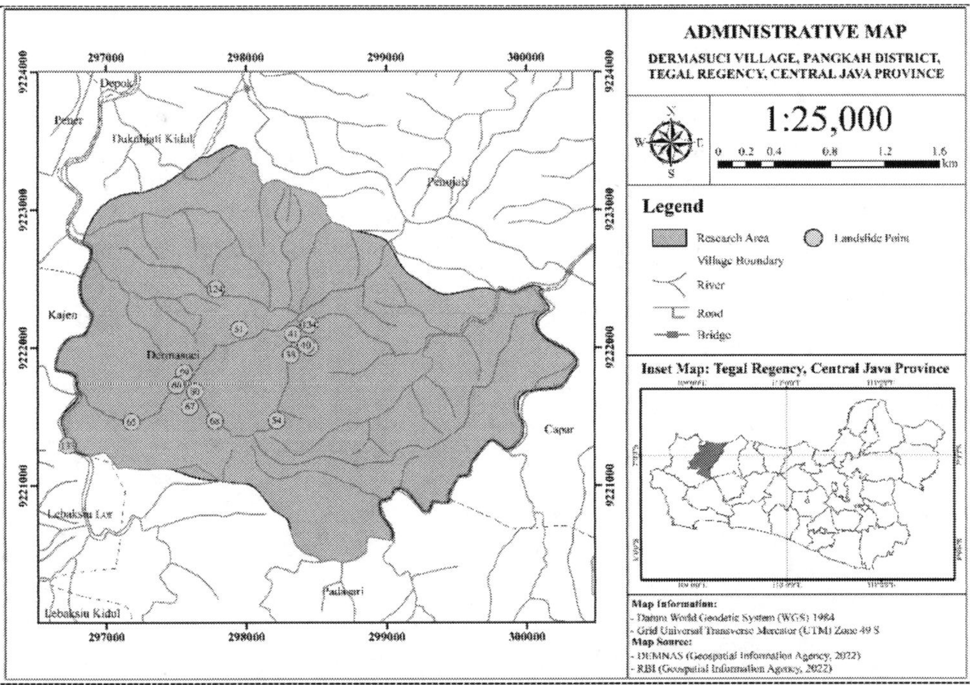

Figure 1. Administrative Map of Research Area
(Source: Geospatial Information Agency/BIG, accessed in 2022)

According to the geological literature study, there are two formation units lied in the research area[5], namely: the Rambatan Formation (Tmr) and the Halang Formation (Tmph). The Rambatan Formation (Tmr) is composed of shale, marlstone, and limed sandstones; and the Halang Formation (Tmph) is composed of andesitic sandstones, tuffaceous conglomerates, and marlstone. Based on regional physiographic conditions, the research area is situated in the North Serayu Anticlinorium zone[2], and based on regional structural geology is located in the Tegal High[16]. These two conditions cause the Dermasuci Village area to have a complex rock bedding inclination and massive geological structure, which can potentially be the factors causing a landslide.

A landslide is the downslope movement of rock or soil or both, that occurs at the ruptured surface. There are two types of rupture surfaces, namely curved rupture/rotational slide, or planar rupture/translational slide, in which much of the material often moves as a coherent or semicoherent mass with little internal deformation[19]. A landslide can occur because it is influenced by two main factors, namely: controlling factors and triggering factors. The controlling factors are factors that make the condition of a slope vulnerable and ready to move, and the triggering factors are factors that change the condition of a slope from a prone condition or ready to move to a critical condition and finally move[7].

This research was carried out with the objective of identifying the controlling and triggering factors that caused landslides in Dermasuci Village. Controlling and triggering factors are determined based on geological and engineering geological conditions obtained from surface mapping results. The controlling factors obtained from surface mapping results include the rock bedding inclination, geological structure in the form of faults, slope inclination, engineering properties of soil and rock, land use, and groundwater level. The triggering factors obtained from secondary data include rainfall in the form of antecedents and intensity, and earthquakes. All these factors will be sorted from the most static to the most dynamic characteristics using the analytical hierarchy process (AHP) method.

2. Data and method

Identification of controlling and triggering factors of landslides in Dermasuci Village was carried out using primary and secondary data. Primary data is obtained from the results of surface geological and engineering geological mapping. The results of surface geological mapping obtained data as follows: strike and dip of rock, distribution of rock types, and distribution of geological structure. Based on the results of the mapping that has been carried out, a total of 134 observation points were obtained, of which 48 observation points obtained strike and dip of rock data. There are four types of rock identified and distributed in the study area, namely marlstone (Tmr), limed sandstone (Tmr), marlstone (Tmph), and andesitic sandstone (Tmph). Geological structures that can be identified include strike-slip faults, normal faults, reverse faults, and indications of anticlines and synclines. The surface geological mapping results are presented in the form of a geological map, as shown in Figure 2.

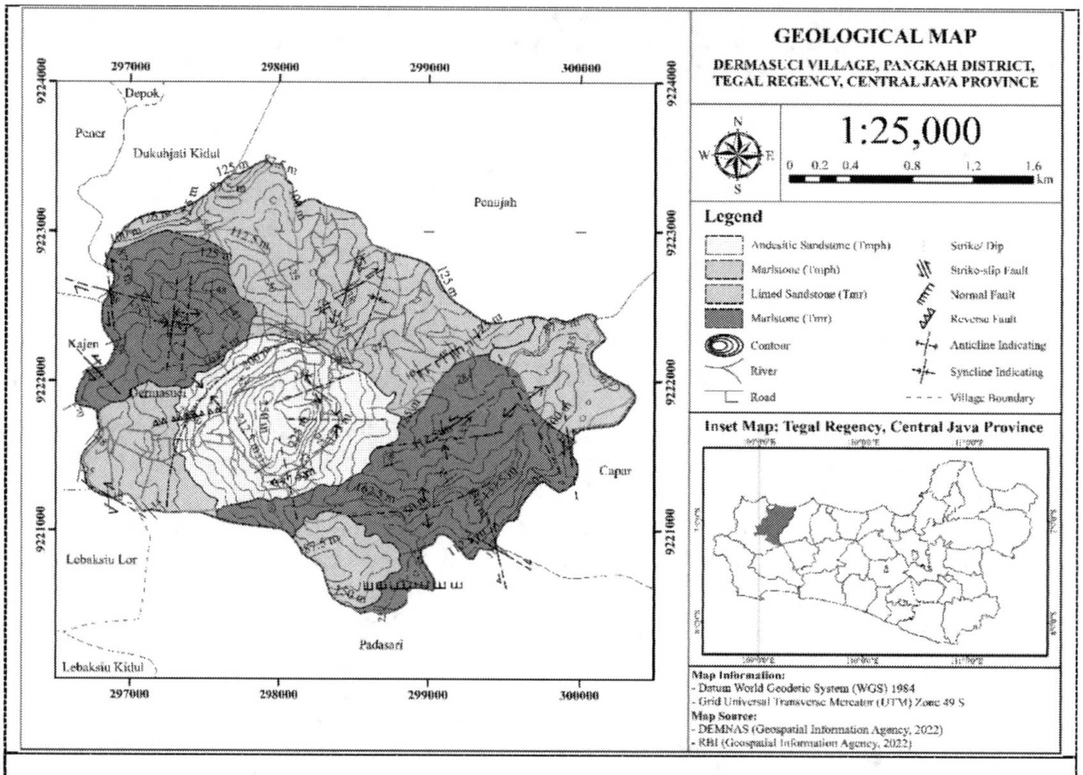

Figure 2. Geological Map of Research Area

Surface engineering geological mapping obtained data in the form of landslide distribution and direction, slope, and soil and rock engineering properties. According to the results of surface engineering geological mapping, there are 15 landslide points spread across Dermasuci Village (Figure 1). Slope data is used to determine the angle of the slope that has the potential to experience landslides, as well as those that have experienced landslides. Data on the engineering properties of soil and rock presented in this paper are limited only to those based on physical tests carried out in the field with reference to the degree of weathering of the rock mass[6]. Based on the results of the degree of rock weathering identification, there are three types of weathering degrees obtained for each rock. Limed sandstone (Tmr) is classified as fresh rock, marlstone (Tmr) and marlstone (Tmph) are classified as highly weathered rocks, and andesitic sandstone (Tmph) is classified as completely weathered rock. The distribution of rocks based on the degree of weathering and the landslide directions are presented in the form of an engineering geological map, as shown in Figure 3.

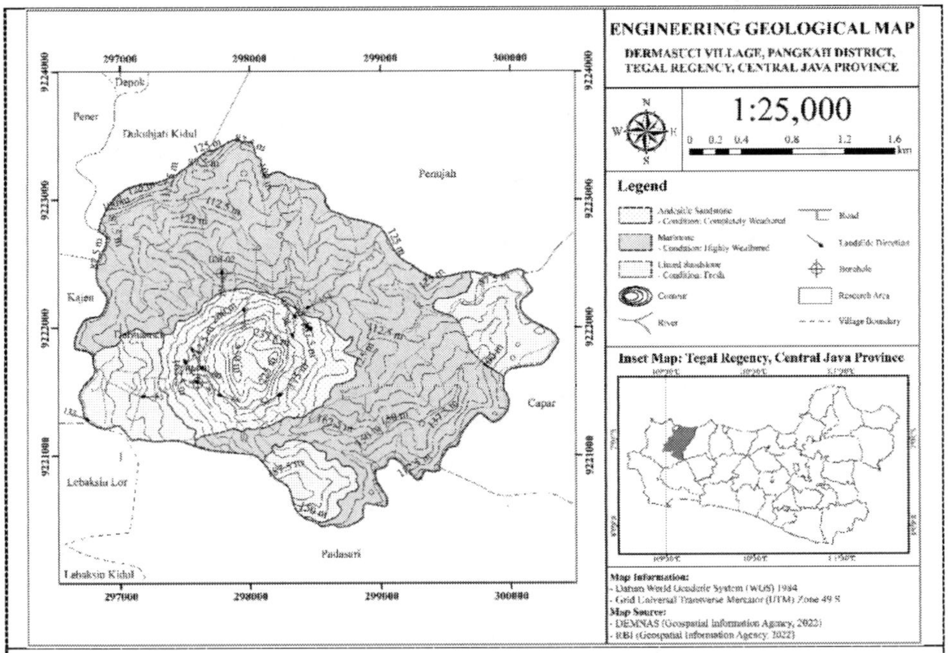

Figure 3. Engineering Geological Map of Research Area

The result of surface geological and engineering geological mapping data in the form of land use distribution combined with secondary data in the form of spatial data on the distribution of residents' houses[4] and satellite image data in the form of Landsat 8-9 OLI/TIRS X2 L2[20], it can be used to identify four types of land use, namely forest, fields, settlements, and vacant land. The land use distribution is presented in the form of a land use map, as shown in Figure 4.

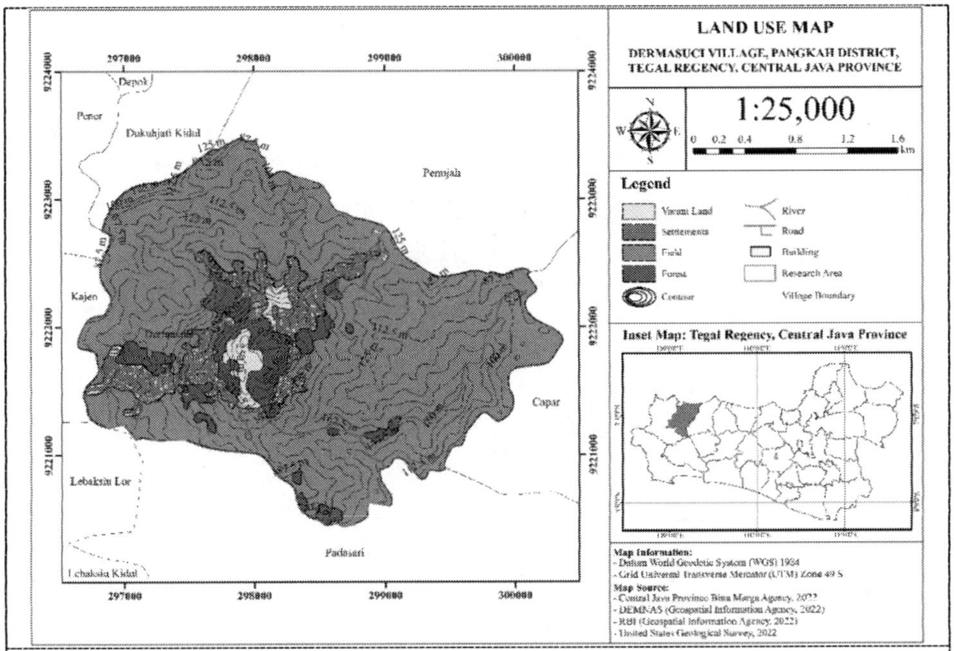

Figure 4. Land Use Map of Research Area

The next set of data is secondary data which includes rainfall data in the form of antecedents and intensities, seismicity data and earthquake potential. Based on rainfall intensity from January 25 to February 14, 2022[8], combined with information on the landslides' occurrence time and groundwater level, an analysis was carried out to identify the highest rainfall intensity before the landslide occurred in Dermasuci Village. The rainfall intensity data is presented in graphical form, as shown in Figure 5. The rainfall intensity data used cannot fully represent the conditions at the study site because the data was obtained outside the boundaries of the research area. The rainfall intensity measurement data is located at the coordinates UTM 49 S 292372 E 9240417 N.

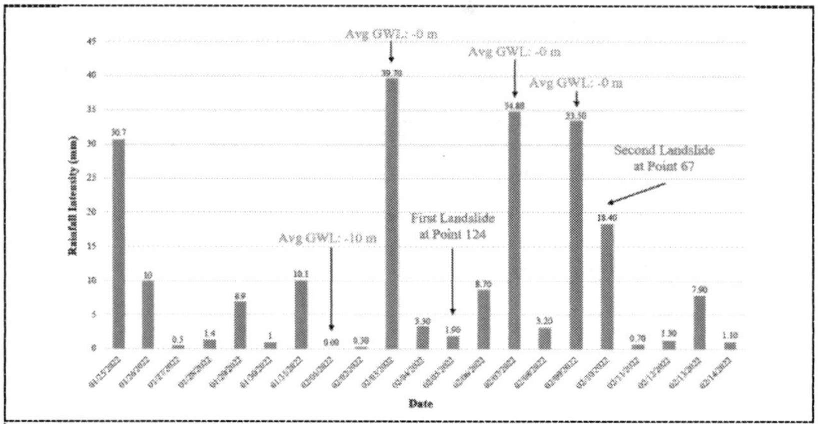

Figure 5. Rainfall Intensity from January 25 to February 14, 2022[8]

Based on historical seismic data and earthquake potential[4], it is known that there are six points of seismicity spread across Tegal Regency and its environs from different time periods, and very low earthquake potential. The distribution of seismic points and earthquake potential in Tegal Regency and its environs is presented in the form of a seismic and earthquake potential map as shown in Figure 6.

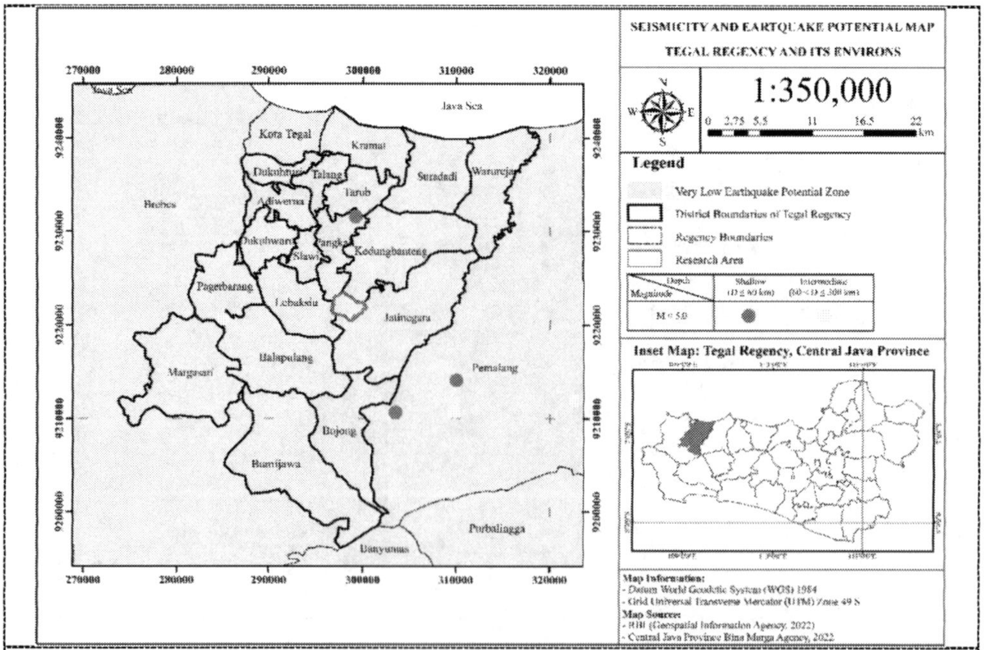

Figure 6. Seismicity and Earthquake Potential Map

IOP Conf. Series: Earth and Environmental Science 1245 (2023) 012005 doi:10.1088/1755-1315/1245/1/012005

Based on the data presented above, the controlling and triggering factors that cause landslides in Dermasuci Village can be determined. The landslide-controlling factors include the rock bedding inclination, geological structure in the form of faults, slope inclination, engineering properties of soil and rock, land use, and groundwater level. The landslide-triggering factors include rainfall in the form of antecedents and intensity, and earthquakes. These factors will then be sorted from the most static to the most dynamic by using the analytical hierarchy process (AHP) method.

The analytical hierarchy process (AHP) is a general theory of measurement used to derive ratio scales from both discrete and continuous paired comparisons. These comparisons may be taken from actual measurements or from a fundamental scale that reflects the relative strength of preferences and feelings. The comparisons are arranged in a pairwise comparison matrix to form the reciprocal matrix needed to get the eigenvalue. Furthermore, the eigenvalue is used to calculate the level of consistency of the comparison results in order to obtain a consistency index and consistency ratio[13].

An assumption is needed to compile a pairwise comparison matrix table before the weighting process begins. The set assumption is the dynamic levels of the factors that cause landslides to begin when it rains with low, moderate, or high intensity, and there is a potential for an earthquake to occur. Weighting is done by comparing a parameter with other parameters. The weighting value given ranges from 1 to 9. A value of 1 is interpreted to have the same level of dynamics. The higher the value given to a factor parameter, the more the factor parameter is considered to have more dynamic properties compared to the comparison factor parameters. Pairwise comparison matrices of the predetermined parameters of the controlling factors and landslide triggers are shown in Table 1.

Table 1. Pairwise comparison matrix.

Y \ X	Rock Bedding Inclination	Geological Structure (Fault)	Slope Inclination	Engineering Properties (Soil & Rock)	Land Use	Groundwater Level	Rainfall Intensity	Eartquakes	Weight Value
Rock Bedding Inclination	1	1.000	1.000	0.333	1.000	0.200	0.125	0.111	0.033
Geological Structure (Fault)	1	1	1.000	0.333	1.000	0.200	0.125	0.111	0.033
Slope Inclination	1	1	1	0.333	1.000	0.200	0.125	0.111	0.033
Engineering Properties (Soil & Rock)	3	3	3	1	3	0.333	0.250	0.200	0.084
Land Use	1	1	1	0.333	1	0.200	0.125	0.111	0.033
Groundwater Level	5	5	5	3	5	1	0.500	0.333	0.161
Rainfall Intensity	8	8	8	4	8	2	1	0.500	0.262
Eartquakes	9	9	9	5	9	3	2	1	0.361
Total	29.000	29.000	29.000	14.333	29.000	7.133	4.250	2.478	1.000

3. Result and discussion

The weighting that has been carried out using the analytical hierarchy process method in Table 1 obtained a principle Eigenvalue (λ_{max}) of 8.181, a ratio index (RI) value of 1.41, a consistency index (CI) value of 0.030, and a consistency ratio (CR) value of 1.84%. Based on the provisions in the analytical hierarchy process calculation, if the weighting results in a CR value of less than 10%, the CR value is consistent and acceptable[13], so the CR value resulting from the weighting of the parameters of the factors that cause and trigger landslides in Dermasuci Village is acceptable. The weighting value results are then sorted from the smallest value to the largest value, where the value represents the level from the most static to the most dynamic factors causing landslides, as shown in Table 2.

IOP Conf. Series: Earth and Environmental Science 1245 (2023) 012005 doi:10.1088/1755-1315/1245/1/012005

Table 2. Controlling and triggering factors of landslide.

No	Landslide Factors	Dynamic Factor Values	Levels
Controlling Factors			**Static**
1	Rock bedding inclination	0.033	↓
2	Geological structure (fault)	0.033	
3	Slope inclination	0.033	
4	Land use	0.033	
5	Engineering properties (soil and rock)	0.084	
6	Groundwater level	0.161	
Triggering Factors			
1	Rainfall intensity	0.262	
2	Earthquakes	0.361	**Dynamic**

According to the weighting results of the dynamic factor values in Table 1, it is known that the controlling factors in the form of the rock bedding inclination, geological structure in the form of faults, slope inclination, and land use have the same dynamic factor value of 0.033 or have the same static properties. This can be interpreted to mean that when the rain conditions start to fall, the four factors do not experience any changes until landslides occur. The next controlling factor in the form of engineering properties (soil and rock) and groundwater level has a dynamic value change that is increasingly dynamic, namely 0.084 and 0.161, which means that when the rain conditions start to fall, these two factors begin to experience a change in conditions. The triggering factor in the form of rainfall intensity has a dynamic factor value of 0.262, and earthquakes have the highest dynamic factor value of 0.361. Each of the factors listed in Table 1 that cause landslides will then be discussed in terms of how they influence the causes of landslides in Dermasuci Village, as follows:

3.1. Rock bedding inclination and geological structure (fault)

The rock bedding inclination and geological structures in the form of faults which are identified from the results of surface geological mapping (Figure 2), can be one of the controlling factors that cause landslides. One example of a landslide case due to the rock bedding inclination that caused a landslide in one of the buildings is located at observation location 134 (Figure 1), as shown in Figure 7.

Figure 7. Influence of Rock Bedding Inclination toward a Landslide (Point 134 in Figure 1)

Fault zone processes control the rainfall-induced landslides in three ways[10]: (1) increasing fracture density near major faults that produce more debris and reduce rock strength, (2) weak minerals growing in fault zones in which clay minerals grow, and (3) topographic features produced by active faulting. Because fault planes and rock bedding planes are both types of discontinuity planes, the processes of rainfall-induced landslides at the fault zone can also be applied to rock bedding. An example of a landslide due to a geological structure (fault) that was identified as causing damage to several buildings on the surface is located at observation point 59 (Figure 1), as shown in Figure 8.

Figure 8. Influence of the Geological Structure Toward a Landslide (Point 59 in Figure 1)

The geological cross-section in this case cannot be used to show how the influence of geological controlling factors (rock bedding inclination, geological structure in the form of fault, and engineering properties of soil and rock) on the triggering factors for landslides. This is due to several reasons, namely; First, geological cross-sections cannot provide detailed rock layering data such as cross-sections made from the correlation results of core drilling. Second, the condition of the rock layers in the study area is composed of rocks with a relatively thin average layer thickness of 3 – 12 cm (Figure 9), so it will be very difficult to draw the detailed rock layers on a geological cross-section with a large-scale.

Figure 9. Relatively Thin Rock Layers Condition on Halang Formation
(Point 37, Not a Landslide Point Area)

3.2. Slope inclination

The slope inclination which is identified from the results of surface engineering geological mapping (Figure 3), can be one of the controlling factors that cause landslides. The slope determines how influential that strength is in ensuring the movement of the object; the steeper the slope, the stronger the gravitational strength that ensures the shear of the slope and allows the object to slide[3]. This statement is in accordance with the results of the mapping of points that experienced landslides, where there were three landslide points (38, 41, and 124) that occurred due to the steep slope of the slope, namely 70°, 71°, and 57°. An example of a landslide due to a slope inclination that was identified as causing damage to several buildings is located at observation point 124 (Figure 1), as shown in Figure 10.

Figure 10. Influence of Slope Inclination toward a Landslide (Point 124 in Figure 1)

3.3. Engineering properties (soil and rock)

Weathering is an essential process that affects the mechanical properties of rock material and mass at shallow depths and on the surface through chemical and physical weathering[1]. The sliding occurs because of strength reduction caused by external influences, such as the propagation of weathering from the surface towards the slope inside, and the progress of weathering through bedding joints and other joints[9]. Those statements are consistent with the results of surface engineering geological mapping (Figure 3), where 11 landslide points are located in rock areas with completely weathered conditions, two landslide points are located in rock areas with highly weathered conditions, and two landslide points with special mechanisms are located in rock areas with fresh conditions.

3.4. Land use

Recent studies focusing on the effect of human-induced land use changes on slope stability have shown that in populated regions, the impact of humans on the environment contributes significantly to the initiation and reactivation of landslides[12]. This statement is consistent with the results of geological and engineering geological surface mapping (Figure 4), where 13 out of 15 landslide points in Dermasuci Village are located in areas with land use in the form of settlements.

3.5. Groundwater level and rainfall intensity

Groundwater level and rainfall intensity, respectively, are controlling and triggering factors that are interrelated to the causes of landslides, as shown in Figure 5, and are followed by their influence on controlling factors in the form of engineering properties (soil and rock). The first highest intensity of rainfall (39.7 mm) on 3 February 2022, is interpreted to be the cause of landslides at point 124 which occurred on 5 February 2022. As a result of this highest rainfall, the average groundwater level has increased from -10 m in the day without rain (1 February 2022) to -0 m. The second landslide at point

67 is interpreted to occur due to prolonged rainfall that occurred from 3 February 2022 until the day the second landslide occurred on 10 February 2022. The second-highest rainfall (7 February 2022) and the third-highest rainfall (9 February 2022) are interpreted to have a significant influence against landslides at point 67, and causing the groundwater level to reach -0 m.

Based on the time sequence of events, it can be known that landslides did not occur immediately at the time of the highest rainfall, but some time after the occurrence of rain with sufficient and/or prolonged intensity until causing the soil condition to become saturated. This is in accordance with the results of a literature study[18], where the intensity of rainfall will take some time to influence the magnitude of the increase in the groundwater level, followed by an increase in pore water pressure and a decrease in shear stress on the soil. When the soil shear stress reaches its critical point, landslides will occur. The length of time required for landslides to occur is interpreted to be highly influenced by the amount of rainfall intensity, and the physical and mechanical conditions of the soil or rocks.

3.6. Earthquakes

Earthquakes are a triggering factor for landslides in the most dynamic research locations because when and where an earthquake will occur cannot be predicted. In addition, the earthquake potential at the study site, as shown in Figure 6, is classified as very low. Based on a literature study from[14,15], in a simple explanation, during the main shock of an earthquake loading will cause a rapid increase in pore water pressure followed by a decrease in soil shear strength in progressive failure motions.

4. Conclusions

Based on the results of the research that has been carried out, it can be concluded as follows:

1) Controlling factors that cause landslides in Dermasuci Village include the rock bedding inclination, geological structure in the form of faults, slope inclination, engineering properties of soil and rock, land use, and groundwater level. Triggering factors that cause landslides in Dermasuci Village include rainfall intensity, and earthquakes.
2) Static properties are factors that do not experience any changes in conditions before the occurrence of landslides. Meanwhile, dynamic properties are factors that experience changes in conditions either individually or as a result of the influence of other factors before the occurrence of landslides.
3) The factors that cause landslides from the most static to the most dynamic based on the results of the analysis using the analytical hierarchy process (AHP) method include the rock bedding inclination, geological structure in the form of faults, slope inclination, and land use that have the same dynamic factor value (0.033), engineering properties of soil and rock (0,084), groundwater level (0.161), rainfall intensity (0.262), and earthquakes (0.361).

The results of this research have limitations and suggestions for further research as follows:

1) This research does not discuss and compare which factors have more influence on the causes of landslides. Further research can be carried out to further explore which factors have the greatest influence on the causes of landslides disaster in Dermasuci Village.
2) This research does not explain in detail how the mechanism of the influence of triggering factors on each of the controlling factors that cause landslides occurs. Further research can be carried out using more detailed data (such as laboratory test data or in-situ testing) to explain the effect of triggering factors on controlling factors in the mechanical analysis.

Acknowledgments

The author would like to thank the financial support that has been provided by the Department of Geological Engineering, Gadjah Mada University, Yogyakarta. The author also would like to thank Mr. Cahyono, the caretaker head of the Pangkah District, and Mr. Mulyanto, the head of Dermasuci Village, for their permission to conduct the site investigation and collect the field data.

IOP Conf. Series: Earth and Environmental Science 1245 (2023) 012005 doi:10.1088/1755-1315/1245/1/012005

References

1. Arikan F, Aydin N. Influence of weathering on the engineering properties of dacites in Northeastern Turkey. *Intern Scholarly Research Network Soil Science* 2012; 1:1
2. Bemmelen Van R W. The geology of Indonesia. Part 1A. General geology of Indonesia and adjacent archipelagos. DenHaag: Government Printing Office Solo Agent; 1949:26-30
3. Cellek S. Effect of the slope angle and its classification on landslide. *Natural Hazards and Earth System Sciences* 2020; 1:4
4. Central Java Province Bina Marga Agency. Peta tematik rawan seismisitas dan peta tematik rawan gempa; 2022 [updated 2023; cited 2022 Dec 12]. Available from: https://webgis.dpubinmarcipka.jatengprov.go.id/
5. Djuri M, Samodra H, Amin T C, Gafoer S. Peta geologi Lembar Purwokerto dan Tegal skala 1:100.0000. Indonesia: Geological Research and Development Center; 1996:1
6. Hack R, Price D. Quantification of wathering. *Proceedings of the Engineering Geology and the Environment*; Rotterdam, Netherlands: Balkema; 1997. p.146.
7. Karnawati D. Mekanisme gerakan massa batuan akibat gempabumi; tinjauan dan analisis geologi teknik. *Dinamika Teknik Sipil* 2007; 7: 1
8. Meteorological Climatological and Geophysical Agency. Probabilistik curah hujan; 2022 [updated 2023; cited 2022 Oct 26]. Available from: https://www.bmkg.go.id/cuaca/
9. Miscevic P, Vlastelica G. Impact of weathering on slope stability in soft rock mass. *Rock Mechanics and Geotechnical Engineering* 2014; 6:243
10. Moinabadi M E. Properties of fault zones and their influences on rainfall-induced landslides, examples from Alborz and Zagros Ranges. *Environmental Earth Sciences* 2022; 81:1
11. National Disaster Management Agency. Kejadian bencana per provinsi tahun 2022; 2022 [updated 2023; cited 2022 Oct 25]. Available from: https://gis.bnpb.go.id/
12. Reichenbach P, Busca C, Mondini C, Rossi M. The influence of land use change on landslide susceptibility zonation: The Briga catchment test site (Messina, Italy). *Environmental Management* 2014; 54:1372
13. Saaty R W. The analytic hierarchy process – what it is and how it is used. *Mathl Modelling* 1987; 9:161-176
14. Setiawan H, Sassa K, Takara K, Miyagi T, Fukuoka H. Initial pore pressure ratio in the earthquake triggered large-scale landslide near Aratozawa Dam in Miyagi Prefecture, Japan. *Procedia Earth and Planetary Science* 2016; 16:61-69
15. Setiawan H, Sassa K, Takara K, Ostric M, Miyagi T, Fukuoka H. TXT-tool 4.081-1.2 mechanism of the Aratozawa large-scale landslide induced by the 2008 Iwate-Miyagi earthquake. *Landslide Dynamics* 2018; 2:807-818
16. Sribudiyani, Muchsin N, Ryacudu R, Kunto T, Astono P, Prasetya I, Sapiie B, Asikin S, Harsolumakso A H, Yulianto I. The collision of the East Java Microplate and its implication for hydrocarbon occurrences in the East Java Basin. *Proceedings of the Indonesian Petroleum Association 29th*; 2003 Oct 14; Indonesia, Jakarta: Indonesian Petroleum Association; 2003. p.11.
17. Tegal Regency Government. 61 Rumah di Desa Dermasuci rusak akibat tanah gerak, lokasi relokasi disiapkan sejak 2017; 2022 [updated 2022; cited 2022 May 11]. Available from: https://utama.tegalkab.go.id/page/view/geografi _20190121082814
18. Tien P V, Sassa K, Takara K, Fukuoka H, Dang K, Shibasaki T, Ha N D, Setiawan H, Loi D H. Formation process of two massive dams following rainfall-induced deep-seated rapid landslide failures in the Kii Peninsula of Japan. *Landslides* 2018; 15:1761-1777
19. United States Geological Survey. The landslide handbook – a guide to understanding landslides. Virginia: U.S. Geological Survey; 2008. p.5.
20. United States Geological Earth Explorer. Landsat 8-9 OLI/TIRS X2 L2; 2022 [updated 2023; cited 2022 Dec 12]. Available from: https://earthexplorer.usgs.gov/
21. World Risk Report. World risk report 2021 focus: social protection. Germany: Ruhr University Bochum; 2021. p.54.

A Statistical review of the dates and patterns of volcanic activity of Lewotolo Volcano, East Nusa Tenggara, Indonesia

T Muliawati[1, a], D G Harbowo[2]

[1] Department of Mathematics, Institut Teknologi Sumatera, 35352, Wayhui, Lampung, Indonesia
[2] Department of Geology, Institut Teknologi Sumatera, 35352, Wayhui, Lampung, Indonesia
[a] Correspondent author: triyana.muliawati@ma.itera.ac.id

Abstract. Lewotolo is a stratovolcano located on the Lembata Island, East Nusa Tenggara (Lesser Sunda Islands), Indonesia. The first geohistory of the Lewotolo volcanic eruption was recorded in 1660 (Volcanic Eruption Index (VEI) 3). Since November 11, 2020, Lewotolo has been included in the list of Indonesian volcanoes with the necessary precautions, comparable to the Anak Krakatau, Merapi, and Semeru volcanoes. We investigated Lewotolo volcanic activity by analyzing the height of the ash column data, maximum seismograph amplitude, and recorded seismic duration from November 29, 2020, to September 23, 2022, which are provided in magma.esdm.go.id. The results showed 191 records of eruption activity data, which formed three clusters for each variable using the Elbow Method in the Non-Hierarchical K-means clustering analysis. These data were plotted on the solar and Hijri lunar calendars. The data plot shows anomalies in the volcanic activity frequency and a significant pattern of activity at specific times. The data plot illustrates that the value of the activity frequency pattern tends to increase at Earth orbital events such as the December-January perihelion (up to 0.124) and June-July aphelion (up to 0.262) and at the new moon phases (up to 0.168). This phenomenon appears as solar and lunar tidal anomalies that commonly occur as gravitational sea tides. Along with the need for more comprehensive data, the results of this study may provide new perspectives for further research on the possible role of gravitational tide phenomena in volcanic activity, at least to explain the volcanic activity in Lewotolo.

Keyword: Volcano, Lewotolo, K-means clustering, Elbow method.

1. Introduction

On November 29, 2020, at 09:45 local time, larger explosive eruptions occurred on the Lewotolo Volcano on Lembata Island, Indonesia [1,2]. Interestingly, this eruption occurred about at the end of the year and coincided with the full Moon date. This condition is the same as that of the Krakatoa destructive eruption followed by the tsunami that occurred at the end of 2018 and during the full moon [3]. These eruption events prompted us to further investigate the data record, the level of activity of Lewotolo volcano, and its periodic eruption based on solar and lunar calendars. This study began with suspicion of the possibility of a link between the position of the Moon and the Sun relative to the Earth with specific volcanic activities, such as Lewotolo in November 2020 and Krakatoa in December 2018. In several similar cases, some hypotheses have proposed that this may be related to the anomalous gravitational forces of the sun and moon [4], which may have affected fluidity-magmatic activity, similar to what occurs with sea tides [5–9]. Although this theory still needs to be proven further, in this study, we attempt to statistically review the dates of eruptions with the orbital events of the Moon and Sun with respect to Earth. This study aims to

IOP Conf. Series: Earth and Environmental Science 1245 (2023) 012006 doi:10.1088/1755-1315/1245/1/012006

review the pattern of eruption events on lunar and solar calendars based solely on the resulting statistical expressions. Furthermore, the process and mechanism of eruptions must be investigated more comprehensively.

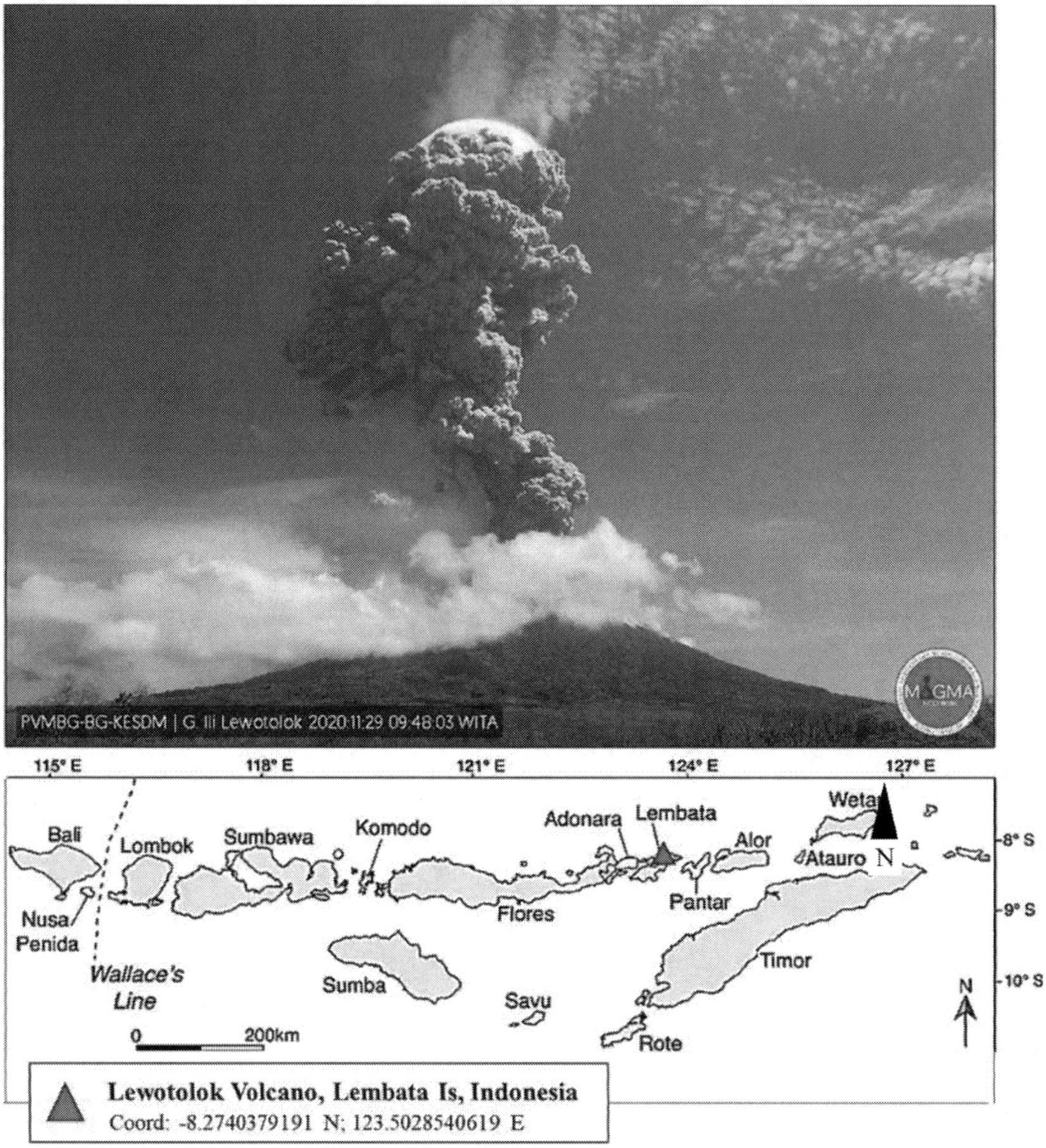

Figure 1. Location map of Lewotolo volcano on Lesser Sunda, East Nusa Tenggara, Indonesia, and photo documentation of the larger explosive eruption [1], date: November 29, 2020.

Lewotolo Volcano (Ili Lewotolo) located di East Nusa Tenggara (Lesser Sunda), Indonesia. It part of Lembata Island in the Flores Sea (coordinates: 8.274°S, 123.508°E; **Figure 1**). The Lewotolo is stratovolcano-type [10]. It mainly has basaltic andesite lava associated with the Eastern Sunda Arc tectonic setting, part of the Adonara-Pantar segment, a regional subduction zone. Lewotolo is likely associated with calc-alkaline magma origins, which have relatively more fluid characteristics, lower viscosity, and lower silica content than rhyolitic magma [11]. These characteristics led to various Lewotolo eruption types, ranging from vulcanian to strombolian. Recent stromobolian eruptions were recorded in November 2020. The fair-confirmed Holocene eruptivity periods occurred at 1660 at VEI 3, then 1819; October 6 1849; October 5-6 1852; 1864; June 2 1899; 1920, December 15 1951; January 2-14 2012, and lastly continuously well-recorded occured since November 27 2020 [1]. Nearly all these periods occurred at about the end of the year (October-Desember). Reports suggest that Lewotolo eruptions are likely to be in the VEI 2.

Several orbital phenomena occur owing to changes in the distance between Earth and the Sun [12]. The distance from the Sun changes every day of the year (365.25 days in the Gregorian calendar) because Earth's orbit around the Sun is elliptical. When the Earth is at its closest distance to the sun, it is called a perihelion. When the Earth is at its farthest distance from the sun, it is called an aphelion. Aphelions occurred every year between July 2 and July 7. Perihelions occurred on January 3. There is also something called the equinox, which is divided into two: the March Equinox and the September Equinox [13,14]. Astronomically, the March equinox is the point of intersection of the ecliptic and celestial equator that the Sun passes through on its quasi-annual journey from the Southern Hemisphere sky to the Northern Hemisphere sky. The March equinox is also known as the vernal equinox in the Northern Hemisphere and autumnal equinox in the Southern Hemisphere. Similar orbital phenomena can also result from changes in the Moon's position relative to the earth and sun. This phenomenon is divided into several moon phases (such as the new moon phase to the full moon phase), which are viewed based on the appearance of the moon's luminescens visible from Earth [6]. This can be described through the lunar calendar, as in hijri month dates [15].

In this study, we aimed to understand the pattern of volcanic eruption activity by statistically reviewing several quantitative parameters, such as the observed height of the ash column and seismograph responses during eruptions, including maximum amplitude and seismograph duration. Data analyses were conducted using *k*-Means Clustering Algorithms. It is a technique for grouping data such that members of the same group are more homogeneous or more similar to each other than members of different groups [16,17]. In addition, the *k*-means algorithm divides the dataset into *k* clusters with a given *k* value, followed by the Elbow Method. It generates specific cluster numbers, which are groups of data that have a smaller distance between data in the group than the distance between data outside the group [18–20]. The advantage of *k*-means clustering is that it is susceptible to implementation and the learning process is complementary and fine for further data interpretation. The *k*-means algorithm represents each cluster with a centroid or a cluster center. The cluster data and its centroid can be interpreted as a pattern in data expression.

2. Method

Data Records. In this study, we used well-observed records of Lewotolo eruption activity from open-access data in *magma.esdm.go.id*, which has officially been reported. The report contains eruption activity, including the date of the event, photos of the volcano when it erupted, height of the visible ash column, and duration and maximum seismograph amplitude when the eruption occurred. The data used in this study were recorded from November 2020 to September 2022 (**Appendix 1**), totaling 191 eruption events. We plotted the eruption date on the Gregorian date as a solar calendar from the date of the first day at the beginning of the year (starting on January 1) and reviewed the position of the Hijri days as the lunar calendar. In the data record, we statistically reviewed the data using numerical data review, the Elbow Method, and *k*-Means Clustering. The data were processed using *Google Collaboratory* and presented in the form of x-y plots.

Elbow Method. This method is used to determine the optimal number of clusters or groups. The Elbow Method considers the Sum of Square Error (*SSE*) value expressed by the estimated *k* value (starting at $k = 2$), which is represented by a line graph. The concept is to start the calculation with $k = 2$ and then increase each step of *k* by one step per *k* unit. The selection of the best *k* value is based on the lowest angle value (sharper) between the two lines, from $k = n$ to $k = n + 1$. The change in the value of k, which increased, was followed by a decrease in the *SSE* value. Thus, the optimal number of clusters can be determined (see **Equation 1**).

$$SSE = \sum_{k=1}^{k} \sum_{x_i \in S_k} \|x_i - \mu_k\|^2 \text{ .... (1)}$$

Where, k is the number of clusters, x_i is the i-th data value, and μ_k is the k-th cluster center point value. The changes in the value of k and *SSE* value are presented in the form of xy line-graphs.

***K*-means Clustering.** The k-means algorithm is a clustering method of unsupervised learning, in which the similarity measure of data in a group is determined based on the smallest distance of each data point to the centroid of each cluster. The objective function of the k-means clustering algorithm is to minimize the within-cluster variance and maximize the variance between different clusters. The k-means clustering algorithm consists of the following steps: A random value of k as the number of clusters to be formed on the data is determined. The centroid value at the initial stage is determined randomly, whereas in the iteration stage conducted continuously, **Equation (2)** is used.

$$\mu_{ij} = \frac{1}{N_i} \sum_{k=1}^{N_i} x_{kj} \text{ ..... (2)}$$

Where, μ_{ij} denotes the average centroid of the i-th cluster for the j-th variable. N_i is the number of members in the i-th cluster. x_{kj} is the k-th data value for the j-th variable in cluster. The proximity of the data to a particular cluster is determined by the distance between the data and cluster center. The centroid distance used is the Euclidean Distance, as shown in **Equation (3)**.

$$D(x_i, \mu_i) = \sqrt{\sum_{j=1}^{p} (x_{ij} - \mu_{ij})^2} \text{ ...... (3),}$$

Where, $D(x_i, \mu_i)$ is the distance between data x_i and centroid μ_i. x_{ij} is the ith data point in the jth data variable. μ_{ij} is the ith cluster centroid of the j-th variable. The data were clustered based on the distance to the nearest centroid. The steps in **Equation (2)** are repeated until the centroid iteration is optimal, that is, if the cluster center no longer changes (converges to single value), then the clustering process is complete.

3. Result

The Elbow Method results in an elbow point that determines the optimal value of k for each parameter (height ash column, duration, and maximum seismograph amplitude). It can be seen that the ash column graph forms an elbow at $k = 4$ with an SSE value = 0.42×10^7, then at $k = 5$ has a decreasing SSE value and decreases very slowly thereafter. Therefore, the value of k used for the observed ash column height above the summit (meter) consisted of four clusters ($k = 4$). In the graph of duration of seismograph activity forming an elbow at $k = 3$ with a value of SSE = 114567.6, at $k = 4$, the SEE value decreases slowly thereafter. Thus, the k value determined for the duration of the seismograph activity was optimum in three clusters ($k = 3$). The maximum seismograph amplitude graph forms an elbow at $k = 4$ with an SSE value of = 1584.0; then, at $k = 5$, it has a decreasing SEE value and decreases very slowly thereafter. Thus, k used as a cluster on the maximum seismograph amplitude was optimum in four clusters ($k = 4$), see **Figure 2**.

Through the results of the k-means clustering analysis with the value of k in each parameter, the data members in each data group/cluster can be determined. Numerically, the centroid of each cluster, the amount of data, the minimum-maximum value, and the standard deviation can be determined. These data groups serve as a reference to distinguish the level of volcanic eruption activity from other eruption events. In this study, we arranged cluster order numbers based on the magnitude of the centroid value in correlation

IOP Conf. Series: Earth and Environmental Science 1245 (2023) 012006 doi:10.1088/1755-1315/1245/1/012006

with the eruption activity level. The greater the cluster order value, the higher the eruption activity level. The descriptive statistics of *k*-means clustering are presented in **Table 1** and **Figure 3**.

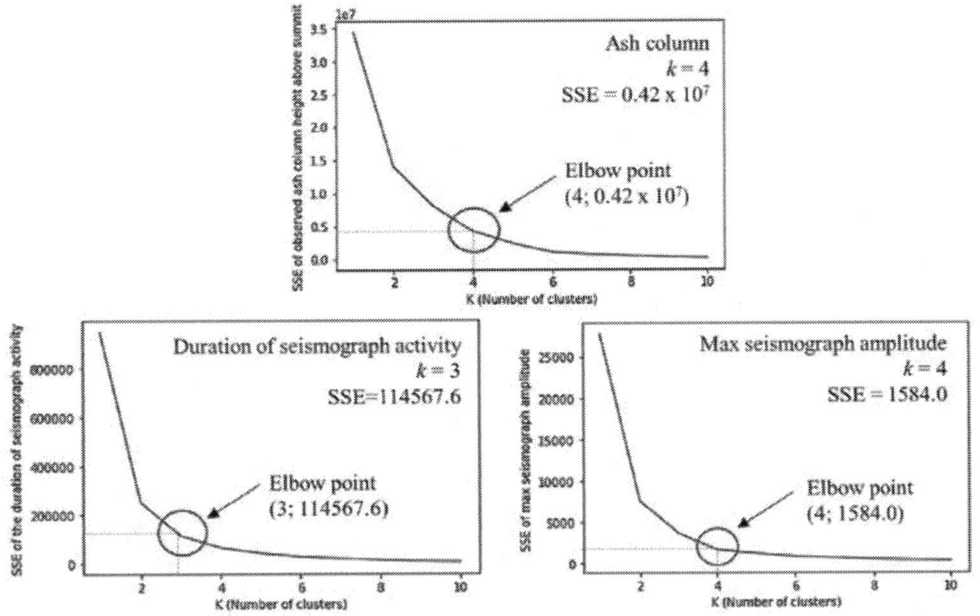

Figure 2. Line graphs of *k*-value (number of cluster) and sum of square error (SSE) each parameter. The graph shows that the elbow point is relatively more sharply bent, which was determined as the selected cluster number chosen in this study.

Table 1. Numerical resume of *k*-means clustering result of each parameter

Parameter	Ash column ($k = 4$)				Seismic duration ($k = 3$)			Max amplitude ($k = 4$)			
Clustering	Clstr1	Clstr 2	Clstr 3	Clstr 3A	Clstr 1	Clstr 2	Clstr 3	Clstr 1	Clstr 2	Clstr 3	Clstr 4
Centroid	583.3	907.4	1557.8	4000	40.7	197.4	500.0	8.1	18.7	29.5	41.8
Data (n)	82	81	27	1	178	10	3	55	51	53	32
Min	200	750	1300	4000	15	128	400	2.5	14	25	36
Max	700	1200	2000	4000	113	300	600	13	24	35.5	8054
St.dev	121.6	121.0	260.1	-	17.6	62.6	81.6	2.9	2.6	2.8	1394.1

Note that the ash column has four clusters; however, in cluster-4, there is only one event (single data). Therefore, we renamed cluster-4 as cluster-3A. In cluster-3A of the ash column, there was only one ash column member with a data value of 4000 m (ID-1). This is because 4000 m is very far from the other data. Thus, these data formed one separate cluster. Seismic duration data formed three clusters. The third cluster of seismic duration only has three members, where the three members are the three longest seismic duration values, namely, 400 s, 500 s, and 600 s. The maximum amplitude data formed 4 clusters. In the fourth cluster, there was the highest value of maximum amplitude, which was 8054 m.

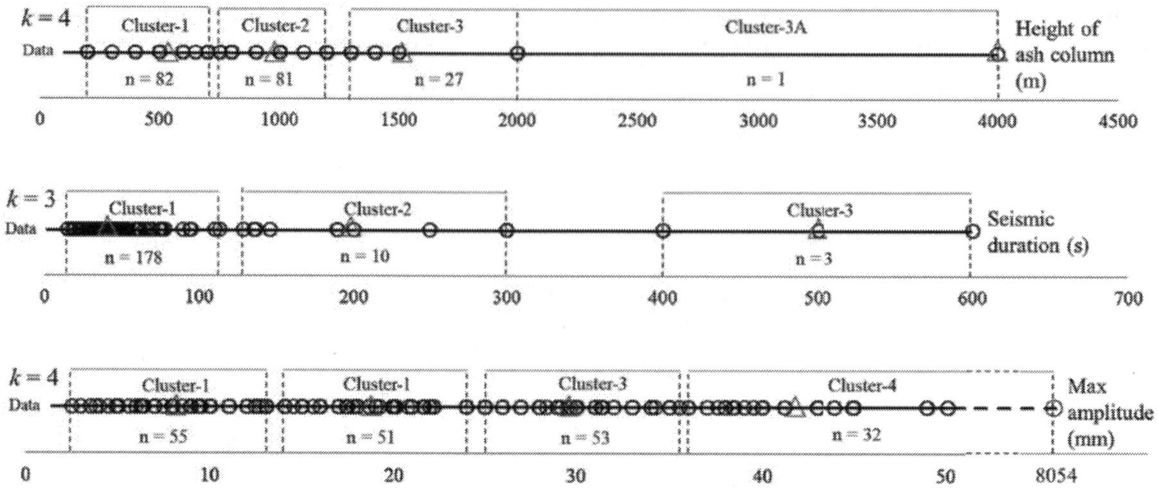

Figure 3. Visualization of the distribution of clustered data (black circles) and centroids (red triangles) for each parameter of the Lewotolo eruption activity.

The clustering results obtained from the calculation were superimposed on a distribution graph that represents volcanic activity events in both the lunar and Gregorian calendars. Each cluster is defined by a range boundary that is determined by the maximum and minimum values of the data within the cluster. To address the gaps between clusters, we assumed that the median value fell within the cluster boundaries. This study illustrates that higher cluster numbers correspond to greater degrees of volcanic activity, whereas smaller cluster numbers indicate lower levels of activity. Consequently, the findings of this study offer a clearer depiction of the pattern-indicating dates with higher volcanic activity in Lewotolok. Furthermore, the clustering results (as **Figure 3**) are re-visualized on the activity vs. date graph, as depicted in **Figures 4-8**.

4. Discussion

In this study, we reviewed the eruption activity based on three observed parameters: height of ash column, duration of seismograph activity (seismic duration), and the maximum recorded seismic amplitude during eruption. We plotted the eruption data together with the Gregorian calendar (**Figure 4** and **Figure 5**), which represents the Earth revolving to the sun, and the Hijri calendar, which represents the Moon revolving to Earth (**Figure 6-8**). Some orbital conditions, called perihelion, aphelion, equinox, and soltice, occur in the Earth's orbit with respect to the sun. This phenomenon was repeated every year. This phenomenon is an implication of the shape of the Earth's orbit of the sun, which is elliptical, and the tilt of the Earth's axis against the orbital trajectory [12]. The Earth periodically revolves around the sun for 365 days, 6 hours, and 9 minutes (or approximately 365.25 days), and one periodic revolution is referred to as one year, which is divided into 12 months in the Gregorian calendar.

On the solar calendar, every early (3rd-4th) January, perihelion always occurs, which is the position of the earth and sun at a relatively close distance [12]. This phenomenon occurred on March 21st and September 23rd. Then, the solstice phenomenon occurs when the sun appears to reach its most northerly or southerly excursion relative to the celestial equator on the celestial sphere. Around the end of June (21st), the Sun travels along its northernmost path in the sky. At the end of December (21st), the sun travels along its southernmost path in the sky. These phenomena are related to the shape of the Earth's elliptical trajectory with respect to the sun and the period of change of the Earth's axis tilt in a one-year cycle [14].

IOP Conf. Series: Earth and Environmental Science 1245 (2023) 012006 doi:10.1088/1755-1315/1245/1/012006

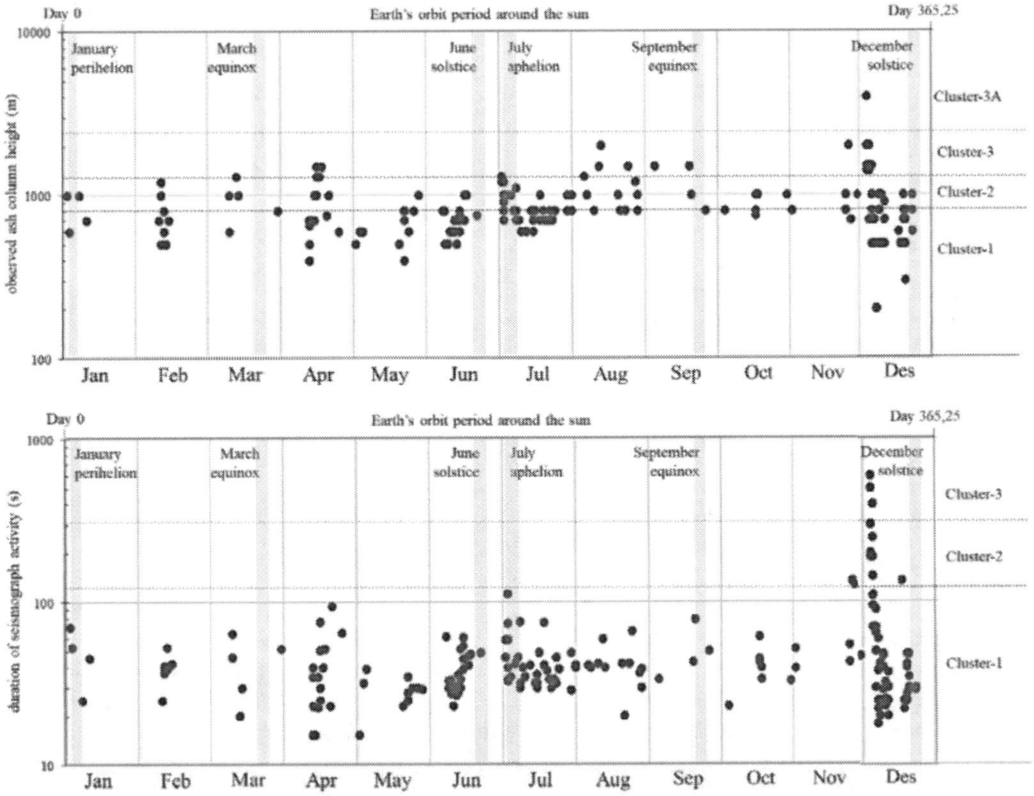

Figure 4. Distribution of clustered data (blue dots) of the parameters of height of ash column (top) and duration of seismic (bottom) activity during eruption. Data are plotted on the Gregorian calendar with Earth's orbital phenomena toward the sun (yellow streak area).

Plotting the eruption data of Lewotolo volcano, it can be seen that the eruption frequency often occurs in the middle of the year (June–July) and towards the end of the year (December). The average ash column generated from 191 recorded Lewotolo eruptions was ± 880 m, and most of the ash columns formed from June to December were relatively higher than this value (**Figure 5**). The maximum recorded value produced by Lewotolo was ± 4000 m in November 2020. In retrospect, the high frequency of eruptions in June and July, the June solstice, and the July aphelion. The frequency of eruptions in December was related to the December solstice and January perihelion (in the following year). Lewotolo eruption activity is relatively small on the equinox, either during the March or September equinox.

The first and second highest volcanic eruptions occurred in December (44/191) and July (36/191), respectively. The first and second highest average seismograph amplitudes occurred in February (32.8 mm) and January (32.2 mm), respectively. Volcanic activity increased from September to December (73/191) (Figure 5). This was followed by several eruptions of ash columns formed at the top of the volcano. The highest ash column height of 4000 m occurred on November 29, 2020, or 14 Rabīʿ al-Awwal 1442, a time relatively close to the end of the year and close to the full moon phase. The lowest volcanic eruption activity occurred from January to March, which was 16/191. On average, the height of the ash column of Lewotolo volcano formed was 876.8 m from the top of the volcano.

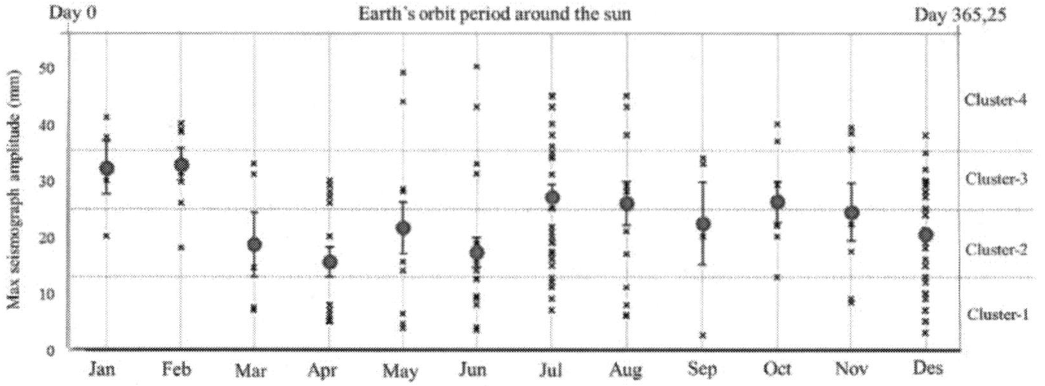

Figure 5. Frequency distribution of eruptions per Gregorian month and maximum amplitude value recorded by the seismograph for each eruption event (black dots). The average maximum amplitude (blue dots) and standard deviation are listed.

From the maximum amplitude on the seismogram, it was observed that the eruptions in January had a higher average maximum amplitude, decreased until June, and increased again in July. The amplitude variations produced in July were relatively more diverse, with higher eruption frequency values. Similarly, the December eruption, although with a smaller average, had a relatively higher frequency of eruptions with a more diverse maximum amplitude variation. The high amplitude generated by the seismograph indicates that the amount of friction between the volcanic material and the volcanic magma channel formed when the volcano erupted. As previously discussed, June-July coincides with the aphelion and mid-year soltice, whereas the December-January eruption coincided with the perihelion, December-soltice, and January-perihelion.

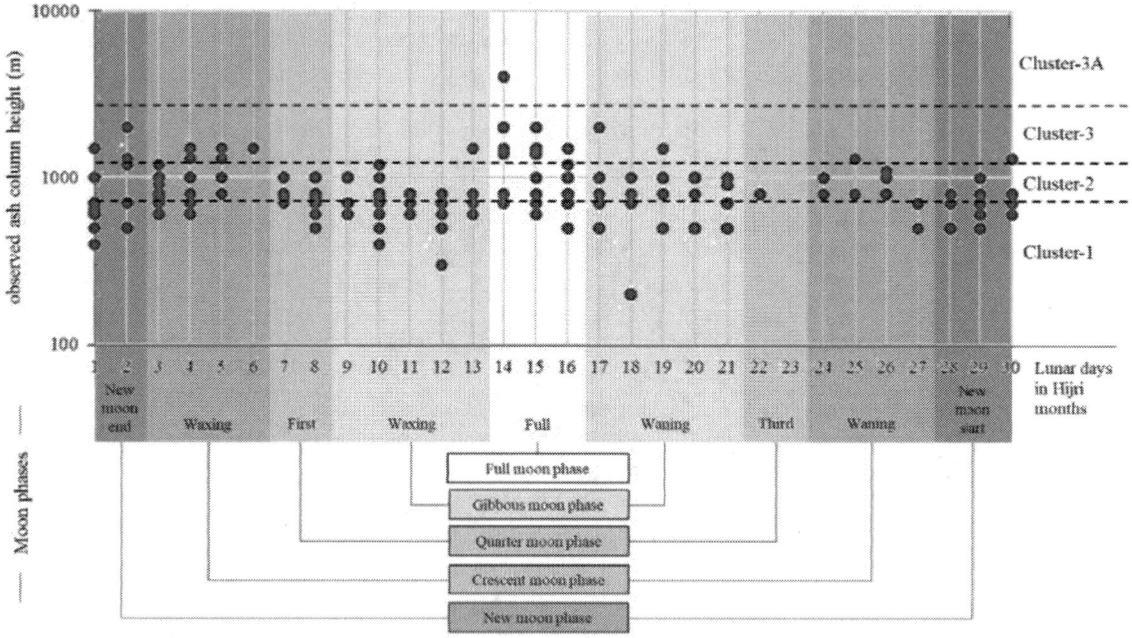

Figure 6. Distribution of ash column height clustered data (blue dots) on the lunar calendar with respect to the moon phase.

We also reviewed the periodic eruptive activity based on the Moon calendar. The Moon calendar was determined based on the phases of the Moon. The moon phase occurs because of certain conditions in the positions of the Earth, Moon, and Sun. The Moon calendar used in this study is the Hijri calendar. The calendar is determined from the appearance of the new moon, the full moon, until it returns to the new moon phase. Traditionally, the lunar calendar is divided into 30 days, which can be divided into several lunar phases, in order, namely the new moon phase, waxing cresent, 1st quarter moon, waxing gibbons, full moon, waning gibbons, 3rd quarter, waning cresent, and back to the new moon phase (**Figure 6-8**). This moon phase can be distinguished by the appearance of the moon, which is an implication of the positions of the sun, moon, and earth.

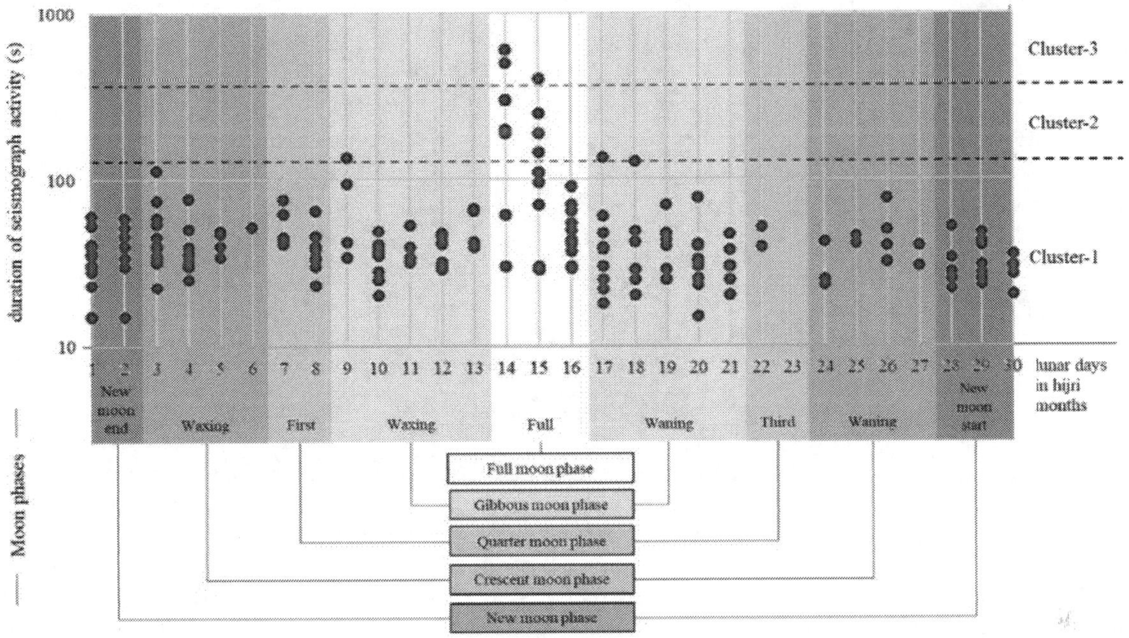

Figure 7. Distribution clustered data of the duration of seismograph activities (blue dots) on the lunar calendar with respect to the moon phase.

When compared with Lewotolo eruption activity, ash columns with a height of>1000 m often occur during the full moon phase and from the new moon phase to the waxing cresent moon phase. The cumulative frequency of eruptions during this phase was also relatively higher than that during the other Moon phases. In addition, it was observed that the amplitude of the eruptions produced during the new moon and full moon phases was relatively more varied. A stark contrast occurred during the full moon phase, where the maximum amplitude recorded on the seismograph was relatively higher than other phases between the two phases. In addition, the graph shows a sinusoidal pattern in the data, which should be further examined in future studies.

Thus far, in the analysis conducted, there is an impression of a special pattern between the eruption parameters used and the orbital moment between the Earth, Moon, and Sun. From the observation span of approximately 2 years, there appears to be a tendency for Lewotolo volcano to have higher eruptive activity at the end of the year. Some of the high eruptive activity is close to the occurrence of the December solstice and January perihelon phenomena. In the middle of the year, eruptions often occur, although not as large as at the end of the year, and the eruption events are close to the phenomena of the june solstice and july aphelion. In the March and September equinox, Lewotolo eruption activity was relatively less frequent. In terms of proximity to the lunar phases, Lewotolo activity appears to increase but is more frequent before and after the new moon.

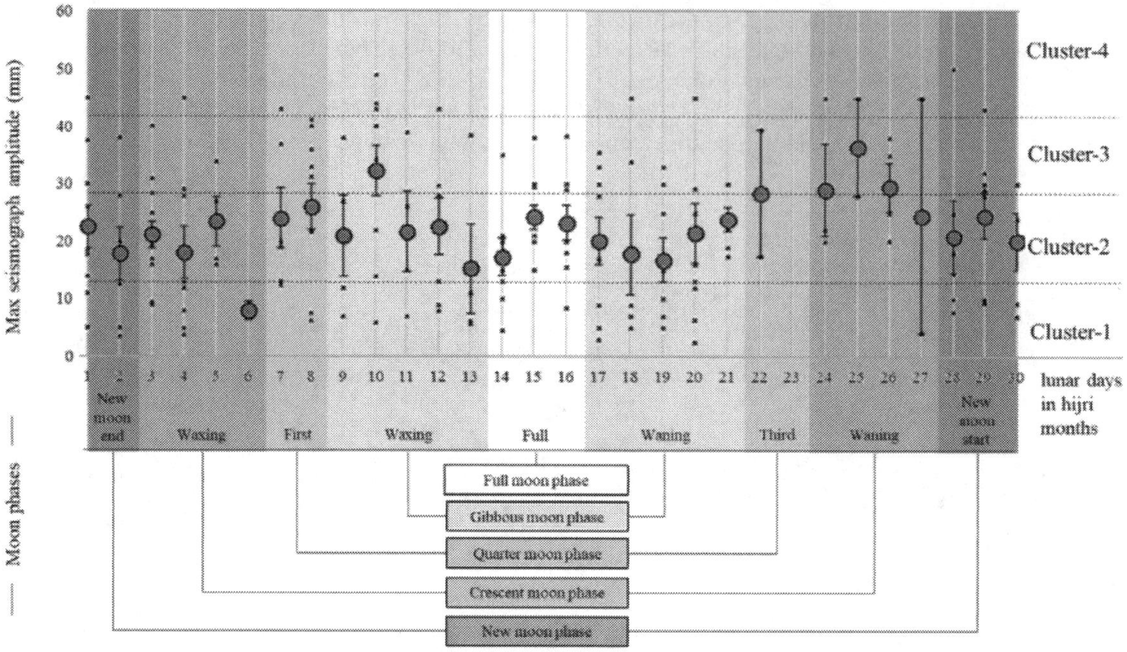

Figure 8. Distribution clustered data of the maximum seismograph amplitude (black dots) on the lunar calendar with respect to the moon phase. Average (blue dots) and its standar deviation are listed.

This study revealed the possible interaction between the position of the Moon and Sun with respect to Earth and the eruptive activity of Lewotolo. This is also related to the distance of the earth from the sun during the buli around the sun and the position of the Moon with respect to the Earth and Moon. In the new Moon and full Moon phases, the Earth, Moon, and Sun are aligned. The difference is that in the new moon phase, the moon is between the position of the earth and the sun, whereas in the full moon phase, the earth is between the position of the moon and the sun. Both produce significantly different gravitational pulls; one of the natural phenomena on earth that can be found is the significance of tidal activity during these moon phases. However, we suggest a more comprehensive study of other volcanoes in Indonesia with relatively similar characteristics. We suspect that the fluidity/viscosity of the magma in volcanic system plays a role in responding to the gravitational pull of the sun and/or moon. Additionally, longer and various observations are required to obtain closer observations.

5. Conclusion

Based on the results and discussion above, the ash column data formed four clusters, the duration of seismograph activity formed three clusters, and the maximum seismograph amplitude formed four clusters. These clusters were formed based on frequency data and patterns of volcanic activity seen in the Gregorian and Lunar Calendar. Activity frequency patterns tend to increase at Earth orbital events such as the December-January perihelion (up to 0.124) and June-July aphelion (up to 0.262) and at the new moon phase (up to 0.168). The results of the statistical analysis suggest a pattern of interaction between certain positions on the Earth, Moon, and Sun that causes an increase in Lewotolo volcanic activity. This interaction is thought to be due to certain gravitational conditions that can affect pressure in the magma chamber.

6. References

[1] PVMBG 2022 Multiplatform Application for Geohazard Mitigation and Assessment in Indonesia

[2] Krotkov N, Realmuto V, Li C, Seftor C, Li J, Brentzel K, Stuefer M, Cable J, Dierking C, Delamere J, Schneider D, Tamminen J, Hassinen S, Ryyppö T, Murray J, Carn S, Osiensky J, Eckstein N, Layne G and Kirkendall J 2021 Day–Night Monitoring of Volcanic SO2 and Ash Clouds for Aviation Avoidance at Northern Polar Latitudes *Remote Sens.* **13** 4003

[3] Pelupessy D, Jibiki Y, Lestari F, Zulys A and Imamura F 2021 Different vantage points amongst different stakeholders in NATECH (NAtural hazard-triggered TECHnological) disasters: A case from the 2018 Mt. Anak Krakatau eruption and tsunami *IOP Conf. Ser. Earth Environ. Sci.* **630** 012024

[4] Gasanalizade A G 1992 Possible semi-annual variation of the Newtonian constant of gravitation *Astrophys. Space Sci.* **195** 463–6

[5] Dinger F, Bredemeyer S, Arellano S, Bobrowski N, Platt U and Wagner T 2019 On the link between Earth tides and volcanic degassing *Solid Earth* **10** 725–40

[6] Myers D E 1995 Gravitational effects of the period of high tides and the new moon on lunacy *J. Emerg. Med.* **13** 529–32

[7] Sottili G, Lambert S and Palladino D M 2021 Tides and Volcanoes: A Historical Perspective *Front. Earth Sci.* **9**

[8] Miguelsanz L, González P J, Tiampo K F and Fernández J 2021 Tidal Influence on Seismic Activity During the 2011–2013 El Hierro Volcanic Unrest *Tectonics* **40**

[9] Dumont S, Silveira G, Custódio S, Lopes F, Le Mouël J-L, Gouhier M and Guéhenneux Y 2021 Response of Fogo volcano (Cape Verde) to lunisolar gravitational forces during the 2014–2015 eruption *Phys. Earth Planet. Inter.* **312** 106659

[10] Venzke E 2016 Report on Lewotolo (Indonesia) *Bull. Glob. Volcanism Netw.* **41**

[11] de Hoog J C M and van Bergen M J 2000 Volatile-induced transport of HFSE, REE, Th and U in arc magmas: evidence from zirconolite-bearing vesicles in potassic lavas of Lewotolo volcano (Indonesia) *Contrib. to Mineral. Petrol.* **139** 485–502

[12] Jenkins A 2013 The Sun's position in the sky *Eur. J. Phys.* **34** 633–52

[13] Reese R L and Chang G Y 1987 The date and time of the vernal equinox: A graphical representation of the Gregorian calendar *Am. J. Phys.* **55** 848–9

[14] Berger A, Loutre M-F and Tricot C 1993 Insolation and Earth's orbital periods *J. Geophys. Res.* **98** 10341

[15] Karray F 1998 A Study on Establishing an International Lunar Calendar *Proceedings of the IIIT Lunar Calendar Conference* (International Institute of Islamic Thought) pp 8-1-8–7

[16] Umargono E, Suseno J E and S. K. V G 2019 K-Means Clustering Optimization using the Elbow Method and Early Centroid Determination Based-on Mean and Median *Proceedings of the International Conferences on Information System and Technology* (SCITEPRESS - Science and Technology Publications) pp 234–40

[17] Lloyd S 1982 Least squares quantization in PCM *IEEE Trans. Inf. Theory* **28** 129–37

[18] Bishop C . 2006 *Pattern Recognition and Machine Learning* (Springer New York)

[19] Sinaga K P and Yang M-S 2020 Unsupervised K-Means Clustering Algorithm *IEEE Access* **8** 80716–27

[20] Muliawati T and Murfi H 2017 Eigenspace-based fuzzy c-means for sensing trending topics in Twitter *AIP Conference Proceedings.* **1862** 030140

Acknowledgments

We gratefully thanks to Hanifah Viralina (Institut Teknologi Sumatera) for compiling Lewotolo eruption data.

IOP Conf. Series: Earth and Environmental Science 1245 (2023) 012006 doi:10.1088/1755-1315/1245/1/012006

Appendix 1. Compilation of Lewotolo eruption date and data records from November 29, 2020, to September 23, 2022.

ID	Date in Gregorian		Date in Hijri		Ash column (m)	Max amplitude (mm)	Seismograph duration (s)
	Date	Doy (th)	Date	Lunar day (th)			
1	29-Nov-20	333	14 Rabīʿ al-Thānī 1442	14	4000	35.0	600.0
2	29-Nov-20	333	14 Rabīʿ al-Thānī 1442	14	2000	20.0	500.0
3	29-Nov-20	333	14 Rabīʿ al-Thānī 1442	14	2000	13.0	200.0
4	29-Nov-20	333	14 Rabīʿ al-Thānī 1442	14	2000	20.0	300.0
5	29-Nov-20	333	14 Rabīʿ al-Thānī 1442	14	1500	21.0	190.0
6	29-Nov-20	333	14 Rabīʿ al-Thānī 1442	14	1400	10.0	300.0
7	30-Nov-20	334	15 Rabīʿ al-Thānī 1442	15	1400	30.0	189.0
8	30-Nov-20	334	15 Rabīʿ al-Thānī 1442	15	1500	21.0	400.0
9	30-Nov-20	334	15 Rabīʿ al-Thānī 1442	15	2000	20.0	250.0
10	30-Nov-20	334	15 Rabīʿ al-Thānī 1442	15	1400	25.0	110.0
11	30-Nov-20	334	15 Rabīʿ al-Thānī 1442	15	2000	15.0	95.0
12	30-Nov-20	334	15 Rabīʿ al-Thānī 1442	15	1500	30.0	70.0
13	30-Nov-20	334	15 Rabīʿ al-Thānī 1442	15	1400	15.0	110.0
14	30-Nov-20	334	15 Rabīʿ al-Thānī 1442	15	700	24.0	145.0
15	01-Dec-20	335	16 Rabīʿ al-Thānī 1442	16	500	30.0	30.0
16	01-Dec-20	335	16 Rabīʿ al-Thānī 1442	16	800	20.0	90.0
17	01-Dec-20	335	16 Rabīʿ al-Thānī 1442	16	700	18.0	70.0
18	01-Dec-20	335	16 Rabīʿ al-Thānī 1442	16	1000	30.0	65.0
19	01-Dec-20	335	16 Rabīʿ al-Thānī 1442	16	1500	20.0	50.0
20	02-Dec-20	336	17 Rabīʿ al-Thānī 1442	17	500	30.0	38.0
21	02-Dec-20	336	17 Rabīʿ al-Thānī 1442	17	500	5.0	60.0
22	02-Dec-20	336	17 Rabīʿ al-Thānī 1442	17	700	3.0	22.0
23	02-Dec-20	336	17 Rabīʿ al-Thānī 1442	17	500	9.0	18.0
24	02-Dec-20	336	17 Rabīʿ al-Thānī 1442	17	700	28.0	25.0
25	03-Dec-20	337	18 Rabīʿ al-Thānī 1442	18	200	7.0	20.0
26	03-Dec-20	337	18 Rabīʿ al-Thānī 1442	18	200	5.0	25.0
27	03-Dec-20	337	18 Rabīʿ al-Thānī 1442	18	200	7.0	42.0
28	04-Dec-20	338	19 Rabīʿ al-Thānī 1442	19	1000	15.0	48.0
29	04-Dec-20	338	19 Rabīʿ al-Thānī 1442	19	800	7.0	25.0
30	04-Dec-20	338	19 Rabīʿ al-Thānī 1442	19	500	5.0	47.0
31	04-Dec-20	338	19 Rabīʿ al-Thānī 1442	19	500	10.0	29.0
32	04-Dec-20	338	19 Rabīʿ al-Thānī 1442	19	800	10.0	40.0
33	04-Dec-20	338	19 Rabīʿ al-Thānī 1442	19	1000	25.0	25.0
34	05-Dec-20	339	20 Rabīʿ al-Thānī 1442	20	500	25.0	32.5
35	05-Dec-20	339	20 Rabīʿ al-Thānī 1442	20	500	16.0	25.5
36	05-Dec-20	339	20 Rabīʿ al-Thānī 1442	20	500	12.0	23.0
37	05-Dec-20	339	20 Rabīʿ al-Thānī 1442	20	1000	13.0	30.0
38	06-Dec-20	340	21 Rabīʿ al-Thānī 1442	21	900	30.0	37.5
39	06-Dec-20	340	21 Rabīʿ al-Thānī 1442	21	500	19.0	20.0
40	06-Dec-20	340	21 Rabīʿ al-Thānī 1442	21	500	30.0	25.0
41	06-Dec-20	340	21 Rabīʿ al-Thānī 1442	21	700	25.0	30.0
42	13-Dec-20	347	28 Rabīʿ al-Thānī 1442	28	500	25.0	25.0
43	13-Dec-20	347	28 Rabīʿ al-Thānī 1442	28	500	10.0	22.0
44	14-Dec-20	348	29 Rabīʿ al-Thānī 1442	29	500	30.0	42.0
45	14-Dec-20	348	29 Rabīʿ al-Thānī 1442	29	1000	10.0	25.0
46	14-Dec-20	348	29 Rabīʿ al-Thānī 1442	29	700	28.0	40.0
47	14-Dec-20	348	29 Rabīʿ al-Thānī 1442	29	500	32.0	25.0
48	14-Dec-20	348	29 Rabīʿ al-Thānī 1442	29	700	29.0	48.0
49	14-Dec-20	348	29 Rabīʿ al-Thānī 1442	29	800	30.0	27.5

ID	Date in Gregorian		Date in Hijri		Ash column (m)	Max amplitude (mm)	Seismograph duration (s)
	Date	Doy (th)	Date	Lunar day (th)			
50	15-Dec-20	349	30 Rabīʿ al-Thānī 1442	30	700	24.0	35.0
51	15-Dec-20	349	30 Rabīʿ al-Thānī 1442	30	800	30.0	30.0
52	15-Dec-20	349	30 Rabīʿ al-Thānī 1442	30	800	30.0	30.0
53	03-Jan-21	2	19 Jumādā al-Awwal 1442	19	1000	30.0	70.0
54	08-Jan-21	7	24 Jumādā al-Awwal 1442	24	1000	20.0	25.0
55	10-Feb-21	40	28 Jumādā al-Thānī 1442	28	700	18.0	25.0
56	14-Mar-21	72	30 Rajab 1442	30	1300	7.0	20.0
57	15-Mar-21	73	01 Shaʿbān 1442	1	1000	31.0	30.0
58	13-Apr-21	102	01 Ramaḍān 1442	1	700	30.0	35.0
59	13-Apr-21	102	01 Ramaḍān 1442	1	400	26.0	23.0
60	13-Apr-21	102	01 Ramaḍān 1442	1	650	5.0	15.0
61	13-Apr-21	102	01 Ramaḍān 1442	1	700	30.0	40.0
62	13-Apr-21	102	01 Ramaḍān 1442	1	500	5.0	15.0
63	14-Apr-21	103	02 Ramaḍān 1442	2	700	5.0	15.0
64	15-Apr-21	104	03 Ramaḍān 1442	3	700	20.0	22.5
65	15-Apr-21	104	03 Ramaḍān 1442	3	1000	20.0	35.0
66	16-Apr-21	105	04 Ramaḍān 1442	4	1000	28.0	25.0
67	16-Apr-21	105	04 Ramaḍān 1442	4	1300	8.0	50.7
68	16-Apr-21	105	04 Ramaḍān 1442	4	1500	29.0	30.0
69	16-Apr-21	105	04 Ramaḍān 1442	4	1500	4.9	76.0
70	17-Apr-21	106	05 Ramaḍān 1442	5	1300	27.0	40.0
71	18-Apr-21	107	06 Ramaḍān 1442	6	1500	8.0	52.0
72	20-Apr-21	109	08 Ramaḍān 1442	8	750	6.2	23.0
73	21-Apr-21	110	09 Ramaḍān 1442	9	1000	7.0	94.0
74	25-Apr-21	114	13 Ramaḍān 1442	13	600	5.6	65.0
75	02-May-21	121	20 Ramaḍān 1442	20	500	6.4	15.0
76	20-May-21	139	08 Shawwāl 1442	8	500	21.6	23.0
77	22-May-21	141	10 Shawwāl 1442	10	700	49.0	28.1
78	22-May-21	141	10 Shawwāl 1442	10	800	22.0	35.0
79	22-May-21	141	10 Shawwāl 1442	10	800	14.0	25.0
80	22-May-21	141	10 Shawwāl 1442	10	400	44.0	25.0
81	24-May-21	143	12 Shawwāl 1442	12	600	28.0	30.0
82	26-May-21	145	14 Shawwāl 1442	14	800	4.5	30.0
83	28-May-21	147	16 Shawwāl 1442	16	1000	15.6	29.2
84	08-Jun-21	158	27 Shawwāl 1442	27	500	4.0	30.0
85	09-Jun-21	159	28 Shawwāl 1442	28	500	7.9	27.6
86	09-Jun-21	159	28 Shawwāl 1442	28	500	50.1	33.6
87	10-Jun-21	160	29 Shawwāl 1442	29	600	9.5	30.7
88	10-Jun-21	160	29 Shawwāl 1442	29	600	9.4	23.0
89	11-Jun-21	161	30 Shawwāl 1442	30	600	9.3	27.0
90	12-Jun-21	162	01 Dhū al-Qaʿdah 1442	1	700	18.6	36.0
91	12-Jun-21	162	01 Dhū al-Qaʿdah 1442	1	600	17.8	31.0
92	12-Jun-21	162	01 Dhū al-Qaʿdah 1442	1	700	14.0	28.0
93	13-Jun-21	163	02 Dhū al-Qaʿdah 1442	2	500	3.5	30.0
94	13-Jun-21	163	02 Dhū al-Qaʿdah 1442	2	500	18.0	34.0
95	13-Jun-21	163	02 Dhū al-Qaʿdah 1442	2	700	12.6	52.0
96	14-Jun-21	164	03 Dhū al-Qaʿdah 1442	3	750	9.0	39.0
97	14-Jun-21	164	03 Dhū al-Qaʿdah 1442	3	800	9.4	54.0
98	14-Jun-21	164	03 Dhū al-Qaʿdah 1442	3	600	19.0	45.0
99	16-Jun-21	166	5 Dhū al-Qaʿdah 1442	5	1000	16.0	47.0
100	21-Jun-21	171	10 Dhū al-Qaʿdah 1442	10	750	43.0	49.0
101	05-Jul-21	185	24 Dhū al-Qaʿdah 1442	24	1000	45.0	42.0

ID	Date in Gregorian		Date in Hijri		Ash column (m)	Max amplitude (mm)	Seismograph duration (s)
	Date	Doy (th)	Date	Lunar day (th)			
102	06-Jul-21	186	25 Dhū al-Qaʿdah 1442	25	800	45.0	46.0
103	07-Jul-21	187	26 Dhū al-Qaʿdah 1442	26	1100	25.0	76.0
104	07-Jul-21	187	26 Dhū al-Qaʿdah 1442	26	800	35.0	32.0
105	08-Jul-21	188	27 Dhū al-Qaʿdah 1442	27	700	45.0	40.0
106	11-Jul-21	191	01 Dhū al-Ḥijjah 1442	1	600	45.0	41.0
107	13-Jul-21	193	03 Dhū al-Ḥijjah 1442	3	800	17.0	32.0
108	14-Jul-21	194	04 Dhū al-Ḥijjah 1442	4	600	45.0	36.0
109	14-Jul-21	194	04 Dhū al-Ḥijjah 1442	4	700	19.0	30.0
110	14-Jul-21	194	04 Dhū al-Ḥijjah 1442	4	800	12.0	32.0
111	15-Jul-21	195	05 Dhū al-Ḥijjah 1442	5	800	17.0	49.0
112	17-Jul-21	197	07 Dhū al-Ḥijjah 1442	7	700	43.0	41.0
113	17-Jul-21	197	07 Dhū al-Ḥijjah 1442	7	1000	19.0	75.0
114	18-Jul-21	198	08 Dhū al-Ḥijjah 1442	8	800	36.0	38.0
115	19-Jul-21	199	09 Dhū al-Ḥijjah 1442	9	700	12.0	34.0
116	21-Jul-21	201	11 Dhū al-Ḥijjah 1442	11	700	7.0	32.0
117	21-Jul-21	201	11 Dhū al-Ḥijjah 1442	11	800	15.0	34.0
118	22-Jul-21	202	12 Dhū al-Ḥijjah 1442	12	700	43.0	46.0
119	22-Jul-21	202	12 Dhū al-Ḥijjah 1442	12	700	9.0	32.0
120	23-Jul-21	203	13 Dhū al-Ḥijjah 1442	13	800	11.0	39.0
121	28-Jul-21	208	18 Dhū al-Ḥijjah 1442	18	1000	45.0	49.0
122	28-Jul-21	208	18 Dhū al-Ḥijjah 1442	18	800	34.0	29.0
123	30-Jul-21	210	20 Dhū al-Ḥijjah 1442	20	800	45.0	41.0
124	30-Jul-21	210	20 Dhū al-Ḥijjah 1442	20	1000	45.0	40.0
125	04-Aug-21	215	25 Dhū al-Ḥijjah 1442	25	1300	28.0	41.0
126	05-Aug-21	216	26 Dhū al-Ḥijjah 1442	26	1000	38.0	40.0
127	08-Aug-21	219	29 Dhū al-Ḥijjah 1442	29	800	43.0	42.0
128	10-Aug-21	221	01 Muḥarram 1443	1	1500	11.0	60.0
129	11-Aug-21	222	02 Muḥarram 1443	2	2000	28.0	40.0
130	18-Aug-21	229	09 Muḥarram 1443	9	1000	38.0	42.3
131	19-Aug-21	230	10 Muḥarram 1443	10	800	6.0	20.0
132	21-Aug-21	232	12 Muḥarram 1443	12	800	8.0	42.0
133	22-Aug-21	233	13 Muḥarram 1443	13	1500	6.3	66.6
134	25-Aug-21	236	16 Muḥarram 1443	16	1200	29.0	37.0
135	26-Aug-21	237	17 Muḥarram 1443	17	800	17.0	39.0
136	26-Aug-21	237	17 Muḥarram 1443	17	1000	21.0	30.0
137	01-Oct-21	273	24 Safar 1443	24	800	22.2	23.1
138	14-Oct-21	286	07 Rabīʿ al-Awwal 1443	7	750	37.0	43.0
139	14-Oct-21	286	07 Rabīʿ al-Awwal 1443	7	800	13.0	62.0
140	14-Oct-21	286	07 Rabīʿ al-Awwal 1443	7	1000	20.0	45.0
141	15-Oct-21	287	08 Rabīʿ al-Awwal 1443	8	1000	40.0	40.0
142	15-Oct-21	287	08 Rabīʿ al-Awwal 1443	8	1000	22.0	34.0
143	27-Oct-21	299	20 Rabīʿ al-Awwal 1443	20	1000	29.2	33.0
144	29-Oct-21	301	22 Rabīʿ al-Awwal 1443	22	800	17.4	39.3
145	29-Oct-21	301	22 Rabīʿ al-Awwal 1443	22	800	39.5	52.0
146	21-Nov-21	324	16 Rabīʿ al-Thānī 1443	16	800	38.4	54.7
147	21-Nov-21	324	16 Rabīʿ al-Thānī 1443	16	1000	8.5	43.0
148	22-Nov-21	325	17 Rabīʿ al-Thānī 1443	17	2000	35.5	136.0
149	23-Nov-21	326	18 Rabīʿ al-Thānī 1443	18	700	9.0	128.0
150	26-Nov-21	329	21 Rabīʿ al-Thānī 1443	21	1000	22.2	47.0
151	13-Dec-21	346	09 Jumādā al-Awwal 1443	9	600	27.0	135.0
152	16-Dec-21	349	12 Jumādā al-Awwal 1443	12	500	28.0	48.0
153	16-Dec-21	349	12 Jumādā al-Awwal 1443	12	300	13.0	28.9

ID	Date in Gregorian		Date in Hijri		Ash column (m)	Max amplitude (mm)	Seismograph duration (s)
	Date	Doy (th)	Date	Lunar day (th)			
154	19-Dec-21	352	15 Jumādā al-Awwal 1443	15	1000	19.9	29.0
155	19-Dec-21	352	15 Jumādā al-Awwal 1443	15	800	38.0	28.9
156	19-Dec-21	352	15 Jumādā al-Awwal 1443	15	600	29.5	30.0
157	04-Jan-22	3	01 Jumādā al-Thānī 1443	1	600	37.6	53.0
158	11-Jan-22	10	08 Jumādā al-Thānī 1443	8	700	41.2	45.6
159	11-Feb-22	41	10 Rajab 1443	10	1000	31.4	40.8
160	11-Feb-22	41	10 Rajab 1443	10	1200	40.0	37.0
161	11-Feb-22	41	10 Rajab 1443	10	500	40.0	39.0
162	12-Feb-22	42	11 Rajab 1443	11	600	26.0	39.0
163	12-Feb-22	42	11 Rajab 1443	11	800	39.0	53.0
164	13-Feb-22	43	12 Rajab 1443	12	500	29.7	41.1
165	14-Feb-22	44	13 Rajab 1443	13	700	38.5	42.0
166	11-Mar-22	69	08 Shaʿbān 1443	8	600	7.5	46.0
167	11-Mar-22	69	08 Shaʿbān 1443	8	1000	33.0	64.5
168	31-Mar-22	89	28 Shaʿbān 1443	28	800	14.5	52.0
169	04-May-22	123	03 Shawwāl 1443	3	600	28.4	32.0
170	05-May-22	124	04 Shawwāl 1443	4	600	3.8	39.0
171	07-Jun-22	157	7 Dhū al-Qaʿdah 1443	7	800	12.4	62.0
172	08-Jun-22	158	8 Dhū al-Qaʿdah 1443	8	800	31.3	33.0
173	14-Jun-22	164	14 Dhū al-Qaʿdah 1443	14	700	15.0	61.0
174	16-Jun-22	166	16 Dhū al-Qaʿdah 1443	16	700	8054.0	41.0
175	17-Jun-22	167	17 Dhū al-Qaʿdah 1443	17	1000	33.0	48.0
176	01-Jul-22	181	02 Dhū al-Ḥijjah 1443	2	1300	20.0	46.0
177	01-Jul-22	181	02 Dhū al-Ḥijjah 1443	2	1200	38.0	59.0
178	02-Jul-22	182	03 Dhū al-Ḥijjah 1443	3	1000	16.0	34.0
179	02-Jul-22	182	03 Dhū al-Ḥijjah 1443	3	1200	40.0	74.0
180	02-Jul-22	182	03 Dhū al-Ḥijjah 1443	3	700	20.8	40.0
181	02-Jul-22	182	03 Dhū al-Ḥijjah 1443	3	800	31.0	33.0
182	02-Jul-22	182	03 Dhū al-Ḥijjah 1443	3	900	25.0	59.0
183	02-Jul-22	182	03 Dhū al-Ḥijjah 1443	3	1000	19.2	113.0
184	03-Jul-22	183	04 Dhū al-Ḥijjah 1443	4	1000	12.8	35.0
185	07-Jul-22	187	08 Dhū al-Ḥijjah 1443	8	700	21.7	30.0
186	09-Jul-22	189	10 Dhū al-Ḥijjah 1443	10	600	34.2	35.0
187	20-Jul-22	200	21 Dhū al-Ḥijjah 1443	21	700	17.5	30.0
188	02-Sep-22	244	05 Shafar 1444	5	1500	34.0	34.0
189	16-Sep-22	258	19 Shafar 1444	19	1500	33.0	43.0
190	17-Sep-22	259	20 Shafar 1444	20	1000	2.5	78.0
191	23-Sep-22	265	26 Shafar 1444	26	800	20.1	50.0

Sustainable Engineered Cementitious Composite using Natural Sisal and Coir Fibers

Suranta Sembiring[1], Patria Kusumaningrum[2]*, Zaki Anwar F[3]

[1,2,3]Civil Engineering Department, Faculty of Civil and Environmental Engineering, The Institute of Technology Bandung, Bandung 40132, Indonesia

*Corresponding author: patria.k@itb.ac.id

Abstract. Engineered Cementitious Concrete (ECC) or bendable concrete is a high-performance concrete (HPC) with high ductility characteristic. Generally, it has been used to strengthen or retrofit existing structures to increase structural resilience against extreme loadings such as earthquakes or blasts. The constituent materials of ECC are fine aggregates, cement, supplementary cementitious composite (silica fume, fly ash, etc.), superplasticizer, water, and the use of fibers up to 2% volume fraction. With the use of fibers and the elimination of coarse aggregate, ECC has deflection-hardening in bending and exhibits fiber-bridging by multiple cracking. However, the development of ECC has so far been limited to using only synthetic and inorganic fibers, which are scarce yet expensive in Indonesia. As Indonesia boasts the second highest level of biodiversity in the world as the source of natural fibers, in this research, the natural fibers are used to replace synthetic and inorganic fibers in making ECC more sustainable, low cost yet locally available. Not only did the natural fibers are locally available at affordable prices, but the fibers also have distinctive characteristics of sustainable materials with high tensile strength and significant elongation. From the experiments, one may be concluded that natural fibers of sisal and coir can improve the ECC flexural strength up to 39.5% of the non-fiber mix. This highlights its ability to replace synthetic and inorganic fibers in ECC mixture.

Keywords. ECC, natural fibers, sisal fiber, coir fiber, sustainable material, low cost, locally available, flexural strength, ductility

1. Introduction

ECC or bendable concrete is a fiber reinforced cementitious material consisting of a cement composite with a random distribution of short fibers with a volume fraction of about 2% with extremely high ductility and toughness. ECC has much higher flexural and tensile strength characteristics than normal concrete, thus reducing structural damage due to extreme tensile and bending events, for example, blast [1] and impact [2]. Engineered Cementitious Composite (ECC) or Bendable Concrete is a Fiber Reinforced Concrete (FRC) development, as ECC contains fiber in its mixture too. However, under tensile loads, FRC produces tension-softening properties after the concrete shows initial cracking, so that the load-bearing capacity of the concrete then decreases. ECC is the result of several constituent materials that can make concrete become ductile. ECC or bendable concrete, is designed with the aim of overcoming the weaknesses of ordinary concrete, low tensile deformation capacity. ECC's stress-strain typically has a similar trend to metallic materials subjected to tensile strain-hardening properties. This thing due to ECC can create multiple cracking after the initial crack is exceeded [3].

Fiber is a small filament of natural or synthetic materials with a minimum aspect ratio of 100 times and has flexible and strong properties. Fibers can be classified as synthetic, semisynthetic, and natural fibers [4]. Synthetic fibers are man-made fibers through chemical synthesis, generally made of polymer or steel, which have a uniform and high mechanical property. However, they are high-priced, scarce especially in remote areas, yet not environmental-friendly. Natural fibers, on the other hand, can easily be found in the remote areas of Indonesia, and have long been used to meet human needs to support a broad type of industries. Nowadays, natural fibers are used in textiles, paper, crafts, accessories, decoration, and bio-composite materials. Natural fibers include cotton, silk, palm, sisal, coir, and empty fruit bunch of palm oil.

Currently, ECC is being developed as a reinforced or retrofit material made of cement paste and synthetic fibers such as polymer and steel fibers. In Indonesia, synthetic fibers, especially polymer fibers, are not manufactured locally. The procurement of synthetic fiber must be done through import, and this causes the cost of ECC production in Indonesia very expensive. Providentially, Indonesia boasts the second highest level of biodiversity worldwide as the source of natural fibers. A Map of the distribution of natural fiber sources in Indonesia is shown in Figure 1. Not only did the natural fibers are locally available with affordable prices, but the fibers also have distinctive characteristics of sustainable materials with high tensile strength and significant elongation. This paper comprises preliminary results of the experimental characterization of mechanical properties of ECC and effect of length and volume fraction of fiber in ECC with natural fibers, namely sisal and coir fiber.

Figure 1. Natural fiber distribution map of Indonesia [4]

2. Research Significance

In the development of ECC, commonly only polymer and/or steel fibers are utilized. In this study, a new set of ECC is developed considering environmental sustainability and domestic availability. Specifically, synthetic fiber is herein substituted by natural fibers namely sisal and coir fiber. These sisal and coir fiber ECCs are expected to promote infrastructure resilience and sustainability through simultaneous enhancements of natural fiber.

3. Material and Mixing Proportion

3.1. Material

The mixtures carried out in this study consist of Ordinary Portland Cement (OPC), Silica Fume, Class F fly ash as cementitious material, quartz sand as fine aggregates, sisal and coir fibers, water, and polycarboxylate-based High Range Water Reducing (HRWR) admixture as shown in Figure 2. The

IOP Conf. Series: Earth and Environmental Science 1245 (2023) 012007 doi:10.1088/1755-1315/1245/1/012007

cement Type 1 OPC which conforms to ASTM C150 standard [5] is used. The chemical composition of OPC, fly ash, and silica fume are evaluated through XRF analysis and presented in Table 1. All the materials are available locally in Indonesia. Sisal and coir fibers were obtained from Blitar, East Java, and Banyumas, Central Java, respectively.

3.2. Natural Fiber Tensile Strength and Elongation Test

Testing of the tensile characteristics of natural fibers is carried out in accordance with SNI 08-1112-1989 [6]. The tensile test was carried out per bundle basis as the dimensions of each single sample vary greatly with the testing setup as shown in Figure 3.

Figure 2. Raw ingredients

Table 1. Composition of OPC, fly ash, and silica fume

Chemical Oxide	OPC (%)	Fly Ash (%)	Silica Fume (%)
Na_2O	0,302	3,87	-
MgO	1,84	3,13	0,619
Al_2O_3	6,4	24,7	0,551
SiO_2	20,4	47,7	96,4
P_2O_5	0,798	0,788	0,187
SO_3	2,44	0,755	1,24
K_2O	0,473	1,81	0,146
CaO	55,4	10,5	0,459
TiO_2	0,464	0,414	-
MnO	0,726	0,089	-
Fe_2O_3	9,84	6,14	0,108
SrO	0.282	0,107	-

3.3. Slump Test

Characteristics of fresh ECC with sisal and coir fibers were evaluated to comprehend its workability. Slump tests conforming to ASTM C143 [7] are measured periodically at 0, 30, 60, 90 and 120 min to obtain the slump loss.

3.4. Mixing Procedures

The mixing was carried out conventionally by using a Hobart mixer as shown in Figure 4. Initially, the dry ingredients of quartz sand, fly ash, silica fume, and OPC, in accordance with the mixture proportion of ECC, were added to the mixing bowl and mixed until homogeneous. Table 2 shows the mixture proportion of the ECC with sisal and coir fibers. Fibers were added gradually while stirring the dry mixture. Water and superplasticizer were then poured into the mixture. After the mix was consistent and homogenous, it was then carefully poured into the specified molds.

Table 2. Mix proportion of ECC using sisal and coir fibers

Mix	ID	OPC	Fly Ash	Silica Fume	Quartz Sand	w/b	SP	Fiber	
								v.f. (%)	Length (mm)
ECC No Fiber	ECC-NF	0.9	-	0.1	0.75	0.30	0.015		-
ECC Sisal Fiber	ECC-SF12-1%	0.9	-	0.1	0.74	0.30	0.015	1	12
	ECC-SF12-2%	0.9	-	0.1	0.73	0.30	0.015	2	12
	ECC-SF30-1%	0.9	-	0.1	0.74	0.30	0.015	1	30
	ECC-SF30-2%	0.9	-	0.1	0.73	0.30	0.015	2	30
	ECC-SF1230-1%	0.9	-	0.1	0.74	0.30	0.015	1	12 and 30
	ECC-SF1230-2%	0.9	-	0.1	0.73	0.30	0.015	2	12 and 30
ECC Coir Fiber	ECC-CF12-1%	0.45	0.55	-	0.53	0.12	0.004	1	12
	ECC-CF12-2%	0.45	0.55	-	0.52	0.12	0.004	2	12
	ECC-CF30-1%	0.45	0.55	-	0.53	0.12	0.004	1	30
	ECC-CF30-2%	0.45	0.55	-	0.52	0.12	0.004	2	30
	ECC-CF1230-1%	0.45	0.55	-	0.53	0.12	0.004	1	12 and 30
	ECC-CF1230-2%	0.45	0.55	-	0.52	0.12	0.004	2	12 and 30

3.5. Curing of Specimens

Upon casting the specimens, they were left in their specified molds. After hardened, the specimens were then placed in a water pond for the required number of days in room temperature water as shown in Figure 5.

3.6. Compressive Strength Test

Casted specimens were tested for 3, 14, and 28 days for their compressive strength. In doing the compressive strength test, ASTM C39 was adhered [8], determining the specimen's maximum compressive load before failure. Figure 6 shows the setup of the compressive strength test with a cylinder specimen.

3.7. Flexural Test

The next key parameter to be studied to reach the objectives of this study is to determine the flexural strength of the ECC mixes. Unlike the compressive strength test, the flexural test will be carried out for the 28-day strength of each mix. For the flexural strength test, a beam mold of 500 mm x 100 mm x 100 mm is used. The beam of length L rests on two roller supports is subjected to a four-point bending test using two concentrated load P in accordance to ASTM C78 as shown in Figure 7 to determine the flexural strength of the ECC [9].

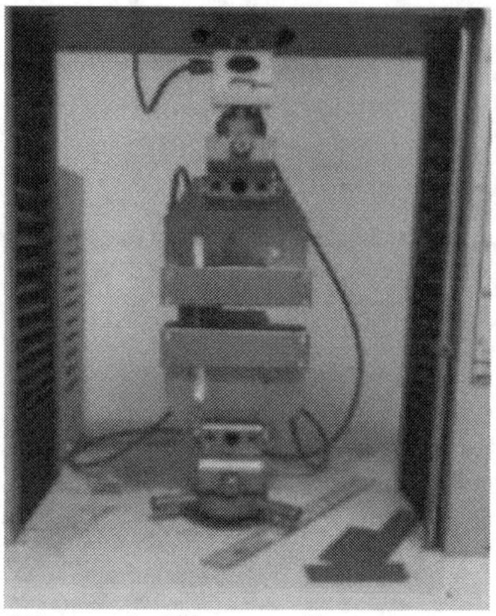

Figure 3. Fiber tensile test setup

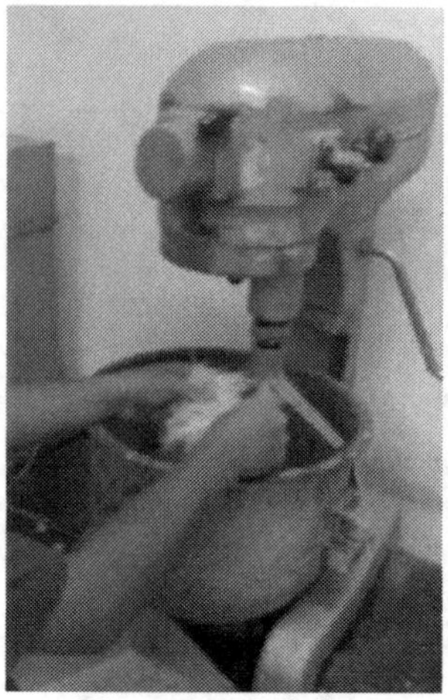

Figure 4. Mixing

Figure 5. Curing

Figure 6. Compressive test setup

Figure 7. Four points bending test setup

4. Result and discussions

4.1. Natural Fiber Tensile Strength and Elongation

The result of tensile strength and elongation of some natural fiber obtained in literature and by experimental observation is represented in Figure 8 and Figure 9. In general, the tensile strength of natural fibers obtained through observation is lower than that of the natural fibers tests carried out by other researchers based on earlier reports [10][11][12][13][14]. This may be due to the nature of the natural fibers. Its strength is strongly influenced by the growth conditions of fiber-producing plants. In Indonesia, natural fibers are generally harvested from fiber-producing plants that were grown with minimal treatment. This may affect the quality of the natural fibers e.g., lower tensile strength. Sisal fiber has the highest tensile strength based on both literature (510-710 MPa) [15] and observation (335.35 MPa). Unlike its tensile strength values, the elongation of natural fibers (Figure 9) that obtained from observations is higher than the elongation based on the literature. The highest elongation of natural fibers is achieved by coir fiber which is 30 % based on literature [10] and 51,84% by observation.

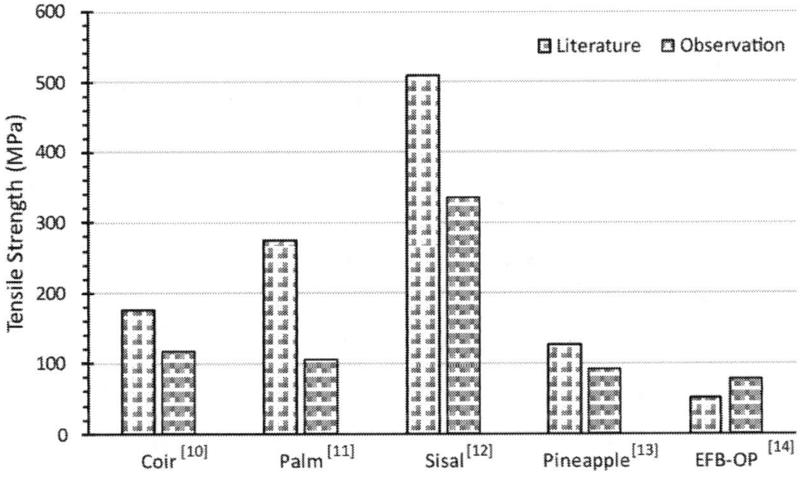

Figure 8. Tensile strength of natural fibers

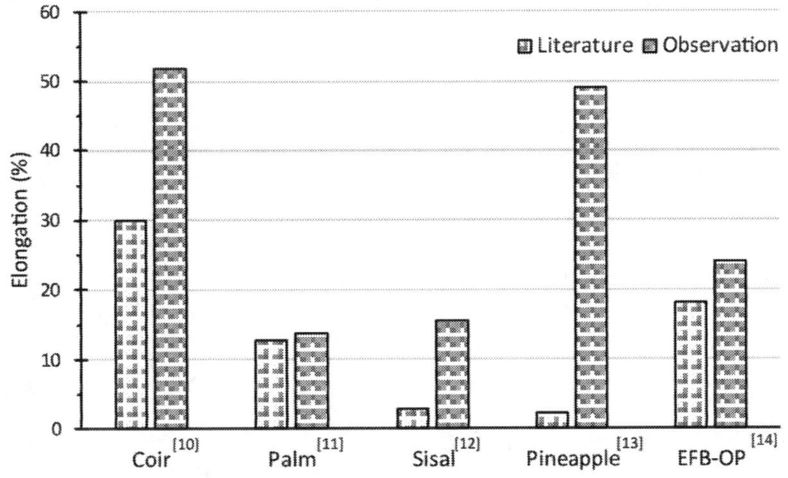

Figure 9. Elongation of natural fibers

4.2. Fresh Concrete Behavior

Characteristics of the fresh mixture of ECC with natural fiber are evaluated with respect to its slump and slump loss. Figure 10 shows the slump values of various ECC proportions. Theoretically, the workability of concrete is only affected by the absorption of water by the aggregate, evaporation, and hydration of the cement. As the ECC mixture contains fibers, the number of fibers, i.e. fiber volume fraction will also affect the workability. Based on the experimental observation, the higher volume fraction of fiber in the concrete, the lower the slump value is. The volume fraction of fiber in the ECC strongly affects its slump value. This is mainly due to the increase in friction between the ECC constituent material and the fiber, which is proportional to the amount of the mixed fiber.

Similarly, slump loss of ECC for both types of natural fibers show a comparable trend over the reviewed time duration. Within two hours of duration, a decrease in the slump of 52.4% for non-fiber mixtures, 60.1% for 1% sisal fiber ECC, and 80.3% for 2% sisal fiber ECC were observed. This shows that the decrease in the slump is considerably affected by the volume fraction of fiber in the ECC mixture.

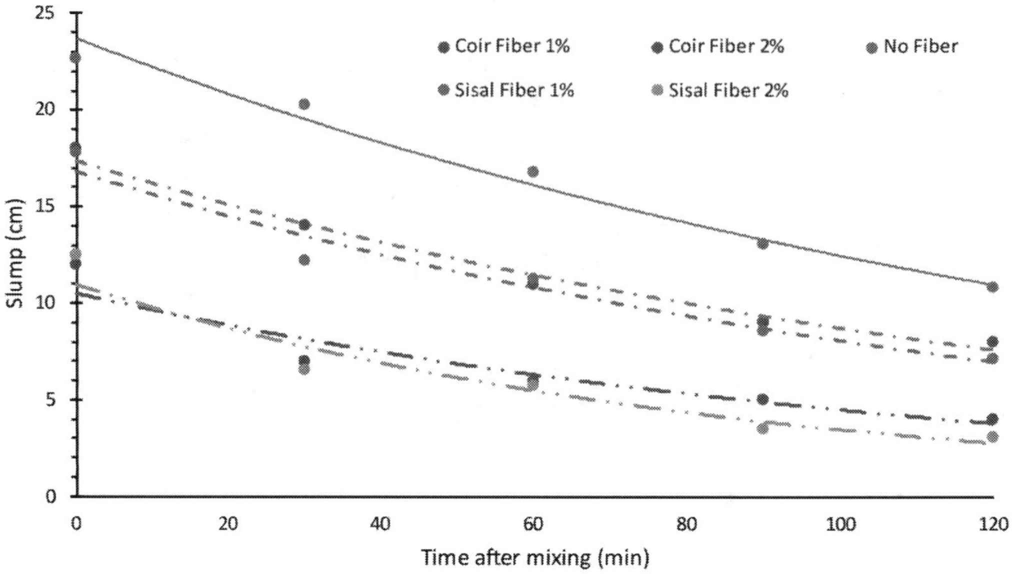

Figure 10. Slump of ECC sisal and coir fiber

4.3. Concrete Compressive Strength

The cylinder specimens were tested to obtain the compressive strength of ECC with sisal and coir fibers as shown in Figure 11(a) and (b) respectively at 14 and 28 days. At 14 days, the compressive strength of ECC non fiber (ECC-NF-S) concrete showed a higher value (41.1 MPa) compared to ECC using a volume fraction of 1% and 2% sisal fiber relatively.

Conversely, the compressive strength of the ECC non fiber (ECC-NF-C) mixture showed a lower value (15.77 MPa). This low strength may be due to the high volume of fly ash as pozzolan used in the ECC coir fiber mixture. For both ECC with sisal and coir fibers, ECC with a volume fraction of 1% has a higher compressive strength value than 2% fiber. The strength variation, however, is rather insignificant for ECC with sisal fiber, in comparison to ECC with coir fiber. Meanwhile, the variation of fiber length is observed to have no effect on the compressive strength of ECC at 14 days.

Until it is fully hydrated, ECC continued to experience an increase in compressive strength. From the experiment, it is observed that the compressive strength of ECC with sisal fiber has a rather similar value with less fluctuation. The non-fiber ECC mixture has a compressive strength of 46.15 MPa, while ECC with sisal fiber has a value between 40.73 to 45.40 MPa. For ECC with a coir fiber mixture, a significant difference in compressive strength is observed as the volume fraction is varied. The non-coir fiber ECC mixture shows an extreme ascent up to 37.27 MPa, while ECC using coir fiber achieves compressive strength from 26 to 36.67 MPa. This may again be due to the high volume of fly ash as pozzolan used in the ECC coir fiber mixture. It may then be concluded that the compressive strength of ECC was not significantly affected by variations in the length and volume fraction of fibers used in the mixture.

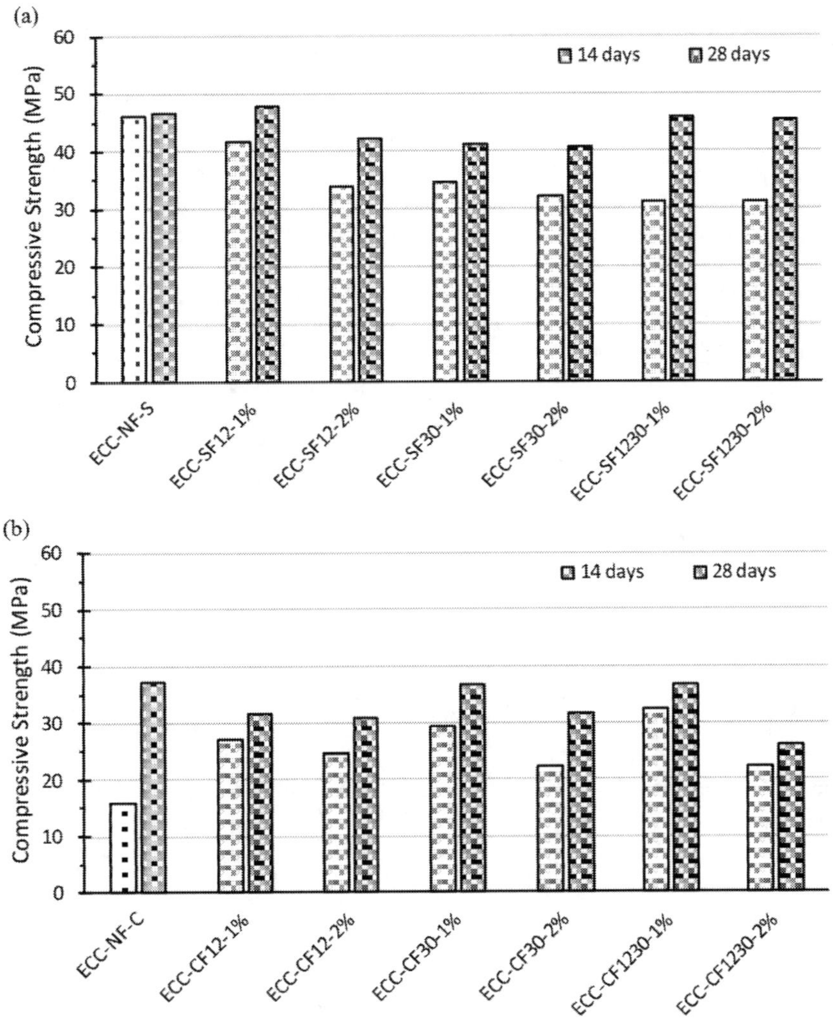

Figure 11. Compressive strength at 14 and 28 days of (a) ECC sisal fiber, and (b) ECC coir fiber

4.4. Flexural Strength

Figure 12 shows the comparison of ECC flexural strength between natural fiber and synthetic fiber obtained from literatures [16][17][18] with respect to fiber volume fraction. It is evident that the addition of natural fibers in the ECC mixture increases the maximum flexural strength of the materials in comparison to the non-fiber sample. In this study, however, this increase is not accompanied by the presence of a deflection hardening zone. The flexural strength of non-fiber ECC mixture is 2,671 MPa. The use of natural fiber in the ECC mixture results in an increase with respect to the flexural strength of around 28.5% to 39.5% for ECC with sisal fiber and 12.3% to 24.7% for ECC with coir fiber in comparison to the flexural strength of non-fiber ECC. The highest flexural strength of ECC with natural fiber was 3.727 MPa i.e., ECC of 2% volume fraction with 12 mm length of sisal fiber and 3.33 MPa for ECC of 1% volume fraction of mix-length (12 and 30 mm length) of coir fiber. The flexural behavior before and after cracking is also reviewed.

The load-deflection curves of three mixtures of different fiber contents are illustrated in Figure 13. The flexural stiffness in the elastic state (first slope) is rather similar for the three curves, while the flexural stiffness after the yielding point varies.

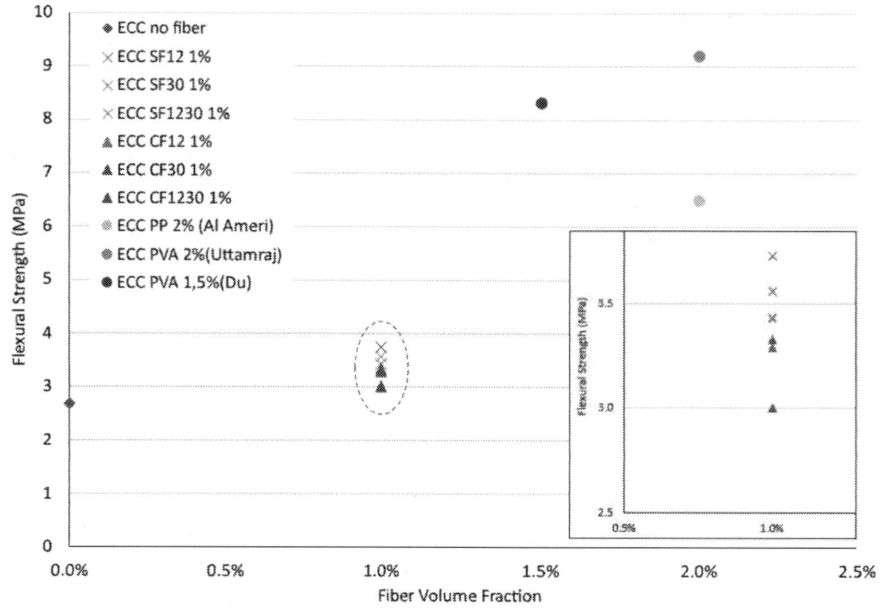

Figure 12. Comparison of ECC flexural strength between natural fiber and synthetic fiber [16][17][18]

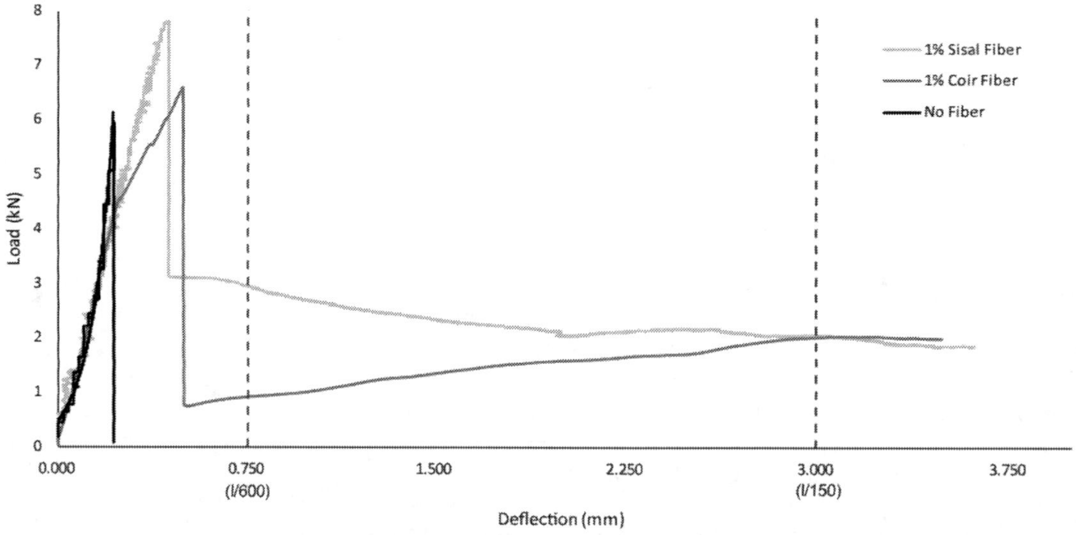

Figure 13. Load-deflection curves of ECC with natural fibers

Conforming to ASTM C1609, equivalent flexural strength ratio, $R_{T.150}^{D}$ which correlates to flexural toughness with the formula shown in Equation (1) is also evaluated for each variation of ECC mixtures.

$$R^{D}_{T.150} = \frac{150 \cdot T^{D}_{150}}{f_1 \cdot b \cdot d^2} \cdot 100\% \quad (1)$$

T^{D}_{150} is toughness, obtained through integration of the load deflection curve, i.e. total area under the curve up to a net deflection of $1/150$ of the span length. f_1 is first-peak strength, whereas b and d are the width and the depth of flexural specimen, respectively. The flexural toughness of ECC with sisal fiber is 10.58 Joule, nearly two times higher than that of the ECC with coir fiber i.e. 5.41 Joule. Equivalent flexural strength ratio, $R^{D}_{T.150}$, of ECC with sisal fiber is 50.4%, about 1.83 times higher than that of the ECC with coir fiber which is 27.4%.

Given Indonesia's archipelagic nature and its susceptibility to geohazards such as earthquakes and landslides, the higher flexural strength and flexural toughness of Engineered Cementitious Composites (ECC) reinforced with natural sisal fibers offer promising prospect. This locally available and cost-effective solution holds great potential as a geohazard retrofit material, thereby contributing to sustainability initiatives aimed at strengthening the housing and infrastructures. The use of ECC with natural sisal fibers as geohazard retrofit material could serve as a ground-breaking suggestion for promoting resilient housing and infrastructure practices.

5. Conclusion

This paper evaluates the mechanical properties of ECC with the use of natural fibers of sisal and coir fibers. Sisal fiber is found to be among the natural fibers with the highest tensile strength, 335 MPa tensile strength. Coir fiber, on the other hand, is among the natural fibers with the highest elongation viz., 51.84% elongation. The volume fraction of natural fibers in the ECC mixture significantly affected its workability, i.e. slump and slump loss. The more volume fraction of ECC, the lower slump, and the higher slump loss is gained. ECC with the use of sisal fiber produces a compressive strength between 40.73 to 45.40 MPa. It was observed that the compressive strength of ECC concrete was not significantly affected by variations in length and volume fraction of fiber. The use of natural fibers increases the flexural strength of ECC around 28.5% to 39.5% and 12.3% to 24.7% for sisal and coir fibers. It has also been analysed that ECC with sisal fiber exhibits higher flexural toughness nearly two times than the ECC with coir fiber. This feature highlights its potential as a retrofit material for geohazard mitigation in housing and infrastructure applications.

Acknowledgement

This paper and the related research are the outcomes of the exceptional support from Institut Teknologi Bandung through the Program of Research, Community Service and Innovation (P2MI) of the Faculty of Civil and Environmental Engineering.

References

[1] Jing Z, M Maalej, and Ser T 2007 *Performance of Hybrid-Fiber ECC Blast/Shelter Panels Subjected to Drop Weight Impact* (Singapore: ASCE-JMCE) pp 855-856
[2] Qi, B., Q. Kong, H. Qian, D. Patil, I. Lim, M. Li, D. Liu, and G. Song 2018 *Study of Impact Damage in PVA-ECC Beam under Low-Velocity Impact Loading Using Piezoceramic Transducers and PVDF Thin-Film Transducer* (Houston: Sensors 18) p 2
[3] Li V 2019 *Engineered Cementitious Composites (ECC) Bendable Concrete for Sustainable and Resilient Infrastructure* (Michigan: Springer) p 12
[4] Suparno O 2020 *The Potency and Future of Indonesia's Natural Fibres as Raw Materials for Various Industries* (Bogor: Jurnal Teknologi Industri Pertanian) pp 221-223
[5] ASTM 2001 ASTM C150: Standard Specification for Portland Cement (Philadelphia: ASTM)
[6] BSN 1989 SNI 08-1112-1989 Cara Uji Kekuatan Tarik dan Mulur Serat Batang per Bundel (Jakarta: BSN)
[7] ASTM 2001 ASTM C143: Standard Test Method for Slump of Hydraulic Cement Concrete (West

IOP Conf. Series: Earth and Environmental Science 1245 (2023) 012007 doi:10.1088/1755-1315/1245/1/012007

Conshohocken: ASTM)
[8] ASTM 2009 ASTM C39: Standard Test Method for Compressive Strength of Cylindrical Concrete Specimens (West Conshohocken: ASTM)
[9] ASTM 2010 ASTM C1609: Standard Test Method for Flexural Performance of Fiber-Reinforced Concrete (Using Beam with Third-Point Loading) (West Conshohocken: ASTM)\
[10] Reddy, P 2010 *Investigation on Tensile and Flexural Properties of Coir Fiber Reinforced Isophthalic Polyester Composites* (Tamil Nadu: Int.J.of Curr. Engg. and Technol. 2) p 223
[11] Mukhtar, I., Leman, Z., Ishak, M. R., and Zainudin, E. S. 2016 *Sugar palm fibre and its composites: A review of recent developments* (BioRes. 11) p 10765
[12] Sathiskumar G 2018 *Vibration and Viscoelastic Characteristics of Sisal Fiber Reinforced Polyester Composite* (Tamil Nadu: TIJEMPERD) p 330
[13] Arib, R.M.N., S.M. Sapuan, M.M.H.M. Ahmad, M.T. Paridah, and H.M.D.K. Zaman 2006 *Mechanical Properties of Pineapple Leaf Fibre Reinforced Polypropylene Composites.* (Selangor: Materials & amp) p 392
[14] Shinoj, S., R. Visvanathan, S. Panigrahi, and M. Kochubabu 2011 *Oil palm fiber (OPF) and its composites: A review* (Andrha Pradesh: ELSEVIER-Industrial Crops and Products) p 10
[15] Mallikarjun B K, Arun S K, Nagaraj, Ravindra A R, Ashok 2017 *Testing Of Sisal Fiber Reinforced Epoxy Composites At Different Orientations For Tensile And Compression* Vol 2 Issue 2 (Telangana: IJARIIE) p 118
[16] Al-Ameri, R.A, Abid, S.R. Özakça 2021 *Mechanical and Impact Properties of Engineered CementitiousComposites Reinforced with PP Fibers at Elevated Temperatures* (Gaziantep: MDPI) pp 7-8
[17] S.Uttamraj, K. Ashwanth, Rafeeq 2016 *An comparative study on Conventional concrete and Engineered Cementitious Composites(ECC-PVA)-REVIEW* (Bandlaguda: IOSR-JMCE) pp 23-24
[18] Maalej M, Li V C 1994 *Flexural/Tensile-Strength Ratio in Engineered Cementitious Composites* (ASCE: JMCE) pp 518-519
[19] Li V 2019 *Engineered Cementitious Composites (ECC) Bendable Concrete for Sustainable and Resilient Infrastructure* (Michigan: Springer) pp 313-367

The Impact of Environmental, Social, And Economic Factors Due to Anak Krakatau Volcano Tsunami Using Remote Sensing Technology

Riantini Virtriana [1,2], Agung Budi Harto[1,2], Pitri Rohayani[3], Tania Septi Anggraini[2,4], and Kalingga Titon Nur Ihsan[4]

[1]Remote Sensing and GIS Research Group, Faculty of Earth Science and Technology, Institut Teknologi Bandung, Indonesia;

[2]Center for Remote Sensing, Institut Teknologi Bandung, Indonesia;

[3]Department of Geodesy and Geomatics Engineering, Faculty of Earth Sciences and Technology, Institut Teknologi Bandung, Indonesia;

[4]Doctoral Program of Geodesy and Geomatics Engineering, Faculty of Earth Sciences and Technology, Institut Teknologi Bandung, Indonesia.

Corresponding author: virtriana@itb.ac.id

Abstract. Tsunamis often occur in the territory of Indonesia because it is widely traversed by encounters between earth plates under the sea. Thus, if an earthquake occurs, it has the potential to generate a tsunami. The Indonesian region is also located in the ring of the fire area, and is the country with the highest number of active volcanoes with more than 127 events, that can encourage volcanic earthquakes. In addition, when volcanoes erupt offshore, the potential for a tsunami will be even greater because the expulsion of material can generate landslides toward the sea, causing a tsunami. The tsunami that happened in the eruption of Anak Krakatau on December 22, 2018, in the Sunda Strait, resulted in a tsunami in the Lampung and Banten areas. This study tries to compare environmental, social, and economic factors before and after the tsunami in the coastal areas of Lampung and Banten. The integration of remote sensing data will be carried out to determine environmental, social, and economic factors such as Land Cover, NDWI, NDVI, Night Light, and NO2 data. Random Forest Machine Learning will also be involved in shaping land cover models. Random Forest is often used because it has high accuracy. The results show a significant change in land cover in coastal areas and the impact of environmental, social, and economic factors. This study is expected to support the mitigation process in the event of a tsunami in the future. In addition, it is expected to be used by policymakers in planning development after the tsunami disaster.

Keywords: Volcano Tsunami, Remote Sensing, Random Forest, Environmental Factor

1. Introduction

On 22 December 2018, Indonesia experienced a tsunami in the Sunda Strait area, which is the strait between Lampung and Banten Provinces (BNPB, 2018). Tsunamis can be caused by shallow earthquakes under the seabed, both tectonic and volcanic earthquakes, which can displace water volumes (Yeh et al., 2022). However, the cause of the tsunami in the Sunda Strait in 2018 was not caused by these two factors. An avalanche of volcanic material caused this tsunami from the eruption of Mount Anak Krakatau (ESDM, 2018). The height of the water waves from the 2018 South Sunda tsunami varied widely, from heights of 6.63 m (Syamsidik et al., 2019) to 12.58 m (Takabatake et al., 2019) in Pandeglang.

The tsunami in the Sunda Strait caused many fatalities. Around 437 people died, and hundreds of others were injured. One of the reasons for the high fatalities is the lack of early warning to the public (Syamsidik et al., 2020). Not only that, but this tsunami also caused a lot of damage to buildings (B. Syamsidik et al., 2020). Previous research conducted by Virtriana et al., (2023) showed that two years after the tsunami occurred in Pandeglang Regency (Banten), there are still traces of damage to buildings due to the tsunami, namely around 10% of buildings have not been repaired, and 19% of buildings are still badly damaged. It was also found that buildings in coastal areas and riverbanks near the estuary suffered more severe damage compared to other buildings(Virtriana et al., 2023).

Based on the previous explanation, a tsunami can harm the surrounding environment, both natural and man-made. In general, tsunamis can cause flooding, building damage, scour and soil instability, strong winds, and fire damage due to oil spills that occur as a result of tsunamis (Yeh et al., 2022). Several previous studies have discussed the impact of the tsunami on an area before and after the tsunami, especially using remote sensing technology and machine learning. The first is research by Sublime & Kalinicheva,(2019) which conducted damage mapping after the tsunami using Deep Learning Remote Sensing with the Tohoku Tsunami case study in 2011. The second is research by Adriano et al., (2019) which integrates various remote sensing data with fusion techniques and machine learning to observe the damage after the earthquake and tsunami in Palu, Sulawesi, in 2018. The third is a study by Virtriana et al., (2023) which tries to observe the impact of the tsunami and focuses on damage to buildings before and after the tsunami using Random Forest with various types of predictors with the 2018 tsunami case study in the Sunda Strait.

Similarly to previous research, this study will also try to observe the tsunami's impact by integrating remote sensing technology and machine learning. However, this research will focus on two major aspects, namely environmental and socioeconomic characteristics before and after the tsunami. Environmental factors will consider land cover before and after the tsunami. Land cover will be processed through the random forest machine learning method. In addition, the environmental impact will also consider the vegetation index and air index. On the other hand, socioeconomic factors will consider air pollution and night light which are used as indicators of economic activities and activities of a community. This research aims to be used as a basis for tsunami disaster management in the future, especially in Indonesia.

2. Methodology

2.1. Study Area

The study area of this research is the provinces of Banten and Lampung. The two regions are separated by the Sunda Strait, which is where Mount Anak Krakatau is located. According to BNPB data (2018), there are 14 sub-districts that cover coastal areas affected by the tsunami, where the fourteen sub-districts are located in Serang, Pandeglang and South Lampung Regencies. Therefore, this study will focus only on these three regencies that were affected by the tsunami. A more detailed look of the study areas can be seen in Figure 1.

IOP Conf. Series: Earth and Environmental Science 1245 (2023) 012008 doi:10.1088/1755-1315/1245/1/012008

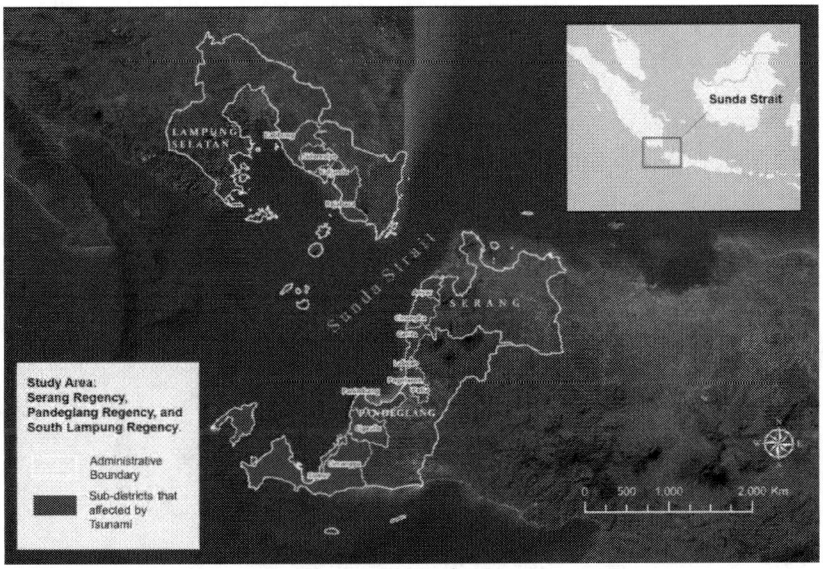

Figure 1. Study Area

2.2. Data

Data used in this study can be seen in Table 1. In general, this study uses five data consisting of administrative data, tsunami-effected region, Landsat-8, VIIRS Nightime, and Sentinel-5P NO_2. The data landsat 8 that used in this study combine some images around 2018 until 2019 to identify the situation before tsunami and after tsunami.

Table 1. Data used in this study

No	Data	Product	Spatial Resolution	Temporal Resolution	Reference
1	Administrative Data	Topographic Map of Indonesia (RBI)	1:25.000	-	(Badan Informasi Geospasial (BIG), 2021)
2	Tsunami-effected Region	Sub-districts affected by the Sunda Strait tsunami, Banten and Lampung provinces	1:25.000	-	(BNPB, 2018)
3	Landsat-8	USGS Landsat 8 Collection 1 Tier 1 TOA Reflectance	30 m	1 Month (Before: November 2018 and After: December 2018)	(USGS, 2018)
4	VIIRS Nighttime Day/Night Band Composites Version 1	Earth Observation Group, Payne Institute for Public Policy, Colorado School of Mines	463.83 m	1 Month (Before: November 2018 and After: December 2018)	(CSM, 2019)
5	Sentinel-5P OFFL NO2: Offline Nitrogen Dioxide	European Space Agency (ESA)	3.5x7 km	1 Day (Before: November 2018 and After: December 2018)	(ESA, 2018)

2.3. Methods

The general methods in this study can be seen in Figure 2. The method can be divided into two sections. The first stage is the environmental factor data processing and the second stage is the socioeconomic data processing.

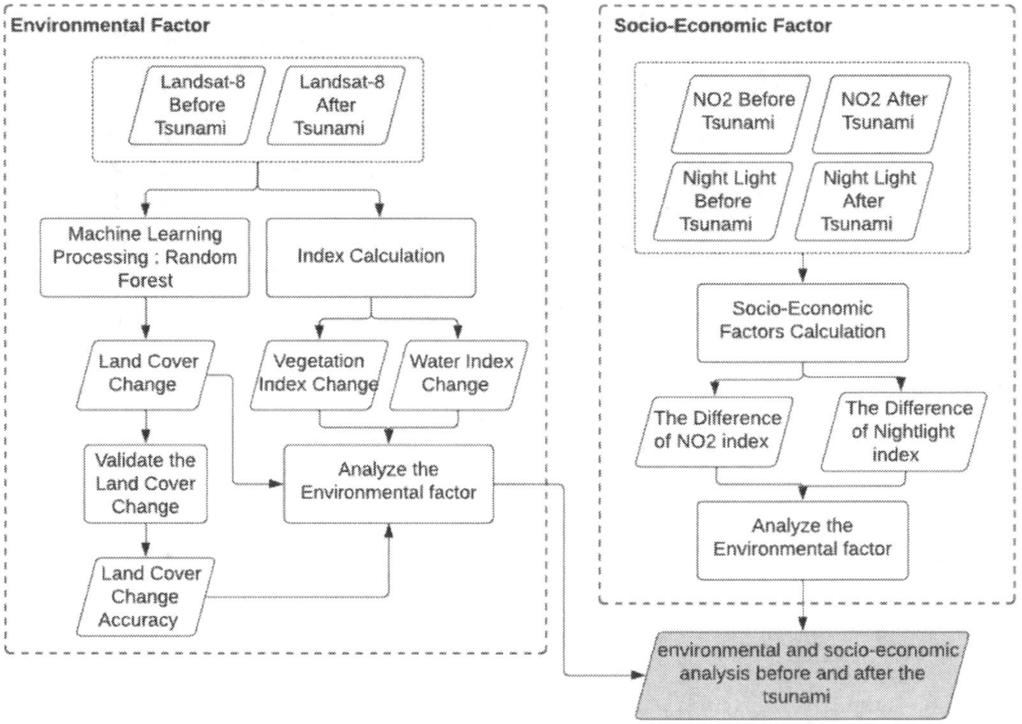

Figure 2. Methodology Flowchart

2.3.1. Environmental Factors

This study looks at changes in land cover that occur due to the tsunami. The changes will be analyzed using Landsat 8 satellite imagery. First, the land cover will be analyzed before the tsunami event using the random forest machine learning classification method. Furthermore, after obtaining changes in land cover before the tsunami, the land cover will be analyzed with the same classification method. After that, changes will be obtained as a result of the tsunami. The land cover classes used are built-up areas, water bodies, vacant lots, and vegetation. In the classification method, 100 points will be taken for each class, where splitting will be performed with 80% of the data used as training data and 20% as testing data.

In this research, the analysis for environmental factors is also represented by The Normalize Difference Vegetation Index (NDVI) and the Normalized Difference Water Index (NDWI). NDVI is an excellent tool for measuring the health and vigor of crop plants and detecting subtle changes in plant health, whereas NDWI is used to highlight open water features in a satellite image, allowing a water body to stand out against the soil and vegetation. With these indexes included in the environmental analysis for differences between after and before the tsunami, it is possible to predict how severely the tsunami affected vegetation and analyse the runoff of water to the mainland so that it changes other land covers into water bodies. Equations 1 and 2 show the NDVI (Weier and Herring, 2018) and NDWI (McFeeters, 1996) algorithms:

$$NDVI = \frac{(NIR - RED)}{(NIR + RED)} \quad (1)$$

$$NDVI = \frac{(GREEN - NIR)}{(GREEN + NIR)} \quad (2)$$

In Landsat 8 imagery, the red reflectance values are obtained from Band 4 (0.630 - 0.680 μm), and the near-infrared (NIR) reflectance values are obtained from Band 5 (0.845 – 0.885 μm). The green reflectance values are obtained from Band 3 (0.525 – 0.600 μm).

2.3.2. Socioeconomic Factors

Socioeconomic data will consider night light data (CSM, 2019) and NO2 air pollution data (ESA, 2018). Night light data is monthly averaged emission data using nighttime data from the Visible Infrared Imaging Radiometer Suite (VIIRS) Day/Night Band (DNB) with units of nanoWatts/cm2/sr (CSM, 2019). While the NO_2 data used is air pollution data in the troposphere (ESA, 2018). Night light and NO2 data can indicate human activity and indicate an urban area (Sakti et al., 2023). Sources of NO2 gas are more dominant in urban areas which can be caused by road activities (motorized vehicles), industry, mining, power plants, and land/forest fires (Sakti et al., 2023). Data processing will compare socioeconomic data at two different temporal points, before and after the tsunami.

3. Result and Discussion

3.1. Environmental Factors

3.1.1. Land Cover Change

This study has determined the changes in land cover due to the tsunami. Figure 3 shows land cover changes before and after the tsunami. Changes from built-up areas and vegetation to bare land often occur due to the tsunami. Some locations found a change of land cover from bare land to water and vice versa. The change of bare land to water could be caused by erosion as an effect of the tsunami. The tsunami's sedimentation can cause the phenomenon of land cover change from water to land.

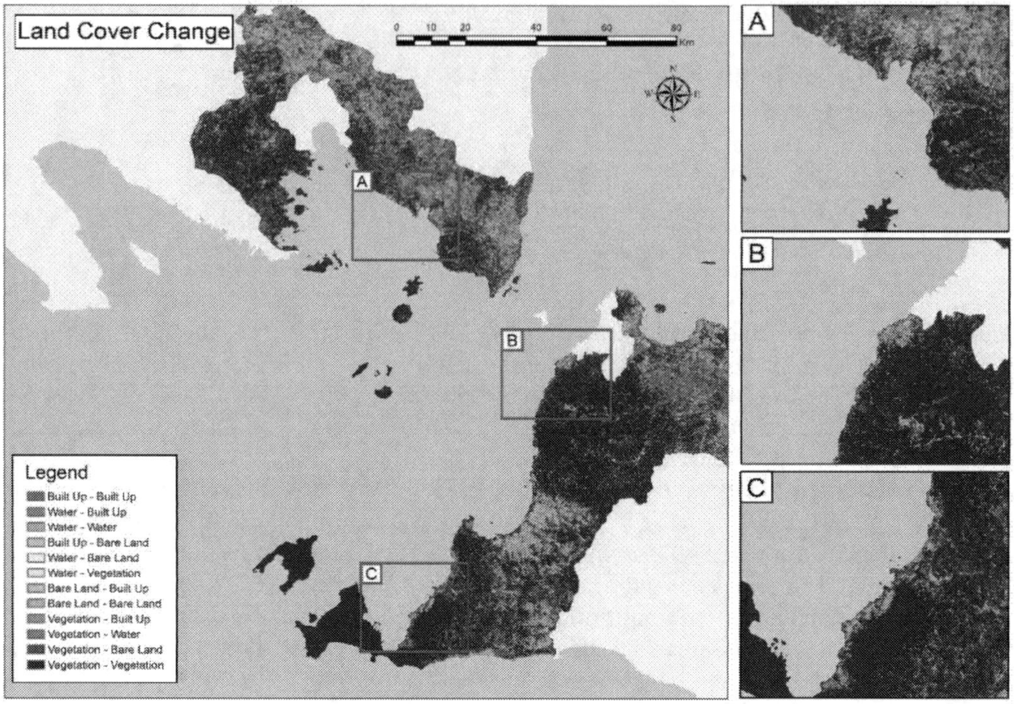

Figure 3. Land Cover Change

3.1.2. NDVI

The impact observation of the tsunami disaster on the environment is analysed through the calculation of the vegetation index (NDVI). The index calculation results are shown in Figure 4, where (a) is the appearance of the vegetation index before the tsunami and (b) after the tsunami. It can be seen in the figure that there has been a decrease in NDVI, from the highest average value of 0.458 before the tsunami to 0.386 after the tsunami. The NDVI values before the tsunami, when viewed based on the affected districts, were as follows: South Lampung = 0.158, Serang = 0.107, and Pandeglang Regencies = 0.171. Meanwhile, after the tsunami, the average NDVI index values in Serang, South Lampung, and Pandeglang Regencies were 0.097, 0.081, and 0.131, respectively.

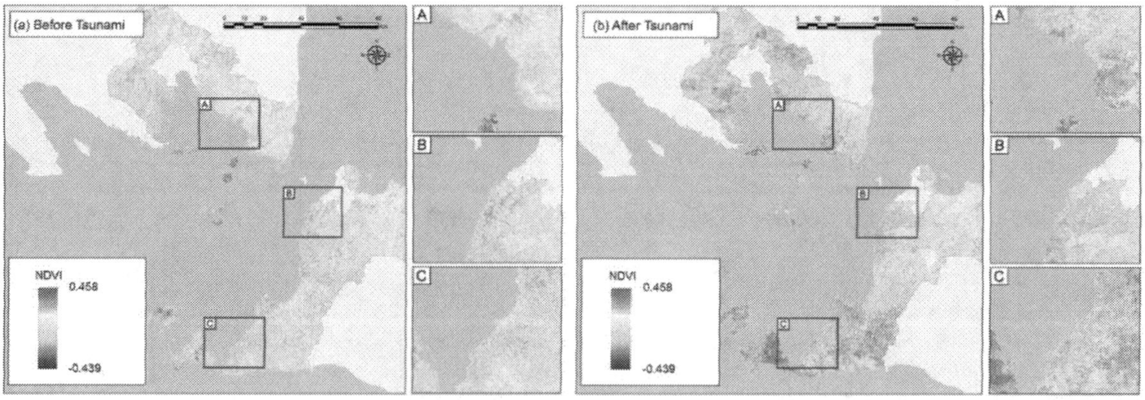

Figure 4. (a) NDVI before tsunami, (b) NDVI after Tsunami

Figure 5 depicts the statistical comparison of the average NDVI value in the 14 tsunami-affected sub-districts. Except for the Labuan, the graph shows that the average NDVI value dropped in all sub-districts. With a lowered value of 0.076, Sumur and Katibung had the largest difference. The sub-districts with substantial decreases in NDVI are also found in three South Lampung Regency sub-districts, namely Rajabasa, Kalianda, and Sidomulyo, with values of 0.071, 0.073, and 0.072, respectively. This suggests that the tsunami impacted the destruction of vegetation in the area. However, in Labuan, the NDVI value appeared to have grown following the tsunami. This is because, first, the Labuan region is smaller in comparison to other sub-districts, while the NDVI value before the tsunami was still somewhat higher than Pagelaran, despite Pagelaran having a larger area. The second possibility is that the tsunami only harmed a limited area of vegetation near the coast, hence the increase in average NDVI occurred in other green areas in Labuan that previously had a relatively high value.

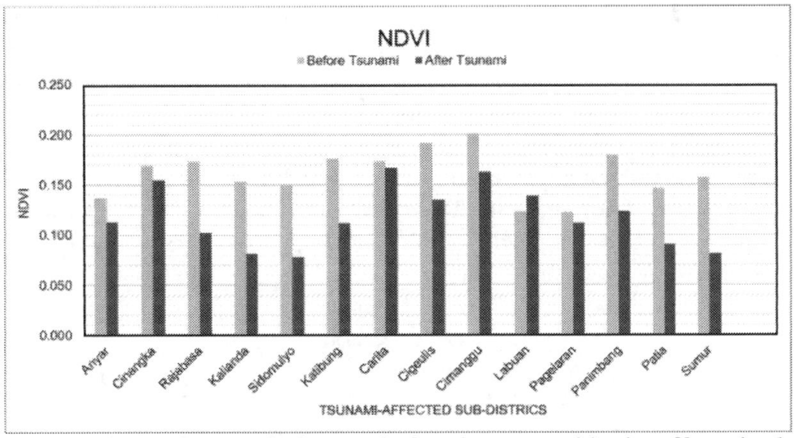

Figure 5. Comparison of NDVI before and after the tsunami in the affected sub-districts

3.1.3 NDWI

This study also makes use of Normalized Difference Water Index (NDWI) observations to assess the environmental impact of the tsunami. Figure 6 depicts the results of the NDWI observations, which show that the index range between before and after the tsunami has increased. The maximum average NDWI score in South Lampung, Serang, and Pandeglang Regencies before the tsunami was 0.689. However, it increased to 0.735 after the tsunami. Furthermore, based on the affected districts, the NDWI value before the tsunami was as follows: South Lampung = 0.321, Serang = 0.376, and Pandeglang = 0.282. Meanwhile, the average NDWI values in Serang, South Lampung, and Pandeglang Regencies after the tsunami were 0.351, 0.301, and 0.319, respectively. According to these results, only Serang had a decrease in the water content index in vegetation, while the other two had a rise. This can be thought to be due to the rainfall factor, which varies by location and thus affects the water content.

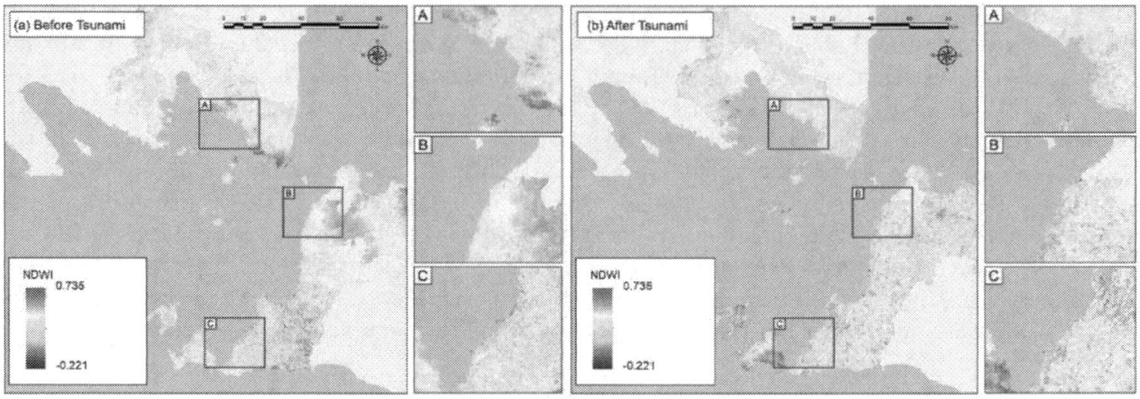

Figure 6. (a) NDWI before Tsunami, (b) NDWI after Tsunami

The average NDWI value was analysed at the regional administration level in more detail; in this case, for the 14 districts affected by the tsunami, it can be seen in Figure 7 that 9 districts experienced an increase in the average water content index. Meanwhile, the other 5 districts experienced a decrease. These districts are Anyar District in Serang Regencies, as well as Rajabasa, Kalianda, Sidomulyo, and Katibung Districts in South Lampung Regencies. Then, all affected districts in Pandeglang experienced an average increase in NDWI after the tsunami.

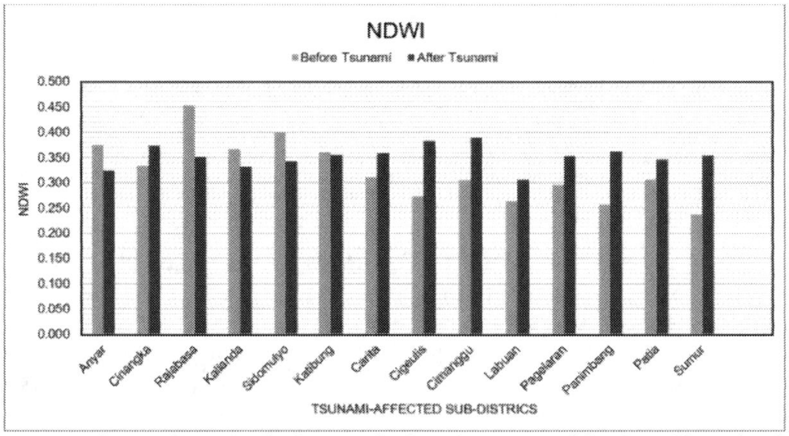

Figure 7. Comparison of NDWI before and after the tsunami in the affected sub-districts

3.2. *Socioeconomic Factors*

3.2.1. Air Pollution

The results of air pollution observations before and after the tsunami can be seen in Figure 6. Based on the results, it can be interpreted that the tsunami affected NO_2 patterns before and after the incident. Before the tsunami, the highest average value in the study area was 148.706 μmol/m^2. Meanwhile, before the tsunami, the highest average value decreased to 76.761 μmol/m2. This indicates that there is a decrease in human activity, especially in motorized vehicles.

This can also be caused because the affected coastal area is also a strategic area for tourism. So that when after the tsunami occurred, the region experienced a temporary stop in its tourism sector and experienced a decrease in air pollution gases. If we look at the air pollution before the tsunami based on the affected districts, South Lampung (Figure 8A), Serang (Figure 8B), and Pandeglang Districts (Figure

8C) respectively have an average NO_2 gas value of 18 (2-70) µmol/m^2, 46 (6-149) µmol/m^2, and 13 (1-36) µmol/m^2. Meanwhile, after the tsunami, the average air pollution values in Serang, South Lampung, and Pandeglang Regencies were 16 (1 - 51) µmol/m2, 31 (10 - 77) µmol/m2, and 15 (0 - 39) µmol/m2. NO2 gas in the Pandeglang area experienced an increase, which could be caused by human activities in areas far from the coast. Thus, this increase was likely not an impact of the tsunami.

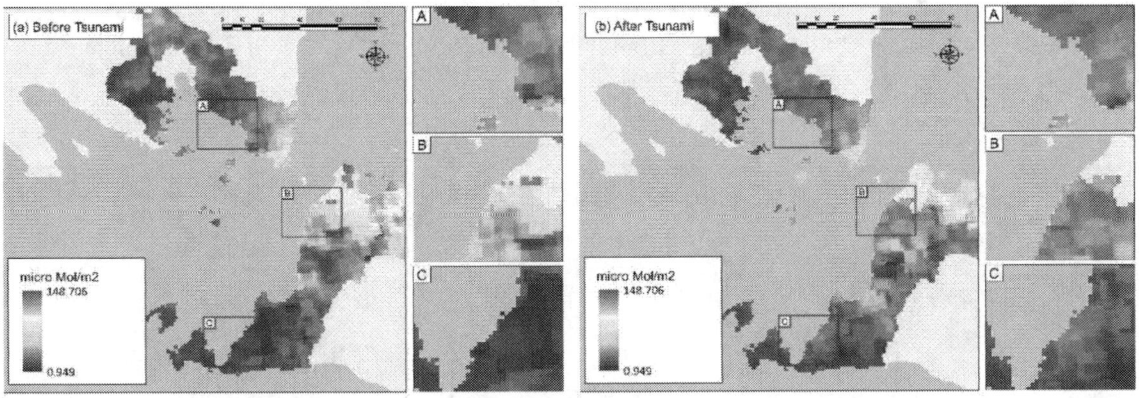

Figure 8. NO_2 air pollution before (a) and after (b) the tsunami.

Based on Figure 9, seven sub-districts experienced a decrease in NO_2 gas after the tsunami occurred. The districts are Anyar, Cinangka, Rajabasa, Kalianda, Labuan and Pagelaran. On the other hand, some sub-districts experienced an increase in NO_2 gas during the post-tsunami period, namely Sidomulyo, Katibung, Carita, Cigeulis, Cimanggu, Panimbang, Patia, and Sumur sub-districts. Either human transportation activities caused this increase in NO_2 gas after the tsunami, or there is a bias from areas far from the coast that do not have a path from the tsunami.

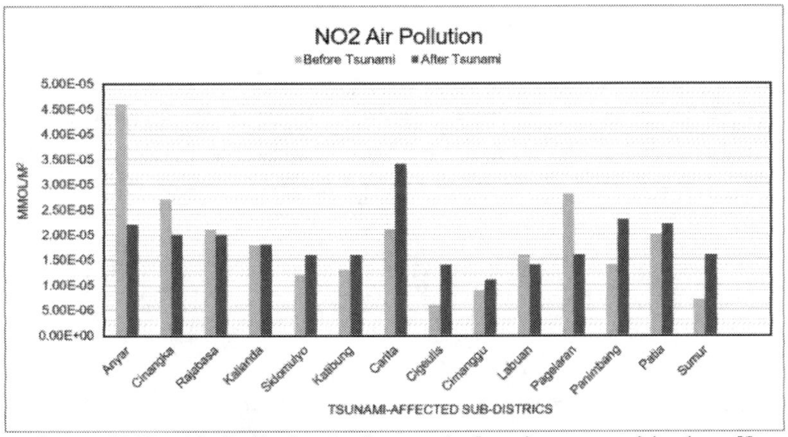

Figure 9. Comparison of NO_2 Air Pollution before and after the tsunami in the affected sub-districts

3.2.2. Night light

Night light indicates human activity at night, which can be marked by light at that location. The night light can also identify the state of an area's economy. If a site has a high night light, the area has a high economic level, such as in a metropolitan city. Conversely, if an area has a low night light, then the area experiences a low financial status.

Based on observations of night light (Figure 7), before the tsunami, the average value of night light in the three districts was 149.775 nanowatts/cm^2/sr and decreased after the tsunami to 88.905 nanowatts/cm^2/sr. When viewed from the affected districts, South Lampung (Figure 10A), Serang

District (Figure 10B), and Pandeglang (Figure 10C), respectively, have average values of light gas at night before the tsunami of 1.16 (0.08 - 34.31) µmol/ m², 2.21 (0.14 - 149.77) µmol/m², and 0.43 (0.01 - 10.50) µmol/m². Meanwhile, after the tsunami, the average values of night light in Serang, South Lampung, and Pandeglang Regencies were 1.06 (0 – 26.89) µmol/m2, 3.15 (0 – 88.91) µmol/m², and 0.45 (0 – 25.51) µmol/m². Similar to NO_2 gas, the night light in the Pandeglang area tends to increase compared to the other two cities, which had decreased values after the tsunami. This can be caused by an increase in night light in areas quite far from the coast (see Figure 10C), so it can be concluded that the rise in night light was not caused by a tsunami.

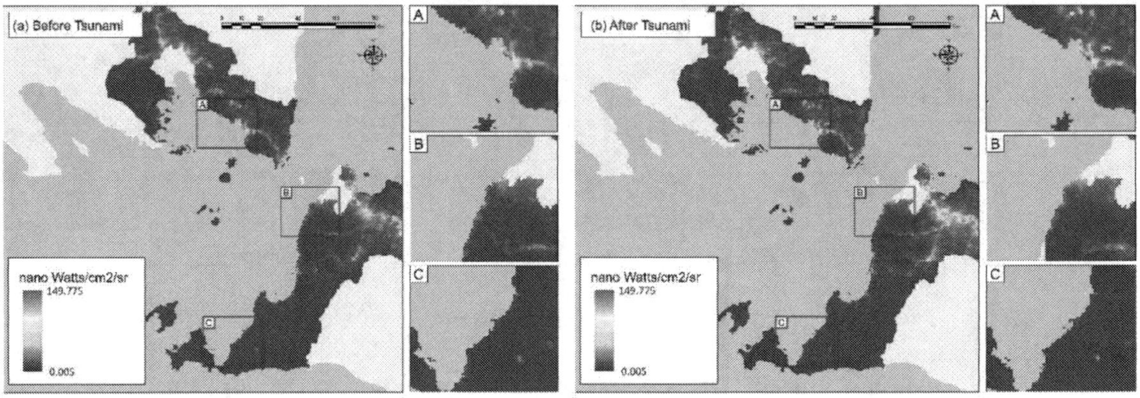

Figure 10. Night light before (a) and after (b) the tsunami.

Figure 11 states the average value of night light in each sub-district in the affected area. Of the fourteen sub-districts observed, nine sub-districts experienced a decrease in the value of night light from before to after the tsunami. The districts are Cinangka, Rajabasa, Katibung, Carita, Cigeulis, Cimanggu, Labuan, Panimbang and Sumur. However, five sub-districts have not experienced a reduction in night light, including Anyar, Kalianda, Sidomulyo, Pangelaran, and Patia, which can be caused by an increase in night light far from the coast. Hence, it is not an impact of the tsunami.

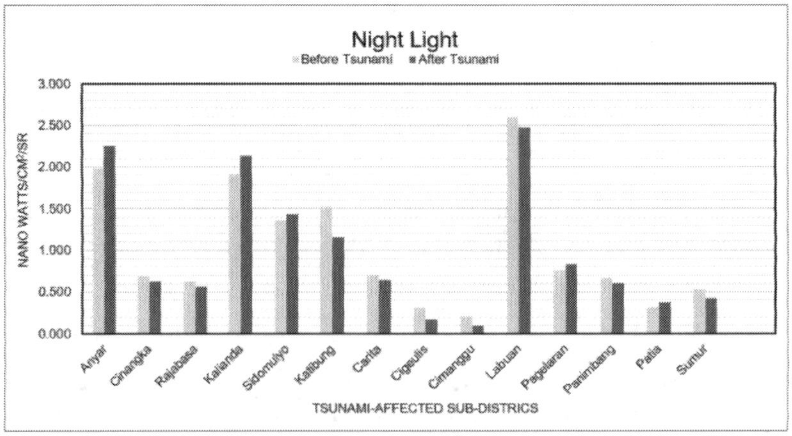

Figure 11. Comparison of Night Light before and after the tsunami in the affected sub-districts

3.3. *Land Cover Validation*

Land cover classification before and after the tsunami event is calculated as the accuracy value of the land cover model. Table 2 and Table 3 are confusion matrices for validation of land cover models made using the random forest method before and after the tsunami. The model that has been made appears to be sufficiently accurate, with an overall accuracy value, at the time before the tsunami reaching 91.089% with a kappa value of 88.083%. The land cover model after the tsunami has an overall accuracy value of 91.176% with a kappa value of 88.144%. However, in modeling land cover with Landsat 8, which has a resolution of 30 meters, there are some shortcomings because it can only detect the affected area, not specific to the object of damage. In addition, the area around the Sunda Strait has quite a lot of cloud cover on each imagery, so it is necessary to carry out cloud masking techniques. The cloud masking technique followed by filling empty images with other images is one solution to eliminate the cloud effect at the time of classification. Although in some locations, a thin layer of cloud was still not successfully removed during cloud masking.

Table 2. Validation Matrix Land Cover Classification Before Tsunami

		Actual			
		Built Up	Water	Bare Land	Vegetation
Predicted	Built Up	21	0	3	1
	Water	0	25	0	0
	Bare Land	4	0	25	1
	Vegetation	0	0	0	21

Table 3. Validation Matrix Land Cover Classification After Tsunami

		Actual			
		Built Up	Water	Bare Land	Vegetation
Predicted	Built Up	19	0	4	0
	Water	0	32	0	0
	Bare Land	3	0	21	0
	Vegetation	0	0	2	21

4. Conclussion

This study identifies changes after the tsunami by considering environmental factors based on land cover conditions, NDVI, NDWI, and socioeconomic factors represented by night light and NO_2. The results of identifying land cover changes using a random forest classification show changes in coastal areas affected by the tsunami. Some locations show a change from vegetation to open land and built-up areas to open land. The overall accuracy value of the land cover model before and after the tsunami was compared with a reference of 91%. Based on the results of NDVI calculations, it can be concluded that there is a decrease in the value of NDVI in the affected area, where the highest average value before the tsunami was 0.458 to 0.386 after the tsunami. One of the causes of the decline in the value of NDVI is the change in vegetation land cover to vacant land. The value of NDWI in the affected areas has increased on average, directly proportional to the change in land cover to the waters due to the tsunami (land erosion). Based on socioeconomic factors, air pollution decreases, indicating a decrease in human activity, especially motor vehicles, due to the tsunami. The same is shown by the night light in the affected areas, where the index after the tsunami is smaller than before the tsunami, which shows a decrease in human activity.

References

Adriano, B., Xia, J., Baier, G., Yokoya, N., Koshimura, S., 2019. Multi-source data fusion based on ensemble learning for rapid building damage mapping during the 2018 sulawesi earthquake and tsunami in Palu, Indonesia. Remote Sens (Basel) 11, 886.

Badan Informasi Geospasial (BIG), 2021. Basemap Rupa Bumi Indonesia (RBI) [WWW Document]. URL http://basemap.big.go.id/rbi/ (accessed 11.2.21).

BNPB, 2018. Benar, Tsunami Menerjang Pantai di Selat Sunda [WWW Document]. URL https://bnpb.go.id/berita/benar-tsunami-menerjang-pantai-di-selat-sunda#:~:text=Dampak%20tsunami%20menyebabkan%20korban%20jiwa,semua%20daerah%20terdampak%20di%20data. (accessed 1.20.23).

CSM, 2019. See The World at Night. [WWW Document]. URL https://eogdata.mines.edu/products/vnl/ (accessed 2.22.23).

ESA, 2018. Sentinel-5P TROPOMI User Guide [WWW Document]. URL https://sentinel.esa.int/web/sentinel/user-guides/sentinel-5p-tropomi (accessed 2.22.23).

ESDM, 2018. Tanggapan Kejadian Tsunami di Selat Sunda, Tanggal 22 Desember 2018. [WWW Document]. URL https://www.esdm.go.id/id/media-center/arsip-berita/tanggapan-kejadian-tsunami-di-selat-sunda-tanggal-22-desember-2018 (accessed 2.22.23).

McFeeters, S.K., 1996. The use of the Normalized Difference Water Index (NDWI) in the delineation of open water features. Int J Remote Sens 17, 1425–1432. https://doi.org/10.1080/01431169608948714

Sakti, A.D., Anggraini, T.S., Ihsan, K.T.N., Misra, P., Trang, N.T.Q., Pradhan, B., Wenten, I.G., Hadi, P.O., Wikantika, K., 2023. Multi-air pollution risk assessment in Southeast Asia region using integrated remote sensing and socio-economic data products. Science of The Total Environment 854, 158825.

Sublime, J., Kalinicheva, E., 2019. Automatic post-disaster damage mapping using deep-learning techniques for change detection: Case study of the Tohoku tsunami. Remote Sens (Basel) 11, 1123.

Syamsidik, B., Luthfi, M., Suppasri, A., Comfort, L.K., 2020. The 22 December 2018 Mount Anak Krakatau volcanogenic tsunami on Sunda Strait coasts, Indonesia: tsunami and<? xmltex\break?> damage characteristics. Natural Hazards and Earth System Sciences 20, 549–565.

Syamsidik, S., Benazir, B., Luthfi, M., 2019. Tsunami flow depths, building damages, and tsunami boulders measured from the December 22, 2018 Sunda Strait tsunami around Western Java and Southern Lampung of Indonesia, Mendeley Data. Mendeley Data [online] Available from: https://data. mendeley. com/datasets/yyyvmxh8vg/1.

Takabatake, T., Shibayama, T., Esteban, M., Achiari, H., Nurisman, N., Gelfi, M., Tarigan, T.A., Kencana, E.R., Fauzi, M.A.R., Panalaran, S., 2019. Field survey and evacuation behaviour during the 2018 Sunda Strait tsunami. Coastal Engineering Journal 61, 423–443.

USGS, 2018. Landsat 8.

Virtriana, R., Harto, A.B., Atmaja, F.W., Meilano, I., Fauzan, K.N., Anggraini, T.S., Ihsan, K.T.N., Mustika, F.C., Suminar, W., 2023. Machine learning remote sensing using the random forest classifier to detect the building damage caused by the Anak Krakatau Volcano tsunami. Geomatics, Natural Hazards and Risk 14, 28–51.

Weier, J., Herring, D., 2018. Measuring Vegetation (NDVI & EVI) : Feature Articles [WWW Document]. 2000.

Yeh, H., Barbosa, A., Mason, B.H., 2022. Tsunamis effects in man-made environment. Complexity in Tsunamis, Volcanoes, and their Hazards 187–211.

Geochemical characteristics of lacustrine source rock: An insight of Banuwati Formation from Sunda and Asri Basins, Indonesia

Bagus Priyanto[a,*], Eddy Ariyono Subroto[b], Asep Heri Patria Kesumajana[b], Very Susanto[b], Agus Dwi Suryanto[c]

[a]Doctoral Program in Geological Engineering, Faculty of Earth Sciences and Technology, Institut Teknologi Bandung, Jl. Ganesha, 10, Bandung, 40132, Jawa Barat, Indonesia
[b]Department of Geological Engineering, Faculty of Earth Sciences and Technology, Institut Teknologi Bandung, Jl. Ganesha, 10, Bandung, 40132, Jawa Barat, Indonesia
[c]Pertamina EP, Standard Chartered Tower, Floor 21-29, Jl. Prof. Dr. Satrio, 164, Jakarta, 12950 Indonesia
*Corresponding author: 32020001@mahasiswa.itb.ac.id

Abstract. The Sunda and Asri Basins are petroliferous in Indonesia. Despite producing billions of barrels of oil equivalent since their initial production, research on the source rock organic geochemistry of these basins remains limited, particularly concerning the sapropelic kerogen of the syn-rift Banuwati Formation. A comprehensive study of the Banuwati Formation in the Sunda and Asri Basins was conducted using kerogen and biomarker data. Various techniques, such as carbon analysis, Rock-Eval pyrolisis, vitrinite reflectance (Ro), gas chromatography (GC), and gas chromatography-mass spectrometry (GC-MS), were employed on samples from six wells. The Banuwati Formation exhibits source rock with good to very good quantities. Biomarker analysis offers a novel interpretation, suggesting that the Banuwati shale source rock was deposited in an oxic-suboxic environment. The formation is primarily composed of Type I organic material, characterized by the presence of biomarkers such as C_{26} tricyclic terpane, C_{30} 4-methylsterane, and a possible botryococcane peak, indicative of a lacustrine setting. Maturity assessments based on Ro and Tmax values show conflicting results. Tmax values imply maturity, while Ro values suggest immaturity. This discrepancy is likely due to Ro suppression, a phenomenon often observed in Type I kerogen samples.

1. Introduction

The Sunda and Asri Basins are back-arc basins located close to each other at southeast of Sundaland (Figure 1). The exploration has been actively conducted since the 1960s in the Sunda Basin, and then several fields have been found to mainly contain oil since the 1970s in the Sunda Basin. A successful exploration in the Sunda Basin was then followed by another one in the Asri Basin.

The Sunda and Asri Basins are prolific basins with numerous hydrocarbon discoveries with total cumulative production of more than 1700 million-barrel oil equivalent (status until the year 2023). Several authors have evaluated the petroleum system in these producing basins [1]–[5]. Nevertheless, none of them conducted an in-depth study of the geochemistry of source rock and simultaneously appraised both the Sunda and Asri Basins. As few exploration wells have penetrated the basin area in Indonesia, this study gives an ideal opportunity to gain knowledge about the organic geochemistry features of the initial sediment basin-fill, the Banuwati Formation. Therefore, the primary objective of this study is to be the first to provide a comprehensive description of the source rock organic geochemistry of Banuwati Formation in the Sunda and Asri Basins, Indonesia.

2. Geological background

The Sunda and Asri Basins are categorized with Cenozoic sediments filled. The geological setting is described by several authors [2]–[10]. Tectonostratigraphically, the formation of the Sunda and Asri Basins could be divided into a period of pre-rift, syn-rift, and post-rift (Figure 1). The pre-rift, basement is pre-Tertiary and consists of various non-sediment rocks such as granodiorite, granite, slate, schist, and marble. In the Eocene/Early Oligocene (Paleogene), due to tensional stresses, rifts were formed by predominantly north-south oriented faults in the eastern part of the basins. In the syn-rift phase, the first sediment filling deposited overlying the basement rock, characterized by sedimentary rock with conglomerate, shale, sandstone, and alluvium fan deposits to the basin margin, known as the Banuwati Formation. This formation is characterized by lacustrine shale deposits containing rich organic material. The maximum thickness of the Banuwati Formation is approximately 1500 and 1000 meters in the Sunda and Asri Basins, respectively. Overlying the Banuwati Formation are Zelda and Gita Members of the Talangakar Formation in the Late Oligocene - Early Miocene. The Zelda and Gita Members are characterized predominantly by the fluvial to the deltaic system with lithology succession of sandstone, siltstone, shale, and interbedded coal. The sandstone of the Talangakar Formation was important as the main reservoir. The Baturaja Formation, predominantly consisting of limestone lithology in marine settings, was deposited overlying the Talangakar Formation due to sea-level rise. The post-rift, during the Middle Miocene, the Gumai Formation deposited and consisted of claystone and minor sandstone. The Gumai Formation served as a regional seal rock. Further, the Airbenakat and Cisubuh Formations were deposited in the Late Miocene-Holocene.

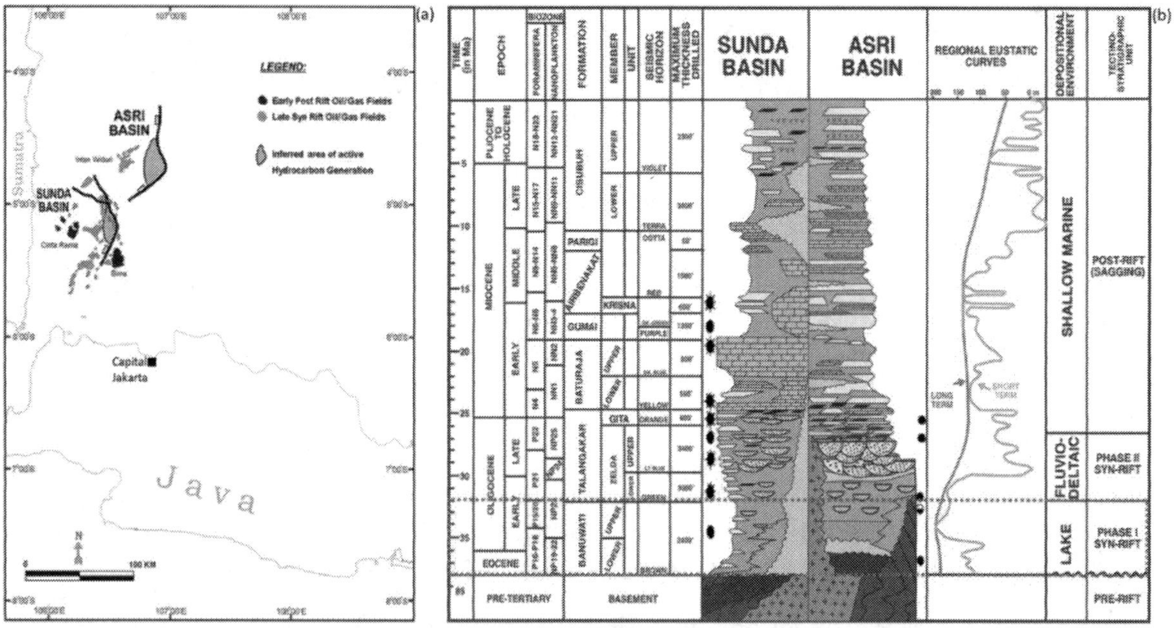

Figure 1. The study area is in the Sunda and Asri Basins, Indonesia. Both basins are characterized by Paleogene half-grabens with a north-south major fault trend in the eastern part (figure 1a, modified after Doust and Noble [5]). The Banuwati Formation was deposited in the Paleogene syn-rift (figure 1b, modified after Prayitno et al., [2]; Sukanto et al., [4]).

3. Samples and methods

For this study, six offshore wells were selected, including SADE-1, SFAN-1, SMON-1, SLAS-1, and SZEL-14 in the Sunda Basin and ADEL-1 in the Asri Basin. The data on source rock samples from those well were obtained and compiled from the current operator. The rock samples were analyzed using conventional geochemical methods. Detail descriptions of the procedures and analyses can be found in the works of several authors [11]–[14]. The laboratory analyses included TOC (Total Organic Carbon), Rock-Eval pyrolysis, vitrinite reflectance (Ro), gas chromatography (GC), and gas chromatography-mass spectrometry (GC-MS) such as m/z 123, m/z 191, and m/z 217. The geochemical data available for this study, focusing on the Banuwati Formation in the Sunda and Asri Basins, are limited and presented in table 1. The simplified flowchart of research methodology shown in Figure 2.

Rock geochemical data is selected and validated to produce an accurate interpretation using geochemical parameters such as TOC and S2 (organic compounds generated by cracking of the kerogen in the Rock-Eval pyrolysis) to maintain the quality and validity of the data [12], [15]. Drilling activities from wells focus of the study do not use oil-based mud (OBM) but water-based mud (WBM). All rock samples used in this research are free of hydrocarbon contamination from the drilling operation. Thus, further evaluation of the source rock is conducted with high confidence.

Table 1. The total numbers of Banuwati Formation data from each geochemical laboratory analysis.

No.	Well Name	TOC	Rock-Eval	Ro	GC	M/Z 123	M/Z 191	M/Z 217
1	SADE-1	2	2	2	-	-	-	-
2	SFAN-1	5	4	1	1	-	-	-
3	SMON-1	6	6	4	1	-	-	-
4	SLAS-1	3	3	3	-	-	-	-
5	SZEL-14	-	-	-	3	3	3	3
6	ADEL-1	3	3	8	10	-	-	-
TOTAL		19	18	18	15	3	3	3

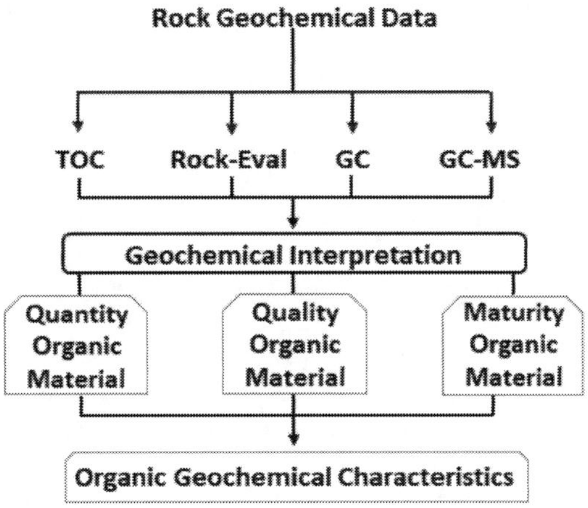

Figure 2. The simplified flowchart of research methodology.

4. Results and discussions

The evaluation of the source rock includes the evaluation of the quantity of organic material, the quality of the organic material, the maturity of the source rock, the biomarkers and interpretation.

The rock geochemical studies are done with good-quality data. Some researchers who have conducted a geochemical study of rocks in the Tertiary basin in western Indonesia, for example [16] in the Central Sumatra Basin use a TOC threshold value (screening) of 1% and S2 of 2.5 mg/g source rock. This study uses a TOC threshold limit value of 1% because this value is the threshold limit for a sedimentary rock categorized as good to be a source rock [12], [17]. The S2 threshold limit used in this study has a higher value, 5 mg/g of source rock because this value is the threshold limit of source rock having a good quantity and quality of organic material [17]. The numbers of data after screening using a minimum TOC value of 1% and S2 minimum value of 5 mg/g are shown in table 1.

4.1. Source rock quantity

The Banuwati Formation core sample is only available from SFAN-1 and SMON-1 wells and shows excellent visual source rock characteristics (Figure 3). Based on classification from Dembicki Jr. [17], the Banuwati Formation in the Sunda Basin has a good to a very good amount of organic material with a TOC value of 1.87% to 7.12% and an S2 value of 5.34 to 94.16 mg/g source rock.

The Banuwati Formation in the Asri Basin exhibit a high quantity of organic material, with a TOC value of 3.29% to 3.83% and an S2 value of 24.44 to 34.55 mg/g source rock. Notably, one sample from ADEL-1 well in the Banuwati Formation reached the coal category, with TOC and S2 values of 58.32% and 183.8 mg/g source rock, respectively (Figure 3).

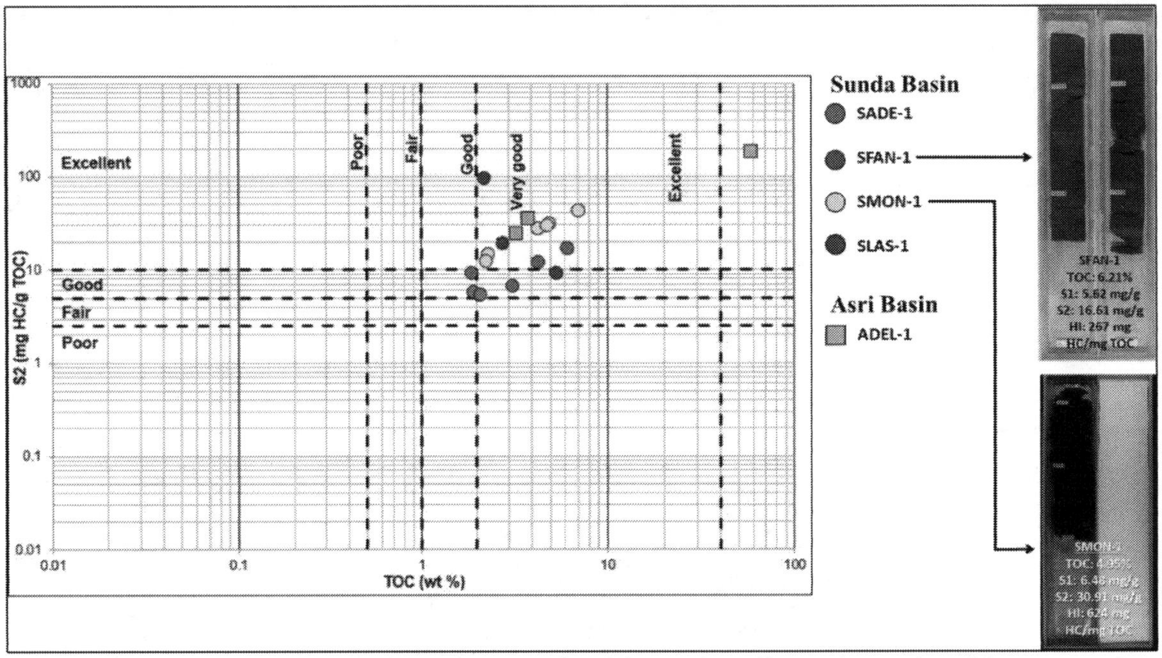

Figure 3. The cross plot between S2 and TOC values shows that all samples analyzed have a good to very good quantity of organic matter (classification based on Dembicki Jr. [17]).

4.2. Source rock quality

The quality of organic material is determined using Rock-Eval pyrolysis data from Banuwati Formation. High hydrogen index (HI = S2/TOC x 100) ranging from 167.96 to 673.84 in the Sunda Basin and 315 to 902 in the Asri Basin, suggest a predominance of Type I kerogen with good oil-prone quality. This classification is based on the work of Dembicki Jr [15] (Figure 4).

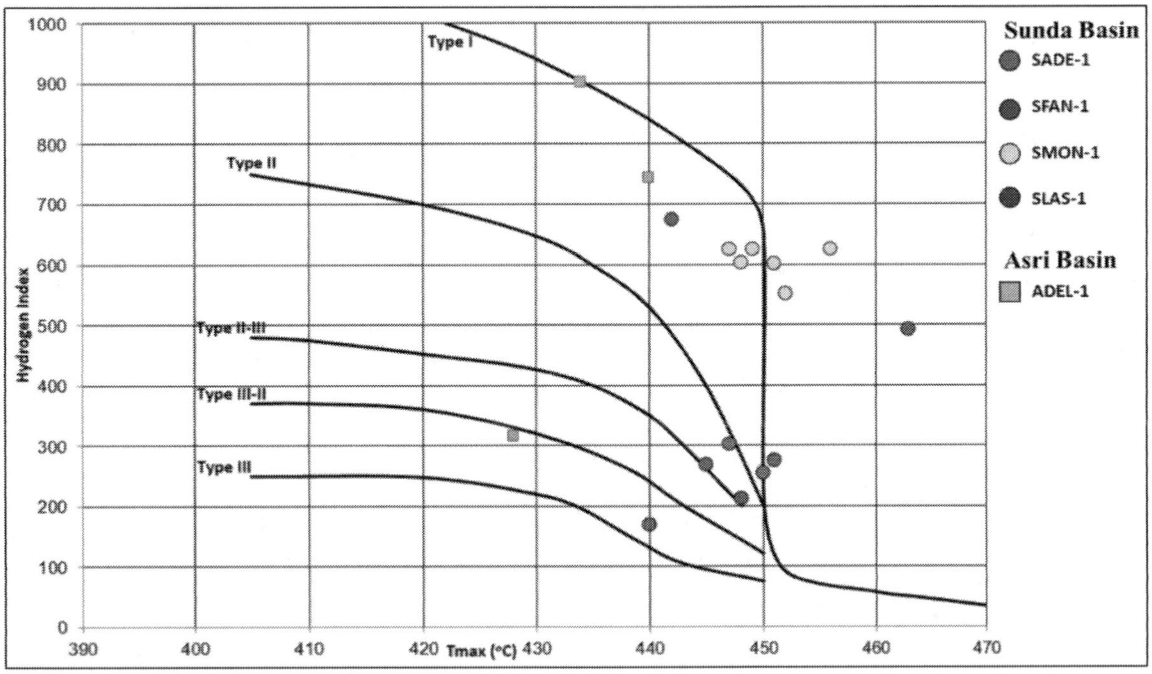

Figure 4. HI and Tmax parameter diagram for determining organic matter quality (classification based on Dembicki Jr. [15]).

4.3. Source rock maturity

The maturity of organic material in sedimentary rocks is determined by vitrinite reflectance (Ro – a parameter for determining the maturation of organic matter in fine-grained rocks using a microscope) and the Tmax of Rock-Eval pyrolysis analysis (Tmax – a thermal maturity parameter based on the temperature at which the maximum amount of pyrolyzed/S2 is generated from the kerogen in a rock sample). The Banuwati Formation has entered the maturity window to the peak of maturity in the Sunda Basin, with a Tmax value ranging between 440 °C and 463 °C. In comparison, this formation has just reached the oil window in the Asri Basin with a Tmax value of 428 °C to 440 °C. However, the Ro values from the Banuwati Formation in the Sunda and Asri Basin indicate that this formation is still immature (Ro < 0.6%), with one sample having reached the peak of maturity (Ro = 0.66%). This difference in maturity can be attribute to two factors. First, the high Tmax value from the Banuwati Formation could result from the samples' low quantity of organic material [12]. Second, vitrinite suppression could cause low Ro values [18]. The first factor is unlikely to occur because the sample used has been geochemically selected, leaving a sample with a good quantity of organic material. Vitrinite suppression is the most likely explanation, given that it occurs in samples with high HI values (HI > 400). Some samples of the Banuwati Formation in the Sunda and Asri Basins have HI values of more than 600 in the SMON-1 and SLAS-1 wells, even reaching 900 in the ADEL-1 well. Therefore, the Ro values of samples with high HI values are corrected on the graph from [18]. After correction, the Banuwati Formation matured, ranging from the peak of maturity to the end of maturity, with Ro values between 0.7% and 1.17% (Figure 5). The Banuwati Formation also exhibits matured source rocks, as shown by the high Ts (Trisnorneohopane) over Tm (Trisnorhopane) from rock sample (Figure 7).

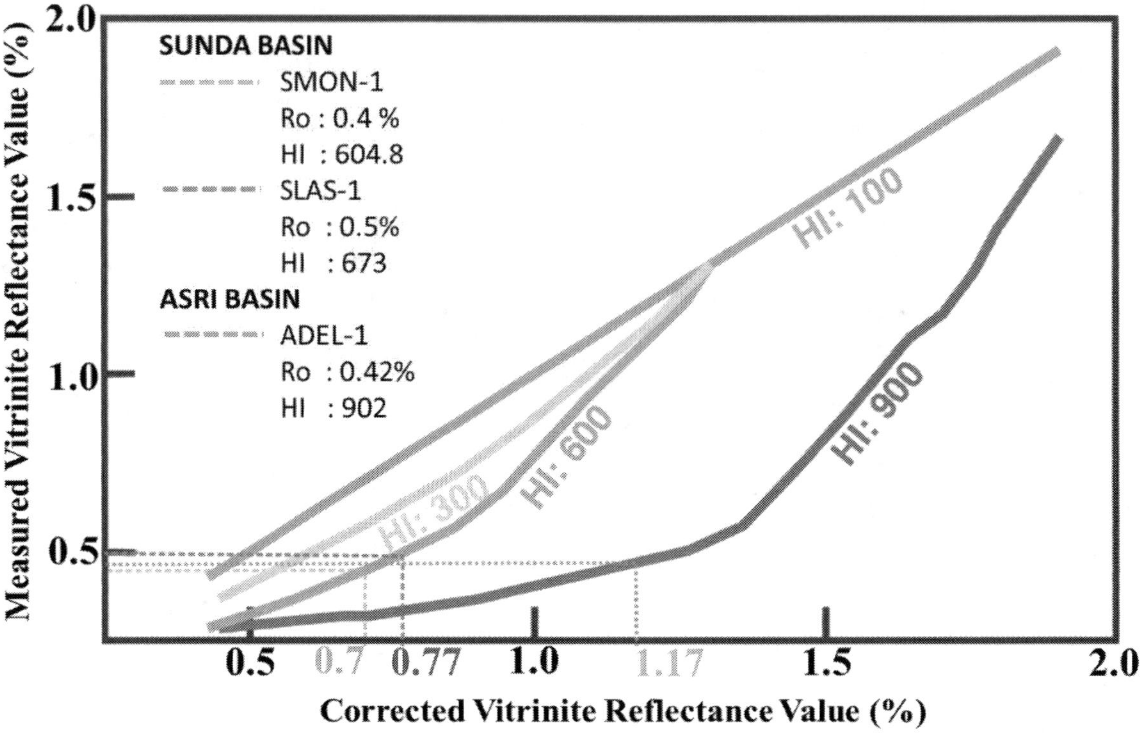

Figure 5. The presence of vitrinite suppression in the Banuwati Formation sample with a high HI value causes a low Ro value. Therefore, the value of Ro is corrected with the re-digitized graph of Lo [18].

4.4. Biomarker and interpretation

The biomarkers were evaluated based on n-alkanes, isoprenoid, bicyclic alkane, terpane, and sterane components. Each saturated biomarker component was identified by comparing the retention times and mass spectra of the monitor ions for bicyclic alkanes (m/z 123), terpanes (m/z 191), and steranes (m/z 217).

Normal alkanes and acyclic isoprenoids (e.g. pristane, Pr; phytane, Phy) are represented by saturated hydrocarbons between nC_5 to nC_{38}. The acyclic isoprenoid composition is used in interpreting redox conditions [19]. All Banuwati rock samples from Sunda and Asri Basins used in this study show that pristane is consistently higher than phytane. The higher pristane content over phytane in all samples indicates that the Banuwati Formation was deposited in a suboxic-oxic environment. However, the ratio of pristane to phytane must be used cautiously because this parameter is influenced by several factors other than the redox conditions, such as organic matter contribution, thermal maturity, and post-expulsion alteration [20], [21]. The oxic-suboxic environment was also demonstrated by the low molecular weight steranes, such as C_{21} sterane (pregnane) and C_{22} sterane (homopregnane), which are lower than normal sterane on the extended m/z 217 fragmentogram [22] (Figure 7) and the presence of C_{29} norneohopane (Cy) on the m/z 191 fragmentogram [20] (Figure 7). Interestingly, Banuwati rock samples have a moderate 17α(H),21β(H),28,30-bisnorhopane (BNH) peak, the origin of which still needs to be understood. The BNH biomarker is frequently found in source rocks deposited in anoxic conditions [20]. Despite this explanation, our interpretation differs and provides novelty interpretation based on biomarkers from the existing interpretation that suggests the crude oils in Asri Basin primarily came from an anoxic environment [4].

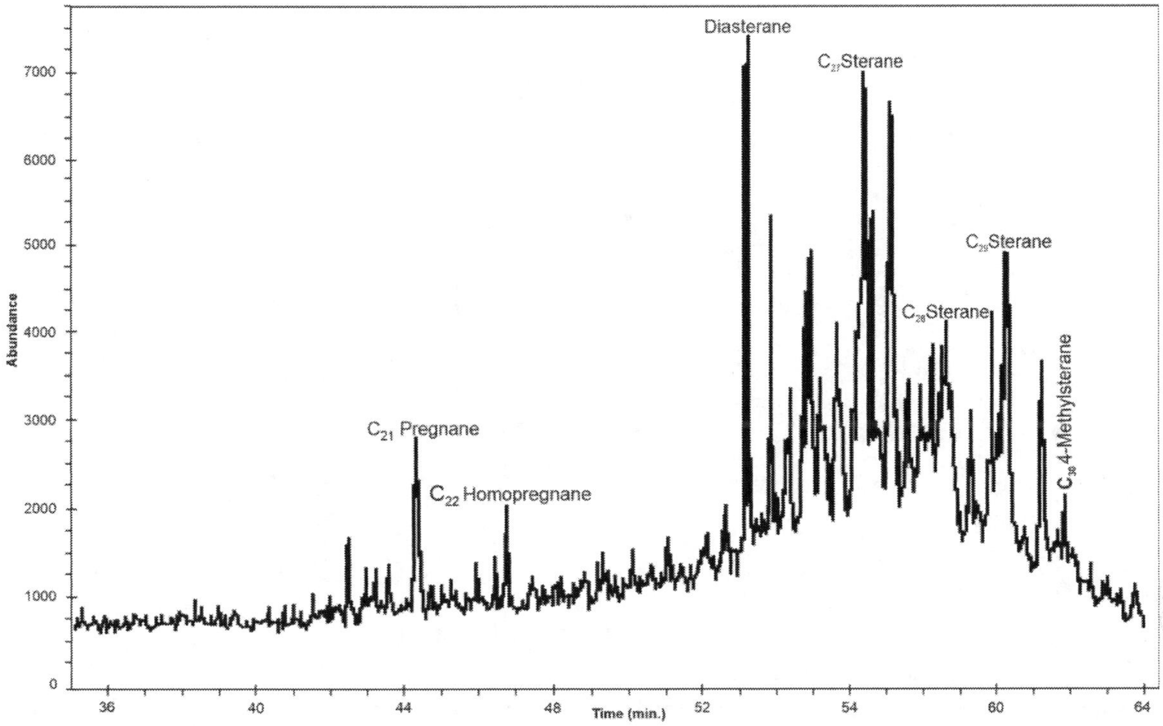

Figure 6. The representative extended sterane fragmentogram m/z 217 from the Banuwati Formation displays several key features: low molecular weight sterane, such as C_{21} sterane (pregnane) and C_{22} sterane (methylpregnane), are lower than normal C_{27}-C_{29} sterane; diasteranes are more abundant than normal steranes; C_{27} sterane predominate over C_{28} and C_{29}; and C_{30} 4-methylsterane is present.

The Banuwati rock samples, dominated by shale sediments, exhibit lower C_{34} homohopane than C_{31} and lower C_{29} hopane than C_{30} hopane [20], [23]. The biomarker that supports the deposition of the Banuwati Formation in a lacustrine environment is the C_{26} tricyclic terpane, which is higher than the C_{25} tricyclic terpane, as shown in figure 7 [20], [24]. The presence of 18α(H)-oleanane (OL) in the m/z 191 fragmentogram indicates the presence of higher plant inputs into the Banuwati Formation. However, the small amount of oleanane present in the Banuwati Formation suggests only minor inputs of higher plants, such as Angiosperms (Figure 7). Other evidence of higher plant material contributing to the organic material in the Banuwati Formation includes the presence of the bicyclic alkanes (e.g. 8β(H)-drimane and 8β(H)-homodrimane) compound in the m/z 123 fragmentogram, which are related to the contribution of higher plants [25] (Figure 8).

Another supporting for the lacustrine environment of the Banuwati Formation is the high the alkane peak after nC_{22} in rock samples, which exhibit relatively high waxy properties [26]. The gas chromatogram data show carotenoid compounds, which are usually associated with algae [27]. One of these samples also displayed an indication of botryococcane, associated with lacustrine algae, in figure 8 [28], [29]. The dominance of algae as a constituent of the Banuwati Formation is also demonstrated by the predominance of C_{27} sterane compared to C_{28} and C_{29} on the m/z 217 fragmentogram [30], [31]. The presence of C_{30} 4-methylsterane further supports the deposition of the Banuwati Formation in a lacustrine environment, as shown in figure 6 [32]. The high diasterane levels also suggest a clay-rich environment, rather than a carbonate one.

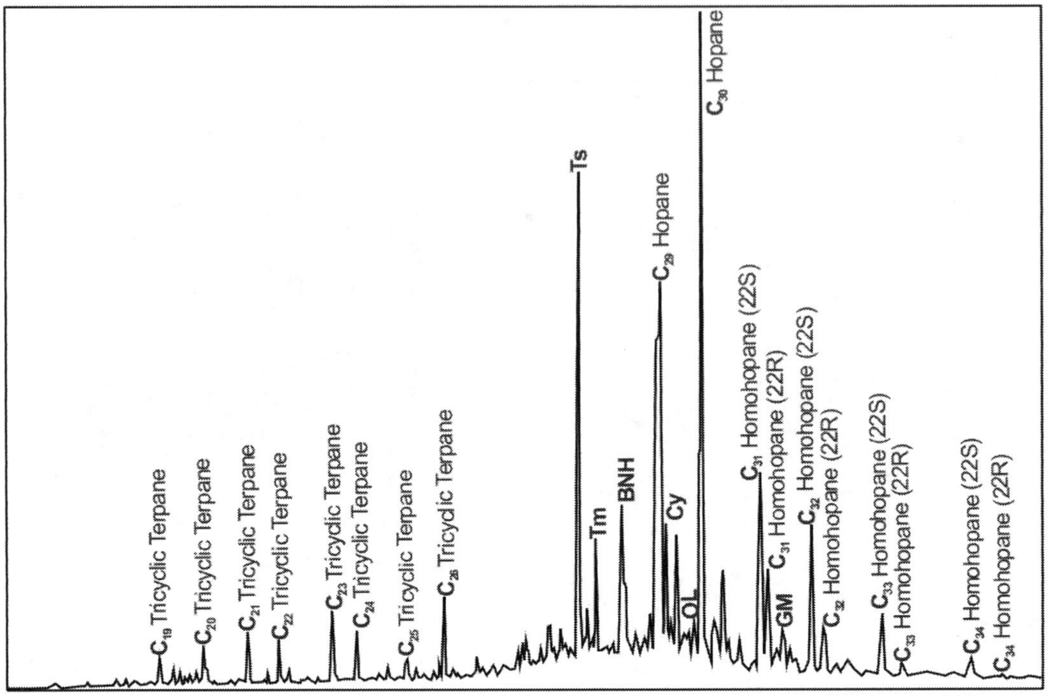

Figure 7. The representative m/z 191 extended terpane fragmentogram of Banuwati Formation shows higher C_{26} over C_{25} tricyclic terpane, higher Ts than Tm, lower C_{29} than C_{30} hopane, and lower C_{34} compared to C_{31} homohopane; the presence of C_{29} norneohopane (Cy), 18α(H)-oleanane (OL), 17α(H),21β(H),28,30-bisnorhopane (BNH).

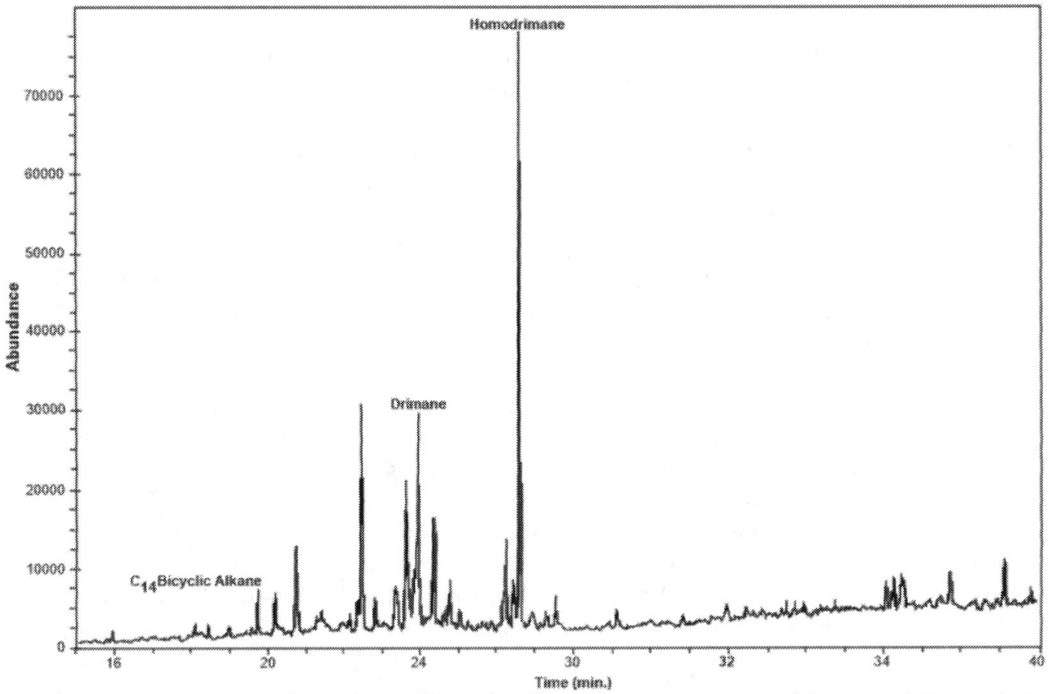

Figure 8. The representative m/z 123 bicyclic alkane fragmentogram of Banuwati Formation shows thepresence of bicyclic alkanes (e.g. 8β(H)-drimane and 8β(H)-homodrimane) compounds.

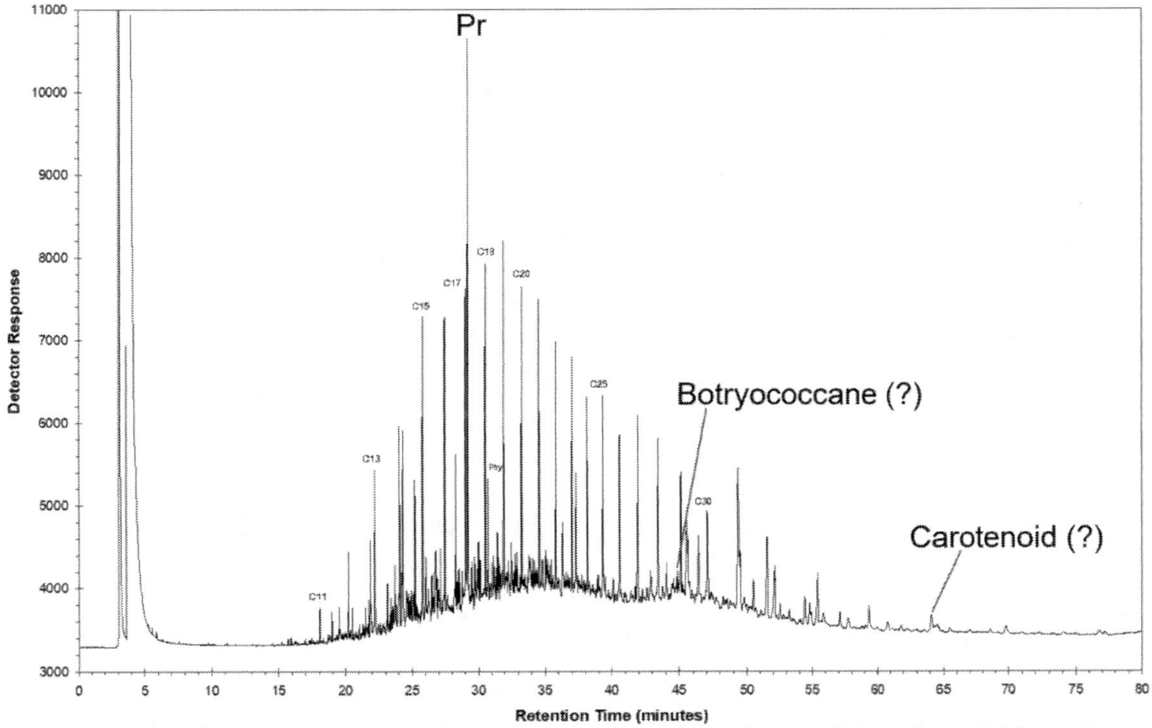

Figure 9. The sediment extract of the Banuwati Formation from the Asri Basin shows higher pristane (Pr) than phytane (Phy), the possible presence of Botryococcane and Carotenoid.

5. Conclusions

This study concludes that the syn-rift packages of the Banuwati Formation in the Sunda and Asri Basins are abundant in source rock, with good to very good quantities. Biomarker analysis reveals a novel interpretation, suggesting that the source rock was clay rich and deposited in an oxic-suboxic environment. This challenged the prevailing notion that anoxic conditions are necessary for substantial hydrocarbon production. The Banuwati Formation primarily consists of algae or Type I organic material and displays excellent kerogen quality and oil-prone characteristics. Distinctive biomarkers indicate deposition in lacustrine environments. However, the mechanism by which oxic-suboxic source rocks produce significant hydrocarbons in these basins was not evaluated in this study. Further research is needed to explore these mechanisms and validate biomarker-based interpretation.

6. Acknowledgement

The authors would like to thank and acknowledge the Management and Subsurface Exploration Team of Pertamina EP for providing the data and allowing this research to be presented and published. The authors also thank to anonymous reviewers for editing the original manuscript and the editors for publishing this paper.

References

[1] Molina J 1985 *Proc. Indo. Pet. Assoc., 14th Ann. Conv.* (Jakarta: IPA) pp. 143–179
[2] Prayitno W, Armon J W, and Haryono S 1992 *Proc. Indo. Pet. Assoc., 21st Ann. Conv.* (Jakarta: IPA) pp. 379–415
[3] Noble R A, Pratomo K H, Nugrahanto K, Ibrahim A M T, Prasetya I, Wu C H, and Howes J V C 1997 *Proc. Indo. Petrol. Assoc., the International Petroleum Systems of SE Asia and Australasia Conference* (Jakarta: IPA) pp. 585–600

[4] Sukanto J, Nunuk F, Aldrich J B, Rinehart G P, and Mitchell J 1998 *Proc. Indo. Pet. Assoc., 26th Ann. Conv.* (Jakarta: IPA) pp. 291–312
[5] Doust H and Noble R A 2008 *J. Mar. Pet. Geol.* **25** (2) pp. 103–129
[6] Bushnell D C and Temansja A D 1986 *Proc. Indo. Pet. Assoc., 15th Ann. Conv.* (Jakarta: IPA) pp. 47–75
[7] Sribudiyani, Muchsin N, Ryacudu R, Kunto T, Astono P, Prasetya I, Sapiie B, Asikin S, Harsolumakso A H, and Yulianto I 2003 *Proc. Indo. Pet. Assoc., 29th Ann. Conv.* (Jakarta: IPA)
[8] Hall R, Clements B, and Smyth H R 2009 *Proc. Indo. Pet. Assoc., 33rd Ann. Conv.* (Jakarta: IPA)
[9] Metcalfe I 2011 *Geol. Soc. London, Spec. Publ.* **355** (1) pp. 7–35
[10] Aldrich J B, Rinehart G P, Ridwan S, and Schuepbach M A 1996 *Proc. Indo. Petrol. Assoc., the International Symposium on Sequence Stratigraphy in SE Asia* (Jakarta: IPA) pp. 261–287
[11] Tissot B P and Welte D H 1984 *Petroleum Formation and Occurrence* (New York: Springer Berlin Heidelberg)
[12] Peters K E 1986 *J. AAPG Bull.* **70** (3) pp. 318–329
[13] Zumberge J E, Russell J A, and Reid S A 2005 *AAPG Bull.* 89(10) pp. 1347–1371
[14] McCarthy K, Rojas K, Niemann M, Palmowski D, Peters K E, and Stankiewicz A 2011 *Oilf. Rev.* 23(2) pp. 32–43
[15] Dembicki Jr. H 2017 *Practical Petroleum Geochemistry for Exploration and Production* (Amsterdam: Elsevier Science)
[16] Katz B J and Mertani B 1989 *Proc. Indo. Pet. Assoc., 18th Ann. Conv.* (Jakarta: IPA)
[17] Dembicki Jr. H 2009 *J. AAPG Bull.* **93** (3) pp. 341–356
[18] Lo H B 1993 *J. Org. Geochem.* 20 (6) pp. 653–657
[19] Didyk B M, Simoneit B R T, Brassell S C, and Eglinton G 1978 *J. Nature* **272** (5650) pp. 216–222
[20] Peters K E, Walters C C, and Moldowan J M 2005 *The Biomarker Guide Vol.2: Biomarker and Isotopes for Petroleum Exploration and Earth History* (Cambridge: Cambridge University Press)
[21] Connan J and Cassou A M 1980 *J. Geochim. Cosmochim. Acta* **44** (1) pp. 1–23
[22] Wang G, Chang X, Wang T G, and Simoneit B R T 2015 *J. Org. Geochem.* **78** pp. 110–120
[23] Waples D W and Machihara T 1991 *Biomarkers for geologists: a practical guide to the applications of steranes and triterpanes in petroleum geology* **9** (Oklahoma: The American Association of Petroleum Geologists)
[24] Zumberge J E 1987 *J. Geochim. Cosmochim. Acta* **51** (6) pp. 1625–1637
[25] Noble R A, Alexander R, Kagi R I , and Knox J 1986 *Org. Geochem.* **10** pp. 825–829
[26] Robinson K M 1987 *Proc. Indo. Pet. Assoc., 16th Ann. Conv.* (Jakarta: IPA) pp. 97–122
[27] Koopmans M P, De Leeuw J W, and Damsté J S S 1997 *J. Org. Geochem.* **26** pp. 451–466
[28] Moldowan J M and Seifert W K 1980 *J. Chem. Comm.* pp. 912–914
[29] McKirdy D M, Cox R E, Volkman J K, and Howell V J 1986 *J. Nature* **320** pp. 57–59
[30] Huang W and Meinschein W G 1979 *J. Geochim. Cosmochim. Acta* **43** (5) pp. 739–745
[31] Volkman J K 2005 *J. Org. Geochem.* **36** (2) pp. 139–159
[32] Longley I M, Barraclough R, Bridden M A, and Brown S 1990 *Proc. Indo. Pet. Assoc., 19th Ann. Conv.* (Jakarta: IPA) pp. 279–297

Comparing Gardner-Knopoff, Gruenthal, and Uhrhammer Earthquake Declustering Methods in Aceh, Indonesia

Juellyan[1,2], Bambang Setiawan[3,*], Muttaqin Hasan[4], Halida Yunita[4], Munira Sungkar[4], Taufiq Saidi[4]

[1] Doctoral Program, School of Engineering, Post Graduate Program, Universitas Syiah Kuala, Banda Aceh 23111, Indonesia
[2] Awardee, Program Riset Unggulan Universitas Syiah Kuala Percepatan Doktor (PRUU-PD) Lembaga Penelitian dan Pengabdian kepada Masyarakat (LPPM), Universitas Syiah Kuala
[3] Program Study of Geological Engineering, Faculty of Engineering, Universitas Syiah Kuala, Indonesia
[4] Department of Civil Engineering, Faculty of Engineering, Universitas Syiah Kuala, Indonesia

*Corresponding author: bambang.setiawan@usk.ac.id

Abstracts. The sudden release of energy from inside the earth, which generates seismic waves, causes earthquakes on the earth's surface. Foreshock, mainshock, and aftershock are terms commonly used in this subject. These phrases refer to the three stages of an earthquake event—before the main event, the primary event, and after the main event. The earthquake data catalogue, however, typically does not distinguish between mainshock data and foreshocks/aftershocks. For seismic-related studies, justifying the mainshock is very crucial. There are several methods for separating the mainshock of earthquake data from the foreshocks/aftershocks. This study employs three declustering techniques, including Gardner-Knopoff (1974), Gruenthal, and Uhrhammer (1986). The present study collected and analyzed 3543 earthquake events from 1971 to 2021 for Aceh region. In order to determine which method provides the highest a-value and b-value outcomes for Aceh's seismicity conditions, this study compares the aforementioned three approaches. The Uhrhammer method (1986) has an a-value of 5.82, a b-value of 1.05, 1893 main shock events, 305 clusters, and 3.3 percent of the total seismic moment generated at that location are the findings from this study. According to the Gruenthal method, the results had an a-value of 4.53, a b-value of 0.823, 776 mainshock events, 304 clusters, and released 1.3 percent of the total seismic moment at that location. a-value of 5.1, b-value of 0.929, 1097 main shock events, 319 clusters, and a release of 1.08 percent of the total seismic moment at that location were the results of the Gardner-Knopoff method (1974). For seismic hazard evaluation with the highest level of risk, the Uhrhammer (1986) technique is appropriate. The Gruenthal approach is suggested for more conservative cases.
Keywords: Aceh, earthquake, mainshock, foreshock, declustering.

1. Introduction

The first step in performing a seismic hazard analysis is to justify the mainshock of the earthquakes. Many declustering techniques, including the Gardner-Knopoff [1], Reasenberg [2], Zhuang [3], Uhrhammer method [4], and others, have been proposed by earlier scholars. Each technique for segregating seismic data has benefits and drawbacks. Stiphout [5] used earthquake data from southern

IOP Conf. Series: Earth and Environmental Science 1245 (2023) 012010 doi:10.1088/1755-1315/1245/1/012010

California and Japan with a time range of 1932–2010 to compare the declustering methods of Reasenberg [2], Zhuang [3], and Gardner–Knopoff [1]. According to this study, using Reasenberg [2] and Zhuang [3], declustering causes a slowdown for long events and a quick drop in time for brief events. Furthermore, Stiphout [5] also suggested that estimations become redundant at close distances; the Gardner-Knopoff technique [1] only eliminates occurrences that are close in space and time and is not possible long-distance estimation. Using earthquake catalogue data from Turkey, Azak [6] attempted to compare the approaches of Gardner-Knopoff [1], Reasenberg [2], and Zaliapin [7]. This study indicated that none of the used approaches accurately identified events from the catalogue. The number of clusters found using the Reasenberg [2] and Gardner-Knopoff [1] approaches differs significantly. The Gardner-Knopoff [1] approach and the Zaliapin [7] method both produced results that were pretty close [6]The present study attempted to compare the declustering approach, which has the same principle as the windows method, as this is the simplest method for figuring out mainshock and foreshocks/aftershocks. The largest earthquakes in a specific window of time and space are known as the main earthquakes or main shocks, whereas the earthquakes that precede the main earthquake are known as foreshocks [8]The aftershocks are earthquakes that follow the main earthquake. The Gardner-Knopoff [1], the Gruenthal [9], and the Uhrhammer [4] methods are a few declustering techniques that take advantage of this fundamental idea [10]

2. Declustering earthquake methods

As aforementioned, the precede shocks will be classified as a foreshock for every earthquake in the catalogue with magnitude *M* if it takes place within a specific window of time and space. The earthquake with the greatest magnitude in the sequence usually becomes the main shock. Within a specific window of time and space, the subsequent shocks are classified as an aftershock for every earthquake in the catalogue. Based on the abovementioned assertion, each declustering technique offers a mathematical formula for the two determining factors, time and distance. The Gardner-Knopoff technique [1] equations (1), (2), and (3) are used to calculate the mainshock and foreshock, whereas equations (4), (5), and (6) are for the Gruenthal approach [9], which the author developed as an alternative to the Gardner-Knopoff method. The final equations are (7) and (8), proposed by Uhrhammer [4] for the relationship between time and distance [10]

Gardner-Knopoff method [1]

$$t = 10^{0.032M+2.7389}\ for\ M \geq 6.5 \tag{1}$$

$$t = 10^{0.5409M-0.547}\ for\ M < 6.5 \tag{2}$$

$$d = 10^{0.1238M+0.983} \tag{3}$$

Gruenthal method[9]

$$t = \left|e^{-3.95+(0.62+17.32M)^2}\right|\ for\ M \geq 6.5 \tag{4}$$

$$t = 10^{2.8+0.024M}\ for\ M < 6.5 \tag{5}$$

$$d = e^{1.77+(0.037+1.02M)^2} \tag{6}$$

Uhrhammer method[4]

$$t = e^{-2.87+1.235M} \tag{7}$$

$$d = e^{-1.024+0.804M} \tag{8}$$

where M is the earthquake magnitude (Mw), t is the time (days), and d is the distance (km).

2.1. *Maximum likelihood model*

In order to obtain the most representative data, the maximum likelihood model can statistically calculate earthquake activity characteristics and can eliminate magnitude gaps at specific intervals. This model's output parameter is called b-value. The fragility of the rock will grow if the b-value is high. The area's shear stress level will also be higher if the b-value is low [11]The equation provided by Utsu [12] can be employed to determine b-values (equation 9) [13]The equation proposed by Wekner [14] can be used to determine a-values (equation 10). The a-value is a parameter that is affected by the volume, time window, and the number of earthquake events. In other words, it can be said that if the a-value is high, the seismic activity is high [15]

$$b = \frac{1}{\bar{M} - M_{min}} \log e \tag{9}$$

$$a = \log N(M \geq M_o) + \log (b \ln 10) + M_o b \tag{10}$$

where, variable $\bar{M}$ is the average magnitude; variable M_{min} or M_o is the minimum magnitude contained in the research data; N is the number of earthquake events that occur; a is the seismicity parameter that depends on the number of earthquake events, volume, and time window; b is the parameter seismicity related to the level of rock fragility or the level of shear stress.

3. Result and discussion

The study is being conducted in the Indonesian province of Aceh. Geographically, the study area lies between latitudes 0.297° – 7.569° N and longitudes 93.208° – 97.822° E. There were 3543 earthquake events recorded in the earthquake data from 1971 to 2021. This data was obtained from the USGS data catalog [16]. The data is earthquake data with a minimum magnitude of 4 Mw. The location of the study and the information used is shown in Figure 1 below.

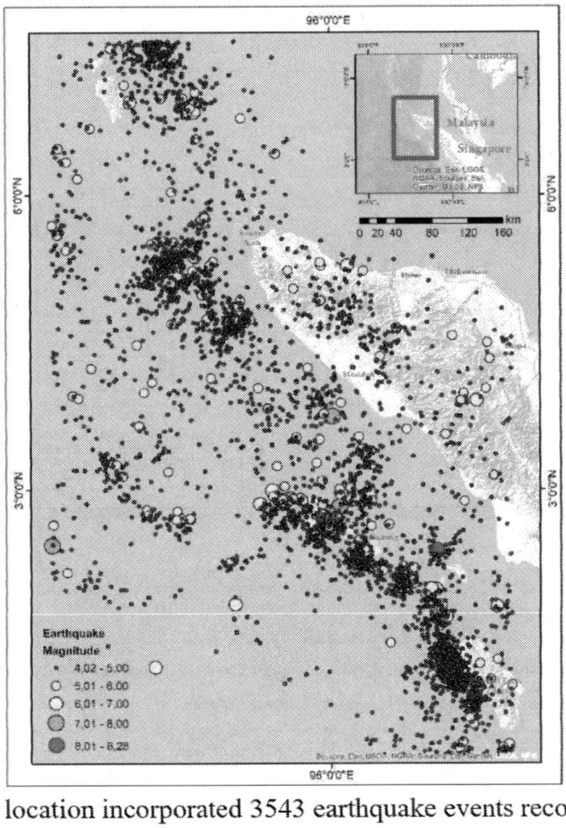

Figure 1. The research location incorporated 3543 earthquake events recorded from 1971 to 2021

3.1. Result

This study separates earthquake data using three techniques: Gruenthal [9], Uhrhammer [4], and Gardner-Knopoff [1]. Table 1 below compares the number of earthquake events falling within a particular scale range. The findings are then modeled with the maximum likelihood model to produce the a-values and b-values, as shown in Table 2. The outcomes of maximum likelihood modeling for the three approaches are shown in Figure 2. The outcomes of the three declustering techniques are shown in Figure 3. The comparison of magnitude-to-time and magnitude-to-distance in each approach is shown in Figure 4.

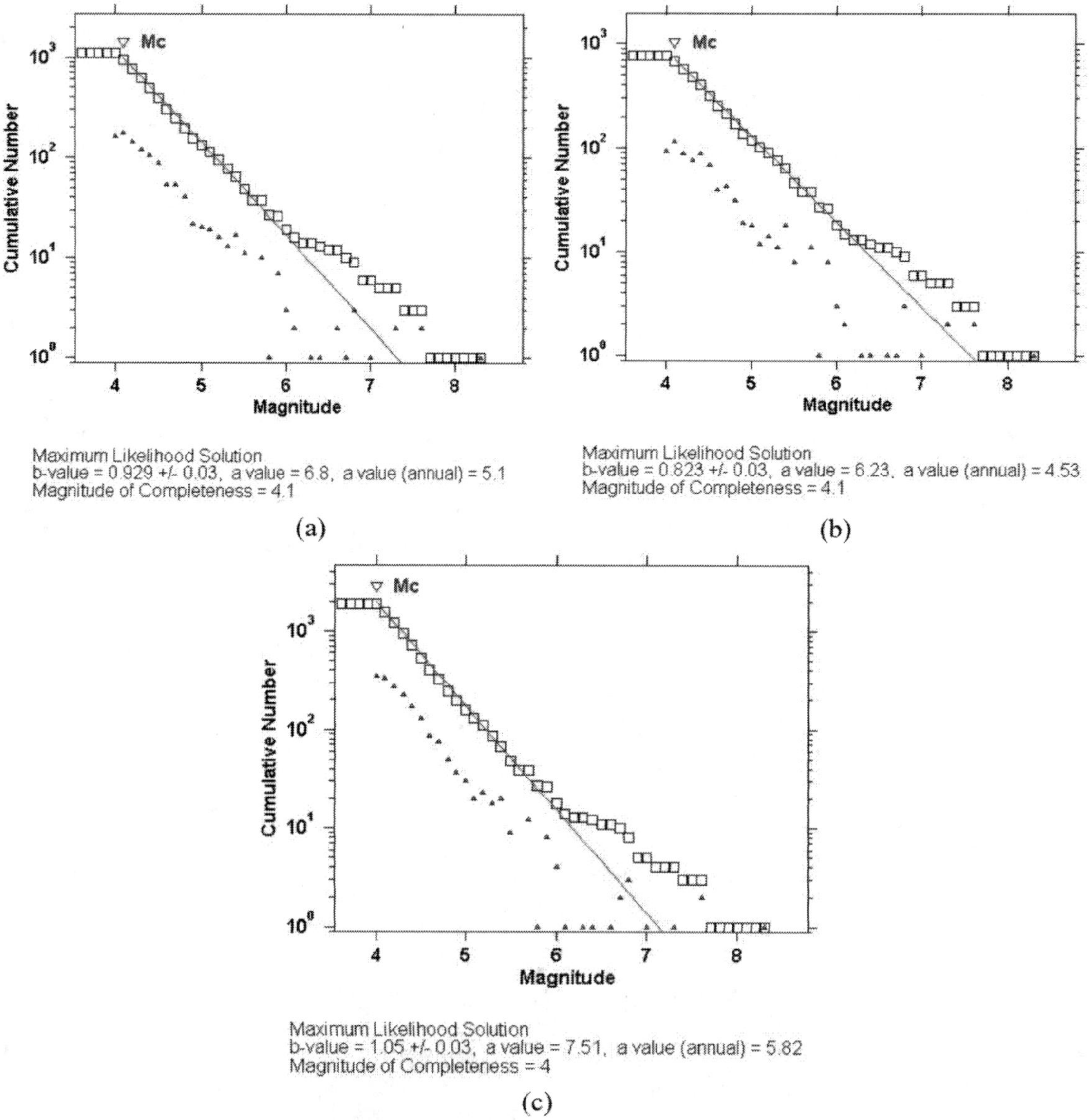

Figure 2. Maximum likelihood recurrence model (a) Gardner-Knopoff method, (b) Gruenthal method, (c) Uhrhammer method.

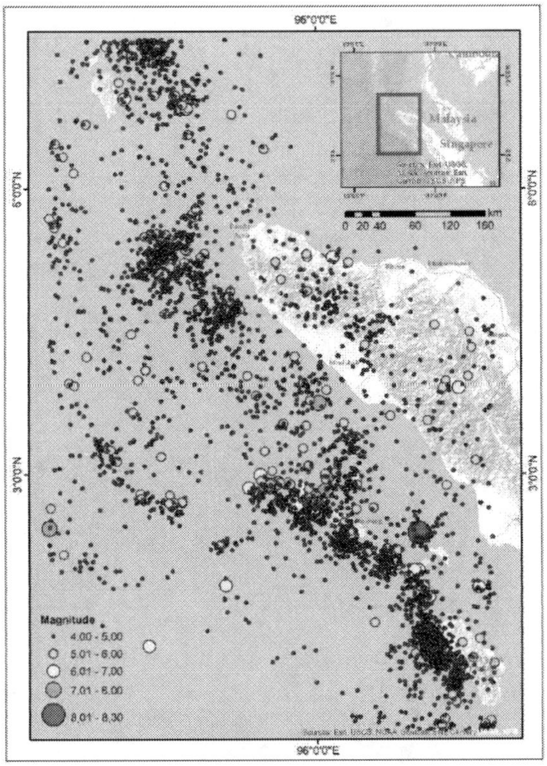

(a)

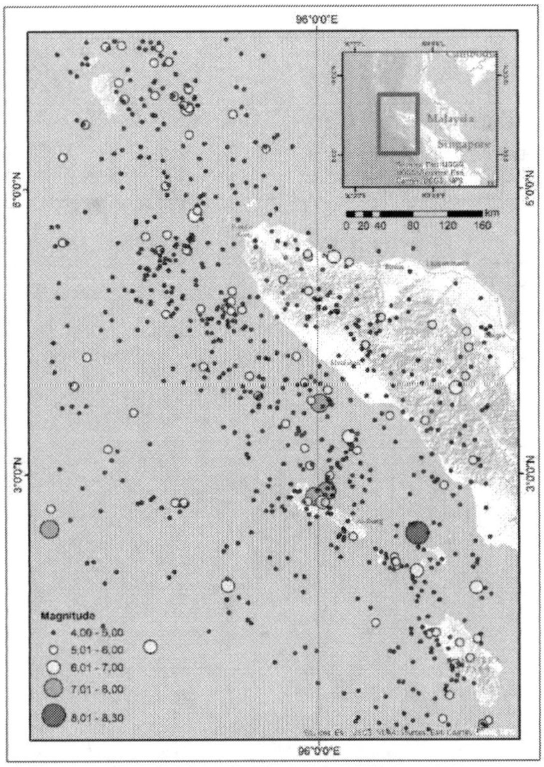

(b)

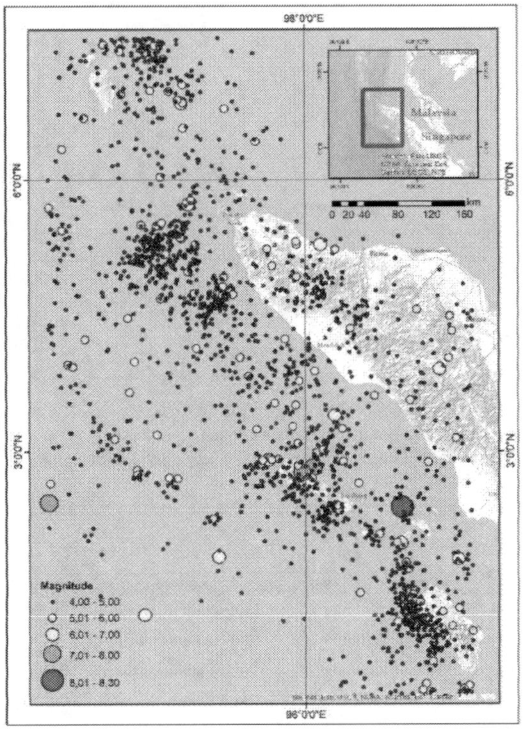

(c)

Figure 3. Declustering earthquake data (a) Gardner-Knopoff method, (b) Gruenthal method, (c) Uhrhammer method.

Table 1. Number of earthquake events in each declustering method

Declustering method	Magnitude (M_w)				
	4.00-5.00	5.01-6.00	6.01-7.00	7.01-8.00	8.00-8.30
Gardner-Knopoff (1974)	984	97	11	4	1
Gruenthal (1985)	675	86	10	4	1
Uhrhammer (1986)	1764	114	10	3	1

Table 2. Parameters a-value and b-value for each declustering method

Declustering method	a	b
Gardner-Knopoff (1974)	5.1	0.929
Gruenthal	4.53	0.823
Uhrhammer (1986)	5.82	1.05

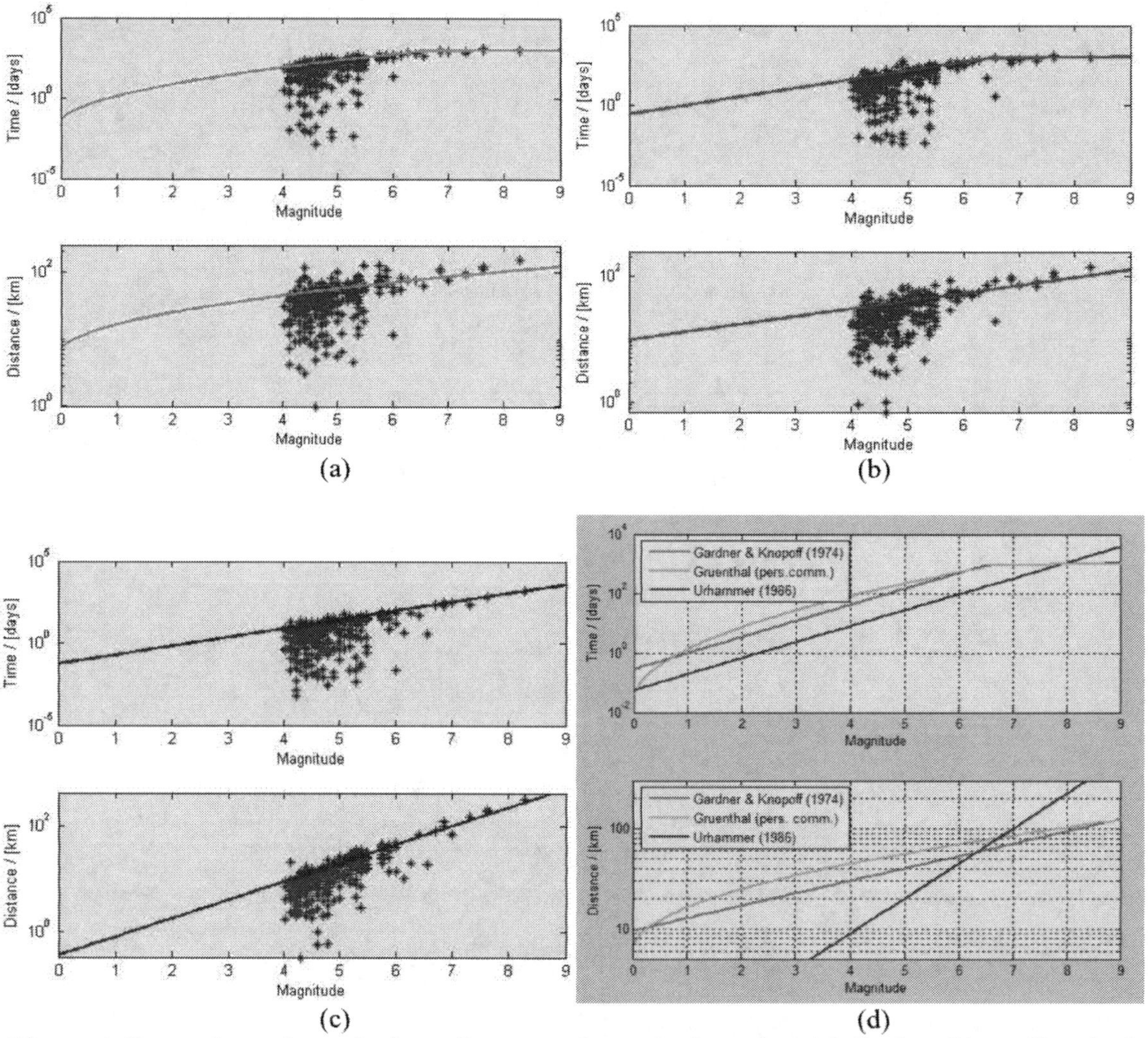

Figure 4. Comparison of magnitude-to-distance and magnitude-to-time (a) Gardner-Knopoff method [1], (b) Gruenthal method[2], (c) Uhrhammer method[3], (d) combined graph of the three methods [1], [4], [9].

3.2. Discussion

The outcomes revealed some variations among the three techniques. The first comparison of how many main shocks each method produced (see Table 1). The Uhrhammer technique [4], which included 1892 earthquake events, identified the highest number of main shocks after the declustering process. The Gardner-Knopoff approach [1], which includes 1097 earthquake events, comes in second. Lastly, Gruenthal approach [9] suggested 776 earthquake events. The outcomes of recurrence modeling using the Gutenberg-Richter [1] approach provide the basis for the second comparison (see Table 2). According to the results of the present study, as shown in Table 2, the Uhrhammer [4] method get the highest a-value 5.82. It is followed by Gardner-Knopoff [1] approach with an a-value of 5.1. The Gruenthal method [9]is the last one with a-value of 4.53. The level of seismic activity in a zone increases with increasing a-value.

Additionally, the b-values for each technique are as follows: 0.929 for the Gardner-Knopoff [1], 0.823 for the Gruenthal method [9], and 1.05 for the Uhrhammer [4]. However, the more brittle the rocks there are, the greater the b-value in a zone denotes a smaller shear stress experienced. Thus, using the Gruenthal approach [9], it may be concluded that there is significant shear stress in the zone. Utilizing the Uhrhammer method [4], the zone exhibits the highest level of rock fragility.

The Gardner-Knopoff method of declustering [1] produced 319 clusters, of which 2446 events (or 69.04 percent) were eliminated from 3543 earthquake occurrences (Figure 4a). As can be observed in Table 1, the total mainshock contained by this method is 1097 occurrences, distributed throughout each magnitude interval. This technique released 1.08 percent of the total seismic moment in the zone as seismic energy. Gruenthal's approach [9] of declustering produced 304 clusters, of which 2767 events (78.1%) were left out of the total of 3543 earthquake occurrences (Figure 4b). The distribution of the 776 total mainshock occurrences included in this manner for each magnitude interval is shown in table 1. The seismic moment that was released using this technique makes up 1.3% of the total seismic moment in the zone. Following that, the Uhrhammer method [4] of declustering produced 305 clusters, of which 1650 occurrences (or 46.57 percent of the total events) were eliminated from the 3543 earthquake events (Figure 4c). This technique contains 1893 mainshock events in total. 3.3 percent of the total seismic moment was discharged using this way.

4. Conclusion

The present study suggests that the Uhrhammer [4] technique produces data with an a-value of 5.82, b-value of 1.05, 1893 mainshock events, 305 clusters, and 3.3% of the total seismic moment released at that location. With an a-value of 4.53, a b-value of 0.823, a total of 776 mainshock events, 304 clusters, and 1.3 percent of the total seismic moment released at that location, the results of the Gruenthal [9] method are presented. With an a-value of 5.1, a b-value of 0.929, 1097 main shock events, 319 clusters, and a release of 1.08 percent of the total seismic moment at that location, the Gardner-Knopoff [1] method produced findings. The Uhrhammer [4] approach is appropriate for assessing seismic hazards with the highest level of risk. The Gruenthal[9] approach is advised in the meanwhile for outcomes that are more conservative.

References

[1] J. K. Gardner and L. Knopoff, "Is the sequence of earthquakes in Southern California, with aftershocks removed, Poissonian?," *Bulletin of the Seismological Society of America*, vol. 64, no. 5, pp. 1363–1367, Oct. 1974, doi: 10.1785/BSSA0640051363.

[2] P. Reasenberg, "Second-order moment of central California seismicity, 1969-1982," *J Geophys Res Solid Earth*, vol. 90, no. B7, pp. 5479–5495, Jun. 1985, doi: 10.1029/JB090iB07p05479.

[3] G. Zhuang, A. S. L. Tsang, N. Zhou, F. Li, and J. A. F. Nicholls, "Impacts of situational factors on buying decisions in shopping malls," *Eur J Mark*, vol. 40, no. 1/2, pp. 17–43, Jan. 2006, doi: 10.1108/03090560610637293.

[4] Uhrhammer R, "Characteristics of northern and southern California seismicity," *Earthquake Notes*, vol. 57, p. 21, 1986.

[5] T. van Stiphout, J. Zhuang, • David Marsan, V. Stiphout, J. Zhuang, and D. Marsan, "Theme V-Models and Techniques for Analyzing Seismicity Seismicity Declustering," 2012, doi: 10.5078/corssa.

[6] T. Eroglu Azak, D. Kalafat, K. Şeşetyan, and M. B. Demircioğlu, "Effects of seismic declustering on seismic hazard assessment: a sensitivity study using the Turkish earthquake catalogue," *Bulletin of Earthquake Engineering*, vol. 16, no. 8, pp. 3339–3366, Aug. 2018, doi: 10.1007/s10518-017-0174-y.

[7] I. Zaliapin, A. Gabrielov, V. Keilis-Borok, and H. Wong, "Clustering Analysis of Seismicity and Aftershock Identification," *Phys Rev Lett*, vol. 101, no. 1, p. 018501, Jun. 2008, doi: 10.1103/PhysRevLett.101.018501.

[8] Y. Zhang *et al.*, "Automatic Inversion of Rupture Processes of the Foreshock and Mainshock and Correlation of the Seismicity during the 2019 Ridgecrest Earthquake Sequence," *Seismological Research Letters*, vol. 91, no. 3, pp. 1556–1566, May 2020, doi: 10.1785/0220190343.

[9] G. Grünthal, "The up-dated earthquake catalogue for the German Democratic Republic and adjacent areas - statistical data characteristics and conclusions for hazard assessment," 1985.

[10] T. Van Stiphout, J. Zhuang, and D. Marsan, "Seismicity declustering," *Community Online Resource for Statistical Seismicity Analysis*, no. February, pp. 1–25, 2012, doi: 10.5078/corssa-52382934.

[11] J. Woessner and S. Wiemer, "Assessing the quality of earthquake catalogues: Estimating the magnitude of completeness and its uncertainty," *Bulletin of the Seismological Society of America*, vol. 95, no. 2, pp. 684–698, 2005, doi: 10.1785/0120040007.

[12] A. Kijko and A. Smit, "Estimation of the frequency-magnitude Gutenberg-Richter b-value without level of completeness Fluid-Induced Seismicity in the Central Basin Area: Ground Motion Prediction and the Development of an Early Warning System for Risk Reduction View project," 2018, doi: 10.5078/corssa.

[13] T. Utsu, "A Statistical Significance Test of the Difference in b-value between Two Earthquake Groups," *Journal of Physics of the Earth*, vol. 14, no. 2, pp. 37–40, 1966, doi: 10.4294/jpe1952.14.37.

[14] C. H. Chen and S. D. Fassois, "Maximum likelihood identification of stochastic Weiner-Hammerstein-type non-linear systems," *Mech Syst Signal Process*, vol. 6, no. 2, pp. 135–153, Mar. 1992, doi: 10.1016/0888-3270(92)90061-M.

[15] S. Rohadi, H. Grandis, and M. A. Ratag, "Studi Variasi Spatial Seismisitas Zona Subduksi Jawa," *Jurnal Meteorologi dan Geofisika*, vol. 8, no. 1, pp. 42–47, 2014, doi: 10.31172/jmg.v8i1.4.

[16] USGS, "USGS Earthquake Catalog," *Earthquake Catalog*, 2021. https://earthquake.usgs.gov/earthquakes/search/ (accessed May 17, 2022).

Preliminary result of dominant frequency and seismic amplification in Penajam Paser Utara and its surrounding regions using the HVSR method

A Mawadah[1], Z Zulfakriza[2*], S Widiyantoro[2,3], P Supendi[4,5], Y M Husni[6], A Lesmana[6], M F Azhari[1]

[1]Earth Science, Faculty of Earth Science and Technology, Bandung Institute of Technology, Indonesia
[2]Global Geophysics Research Group, Faculty of Mining and Petroleum Engineering, Bandung Institute of Technology, Indonesia
[3]Faculty of Engineering, Maranatha Christian University, Bandung 40164, Indonesia
[4]Agency for Meteorology, Climatology, and Geophysics (BMKG), Jakarta 10720, Indonesia
[5]Department of Earth Sciences, Bullard Labs, University of Cambridge, Cambridge CB3 0EZ, UK
[6]Geophysical Engineering, Faculty of Mining and Petroleum Engineering, Bandung Institute of Technology, Indonesia

*E-mail: zulfakriza@itb.ac.id

Abstract. East Kalimantan has been designated as the new capital city of Indonesia, precisely in Penajam Paser Utara and Kutai Kartanegara Regency. According to the historical records of seismic activity, there were several destructive earthquakes in East Kalimantan. Therefore, it is necessary to study the characteristics of the local site to assess the seismic hazard potential in the new capital city area and its surroundings. In this research, the Horizontal to Vertical Spectral Ratio (HVSR) method was applied by utilizing microtremor data from 21 seismometers installed in the research area. The results of the HVSR analysis obtained the dominant frequency (f_0) and amplification factor (A_0) from the H/V curve. Based on the H/V calculation, the highest f_0 is 13.04 Hz and the lowest value is 0.58 Hz. Meanwhile, the highest value for A_0 is 6.41, and the lowest value is 1.41. The results of the study show that the research area has a low to medium level of seismic hazard.

1. Introduction

East Kalimantan Province, specifically in Penajam Paser Utara and Kutai Kartanegara Regency, has been designated as the new capital city of Indonesia, or IKN. One of the considerations for selecting these regions was that the potential for natural disasters is relatively small, especially earthquakes. Based on USGS data, there were about 60 earthquake events with M 3–6 from 1921–2022 (Figure 1), which is a small number when compared to earthquake events that occurred in other regions in Indonesia. However, according to the historical seismic record, several destructive earthquakes have occurred in East Kalimantan, as can be seen in Figure 1.

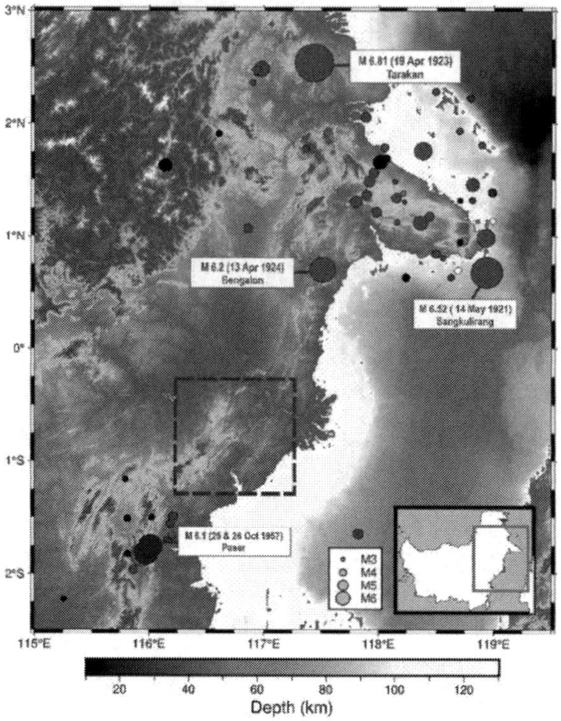

Figure 1. Historical of seismic event in East Kalimantan based on USGS catalogue 1921 – 2022. The dashed line indicated the research area. The insert map shows the Kalimantan region and the eastern Kalimantan marked by red square

Tectonically, there are three main faults in East Kalimantan that are still active today, namely the Maratua Fault, the Paternoster Fault, and the Mangkalihat Fault. A study revealed that these faults have a potential magnitude of up to 7.0 and can impact levels VI–VII on the MMI scale [1]. Therefore, it is important to study the subsurface profile and characteristics of the new capital city and its surroundings in order to know the potential for seismic hazards in this region, considering that research in this area has not been widely carried out. This research was conducted to calculate the dominant frequency (f_0) value and amplification factor (A_0) to determine the subsurface characteristics by utilizing single-station microtremor data and analyzing it using the HVSR method.

2. Geological Setting

Kalimantan Island is part of Sundaland, which is part of the Eurasian continental plate [2]. The Indo-Australian plate is moving northward with a movement velocity of 7 cm/year and colliding with the Eurasian plate at a velocity of about 0.4 cm/year, moving southeast [3].

The research area is included in the Kutai Basin, which has a thick layer of sediment with a thickness of > 3 km [13]. Physiographically, to the north of the Kutai Basin, it is bordered by Sangkulirang, the Bengalon Fault Zone, and the Mangkalihat High. Whereas in the south it is bordered by the Adang Fault Zone, which has served as the axis zone of the basin since the end of the Paleogene until now [14]. The tectonic framework in eastern Kalimantan is influenced by regional tectonic developments through interactions between the Pacific Plate, the Eurasian Plate, and the India-Australia Plate, as well as regional tectonics in southeastern Asia [15].

The Kalimantan area is surrounded by several faults, namely the Adang Fault in West Kalimantan, the Tinjia Fault in Serawak, the Mangkalihat Fault, which is located on the east coast of Kalimantan, the Sangkulirang Fault in East Kalimantan, and the Paternoster Fault in the Makassar Strait, which is trending west-east through Paser Regency [4].

In general, the geological conditions of the research area consist of six rock formations (Figure 2), which are composed of sedimentary rocks that contain a fairly high mixture of organic materials such as carbonate, lignite, and coal [5]. These six formations are:

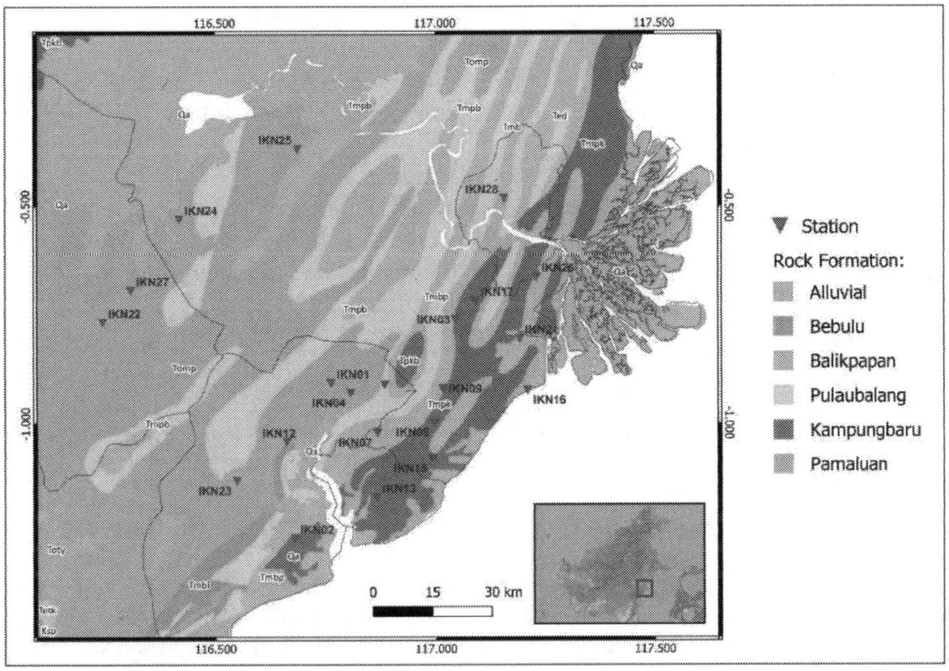

Figure 2. The regional geological map of the research area [16]. Inverted triangles represent the measurement stations and color bars represent the rock formations

a. The Pamaluan Formation is an Oligocene-Early Miocene rock composed of sandstone with intercalated claystone, shale, siltstone, and limestone.
b. The Bebulu Formation dates from the Early Miocene to the Middle Miocene. The rocks formed in this formation are reef limestone intercalated with sandy limestone and shale.
c. The Pamaluan Formation is an Oligocene-Early Miocene rock composed of sandstone with intercalated claystone, shale, siltstone, and limestone.
d. The Bebulu Formation dates from the Early Miocene to the Middle Miocene. The rocks formed in this formation are reef limestone intercalated with sandy limestone and shale.
e. The Pulau Balang Formation is Middle Miocene composed of alternations of greywacke and quartz sandstones with intercalated limestone, coal, claystone, and dacite tuff that formed in a shallow marine depositional environment.
f. The Balikpapan Formation is a Middle Miocene-Late Miocene formation, composed of alternations of sandstones and clays with intercalated silt, shale, limestone, and coal.
g. The Kampungbaru Formation is a Late Miocene-Plio Pleistocene formation composed of quartz sandstones intercalated with clay, shale, silt, and lignite, which are generally soft and easily crushed.
h. Alluvium deposits consist of gravel, sand, and silt that have been deposited in the river, swamp, delta, and beach environments. Formed during the Holocene and is the youngest formation.

The geological structure of this area is dominated by anticlinorium folds and faults, which have three directions, namely north-south, northwest-southeast, and southwest-northeast. Three types of faults are found in this area: reverse fault, normal fault, and strike-slip fault. In the late Miocene, reverse faults formed, which were then cut by a strike-slip fault. Meanwhile, normal faults formed during the Pliocene [5].

3. Data and Method

3.1. Data

This research used microtremor data from 21 seismometers installed in the IKN area and its surroundings, covering several districts and cities, including Penajam Paser Utara, Kutai Kartanegara, Balikpapan, Samarinda, and Kutai Barat, which can be seen in Figure 2. The recording data consists of three broadband components: north-south horizontal components, east-west horizontal components, and vertical components. The data was recorded in 2021, with a recording duration of 3 days for each station.

3.2. Method

Data processing was carried out using Geopsy software to obtain the HVSR curve. The HVSR method was first introduced by Nogoshi and Iragashi (1971) and made widespread by Nakamura. The HVSR method is a simple method that can describe soil characteristics by calculating the spectral ratio of the horizontal component to the vertical component [6]. The HVSR method can also explain the phenomenon of the surface layer resonance frequency. The HVSR calculation can be expressed in the following equation:

$$R(T) = \frac{\sqrt{F_{NS}(T)^2 + F_{EW}(T)^2}}{F_Z(T)} \tag{1}$$

where F_{NS}, F_{EW}, and Fz are the components of the Fourier spectra in the north-south, east-west, and vertical, respectively.

The recorded microtremor data in this study is continuous, with a recording time of 3 days. In data processing, all microtremor signals were plotted to see the comparison and consistency of the data by making comparisons between 24-hour data and 1-hour data from each station. In the selection of 1-hour data, the data was cut at a quiet time, in the range of 23.00 p.m. to 02.00 a.m. This is done to reduce the noise that may be recorded on seismometers caused by human activities, animals, vehicles, etc. In general, the H/V curves obtained from 24-hour data and 1-hour data are quite similar.

The result of the HVSR calculation is a H/V curve that contains the dominant frequency value (f_0) and amplification factor (A_0) of each station. Based on these two parameters, the seismic hazard potential in the research area can then be determined. To extract the microtremor signal, a bandpass filter with a frequency range of 0.5–15 Hz is used, with a time window length of 60 seconds, an STA/LTA value range of 0.2–2.5, and a measurement duration of 1 hour.

4. Results and Discussion

4.1 H/V Curve

The results of the HVSR calculation obtained two types of H/V curves. These types of curves are characterized based on [7], namely clear peak (a), and multiple peaks (b). The variations in the H/V curve can be caused by differences in impedance contrast between sediment layers and bedrock, subsurface geological conditions, and rock compactness.

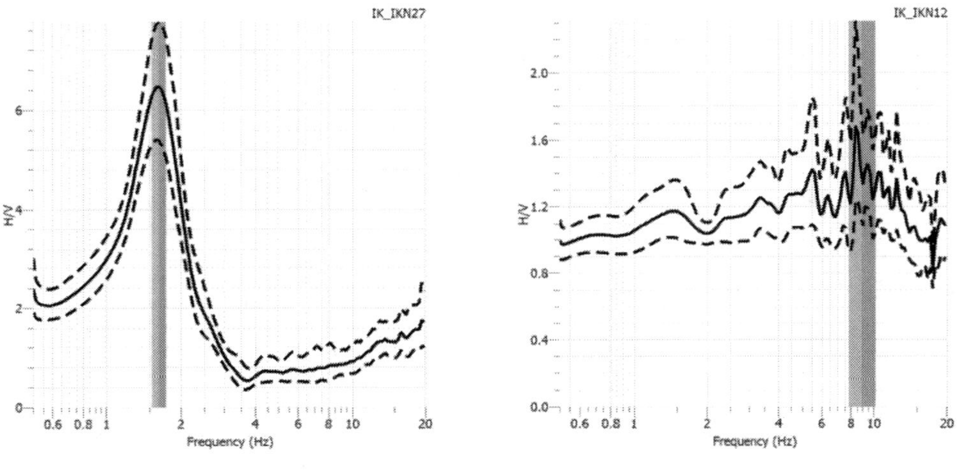

(a) (b)

Figure 3. Two types of H/V curves. (a) clear peak, and (b) multiple peaks. The solid line indicates the average HVSR value and the dashed line indicates the deviation standard

4.2 Dominant Frequency (f_0) and Amplification Factor (A_0)

The distribution of dominant frequency (f_0) and amplification factor (A_0) can be seen in Figure 4 and Figure 5, respectively.

As shown in Figure 4, it can be seen that the research area has a dominant frequency value ranging from 0.58 to 13.04 Hz. In the eastern part of the research area, most areas have a relatively low dominant frequency of < 2.5 Hz, especially in Samboja District, Kutai Kartanegara Regency. Meanwhile, the dominant frequency value tends to increase to the west of the research area. The highest value is up to 13.04 Hz at IKN22 station, precisely in Bongan District, Kutai Barat Regency. At other stations, namely IKN12, IKN05, and IKN01 in Sepaku District, Penajam Paser Utara Regency, the dominant frequency value is relatively medium-high, ranging from 8–12 Hz.

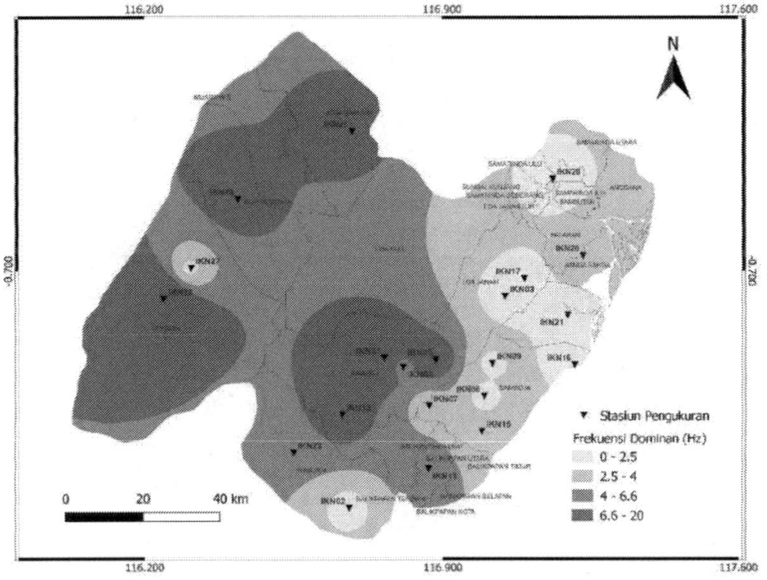

Figure 4. Dominant frequency map of IKN area and its surroundings

IOP Conf. Series: Earth and Environmental Science 1245 (2023) 012011 doi:10.1088/1755-1315/1245/1/012011

Table 1. The classification of soil types [9]

Frequency (Hz)	Type	Soil Condition	Description
6,6-20	I	Tertiary or older rocks (hard sandy, gravel)	Sediment thickness is very thin, dominated by hard rock
4-10	II	Alluvial rocks (sandy gravel, sandy hard, clay, loam)	Sediment layer with a thickness of about 5-10 m
2,5-4	III	Alluvial rocks (sandy gravel, sandy hard, clay, loam)	Sediment layer with a thickness of about 10-30 m
< 2,5	IV	Consists of alluvial sediment delta, top soil, and mud	Sediment thickness is very thick (> 30 m)

The dominant frequency is associated with sediment thickness [8], and it is also supported by the research of [12]. Based on the classification of soil types by [9] (Table 1), the low dominant frequency value shows that the area has relatively deep bedrock depth. On the other hand, the areas with high dominant frequency values have relatively thin bedrock depth.

On the geological map, the eastern part of the research area is located in the Kampungbaru formation, which is the youngest formation after alluvium deposits and also located near the Mahakam Delta. This formation is composed of quartz sandstone intercalated with clay, shale, silt, and lignite, which are generally soft and easily crushed. This is presumably what causes the dominant frequency value to tend to be low. While the high dominant frequency values that dominate the western part of the research area are sedimentary layers composed of older tertiary rocks with relatively thin sediment thickness (hard sandy, hard gravel, clay, loam). This area is included in the Pamaluan formation, which is the oldest formation in the research area. Composed of sandstone with intercalations of claystone, shale, siltstone, and limestone.

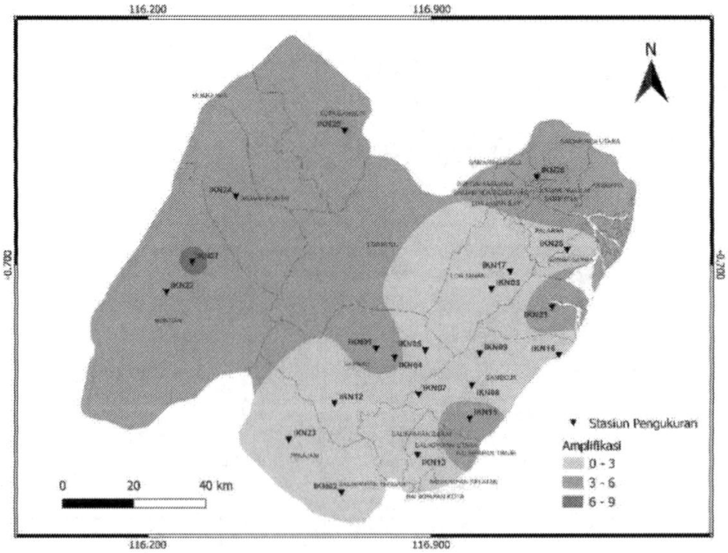

Figure 5. Amplification map of IKN area and its surroundings

The amplification factor obtained at a location can be affected by differences in wave velocity associated with soil compaction. Damage to a building can be influenced by soil characteristics and building conditions [11]. Typically, rocks with low compaction or soft soil will have a high amplification value. It is because the soft sediment layer can slow down the wave propagation from bedrock to the surface, so the amplification becomes high and has the potential to damage buildings during shocks like earthquakes.

The soil amplification zone is divided into 4 risk categories based on [10], such as low risk with amplification ($0 < A_0 < 3$ times), medium risk with amplification ($3 < A_0 < 6$ times), high risk with amplification ($6 < A_0 < 9$ times), and very high risk with amplification (> 9 times).

Based on the amplification map above (Figure 5), it can be seen that the research area has amplification that varies from low to high levels, ranging from 1.41 to 6.41. In the south-northeast, the amplification value is in the range of 1.41–4.5, which is estimated to have a low–medium risk level. Meanwhile, the amplification value in the southwest to the northern part of the research area is at a medium–high level, ranging from 4.5 to 6.41. The amplification value is quite high at the IKN27 station, which is 6.41 Hz. In general, based on the amplification value, the seismic hazard in the research area is at a low–medium risk level.

5. Conclusion

The dominant frequency value in the IKN area and its surroundings varies from 0.58 to 13.04 Hz. In the western part, precisely in Samboja District, Kutai Kartanegara Regency, which will be one of the IKN locations, has a low dominant frequency value which can be assumed to have a very thick sediment layer (> 30 m) and is composed of delta sedimentation, top soil, and mud. Meanwhile, in Sepaku District, Penajam Paser Utara Regency, which will become the central government area, has a high dominant frequency. It can be assumed that the sediment layer in this area is quite thin, composed of hard rock (sandy gravel, sandy hard, clay, loam). The amplification value in the research area varies from 1.41 to 6.41. In the two districts that will become the IKN location, the amplification varies from 1.41 to 4.49. This indicates that these areas have a low to medium risk level. In other words, the IKN area is an area with low to medium potential for seismic hazards.

References

[1] PUSGEN 2017 *Peta Sumber dan Bahaya Gempa Indonesia Tahun 2017* (Bandung: Badan Penelitian dan Pengembangan Kementerian Pekerjaan Umum dan Perumahan Rakyat)

[2] Hall R 2002 *J. Asian Earth Sci.* **20** pp 353 – 431

[3] Yeats R.S, Sieh K and Allen C.R 1997 *The Geology of Earthquakes* (Oxford: Oxford University Press)

[4] Supartoyo 2017 Ancaman dan Potensi Gempabumi di Kalimantan *Prosiding Pertemuan Ilmiah Tahunan Riset Kebencanaan ke 4 – 2017*, Universitas Indonesia, Depok

[5] Supriyatna S, Sukardi and Rustandi 1995 *Peta Geologi Lembar Samarinda* (Bandung: Pusat Penelitian dan Pengembangan Geologi)

[6] Nakamura Y 1989 A Method for Dynamic Characteristic Estimation of Subsurface Using Microtremor on the Ground Surface (Japan: *Quarterly Report of the Railway Technical Research Institute*) **30** pp 25 – 33

[7] The SESAME Team 2004 Guidelines for the implementation of the H/V spectral ratio technique on ambient vibrations: measurements, processing, and interpretation *SESAME European Research Project WP12* pp 1 – 62

[8] Nakamura Y 2008 On the H/V spectrum *The 4th World Conference on Earthquake Engineering* Beijing, China

[9] Kanai K 1983 *Engineering Seismology* (Tokyo: University of Tokyo Press)

[10] Marjiyono 2010 Estimasi karakteristik dinamika tanah dari data mikrotremor wilayah Bandung *Msc Thesis* Bandung Institute of Technology

[11] Gosar A 2009 A microtremor HVSR study of the seismic site effects in the area of the town of Brežice (SE Slovenia) *Acta Geotech Slov.* **6** pp 31-45
[12] Parolai S, Bormann P, Milkereit C 2001 Assesment of the natural frequency of the sedimentary cover in the Cologne area (Germany) using noise measurements *J. Earthq. Eng.* **5** pp 541-564
[13] Syuhada S, Pranata B, Anggono T, Ramdhan M, Zulfakriza Z, Febriani F, Prasetio A D, Dewi C N, Hasib M, Sulaiman A 2022 Crustal velocity structure in Borneo Islan using receiver function inversion *Acta Geophysica*, **70** pp 2529-2533
[14] Moss S J and Chambers J L C 1999 Depositional modelling and facies architecture of rift and inversion in The Kutai Basin, Kalimantan, Indonesia *Proceeding of 27th Annual Convention of Indonesian Petroleum Association*, Jakarta
[15] Biantoro E, Muritno B P, Mamuaya J M B 1992 Inversion faults as the major structural control in the northern part of The Kutai Basin, East Kalimantan *Proceeding of 21st Annual Convention of Indonesian Petroleum Association*, Jakarta
[16] Indonesia Geospasial 2019 Download kumpulan shapefile (SHP) terlengkap dan terbaru seluruh Indonesia https://www.indonesia-geospasial.com/

Acknowledgment
The authors are grateful to Global Geophysics Research Group of Bandung Institute of Technology, Agency of Meteorological, Climatological, and Geophysical (BMKG), and Cambridge University.

Application of modified Segment Tracing Algorithm (mSTA) Method to Identify Landslide Susceptibility Zones Around Mt. Sinabung, Indonesia

Zahratunnisa[1], Asep Saepuloh[1*], Estu Kriswati[2], Ahmad Basuki[3].

[1]Geological Engineering Study Program, Faculty of Earth Sciences and Technology, Bandung Institute of Technology, Indonesia
[2]National Research and Innovation Agency, Jl. Sangkuriang, Coblong, Bandung, West Java, Indonesia.
[3]Center for Volcanology and Geological Hazard Mitigation (CVGHM), Geological Agency, Jl. Diponegoro No.57, Bandung, West Java, Indonesia.

Email: zahratunnisa57@gmail.com
[*]Corresponding author e-mail: saepuloh@itb.ac.id

Abstract. Lineament could present the fractured zones and increase the landslides occurrence probabilities. This study aims to assess landslides susceptibility zones based on recorded landslide events corresponded to the detected lineaments around Mt. Sinabung, Indonesia. In this study, we use Synthetic Aperture Radar (SAR) data to increase the extraction accuracy of fracture-related lineament at Mt. Sinabung. The lineament densities were calculated to identify the landslide susceptibility zones including frequency, length, and intersection using modified Segment Tracing Algorithm (mSTA) method for area about 10.9 × 9 km. Based on visual observations and technical report, a total of 61 landslides were occurred around Mt. Sinabung from 2016, 2018, and 2021. An Ordinary Kriging density map of the lineament frequency using 800 × 800 m cells showed agreement between the lineament density and landslide distribution. Comparing landslide location and lineament density, landslide occurrences were located mainly at high and moderate lineament densities. We have classified the landslide density for three years into low, moderate, and high based on histogram quartile analyses. This classification could be used to mitigate the landslide hazard in a large area.

1. Introduction

Indonesia has a humid tropical climate with an average annual rainfall of 2500 mm and average temperatures of 22 - 28°C. The high intensity tropical rainfall coupled with hilly areas and a deeply weathered rock profile comprise near-ideal conditions for landslide development [1,2]. Landslide is one of the most common geomorphological hazards caused by the acceleration of slopes dynamics in the river basins leading to long-term irreversible changes affecting the natural environment, human life properties, and infrastructure facilities [3]. The severity of the landslide impact is related to the number of exposed elements and their linked vulnerabilities, the intensity of the landslide event [4,5]. Among Indonesia's physiographic divisions, the volcano region, which consist of mountain belts is particularly vulnerable to landslide due to its high seismicity, young and fragile geology, geomorphology, and heavy precipitation. In this regard, landslide susceptibility zonation to categorizing the land surface into

IOP Conf. Series: Earth and Environmental Science 1245 (2023) 012012 doi:10.1088/1755-1315/1245/1/012012

different zones of landslide possibility based on the unstable slope, and field measurements is important, but not feasible due to rough terrain and hazardous volcanic materials. Therefore, high-accuracy and low-cost advanced techniques to identification landslide susceptibility zones are important for achieving landslide disaster risk reduction goals.

Landslides are reported in many cases to be controlled mostly by structural trends such as discontinuities [6,7]. These discontinuities can be identified and measured by satellite images; therefore, lineament extraction and analysis are valuable tool to identified discontinuities in landslides studies. A discontinuity plane can be defined as separation or fracture in geological formation which divides the rock into two or more pieces. This structural weakness originated mostly around physical discontinuities such as faults, joint sets, or dykes are important elements to controlled landslide event [8,9]. However, lineament extraction from optical sensor images as hindered by low contrast due to the sun illumination effect and high humidity, cloud cover, and demonstrate poor performance in the detection of outcrop or geological lineaments in areas of vegetation [10]. To avoid such non fractures related lineaments, this study uses images of the radar Sentinel 1. SAR imagery can be acquired under cloud cover with all weather conditions and can identify small topographic relief with relatively high spatial resolution by using oblique microwave radiation [11].

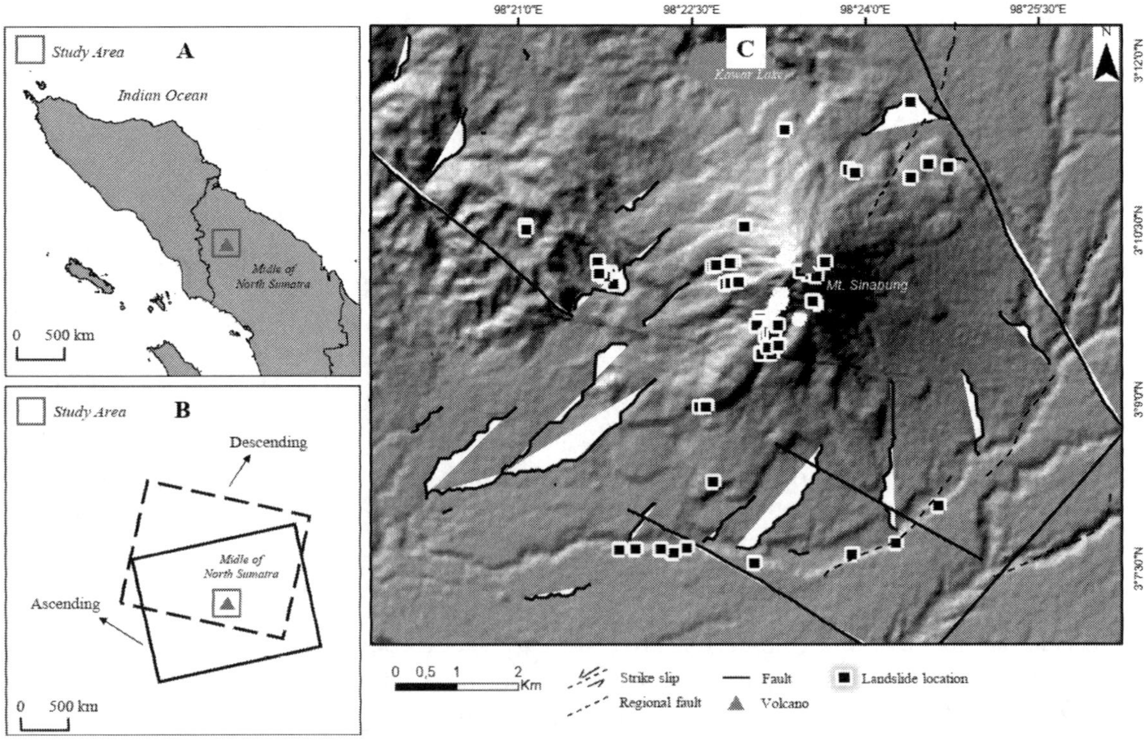

Figure.1. Study area located at Mt. Sinabung in the middle of North Sumatra (A), showing footprints of Sentinel-1 in ascending and ascending orbits (B), and shaded topographic map using DEMNAS 9 m data showing the eruption centre, the distribution of faults, and landslide locations (C).

Mt. Sinabung is an andesitic stratovolcano elevated about 2460 m above sea level [12]. This volcano is located in North Sumatra, 40 km northwest of lake Toba formed by catastrophic caldera eruptions [13]. Volcanic and tectonic activities around Mt. Sinabung cause potential secondary hazards such as landslides and mass movements around the edifice. Volcanic landslide is a catastrophic feature and represents an enormous danger to the neighbouring population [14]. Landslides following the eruption affect the stability of the slopes around Mt. Sinabung and produce hazards in certain areas [14]. The prime purpose of this study is to develop a method that automatically extracts lineaments from SAR images and identifies landslide susceptibilities zone based on lineament distribution at Mt. Sinabung, Indonesia.

2. Data Collection

We have used Sentinel-1 data, and landslide inventory from the Center for Volcanology and Geological Hazard Mitigation (CVGHM) and visual images from Google Earth. The Sentinel-1 is equipped with twin polar-orbiting satellites operating up to 25-min per orbit with a repeat cycle of 12 days. The satellites are to operate day and night and perform a synthetic aperture with radar imaging. Sentinel-1 bands allow us to get imagery in all weather conditions. Sentinel-1 was composed of three data products, and we use level-1 SLC product focused on the slant range by azimuth imaging plane, single looks, which contain phase and amplitude information. Sentinel-1 has azimuth and range. Azimuth is the satellite track or so-called orbit. Both orbits have opposite ranges. It can provide images of the opposite line of sight (LOS) utilizing two orbit modes: the ascending mode in which the satellite moves from south to north direction, and the descending is when the satellite moves from north to south direction. We combined a pair of Sentinel images of the ascending and descending orbit modes to reduces the effect of geometrical distortion. Then, the combination can equalize the contrast of both flanks, because the back slope at one mode changes to the fore slope at the other mode [15]. We used six scenes level-1 dual-polarization (VV and VH) C-band Sentinel-1 Single Look Complex (SLC) for each ascending and descending orbit. The detail of SAR data used in this study was listed in Table 1.

Table 1. Details of Sentinel 1 SAR data used in this study.

No	Scene ID	Acquisition date	Product Type	Orbit	Polarisation
1	S1B_IW_SLC_1SDV_20161209T114114_20161209T114144_003319_005A9D_11D8	09 December 2016	SLC	Ascending	VV/VH
2	S1B_IW_SLC_1SDV_20180427T114127_20180427T114157_010669_0137AB_4375	27 April 2018	SLC	Ascending	VV/VH
3	S1A_IW_SLC_1SDV_20211213T114233_20211213T114300_040990_04DE62_E9DA	13 December 2021	SLC	Ascending	VV/VH
4	SIA_IW_SLC_1SDV_20160531T231154_20160531T231222_011509_0118E5_1757	31 May 2016	SLC	Descending	VV/VH
5	S1A_IW_SLC_1SDV_20180427T231153_20180427T231220_021659_02559D_2D3A	27 April 2018	SLC	Descending	VV/VH
6	SIB_IW_SLC_1SDV_20211213T231145_20211213T231215_030013_03955F_0DE1	13 December 2021	SLC	Descending	VV/VH

for this research, only the 2016, 2018 and 2021 landslides were used for the analysis and these landslides were also digitized in a point format (see Fig 1). The 61 landslide locations have been identified well visually using Google Earth engine and soil movements inspection reports from CVGHM. The identified landslide locations are not primary hazards from volcanic products such as debris avalanche or pyroclastic flows but originated to the secondary hazards following the eruption. A volcano-triggered landslide is a landslide occurrence following the eruption period passing a steep slope.

3. Methodology

3.1 Lineament extraction

Linear features represent the geological structure with a different hue contrast and relief in the images. analysis using satellites for lineament detection is currently an effective option, but cannot be performed for large areas [16]. Lineament detection is automatically referred to extraction which is carried out based on the examination of digital numbers from satellite images with certain criteria. An effective method of automatic lineament detection and extraction is mSTA (modified Segment Tracing Algorithm). The mSTA works based on the STA principle [17]. The STA principle detects lines from pixels that are read as vector elements by examining local variations in the gray level of digital numbers [18]. The program used for the linear features extraction is developed using MATLAB syntax.

In this study, the automatic lineaments extraction was carried out on the basis of two processing: first, preparation images of Sentinel-1 were performed through three main steps, multilooking, speckle filtering, and geometric correction. Multilooking was increasing focus without losing resolution, speckle filtering was minimizing the noise from surrounding objects, and the last step was the correction of the geometrical distortions attributed to the topographic variation in the scene as well as the inclination of the radar. Consequently, geometric corrections task is especially necessary so that the image will be represented geometrically as close as possible to the reality, and subsequently be ready for use. These pre-processing is indispensable to mSTA application. Next, a line element is identified by examining the local variation along 16 directions at 11.25° intervals within a small window size of 11 × 11 pixels, following the principle of STA. In the preceding STA applications [17], the direction that minimizes the variation is assumed the valley direction and expressed by a symbol k_{min}. k_{min} is also defined to minimize εj that represents the difference between the values at the central pixel in the window. Since the image may produce more than one k_{min} due to remaining speckle noises [18] average the three neighbouring εj values in the window as:

$$E_j = \frac{E_{j-1} + E_j + E_{j+1}}{3} \text{ for } J \in K_{min} \tag{1}$$

Where E_j is the average value of three εj to obtain k_{min} value. The direction J that minimizes E_j was assigned to k_{min}. Next, we transform the segments into a lineament. A search ellipsoid window is set along a target segment whose start and middle points are inside the window with angle differences from the segment less than 11.25° are selected. Segments that only have endpoints inside the window are excluded. The window size was determined by considering the agreement of the lineaments and major faults length in the study area. Then, one line is defined as lineament using the least squares method, which approximates the coordinates of the start and end points of the selected segments. For the final step, the lineaments from the ascending and descending mode images were merged and the overlaps were grouped into one line using the least square method.

Therefore, to quantify the spatial concentration of the extracted lineaments, the study area was covered by a grid cell size 800×800-m and the three indices related to the density: the lineaments intersection (L_i) that's the number of the intersected lineaments which may extend the estimated damage zone by the multiple effect of fracturing, lineament frequency (L_f) that's the number of extracted lineaments, and lineament length (L_l) that's the total length of lineaments in the cell per unit area were calculated.

3.2 Landslide distribution and identification

Landslide occurrences were collected visually from the Google Earth engine explained as follows. The locations of evidence were identified based on image color, tone, surface roughness, and texture. These image parameters depend on the light reflected by the surface which useful to distinguish rock type, soil, and vegetation. Landslides are easily recognizable immediately after a landslide event. Their boundaries are usually recognizable easily in the image that the new occurrence of the landslides is less vegetation covers or opened area. As time passes, vegetation starts to cover the landslides and humans might work the land thus destroying the boundary and the landslides might be reactivated and move as time progresses. In addition, we also used the soil movement inspection reports from CVGHM for 2016 – 2021 to ensure that landslide locations are correctly digitized. The areas of these landslides were also digitized in a point format. Therefore, the landslide density map was calculated based on the number of landslides in every 800 m × 800 m grid and using the fishnet tools to create a grid set that was later overlaid with a lineament map. Until now, there is no standard landslide density classification to indicate an area's high or low landslide density. For that reason, this study uses three density classes are low (for one landslide/640000 m^2), moderate (2 - 3 landslides/640000 m^2), and high ($\geq$ four landslides/640000 m^2).

Next, the landslide susceptibility map is produced by combining both lineament and landslide density maps. The overlapping areas indicated as high lineament density and high landslide density could be considered highly susceptible areas for the landslide. This susceptibility map is validated by calculating the numbers and percentage of landslides in the high susceptibility zone. The level of susceptibility indicated by the Australian Geomechanics Society [19] is used to indicate the level of susceptibility based on the proportion of landslides in the high susceptibility class with the total landslides in the study area. The hazard level was also calculated for the study area by calculating the number of landslides in the high susceptibility zone for the assessment year. The levels of hazard and susceptibility based on the guidelines are given in Table 1. Susceptibility is the proportion of the total landslide population in the study, and area hazard is the number of landslides/year/800m^2.

Table 2. The descriptor for the susceptibility and hazard degrees used in this study [19]

Descriptor	Relative Susceptibility	Hazard
Very High	-	>10
High	>0.5	1 – 10
Moderate	0.1 – 0.5	0.1 – 1
Low	0.01 – 0.1	0.01 – 0.1
Very Low	0 – 0.1	<0.01

4. Result and Discussion

4.1 Landslide and lineament density

The interpretation of visual observations using the Google earth engine resulted in 11 landslides in 2016, 22 landslides in 2018, and 28 landslides in 2021. The total landslides recorded for these 3 years are 61 landslides that spanned along the 119 km Mt. Sinabung area. The distribution of landslides based on each assessment years are presented in Fig 4. The landslide was classified to examine their density in every 800 m^2 of the study area.

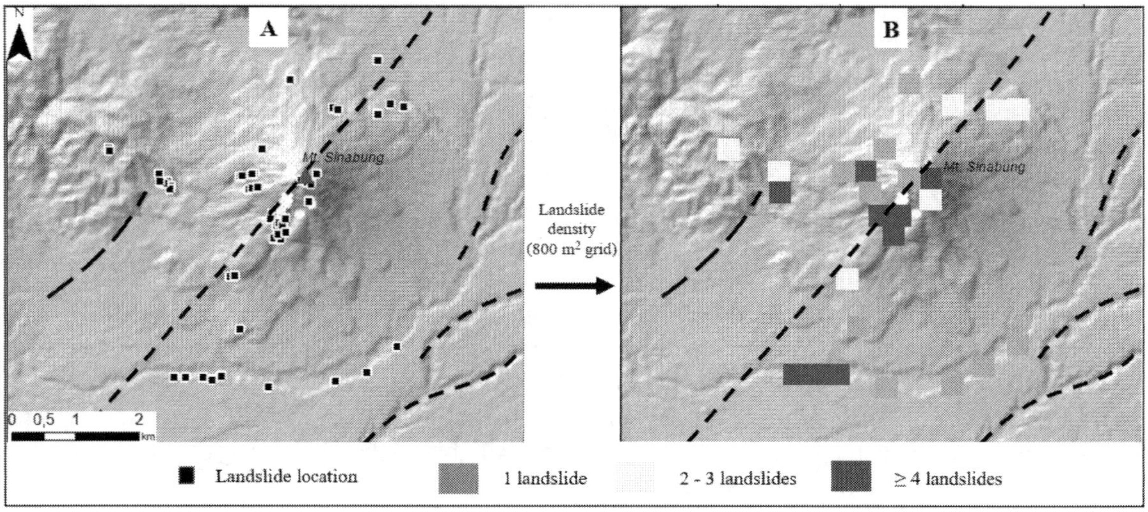

Figure 2. The occurrences map of landslide distribution at study area in 2016, 2018, and 2021 (A) and landslide density map calculated from landslide occurrences (B).

The classification was divided into three classes representing low landslide occurrence, moderate occurrences, and high occurrences. Although the study area is dominantly characterized by low landslide occurrences all three assessment years, there are several of the area that were classified as high landslide density area. The indicated high density sections of the area shown with red pixel at Fig 2 in three different locations, on around the crater of Mt. Sinabung, at the west of Mt. Sinabung, and in the southwest aligned to the Lou Borus River close to community settlements. Although there were just few high landslide density zones in each of the assessment year, it can be observed that at least 1 landslide occurred in every 2 – 3 km of the area Mt. Sinabung. To examine the landslide density as a total density for the three assessment years, landslides recorded in those three years were combined and analysed together to produce the final landslide density map.

Modified Segment Tracing Algorithm (mSTA) was used to extract the linear feature, which was interpreted as fault or fractured based on Sentinel-1A data with Ascending and Descending modes. The improvement of mSTA was achieved by excluding short lineaments of mSTA, which originated from speckle noises in the black slope of LOS with low backscattering intensities [18]. The mSTA data is gained by combining two orbit mode lineaments, grouping them, and eliminating short lineaments. In the rossete diagram, ratios of the counts of individual linear feature and the total amount of all linear features for ascending and descending images along each direction was calculated. There were 1152 on ascending and 1186 on descending detected lineaments in total using mSTA method. In the ascending and descending image, since 2016 and 2018, more variated in all directions but the most dominant is NW-SE. However, the linear features direction at 2021 NW-SE direction is the most dominant direction than the others corresponding to geologi structure map of Sinabung [20] (See Fig 3). The number of extracted linear features in descending images was more significant than in descending images, where counts of individual linear features to the total amount of all linear features for ascending and descending images along each direction were also calculated.

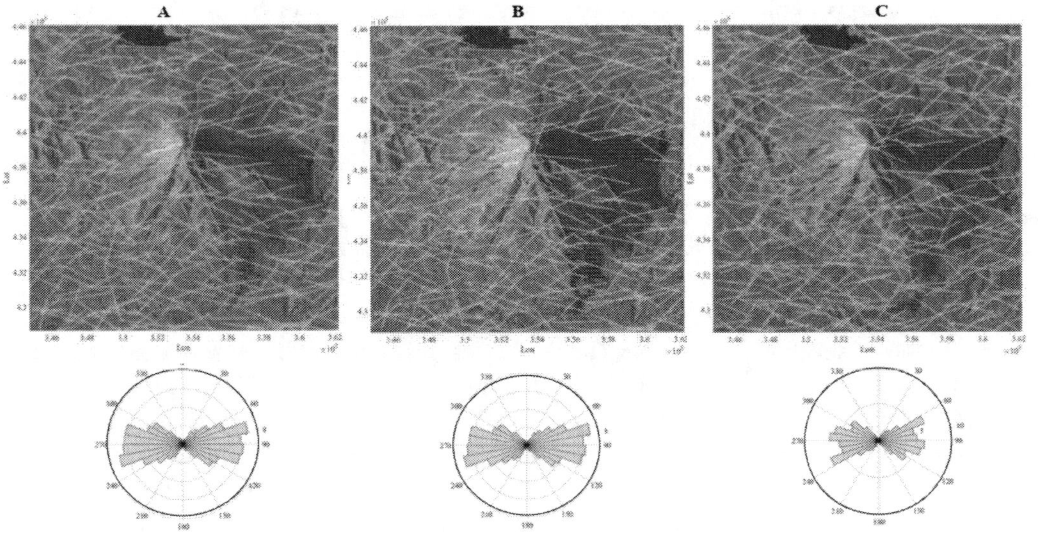

Figure 3. The final lineament map produced by grouping the lineaments from the two observation modes in 2016 (A), 2018 (B), and 2021 (C). The rose diagrams (lower panels) show the frequency of lineament directions.

Therefore, to map the three indices of lineament density; L_i, L_f, and L_l, their semivariograms were calculated omnidirectionally and all the data were well-fitted to the spherical or exponential model. Furthermore, the density of intersections between lineaments has to 0 to 7 km/km^2. While the lineament length density is spread from 0 to 8 counts/km^2. The number of lineament with density has values from 0 to 8 counts/km^2. The lineament results could be seen in Fig 4.

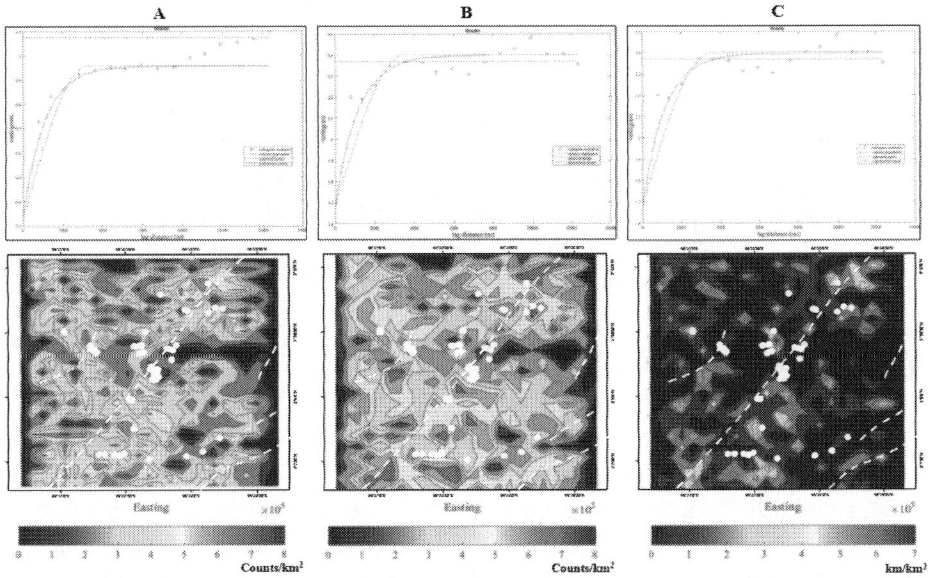

Figure 4. Ordinary kriging maps of the lineament density indices for length L_l (A), frequency L_f (B), and intersection L_i (C) in 2021 indicate the spatial concentration of lineaments overlaid by landslide distribution (white dots) and faults (white dashed lines).

The three ordinary kriging maps show large and small index values by red-toned and blue-toned colours, respectively. The good match between the major faults and the lineament densities for the three indices verified the ability of mSTA to detect faults and the effectiveness of the ordinary kriging mapping. However, from the three indices of lineament density, L_i, L_f, and L_l, we found L_f has a good indicator that caused a good match between the distribution of landslides with high-density values compared to the others. Comparison between landslide occurences and the lineament density map showed that most landslide occured in high and moderate lineament density area. Thus, from the result, we conclude that L_f is a good indicator of the degree of near-surface fracturing (See Fig 4).

A lineament map that was produced from the interpretation using mSTA is used to observe the association between landslide occurrences (landslide density) with the lineament density in the study area. The corrected lineament map that was transformed into a lineament density map is shown in Fig 3. Distribution of landslides have been known to located in highly fractured zones with high permeability over the shallow to deep depth range. To quantify the spatial concentration of the extracted lineaments, the conversion on the lineament map from vector to density in a grid format is based on the total length of lineament in each of the 800 m^2. Subsequently, the lineament density map was classified into three density classes, namely: low, moderate, and high.

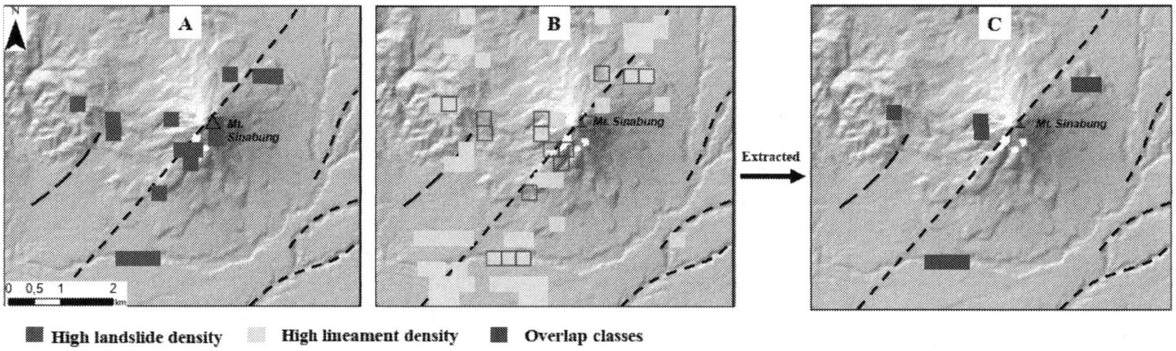

Figure 5. High landslide density maps around Mt. Sinabung (A), overlaid with high lineament density (B), and overlap of high landslide and lineament densities zones (C) with the black dashed lines are faults at Mt. Sinabung [20].

Table 3. Description on the degree of susceptibility, hazard, and landslide density of the study area

Degree	Quantification	Value	Classify
Susceptibility	36 landslides / 61 total landslides	0.59	High
Hazard	36 landslides / 6 years / 6.4 km	0.9	Moderate
Landslide Density (per km^2)	36 landslides / 6.4 km^2	5.6	High

The degree of susceptibility and hazard can be quantified and classified based on the method and proposed classes of susceptibility and hazard by AGS (Table 1). To proceed with the quantification process, the total landslide from the three assessment years were intersected with 8 zones of high susceptibility classes shown in Fig 4. From the total 61 landslides acquired from the three assessment years, 36 landslides are within the 8 high susceptibility zones, which indicates that if the degree of susceptibility and hazard are calculated based on the guidelines by AGS [19] the area at Mt. Sinabung can be classified as highly susceptible (0.59) but moderate in terms of hazard (0.9). In terms of landslides that occurred in every 800 m^2, which is also can be indicated as high based on the classification of landslide density used in this study (Fig 4). Table 3 shows the degree of susceptibility, hazard, and landslide density (per km^2) that are based on the 36 landslides in the 8 high susceptibility zones at the Mt. Sinabung.

4.2 Discussion

In landslide susceptibility mapping, the selection of landslide factors is one of the most important elements. However, there is no well-defined standard to select the most significant landslides factors. The factors that initiate the landslide incidence in the study area selected based on data availability, literature review and field evaluation. On this study, we use manual land use classification of the Google Earth imagery and found that be effective as it has a high spatial resolution to identify landslide location. Methodologies for producing landslide susceptibility maps are mainly based on GIS technology. Previous published works has been explained to solve the deficiencies and difficulties in landslide susceptibility assessment [19,21–23]. Practically, the procedure for preparing landslide susceptibility maps should be applicable and accurate to the field scale. To validate the high susceptibility zones generated for the study area, 15 landslides in 2022 that occurred close to the area Mt. Sinabung were used to validate the map (Fig 6). However, due to a lack of data from google earth and a report from the Centre for Volcanology and Geological Hazard Mitigation (CVGHM), the landslide interpreted from the 2022 dataset only covered the area and only 15 landslides were recorded based on the availability of the satellite photographs. Nevertheless, the landslides from 2022 are adequate to validate the landslide susceptibility map (see Fig 6). As a result, from the intersection between the high susceptibility zone and the 2022 landslides, 15 out of 21 landslides are located within the high susceptibility zones. This means that the susceptibility map accurately captures almost most of the landslides in 2022 at Mt. Sinabung. This also demonstrates that the combined high lineament and landslide densities maps can be used to indicate locations that are highly susceptible to landslides in the future.

Regarding the landslide occurrences in 2022 and landslide zone susceptibility using lineament density results, we obtained 15 landslides based on visual observations using the Google Earth engine match with the predicted landslide zone. Generally, the landslide occurred in several zones (see Fig 6) as shown in L14 located northeast of Mt. Sinabung. Based on the geological map of Mt. Sinabung, this type of landslide is in pyroclastic flow deposits [24]. The landslides are reported to be mostly controlled by structural trends and made low shear strength effects of infill materials and decrease the slope strength thus causing landslides to occur. The rock types played a significant role in the rainfall triggered landslides, as they underwent intensive weathering to form thick soil profiles observed at landslide ridges and flanks/headwalls. The geological conditions, including physical and weathering characteristics can control the triggering mechanism, type, and pattern of landslides. Bartelliti et al. [25] reported that bedrock lithology seems to be an important factor that influences the localization of shallow landslides. The effect of geological conditions on landslides may be related to the mechanical properties of the soil present in the slope and originating from parent rock. The physical and mechanical properties of soil vary depending on geological conditions because they are the results of the physical and chemical weathering of bedrock. Soil properties are also affected by the composition of minerals present, which depend on the geological conditions of the site [26,27]. The content and geochemical characteristics of the minerals constituting the soils are important factors in landslide susceptibility and size [27]. Based on the geological map, the bedrock in Mt. Sinabung in age and consist of layers of lava andesite-dacite, pyroclastic flow deposits, and alluvium [24]. After weathering, andesite also has characteristics of clay on the surface. The bedrocks of the landslides and their constituent mineral types are prone to rapid and deep chemical weathering by hydrolysis and produce weakened clay-rich fractured rocks and soils. Clay, a weathering product of rock mass, contributes to landslide occurrence because of their chemical and physical properties. The presence of clay minerals in the altered andesitic can be considered one of the factors that induced Landslides at Mt. Sinabung.

Moreover, a high landslide also occurred west of Mt. Sinabung located around the volcanic crater and that occurrence is triggered by the eruption of Mt. Sinabung (see Fig 6B). Based on the geological map of Mt. Sinabung, this type of landslide developed in a thick layer of lava andesit-dasit where erosion is experienced. This condition makes landslides on volcanic cones common due to the weakening of the volcanic rock hydrothermal caused alterations as well as eruptions. The intercalation of lava unconsolidated pyroclastic rocks can also lead to the weakening of slope stability, which triggers landslides. We also found the landslide located near the river Lou Borus, Mt. Sinabung (See Fig 6C). Based on the geological maps of Mt. Sinabung, this type of landslide is in alluvium deposited [24]. The area is predicted to be a zone with high landslide intensity and close to residential areas that were about 92-m away from the landslide location. This is of greater concern to the local community and government and makes this information useful for preventive or mitigating actions for the surrounding community.

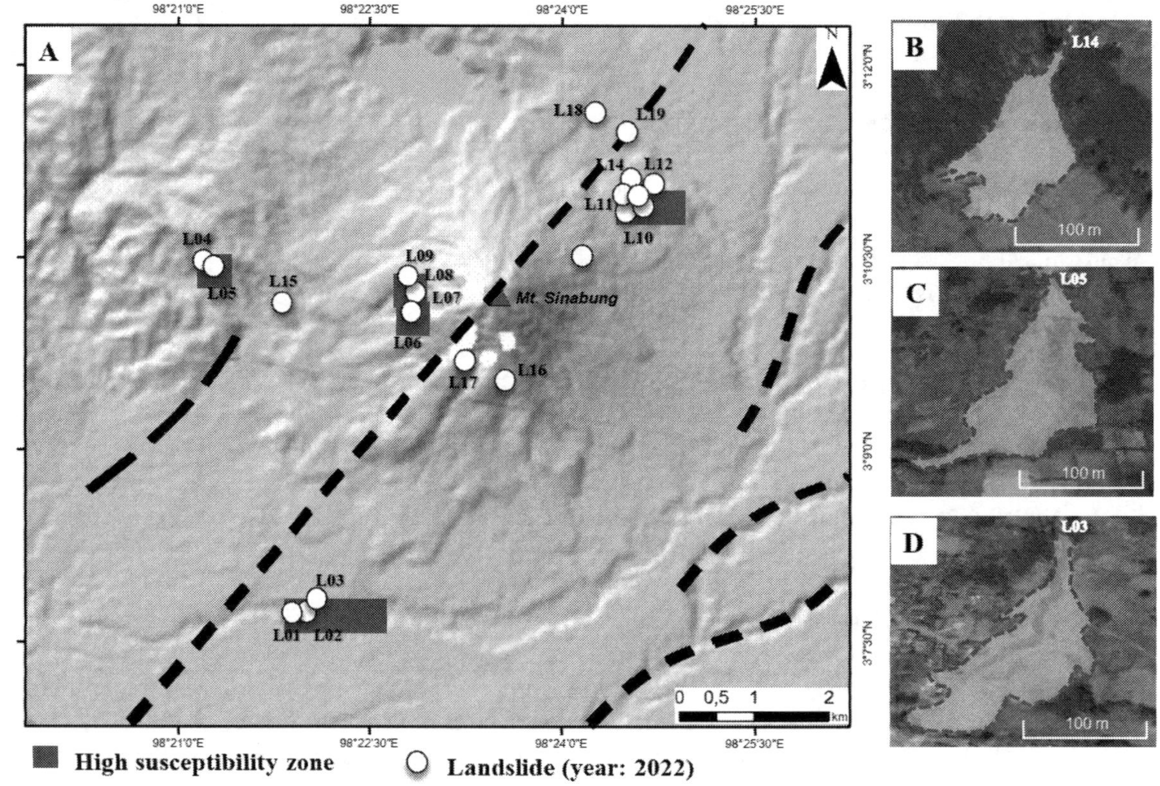

Figure 6. The landslide susceptibility map presents the distributions of landslide occurrences in 2022 overlaid by faults in the black dashed lines (A) and verification of the high susceptibility zone at Mt. Sinabung presented visually by Google Earth Engine (B, C, D).

5. Conclusion

In this study, a simple but useful technique using the landslide density method was used to carried out a temporal landslide assessment at the Mt. Sinabung. We used the mSTA method to extract linear features derived from ascending and descending images producing 1152 and 1186 lineaments, respectively. Lineaments related to geological structures were presented by continuous lineaments associated with morphology. The detected lineaments using the mSTA method are more effective to identify geological structures than visual method especially for large coverage area. The agreement between statistics of lineament direction with confirmed geological structure was used as verification approach. The major lineaments directions agreed to the fault directions of the NW-SE corresponding to geologi structure map of Sinabung. The densities of the extracted lineaments were mapped by ordinary kriging. The high density of lineaments frequency (L_f) is concordance to the 48 out from 61 landslide location. We have also obtained that the 36 landslides located at the high and moderate potential hazard at Mt. Sinabung. The landslide occurrences in this study were associated mainly with high lineament density potential to the instability. Integrating the landslide and lineament densities, we calculated the accuracy of the produced susceptibility map is about 75%. This map could be used to mitigate landslides potential and hazard mitigation purposes.

Acknowledgement

This study is supported fully by LPPM ITB under "Riset Unggulan 2022" scheme.

Credit author statement

Zahratunnisa: Making first draft and revising the draft. **Asep Saepuloh:** Revising substantial context of the paper and designing of the work. **Estu Kriswati:** Analysing the data and revising the important intellectual content. **Ahmad Basuki:** Interpreting the results and revising the important intellectual content.

Reference

[1] BMKG 2022 Indonesia online climate database *http://dataonline.bmkg.go.id/data_iklim*

[2] Dunning S A, Massey C I and Rosser N J 2009 Structural and geomorphological features of landslides in the Bhutan Himalaya derived from Terrestrial Laser Scanning *Geomorphology* **103** 17–29

[3] Reichenbach P, Rossi M, Malamud B D, Mihir M and Guzzetti F 2018 A review of statistically-based landslide susceptibility models *Earth-science reviews* **180** 60–91

[4] Froude M J and Petley D N 2018 Global fatal landslide occurrence from 2004 to 2016 *Natural Hazards and Earth System Sciences* **18** 2161–81

[5] Glade T and Crozier M J 2005 The nature of landslide hazard impact *Landslide hazard and risk* 41–74

[6] Nagarajan R and Khire M V 1998 Debris slides of Varandh Ghat, west coast of India *Bulletin of Engineering Geology and the Environment* **57** 59–63

[7] Pradhan A M S and Kim Y T 2016 Evaluation of a combined spatial multi-criteria evaluation model and deterministic model for landslide susceptibility mapping *Catena* **140** 125–39

[8] Clark C D and Wilson C 1994 Spatial analysis of lineaments *Computers & Geosciences* **20** 1237–58

[9] Davis G H, Reynolds S J and Kluth C F 2011 *Structural geology of rocks and regions* (John Wiley & Sons)

[10] Mwaniki M W, Moeller M S and Schellmann G 2015 A comparison of Landsat 8 (OLI) and Landsat 7 (ETM+) in mapping geology and visualising lineaments: A case study of central region Kenya. *International Archives of the Photogrammetry, Remote Sensing & Spatial Information Sciences*

[11] Yamaguchi Y 1985 Image-scale and look-direction effects on the detectability of lineaments in radar images *Remote Sensing of Environment* **17** 117–27

[12] Syarifuddin M, Oishi S, Hapsari R I, Shiokawa J, Mawandha H G and Iguchi M 2019 Estimating the volcanic ash fall rate from the Mount Sinabung eruption on February 19, 2018 using weather radar *Journal of Disaster Research* **14** 135–50

[13] Chesner C A and Rose W I 1991 Stratigraphy of the Toba tuffs and the evolution of the Toba caldera complex, Sumatra, Indonesia *Bulletin of Volcanology* **53** 343–56

[14] Hürlimann M, Garcia-Piera J O and Ledesma A 2000 Causes and mobility of large volcanic landslides: application to Tenerife, Canary Islands *Journal of Volcanology and Geothermal Research* **103** 121–34

[15] Saepuloh A, Haeruddin H, Heriawan M N, Kubo T, Koike K and Malik D 2018 Application of lineament density extracted from dual orbit of synthetic aperture radar (SAR) images to detecting fluids paths in the Wayang Windu geothermal field (West Java, Indonesia) *Geothermics* **72** 145–55

[16] Saepuloh A, Urai M, Aisyah N, Widiwijayanti C and Jousset P 2013 Interpretation of ground surface changes prior to the 2010 large eruption of Merapi volcano using ALOS/PALSAR, ASTER TIR and gas emission data *Journal of Volcanology and Geothermal Research* **261** 130–43

[17] Koike K, Nagano S and Ohmi M 1995 Lineament analysis of satellite images using a Segment Tracing Algorithm (STA) *Computers & Geosciences* **21** 1091–104

[18] Saepuloh A, Army E K and Hilman Z 2020 Pemetaan Permeabilitas Magnetik Permukaan Berbasis Citra SAR Polarimetrik dengan Pengukuran In Situ di Lapangan Gunung Api *RISET Geologi dan Pertambangan* **30** 131–41

[19] Society A G 2007 Guideline for landslide susceptibility, hazard and risk zoning for land use management. Australian Geomechanics Society Landslide Taskforce Landslide Zoning Working Group *Aust Geomechanics Soc* **42** 13–36

[20] Kriswati E, Meilano I, Iguchi M and Abidin H Z 2019 An evaluation of the possibility of tectonic triggering of the Sinabung eruption *Journal of Volcanology and Geothermal Research* **382** 224–32

[21] Dong S, Yin H, Yao S and Zhang F 2013 Detecting surface subsidence in coal mining area based on DInSAR technique *Journal of Earth Science* **24** 449–56

[22] Kriswati E, Frederik M, Saepuloh A, Darmawan S and Alfianti H 2021 Long Term Ground Deformation of Mount Raung as Inferred by InSAR and GPS Data *2021 7th Asia-Pacific Conference on Synthetic Aperture Radar (APSAR)* (IEEE) pp 1–4

[23] Beccaro L, Tolomei C, Gianardi R, Sepe V, Bisson M, Colini L, De Ritis R and Spinetti C 2021 Multitemporal and Multisensor InSAR Analysis for Ground Displacement Field Assessment at Ischia Volcanic Island (Italy) *Remote Sensing* **13** 4253

[24] Prambada O, Zaennudin A, Iryanto S, Nakada S and Yoshimoto M 2011 Geologic map of Sinabung Volcano, North Sumatra Province (1: 25,000) *Center for Volcanology and Geological Hazards Mitigation. Geological Agency*

[25] Bartelletti C, Giannecchini R, D'Amato Avanzi G, Galanti Y and Mazzali A 2017 The influence of geological–morphological and land use settings on shallow landslides in the Pogliaschina T. basin (northern Apennines, Italy) *Journal of Maps* **13** 142–52

[26] Kim K-S, Lee M-S, Cho Y-C, Chae B-G and Lee C-O 2004 Engineering characteristics of soil slopes dependent on geology-Hwangryeong Mt. District, Busan *The Journal of Engineering Geology* **14** 487–98

[27] Kim K-S, Choo C-O, Booh S-A and Jeong G-C 2005 Importance of microtextural and geochemical characterizations of soils on landslide sites *The Journal of Engineering Geology* **15** 447–62

Water Quality Assessment of Lake Maninjau After the Mass Fish Kill Event

Taofik Jasalesmana[1,4*], Mutiara Rachmat Putri[2], Mirzam Abdurrachman[3], Cynthia Henny[4], Sulung Nomosatryo[4], Ira Akhdiana[4],

[1]Earth Science Doctor Program, Faculty of Earth Sciences and Technology, Bandung Institute of Technology, Bandung, Indonesia

[2]Research Group of Oceanography, Faculty of Earth Sciences and Technology, Bandung Institute of Technology, Bandung, Indonesia

[3]Geological Engineering Study Program, Faculty of Earth Sciences and Technology, Bandung Institute of Technology, Bandung, Indonesia

[4]Research Center for Limnology and Water Resources, National Research and Innovation Agency (BRIN), Cibinong 16911, West Java, Indonesia

*Email: t.jasalesmana@gmail.com

Abstract. A mass fish kill is often occurring in Lake Maninjau. A lack of oxygen has reportedly resulted in numerous fish kills, including the recent lake Maninjau condition. Deoxygenation of lake water is a natural phenomenon that often occurs after heavy rain events and low sunlight intensity. Additionally, during the strong wind for a long period, the decrease in oxygen could be chemically triggered by the diffusion of sulfide from the bottom layer of the water column. Furthermore, this study examines water quality parameters after the mass fish mortality occurred across locations in Lake Maninjau. Field surveys were conducted by measuring physiochemical parameters and total sulfide at the Fifteen sampling points (five points in the middle and ten points around the lake) from November 26 to December 2, 2022. They were measured directly using the Horriba©U 52 water quality checker (WQC), DO meter YSI ProDO© International and the HACH spectrophotometer DR3900 method. The results show that the average DO levels at each observation point were close to 0 mg/L while the average sulfide level on the surface was 9.5 micro g/L. The average Oxidation Reduction Potential (ORP) of -79.26 mV indicates that the lake is in a reduced state and signifies that much of the dead and decaying material in the water would be slowly broken down and decomposed. Low ORP values indicate that the compound in the water is available in reduced form, which is more toxic. Our diel DO measurement show that sunlight intensity was effective in increasing oxygen concentration in the lake.

Keywords: Lake Maninjau, dissolved oxygen, fish kill, sulfide.

1. Introduction

Lake Maninjau is one of 15 lakes that are prioritized for management (Perpres No. 60). Recently, the decrease in water quality due to eutrophication appears to be serious. Indications of the eutrophication are a decrease in the water clarity (Secchi Depth) and frequent occurrence of a mass fish kill. This could be due to a much lower oxygen level in the epilimnion zone now (Fukushima et al., 2017) than that of the condition of Lake Maninjau in 1929 (Rutner, 1931). According to previous studies, the input of organic matter overloading from the intensive activity of floating net cages (FNC) has caused the deterioration of the water quality (Henny, 2009; Henny & Nomosatryo, 2012; Syandri et al., 2014; Syandri, 2016; Anggraini, 2019). The enhancement of organic matter in the water column could accelerate the rate of aerobic decomposition of organic matter by consuming a lot of oxygen in the oxic layer of the water column, which subsequently impacts the oxygen deficit in lake surface water (Komala et al., 2019; Santoso & Triwisesa, 2020). Additionally, organic matter triggers the generation of sulfides via sulfate reduction by anaerobic sulfate-reducing bacteria (Holmer & Storkholm, 2001; Sakai et al., 2013; Luek et al., 2017).

Changes in Dissolved Oxygen (DO) concentration in Lake Maninjau occur on a daily and seasonal scale. DO concentrations drop dramatically at night. In contrast, the DO concentration increases during the day when the sun is shining, moving in tandem with the increasing photosynthesis process. (Romaire & Boyd, 1979; Lukman et al., 2014; Ajie & Setiadewi, 2020). The seasonal factors that influence DO fluctuations in the lake included the rainy, cold season, and strong winds event (Araoye, 2009; Mihu-Pintilie et al., 2014; Siriwardana et al., 2019). Increased precipitation and minor sunlight prevent photosynthesis activity by phytoplankton and lower the temperature gradient in the water column (Liu et al., 2020). The water column's ability to stratify is weakened by a decreasing temperature gradient, which triggers the movement of contaminants from the anoxic layer that act as oxygen-reducing agents (Santoso et al., 2018). Furthermore, a period of strong winds can cause mixing to transport various reducing agents including sulfide, up to the oxic layer through mixing. Sulfide that diffuses to the surface will reduce oxygen to form sulfate (Henny, 2009; Couture et al., 2016).

The elevate of sulfide to the surface is a phenomenon that frequently occurs in Lake Maninjau. The local people called it *tubo belerang*. *Tubo* is a dialect for noxious while *belerang* is sulfur. *Tubo belerang* often occurs during extreme weather, such as prolonged periods of cloudy days and heavy rain accompanied by strong winds (Henny, 2009; Lukman et al., 2013; Anggraini, 2019). Typically, *tubo belerang* takes place between the end and the beginning of the year. The *tubo belerang* occurrence causes a significant increase in fish mortality and deteriorating water quality.

Even though the mass fish kill in this lake has been the subject of numerous studies, none of them specifically described the lake water quality conditions following the incident. Knowing the lake water quality condition after the incident is crucial to understand the root cause of the mass fish kill. Therefore, this study will reveal the water quality conditions specifically after the mass fish kill that occurred on November 2, 2022. The goal of this study is to provide the fundamental knowledge needed for effective lake management.

Method

1.1. Study site

Lake Maninjau is located in West Sumatra Province, Indonesia (0°19′S, 100°12′E). This tropical basin was originally formed from a volcanic eruption around 51000 ± 6000 years ago by ejecting volcanic material as much as 220-250 km^3 and spreading as far as 75 km (Alloway et al., 2004; Pribadi et al., 2007). Lake Maninjau has a surface area of 99.5 km^2, a volume of 10.1 km^3, an average depth of 105 m, and a maximum depth of 165 m (Santoso et al., 2018). The height of the Maninjau cliff wall is about 459 m above lake level, with the south side being the steepest part. This lake stretches from north to south with a length of 17 km and a width of about 8 km (Pribadi et al., 2007; Nontji, 2016). Lake Maninjau has been exploited by floating net cages since 1992. These fishery activities had deteriorated the water quality of Lake Maninjau, which is indicated by the frequency of mass fish kills every year and caused enormous economic losses. Until 2017, 11 times fish were killed happened, which increased in 2016 (Makmur et al., 2020).

1.2. Water Quality Measurement and Sampling

The survey was conducted in November 2022 at 15 sampling points. The temperature, DO, and conductivity parameters were measured with the CTD Rinko Profiler and DO meter ProDO YSI© International, while pH and Oxidation Reduction Potential (ORP) were measured by the Horriba©U 52 water quality checker. Each parameter is measured in strata to see the variation in the water column. Sulfide was analyzed by taking a water sample using a syringe to prevent it from oxidizing by oxygen in the air. The water sample was quickly transferred into the screw tube, which had previously been added with Sulfide 1 Reagent. Sulfide 2 Reagent then was added quickly to the tube and measured with a DR3900 spectrophotometer (DR 2800 Spectrophotometer, 2007).

The diel pattern of DO was estimated by measuring the water surface every hour at Tan Dirih station using DO meter ProDO YSI© International. To identify the weather condition around Lake Maninjau, we analyzed the meteorological data from Padang Panjang Geophysics station, which can be freely accessed from the Meteorological, Climatological, and Geophysical Agency (BMKG) website (https://dataonline.bmkg.go.id/data_iklim).

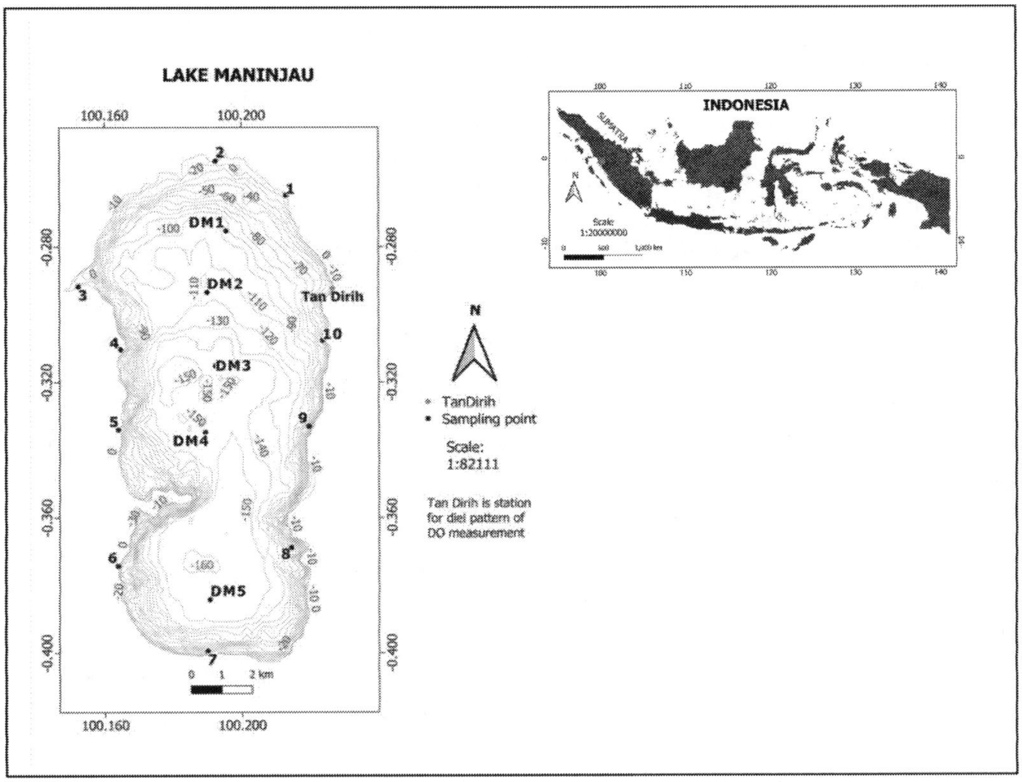

Figure 1. Bathimtery map of Lake Maninjau. DM1, DM2, DM3, DM4, DM5, 1, ..., 10 are sampling points. Tan Dirih is station for diel pattern of DO measurement. (Lake Maninjau and Indonesian Map are created with QGIS Software 3.16.15-Hannover based on maps provided by GIS Lab. RC for Limnology-LIPI and https://tanahair.indonesia.go.id/portal-web, retrieved on 27 February 2023).

2. Result and Discussion

2.1. *The water quality condition*

Lake Maninjau's condition after the mass fish kill event is shown in Figure 2. The lake water is whitish green with a foul odor. The stench comes from fish carcasses inside the cage of several FNC or thrown outside the FNC. The DO concentration at the lake's periphery on the first day of the survey was 0.52 mgL^{-1}, and the temperature was 27.2 °C. The critical DO level indicated that the water quality of Lake Maninjau after the mass kill was still very low.

Figure 3 depicts the profile of the temperature and thermocline depth in the middle of the lake in November 2022 measurements. The vertical structure of the temperature at each station looks the same. This condition is caused because the amount of heat energy absorbed from sunlight by each station is minimal. Based on the BMKG station measurements, the sun duration (SD) when measuring the temperature of the water column is short (SD = 3 hours). In addition, because the lake is narrow and there is no other heat source outside sunlight, the variation in sampling points does not produce a substantial difference in temperature.

The average surface temperature of the lake is 28.07 °C, with the temperature difference between the bottom and surface of 1.16 °C. The temperature difference in this period is smaller than the measurement in February 2017, which is 3 °C (Santoso et al., 2018). Furthermore, the water column did not have a thermocline layer, except at station DM5 with a thermocline layer depth of 0.46 m. The disappearing thermocline layer accompanied by a homogeneous water column indicates that the convective process occurred in Lake Maninjau that mixes each layer of the water column. The mixing process can be identified not only by the vertical temperature of the water column, but also by the vertical structure of homogenous conductivity and DO (Fukushima et al., 2021).

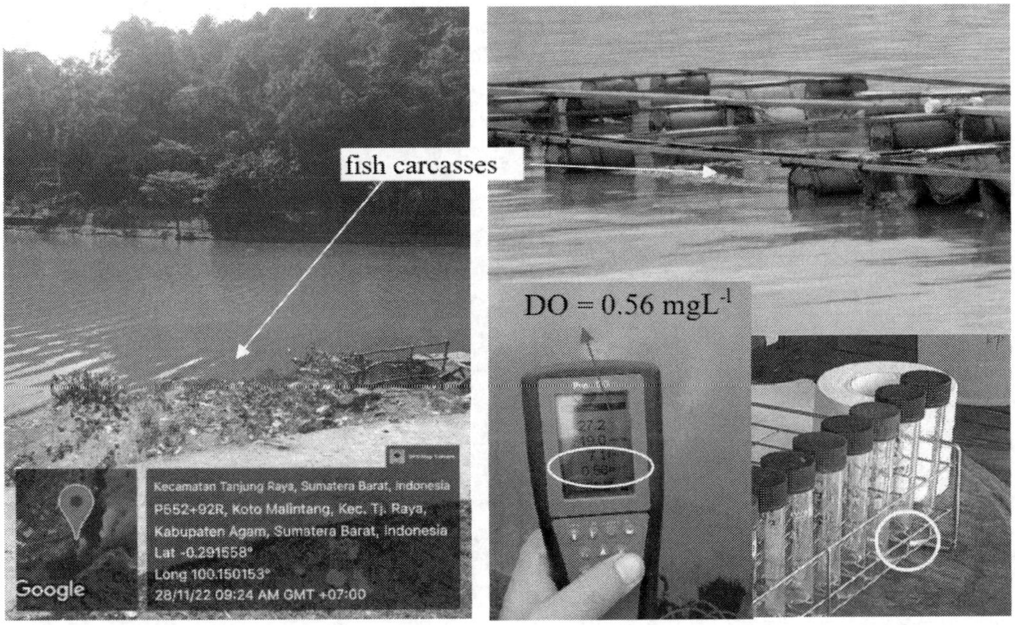

Figure 2. The water quality condition of Lake Maninjau after the mass fish kill event. Some fish carcasses are still inside and outside FNC. The water in the screw tube was taken from the surface. The green color indicates water containing sulfide

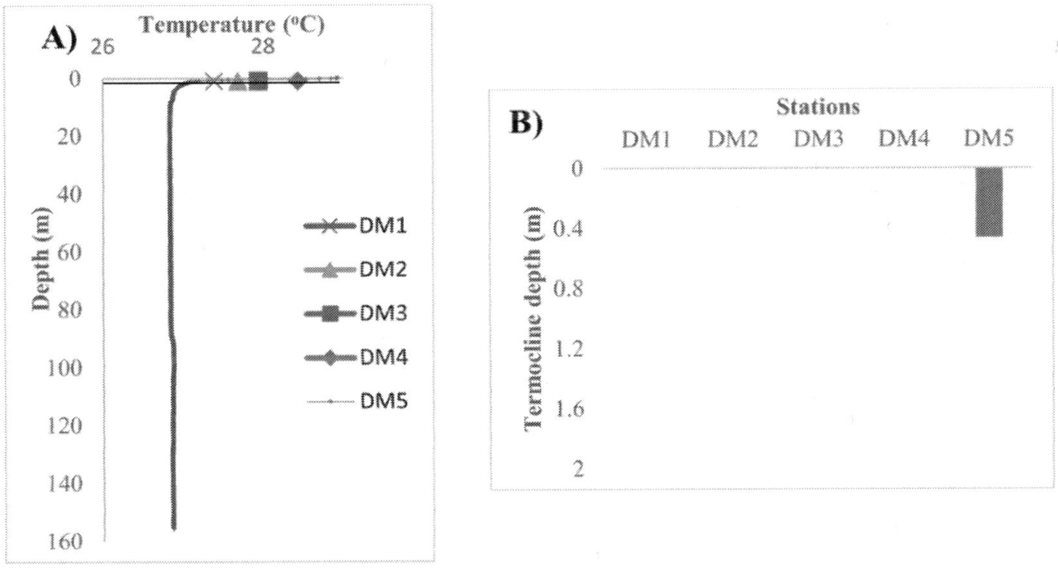

Figure 3. A) Vertical profile of water column temperature and B) thermocline depth. The water column is homogenous in most stations, except in DM5 with the shallow thermocline (0.46 m). Black line in Figure A indicated the thermocline depth in DM5.

IOP Conf. Series: Earth and Environmental Science 1245 (2023) 012013 doi:10.1088/1755-1315/1245/1/012013

Mixing in the water column was indicated by the DO profile in the water column (Figure 4). When the water column is not stratified, the thermocline layer as a barrier to the mixing process between the epilimnion and hypolimnion layers disappears, causing pollutants and reducing oxygen to easily rise from the anoxic hypolimnion to the oxic epilimnion. This condition causes the oxygen level in the epilimnion to decrease. If this occurs continuously, the oxygen level in the epilimnion layer will balance with the hypolimnion layer to form a homogeneous layer.

A straight-line profile with DO level close to zero from the surface to the bottom is also due to the increase in rainfall. Local people have observed constant rain since the beginning of November, which inhibited the photosynthesis process. On the other hand, respiration by autotrophs and heterotrophs (the decomposition of organic and inorganic matter) continued to increase, causing DO concentrations to decrease drastically (Santoso & Triwisesa, 2020).

The effect of photosynthesis and respiration on DO dynamics in Lake Maninjau is demonstrated by surface DO measurements for 24 hours. Figure 4. B shows that at 8:00, the DO content began to increase with the appearance of sunlight. DO surface reached its peak at 18.00. Along with the loss of sunlight, the process of photosynthesis stopped, as indicated by the decreased DO content, and reached a critical condition at 05:00. Based on Figure 4. B, it can be seen that although that sunlight intensity was effective in increasing oxygen concentration in the lake, the DO concentration shown was still below the standard quality set for fisheries, (4 mg L^{-1}) (Andria & Rahmaningsih, 2018; Paputungan et al., 2022). This condition occurs because the lake needs time to purify itself (Komala et al., 2019). Because of the low DO levels, the fish carcasses monitored were dispersed around Lake Maninjau.

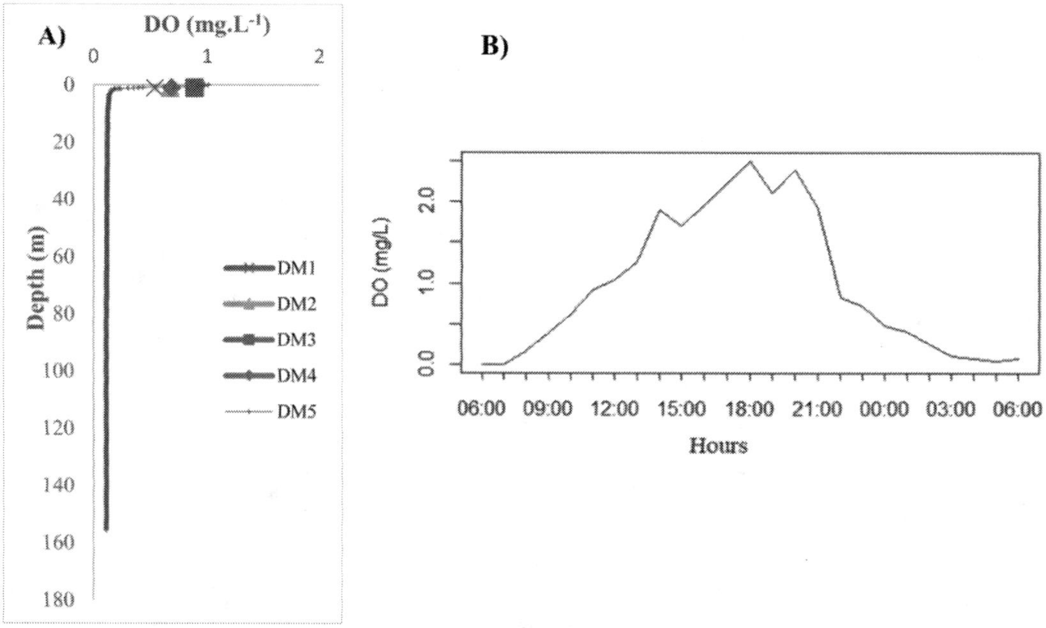

Figure 4. A) DO vertical profile of water column and B) Diel of pattern DO at the surface water

Biogeochemical processes in waters often involve oxidation-reduction processes (Wetzel, 2001). The state of the ongoing reduction or oxidation process in a body of water can be tested by measuring the ORP. By measuring the ORP, it can be seen the ability of water to recover itself from the pollutants that enter it. A high ORP value shows much available oxygen, so bacteria effectively break down organic matter. Conversely, if the ORP value is low, it indicates low oxygen levels, and the water is in the reduction process.

Figure 5 shows that the water column at almost all sampling points had a negative ORP, except at surface and 2 m depth in DM3. A negative ORP indicates that a reduction process of various ions and molecules has occurred in the water column (Sulastri et al., 2016). Wetzel. (2001) stated that an ORP of less than 100 mV, the sulfate-to-sulfide reduction would occur in eutrophic lakes. As a result, a negative ORP (Figure 5. A) indicated that Lake Maninjau was in a reduced state, including sulfate reduction to sulfide.

The ORP value of the water column can be used as a proxy in identifying the response of the water column to seasonal conditions. The ORP of the water column measured during the dry season will differ from the rainy season. Figure 5 B shows the ORP measured during the transition from rainy to dry seasons. The ORP remains positive to a depth of 10 m at each sampling point, indicating that oxygen is still available at that depth (Sulastri et al., 2016). This condition can be caused by the intensity of the sun starting to increase, thereby increasing the photosynthetic process. In addition, the water column stratification has formed forms of thermocline layers that act as a barrier to the mixing process.

2.2. Spatial distribution of sulfide in lake surface November 2022

The weakening of thermal stratification and the oxic layer of the water column triggers the transport process from the hypolimnion to the epilimnion. Sulfides that accumulates in the hypolimnion layer (Henny & Nomosatryo, 2016) could quickly diffuse to the surface and become toxic to fish in the FNC.

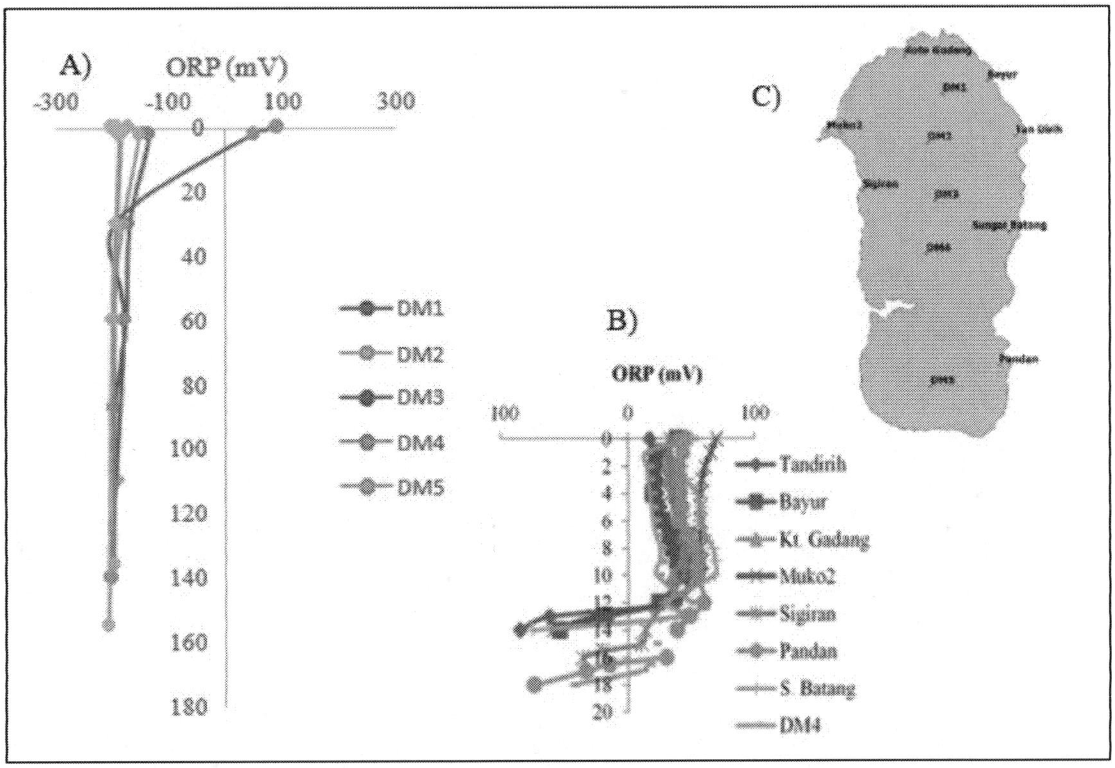

Figure 5. A) Vertical profile of ORP in November 2022, B) April 2014 (Sulastri et al., 2016), and C) Sampling points in November 2022 and April 2014.

We found that dissolved sulfides were detected over the entire surface of the lake at an average of 9.5 μgL^{-1} (Figure 6). The sulfide concentration in the lake exceeded the class III water quality standard set at 2.0 μgL^{-1} (Piranti et al., 2018). Based on Figure 5, the lake's periphery (points 1, 7, and 8) had a higher sulfide concentration than in the center. With low sunlight intensity and high organic matter of dead fishes, we estimate the recovery of the lake will take a long time.

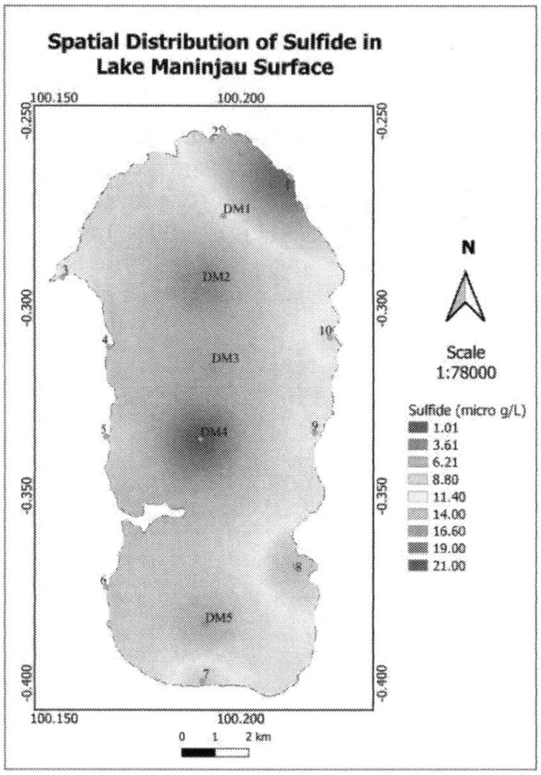

Figure 6. Distribution of sulfides on the surface November 2022.

2.3. The weather patterns in 2022 related to changes in the water quality of Lake Maninjau

Biogeochemical processes in the waters are strongly influenced by physical factors, such as the stratification of the water column and the mixing process between the epilimnion and hypolimnion (Katsev et al., 2010). An unstratified water column will trigger the exchange of several molecules, ions, or compounds between the epilimnion and hypolimnion (Henny, 2009). The strength of the stratification and mixing of the lake water column is greatly influenced by the weather conditions above (Yang et al., 2018; Kunz & Wildman, 2019). Therefore, the DO decrease and the rise of sulfide to the surface during this observation period are strongly influenced by weather variabilities, such as rainfall and air temperature.

Figure 7 show the daily rainfall and maximum air temperature during 2022. The rainfall has increased since June and reached its maximum in November (average rainfall is 21.34 mm/day) (Figure 7. A). This follows the report of Fakhrudin et al. (2002), which states that rain in Lake Maninjau reaches its maximum in November. The rainfall increase since May corresponds to the air temperature decrease from May until the end of the year (Figure 7. B). Rain is a weather parameter that significantly influences heat distribution in the lake water column, such as lowering the lake's surface temperature (Liu et al.,

2020). Therefore, decreasing the lake's surface temperature will reduce the temperature gradient between the surface and the bottom, resulting in the loss of lake stratification. In addition, a decrease in air temperature on the surface layer will trigger a flux of heat energy from the surface to the air, resulting in the mixing and deepening of the surface layers (Bouffard & Wüest, 2019). Fukushima et al. (2021) reported that large convection in Lake Maninjau corresponds to a decreased air temperature. Therefore, rising sulfides to the surface and decreasing DO in Lake Maninjau are closely related to significant weather changes.

Changes in lake ecosystem parameters due to changes in meteorological parameters depend on the magnitude of each parameter. Due to the short research time and limited data on physicochemical parameters, in this study, we did not quantify in detail how long it took for ecosystem parameters to change as a result of their response to climatological parameters. However, for a rough estimate, November 2022's heaviest rainfall occurred on the 14th (75 mm), and local media reported that on the 17th, there was a mass fish kill (Nugroho, 2022). The mass fish kill in that period was probably preceded by a decrease in oxygen levels due to inhibition of the photosynthetic process or sulfides rising from below reducing oxygen on the surface. Therefore, we hypothesize that in these conditions with extreme weather, ecosystem parameters can respond immediately to climatological parameters.

A)

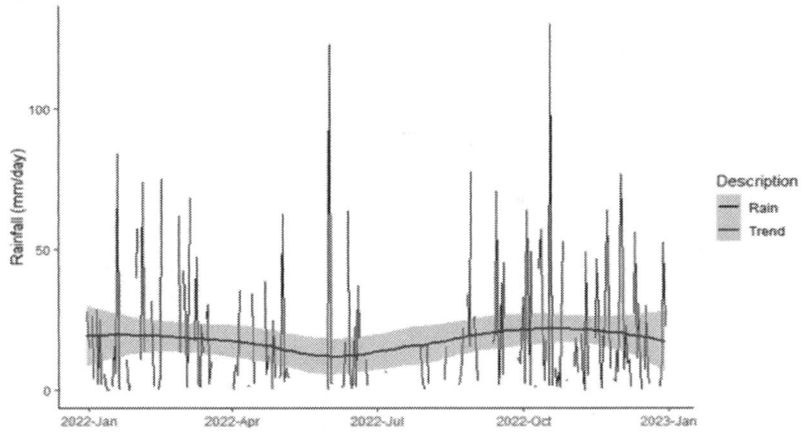

B)

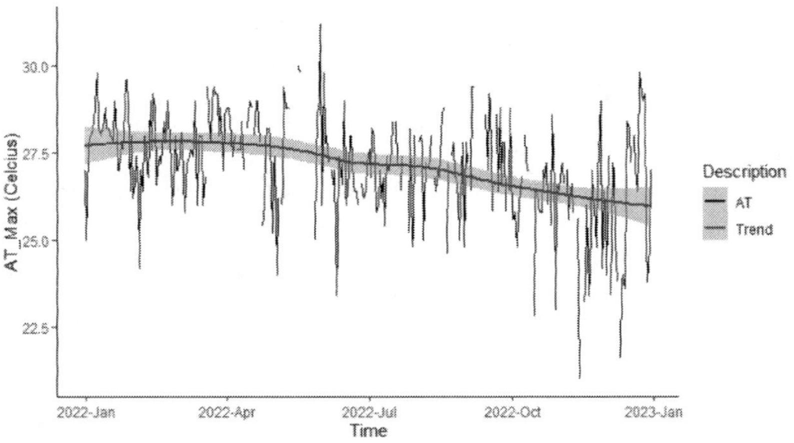

Figure 7. A) Daily rainfall and B) Maximum air temperature (AT_Max) for 2022 (data from Meteorological, Climatological, and Geophysical Agency (BMKG))

3. Conclusions

A few days after the mass fish kill, the water column temperature appears to be homogeneous with a very low temperature gradient of 1.16 °C. In addition, the water quality of Lake Maninjau was still poor, as indicated by deficient DO levels on the surface. Although the DO increased when sunlight reached the lake's surface, DO levels were still too low to provide the aquatic biota demand. High levels of detectable sulfide (an average 9.5 μgL^{-1}) on the surface also indicated poor water quality. The decrease in DO and the presence of sulfide on the lake surface water is an implication of increased rainfall and decreased air temperature, which are directly related to primary production and the regulation of water column stratification patterns.

Acknowledgment

This work was supported by Saintek Scholarship Program, Ministry of Research and Technology/National Research and Innovation Agency (BRIN).

References

Alloway, B. V., Pribadi, A., Westgate, J. A., Bird, M., Fifield, L. K., Hogg, A., & Smith, I. (2004). Correspondence between glass-FT and 14C ages of silicic pyroclastic flow deposits sourced from Maninjau caldera, west-central Sumatra. *Earth and Planetary Science Letters*, *227*(1–2), 121–133. https://doi.org/10.1016/j.epsl.2004.08.014

Andria, A. F., & Rahmaningsih, S. (2018). Technical Study of Abiotic Factors in Clay Embankment Used at PT. Semen Indonesia Tbk. for Utilization of Fish Cultivation with KJA Technology. *Jurnal Ilmiah Perikanan Dan Kelautan*, *10*(2), 95–105. https://doi.org/10.20473/jipk.v10i2.9825

Anggraini, Y. (2019). Ancaman dan Peluang Keramba Terhadap Masyarakat Salingka Danau Maninjau. *Jurnal Kapita Selekta Geografi*, *2*(2), 21–29.

Araoye, P. A. (2009). The seasonal variation of pH and dissolved oxygen (DO2) concentration in Asa lake Ilorin, Nigeria. *International Journal of Physical Sciences*, *4*(5), 271–274.

Bouffard, D., & Wüest, A. (2019). Convection in Lakes. *Annual Review of Fluid Mechanics*, *51*, 189–215. https://doi.org/10.1146/annurev-fluid-010518-040506

Couture, R. M., Fischer, R., Van Cappellen, P., & Gobeil, C. (2016). Non-steady state diagenesis of organic and inorganic sulfur in lake sediments. *Geochimica et Cosmochimica Acta*, *194*, 15–33. https://doi.org/10.1016/j.gca.2016.08.029

Fakhrudin, M., Wibowo, H., Subehi, L., & Ridwansyah, I. (2002). *Karakterisasi hidrologi danau maninjau sumbar*. *January*.

Fukushima, T., Matsushita, B., Subehi, L., Setiawan, F., & Wibowo, H. (2017). Will hypolimnetic waters become anoxic in all deep tropical lakes? *Scientific Reports*, *7*(March), 1–8. https://doi.org/10.1038/srep45320

Fukushima, T., Setiawan, F., Subehi, L., Fakhrudin, M., Triwisesa, E., Dianto, A., & Matsushita, B. (2021). Convection of waters in Lakes Maninjau and Singkarak, tropical oligomictic lakes. *Limnology*, *0123456789*. https://doi.org/10.1007/s10201-021-00686-8

Henny, C. (2009). Dynamics of Biogeochemisrty of Sulfur in Lake Maninjau. *Limnotek*, *XVI*(2), 74–87.

Henny, C., & Nomosatryo, S. (2012). Sulfide Dynamics in Lake Maninjau: The Implication on Phosphate Release in the Hypolimnion. *Limnotek : Perairan Darat Tropis Di Indonesia*, *19*(2),

102–112.

Henny, C., & Nomosatryo, S. (2016). Changes in water quality and trophic status associated with cage aquaculture in Lake Maninjau, Indonesia. *IOP Conference Series: Earth and Environmental Science*, *31*(1). https://doi.org/10.1088/1755-1315/31/1/012027

Holmer, M., & Storkholm, P. (2001). Sulphate reduction and sulphur cycling in lake sediments: A review. *Freshwater Biology*, *46*(4), 431–451. https://doi.org/10.1046/j.1365-2427.2001.00687.x

Katsev, S., Crowe, S. A., Mucci, A., Sundby, B., Nomosatryo, S., Douglas Haffner, G., & Fowle, D. A. (2010). Mixing and its effects on biogeochemistry in the persistently stratified, deep, tropical Lake Matano, Indonesia. *Limnology and Oceanography*, *55*(2), 763–776. https://doi.org/10.4319/lo.2009.55.2.0763

Komala, P. S., Afrianita, R., & Prima, R. (2019). Determination of purification ratio of Lake Maninjau in epilimnion zone. *MATEC Web of Conferences*, *276*, 06008. https://doi.org/10.1051/matecconf/201927606008

Kunz, K., & Wildman, R. (2019). *Effects of Solar Radiation and Air Temp on Thermal Stratification of Lakes*. 0–3. https://www.nalms.org/wp-content/uploads/2019/04/39-1-6.pdf

Liu, M., Zhang, Y., Shi, K., Zhang, Y., Zhou, Y., Zhu, M., Zhu, G., Wu, Z., & Liu, M. (2020). Effects of rainfall on thermal stratification and dissolved oxygen in a deep drinking water reservoir. *Hydrological Processes*, *34*(15), 3387–3399. https://doi.org/10.1002/hyp.13826

Lukman, Sutrisno, & Hamdani, A. (2013). PENGAMATAN POLA STRATIFIKASI DI DANAU MANINJAU SEBAGAI POTENSI TUBO BELERANG PENDAHULUAN. *Limnotek*, *20*(2), 129–140.

Lukman, Sutrisno, & Hamdani, A. (2014). Fluktuasi Oksigen Terlarut di Kawasan Karamba Jaring Apung di Danau Maninjau dan Hubungannya Dengan Ketersediaan Klorofil dan Bahan Organik. *Limnotek*, *21*(1), 30–40.

Makmur, S., Muthmainnah, D., & Subagdja. (2020). Fishery activities and environmental condition of Maninjau Lake, West Sumatra. *IOP Conference Series: Earth and Environmental Science*, *564*(1). https://doi.org/10.1088/1755-1315/564/1/012025

DR 2800 Spectrophotometer, (2007).

Mihu-Pintilie, A., Romanescu, G., & Stoleriu, C. (2014). The seasonal changes of the temperature, pH and dissolved oxygen in the Cuejdel Lake, Romania. *Carpathian Journal of Earth and Environmental Sciences*, *9*(2), 113–123.

Nontji, A. (2016). *Danau-Danau Alami Nusantara*.

Nugroho (Ed.). (2022, November 19). Jumat, ikan mati di Danau Maninjau capai 25 ton. Anatara Sumbar. Retrieved May 24, 2023, from https://sumbar.antaranews.com/berita/540585/jumat-ikan-mati-di-danau-maninjau-capai-25-ton

Paputungan, F., Pangemanan, N. P., Tumbl, R. A., Undap, S. L., Tumembuw, S. S., & Rantung, S. (2022). A study of water quality to support aquaculture at Moaat Lake, North Sulawesi. *Budidaya Perairan*, *2*(8.5.2017), 134–143.

Peraturan Presiden Republik Indonesia No. 60 Tahun 2021 tentang Penyelamatan Danau Prioritas Nasional, Peraturan Pemerintah Republik Indonesia Nomor 60 Tahun 2021 1 (2021). https://www.google.com/url?sa=t&rct=j&q=&esrc=s&source=web&cd=1&cad=rja&uact=8&ved=2ahUKEwjWxrKeif7eAhVYfysKHcHWAOwQFjAAegQICRAC&url=https%3A%2F%2Fwww.ojk.go.id%2Fid%2Fkanal%2Fpasar-modal%2Fregulasi%2Fundang-undang%2FDocuments%2FPages%2Fundang-undang-nomo

Piranti, A. S., Rahayu, D. R. U. S., & Waluyo, G. (2018). Evaluasi Status Mutu Air Danau Rawapening. *Jurnal Pengelolaan Sumberdaya Alam Dan Lingkungan (Journal of Natural Resources and Environmental Management)*, *8*(2), 151–160. https://doi.org/10.29244/jpsl.8.2.151-160

Pribadi, A., Mulyadi, E., & Pratomo, I. (2007). Mekanisme erupsi ignimbrit Kaldera Maninjau, Sumatera Barat. *Indonesian Journal on Geoscience*, *2*(1), 31–41. https://doi.org/10.17014/ijog.vol2no1.20073

Romaire, R. P., & Boyd, C. E. (1979). Effects of Solar Radiation on the Dynamics of Dissolved Oxygen in Channel Catfish Ponds. *Transactions of the American Fisheries Society*, *108*(5), 473–478. https://doi.org/10.1577/1548-8659(1979)108<473:eosrot>2.0.co;2

Rutner, F. (1931). *Hydrographische und hydrochemische Beobachtungen auf Java, Sumatra und Bali.* Arch. f. Hydrobiol.

Santoso, A. B., & Triwisesa, E. (2020). Ecosystem Metabolism and Oxygen Deficit in Lake Maninjau: Insight From High-Frequency Measurement. *Limnotek : Perairan Darat Tropis Di Indonesia*, *27*(2), 93–102. https://doi.org/10.14203/limnotek.v27i2.306

Santoso, A. B., Triwisesa, E., Fakhrudin, M., Harsono, E., & Rustini, H. A. (2018). What do we know about Indonesian tropical lakes? Insights from high frequency measurement. *IOP Conference Series: Earth and Environmental Science*, *118*(1), 0–5. https://doi.org/10.1088/1755-1315/118/1/012024

Siriwardana, C., Cooray, A. T., Liyanage, S. S., & Koliyabandara, S. M. P. A. (2019). Seasonal and Spatial Variation of Dissolved Oxygen and Nutrients in Padaviya Reservoir, Sri Lanka. *Journal of Chemistry*, *2019*. https://doi.org/10.1155/2019/5405016

Sulastri, Nomosatriyo, S., & Hamdani, A. (2016). Kondisi Lingkungan Perairan dan Keanekaragaman Sumber Daya Ikan di Danau Maninjau, Sumatra Barat. *Bawal*, *8*(1), 1–12.

Syandri, H., Junaidi, Azrita, & T, Y. (2014). State of aquatic resources Maninjau Lake West Sumatra Province , Indonesia. *International Journal of Ecology and Environmental Sciences*, *5*(1), 109–113. http://www.bioinfopublication.org/jouarchive.php?opt=&jouid=BPJ0000261

Wetzel, R. G. (2001). Limology Lake and River Ecosystems 3rd ed. In *Low Temperature Physics: Vol. Academic P* (Issue California). Academic Press.

Yang, Y., Wang, Y., Zhang, Z., Wang, W., Ren, X., Gao, Y., Liu, S., & Lee, X. (2018). Diurnal and Seasonal Variations of Thermal Stratification and Vertical Mixing in a Shallow Fresh Water Lake. *Journal of Meteorological Research*, *32*(2), 219–232. https://doi.org/10.1007/s13351-018-7099-5

An Enhanced Earthquake Risk Analysis using H3 Spatial Indexing

A N Aini[1,2*], O A Dewantari[1], D P A Mandala[1], M B F Bisri[1,3]

[1] Cerdas Antisipasi Risiko bencana Indonesia (CARI!), Jalan Sepak Bola 5, Bandung 40293, Indonesia
[2] Faculty of Earth Sciences and Technology, Bandung Institute of Technology, Jalan Ganesha 10, Bandung 40132, Indonesia
[3] GSICS, Kobe University, 2-1 Rokkodai-cho, Nada-ku, Kobe 657-8501, Japan

*E-mail: atiknurulaini@students.itb.ac.id

Abstract. Risk assessment is the first step in disaster risk reduction and crucial for types of disasters with potentially significant impacts, such as earthquakes. Besides the direct impact of very devastating shocks, earthquakes can also trigger cascading events that are no less damaging such as landslides, liquefaction, and tsunamis. This study aims to apply the H3 spatial indexing system to improve earthquake risk assessment using Sorong City as a case study, and compare the result of H3-based risk assessment with the normative risk assessment in Indonesia as defined by relevant *Perka* BNPB (Badan Nasional Penanggulangan Bencana/ National Disaster Management Authority) 2/2012. H3 is an open-source geospatial indexing system developed by Uber, which utilizes a hierarchical hexagonal grid for indexing, calculations, and visualization. The hexagonal cells in the grid have equidistant neighbors and possess the property of expanding neighboring rings, approximating circles and optimizing space filling. The H3 spatial index finds extensive applications in earthquake hazard, vulnerability, capacity, and risk assessment systems. The value of each parameter is entered into each hexagon so that in the end each hexagon has its risk index value. The study shows that the use of H3 in earthquake risk assessment can provide a more detailed level of analysis and indirectly does not depend on regional administrative boundaries. Technically, this advantage provides a more efficient calculation method and is also easier to modify partially both spatially and in terms of the parameters used. Based on the performance evaluation results on the spatial join aspect, H3 provides much faster performance compared to traditional indexing systems. Hence, the H3 spatial index will be very useful to meet the need for an earthquake risk assessment on a larger scale with a higher level of resolution.
Keywords: earthquake, risk assessment, H3 spatial indexing, Perka BNPB 2/2012, Sorong

1. Introduction

Earthquake risk assessment is an important component of disaster management. It helps identify areas and populations most at risk of earthquakes and guides the development and implementation of effective mitigation and response strategies. Earthquakes are one of the most destructive and frequent natural disasters, causing heavy casualties and damage to infrastructure and the built environment.

Carrying out regular earthquake risk assessments allows disaster management authorities and municipalities to identify areas of high risk and prioritize the allocation of resources for risk reduction actions. Such measures may include retrofitting existing buildings to make them more earthquake-resistant, strengthening critical infrastructure such as hospitals and emergency services, and establishing evacuation plans to ensure the safety of residents in the event of an earthquake. [1]

Risk assessment plays a crucial role in identifying vulnerable areas and populations, as well as discerning the diverse needs and required resources in each region to effectively mitigate the impacts of earthquake disasters. It serves as an essential tool for strategic planning, capable of both saving lives and minimizing economic losses resulting from earthquake events. Conducting thorough earthquake assessments also proves instrumental in the formulation and implementation of effective risk reduction measures. [1]

Various methods have been developed to evaluate the potential impact of earthquakes on the human population and the environment. Two methods commonly used for seismic risk assessment are based on raster data and administrative polygons. The raster-based method represents the study area as a grid of cells, each with an assigned value representing a specific attribute, such as population density or building type. This approach allows the integration of multiple data sources, such as remote sensing and Geographic Information System (GIS), to create a detailed and accurate representation of the study area [2]. Raster-based methods have been used for hazard and vulnerability assessments, as well as for damage and loss prediction [3]. The administrative polygon-based method represents the study area as a set of geographically defined areas, which are then assigned values that represent certain attributes. This approach enables the integration of socio-economic data, such as population demographics and land use, with geospatial data to understand the potential impact of earthquakes on certain areas or population groups [4].

Grid systems in vector format are gaining popularity for use in earthquake risk assessment because of their ability to efficiently organize and analyze large amounts of data. The hexagon grid, in particular, is advantageous for earthquake risk assessment as it preserves the relationships of adjacent cells while reducing the “points-in-polygon” problem thereby increasing spatial accuracy compared to traditional square grids. [5]

A specific hexagon grid system that is increasingly being used is the Uber H3 system. H3 is a hierarchical hexagonal grid system that divides the world into equal-sized hexagons, providing a stable and efficient way of indexing and querying large-scale geospatial data. It can be used for hazard and vulnerability assessments, as well as for damage and loss prediction. In addition, it is also possible to evaluate the accessibility of emergency services and critical infrastructure, and to simulate evacuation scenarios. [6]

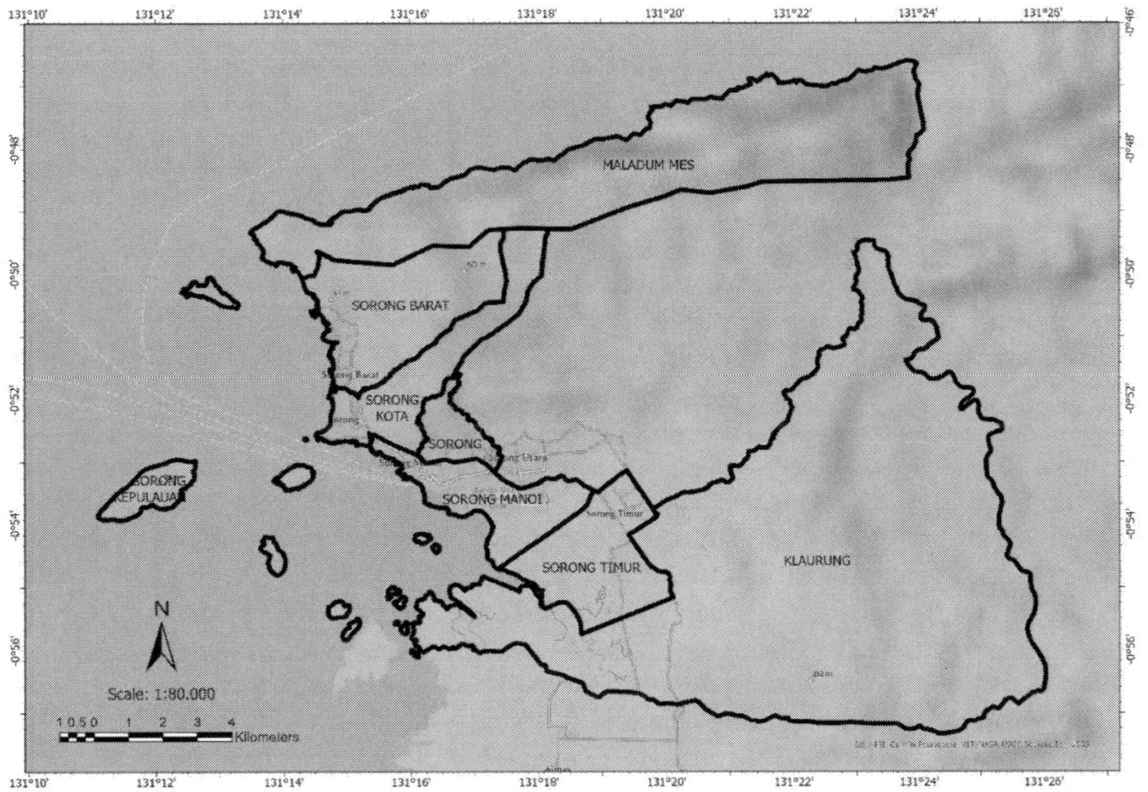

Figure 1. Sorong City

To demonstrate the potential application of H3, Sorong City (Figure 1) is chosen as a case study for comparing the process and results of earthquake risk assessment using H3 and the ones based on Head of BNPB (Badan Nasional Penanggulangan Bencana/National Disaster Management Authority) Regulation 2/2012 on disater risk assessment. Sorong City is located in the western part of Papua Province, Indonesia, is an area at high risk of earthquakes due to its proximity to several active faults. The most significant fault is the Sorong Fault, which crosses the city and has the potential to cause moderate to large earthquakes. In its history, the city of Sorong has experienced several earthquakes, the last and most significant being the 2009 Sorong earthquake with a magnitude of 7.2 which caused damage to buildings, and infrastructure, and resulted in casualties [7]. The city is also located near the boundary of the Bird's Head plate, where the Australian plate is subducting under the Sunda plate, which can cause tsunamis.

Given the high risk of earthquakes, it is very important to carry out regular and comprehensive earthquake risk assessments for Sorong City. Thus, the H3 hexagonal grid system will be applied in this study to help assess earthquake risk in Sorong City, and the results will be compared with the results of risk assessments that are common in Indonesia. The use of H3 is expected to support disaster management authorities in mitigating the negative impact of earthquakes, especially by identifying high-risk areas and populations and providing appropriate actions for those areas.

2. Literature Review

Earthquake risk assessment (ERA) is an important tool for identifying, analyzing, and mitigating the potential impacts of seismic events on human and natural systems. The goal of ERA is to determine the likelihood and consequences of earthquakes occurring in a given area. It is done to inform decision making related to disaster risk reduction and management. [3]

One key aspect of ERA is the unit of analysis, or the level at which risk is being assessed. Common units of analysis include individual buildings, communities, cities, and regions. The choice of unit of analysis will depend on the specific goals and objectives of the risk assessment, as well as the available data and resources. [8]

Another important aspect of ERA is the type of data used in the assessment. Data can be divided into two main categories: hazard data, which relates to the likelihood of an earthquake occurring, and vulnerability data, which relates to the potential impacts of an earthquake on human and natural systems. Hazard data can include information on historical earthquake activity, seismic hazard maps, and probabilistic seismic hazard analyses. Vulnerability data can include information on building codes and construction practices, population density and demographics, and critical infrastructure. [3]

There are several different methods used for earthquake risk assessment, depending on the unit of analysis and the data type used. Some common methods include:

a. Building-specific methods: These methods focus on the risk to individual buildings and structures. They typically involve detailed engineering analyses and inspections to assess the seismic performance of the building and identify potential vulnerabilities. Examples include performance-based engineering and fragility analysis. [9]
b. Community-based methods: These methods focus on the risk to neighborhoods, cities, or regions. They typically involve the use of GIS data and spatial analysis techniques to map the distribution of hazards and vulnerabilities across the community. Examples of this method include its usage in the HAZUS and RAPID. [10]
c. Probabilistic methods: These methods use statistical and probabilistic techniques to estimate the likelihood of earthquakes occurring and the potential impacts on human and natural systems. Examples include probabilistic seismic hazard analysis (PSHA) and seismic loss estimation. [11]

The grid system is a method of dividing a geographical area into a regular grid of cells, typically in a square or hexagonal shape, for the purpose of spatial analysis and mapping. The grid system has a long history in geographical information science and has been widely used in a variety of applications, including land-use planning, environmental modeling, and risk assessment. [6]

One recent development in the grid system is the implementation of the H3 hexagonal grid system. H3 was developed by Uber Technologies, Inc. as a way to index and spatially partition geographical data in a hierarchical manner. The hexagonal shape of the H3 grid allows for more efficient spatial analysis and mapping than traditional square grid systems. [6]

In recent years, the application of the grid system and H3 to earthquake risk assessment has gained significant attention. This is due to the ability of the grid system and H3 to efficiently and accurately map and analyze the spatial distribution of seismic hazards and vulnerabilities. The use of the grid system and H3 in earthquake risk assessment allows for the development of detailed and accurate maps that can be used to identify areas of high risk and inform decision making related to disaster risk reduction and management. [6]

One of the advantages of using the H3 grid system in earthquake risk assessment is that it allows for more precise and accurate mapping of seismic hazard and vulnerability data. The hexagonal shape of the H3 grid allows for better representation of areas with complex shapes and irregular boundaries. Additionally, the hierarchical nature of the H3 grid allows for more efficient storage and processing of large datasets. [6]

Another advantage of using the H3 grid system in earthquake risk assessment is that it allows for more flexible and adaptable spatial analysis. The hierarchical structure of the H3 grid allows for the analysis of data at different levels of resolution, from a global scale down to a local scale. This can be useful in identifying patterns and trends in seismic hazard and vulnerability data that may not be visible at a single scale. [6]

3. Materials and Methods

3.1. Data

The data used includes physical and socio-economic data, which are collected from various official sources, namely government agencies. The seismic hazard index data was taken from InaRisk BNPB. The obtained data is in raster format. The data used to assess vulnerability includes demographic, infrastructure, and economic characteristic data in the study area. The capacity index was taken from the latest IKDn (Indeks Ketahanan Daerah/Regional Resilience Index) assessment results by BNPB. Supporting data was collected from observations via Google Earth.

Specifically, vulnerability assessment includes the assessment of social, physical, environmental, and economic vulnerability. In assessing social vulnerability, the data used includes vulnerable groups, population density, and demographic structure. The assessment of physical vulnerability uses data on critical facilities, support facilities, roads, and buildings. The assessment of environmental vulnerability uses data on built and non-built areas. Meanwhile, the assessment of economic vulnerability includes data on productive land, productive land GDP, and economic activity centers.

3.2. H3 Spatial Indexing

H3 Spatial Indexing is a hierarchical geospatial indexing system that partitions the world into hexagonal cells. This system, which is open source under the Apache 2 license, was specifically developed to address the challenges faced by Uber's data science needs. One of the key advantages of H3 is its ability to join disparate data sets, making it a versatile tool for analysis. [6]

The hexagonal shape of the grid employed by H3 offers several benefits. Firstly, it provides smooth gradients, allowing for accurate measurement of differences between cells. Additionally, the equidistant relationship between hexagonal neighbors simplifies the analysis of movement, as hexagons exhibit expanding rings of equidistant neighbors that approximate circles. Moreover, hexagons are optimally space-filling, allowing for a more precise tiling of polygons compared to other shapes like squares. [6]

As a standard unit of analysis, H3 serves as a basis for joining various data sets. The H3 library supports the indexing of points, lines, and regions into the grid. This means that different data formats, such as rasters, can be easily indexed into H3 using combinations of these basic operations. Consequently, once data is indexed, it can be seamlessly joined with other datasets based on the H3 index. [6]

Beyond its indexing capabilities, H3 offers additional features for modeling flow. It can create indexes that represent movement from one cell to a neighboring cell, allowing for the association of weights with specific movements. This facilitates the analysis of movement patterns within the grid. [6]

Furthermore, H3 is well-suited for the application of machine learning techniques to geospatial data. The hexagonal grid defined by H3 can be treated as a pixel grid, enabling the use of computer vision techniques like convolution. The H3 library includes functions for finding neighbors (kRing) and transforming indexes into a two-dimensional coordinate space, enabling the implementation of various computer vision algorithms. [6]

The hierarchical nature of H3 indexing is an essential characteristic of the system. Each hexagonal cell in the grid, up to the maximum supported resolution, has seven child cells below it. This hierarchical subdivision, known as aperture 7, enables efficient traversal and organization of data within the grid. Although hexagons do not perfectly subdivide into seven finer hexagons, the alternate orientation of grids approximates this subdivision, allowing for precision truncation and determination of children cells. [6]

While geographic containment is approximate, logical containment in the index is exact. This means that H3 can be used as an exact logical index on top of data indexed at a specific resolution. The hierarchical structure of H3 also facilitates the efficient relation of datasets indexed at different resolutions, as functions for changing precision (h3ToParent, h3ToChildren) are implemented with minimal computational overhead. Furthermore, due to the numerical proximity of geographically close locations, the H3 index structure enables fast operations on spatially related data points. [6]

In analysis scenarios where precision or uncertainty of a location needs to be encoded in the spatial index, the hierarchical structure of H3 can be leveraged. For instance, a GPS receiver's point can be indexed at a coarser resolution when the precision of the signal is lower, or cells can be aggregated to a parent cell when there are insufficient data points in each finer cell. This flexibility allows for adaptable and contextually appropriate indexing based on the specific analysis requirements. [6]

3.3. Earthquake Risk Assessment with IRBI Method

IRBI (Indeks Risiko Bencana Indonesia/Natural Disaster Risk Index) is a method developed by the Indonesian National Disaster Management Authority (BNPB) to assess earthquake disaster risks in Indonesia. The method uses a risk index that combines several factors, including hazard, exposure, and vulnerability. The IRBI method considers several factors that may affect earthquake risk, including the seismicity rate, ground motion intensity, population density, and building quality.

The IRBI method is a multilevel index that assesses risks on a scale of 0 to 7, with higher numbers indicating higher risks. The index is based on three main components: hazard, exposure, and vulnerability. The hazard component assesses the potential for earthquakes in a given area. The exposure component assesses the number of people and buildings that may be affected by an earthquake. The vulnerability component assesses the susceptibility of people and buildings to damage from an earthquake.

To calculate the hazard component of the IRBI index, BNPB Indonesia considers several factors, including seismicity rates, ground motion intensity, and geological characteristics of the region. The exposure component considers population density, building density, and other demographic factors. Finally, the vulnerability component assesses the quality of buildings, infrastructure, and other factors that may impact the ability of people and communities to cope with an earthquake.

The IRBI method also takes into account the potential consequences of an earthquake disaster. For example, it considers the potential impact on critical infrastructure such as hospitals, schools, and transportation systems. The method also considers the potential for secondary disasters such as landslides, fires, and tsunamis.

3.4. Earthquake Risk Assessment with H3

Earthquake risk assessment with H3 spatial index involves several steps. Firstly, the study area is divided into a grid system using the H3 spatial index. Each grid cell contains a unique H3 index, which is used to represent the location of the cell. The H3 resolution used is 11, following Tobler's principle at a 1:50,000 administrative boundary scale.

The seismic hazard index is then calculated for each grid cell. The seismic hazard index provides information about the probability and intensity of earthquakes. The obtained data is in raster format, and subsequently, zonal statistics operation with the mean measure is applied to acquire index values at each hexagon point (resulting from the conversion of polygons to points). Furthermore, a spatial join is conducted to incorporate hazard values into the attributes of the hexagons (polygons).

The vulnerability index is obtained from social, physical, environmental, and economic vulnerability indices. The weight values are adjusted using the IRBI method of BNPB. Mathematically, the formula to calculate the vulnerability index can be seen as follows:

$$VI = 0.4S \times 0.25F \times 0.15L \times 0.2E$$

where VI represents the Vulnerability Index, S represents the Social Vulnerability Index, F represents the Physical Vulnerability Index, L represents the Environmental Vulnerability Index, and E represents the Economic Vulnerability Index.

IOP Conf. Series: Earth and Environmental Science 1245 (2023) 012014 doi:10.1088/1755-1315/1245/1/012014

Each type of vulnerability includes components and sub-components, which can be seen in detail in Table 1. For data analyzed at the district level, the methodology employed is akin to the IRBI method. For data pertaining to sub-components subjected to spatial analysis at the hexagonal grid level, each distance-related sub-component is computed utilizing the Euclidean distance.

Table 1. Component and Sub-Components of Vulnerability

Component	Sub-Components	Remarks
Social Vulnerability Index (0.4)		
Vulnerable Group (0.5)	Elderly and Underage Group (0.25)	Data is analyzed at the district level.
	Disability (0.38)	
	Poor Population (0.38)	
Population Density (0.25)	Population Density	Data is analyzed at the hexagonal grid level.
	Hexagon Area	
Demographic Structure (0.25)	Population Count (0.6)	
	Gender Ratio Value (0.4)	
Physical Vulnerability Index (0.25)		
Critical Facilities (0.35)	Accessibility (0.3)	Data is analyzed at the hexagonal grid level.
	Energy (0.3)	
	Water (0.25)	
	Telecommunication (0.15)	
Supporting Facilities (0.2)	Health Facilities (0.55)	
	Education (0.45)	
Roads (0.2)	Road Length (km)	
Buildings (0.25)	Building Area (0.55)	
	Number of Buildings (0.45)	
Environmental Vulnerability Index (0.15)		
Developed (0.45)	Residential Area (0.6)	Data is analyzed at the hexagonal grid level.
	Reclamation (0.4)	
Undeveloped (0.55)	Forest (0.25)	
	Swamp and Mangrove (0.3)	
	Coral Reefs (0.3)	
	Lakes and Ponds (0.15)	
Economic Vulnerability Index (0.2)		
Productive Land (0.35)	Agricultural Land Area (0.30)	Data is analyzed at the hexagonal grid level.
	Number of Mining Activities (0.20)	
	Plantation Area (0.10)	
	Fishery Area (0.25)	
	Seasonal Agriculture (0.15)	
GDP of Productive Land (0.35)	GDP Score (0.30)	Data is analyzed at the district level.
	Value of Gross Regional Domestic Product in Wholesale and Retail Trade Sector (0.35)	
	Value of Gross Regional Domestic Product in Construction Sector (0.35)	
Economic Activity Center (0.30)	Number of Markets/Malls (0.45)	Data is analyzed at the hexagonal grid level.
	Number of Tourist Attractions (0.55)	

The capacity index is taken from the IKDn data. This involves evaluating the availability and effectiveness of emergency services, evacuation routes, and other critical infrastructure. The capacity index value is given the same value for each grid cell as the IKDn assessment is done at the district/city level.

The results of seismic hazard, vulnerability, and capacity assessments are combined to generate a comprehensive earthquake risk index for each grid cell. This index provides a comprehensive measure of the risk of damage due to earthquakes in each study area location. Mathematically, the formula to calculate the risk index can be seen as follows:

$$RI = HI \times \sqrt{\frac{VI}{CI}}$$

where *RI* represents the Risk Index, *HI* represents the Hazard Index, *VI* represents the Vulnerability Index, and *CI* represents the Capacity Index.

3.5. Comparative Analysis

Comparative analysis is a valuable method for comparing the results of different risk assessment methods. One approach involves using statistical analysis to compare the distribution of risk across various domains. Statistical analysis for comparing the distribution of risk across different domains using descriptive statistics is an approach used to delineate and present the fundamental characteristics of risk distribution within each domain under consideration [12]. This method aids in gaining initial insights into how risk data is spread across all the relevant domains.

In addition to statistical methods, visual methods are also employed for comparative analysis [13]. The approach entails creating maps that display the results of both methods side by side. This provides a visual representation of the disparities in risk scores between the two methods and can be instrumental in identifying regions that consistently exhibit high or low risks in both methods, or areas where the methods diverge.

Another analysis method that can be utilized is SWOT analysis (Strengths, Weaknesses, Opportunities, and Threats) [14]. This method involves identifying and analyzing the strengths and weaknesses of each risk assessment method, as well as the opportunities and threats associated with the implementation of each method. This allows for a comprehensive evaluation of each method, considering not only technical aspects but also social, economic, and political factors that may impact the implementation and effectiveness of the methods. This study will present a SWOT analysis of the H3 application for risk assessment in the context of disaster management in Indonesia.

4. Results and Discussion

4.1. Results of Hazard, Vulnerability, and Risk Analysis using the IRBI Method based on Administrative Boundaries

Based on the 2020 IRBI, Sorong City is classified as high risk. The highest risk in Sorong City comes from the earthquakes, forest and land fires, and floods. Looking at the results of disaster incidence analysis, it is known that Sorong City has experienced 18 natural disasters since 2009 to 2021. Earthquakes are one of the disasters with significant impacts occurring in 2009, 2015, and 2021.

The maps of hazard, vulnerability, and capacity indices for Sorong City can be seen in Figure 2. The hazard index assessment is based on data from disaster incidents in Sorong City. Based on the hazard index assessment results, the Sorong Manoi District has the highest level of natural hazard compared to other districts and is classified as a high hazard class. The high level of hazard in the Sorong Manoi District is influenced by a combination of natural and anthropogenic hazards with both values considered high. The Sorong District is classified as a moderate hazard class. In addition, the Sorong Kepulauan and Sorong districts have a low hazard index value.

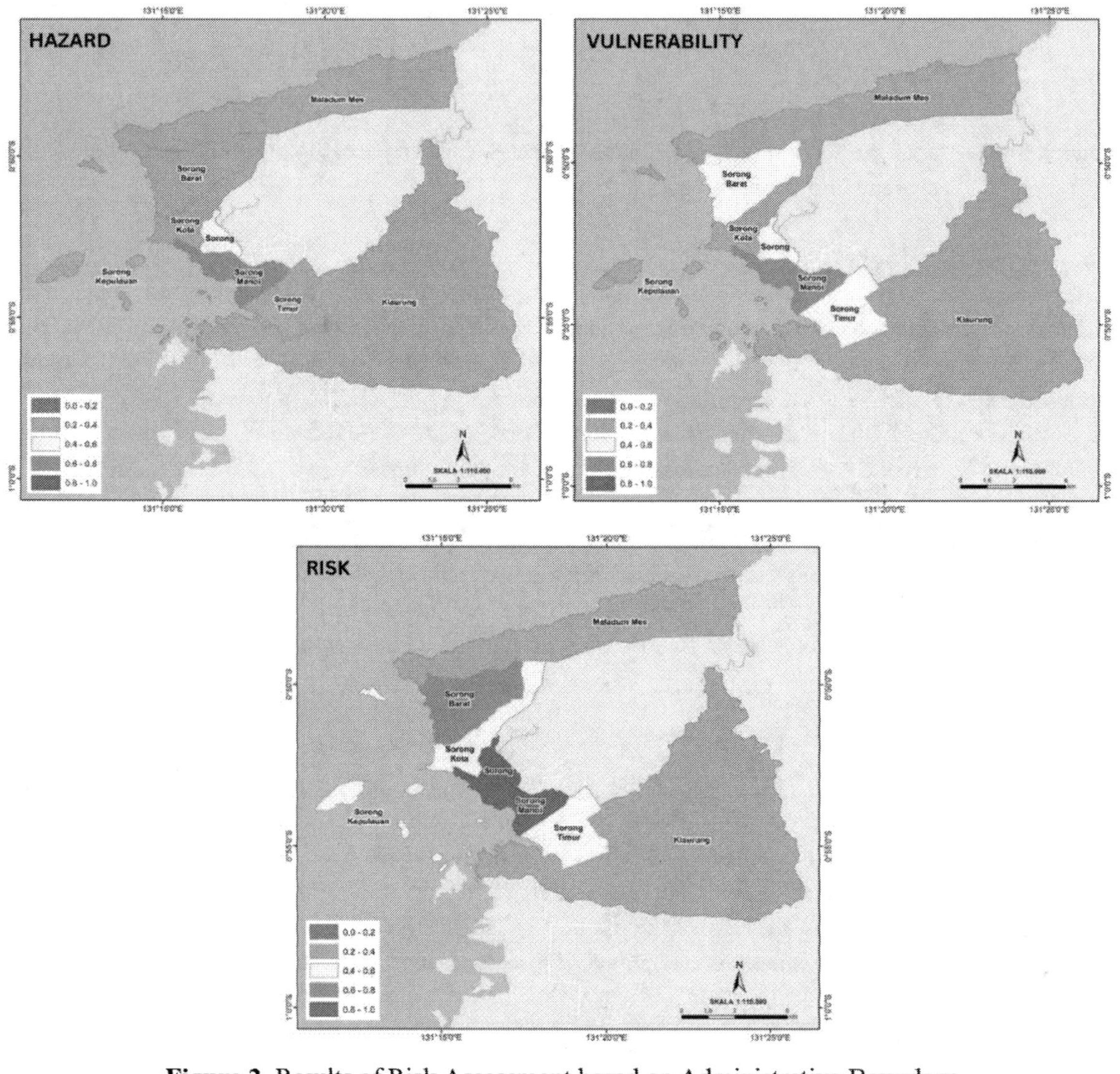

Figure 2. Results of Risk Assessment based on Administrative Boundary

Vulnerability assessment is a comprehensive evaluation of the condition and characteristics of a society and its living location to determine the factors that are susceptible to a disaster as well as the factors that can reduce it. The vulnerability index assessment in this research uses an approach of vulnerability assessment on the administrative scale unit aspect that combines exposure, sensitivity, and capacity aspects as a potential value of a disaster. Vulnerability studies in Indonesia are generally separated into social, physical, environmental, and economic components as regulated in BNPB Regulation No. 2 of 2012. Vulnerability assessment of the coastal areas of Sorong City is conducted by analyzing the characteristics of Sorong City from various government data sources and field results. Based on the vulnerability assessment results in Sorong City, the most vulnerable district is the Sorong Manoi District with an index value of 0.64, while the least vulnerable district is the Sorong Kepulauan District with an index value of 0.25. The average vulnerability value is 0.44, so the vulnerability level of Sorong City is in the medium range.

The physical vulnerability profile of coastal district in Sorong City is found in the Sorong Manoi District, which has an index value of 0.80. This is due to the location of the Sorong Manoi District, which has more supporting facilities and residential buildings compared to other districts. The physical aspect is assessed based on the number of critical and supporting infrastructure facilities in the study area, including accessibility facilities consisting of public facilities and airports. Then, supporting facilities are obtained from health and education facilities.

The environmental vulnerability profile is obtained from the calculation of the built-up area and the unbuilt-up area. The built-up area is viewed from residential and reclamation areas, while the unbuilt-up area is calculated from the land-use area value such as forests, swamps, mangroves, and coral reefs. The environmental vulnerability profile of coastal district in Sorong City is found in the Klaurung District with an index value of 0.48. This is due to the fact that this district has the highest forest area compared to other districts, which is 80.09 km2. In addition, the built-up area score in this district is also among the highest with a value of 0.60. On the other hand, the coastal district with the lowest environmental vulnerability index value is the Sorong Kota District. This is because this district does not have swamp and mangrove areas and has the third-lowest built-up area score after the Sorong Kepulauan and Maladum Mes districts.

The calculation of the Economic Vulnerability Index (EVI) involves the computation of the Gross Regional Domestic Product (GRDP) of productive areas and the number of economic centers within the study location. The GRDP score for each regency/city is based on the total value of the GRDP within the area divided by its district area. The economic aspect also encompasses the calculation of the highest GRDP within the economic index of each regency/city. The economic vulnerability profile of Sorong coastal district is found in the West Sorong district, with an index value of 0.58. This is because the GRDP value in this district is known to be the third-highest after Klaurung and Maladum Mes districts, at Rp19.13679 billions. West Sorong district also has the largest number of tourist spots, amounting to 13 points, and the second-largest agricultural land area, at 11.04 km^2.

The risk index value is obtained by considering the evaluation of hazards and vulnerabilities. Based on the calculation results, it is known that the Sorong Manoi district has the highest risk index value compared to other districts. The high risk index value needs to be considered because the population and settlements in this area are the highest, thus in the event of a disaster, the resulting losses will also be greater compared to other districts.

4.2. Results of Hazard, Vulnerability, and Risk Analysis using the H3 Method

The earthquake disaster risk assessment using H3 resulted in an index ranging from 0 to 1, indicating the level of risk ranging from low to high. A value of 0 indicates the lowest level of risk, while a value of 1 indicates the highest level of risk. The map above shows that this result can be used to identify areas with the highest risk of earthquake disasters and help prioritize disaster preparedness and response measures.

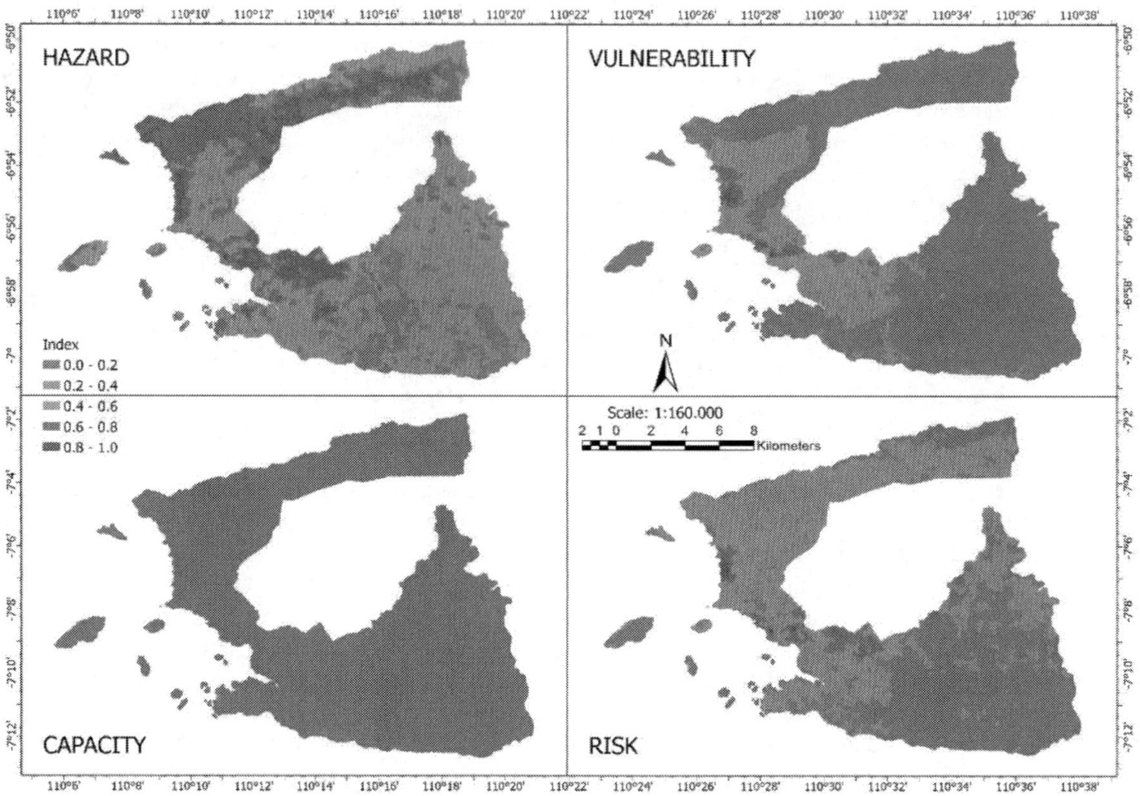

Figure 3. Results of Risk Assessment based on H3

Looking at the spatial distribution of risk indexes in Sorong City (Figure 3), the pattern and cluster of high-risk areas are found in Sorong Kota, Sorong Manoi, and Sorong Timur districts. Meanwhile, low-risk areas are found in Klaurung district. These results can be used to inform specific and targeted risk management decisions and actions. This includes identifying priority areas for disaster preparedness and response measures, such as improving buildings, conducting public awareness campaigns, and strengthening emergency response capacity.

4.3. Comparative Analysis

To carry out this comparative analysis, we used three types of analysis: statistical, visual, and SWOT analysis. The statistical analysis involved calculating the mean, standard deviation, and correlation coefficient of the risk values obtained from each method. The visual analysis involved creating maps to identify similarities and differences in the spatial distribution of risk. The SWOT analysis involved identifying the strengths, weaknesses, opportunities, and threats associated with each method.

4.3.1 Statistical Analysis

When comparing earthquake risk assessment results using administrative boundary-based and H3-based methods, descriptive statistical analysis can be conducted to obtain further information about the distribution of values in both methods. Several descriptive statistical components that can be used include mean, median, standard deviation, skewness, kurtosis, min, and max.

In the administrative boundary-based method, the mean value shows the average risk value obtained from all measured areas, while the median indicates the middle value of the data. Standard deviation indicates how far the data is spread from the mean, while skewness and kurtosis inform the shape of the value distribution. The minimum and maximum values indicate the range of risk values from the measured areas.

Meanwhile, in the H3-based method, the descriptive statistical values will be obtained from each hexagonal cell. The mean, median, and standard deviation values can provide information about the spread of risk values across all hexagonal cells. Skewness and kurtosis can also provide information about the shape of the risk value distribution in each hexagonal cell. The minimum and maximum values in each hexagonal cell can indicate the range of risk values in that area.

Table 2. Statistical Analysis

Statistics	Administrative Boundary-based	H3-based
Mean	0.346	0.24
Median	0.2	0.21
Standard Deviation	0.35	0.153
Minimum	0	0
Maximum	1	1
Skewness	0.89	1.32
Kurtosis	2.4	4.8

A comprehensive analysis of the statistical results presented in the Table 2 reveals several key findings when comparing the risk assessment based on administrative boundaries and the H3-based approach. In terms of the mean, the administrative boundary-based risk assessment yielded a value of 0.346, while the H3-based assessment resulted in a lower mean of 0.24. This suggests that, on average, the H3-based approach tends to estimate lower levels of risk compared to the administrative boundary-based method.

Looking at the median values, we observe that the administrative boundary-based assessment has a median of 0.2, while the H3-based assessment has a slightly higher median of 0.21. This indicates that the central tendency of the risk distribution is similar between the two approaches, with the H3-based approach showing a slightly higher median value.

In terms of variability, the standard deviation provides insights into the dispersion of the risk values. The administrative boundary-based assessment exhibits a higher standard deviation of 0.35, indicating a greater spread of risk values across the study area. On the other hand, the H3-based assessment demonstrates a lower standard deviation of 0.153, suggesting a more concentrated and less variable risk distribution.

Examining the minimum and maximum values, we find that both approaches yield the same range of risk values, ranging from 0 (indicating no risk) to 1 (indicating the highest level of risk). This indicates that both methods capture the full spectrum of risk levels present in the study area. Considering the skewness values, we observe that the administrative boundary-based assessment has a skewness of 0.89, indicating a moderate positive skew in the risk distribution. In contrast, the H3-based assessment exhibits a higher skewness of 1.32, indicating a more pronounced positive skewness and a potential presence of extreme values or a longer tail in the risk distribution.

The H3-based kurtosis being twice as large as the administrative boundary-based can be attributed to several factors. Firstly, the H3-based analysis utilizes a finer-grained spatial partitioning system, resulting in a higher number of data points and a more detailed representation of the risk distribution. This increased granularity can lead to greater variability and higher kurtosis values in the H3-based assessment. Secondly, the H3-based analysis captures the inherent spatial heterogeneity and clustering patterns within the data, which can contribute to a wider distribution and heavier tails in the risk assessment. In contrast, the administrative boundary-based analysis tends to aggregate data within larger administrative units, potentially smoothing out localized variations and reducing the overall kurtosis of the risk distribution. Additionally, the choice of statistical measure and calculation methodology can also influence the kurtosis values. It is possible that the specific calculation method used for the H3-based analysis is inherently more sensitive to extreme values or exhibits a different skewness-kurtosis relationship compared to the administrative boundary-based approach.

These statistical results demonstrate that the H3-based approach tends to yield lower mean values, a slightly higher median, a lower standard deviation, and a more positively skewed risk distribution compared to the administrative boundary-based approach.

4.3.2. Visual Analysis

When comparing the results of earthquake risk assessments using administrative boundary-based and H3-based methods, visual analysis can provide more easily understandable information and help identify certain patterns or trends in the risk data.

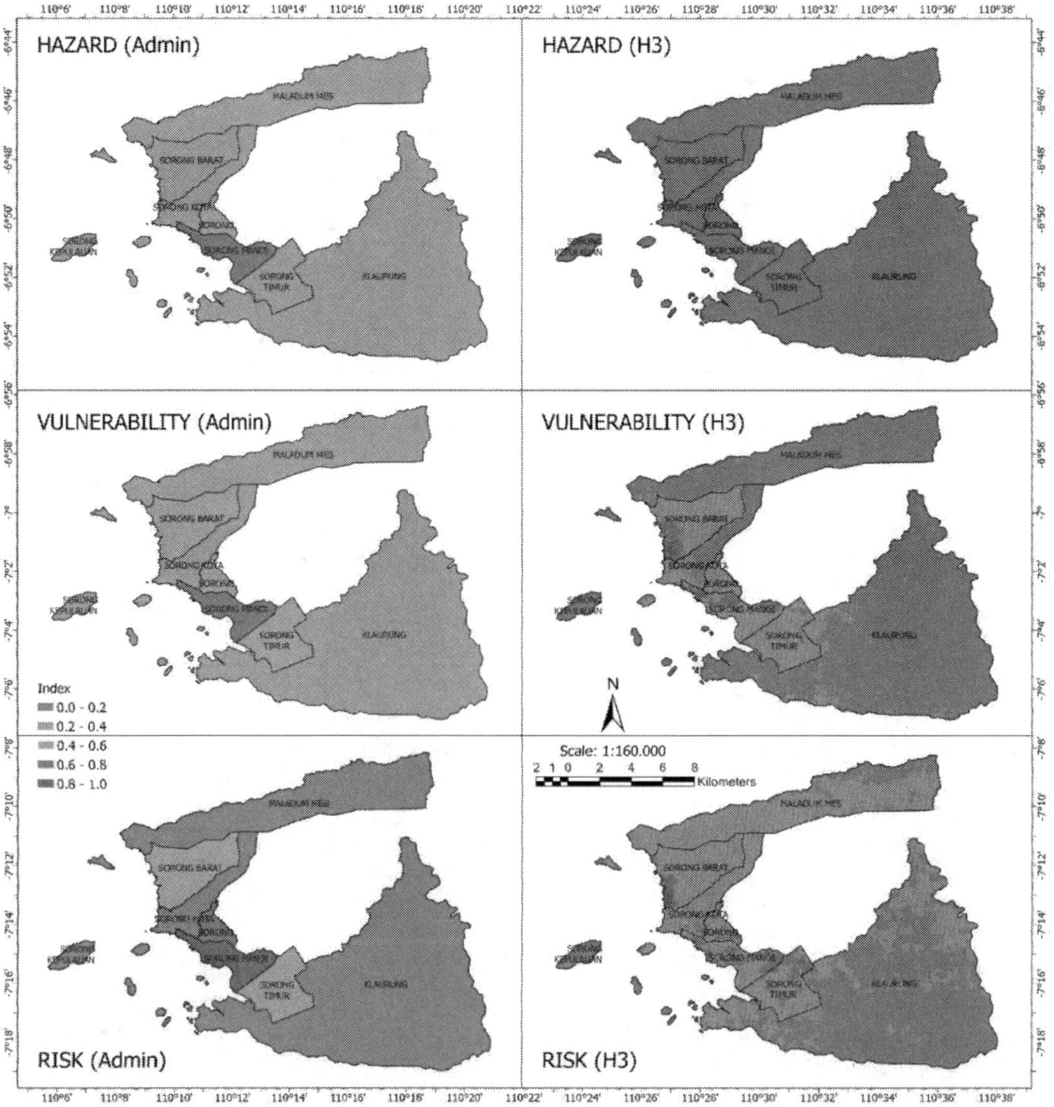

Figure 4. Comparison of Risk Map with Perka BNPB Guideline (left) and H3 (right)

From the map of the risk assessment results based on administrative boundaries, it can be seen that the area with the highest risk is located in the Sorong Manoi District, while other areas have relatively low levels of risk. From the map of the risk assessment results based on H3, it can be seen that the highest risk values are located in the Sorong Manoi, Sorong Kota, and Sorong Timur Districts.

From the visual analysis (Figure 4), it can be seen that both methods yield similar results by showing that the area with the highest risk is located in the Sorong Manoi District. However, the H3-based method provides more specific information at the hexagonal cell level, which successfully identifies Sorong Kota and Sorong Timur Districts as high-risk areas. H3 can show finer risk patterns at the cell level.

4.3.3. SWOT Analysis

In this section, we conduct a SWOT analysis of the H3 method with the aim of comprehensively identifying the intrinsic strengths and weaknesses of this method, as well as recognizing the opportunities and threats that may be associated with its use in the context of risk analysis. This analysis will provide profound insights into the potential for the development and improvement of the H3 method in disaster risk modeling.

4.3.3.1 Strengths

The utilization of H3 for earthquake risk assessment offers several advantages and positive value compared to the traditional administrative boundary-based method. One of the significant advantages is that H3 can provide a higher resolution spatial unit for risk assessment compared to administrative boundaries. With the H3 method, it is possible to create a more precise spatial unit, allowing for more accurate and detailed risk assessment. The H3 method also enables the combination of various types of data and models, making it a useful tool for multi-hazard risk assessment.

Another advantage of H3 is that it can overcome the issue of spatial heterogeneity in risk assessment, which is a common challenge in administrative boundary-based methods. This is because H3 units have more uniform spatial distribution and size, allowing for a more standardized risk assessment approach. Additionally, H3 can also facilitate the integration of community-level data, which can improve the accuracy of risk assessment and increase community engagement in disaster risk reduction efforts.

4.3.3.2 Weaknesses

Despite its advantages, the utilization of H3 for risk assessment also has several limitations and weaknesses. One of the primary limitations is the potential for data inconsistency and inaccuracy due to the complexity of the H3 method. This is because H3 involves the integration of various types of data and models, which can lead to errors in data processing and analysis.

Another limitation is the need for significant computational resources and technical expertise to perform H3-based risk assessment. This can be challenging for many local and regional disaster management agencies, which may not have access to the necessary resources or technical expertise.

Additionally, the H3 method requires careful consideration of the choice of resolution level to balance the accuracy and efficiency of risk assessment. Choosing too high of a resolution level can result in excessive computational requirements and potential data inconsistency, while too low of a resolution level may lead to inaccurate risk assessment results.

4.3.3.3 Opportunities

In terms of opportunities, the H3-based risk assessment offers several promising prospects. Firstly, it provides an enhanced spatial resolution compared to the administrative boundary-based approach. By utilizing hexagonal grids, the H3 method allows for a more detailed analysis and identification of localized risk patterns and hotspots within the study area. This finer granularity can greatly improve risk management strategies by targeting specific areas with higher vulnerability.

The H3-based approach presents an opportunity for improved accuracy in risk assessment. Leveraging geospatial algorithms and techniques specifically designed for hexagonal grid systems, this method has the potential to yield more precise estimations of risk levels. The incorporation of these techniques can enhance the reliability of the results and support better-informed decision-making processes. Additionally, the H3-based approach offers scalability and flexibility in terms of data integration and analysis. The hexagonal grid structure enables seamless aggregation and disaggregation of data, allowing for comparisons across various scales. This flexibility facilitates the integration of different datasets or models, further enhancing the comprehensiveness of the risk assessment.

4.3.3.4 Threats

However, the H3-based approach also presents certain threats that need to be considered. One such threat is the compatibility and availability of data in the hexagonal grid format. Adapting existing datasets to the H3 framework may require additional efforts and resources, which could introduce challenges and potential inaccuracies in the analysis. Furthermore, the interpretation and understanding of the results derived from the H3-based approach may pose challenges for stakeholders and decision-makers. Familiarity with the new methodology and potential training may be necessary to fully comprehend and interpret the outcomes generated from the hexagonal grid system.

Lastly, the successful implementation of the H3-based risk assessment relies on stakeholder acceptance and adoption. Resistance to change or a lack of familiarity with the new approach may hinder its widespread acceptance and utilization, potentially limiting its impact and effectiveness. Considering these opportunities and threats is crucial when evaluating and selecting the most suitable risk assessment methodology. Proper assessment and mitigation of the identified threats, along with leveraging the opportunities, can contribute to the successful implementation of the chosen approach and ultimately enhance risk management and decision-making processes.

The results of this comparative analysis provide insights into the effectiveness of the administrative boundary-based method and the H3 method in identifying areas of risk. The statistical analysis showed that the mean risk values obtained from the two methods were significantly different, with the H3 method generally resulting in higher risk values. The visual analysis showed that the spatial distribution of risk was also different between the two methods, with the H3 method identifying more areas of high risk. The SWOT analysis identified the strengths and weaknesses of the H3 method, as well as the opportunities and threats associated with its implementation.

5. Conclusions

Risk assessment is a crucial stage in disaster risk management. The use of H3 for earthquake risk assessment offers several advantages and positive values compared to traditional administrative boundary-based methods, namely providing spatial units with higher resolution, creating more precise spatial units, allowing for more accurate and detailed risk assessment, and addressing spatial heterogeneity issues in risk assessment. However, the use of H3 also has some limitations, including potential inconsistency and inaccuracy of data, and the need for significant computational resources and technical expertise. Further development of the H3 method can be done using machine learning and big data. With these technologies, H3 can be used to analyze big data in a shorter time and provide more accurate results in earthquake risk assessment. Further development of H3 can also involve community participation in data collection and risk assessment. By involving communities, H3 can obtain more data and increase community awareness of earthquake disaster risks in their area.

Acknowledgement
This article is partially associated to CARI! Support for Yayasan Konservasi Alam Nusantara on "Baseline Risk Assessment of Selected Coastal Areas in Indonesia".

References

[1] V. Silva, C. Yepes-Estrada dan G. Weatherill, "Words into Action Guidelines: National Disaster Risk Assessment," *UNISDR,* 2017.

[2] F. H. Shah, A. Ali dan M. M. Iqbal, "A Review on Hazard Risk Assessment using Remote Sensing and GIS," *Central Asian Journal of Environmental Science and Technology Innovation,* p. 252–261, 2020.

[3] V. Silva, D. Amo-Oduro, A. Calderon, C. Costa, J. Dabbeek, V. Despotaki, L. Martins, M. Pagani, A. Rao, M. Simionato, D. Vigano, C. Yepes-Estrada, A. Acevedo, H. Crowley dan N. Horspool, "Development of A Global Seismic Risk Model," *Earthquake Spectra,* pp. 1-23, 2020.

[4] C. J. v. Westen, L. Montoya, L. Boerboom dan E. B. Coto, "Multi-Hazard Risk Assessment using GIS in Urban Areas: A Case Study for the City of Turrialba, Costa-Rica," *Hazard Mapping and Risk Assessment,* pp. 53-72, 2002.

[5] S. Woźniak dan P. Szymański, "hex2vec - Context-Aware Embedding H3 Hexagons with OpenStreetMap Tags," *GEOAI,* 2021.

[6] Uber Technologies, Inc., "H3: Hexagonal Hierarchical Spatial Indexing," *Uber Technologies, Inc.,* 2019.

[7] BNPB, "Data Informasi Bencana Indonesia," *DIBI BNPB,* 2009.

[8] R. Jena, B. Pradhan, G. Beydoun, A. Al-Amri dan H. Sofyan, "Seismic Hazard and Risk Assessment: A Review of State-of-the-Art Traditional and GIS Models," *Arabian Journal of Geosciences,* 2020.

[9] G. Andreotti dan C. Lai, "Use of Fragility Curves to Assess the Seismic Vulnerability in the Risk Analysis of Mountain Tunnels," *Tunnelling and Underground Space Technology,* 2019.

[10] B. Gulati, "Earthquake Risk Assessment of Buildings: Applicability of HAZUS in Dehradun, India," *Indian Institute of Remote Sensing,* 2009.

[11] A. Guardiola-Víllora, S. Molina dan D. D'Ayala, "Performance based Probabilistic Seismic Risk Assessment for Urban Heritage: An Example in Pla del Remei Area (Valencia)," *Bulletin of Earthquake Engineering,* vol. 21, p. 4951–4991, 2023.

[12] I. Krnjić dan V. M. Cvetković, "Investigating Students Attitudes and Preferences Towards Disaster Learning Multimedia to Enhance Preparedness," *Research Square,* 2021.

[13] J.-F. Girres, "Hazard-Based Images Comparison Methods for Damage Assessment in Emergency Mapping," *International Journal of Cartography,* Vol. %1 dari %22-3, no. 5, pp. 332-348, 2019.

[14] T. A. Rachmawati, "Comparison Studies on Integrating of Disaster Risk Reduction (DRR) in Spatial Planning Systems in Indonesia, Japan, and European Countries," *Jurnal Dialog Penanggulangan Bencana,* vol. 4, pp. 13-22, 2013.

Distributions of Chromium, Copper, and Cadmium in the estuary-coastal surface sediment of Tin-Rich Island

A U Nurhidayati[1], S Y Cahyarini[1] and D A Utami[1]

[1] Paleoclimate and Paleoenvironment Research Group, Research Centre for Climate and Atmosphere, Research Organization for Earth Science and Maritime, National Research and Innovation Agency (BRIN), KST Samaun Samadikun, Jl Sangkuriang, Coblong Bandung, Indonesia 40135

ayuutaminurhidayati@gmail.com

Abstract. Belitung Island known as one of the tin producer in the world, as it lied in the South East Asian tin belt. The study was conducted in the North Belitung Island to investigate some metal elements concentration of sediment along the estuary river and coastal area near to carbonate reef environment system. Thirteen sediment samples were collected from estuarine riverbed, shoreline, and offshore of North Belitung. The concentrations of metal elements were analysed to assess the environmental pollution characteristics by using flame AAS (Atomic Absorption Spectrophotometry) method. The result indicated that the average concentrations of chromium (Cr), copper (Cu), and cadmium (Cd) were 1.62-18.26, 5.72-80.07, and 1.4-4.59 mg/kg dry weight, respectively. The calculation of geo-accumulation index (I_{geo}) was adopted in this study to evaluate the heavy metal pollution. The I_{geo} values revealed that this research area has been moderately polluted by Cr and Cu, and unpolluted-moderately polluted by Cd. Based on the contamination level to adverse biological effect, the heavy metals concentration in this research area has the low potential to adverse biological effect by Cr and Cu, and low-moderately potential by Cd. This information is important to lead the study about any other elements and heavy metals to evaluate the risk assessment in the tropical area.

Keywords: Heavy metals, surface sediment, reef carbonate, tin mining, AAS

1. Introduction

Tin (Sn) is a precious metal that resists corrosion and is non-toxic, which is why it is often used as a protective coating for other metals. As a result, tin is a key industrial metal in global markets [1], [2]. Due to its high demand, some production countries of tin, mine the metal extensively. Around 38% of the tin offered worldwide comes from the region known as the 'Southeast Asian Tin Belt' [3]. The Southeast Asian Tin Belt is a north-south elongate zone 2,800 km long and 400 km wide, extending from Burma (Myanmar) and Thailand to Peninsular Malaysia and the Indonesian Tin Islands (Fig. 1) [3]. Belitung island, with a land surface area of about 5000 km^2, forms the easternmost expression of the Southeast Asian tin belt [4] and known as the Indonesian Tin Island altogether with Bangka Island

(Fig. 1). While global demand for tin is rising, the mines in Belitung are becoming increasingly exhausted, increasing the pressure on the environment.

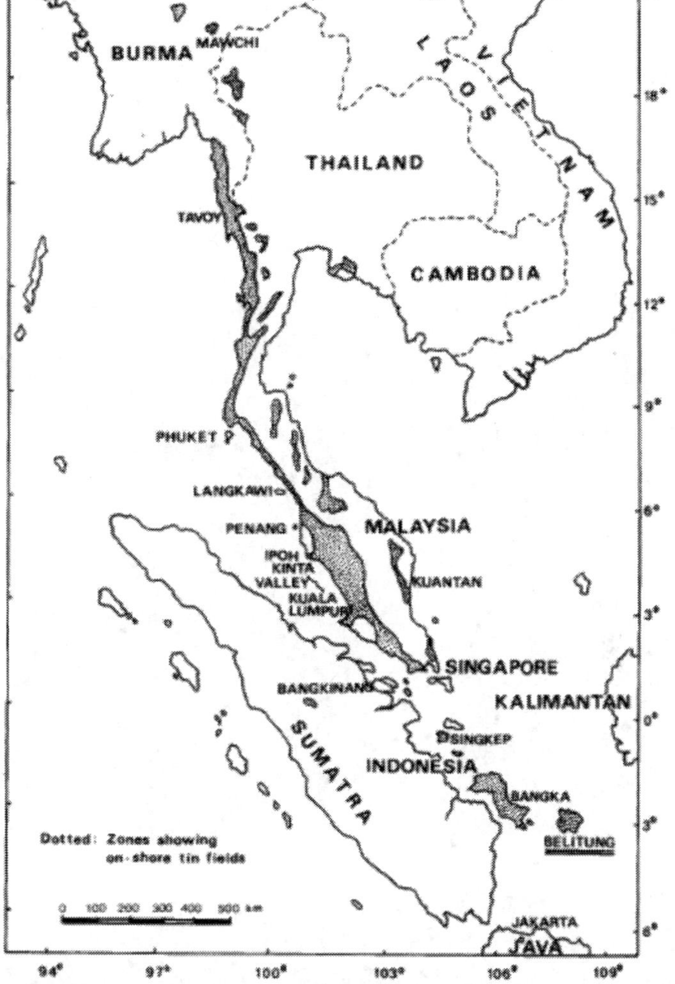

Figure 1. South-East Asian Tin Belt, and Belitung Island is at the end of it [5] cited in[4]

The excessive accumulation of heavy metals in mining areas may threaten the surrounding ecological environment and the health of surrounding residents [6] [7]. In addition, the excessive amounts of heavy metals are introduced to estuarine and coastal environments through river, runoff, and land-based point sources [8]. Many metals can be incorporated into marine ecosystems as the result of intense industrial development along the coastline [8]. Marine sediments have often been regarded as the ultimate reservoir for heavy metals in the coastal environment [9], [10]. These metals are associated with the sediment via adsorption in surface particles, precipitation, and incorporation into biogenic material [11], [12]. As a result, sediment contamination poses one of the worst environmental problems in marine ecosystems, acting as sinks and sources of contaminants in aquatic systems [13], [14], [15].

Belitung Island is not only as Indonesia's Tin Island, but also an island with high diversity coastal ecosystem, including 244 different coral genus from 15 families [16]. As one of the carbonate reef Island in Indonesia, Belitung's reef is strategically connected to the South China Sea and Java Sea, due to seasonally reversing monsoon currents [17]. In addition, Belitung is less exposed by anthropogenic warming of sea surface temperatures, marine heat waves and destructive cyclones [17]. As a result, Belitung Island could potentially provide coral larvae to replenish other reef areas in Indonesia following disturbances such as mass bleaching events [17]. Therefore, Belitung Island is important to be protected.

Heavy metals gained importance due to their known detrimental environmental effects [18]. But, the study of heavy metal accumulation in Belitung Island is still rarely done previously, especially around marine/reef sediment. The analysis of marine sediment is an important area of environmental analytical chemistry [18]. Based on the ecological risks that may occur due to mining activity in Belitung Island, it is crucial to provide some information about heavy metals distribution in this island. The objective of this study is to provide the information of the heavy metal distribution in the Belitung Island affected by tin mining activity, and to evaluate the contamination level of the heavy metals towards the marine ecosystems.

2. Material and Methods

2.1. Sample Collection and Analysis

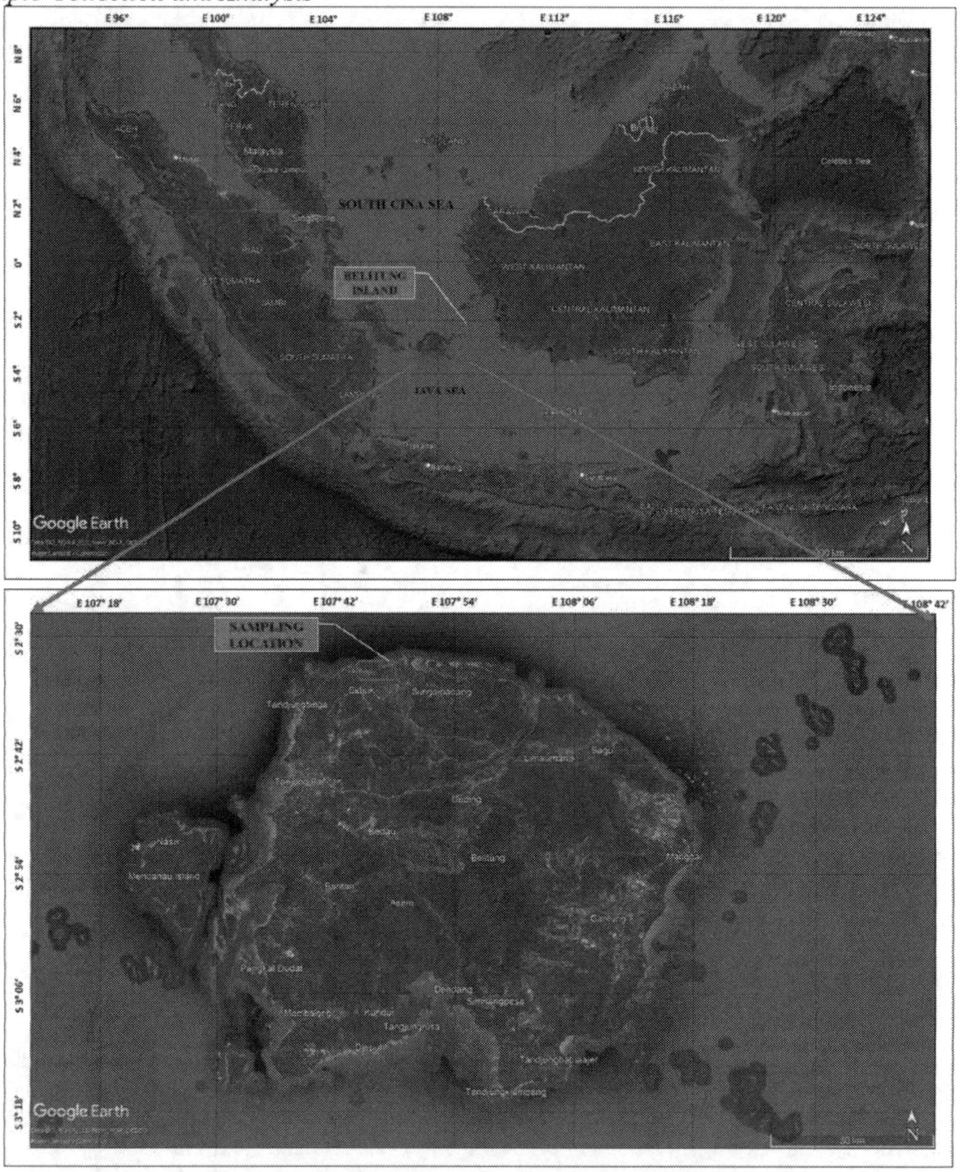

Figure 2. Study area, marked with the red square

This study investigates the distribution of metal concentrations in surface sediments of the eastuary-coastal in Belitung Island, close to reef environment. The study area lies in the north of Belitung Island

(Fig. 1). The samples were collected from the estuarine river (n=6), shoreline (n=5), and offshore (n=2) (Fig. 2). The river is natural (Penyairan River) and artificial (Tambak Udang River) ones, ex mining dredging activity. The sediment samples were selected to represent both proximal and distal sites. Sampling was conducted in August 2022, and 13 samples from different sites were collected using plastic shovel and stored in polyethylene plastic bag. The sediment samples collected were undisturbed with 0-5 cm depth. The sampling locations were recorded by Avenza Maps Apps.

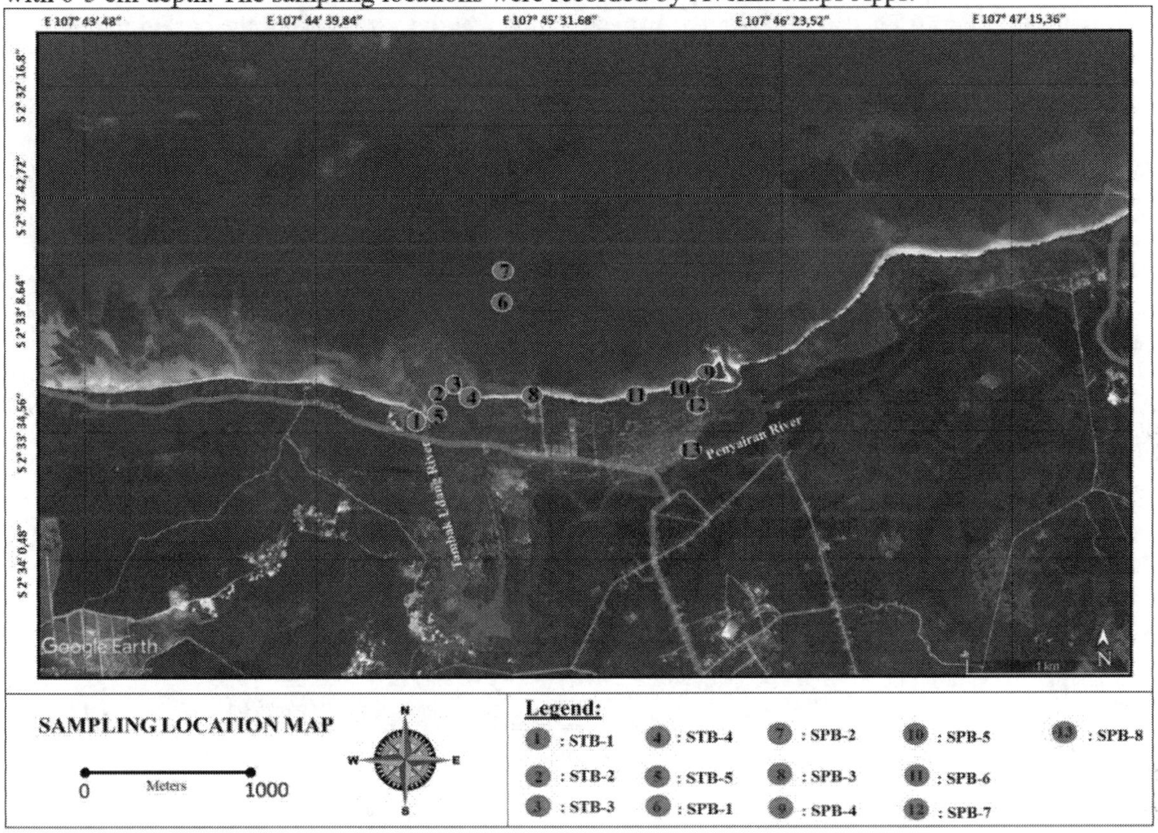

Figure 3. Sampling location of surface sediment in North Belitung

In the laboratory, the sediments were dried using a Memmert™ oven set to 80° C for 8 hours. Afterward, the sediment samples were sieved through a 125 µm sieve using a Durham Geo Slope Indicator (DGIS) sieve shaker. The remaining portion of the coarse fraction was then sieved through 250 µm, 500 µm, 1 mm, and 2 mm sieves. Each individual grain-size fraction was digitally weighed. The sediment from the respective location was digitally weighed around 0.5000-0.5010 gram in 100 mesh (0,147 mm) grain size of sediment.

For the analysis of solid samples such as sediment, a dissolution step must precede the analysis step, as solutions are most easily analyzed by atomic absorption [18]. The digestion process in this study following the procedure from Ozkan et al., 1980 for heavy metal extraction from marine sediment for AAS, with a little modifications. The mixture acids used to dissolve the undesired material were HNO_3, H_2SO_4, HF, and HCl. The dissolved sediment was then diluted with ultrapure deionized water and filtered with filter paper Whatman 41, 100 cycles, diameter 125 mm (CAT No. 1441-125) and made up to 100 ml with ultrapure deionized water. The duplicate sample was performed for every samples respectively.

Flame Atomic Absorption Spectrophotometry (AAS) analysis was performed on all samples to obtain the concentration of heavy metals (Cr, Cu, and Cd), using the AAS machine AA-7000 by Shimadzu (Serial No. A30694901634) and an auto sampler machine ACS-7000 by Shimadzu (Serial No. A30694901446).

2.2. *Geo Accumulation Index (I_{geo})*

The geo-accumulation index (I_{geo}) has been adopted to evaluate the contamination levels of heavy metals in sediments as proposed by Muller, 1969. The I_{geo} determines the pollution grade by comparing current concentrations with background concentrations [19]. The Igeo is calculated by Eq. (1):

$$I_{geo} = \log2 \, (Ci*1.5*Bi)$$

Where Ci is the measured concentration of heavy metal i, Bi is the background value of heavy metal i in global mean crust [20] and 1.5 is the coefficient to correct the variations in the background values caused by lithogenic effects [19]. The I_{geo} values are classified as follows: no contamination ($I_{geo} < 0$); low to medium contamination ($0 \leq I_{geo} < 1$); medium contamination ($1 \leq I_{geo} < 2$); medium to strong contamination ($2 \leq I_{geo} < 3$); serious contamination ($3 \leq I_{geo} < 4$); serious to severe contamination ($4 \leq I_{geo} < 5$); and severe contamination ($I_{geo} > 5$).

3. Result and Discussion

All sediment samples in the study area are grain supported. Mean grain size sediments are mostly fine (n=5) to medium (n=4) sand, although very coarse sand (n=3) and coarse sand (n=1) also occurred. The sediment from the estuarine river zone (STB.1, STB.2, STB.5, SPB.4, SPB.7, and SPB.8) and coastal area (STB.3, STB.4, SPB.4, SPB.5, and SPB.6) was vary, ranging from very coarse – fine sand, while the sediment from the offshore (SPB.1 and SPB.2) area was very coarse - medium sand. The range of the grain size in this study area might be due to the different energy of the wave around this water body from river, coastal to the offshore.

The sediment in the North Belitung are primarily from runoff by around small streams (Penyairan Beach and Tambak Udang Beach), and transported of suspend particulates from open sea by tidal current. Grain size analysis indicates that composition of the surface sediments in North Belitung is relatively very vary from fine-very coarse sand.

Table 1. Summary of heavy metal concentrations in dry weight, from surface sediment in North of Belitung analyzed by AAS.

Statistics	Cr (mg/kg)	Cu (mg/kg)	Cd (mg/kg)
Max	18.26	20.07	4.59
Min	1.62	5.72	1.4
Average	7.232	8.612	2.415
Standard Deviations	5.05	4.21	0.95
Samples Number	13	13	13

The range and mean concentrations of heavy metals in surface sediments from North Belitung Island are summarized in Table 1. The concentrations of Cr ranges from 1.62-18.26 mg/kg, Cu ranges from 5.72-20.07 mg/kg, and Cd ranges from 1.4-4.59 mg/kg in dry weight, respectively. The average concentrations for Cr, Cu, and Cd were 7.232, 8.612, and 2.415 mg/kg, respectively. The highest heavy metal concentration is Cu, and the lowest is Cd in North Belitung Island.

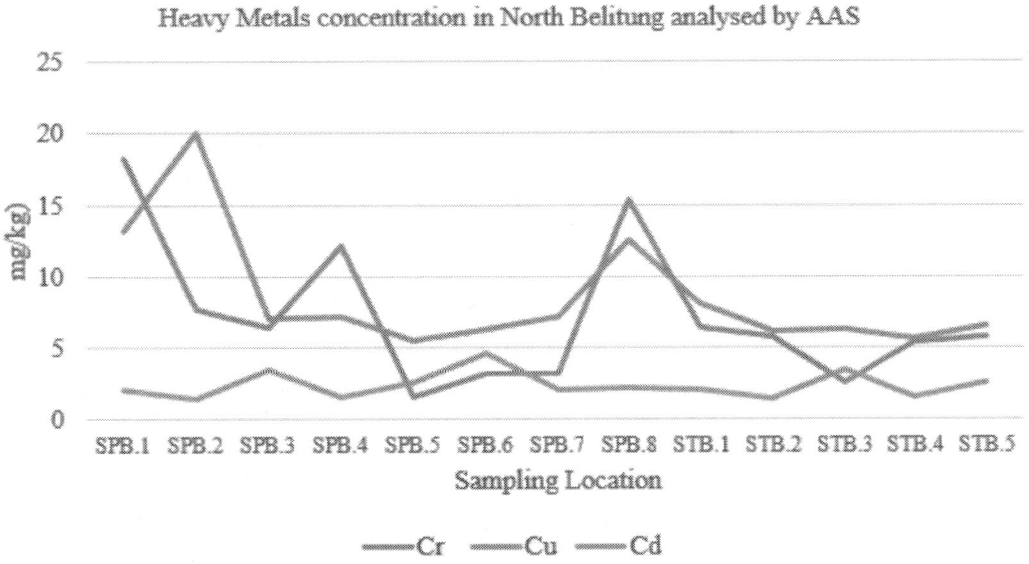

Figure 4. Cr, Cu, and Cd concentrations respectively from estuary river-coastal surface sediment in North of Belitung.

The Cd concentrations that has been analysed from surface sediment in the North Belitung is relatively low and stable in each particular sampling site. But, Cu and Cr have different patterns from Cd concentrations. They have vary concentrations in each particular sampling site. The graphic shown that the highest concentrations value for Cu is located in the offshore area (SPB.1 and SPB.2) and in the estuarine river zone (SPB.1, SPB.8, and STB.1). The concentration of Cr is also vary in each particular sampling site. Likewise concentrations of Cu, Cr also has the highest concentrations value in offshore (SPB.1) and estuarine river area (SPB.4, SPB.8, and STB.1). The highest concentrations value of Cu is in the very coarse sand and the highest concentration value for Cr is in the medium sand.

Table 2. Environmental Assessment of Heavy Metals (Cr, Cu, and Cd) from estuary-coastal surface sediment in North Belitung

Environmental Assessment	Cr	Cu	Cd
I_{geo}	0.163-1.83	0.287-1.00	0.002-0.005
I_{geo} Class	Class 1-2	Class 1-2	Class 1
Sediment Quality	Moderately Polluted	Moderately Polluted	Unpolluted-Moderately Polluted
ANZECC Classification	Low Risk	Low Risk	Low - Moderate Risk

The average I_{geo} values of heavy metals in surrounding area in North Belitung were in order of Cd<Cu<Cr. The I_{geo} class of this study area is class 1-2 for Cr and Cu, and class 1 for Cd. Based on the I_{geo} class assigned in this study, the sediment quality in this study area can be determined. Sediment quality is the ability of sediment to maintain a healthy benthic population[21]. This area is moderately polluted by Cr and Cu, and unpolluted – moderately polluted by Cd. The ANZECC classification [22] following Default Guideline Value (DGV) for toxicant sediment was also assigned in this study to know the contamination level to adverse biological effect around the study area. Based on the DGV, the Cr and Cu has the low risk contamination level to adverse biological effect, but Cd has low-moderate risk contamination level to adverse biological effect.

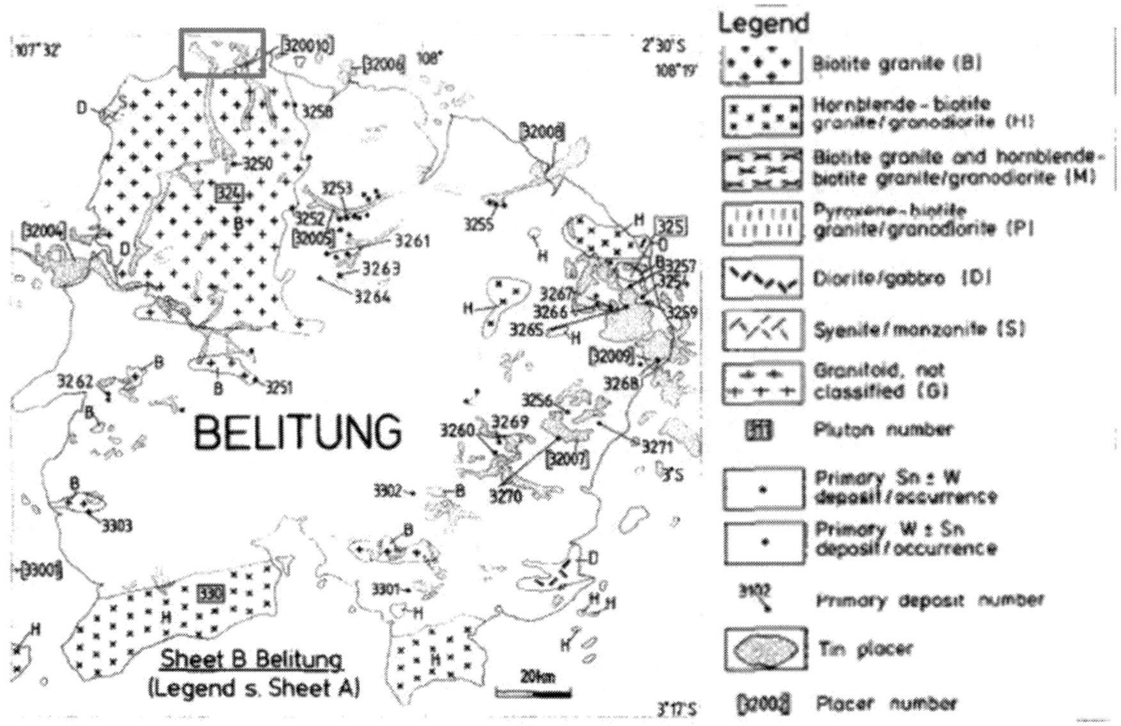

Figure 5. Distribution of tin-tungsten deposits and granitoid in Belitung Island [3], red square indicated the study area.

Tin mining activity has developed in Belitung since 1850 until today (Fig. 5) [23]. Nowadays, up to 80% of Indonesian tin exports are produced by artisanal and small-scale mining (ASM) [24], [1], [2]. ASM has a large share in Belitung's tin production and is of great importance for the economic and social development on the island [1]. However, while ASM mining increases the wealth of Belitung's population, good mining practices, safety standards and land reclamation are neglected [24], [1]. Therefore ASM has serious consequences for the environment [24], [25], [26], [27]. Tin mining in Belitung has been extensively performed that remains some tailings for the environment. The tailings generated from tin mining activity as such Pb, Cd, Zn, Cu, and Cr as well.

In the study area, the source of tin deposit was from biotite granite and tin placer deposits (Fig. 5). According to extensive distribution of tin deposits and the main economic activity in this island, the major pollution generated to environment mostly possible from the tin mining itself rather than other anthropogenic activities.

One of the estuary river in the study area, which is Tambak Udang River was generated due to tin mining dredging activity. It also can be distinguished from the river's meandering among these two rivers. The natural river will commonly have meanders in its river body, and vice versa. In this case, Penyairan River has some meanders in its body, while Tambak Udang River does not. It also enhanced the argument that in the study area, the tin mining activity was extensive.

Toxicants adsorb to fine-grain organic and inorganic materials and thus sediments play a major role in the transport of contaminants as suspended particles in the water column [21]. This finely-disseminated matter influences the feeding habits of benthic and pelagic animals and is part of the uptake process [21]. Since the study area is located near to carbonate reef environment, the sediment quality therefore has a strong influence to the biodiversity and ecological health of marine ecosystems [21] including the coral and reef ecosystems.

The sampling location is spotted around the estuary river zone, shoreline and offshore area that close to the marine environment where carbonate reef and marine aquatic lied (Fig. 2). Carbonate reef ecosystem where coral reefs live, protect coastlines from the damaging effects of wave action and tropical storms; provide habitats and shelter for many marine organisms; and are vital for fisheries and tourism. Many factors may be detrimental to coral health, but in East Belitung the main factor compromising coral health is sedimentation [28].

Mining leads to the exposure of soils to the erosive forces of rain and runoff [29], therefore, the sedimentation rate may increase due to intensive mining. Sedimentation will cause a decrease in live coral cover, growth, and recruitment [30], [31], [32], and lead to a decline in the condition of coral reefs particularly in inshore or near-shore settings. In addition, suspended sediment inhibits the sun's energy from being used for photosynthesis. Therefore, due to the intensive mining, the near-shore coral reefs of Belitung Island are under pressure from sediment runoff [33], [34].

4. Conclusions

The sediments in the North Belitung consisted mainly of fine sand and minor part of coarse sand that indicated similar feature among different sites, showing a stable hydrodynamic setting. Based on the concentrations Cu, Cd, and Cr in the surface sediments in the North Belitung, the highest heavy metal concentrations as follow Cu>Cr>Cd. The highest heavy metal concentrations is located in the offshore and estuarine river zone compare to shoreline area. Sediments quality at most of sites in this study area met I_{geo} class 1-2 for Cr and Cu, and class 1 for Cd. It suggested that the sediment quality in this surface sediment in North Belitung is moderately polluted by Cr and Cu, and unpolluted to moderately polluted by Cd. The contamination level of Cr and Cu is in low level to adverse biological effect, but for Cd it has low-moderate contamination level to adverse biological effect.

Based on the contamination level and the environmental assessment, North Belitung started to have the ecological risk due to the tin mining activity according to the development of tin mining activity in Belitung since 1850. That is why the further and more detail study in this area should be paid more attention on monitoring of heavy metals variation in the future.

References

[1] Vasters J and Franken G 2020 Zinn - Informationen zur Nachhaltigkeit. *Bundesanstalt für Geowissenschaften und Rohsoffe.* https://www.bgr.bund.de/DE/Gemeinsames/Produkte/Downloads/Informationen_Nachhaltigkeit/zinn.pdf?__blob=publicationFile&v=3

[2] Mischler G 2021 Kritische Rohstoffe. *Technik Einkauf.* https://www.technik-einkauf.de/rohstoffe/kritische-rohstoffe/rohstoff-einkauf-zinn-118.html

[3] Schwartz M O, Rajah, S S, Askury A K, Putthapiban P and Djaswadi S 1995 The Southeast Asian tin belt *Earth Sci. Rev.* **vol. 38** no. 2–4, pp. 95–293, doi: 10.1016/0012-8252(95)00004-T.

[4] Johari M 1986 Relationship between Sn mineralization and geochemical anomalies in Non-Residual overburden at Tebrong Area, Belitung, Indonesia. *Journal of Geochemical Exploration* Vol. **28**, 219-234. DOI 10.1016/0375-6742(87)90049-5

[5] Hosking K F G 1970 Aspects of the geology of the tin fields of Southeast Asia. *2nd Tech. Conf. on Tin Bangkok, Int. Tin Commission* pp. 39-79

[6] Cui L, Bai J, Shi Y, Yan S, Huang W and Tang X 2004 Heavy metals in soil contaminated by coal mining activity. *Acta Pedol. Sin.* 41 (**6**), 896–903 (In Chinese) DOI 10.11766/trxb200401070609.

[7] Li Z, Ma Z, Kuijp T J v d, Yuan Z and Huang L 2014 A review of soil heavy metal pollution from mines in China: pollution and health risk assessment. *Sci. Total Environ.* **468**, 843–853. https://doi.org/10.1016/j.scitotenv.2013.08.090.

[8] Gargouri Z, Soussi M, Souissi R, Abdeljaoued S and Turki D 2006 Diagnostic bio-sédimentaire et géochimique du littoral de la frange littorale de Nakta (Sud de Sfax, Tunisie). *Revue Méditerranéenne de l'Environnement*, **1**, 170–184.

[9] Sin S N, Chua H, Lo W and Ng L M 2001 Assessment of heavy metal cations in sediments of

Shing Mun River Hong Kong. *Environmental International,* **26**, 297–301.
[10] Santos I R, Silva-Filho E V, Schaefer C E G R, Albuquerque-Filho M R and Campos L S 2005 Heavy metal contamination in coastal sediments and soils near the Brazilian Antarctic Station, King George Island. *Marine Pollution Bulletin,* ***50***, 185–194.
[11] Szefer P, Szefer K, Glasby G P, Pempkowiak J and Kaliszan R 1996 Heavy metal pollution in surficial sediments from the southern Baltic Sea off Poland. *Journal of Environmental Science and Health, A31*(**10**), 2723–2754.
[12] Cho Y, Lee C and Choi M 1999 Geochemistry of surface sediments off the southern and western coast of Korea. *Marine Geology*, **159**, 111–129.
[13] Adams W J, Kimerle R R and Barnett J W 1992 Sediment quality and aquatic life assessment. *Environmental Science and Technology*, **26** (10), 1864–1875.
[14] Rowlatt S M and Lovell D R 1994 Lead, zinc and chromium in sediments around England and Wales. *Marine Pollution Bulletin,* ***28***(5), 324–329.
[15] Mucha A P, Vasconcelos M T S D and Bordalo A A 2003 Macro benthic community in the Douro Estuary: Relations with trace metals and natural sediment characteristics. *Environmental Pollution,* ***121,*** 169–180.
[16] Hadi T A, Giyanto, Prayudha B, Hafizt M, Budiyanto A and Suharsono 2019 Status Terumbu Karang Indonesia 2018 (In Bahasa Indonesia) Jakarta: Puslit Oseanografi-LIPI p 25.
[17] Beyer H L, et al 2018 Risk-sensitive planning for conserving coral reefs under rapid climate change. *Conservation Letters*, **11**, e12587. Doi: 10.1111/conl.12587.
[18] Özkan M A, Tunçel G, Tugrul S, Ramelow G and Balka T I 1980 Extraction of heavy metals from marine sediments for Analysis by atomic absorption spectrometry - some factors affecting extraction efficiency. *Analytical techniques in enviromental chemistry. Proc. International Congress*, Barcelona, Spain, p. 615-619
[19] Müller G 1969 Index of geoaccumulation in sediments of the Rhine River. *Geojournal* **2**, 108–118.
[20] Bowen H J M 1979 Environmental chemistry of the elements *Academic Press* London
[21] Birch G F 2017 Determination of sediment metal background concentrations and enrichment in marine environments – A critical review, *Sci. Total Environ.*, **580**, pp. 813–831, doi: 10.1016/J.SCITOTENV.2016.12.028.
[22] ANZECC/ARMCANZ 2000 Australian and New Zealand Guidelines for fresh and marine water quality. Volume **1**, Section 3.5 – Sediment Quality Guidelines. *Australian and New Zealand Environment and Conservation Council, and Agriculture and Resource Management Council of Australia and New Zealand* Canberra, Australia.
[23] Heidhues M F S 1991 Company Island: A note on the histoy of Belitung. Indonesia , No. **51**, pp. 1-20, Cornell University.
[24] Nurtjahya, Franklin E J, Umroh and Agustina F 2017 The Impact of tin mining in Bangka Belitung and its reclamation studies *MATEC* **101** 04010 DOI: 10.1051/matecconf/201710104010.
[25] Hermon D, Iskarni P, Oktorie O and Wilis R 2017 The model of land cover change into settlement area, tin mining, and its affecting factors in Belitung island, Indonesia *J. Environ. Earth Sci.* 7 (**6**): 32–39.
[26] Sari S P and Rosalina D 2016 Mapping and monitoring of mangrove density changes on tin mining area. *Procedia Environmental Sciences* **33** 436–442.
[27] Erwana F, Dewi K and Rahardyan B 2015 Study of socio-economic and environment impacts of inconventional tin mining (a case study: Bangka Barat District of Bangka Belitung Province) *ESDP* **3** 147–160.
[28] Johan O, Ginanjar R and Priyadi A 2015 Coral health levels of wild ornamental coral in East Belitung waters, Indonesia. *Nusantara Bioscience* **7**, No. 2, pp. 127-132. DOI: 10.13057/nusbiosci/n070212.
[29] Fallon S J, White J C and McCulloch M T 2002 Porites corals as recorders of mining and

environmental impacts: Misima Island, Papua New Guinea. *Geochim. Cosmochim. Acta* **66**, 45-62.

[30] Idris, Putri A R, Adiwijaya C, Gilang M, Santoso P, Prabowo B, Muhammad F, Lestariningsih W A, Lestari D F, Setyaningsih W A and Zamani N P 2020 Assessment of coral reefs health in Nature Recreation Park (TWA=Taman Wisata Alam) Sangiang Island, Banten. *IOP Conf. Ser. Earth and Environ Sci.* **429** 012020.

[31] Segal B and Castro B 2011 Coral community structure and sedimentation at different distances from the coast of the Abrolhos Bank, Brazil. *Brazilian. Journal of Oceanography* 59 (**2**): 119–129 DOI: 10.1590/S1679-87592011000200001.

[32] Birrell C, Laurence J, McCook L J and Bette L W 2005 Effects of algal turfs and sediment on coral settlement. *Marine Pollution Bulletin* 51 (**1-4**): 408–414. DOI: 10.1016/j.marpolbul.2004.10.022.

[33] Siringoringo R M and Hadi T A 2013 Condition and distribution of coral reefs (Scleractinia corals) in Bangka Waters. *Jurnal Ilmu Dan Teknologi Kelautan Tropis* 5(**2**):273-286.

[34] Putra S A, Syari I A, Akbar H, Suyatna I and Setiamarga D H E 2018 Human activities and persistent coral reef degradation in Gaspar Strait, Bangka Belitung Islands, Indonesia. *F1000 Research* **7**:1962.

Acknowledgement

This research was supported by the research collaboration between Kiel University and BRIN. We would like to acknowledge the laboratory of Advance Characterization, BRIN-KST Samaun Samadikun, Google Earth Pro, and Avenza Maps Apps that support our fieldwork. Salsabila Fathona, student of ITB and our research assistant ITB-BRIN, Ahmad Al Fauzan for the assistance during the sample preparation.

Cianjur Earthquake- Roles of Social Media and the Distribution of Sciences

A Rudyawan[1]* , A Pamumpuni[1], M R Puspaningrum[1], I Gunawan[1], B Sapiie[1], A P Handayani[1], M Abdurrachman[1]

[1] **Faculty of Earth Sciences and Tecnhology, Institut Teknologi Bandung, Jalan Ganesha No 10 Bandung.**
*alfend@geodin.net; alfend@itb.ac.id

Abstract. A magnitude 5.2 earthquake hit Cianjur Regency, West Java, Indonesia on November 21, 2022, killing more than 300 people, leaving many more missing, and damaging more than 60,000 houses which caused thousands of people to be displaced from their homes. Significant seismic activity was recognized in 1699, 1834, 1879, and 1900 near Cianjur in West Java. The news quickly hit social media, with some responding in group chats, feeds, reels, and literature reviews shared via other social media claiming that the Cimandiri Fault had caused the event. Moreover, discussions were featured in webinars and on national television. However, this information was based solely on limited information and prior knowledge of the area without further analysis of the actual data.

New satellite data was uploaded via social media a few days later. This update soon reduced online discussion about the Cimandiri Fault. There was further evidence of a previously buried fault between the Cimandiri Fault and the Gede Pangrango Range, 12 km north of the estimated location of the Cimandiri Fault. Our field observations indicate that this fault caused most of the rupture in this region.

Socially, the earthquake event also caused immediate problems in the local community. There have been incidents of earthquake survivors, donors, and people taking advantage of the situation. Blockades were everywhere, and those claiming to be survivors demanded that oncoming donation fleets be stopped, and all donations abandoned. These incidents have hampered efforts to distribute aid to the most remote areas. As a result, survivors who had taken refuge far from major logistical lines were left stranded. Heartless selfies and video content created in the face of grieving communities are everywhere and create a rift between survivors and donors.

The recent Cianjur earthquake has taught us how important it is to understand the geological phenomena behind our homes and the effect of information and misinformation via social media. While social media serves as a valuable tool for rapid information dissemination during disasters, verifying information from reliable sources and encouraging responsible usage of social media platforms is essential. Effective filtering of information becomes pivotal to combat misinformation in developing more resilient communities, safeguarding lives, and minimising the impact of the disaster.

Keywords: Cianjur Earthquake, Science Communications, Social Media

1. Introduction

Indonesia is a country located in Southeast Asia, comprising thousands of islands that are home to diverse cultures and languages. It is also a country that is prone to natural disasters, including earthquakes, volcanic eruptions, and tsunamis. On November 21, 2022, a magnitude 5.2 earthquake hit Cianjur Regency, West Java, Indonesia (Figure 1), causing significant damage and loss of life.

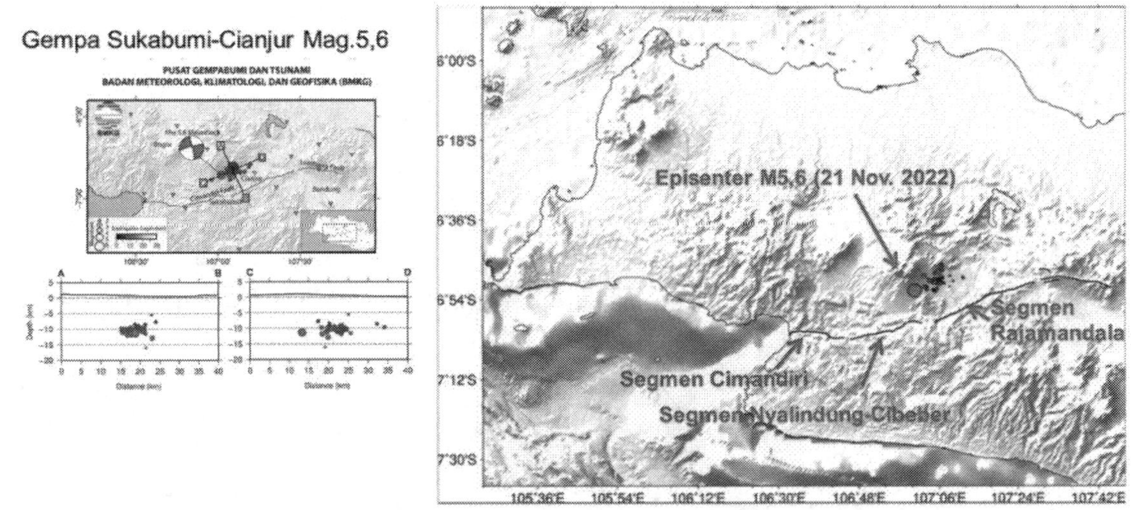

Figure 1. The location of Cianjur Earthquake epicentre (BMKG)

After the 2004 Indian Ocean earthquake and tsunami, the 2018 Lombok earthquakes, and the 2018 Palu earthquake and tsunami, there were reports of aid distribution problems, accusations of corruption and favoritism, and confusion caused by government officials' statements. To address this issue, the Indonesian National Board for Disaster Management (BNPB) has developed guidelines for communicating during disasters, emphasizing the importance of accurate and timely information, clear messaging, and a focus on the needs and concerns of affected communities [1]. The BNPB has also established a network of communication centers across the country staffed by trained personnel to provide information and support to those affected by disasters.

2. Cianjur Earthquake 2022

Historically, Indonesia has been susceptible to earthquakes due to its location on the Pacific Ring of Fire, a region characterized by intense seismic and volcanic activity. In the late 17th century, earthquakes in the region began to be catalogued, including significant seismic activity recognized in 1699, 1834, 1879, and 1900 near Cianjur in West Java. Over time, scientists and geologists have continued to study the geological phenomena of the region to understand the cause and impact of earthquakes in the area.

On November 21, 2022, a magnitude 5.2 earthquake hit Cianjur Regency, West Java, Indonesia, killing more than 300 people, leaving many more missing, and damaging more than 60,000 houses. The earthquake's epicentre was about 12 km north of the estimated location of the Cimandiri Fault, a previously identified fault initially thought to have caused the quake. However, new satellite data was uploaded via social media a few days later, indicating a previously buried fault between the Cimandiri Fault and the Gede Pangrango Range. Field observations suggest that this fault caused most of the rupture in the region. The discovery of the previously buried fault highlights the importance of continually monitoring geological activity in seismically active regions. With the advent of new technologies, such as satellite data, we can gather more accurate data and make more informed decisions about the potential risks posed by geological phenomena.

The immediate aftermath of the Cianjur earthquake was marred by incidents of earthquake survivors, donors, and people taking advantage of the situation, causing blockades and hampering aid distribution to the most remote areas [2]. Having failed and knowing they were being recorded, they forced the fleet to turn back, and people in the most remote area were left stranded for days, preventing others from accessing essential supplies. There were traffic jams everywhere as everyone wanted to make their donations. Additionally, people posted so much content on social media very quickly, including heartless selfies and video content created in the face of grieving communities (Figure 2). The lack of trust between survivors and donors led to a rift between the two groups, and social media played a significant role in exacerbating this division. Even after the government set up a disaster relief base, the problem continued. Logistics were in excess, but they needed to reach the most remote communities.

Figure 2. Response from survivors to the social media content

2.1. Buried faults, surface deformation and relocated homes

After the news spread, several analyses claimed that the Cimandiri fault must have caused the quake. However, the epicenter was about 12 km north of the estimated location of the Cimandiri Fault. There was a need to reassess the position of the fault that caused the earthquake since it is a significant distance from the Cimandiri Fault. Before more analyses were carried out, the news had already hit social media, with some responding in group chats, feeds, reels, and literature reviews shared via other social media. Moreover, others were featured in webinars and on national television. However, this information was based solely on limited information and prior knowledge of the area without further analysis of the actual data [3]. This is consistent with the cognitive needs of communities when such as information seeking, the will to get timely and unfiltered information and to estimate disaster magnitude [4].

New satellite data was uploaded via social media a few days later. This update soon reduced online discussion about the Cimandiri Fault. There was further evidence of a previously buried fault between the Cimandiri Fault and the Gede Pangrango Mountain [5]. Our field observations (Figure 3) indicate that this previously unknown fault may have caused the most surface deformation in the area.

Figure 3. Field observation of surface deformation caused by mass movement (left), and potential ruptures (centre and right).

New observations from the heavily affected area suggest that most damages resulted from movements on uncompacted sediments sitting on the inclined topography of the mountain foothill triggered by the earthquake. Evidence of surface ruptures is limited in certain areas and does not manifest in significant surface deformation aside from mass movements. However, reports published online were based on limited observation stating that an 8-km-long active fault, namely Cugenang Fault, cross-cut the area. This has sparked more discussion as 1.800 homes and their inhabitants were suggested to be relocated [6] (Figure 4). Some other scientists from different institutions suggest that it is too early to determine which fault is responsible for earthquake since there was only limited evidence that suggests surface deformation [7].

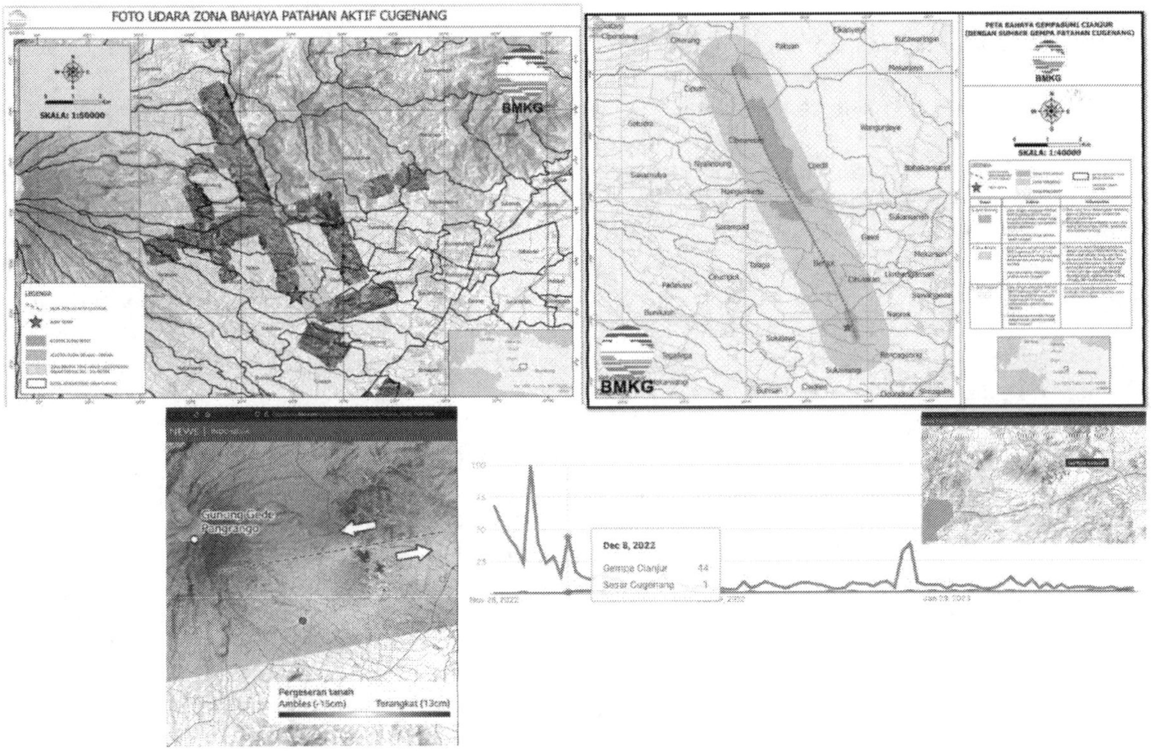

Figure 4. Maps and information in electronic and social media published one after the other with conflicting different interpretations. Information from Google Trends (lower right) shows sparks of interest and conversation triggered at the event of the earthquake and on 8/12/22 when the Cugenang Fault was introduced.

3. Science Communication and Distribution via Social Media

In each of these cases, effective communication and coordination between survivors and aid organisations could have helped to prevent or mitigate tensions and conflicts. Aid organisations must involve local communities in the distribution of aid and ensure that resources are distributed fairly and transparently. This can help to build trust and reduce tension between survivors and donors, ultimately improving the overall response to disasters. It is important to note that effective communication and coordination can be critical in ensuring that aid reaches those who need it most. In the aftermath of a disaster, there is often chaos and confusion, making it difficult for aid organisations to distribute resources fairly and equitably.

Science communication can play a role in addressing these challenges by providing accurate and timely information to survivors and aid organisations about the nature and extent of the disaster, as well as the specific needs of affected communities. This helps prevent misunderstandings and ensures that resources are distributed in a way that is responsive to the needs of survivors. Collaboration of all parties connected by the disaster event should come above all. Scientific communities, governments, NGOs, and volunteers need to put aside their sectoral egos and work together to build comprehensive, elaborate, and testable public information that holds a strong scientific prowess and social sensitivity. Effective aid distribution also relies on clear communication and coordination among all parties, including local communities. This requires a deep understanding of the local context, including the specific needs and vulnerabilities of different communities and the logistical challenges involved in delivering aid to remote or hard-to-reach areas. Scientific discussion often shows differences, and that is normal. However, when it comes to public policy, there must be some form of unity in information and the action plan synchronising resources from all parties.

By leveraging scientific knowledge and expertise, all parties can develop more effective strategies for distributing aid, such as using data and mapping tools to identify areas of greatest need and developing innovative technologies for delivering aid to remote or inaccessible areas. Unnecessary traffic can be avoided and helps can reach target in time. It is also important to involve local communities in the distribution of aid and to ensure that resources are distributed fairly and transparently. This requires ongoing engagement and dialogue with survivors, their families, community leaders, and other stakeholders. Effective science communication and aid distribution are critical components of a comprehensive disaster response strategy. By working together and leveraging the latest tools and technologies, we can improve the response to disasters and help communities become more resilient in the face of future challenges.

To measure community resilience by looking at how it communicates during hazard event, A multi-faceted approach is necessary to promote community resilience and preparedness [8]. First, educating the communities on how to prepare for and respond to natural disasters is essential. Disaster training and education should be widely implemented in communities to make them more resilient. It should include accurate information on the types of natural disasters that may occur in the area, how to prepare for them, and what to do in an emergency. Such training can be delivered through community workshops, public education campaigns, and school programs.

Second, a concerted effort should be made to build more resilient infrastructure in seismically active regions that would benefit from better disaster response plans. It should consider the needs of the community's most vulnerable members, such as the elderly, disabled, and low-income individuals. The plans should also be designed to be flexible and adaptable to changing circumstances. It includes having a clear chain of command, identifying roles and responsibilities, and establishing communication channels between agencies and organizations involved in the response effort.

Third, targeting the right audience seems to be a good starting strategy. When compared to the individual, psychological, or mass, societal levels, communities are rarely seen as the ideal level at which to focus science communication efforts. However, evidence from other fields suggests that developing interpersonal relationships with specific communities over time is essential for meaningful engagement, therefore orienting science communication toward communities is warranted [9].

In addition to these measures, effective science communication is crucial for promoting community resilience. Scientists and researchers should work closely with government agencies and NGOs to communicate their findings in a way that is accessible and understandable to the general public. It includes using plain language, visual aids, and other tools to help people understand the risks associated with natural disasters and what they can do to prepare for them. By promoting science literacy and public understanding of natural hazards, communities can become better equipped to respond to disasters and more resilient in the face of future challenges.

This is aligned with the priorities that need to be addressed as identified in the Sendai Framework for Disaster Risk Reduction, Sustainable Development Goals and the Paris Climate Agreement. Understanding Disaster Risk Analysing, collecting, and managing disaster data (post-disaster and potential losses and probabilities) is essential for achieving a comprehensive understanding of disasters. Strengthening disaster risk governance to manage disaster risk in recognition of the need to manage disaster risks and respond from a holistic perspective, all parties need to contribute to the establishment of a robust mechanism for disaster risk management, supported by a comprehensive policy framework. Investing in disaster risk reduction for resilience and robust funding mechanisms for all disaster and climate risk reduction activities is elemental for facilitating development as envisaged in any given policy in order to arrive in enhancing disaster preparedness for effective response to "Build Back Better" in recovery, rehabilitation and reconstruction (Disaster Risk Reduction in Indonesia; Status Report 2020; [10].

Social media certainly holds important roles in creating the environment of a resilient community. Its ability to convey messages rapidly allows social media to become the main source of information in real-time. However, lack of accuracy can be its Achilles heels, as unfiltered news can lead to further chaos. In social media, information is the currency. Everyone can be the source of information as social media enables individuals to become citizen journalists who can share photographs, videos, and eyewitness accounts from areas where no formal media have information from or where access was limited. This can give a broader perspective and a more realistic situation on the ground so the first responders and volunteers can act based on priority. On the other hand, people who were affected by the disaster can also use social media to ask for support and assistance. However, there were incidents where people allegedly used social media to track for donation fleet, intercept the routes by setting up blockades, and subsequently cut off the donation line [2]. This shows the improper use of social media.

Following the disaster event, the roles of social media is even greater. Generating situational awareness on the ground to identify areas of concern, assess the magnitude of events, and monitor the dynamics of the community are the most important roles of social media as they will give valuable insight into the needs of the community and people can react accordingly. Often after a disaster occurs, information is everywhere and tends to spiral out of control. It is difficult to determine whether the information is real and scientific, or bogus and click-baited. Users can filter the information by double-checking the facts or at least trying to find official information released by the official institution. Governmental institutions can use social media to promote accurate, updated, official information to minimize and, whenever possible, dispel the rumors. This can alleviate further anxiety and fear in the community and will certainly help the rebuilding process where social media will be a reliable source for information in mobilizing resources, volunteering opportunities, and fostering a sense of solidarity.

4. Conclusions

In conclusion, the recent earthquake event serves as a stark reminder of the importance of understanding geological phenomena and implementing effective strategies to build resilient communities. Communication plays a vital role in disaster situations, as it can either foster trust or create tension between survivors and aid organizations. It is crucial to prioritize education and disaster training programs to enhance community preparedness and response capabilities.

The incidents of hoarding and blockades witnessed during the aftermath of the earthquake underscore the need for improved disaster preparedness and education. Governments, supported by all parties involved, must prioritise educating communities on how to effectively prepare for and respond to natural disasters. Low-tech solutions, such as disaster training and education initiatives, can play a significant role in equipping communities with the necessary knowledge and skills to mitigate the impact of future catastrophes.

Accurate scientific information is paramount for effective communication during disaster management [11]. Disasters have profound consequences on people's lives, health, and well-being, making reliable and up-to-date scientific information crucial for informed decision-making. By relying on credible sources and ensuring clear and timely communication, we can dispel misinformation and rumours that can lead to panic and further harm.

While social media serves as a valuable tool for rapid information dissemination during disasters, caution must be exercised due to the potential for misinformation. It is essential to verify information from reliable sources and encourage responsible usage of social media platforms. Effective filtering of information becomes crucial to combat misinformation and maintain clarity. Official institutions can leverage social media to provide accurate updates that can help alleviate anxiety and support the rebuilding process effectively and efficiently.

By harnessing the power of social media responsibly, we can foster solidarity and resilience within communities affected by disasters. By prioritising education, promoting accurate information, and leveraging technology responsibly, we can enhance our preparedness, response, and recovery efforts, ultimately building more resilient communities, safeguarding lives and minimising the impact of disaster.

5. References

1. https://bnpb.go.id/storage/app/media/uploads/migration/pubs/451.pdf (accessed in 20/02/22)
2. https://www.merdeka.com/jateng/5-fakta-warga-cianjur-stop-mobil-bantuan-gempa-kini-minta-maaf.html (published on 24/11/22)
3. https://kumparan.com/kumparansains/beda-kata-ahli-soal-penyebab-gempa-cianjur-bukan-sesar-cimandiri-1zIqczm6rTV/4 published on 22/11/22
4. Cerase, A. (2018). Re-Assessing the Role of Communication in the Aftermath of A Disaster: Case Studies And Lesson Learned in Antronico dan Marincioni. Natural Hazards and Disaster Risk Reduction Policies. Il Sileno Ed. Geographies of The Antropocene., 1(2), 213–242.
5. https://www.bbc.com/indonesia/extra/mj3nzjl1cx/earthquake_visual_indonesia published on 6/12/22
6. https://jabar.antaranews.com/berita/422479/bmkg-sarankan-1800-rumah-direlokasi-dari-patahan-cugenang-cianjur published on 8/12/22
7. https://www.detik.com/jabar/berita/d-6465087/pvmbg-sebut-sesar-cugenang-perlu-diteliti-lebih-detail published 16/12/22

8. Alexander, DE; (2015) Disaster and Emergency Planning for Preparedness, Response, and Recovery. In: **Oxford Research Encyclopedia of Natural Hazard Science.** (pp. 1-20). Oxford University Press: Oxford.
9. Orthia, Lindy & McKinnon, Merryn & Viaña, John Noel & Walker, Graham. (2021). Reorienting science communication towards communities. Journal of Science Communication. 20. A12. 10.22323/2.20030212.
10. https://www.undrr.org/media/48530/download accessed on 20/02/22
11. Khumairoh, Z., Widana, I.D.K., and Sumantri, S.H., 2021, The role of communication as the disaster risk reduction in Indonesia capital city transference policy, IOP Conference Series: Earth And Environmental Science, Vol. 708, **DOI** 10.1088/1755-1315/708/1/012101

Acknowledgements
The authors wish to acknowledge assistance from Geodynamics Research Group – ITB, The Faculty of Earth Sciences and Technology – ITB, IAGI, and LPPM ITB for their helps in collaborating resources for the rupture mapping studies.

Author roles and responsibility

The following paragraph is to explain the roles and responsibilities of the authors. A Rudyawan is responsible for conceptualisation, acquiring field data, and putting together the early manuscript and discussion. A. Pamumpuni, M. R. Puspaningrum, and A. P. Handayani are responsible for fieldwork and organising surveys and analyses. I. Gunawan, B. Sapiie and M. Abdurrachman are responsible for evaluating the manuscript, interpretation and discussion.

Utilization of By-Products and Waste GGBFS Materials from Steel Processing Industry to Produce Sustainable Green Pervious Concrete in Indonesia

Patria Kusumaningrum[1]*, Muhammad Risnan[2]

[1,2] Civil Engineering Department, Faculty of Civil and Environmental Engineering, The Institute of Technology Bandung, Bandung 40132, Indonesia

*Corresponding author: patria.k@itb.ac.id

Abstract. Pervious concrete is a reliable solution for environmentally friendly concrete pavements, especially for densely populated urban areas. Some of the benefits of pervious concrete to the environment are mainly increasing the absorption of water into the soil, so that it can quickly purify rainwater runoff, recharge groundwater, reduce rainwater runoff, and when combined with well-designed drainage system it can minimize the potential of flooding. The uniqueness of this concrete is that it uses few or no fine aggregate at all, with the main composition consisting of coarse aggregate, binder (cementitious/pozzolanic materials), water and admixture. The development of pervious concrete with steel slag as raw material as coarse aggregate and Ground Granulated Blast Furnace Slag (GGBFS) as a partial replacement for cement is carried out to address environmental problems caused by impervious pavement that is commonly used in urban areas, as well as to increase the added value of material by-products of the steel processing industry. The aim of this research is to obtain a pervious concrete mixture with an optimum compressive strength and porosity utilizing locally available by-products and waste materials from steel processing industry to produce a more sustainable construction material. Evaluation based on the experimental results shows that the specimen with well graded or blended-size of steel slag coarse aggregates, with a maximum of 10% steel slag fine aggregate, and GGBFS as 32% cement substitute material (CSM) is found to be an optimized mixture in achieving high compressive strength of 20.24 MPa at which both density and porosity values complying ACI 522R-10.

Keywords. green pervious concrete, low cost, low carbon, high strength, density, void content

1. Introduction

Pervious concrete is a reliable solution for environmentally friendly concrete pavements, especially for densely populated urban areas. The benefits of pervious concrete for the environment are permeable at which it increases the absorption of water into the soil. By permeable layer of pervious concrete in the pavement, it may quickly purify rainwater runoff, replenish groundwater, reduce rainwater runoff [1]. When such pervious concrete pavement is combined with a properly designed drainage system, it may further minimize the potential for flooding. The pervious concrete may also be advantageous in terms of sound absorption and noise reduction of a busy highway [2].

Pervious concrete has its own uniqueness. First, it uses only a small portion or even without fine aggregate. Its main composition consists of coarse aggregate, binder (cementitious/pozzolanic materials), water and admixture. In this study, with the aim to provide society with a more sustainable construction material, as well as to increase the added value of material by-products of the steel processing industry, the development of pervious concrete with steel slag as raw material as coarse aggregate and GGBFS as a partial replacement for cement is carried out.

Steel slag and GGBFS, by-products of steel production, have been considered as industrial wastes due to the challenges associated with its accumulation and disposal. In Indonesia, there are at least three prominent steel industries, namely PT Krakatau Steel, PT Krakatau Posco in Cilegon, Banten, and PT DSI in Morowali, Central Sulawesi, that produce steel slag and GGBFS through the smelting of iron ore into pig iron using blast furnaces. Extensive research has been conducted on the potential utilization of steel slag as a substitute for natural aggregates [3]. However, the exploration and utilization of steel slag as a construction material especially in the surroundings of these steel industries in Indonesia, particularly in the context of pervious pavement, have not been widely explored and implemented.

ACI 522R-10 guideline regulated minimum compressive strength' requirement for the pervious compressive strength of concrete ranging from 2.8 to 28 MPa with a range of void content from 15% to 35% as an effective permeability measure of the pervious concrete. The properties of pervious concrete are primarily dependent on its porosity (void content), cementitious content, ratio of water to binder (w/cm), compaction level, and aggregate gradation and quality [4]. The relatively high compressive strengths of pervious concrete mixtures are possible to achieve. However, such a high strength of pervious concrete is achieved only by the reduction of void content. Therefore, in order to obtain an optimum mixture of pervious concrete, efforts need to be made in balancing the compressive strength and porosity in regard to void content. Previous study had shown that a minimum porosity of approximately 15% was required to achieve significant percolation. To generate larger-sized pores in the material, larger aggregate sizes are recommended [4] as they may reduce chances of pore-clogging [5].

2. Experimental Work

2.1 Material

The main ingredient of this research was steel slag obtained from PT. Krakatau Steel, Cilegon that was used as coarse aggregate as a replacement of natural aggregate. The steel slag that has been previously separated from the metal impurities and shown in Figure 1(a). Variation of sizes of the steel slag aggregate (SSA) studied herein may either be uniformly single-size of either 4.75-9.5 mm or 9.5-19 mm, or blended-size coarse aggregate system with both sizes, with or without the addition of steel slag fine aggregate (0.15-2.36 mm). In this study, a blended size of aggregates was carried out, as replacing smaller-sized aggregates with a higher percentage of larger-sized ones increases the pore size. This is mainly due to the coarser particle may not be able to fit in the void left by the removed finer particle [6, 7]. The main components of steel slag are iron, magnesium, aluminium, calcium, and other oxides [2]. The content of oxides and elements of SSA in this study are shown in Table 1.

The second material was Ordinary Portland Cement (OPC) Type 1, used as the main binder in the pervious concrete mix. OPC obtained from PT. Semen Jakarta as shown in Figure 1(b). The oxide content and elements of the OPC cement used are shown in Table 1.

The third raw material incorporated into the mix was Ground Granulated Blast Furnace Slag (GGBFS) from PT. Krakatau Steel, Cilegon, Indonesia. GGBFS as a by-product of iron production in blast-furnace which mainly consists of silicate, aluminosilicate and calcium oxide of melted calcium (Table 1) which possess both pozzolan and cementitious properties, thus can work as cement substitute material (CSM). Its chemical compositions depend on the raw materials used in the production of iron while the physical characteristics of Granulated Blast Furnace Slag (GBFS) depend on the cooling process of the molten iron. To work as cementitious material, GBFS material needs to be grind until it reached a fineness the size of cement grains as shown in Figure 1(c). Fineness of GGBFS cement is considered an important parameter affecting the strength development of pervious concrete. However, the percentage of use of GGBFS needs to be limited to less than 50% OPC replacement to ensure the effectiveness of the resulting compressive strength of pervious concrete.

(a) (b) (c)

Figure 1: (a) Steel Slag, (b) Semen OPC, (c) GGBFS

Table 1. Oxide Content of SSA, OPC and GGBFS

Oxide	Steel Slag PT. Krakatau Steel	OPC Cement PT. Semen Jakarta	GGBFS PT. Krakatau Steel
		% mass	
CaO	29.90	55.40	29.20
SiO2	18.80	20.40	35.30
Al2O3	11.90	6.40	22.30
SO3	0.76	2.44	1.76
Fe2O3	28.10	9.84	1.23
MgO	6.23	1.84	8.22
K2O	0.21	0.47	0.59
SrO	0.04	0.28	0.06
TiO2	0.76	0.46	0.52
Cl	-	0.62	-
P2O5	0.94	0.80	0.57
MnO	1.38	0.73	0.28

IOP Conf. Series: Earth and Environmental Science 1245 (2023) 012017 doi:10.1088/1755-1315/1245/1/012017

Other materials that were considered key ingredients were water and stabilizer. The ratio of w/c in this study was limited to 0.28, corresponding to the range of 0.26-0.4 recommended by [4]. An excessive amount of cement paste content, and/or fine aggregate may result in a 'clogged' void of the concrete and, consequently, the porosity is expected to be reduced.

All the raw materials were then identified with regards to the physical characteristics, i.e. specific gravity, water absorption, density, air content, water content, fineness modulus, abrasion and soundness, content of friable particles and organic material, as well as physical gradation. These data are necessary to decide and proportionate the mixture of pervious concrete.

2.2 Mix Proportion

In this study, three groups of specimens as given in Table 2 are varied by the aggregate grading, of single and blended coarse aggregate effects, use of steel slag fine aggregate and increased percentage of GGBFS as a partial replacement for cement. The proportion of pervious concrete mixture is shown in Table 3.

Table 2. Specimen Nomenclature

Specimen Variations	Steel Slag Aggregates (SSA)			Binder	
	Coarse Aggregate (SSCA)		Fine Aggregate (SSFA)	OPC	GGBFS
	4.75 - 9.5 mm	9.5 - 19 mm			
SSPC I	100%	-	-	100%	0%
SSPC II	-	100%	0%	100%	0%
SSPC III	50%	50%	0%	100%	0%
SSPC IV	45%	45%	10%	100%	0%
SSPC V	45%	45%	10%	16%	84%
SSPC VI	45%	45%	10%	68%	32%

Table 3. Mix Proportion per m^3

Specimen Variation	SSCA 4.75-9.5 mm	SSCA 9.5-19 mm	SSFA	OPC	GGBFS	w/c	Stabilizer
SSPC I	4.50	-	-	1.00	-	0.28	0.33%
SSPC II	-	4.50	-	1.00	-	0.28	0.33%
SSPC III	2.25	2.25	-	1.00	-	0.28	0.33%
SSPC IV	2.07	2.07	0.46	1.00	-	0.28	0.33%
SSPC V	2.07	2.07	0.46	1.00	0.56	0.28	0.33%
SSPC VI	2.07	2.07	0.46	1.00	1.13	0.28	0.33%

Steel slag coarse aggregate (SSA), Steel Slag Fine Aggregate (SSFA), Ordinary Portland Cement (OPC), Ground Granulated Blast Furnace Slag (GGBFS), water to cement ratio (w/c)

2.3 Mixing, samples, and testing procedures

To obtain a consistent mix, dry mix method was first carried out to blend the aggregates with the binder. Water and stabilizer that had been mixed with were then poured into the aggregate and binder mixture gradually until homogenously mixed. This method is carried out to prevent the accumulation of paste in certain parts only and to ensure the paste smears aggregates thoroughly. Two indications for right mixture are the slump value of fresh pervious concrete that is close to zero and the mixture can be made as lump as shown in Figure 2. These indicate that the pervious concrete mixture contains sufficient water (neither dry nor too wet) and is homogenous. Homogeneity greatly affects the formation of pores and the rate percolation of the hardened pervious concrete.

Cylinder specimens of 100 mm diameter and 200 mm height were casted. To achieve consistent compaction on all specimens, compaction was carried out using a Standard Proctor Hammer. ACI 522R-10 [4] stated that the compressive strength of pervious concrete is strongly influenced by the mix proportion and compaction method during concrete casting.

After casting, the specimens were then cured by air curing of 20 ± 2°C and 95 ± 5% relative humidity for a day, demoulded and water cured to maintain the hydration process. The specimens were tested for its compressive strength conforming ASTM C39 requirements, and for density as well as porosity according to ASTM C1754 requirements.

Figure 2. Mixing of Pervious Concrete Materials

3. Results and Discussion

The experimental results are summarized in Table 4 as follows. Variation of specimens following Tables 2 and 3 gives rise in compressive strength variation between 7.70 to 20.24 MPa, density between 2191.44 to 2540.42 kg/m^3, and porosity between 18.37 to 31.18%.

Table 4. Summary of Experimental Results

Specimen	Compressive Strength (MPa)	Density (kg/m^3)	Porosity (%)
SSPC I	11.28	2191.44	31.18
SSPC II	7.70	2219.32	32.14
SSPC III	14.43	2317.94	26.90
SSPC IV	14.66	2353.42	25.07
SSPC V	19.89	2353.38	25.16
SSPC VI	20.24	2540.42	18.37

3.1 Effect of aggregate grading

Compressive strength of pervious concrete is greatly influenced by aggregate grading as illustrate in Figure 3. Well graded coarse aggregate in a form of blended sizes of both 4.75-9.5 mm and 9.5-19 mm enhances the compressive strength of the specimen by 27.92% to 87.40%. Similarly, Figure 4 shows well graded coarse aggregate slightly increases the density of specimen by 4.44% to 5.77%. Consequently, it hence decreases the porosity 13.73% to 16.30%. This is in accordance to ACI 522R-10 [4].

The replacement of coarse aggregate by 10% fine aggregate to the well graded coarse aggregate mixture (SSPC IV) results in slightly higher compressive strength and density compared to SSPC III. Small amount of fine aggregate in the mixture maintains the porosity of the pervious concrete specimen.

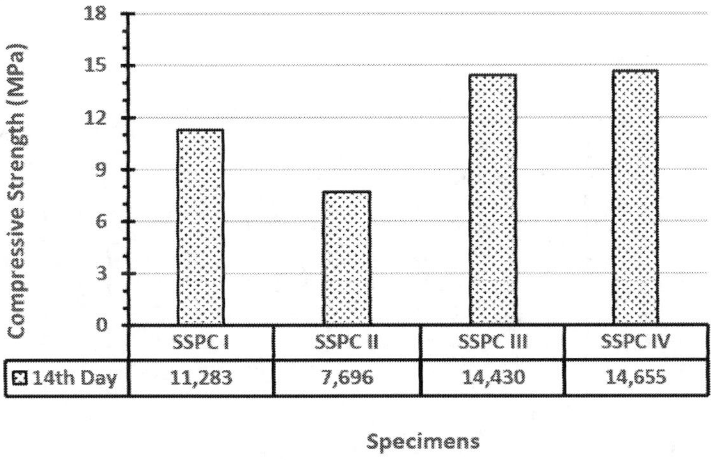

Figure 3. Effect of the aggregate grading on the compressive strength of pervious concrete

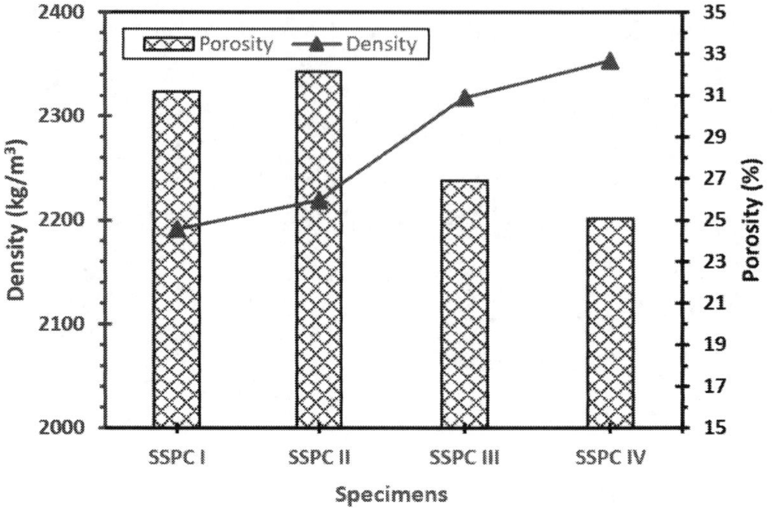

Figure 4. Effect of the aggregate grading on density and porosity in pervious concrete

3.2 Effect of GGBFS as CSM

GGBFS is incorporated into the mixture as cement substitute material as given in Tables 2 and 3. Figure 5 illustrates the increase of compressive strength as percentage of GGBFS as CSM in the mix increases. Similarly, the density of pervious concrete increases as the percentage of GGBFS as CSM is higher as illustrated in Figure 6. Conversely does the porosity, as the percentage of GGBFS as CSM increases, the porosity is reduced respectively.

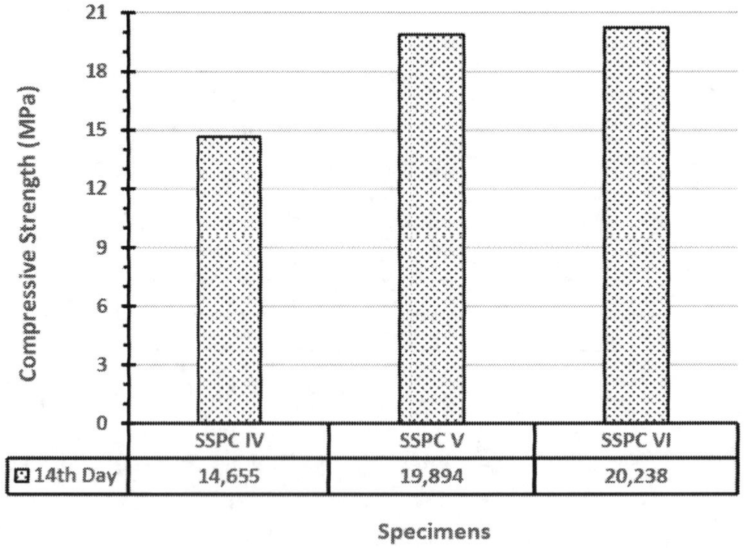

Figure 5. The effect of GGBFS as CSM on the compressive strength of pervious concrete

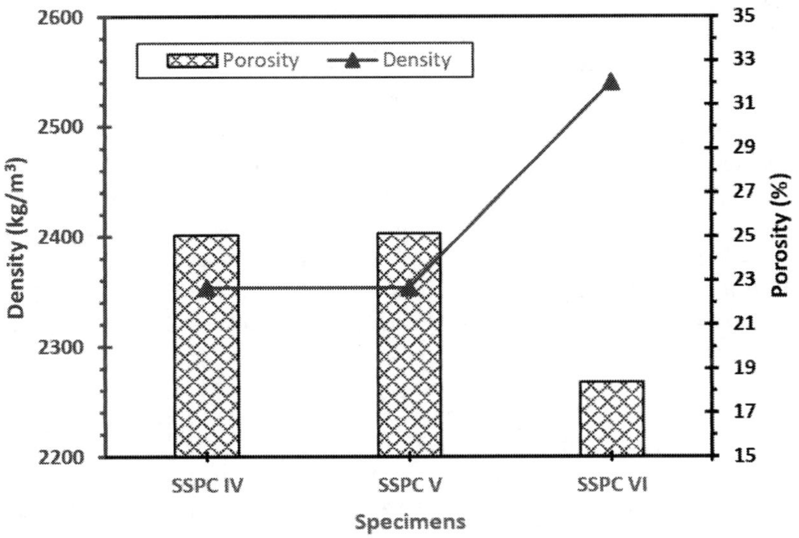

Figure 6. The effect of GGBFS as CSM to density and porosity in pervious concrete

4. Conclusion and Further Study

From the experiment it may therefore be concluded that to obtain an optimum mix of sustainable green pervious concrete, there are at least five important factors to consider, i.e., utilization of well blended coarse aggregate, replacement of small amount of coarse aggregate with fine aggregate, optimization of the achieved compressive strength in regard to the porosity or void content, ensuring a consistent compaction work and also the optimization of GGBFS as CSM. In this study, the effect of well graded or blended-size steel slag coarse aggregate and inclusion of steel slag fine aggregate as much as 10% significantly increases the compressive strength and density of pervious concrete by conforming porosity as per ACI 522R. Finally, replacement of OPC by 32% GGBFS as CSM is found to be an optimized mix proportion to produce high compressive strength, high density and required porosity of a more sustainable yet green pervious concrete made by the by-products of steel processing industry. Further study on the feasibility of the use of steel slag and GGBFS in producing green pervious concrete is currently carried out by evaluating the effectiveness and cost efficient of the mix.

Acknowledgements

This paper was funded by Institut Teknologi Bandung under the Program of Research, Community Service, and Innovation (P2MI) and had been carried out in Structural Engineering Laboratory of the Faculty of Civil and Environmental Engineering, Institut Teknologi Bandung. Exceptional supports from PT. Krakatau Steel by providing the raw materials and PT. Sika Indonesia by providing the chemical admixtures are greatly appreciated.

References

[1] Öz, H. Ö. (2018). Properties of pervious concretes partially incorporating acidic pumice as coarse aggregate. *Construction and Building Materials, 166,* 601–609. https://doi.org/10.1016/j.conbuildmat.2018.02.010

[2] Meng, X., Chi, Y., Jiang, Q., Liu, R., Wu, K., & Li, S. (2019). Experimental investigation on the flexural behavior of pervious concrete beams reinforced with geogrids. Construction and Building Materials, 215, 275–284. https://doi.org/10.1016/j.conbuildmat.2019.04.217

[3] Xu, W., Chen, B., Chen, X., & Chen, C. (2021). Influence of aggregate size and notch depth ratio on fracture performance of steel slag pervious concrete. Construction and Building Materials, 273, 122036. https://doi.org/10.1016/j.conbuildmat.2020.122036

[4] ACI Committee 522. (2010). Report on Pervious Concrete. American Concrete Institute.

[5] Nelson, P. M., and Phillips, S., 1994, "Designing Porous Road Surfaces to Reduce Traffic Noise," Transportation Research Laboratory Annual Review, TRL, Crow Thorne, England, 58 pp.

[6] Neithalath, N., 2004, "Development and Characterization of Acoustically Efficient Cementitious Materials," PhD thesis, Purdue University, West Lafayette, IN, 269 pp.

[7] Neithalath, N.; Weiss, W. J.; and Olek, J., 2003, "Development of Quiet and Durable Porous Portland Cement Concrete Paving Materials," Final Report, The Institute for Safe, Quiet, and Durable Highways, 179 pp.

IOP Conf. Series: Earth and Environmental Science 1245 (2023) 012018 doi:10.1088/1755-1315/1245/1/012018

Assessing Geotourism Aspects of Ungaran Geothermal Areas for Promoting Sustainable Energy through Volcano-Geothermal Tourist Destination

J Marin[1], A Paskarana[1], E B S Sandy[1] and J Agustian[1]

1 Department of Geological Engineering, Faculty of Engineering, Universitas Diponegoro, Jl. Prof. Soedarto, SH, Tembalang, Semarang, Jawa Tengah, Indonesia 50275

Corresponding author: jenian.marin@live.undip.ac.id

Abstract: The indigenous nature of the Ungaran volcano-geothermal area allows us to bring the benefit to its nearest community. Geotourism can be an opportunity to introduce geological knowledge and foster the socio-cultural acceptance of society towards geothermal. This study aims to assess the potential of the study area as tourist destination. The research method is collecting data on geothermal manifestations and volcanic feature sites, quantitative assessment of geosites and geomorphosites, and preliminary hazard risk analysis. Study results identify six geosites: Gedong Songo area, Nglimut hot spring, Diwak-Derekan-Kaliulo warm spring, and Mawar camping area which features a volcanic landscape. Five of six geosites have high geotourism value of more than 0.5 in average, except for Kaliulo site. The recommendation of activities to do are leisure activities in nature, hiking, camping, thermal bathing, horseback riding, and geological excursion.Ttourists should be cautious of potential hazards risk from volcanic gases, hot and acid waters, slippery and unstable areas, landslides, rugged terrain, and extreme weather. A geotourism model is constructed to highlight each site and their attributes to Ungaran volcano-geothermal setting, tourism activities, and possible risks. By exploring the natural and geological richness offered by this destination, tourists will discover the uniqueness of volcano-geothermal tourism experience.

1. Introduction

Geotourism is a new concept in nature-based tourism that has grown rapidly, simultaneously with educational and sustainable tourism. It focuses on geological and geomorphological elements of the geosites to attract visitors [1, 2]. Natural conditions in volcano-geothermal areas open up opportunities for various recreational activities [3]. The challenging climate and rugged mountainous terrain are suitable for adventure tourism. The existence of hot and warm springs is an attraction as a location for natural hot springs which are beneficial for relaxation and health. Over time, this interest has become increasingly evident with the development of tourism in volcanic and geothermal areas. Even so, people associate it more with nature tourism, not geotourism or ecotourism, especially education-based ones [4].

Education-based geotourism aims to foster an understanding of geological phenomena, their benefits and implications for our life. This geotourism sub-sector is part of education which creates holistic access to geoscience knowledge [5]. The development of tourism in geothermal-volcano areas can also be used as a means to build broad public acceptance of geothermal energy as a sustainable energy source [6]. Although health and safety aspects are important factors in determining tourist destinations, visitors to geothermal-volcano areas are aware that they are

approaching a dangerous area. It should be realized that volcanic and geothermal activities have risks of hazards such as eruptions, volcanic gases, steam and hot water, ground motions, and so on. In order for the development of a site to become a tourist destination to remain safe, the potential for hazards must be identified and mitigated, especially those related to volcanic-geothermal conditions in the study area [3, 7].
Ungaran is an andesitic volcano, part of the Quaternary Java volcanic arc which is formed from the subduction between the Indo Australian Plate to Eurasian Plate. Lithology of Ungaran volcano and its surroundings (Figure 1) consists of Neogene sedimentary rocks (Tmk), andesite intrusion (Tma), volcaniclastic rocks (Qpkg), Old Ungaran volcanic rocks (Qpj, Qpk), and Younger Ungaran volcanic rocks (Qhg, Qls) [8, 9]. The remnant of volcanic activity of Ungaran is manifested as fumaroles, altered rocks, steaming ground, and hot springs [10, 11].

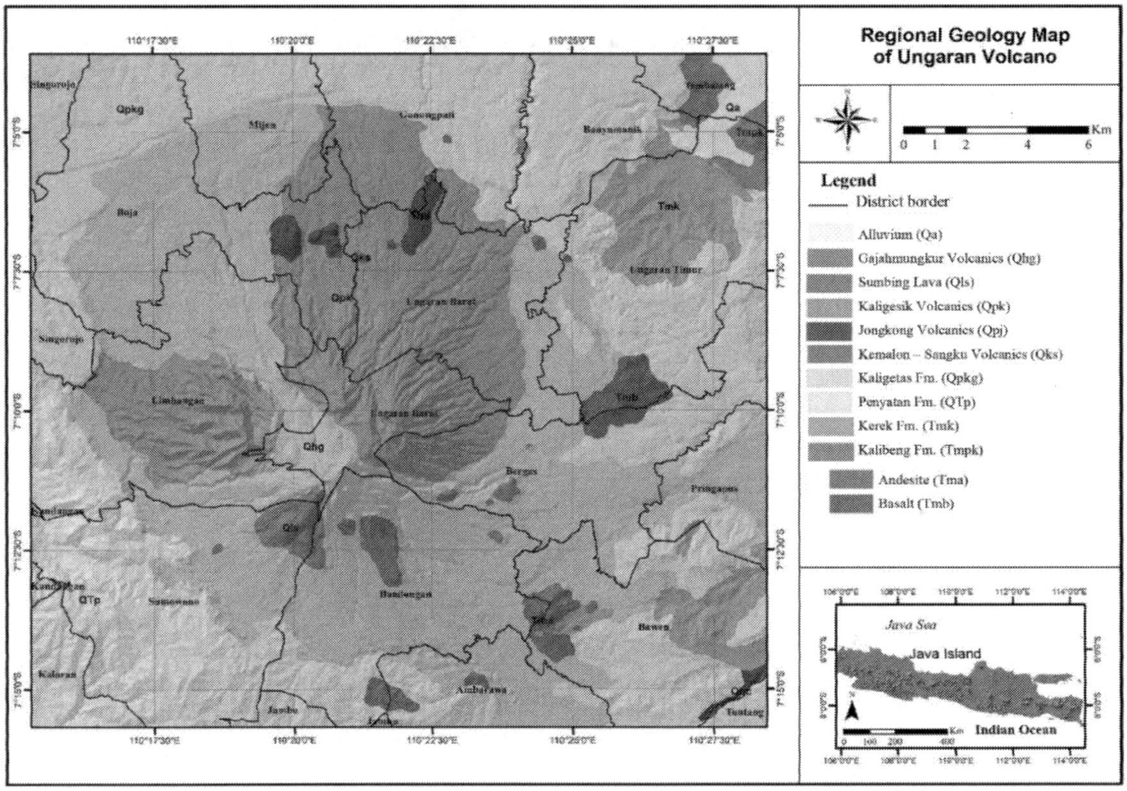

Figure 1. Geological map of the research area according to [9]

The unique volcanic features, beautiful volcanic landforms, and geothermal manifestations are a good opportunity to be developed as a geotourism site. This research aims to assess geosite and geomorphosite of the study area, evaluate and identify potential hazards. The proximity to various Geological Engineering campuses in Central Java and Yogyakarta makes the Gedong Songo area, Ungaran a good choice of location for studies related to volcanoes – geothermal. The purpose of this study is expected to provide further insight to attract students and geoscientists in developing geotourism models that can be applied in volcano-geothermal areas.

2. Methods

Geosite mapping and field survey were conducted to assess geological and geomorphological conditions related to the volcano and geothermal activity at the Ungaran volcano area. Geosite selection was based on public information and previous study so that it represents volcano-geothermal diversity, mostly geothermal manifestations as well as geomorphosites with specific landscapes. A field survey is essential to obtain descriptive data and documentation of each site. The geological and geomorphological aspects of each site are based on previous study, map analysis and direct field observation. Documentation consists of coordinate location and taking photos to capture geomorphological conditions, rock/ outcrop condition, and thermal features. Other than geological and geomorphological conditions, information related to the potential hazards was collected.

Geosite and geomorphosite assessment used in this research is a quantitative method using five values that define geoconservation and geotourism purposes [12, 13]. The five values are (1) scientific and intrinsic value, (2) educational value, (3) economic value, (4) conservation value, and (5) added value. Furthermore, each value is divided into 3-5 criteria and then for every criterion, a 0 – 1 numerical value is added (Table 1).

Table 1. Geosite and geomorphosite assessment based on [12]

Intrinsic and scientific value		
A	Integrity	0 – totally destroyed site 0.5 – disturbed site, but with visible abiotic features 1 – site without any destruction
B	Rarity	0 – more than 5 sites 0.5 – there are 2-5 similar sites 1 – the only site within the area of interest
C	Diversity	0 – only one visible feature/processes 0.5 – there are 2-4 visible feature/processes 1 – more than 5 visible feature/processes
D	Scientific knowledge	0 – unknown site 0.5 – scientific papers on national level 1 – high knowledge of the site, monographic studies
Educational value		
A	Representativeness and visibility/ clarity of the features/ processes	0 – low representativeness/clarity 0.5 – medium representativeness, especially for scientists 1 – high representativeness, also for the public
B	Exemplarity, pedagogical use	0 – very low exemplarity and pedagogical use 0.5 – existing exemplarity, but with limited pedagogical use 1 – high exemplarity and potential for pedagogical use
C	Existing educational product	0 – no products 0.5 – leaflets, maps, web pages 1 – info panel, information at the site
D	Actual use for educational purposes	0 – no educative use of the site 0.5 – sites as a part of specialized excursions 1 – guided tours for public
Economic value		
A	Accessibility	0 – more than 1000 m from the parking place 0.5 – less than 1000 m from the parking place 1 – more than 1000 m from the stop of public transport
B	Presence of tourist infrastructure	0 – more than 10 km ftom the site existing tourist facilities 0.5 – 5-10 km tourist facilities 1 – less than 5 km tourist facilities
C	Local products	0 – more than 10 km ftom the site existing tourist facilities 0.5 – 5-10 km tourist facilities 1 – less than 5 km tourist facilities
Conservation value		
A	Actual threats and risks	0 – high on both natural and atrophic risks 0.5 – existing risks that can disturb the site 1 – low risks and almost no threats
B	Potential threats and risks	0 – high on both natural and atrophic risks 0.5 – existing risks that can disturb the site 1 – low risks and almost no threats
C	Current status of a site	0 – continuing destruction of the site 0.5 – the site destroyed, but now with management 1 – no destruction
D	Legislative protection	0 – no legislative protection 0.5 – existing proposal for legislative protection 1 – existing legislative protection
Added value		
A	Cultural value, presence of historical/ archaeological/ religious aspects related to the site	0 – no cultural features 0.5 – existing cultural features but without strong relation to abiotic features

		1 – existing cultural features with the strong relations to abiotic features
B	Ecological values	0 – not important 0.5 – existing influence but not so important 1 – important influence of the geomorphologic feature on the ecologic feature
C	Aesthetic values, number of colours, structure of the space, viewpoints	0 – one color 0.25 – 2-3 colors 0.5 – more than 3 colors 0 – one pattern 0.25 – 2-3 patterns clearly distinguishable 0.5 – more than 3 patterns 0 – none 0.25 – 1-2 0.5 – 3 and more

A geotourism map is prepared to show the location of each site with its uniqueness as geosite and geomorphosites. To provide a more comprehensive model of volcano-geothermal based geotourism, a cross section model is constructed with the incorporation of simple geothermal conceptual model, visitor activities, and possible hazard risks along the volcano area.

3. Results and Discussion

3.1. Geosites Distribution

Based on the field survey as illustrated in Figure 2, six geosites and geomorphosites are selected in the Ungaran volcano-geothermal area. Most of them are geothermal manifestations on a volcanic landscape: hot springs, warm springs, steaming ground, fumarole, and altered rocks. One of the sites is a geomorphosite which represents the volcanic landscape of Ungaran and its surrounding volcano as a part of Java volcanic arc.

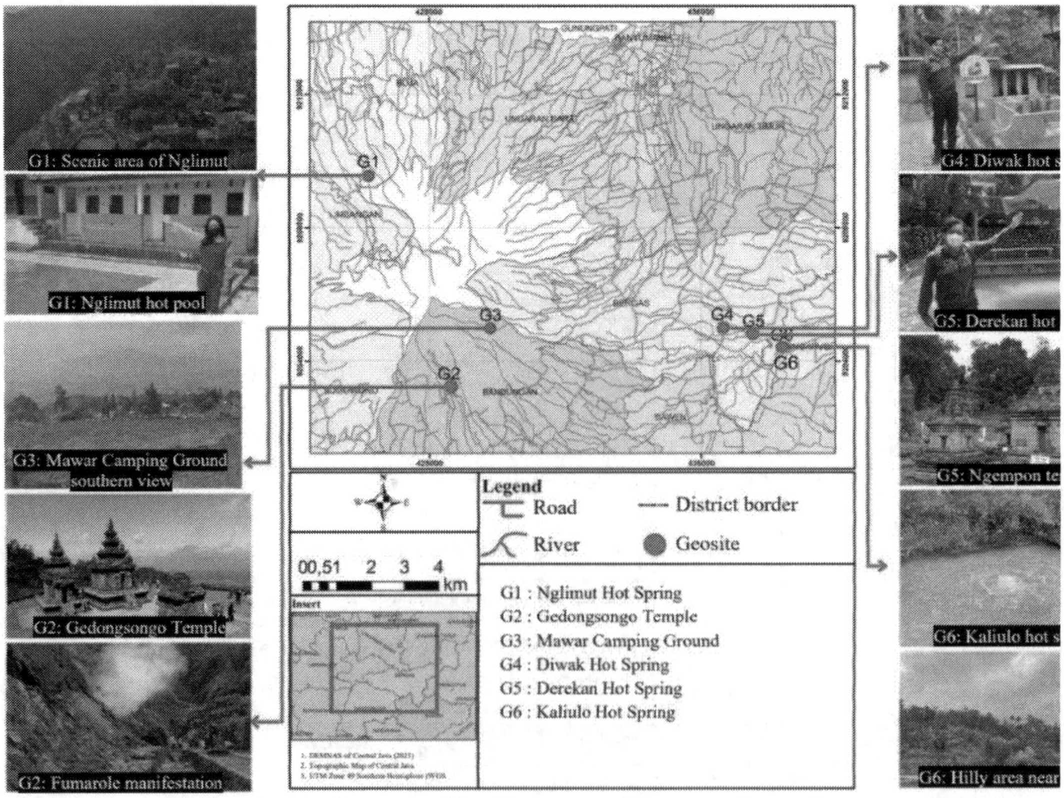

Figure 2. Geosite map of the Ungaran volcano-geothermal area

3.1.1. Gedong Songo Temple Geothermal Area

Gedongsongo Temple is located in Candi Village, Bandungan District, Semarang Regency at the southern flank of Ungaran volcano. This is a cultural site of a historical temple that coexist with several geothermal manifestations such as fumaroles, hot springs, mud pools, steaming ground, and altered rocks (Figure 3). Gedongsongo Temple is located at 1,200 msl elevation. It can be accessed easily thanks to the fairly good road infrastructure with an estimated journey of approximately 1.5-hour drive from the nearest city, Semarang. This location is well facilitated and managed by the local culture and forestry service. The morphology of the Gedongsongo area is controlled by the rocks and geological structures present at the site. The lithology of Gedongsongo area is dominated by volcanic breccia and andesitic lava. The northeast-southwest trending scarp can be interpreted as a fault and fracture zone that resulted in geothermal activity on the surface. Gedongsongo area is an active fumarole field, emitting volcanic steam and gases such as SO_2 and CO_2 through fissures. Sulphate hot springs, mud pools, and acid alteration rocks are an indication of upflow geothermal area [11]. Apart from the geothermal manifestations, this site was a religious monument from the Ancient Mataram Kingdom which is built in the 8th Century. This temple at its time was used as a place to worship their gods, as indicated by the presence of the statues of Shiva, Mahakala, Nandiswara, and Ganesha. Its location in the highlands is also related to the belief of the Hindu community who believe that the mountain is the abode of the gods and worship at a high place will be closer to heaven [14].

(c) (d)

Figure 3. Gedongsongo Temple geothermal area, (a) the entrance; (b) Complex III Gedongsongo Temple; (c) fumarole; (d) acid hot spring and altered rocks

3.1.2. Nglimut Hot Spring

The Nglimut Hot Spring is located in Nglimut Hamlet, Gonoharjo Village, Limbangan District, Kendal Regency. The geosite that was found was the location of a pool with hot springs that had been managed and also used as a tourist spot. It is situated at an elevation of 700 msl and has a relaxing view of forest in a hilly topography. Access to this location is also quite easy, it can be reached by motorcycle and car, with slightly damaged and rocky road access. From the center of Semarang City, the trip will take around 1.5 hours' drive. The uniqueness of this geosite is the discovery of several hot springs that have been used as warm bathing pools. The water temperature in this hot spring ranges from 30°C - 45°C. Based on [15], the chemical content of water is a mixture of chloride and bicarbonate water originating from reservoir fluids that underwent a mixing process with meteoric water.

3.1.3. Diwak Hot Spring

Diwak hot spring is located in the residential area of Diwak Village, Bergas District, Semarang Regency. With an hour distance trip from Semarang City, its location is accessible using a car or motorcycle through a relatively decent road access. This thermal manifestation has been used as a hot bathing pool with the only facilities are parking lot, toilet, and bathroom. Diwak hot springs have two separate pools, for men and women. Diwak area is situated on weathered andesitic lava and breccia associated with the Gajahmungkur Volcanics Formation. The chloride water spring indicates the presence of a high permeability zone, which is formed by faults and fractures on volcanic rock formation [16].

3.1.4. Derekan Hot Spring

Derekan hot spring is located near Diwak along the same river valley. It only takes about 15 minutes to get here from the Diwak hot spring. Historically, Derekan hot spring area is an ancient bathhouse located near a Hindu temple called Ngempon Temple (Figure 4). The temple was built around 8th - 9th century with an architecture similar to Gedongsongo Temple [17]. The lithology of the study area is composed of andesitic breccia and tuffaceous sandstone from the Ungaran volcano. It is indicated that the hot spring came naturally due to structural control [18]. The geothermal water of this site is characterized by bicarbonate type and the occurrence of travertine layers. This is a result of the interaction between condensate water and shallow groundwater which accumulate CO_2 while water flows laterally to the outflow zone [16].

Figure 4. Tourists' attraction at the research area, (a) Tirta Mulya cave at Mawar camping ground; (b) Ngempon Temple; (c) Travertine at Kaliulo area; (d) Derekan hot pool

3.1.5. *Kaliulo Hot Spring*

Kaliulo hot spring is located in the Kaliulo Village, Pringapus District, Semarang Regency. It can be accessed from Semarang-Yogyakarta main road and then through the local road which are in rather poor condition. Several hot spring spots appear on the ground near local rice fields, some of them have been accommodated in a pond. This hot spring is quite unique due to its salty taste, where the locals use its water as a thickening agent for making *gendar*, some kind of rice cake. Travertine deposit is also found as one of the manifestations. The lithology of the surrounding area is dominated by tuffaceous sandstone and claystone units of the Kaligetas Formation. A deep subsurface structural geology inferred by gravity study controlled the hot spring emergence [19]. Water geochemistry shows a bicarbonate water type with an anomalously high concentration of sodium chloride and boron from connate water in the Neogene marine sedimentary rock [19].

3.1.6. *Mawar Camping Ground*

Mawar Camping Ground is a location in the Umbul Sidomukti Tourism area, Bandungan District, Semarang Regency, which has been developed into a tourist location with various facilities. Mawar Camping Ground can be reached in about 1 hour if departing from the center of Semarang City. The price of the entry ticket is Rp. 5,000, - for each person, but not including parking fees and specific attraction fees. The location of Mawar Camping ground is at an altitude of 1,265 msl. This site is also considered fascinating because it faces directly towards the beautiful scenery at the volcanic complex nearby where we can view Merbabu volcano, Merapi volcano, Telomoyo volcano, and Pening Lake at the southern view. In the Mawar Camping Ground area, visitors will not only find locations for camping, but there are several facilities for other activities, such as cafes, restaurants, villas, playgrounds, and also horse-riding areas. In this area there is also an artificial cave, namely the Tirta Mulya Cave which was made in 2009. At the end of the cave, there is a small place which is limited by a fence as a place to enjoy the view of Ungaran volcano up close.

3.2. *Geosite and Geomorphosite Assessment*

Each geosite that has been selected is then analyzed by quantitative analysis with an assessment based on five parameters [13]. After the assessment was carried out (Table 2), the 5 out of 6 geosites have an average value above 0.5 and later can be prioritized to become a geotourism area around Mount Ungaran. One of the geosites does not contain most of tourism elements, hence the low assessment value.

Table 2. Geosite and geomorphosite assessment of Ungaran volcano-geothermal area

Parameter	Nglimut Hot Spring	Geodongsongo Temple Area	Mawar Camping Ground	Diwak Hot Spring	Derekan Hot Spring	Kaliulo Hot Spring
Intrinsic and scientific value						
A	1	1	1	1	0.5	1
B	0.5	1	0.5	0.5	0.5	1
C	0.5	1	0.5	0.5	0.5	0.5
D	0.5	1	0.5	0.5	1	0.5
Educational value						
A	1	1	1	1	1	1
B	0.5	1	0.5	0.5	1	0.5
C	1	1	1	0	1	0
D	1	1	1	1	1	0
Economical value						
A	0.5	1	1	0.5	0.5	0.5
B	1	1	1	1	1	0
C	0	1	0.5	0	0	0

Conservation value						
A	0.5	0.5	0.5	1	1	1
B	1	0.5	0.5	1	1	1
C	1	1	1	1	0.5	0.5
D	0.5	1	1	0.5	1	0
Added value						
A	0	1	1	0.5	1	0
B	0.5	0.5	0.5	0	0.5	0
C	0.5	0.5	0.5	0.5	0.5	0.25
D	0.5	0.5	0.5	0.5	0.5	0.25
E	0.5	0.5	0.5	0.5	0.5	0.25
Total	12.5	17	14.5	12	14.5	8.25
Average value	0.63	0.85	0.73	0.6	0.73	0.41

3.3. Geotourism Activities

3.3.1. Leisure Hiking

This activity can be done at the Gedongsongo Temple area with a specific route facilitated by the site manager. Starting from the entrance, with an uphill path and through several traditional pavilions, equestrian areas, stage areas, lower temple areas (complex I-II), and hot spring pools. Then proceed with hiking through higher temple areas (complex III – V) past several vendor stalls selling food and drinks. Hiking activities carried out in the Gedongsongo Temple area take approximately 2 hours starting from the entrance to the exit. Other than this site, the Nglimut area is also suitable for hiking and sightseeing. The natural ambience of pine forest and waterfall with a serene environment is a relaxing experience for many people.

3.3.2. Cultural Tourism

There are two temple complexes in the research location, namely Gedongsongo temple and Ngempon temple. Both are historical relics from the Hindu culture of the Sanjaya dynasty in the 8th-9th centuries. The management of the temple is under the cultural heritage preservation center. The gedongsongo area offers beautiful views of the slopes of Ungaran Volcano, breezy air and geothermal manifestations. Meanwhile, the Ngempon temple is a former ancient bathhouse which is still being used today as Derekan hot pool.

3.3.3. Thermal Bathing

Several hot springs can be found in the study area: Gedongsongo area, Diwak hot spring, Derekan hot spring, and Nglimut hot spring. All of them are equipped with hot pool facilities from the natural hot spring water. Many cultures all over the world have traditions around hot springs and their utilization [20]. Thermal bathing has therapeutic effects that helps us to relieve pain, relax our muscles, rejuvenate and detoxify the body. Its therapeutical benefits are due to the mineral composition of geothermal water. Natural hot springs usually has common soluble elements such as sodium, calcium, magnesium, iron, sulphur, and other trace elements [21]. Sulphur-rich water can help with skin problems due to its antiseptic properties. Its healing benefit include treating skin conditions such as dermatitis, psoriasis, and fungal diseases.

3.3.4. Horseback Riding

Gedongsongo temple and Mawar camping ground offer horseback riding experiences for adults and kids. It can be an alternative for those who want to explore the area and are too exhausted to hike. The horse stable can be found near the entrance of Gedongsongo temple. Each horse will be accompanied by an equestrian. Horseback riding usually cost about Rp. 50,000 – Rp.100,000 depends on the package and distance.

IOP Conf. Series: Earth and Environmental Science 1245 (2023) 012018 doi:10.1088/1755-1315/1245/1/012018

3.3.5. Camping

Camping can be an adventurous or family friendly activity. In several sites of the study area, location such as Nglimut Hot Springs area and Mawar Camping Ground also provides an allocated spot for tourists who want to camp. The manager also facilitates bathroom and toilet, water supply, and camping equipment rental to make it convenient for visitors.

3.3.6. Volcano-Geothermal Excursion

The research area is a volcano-geothermal complex from Ungaran volcano activity. Based on the assessment, it meets the parameter of educational aspects of general geoscience, volcanology, and geothermal studies. For the volcanology fieldtrip, it can be suggested to visit and observe the location of volcanic rocks outcrop in the Gedongsongo area. Mawar Camping area is a geomorphosite to witness the development of the geomorphological evolution of Ungaran volcano and its surrounding volcanoes. Geothermal excursions are carried out by observing the naturally occurring manifestations and estimating the potential of geothermal energy resources in this area. In the field of geophysics, the relationship between volcanic features, geothermal manifestation, lithology, and structural geology can be studied in the Ungaran volcano-geothermal area.

3.4. Geotourism Model of Ungaran Geothermal Area and Its Possible Hazard

Volcano – geothermal based geotourism in Ungaran area offers many touristic activities to the visitors. This tourism prospect carries geoheritage values and geological education values, is considered relatively new at this time. Based on geothermal conceptual model, this area can be divided into upflow zone and outflow zone (Figure 5). Upflow zone is characterized by higher terrain, fumarole, steaming ground, and acid sulphate water. Meanwhile, outflow zone is located in the lower area with travertine and bicarbonate-chloride spring manifestation. Using this approach, several possible hazards and risks are identified specifically in every zones.

Geothermal Zone	*Upflow zone*	*Lateral upflow zone*
Touristic Activities	*Hiking, camping, horseback riding, thermal bathing, cultural tourism, volcano-geothermal excursion*	*Thermal bathing, volcano-geothermal excursion, cultural tourism*
Possible Hazards and Risks	*Volcanic gases, steam and gas explosion, hot and acid water, slippery and unstable area, rock fall, landslide, rugged terrain, extreme weather*	*Hot water, slippery and unstable area*

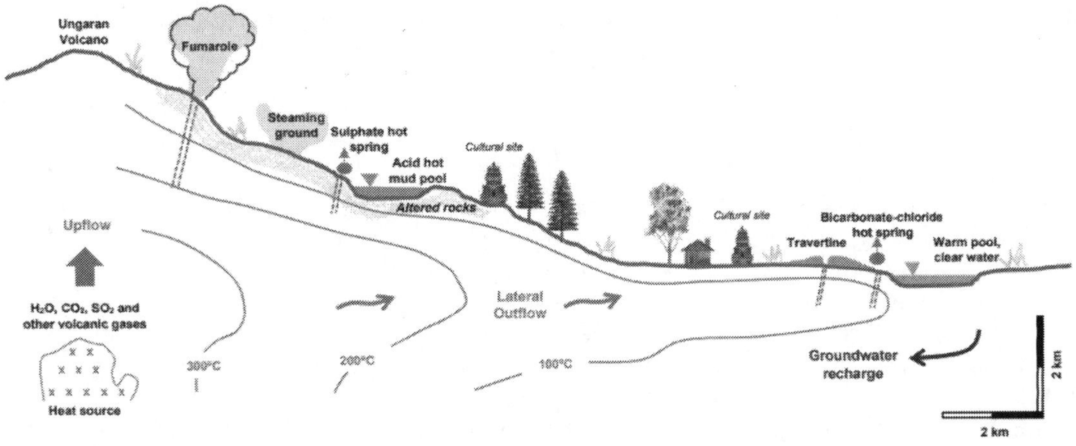

Figure 5. Volcano-geothermal based teourism of Ungaran volcano

Exploring geothermal related environments is not always completely safe as there are a number of hazards that can occur, some of which are even quite unexpected. In conducting tours to locations with geothermal potential, there are many things that must be considered. The tourists will usually be very curious about geothermal manifestations such as fumaroles, steaming ground, and hot spring. However, the thing that also needs to be considered is the safety of tourists who will be threatened if unwanted things happen at the geothermal location. There are several hazards caused by geothermal manifestations and processes that can threaten the health and safety of tourists. Apart from the risk of serious burns from erupting hot springs, fumaroles which release extremely hot gases and vapors can also cause serious burns. Unexpected hydrothermal eruptions can also occur in any active geothermal field. Hot mud eruptions and steam eruptions can sometimes occur without any previous signs. This is caused by tourist sites located in geothermal fields with active seismic or volcanic activity. Other geothermal geotourism hazards also include the emission of gases such as H_2S and CO_2 which can pose a serious threat, as these gases are potentially lethal within a few moments at higher concentrations.

Unfortunately, Ungaran volcano area is still lack of measure regarding of its possible hazards and risks. It is important to assure the sense of security of tourists when traveling. Geosite managers can add geographic paths, tour guides, and interpretive signs to increase tourist awareness of the potential hazards that exist in tourist sites. Ordinary visitors, who are not familiar with existing tourist sites will be greatly helped by the basic information provided by the manager. It also helps tourists navigate safely through the right zones. Managers can also provide pamphlets and brochures that can help tourists get to know tourist areas and locations that are quite important, such as evacuation routes, shelters, and also the nearest health facility. Tourists must have initial knowledge about the tourist environment in preparation for travel in a geothermal environment and at least visitors must also be prepared to deal with unexpected hazards that can occur at geothermal field locations.

4. Conclusions

Ungaran volcano and its surroundings have the potential to be developed as a volcano-geothermal based geotourism and geoedutourism. Total of six selected geosites which include geothermal manifestation sites and geomorphosite to view the volcanic landscape. After conducting the assessment using scientific and intrinsic approaches, educational value, economic value, conservation value and additional value, the results showed that five geosites met the assessment criteria. A constructed volcano-geothermal tourism model proposes several activities in this area based on geothermal upflow-outflow zone. Regarding to the possible hazards and risks, tourists must have proper understanding as well as appropriate management from the geosite managers.

5. References

[1] Dowling R K 2013 Global Geotourism – An Emerging Form of Sustainable Tourism Czech J. Tour. 2
[2] Newsome D, Dowling R 2018 Geotourism and Geoheritage in Renard E, Brilha J *Geoheritage: assessment, protection, and management* 305-321
[3] Erfurt-Cooper P, Cooper M 2010 *Volcano and Geothermal Tourism*
[4] Erfurt-Cooper P 2016 *Volcanic Tourist Destination*
[5] Stolz J, Megerle H E 2022 Geotrails as a Medium for Education and Geotourism: Recommendations for Quality Improvement Based on the Results of a Research Project in the Swabian Alb UNESCO Global Geopark *Land* **11**
[6] Pavlakovic B, Demir M R, Pozvek N, Turnsek M 2021 Role of Tourism in Promoting Geothermal Energy: Public Interest and Motivation for Geothermal Energy Tourism in Slovenia *Sustainability* **13**

[7] Heggie T W 2009 Geotourism and volcanoes: Health hazards facing tourists at volcanic and geothermal destination *Travel Medicine and Infectious Disease* **7** 257-261
[8] Marin J, Winarno T, Mindasari D 2019 Mineralogy and Geochemistry Variation of Igneous Rocks from Ungaran and Muria Volcano and Its Processes related to Subduction Zone Magmatism of Sunda Arc *IOP Conference Series Earth and Environmental Science* **279**
[9] Thanden R E, Sumadirdja H, Richard P W, Sutisna K, Amin T C 1996 Peta Geologi Lembar Magelang dan Semarang Skala 1:100.000
[10] Zarkasyi A, Rezky Y, Nurhadi M 2011 Keprospekan Panas Bumi Gunung Ungaran berdasarkan Analisis Geosain Terpadu *Buletin Sumber Daya Geologi* **6**
[11] Rezky Y, Zarkasyi A, Risdianto D 2012 Sistem Panas Bumi dan Model Konseptual Daerah Panas Bumi Gunung Ungaran, Jawa Tengah *Buletin Sumber Daya Geologi* **7** 109-117
[12] Kubalikova L 2013 Geomorphosite assessment for geotourism purposes *Czech Journal of Tourism* **2** 81-104
[13] Kubalikova L, Kirchner K 2016 Geosite and Geomorphosite Assessment as a Tool for Geoconservation and Geotourism Purposes: a Case Study from Vizovická vrchovina Highland (Eastern Part of the Czech Republic) *Geoheritage* **8** 5-14
[14] BPCB Jateng 2020 *Candi Gedongsongo* https://bpcbjateng.id/berita/candi-gedongsongo
[15] Emianto Y B, Aribowo Y 2011 Studi Geokimia Fluida Panasbumi Daerah Prospek Panasbumi Nglimut, G Ungaran, Kecamatan Limbangan, Kabupaten Kendal, Jawa Tengah *Jurnal Teknik* **32**
[16] Trisnaning P T, Fatimah 2021 Tipe Fluida Hidrotermal Sepanjang Manifestasi Panas Bumi Ungaran dan Sekitarnya *Angkasa Jurnal Ilmiah Bidang Teknologi* **13** 95-102
[17] Yuliati D 2020 Pelestarian Candi Ngempon dan Pemanfaatannya untuk Atraksi Pariwisata *Project: Studying the Admiration Movement on the Indonesian Maritime Culture for Strengthening the Indonesian National Character and Creative Economy*
[18] Falah M F, Marin J, Winarno T 2021 Geological Aspect to Enhance Cultural Heritage at Ngempon Temple Geotourism Site, Semarang Regency, Central Java *E3S Web of Conference* **317**
[19] Saibi H, Setyawan A, Aboud E, Barianto D H 2008 Relationship between the locations of the hot springs and the deduced subsurface structure from gravity data analysis in Ungaran volcano (Indonesia) *Proceeding of the 30th Annual New Zealand Geothermal Workshop* 167-173
[20] Erfurt-Cooper 2010 The Importance of Natural Geothermal Resources in Tourism *Proceedings World Geothermal Congress* 1-10
[21] Simon N, Unjah T, Yusry M, Dzulkafli M A 2019 Physico-chemical Characterization and Potential Health Benefit of the Hulu Langat Hot Spring in Selangor, Malaysia *Sains Malaysiana* **48** 2451-2462

Acknowledgments
We would like to thank our students: Ulfa Dian Ratnasari, Kristanti Wulandari, and Aji Bayu Pamungkas from Geological Engineering Department, Diponegoro University for their assistance and support during field and laboratory work. This research is partly funded by RKAT 2022 from Faculty of Engineering, Diponegoro University.

Surface Geology Analysis on the Relationship between Fault Creep and Overpressure in Grobogan, Central Java, Indonesia

Eri S. Sitinjak[1*], Benyamin Sapiie[2], Agus M. Ramdhan[2], Ana N. Hidayati[3], Yanuar A. Azhari[3], Nanda F. Adriyansyah[3]

[1]Doctoral Program at Department of Geological Engineering, Faculty of Earth Sciences and Technology, Institut Teknologi Bandung, Jl.Ganesha, 10, Bandung, 40132, Jawa Barat, Indonesia.
[2]Department of Geological Engineering, Faculty of Earth Sciences and Technology, Institut Teknologi Bandung, Jl.Ganesha, 10, Bandung, 40132, Jawa Barat, Indonesia.
[3]Undergraduate Program at Department of Geological Engineering, Faculty of Earth Sciences and Technology, Institut Teknologi Bandung, Jl.Ganesha, 10, Bandung, 40132, Jawa Barat, Indonesia

*Corresponding Author: erygea@yahoo.com or 32020003@mahasiswa.itb.ac.id

Abstract. The Grobogan Regency is a significant location for oil and gas exploration due to its proximity to active oil and gas fields. Geologically, this area is intersected by the Kendeng thrust fault and exhibits several surface manifestations of mud volcanoes near the fault. However, there have been limited studies examining the activity of the Kendeng thrust fault and its relationship to overpressure and earthquakes, which could affect oil and gas exploration and natural disaster mitigation in the region. This paper focuses on utilizing surface geological mapping data and Interferometric Synthetic Aperture Radar (InSAR) data to analyze the relationship between fault creep and overpressure manifestations in Grobogan, Central Java. While continuous research on fault creep has been conducted since the 1960s on the San Andreas fault in the United States, there has been a lack of continuous research on onshore fault creep in the Northeast Java Basin or Kendeng Basin, particularly in the Grobogan area. This region, known for its abundance of active mud volcano manifestations, is predominantly characterized by marl and clay lithology. It represents a warm basin environment that is prone to overpressure and ideal for fault creep occurrences.To investigate further, fluid, gas, and mud samples were collected from the mud volcanoes, along with rock samples from the surrounding area, for comprehensive laboratory analysis. The integration of these sample analyses aimed to determine the correlation between the activity of the Kendeng thrust fault creep and the subsurface overpressure conditions. While this study observed several indications of creeping in the vicinity of InSAR anomalies, the exact location of the creeping fault plane could not be definitively determined. The Kendeng thrust fault exhibits the three main factors necessary for initiating fault creeping. The fault segment's extremely slow creeping may be linked to the activity of the mud volcanoes, as indicated by the results of fluid, gas, and rock sample analysis. The findings of this research hold potential for informing the development of oil and gas fields, as well as enhancing the mitigation of natural disasters associated with oil and gas exploration and earthquakes in the region.

Keywords: *fault creep, aseismic fault, overpressure, mud volcano, earthquake*

IOP Conf. Series: Earth and Environmental Science 1245 (2023) 012019 doi:10.1088/1755-1315/1245/1/012019

1. Introduction

The Onshore Northeast Java Basin, also known as the East Java Basin, has experienced several past earthquakes. Within this region, the Kendeng thrust fault plays a significant role as it separates the Kendeng Zone from the Rembang Zone [1]. In fact, there have been at least 10 major earthquakes in the vicinity of the Kendeng thrust fault zone (Figure 1), resulting in numerous casualties and substantial damage to infrastructure [2]. Recognizing the seismic potential of this area, it is crucial to investigate the activity of the Kendeng thrust fault, particularly its surface and subsurface manifestations of fault creeping. To undertake this study, the Grobogan area was chosen due to its intricate geological characteristics and the presence of mud volcano manifestations on the surface. The research primarily focused on observing the surface geology surrounding the Kendeng thrust fault in Grobogan and evaluating its activity (refer to Figure 1).

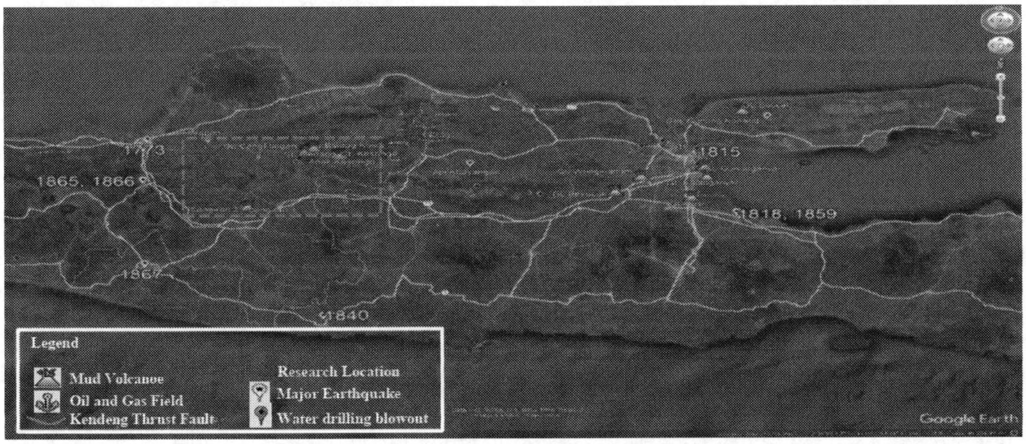

Figure 1. Old documented major earthquakes in Java Island and location of research.

The onshore region of East Java can be categorized into four distinct zones, characterized by their morphology, physiography, and major structural domains. Starting from the northernmost part, we have the Rembang zone, followed by the Kendeng zone, the Central Volcanic Mountains zone, and finally, the Southern Mountain zone [1]. The Kendeng Thrust fault marks the boundary between the Kendeng zone and the Rembang zone (refer to Figure 2).

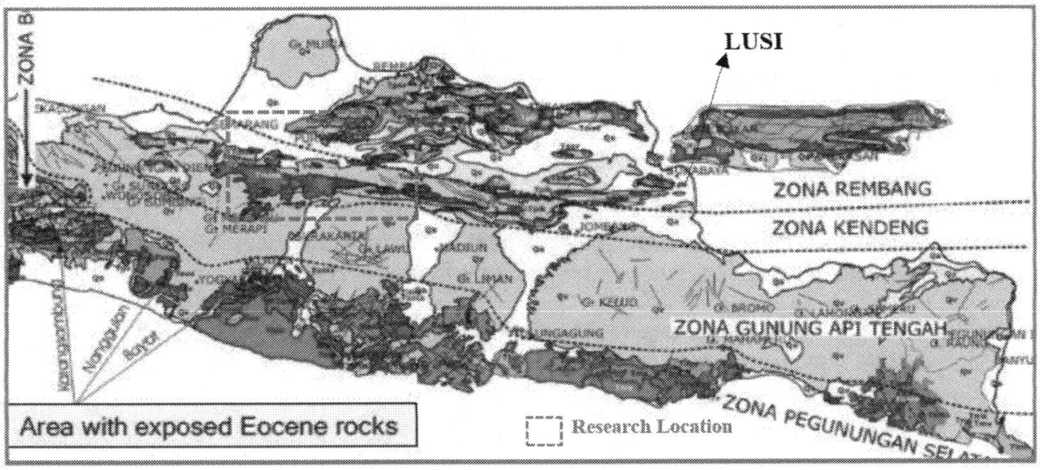

Figure 2. Physiography and major structural domains of East Java [1].

Java exhibits three major structural patterns that govern its structural behavior: the NE-SW Meratus pattern, the E-W Java/Sakala pattern, and the N-S Sunda pattern. The Kendeng thrust fault, situated in Central and East Java, aligns with the East-West (E-W) trend associated with the Java/Sakala pattern. Surface physiography reveals a transition in lithology from the southern to the northern part of Java. The southern region (Southern Mountains Zone and Central Mountains Zone) is predominantly composed of volcanic rocks, while the Kendeng Zone is characterized by clay, and the Rembang Zone is dominated by sandstone. Furthermore, the onshore section of the Northeast Java Basin is recognized as an overpressure zone [3-6]. To provide a comprehensive overview, a cross-sectional and regional stratigraphic column was constructed, illustrating that the Kendeng Zone's lithology consists of clay and marl from the Kalibeng formation, the Kerek formation, and the Pelang formation [7].

Fault creep refers to the gradual movement of an aseismic fault in the upper part of the Earth's crust between two earthquakes, relieving significant stresses on the fault. It can also occur as an after-slip event, involving daily or yearly shifts following an earthquake. The study of earthquake creep originated around 1960 in the United States, specifically focusing on the San Andreas fault and its observable effects on physical infrastructure [8-10]. Changes in structures such as buildings, roads, or exposed rock formations were measured using basic creep meter tools, and these methods have since undergone further refinement [11-13]. This assessment has been applied to actively creeping faults worldwide, including locations like Italy, Taiwan, and other countries [14-20].

Fault creep can be attributed to three primary factors: the geological characteristics of the region, the existence of overpressure fluids, and the geometry of the fault [21-29]. To monitor the speed of fault plane creep, a commonly employed tool is the creep meter. Initially, this instrument was deployed to observe the activity of creep on the San Andreas fault in the United States [30,31]. Presently, there exist various models of creep meter tools, each utilized based on the specific objectives of individual studies. Typically, this tool consists of two iron stakes, an invar rod, a protective cover, and a centrally positioned sensor [18,19].

In the field of geology, three primary types of major faults are recognized: normal faults, strike-slip faults, and reverse or thrust faults, classified according to the Anderson fault classification system. Notable examples of fault creep can be observed in major faults worldwide, such as the Alto Tiberina Normal Fault in Italy, the strike-slip San Andreas Fault in the United States, and the Chishang Fault in Taiwan [14,16-17, 28, 32, 33]. By employing a creep meter, it is possible to monitor the creep behavior of a measured fault [39]. Generally, three types of creep patterns are observed across most sites: steady-state creep, episodic creep, and episodic creep accompanied by a steady-state background creep. As fault creep represents the movement of an aseismic fault between significant earthquakes, the creep meter may provide insights into the cyclic occurrence of major earthquake events, allowing for prediction or approximation of earthquake cycles. The study of the creeping activity of the Kendeng Thrust fault has not been thoroughly explored, particularly through surface geological observations. Similar to other well-known creeping faults, the presence of mud volcanoes in close proximity to the Kendeng Thrust fault indicates an obvious manifestation of overpressure conditions in the subsurface. This research aims to elucidate the relationship between the overpressure conditions in the region and the creeping activity of the Kendeng Thrust fault, potentially associated with a series of significant earthquakes in the past. Given that a significant portion of the Indonesian population resides in Java, particularly in East Java and Central Java, this research focuses on the Kendeng Fault. According to the Central Bureau of Statistics, Java Island alone accommodates at least 56.1% of the Indonesian population [40]. Therefore, the impact of this research holds considerable importance for Indonesia.

2. Methodology and Research Procedures

The methodology and research procedures primarily focus on surface geological mapping and observation. Field mapping was conducted in the Grobogan region, specifically targeting areas near the Kendeng Thrust fault zone. This region is well-known for its complex of mud volcanoes and its proximity to the Kendeng Thrust fault zone. The Google Earth application was also utilized to identify potential geological features of interest on the surface. According to the regional geological map, the Kendeng thrust fault runs through the southern part of Grobogan.Various samples were collected in this area, including rock, mud, fluid, and gas, and were carefully preserved before being sent to laboratories. The rock samples obtained from the field were cut and prepared as thin sections for further analysis at the Geology Department of the Bandung Institute of Technology (ITB). Thin section analysis enables the examination of mineral composition, texture, and fabric of rocks and minerals in thin slices. This provides valuable information about rock and mineral properties, facilitating the interpretation of geological history and processes.The gas, fluid, and mud samples from each mud volcano were also collected and prepared for laboratory analysis. The gas samples were sent to the Coal and Geothermal Mineral Resource Center Laboratory in Bandung. Gas analysis involved methods such as Titrimetric Analysis and Gas Chromatography Analysis. These analyses measure parameters including H2, He, O2, Ar, N2, CH4, CO, and CO2. Titrimetric analysis is a quantitative analysis technique based on volume measurement, while Gas Chromatography Analysis is commonly used for the separation and analysis of volatile compounds.The fluid samples were sent to the Water Quality Laboratory at the Faculty of Civil and Environmental Engineering (Bandung Institute of Technology) in Bandung. The fluid analysis followed the guidelines outlined in the *Standard Methods for The Examination of Water and Wastewater 23rd Edition 2017* (APHA). The fluid samples are pure water samples extracted from the mud. The analysis covers approximately 11 elements, including Boron (B), Calcium (Ca), Magnesium (Mg), Potassium (K), Sodium (Na), Chloride (Cl), Sulfate (SO4), Lithium (Li), Strontium (Sr), Barium (Ba), and Iron (Fe). Additionally, the laboratory measures the pH level of the water. The results of the gas and fluid analysis from the subsurface (mud volcanoes) will be utilized to determine the impact of dominant elements on lithology and their relationship to the activation of the Kendeng thrust fault creeping.

Furthermore, the physical properties of the mud particles from each mud volcano were analyzed. The mud samples were sent to the Soil Mechanics Laboratory at the Faculty of Civil and Environmental Engineering, Bandung Institute of Technology. Some of the mud samples were also sent to Geological Survey Center laboratory in Bandung for Scanning Electron Microscope (SEM) analysis, the result will be in discussion (refer to Figure 9). Several methods were employed for the analysis, including Specific Gravity Analysis, Sieve Analysis, and Hydrometer Analysis. Specific Gravity Analysis helped determine the specific gravity value of the grains present in the mud samples. Sieve Analysis and Hydrometer Analysis were utilized to assess the grain size of the mud samples. This information will be crucial in establishing correlations with subsurface lithology and will be integrated with surface geological observations. It is widely recognized that lithology is one of the factors that can influence fault creep activity [27, 28].

The Kendeng Thrust fault exhibits a very slow creep rate of 2-3 mm per year, which is significantly slower compared to most other active creeping faults. This slow movement makes it challenging to observe the surface manifestations of creep. Relying solely on the regional geological map would prove difficult in identifying the creep planes on the surface. Hence, additional methods are required to support surface geological mapping efforts. One of the chosen methods is Interferometric Synthetic Aperture Radar (InSAR) modeling. InSAR is an established technology that evolved from Synthetic Aperture Radar (SAR) invented by Carl Wiley in 1951 [34]. InSAR modeling serves as a valuable tool to aid in defining the possible surface manifestations of the Kendeng Thrust fault and complement surface geological mapping efforts. InSAR operates as a remote sensing technique that utilizes radar satellite imagery captured by satellites orbiting the Earth. Radar waves emitted by the satellites are reflected off the Earth's surface and captured by sensors on the satellite [35]. The fundamental principle of InSAR involves combining two or more SAR images acquired from different positions to generate three-dimensional (3D) positional information. Through SAR processing, multiple interferometric pairs are formed using SAR imagery, which allows for the extraction of distance and elevation information of objects on the Earth's surface based on phase differences. These phase differences can be converted into Line-of-Sight (LOS) shift values, providing an indication of surface shifts or deformations [36].

In a specific direction, the Line-of-Sight (LOS) in InSAR can detect displacements of points in particular locations (refer to Figure 3). InSAR measurement and modeling techniques are valuable tools for identifying displacements caused by various factors, including tectonic activity, earthquakes, volcanic eruptions, glaciology, oil and gas extraction, hydrology, and infrastructure monitoring [37]. In the case of the Kendeng Thrust fault, its movement primarily involves vertical displacements. Theoretically, InSAR measurements can detect these vertical movements or displacements. Therefore, combining the results from laboratory tests, surface geological mapping, and InSAR modeling will allow for a comprehensive understanding of the relationship between Kendeng Thrust fault creeping and subsurface overpressure conditions.

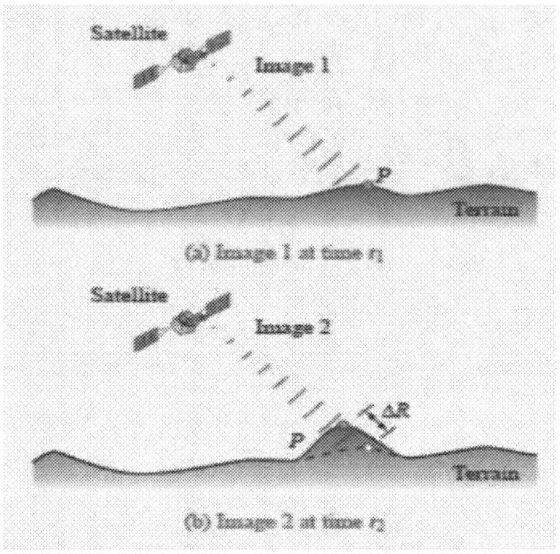

Figure 3. Displacement of surface deformation due to tectonic on the point of P by a ΔR along the direction of LOS [38].

3. Result

The Grobogan region is home to at least 8 main mud volcanoes, namely Kesongo, Anak Kesongo, Crewek, Banjarlor, Bledug Kuwu, Detil, Cakringan, and Medang (refer to Figure 6D). These mud volcanoes are situated at distances ranging from 5 to 11 kilometers from the Kendeng Thrust fault. Among them, Crewek mud volcano stands as the closest to the Kendeng Thrust fault zone, at approximately 5 kilometers away. Samples were collected from each of these mud volcanoes and subsequently sent to specialized laboratories for analysis. The fluid samples extracted from the mud volcanoes were dispatched to the Water Quality Laboratory at the Faculty of Civil and Environmental Engineering, Bandung Institute of Technology. It was observed that all the fluid samples possessed a salty taste, and this was reflected by the significantly high levels of natrium and chloride content (as shown in Table 1). According to the Regulation of the Minister of Health Number 492 (2010), the chloride content should not exceed 250 mg/L, and the natrium content should be below 200 mg/L.

Table 1. Fluid analysis from each main mud volcanoes in Grobogan.

ID	Mud Volcano	Parameter Analysis (Mg/L)										pH
		Cl	Na	K	Mg	Ca	SO_4	Sr	Li	Fe	B	
S24	Kesongo	20094	4147	126.2	1104	210.6	238	34.69	6.06	0.001	0.001	7.81
S25	Anak Kesongo	19485	4743	175.6	708.7	298.9	510.4	39.17	0.96	0.05	0.001	7.86
S18	Crewek	31663	7213	2136	1263	672.6	187.1	62.86	23.01	3.46	0.001	7.29
S19	Banjarlor	25878	7209	1921	939.5	635.3	55.24	5.13	28.92	0.001	0.001	7.7
S20	Bledug Kuwu	74285	12908	4031	5335	2928	66.44	162.3	0.01	0.001	0.001	7.13
T10	Detil	36534	56.41	537.6	20.38	160.82	261	5.242	0.848	0.01	0.01	6.48
S33	Cakringan	32575	64S3	2122	1442	774.5	640.2	68.86	22.58	0.01	0.001	7.07
S34	Medang	13700	4256	17.2	382.7	312.5	72.04	2.46	0.09	0.13	0.001	7.48

Gas samples were also collected from each of the mud volcanoes and sent to the Coal and Geothermal Mineral Resource Center Laboratory in Bandung for analysis (refer to Table 2). The gas samples were obtained using a customized device, consisting of a funnel pipa connected to a hose, which was further connected to a plastic bottle filled with water. As the gas entered the bottle, it displaced the water, causing it to be pushed out. Upon analysis in the laboratory, it was revealed that carbon dioxide (CO_2) was the dominant gas component across all the samples, followed by nitrogen (N_2) and methane (CH_4).

Table 2. Gas analysis from each main mud volcanoes in Grobogan.

ID	Location	Non-Condensable Gas Parameters (%mmol)						
		CO_2	N_2	CH_4	O_2/Ar	He	H_2	CO
S24	Kesongo	23.326	61.293	8.825	6.547	0.005	0.004	0
S25	Anak Kesongo	81.71	14.219	2.455	1.604	0.011	0.002	0
S18	Crewek	90.824	7.587	0.646	0.942	0	0	0
S19	Banjarlor	17.487	73.86	0	8.65	0	0.003	0
S20	Bledug Kuwu	81.71	14.219	2.455	1.604	0.011	0.002	0
T10	Detil	92.519	6.788	0.088	0.598	0.003	0.003	0
S33	Cakringan	72.464	25.275	0.05	2.21	0.002	0	0
S34	Medang	95.714	2.258	1.841	0.183	0.004	0	0

This research primarily focused on conducting geological field observations within the onshore Central Java province. In order to identify and delineate the creeping fault segments of the Kendeng thrust fault, InSAR modeling was utilized, covering the research area (refer to Figure 5 and 6). Both ascending and descending InSAR models were generated to assess potential ground movement associated with fault creep. Remarkably, these models produced similar anomaly results in terms of ground movement. The research clearly highlights the trend and continuity of the Kendeng thrust fault within the designated study area. Furthermore, it was noted that several mud volcanoes are located in the Grobogan area, in close proximity to the suspected fault plane.

The predominant lithology in the area was found to be massive marl, which has undergone significant structural deformation (refer to Figure 4 and 6A). Based on thin section analysis, the rock was classified as mudstone according to Dunham's classification system from 1962 (refer to Figure 7B and 7C). The physical properties analysis of the mud samples collected from the mud volcanoes in Grobogan revealed that the specific gravity of the mud ranged from 2.65 SG to 2.72 SG. Additionally, the grain size analysis of the mud samples indicated a predominance of clay particles. This suggests that the mud source in the subsurface is also predominantly composed of clay lithology. Consequently, it can be concluded that the Grobogan region is characterized by a clay-dominated lithology extending from the surface down to the subsurface. This clay lithology plays a significant role in the occurrence of fault creeping in the area.

Figure 4. Geological observation in S14 location (see Figure 5-6). (A) Lithology marl domination in S14 location (insert is the brittle marl/clay due to structural deformation from Kendeng fault). (B) Rock sample for thin section analysis.

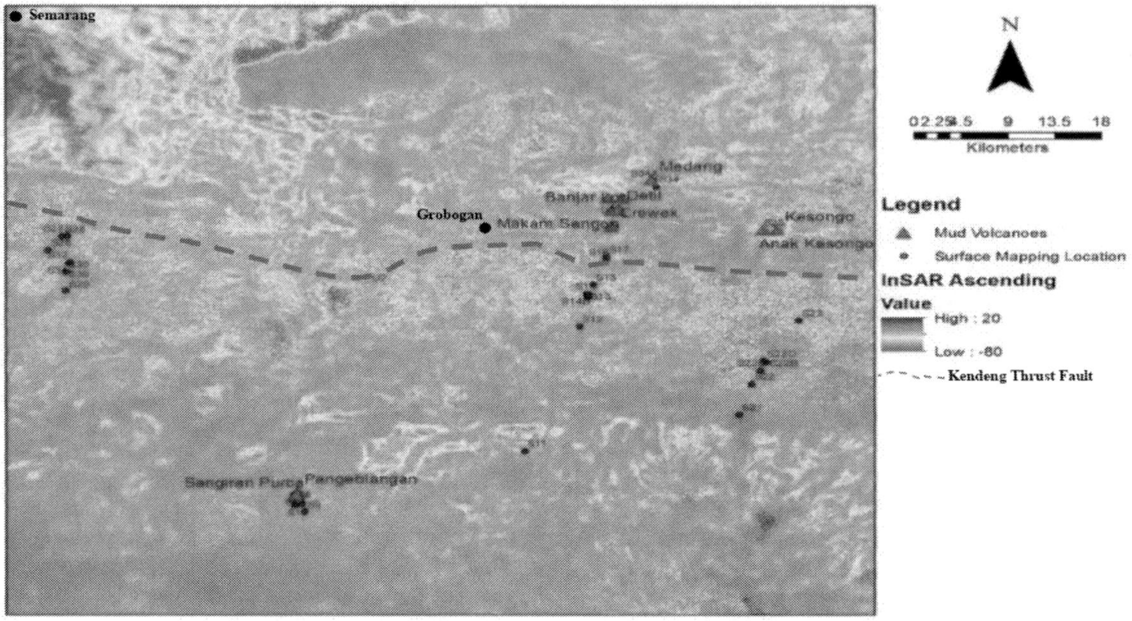

Figure 5. Ascending InSAR modelling result in Grobogan area

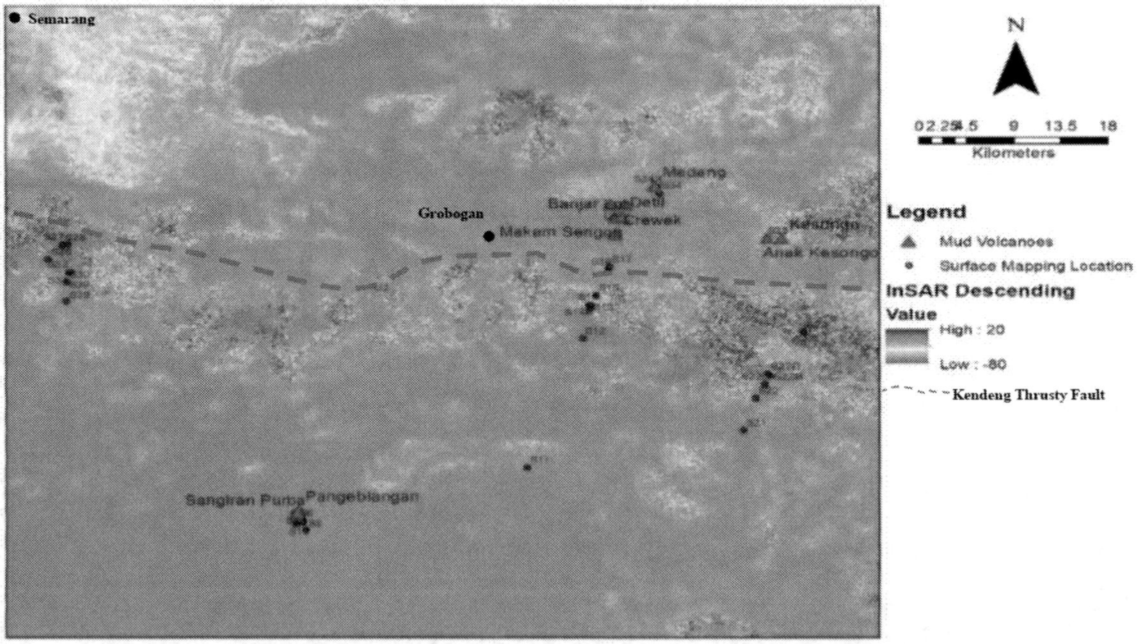

Figure 6. Descending InSAR modelling result in Grobogan area

Figure 7. Geology of the Grobogan area. (A) The presence of fault escarpments in S15, indicated by the red arrow, makes this area highly susceptible to sliding. These escarpments are primarily composed of massive brittle marl lithology (the location of S15 can be found in Figure 4-5). (B) A parallel polarized thin section of the S14 rock sample, located as shown in Figure 4-5, was classified as foraminifera mudstone according to Dunham's classification from 1960. (C) Similarly, a cross-polarized thin section of the S14 rock sample, also located as shown in Figure 4-5, was classified as foraminifera mudstone according to Dunham's classification. (D) The geometry of the Anak Kesongo Mud volcano in the Grobogan area is depicted in this image (location can be found in Figure 4-5).

4. Discussion

One of the common characteristics observed in the example of three main creeping faults worldwide is the presence of mud volcanoes on the surface, indicating subsurface overpressure conditions. Each of these faults exhibits mud volcanoes in close proximity to the fault segment. Overpressure conditions or elevated pressure in the subsurface are among the factors contributing to fault creep. For instance, the Alto Tiberina Normal fault is surrounded by mud volcanoes, with the closest ones located in the northern part of the fault [41]. The Pieve Santo Stefano mud volcano complex, connected to a deep CO_2 reservoir, is particularly notable [42]. The Alto Tiberina Normal fault experiences creep at a rate of approximately 1.7 ± 0.3 mm/yr from the surface to a depth of 5 km. At a depth of around 7 km, the fault becomes locked, which correlates with earthquakes of magnitudes 6.5 and 6.7 [43]. Similarly, the San Andreas Strike Slip (SAF) fault has mud volcanoes located in the southern part of the fault. The southern section of the SAF creeps at a rate of 2-4 mm/yr, while the central and northern parts exhibit higher rates ranging from 10 to 34 mm/yr [45]. Notably, the central part of the SAF, particularly the Parkfield area, demonstrates the highest creep rate at 34 mm/yr [46].

To accurately map the displacement velocity along the San Andreas Fault, a combination of InSAR (Interferometric Synthetic Aperture Radar) and Global Navigation Satellite System (GNSS) time series was utilized [47]. This modeling approach successfully delineated the fault plane and provided precise information on displacement velocity. Notably, the central section of the San Andreas Fault, particularly in the Parkway area, exhibited the highest velocity displacement ranging from 10 to 30 mm/yr.

Menawhile, the Chishang Thrust fault, on the other hand, demonstrated a creep rate of 3 cm/yr [48]. The Chishang and Kendeng are both thrust fault and has a few similarities. InSAR modeling was employed to visualize the fault displacement, revealing an approximate displacement rate of 30 mm/yr along the Chishang Thrust fault. Additionally, creepmeter measurements yielded varied results ranging from 3 to 16 mm/yr. It is worth noting that there are also mud volcanoes in close proximity to the Chishang area. The laboratory analysis confirmed that the fluid from the mud volcanoes had a high salinity content, indicating its highly salty nature. The gas content of the samples was primarily dominated by CO_2, followed by Nitrogen (N_2) and Methane (CH_4). The interaction between CO_2 and the brine fluid with the reservoir rock can cause deformation, potentially contributing to the movement of nearby fault planes [49, 50]. Experimental studies on the CO_2-brine interaction with Knox sandstone rocks from Indiana in the United States demonstrated that after 6 months, it resulted in the development of wider cracks in the samples (refer to Figure 8A and 8B). This interaction also led to a 50% decrease in permeability in the sandstone rock samples due to dissolution, migration, and secondary mineral precipitation [49]. Similarly, the brine-CO_2 interaction with Eau Claire shale rocks from Wisconsin in the United States resulted in the formation of additional secondary clay layers on the surface of quartz (refer to Figure 8C and 8D). The alteration process in the shale rock sample led to the generation of flaky clay minerals like illite [50]. The fluid samples collected from the mud volcanoes contained a high salt content. The salinity and CO_2 content played a role in creating more cracks and additional clay minerals, as clearly shown in the SEM (Scanning Electron Microscope) analysis results (refer to Figure 9).

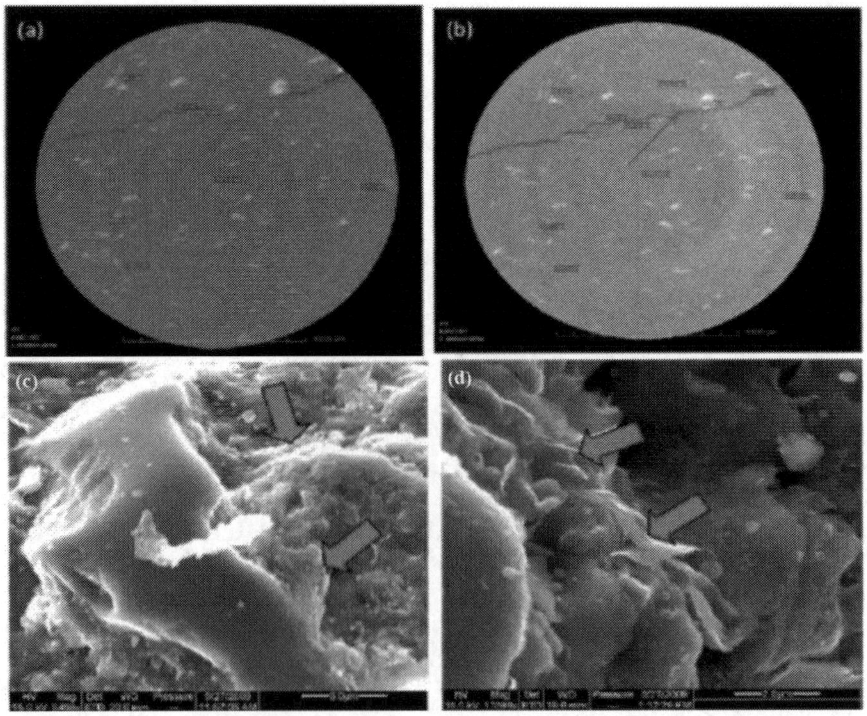

Figure 8. Brine-CO2 interaction with rock samples [49,50]: (a) Interaction between CO_2-Brine and Knox sandstone before 6 months [49]. (b) Interaction between CO_2-Brine and Knox sandstone before 6 months, showing wider cracks [49]. (c) Interaction between CO_2-Brine and Eau Claire shale after 60 days, resulting in secondary illite mats [50]. (d) Interaction between CO_2-Brine and Eau Claire shale after 60 days, showing flaky clay minerals on the surface of quartz [50].

The flaky clay minerals (illite-smectite) are also present in the mud sample from each mud volcanoes (refer to Figure 9). The combination of brine and CO2 from the mud volcanoes material motivate the generation of more flaky clay minerals in subsurface. Additional clay minerals will enhance the possibility of the faulted area to creep. The increased clay content around the fault zone facilitated fault creeping. The significant presence of CO_2 and brine fluid from the subsurface of the mud volcanoes in the Grobogan region may potentially influence the behavior of the surrounding Kendeng thrust fault. This factor strengthens the relationship between the overpressure manifestation in the study area and the activity of the Kendeng Thrust fault. Similarities can also be observed in the gas and fluid content of mud volcanoes in close proximity to the Chishang Thrust fault in Taiwan [51]. The Chishang Thrust fault is known as the fastest creeping thrust fault in the world. It is recommended to conduct further field mapping to determine suitable locations for installing creep meters. Installing creep meters will enable continuous monitoring of the creeping behavior.

Figure 9. SEM analysis from the mud sample of Crewek mud volcano. The flaky clay minerals of smectite-illite dominated the sample.

5. Conclusions

The geological characteristics of the Kendeng Thrust fault in the Grobogan area closely resemble those of other active creeping faults discussed earlier in this paper. This indicates that the Kendeng Thrust fault has the potential to undergo creep, which is supported by laboratory findings. It is likely that the overpressure conditions in the subsurface are influencing the behavior of the creeping activity along the Kendeng thrust fault. Therefore, it is highly recommended to conduct a more comprehensive field mapping initiative to identify the specific segment where creep is occurring.

References

[1] Prasetyadi, C., Suparka, E.R., Harsolumakso, A.H., and Sapiie, B. 2005. Eastern Java basement rock study: preliminary results of recent field study in Karangsambung and Bayat areas. Proceedings of the 34th Annual Convention of the Association of Indonesian Geologists, Surabaya, 310-321.

[2] Harris, R. and Major, J. 2016. Waves of destruction in the East Indies: The Wichmann catalogue of earthquakes and tsunami in the Indonesian region from 1538 to 1877. Geological Society of London, Special Publications, **441**, 9-46.

[3] Ramdhan, A.M., Hakim, F., Hutasoit, L.M., Goulty, N.R., Sadirsan, W., Arifin, F., Endarmoyo,

K., Firmansyah, R., Zainal, R.M., Gulo, M.Y., SihmN, m., Suseno, P.H. and Purwanto, A.H. 2013. Importance of Understanding Geology in Overpressure Prediction: The Example of The East Java Basin. Proceedings of 37th Annual Convention and Exhibition, Indonesian Petroleum Association, Jakarta.IPA13-G-152.

[4] Burhannudinnur, M. 2013. Pengaruh Tektonik dan Laju Sedimentasi dalam Pembentukan Gunung Lumpur di Zona Kendeng dan Rembang Cekungan Jawa Timur. Doctoral Program Dissertation, Institut Teknologi Bandung, 55-78.

[5] Sitinjak, E., Sapiie, B., Ramdhan, A.M., Harianto, M., Somantri, W., and Nurcahyo, M. 2016. Wellbore Stability in Soft Sediment and Overpressured Environment, Study Case in Tanggulangin, Northeast Java Basin. EAGE/SPE Workshop on Integrated Geomechanics in Exploration and Production, Volume 2016, 1- 5.

[6] Sitinjak, E., Sapiie, B., Ramdhan, A., Harianto, M., Soemantri, W., Nurcahyo, M., Juriansyah, M., 2017. Implication of Overpressure Zone to Wellbore Stability for Field Development in Onshore Northeast Java Basin. HAGI-IAGI-IAFMI-IATMI Joint Convention Malang (JCM).

[7] Smyth, H., Hall, R., Hamilton, J., and Kinny, P. 2005. East Java: Cenozoic Basins, Volcanoes and Ancient Basement. Proceedings of 30th Annual and Exhibition of Indonesian Petroleum Association, IPA05-G-045.

[8] Steinbrugge, K.V., and Zacher, E.G. 1960. Creep on the San Andreas Fault: fault creep and property damage. Seismological Society of America Bulletin, 50, 389-396.

[9] Tocher, D. 1960. Creep on the San Andreas Fault: related measurements at Vineyard, California. Seismological Society of America Bulletin, 50, 396-404.

[10] Tocher, D. and Nason, R.D. 1967. Fault creep at the Almaden-Cienega winery, San Benito County, Guidebook to the Gabilan Range and Adjacent San Andreas Fault, PacificSection. American Association Petroleum Geology.

[11] Yamashita, P. A., and Burford, R.O. 1973. Catalog of preliminary results from an 18-station creepmeter network along the San Andreas fault system in central California for the time interval June 1969 to June 1973. USGS Open File Report.

[12] Nason, R. D., F. R. Phillippsborn, F.R., and P. A. Yamashita, P.A. 1974. Catalog of creepmeter measurements in central California from 1968 to 1972. United States Geological Survey, Open File Report, 74-31.

[13] Schultz, R.A. 1989. Strike-slip faulting of ridged plains near Valles Marineris, Mars. Nature, 341, 424-426.

[14] Anderlini, L., E. Serpelloni, and M. E. Belardinelli. 2016. Creep and locking of a low-angle normal fault: Insights from the Altotiberina fault in the Northern Apennines (Italy). Geophysical Research Letters, 43, 4321–4329.

[15] Hreinsdottir, S. and Bennett, R.A. 2009. Active aseismic creep on the Alto Tiberina low-angle normal fault, Italy. Geology, 37 (8), 683-686.

[16] Gori, S., Falcucci, E., Kayen, R.E., Zimmaro, P., Lingwall, B., Galadini, F., and Stewart, J.P. 2018. Surface fault rupture from 2016 Central Italy Earthquake Sequence. Proceedings of the 11th National Conference in Earthquake Engineering, Earthquake Engineering Research Institute, Los Angeles, California.

[17] Blumetti, A.M., Grützner, C., Guerrieri, L., and Livio, F. 2017. Quaternary earthquakes: Geology and palaeoseismology for seismic hazard assessment. Quaternary International, 451, 1-10.

[18] Lee, J. C., Jeng, F.S., Chu, H.T., Angelier, J., and Hu, J.C. 2000. A rod-type creepmeter for measurement of displacement in active fault zone. Earth Planets Space, 52, 321-328.

[19] Lee, J.C., Angelier, J., Chu, H. T., Hu, J.C., Jeng, F.S. 2005. Monitoring active fault creep as a tool in seismic hazard mitigation, Insights from creepmeter study at Chihshang, Taiwan. Comptes Rendus Geoscience, 337, 1200-1207.

[20] Chen, K. H., Nadeau, R.M., and Rau, R.J. 2008. Characteristic repeating earthquakes in an arc-continent collision boundary zone: The Chihshang fault of eastern Taiwan. Earth and Planetary Science Letters, 276, 3-4, 262-272.

[21] Simpson, R. W., Schulz, S. S., Dietz, L. D., and Burford, R. O. (1988): The response of creeping parts of the San Andreas Fault to earthquakes on nearby faults: Two examples. Pure Applied Geophysics, 126(2), 665-685.

[22] Lienkaemper, J., Galehouse, J.S., and Simpson, R.W. 2001. Long-term monitoring of creep rate along the Hayward fault and evidence for a lasting creep response to 1989 Loma Prieta earthquake. Geophysical Research Letters, 28(11), 2265-2268.

[23] Harris, R.A. dan Simpson, R.W. 1998. Suppression of large earthquakes by stress shadows: A comparison of Coulomb and rate-and-state failure. Journal of Geophisical Research, 103(B10), 24439-24451.

[24] Simpson, R. W., Lienkaemper, J.J., and Galehouse, J.S. 2001. Variations in creep rate along the Hayward Fault California interpreted as changes in depth of creep. Geophysical Research Letters, 28(11), 2269-2272.

[25] Lienkaemper, J., McFarland, F., Simpson, R., Bilham, R., Ponce, D., Boatwright, J., Caskey, John. 2012. Long-Term Creep Rates on the Hayward Fault: Evidence for Controls on the Size and Frequency of Large Earthquakes. The Bulletin of the Seismological Society of America, 102, 31-41.

[26] Avouac, J. P. 2015). From geodetic imaging of seismic and aseismic fault slip to dynamic modeling of the seismic cycle. Annual Reviews Earth Planetary Sciences, 43, 233-271.

[27] Harris, R. A. 2017. Large earthquakes and creeping faults. Reviews of Geophysics, 55, 169-198.

[28] Chen, K.H. and Burgmann, R. 2017. Creeping faults: Good news, bad news?. Reviews of Geophysics, 55, 282-286.

[29] Burgmann, R. 2018. The geophisics, geology and mechanics of slow fault slip. Earth and Planetary Science Letters, 495, 112-134.

[30] Nason, R. D. 1971. Measurements and theory of fault creep slippage in central California. Bulletin of Royal Society of New Zealand, 9, 181–187.

[31] Goulty, N.R. and Gilman, G. 1978. Repeated creep events on the San Andreas fault near Parkfield, California, recorded by a strainmeter array. Journal of Geophysics Research, 83, 5415-5419.

[32] Hreinsdóttir, S., and Bennett, R.A. 2009. Active aseismic creep on the Alto Tiberina low-angle normal fault, Italy. Geology, 37 (8), 683–686.

[33] Rymer, M.J., Hickman S.H., and Stoffer, P.W. 2006. A field guide to the central, creeping section of the San Andreas Fault and the San Andreas Fault Observatory at Depth, in Prentice. C.S., Scotchmoor, J.G., Moores, E.M, and Kiland, J.P., eds., 1906 San Francisco Earthquake Centennial Field Guide. Geological Society of America Field Guide, 7, 237-272.

[34] Lasswell, S.W. 2005. History of SAR at Lockheed Martin (previously Goodyear Aerospace)". Proceeding of SPIE, Radar Sensor Technology IX, 5788, 1-12.

[35] Teknologi InSAR untuk Monitoring Gempa Bumi. Pusat Penelitian Fisika (LIPI). http://www.fisika.lipi.go.id/news/read/teknologi-insar-untuk-monitoring-gempa-bumi#:~:text=InSAR%20(Interferometric%20Synthetic%20Aperture%20Radar,dari%20satelit%20yang%20mengorbit%20bumi (accessed on February 26th, 2023 at 22.00).

[36] InSAR (Interferometry Synthetic Apature Radar). Magma Indonesia. Kementrian Energi dan Sumber Daya Mineral. https://magma.esdm.go.id/v1/edukasi/glossary/insar-interferometry-synthetic-apature-radar (accessed on February 26th, 2023 at 22.30).

[37] Hooper, A. 2016. InSAR theory. ESA Advance Training Course. https://eo4society.esa.int/wp-content/uploads/2021/02/2016Cryosphere_2_Hooper_InSAR-Theory.pdf (accessed on February 26th, 2023 at 23.00).

[38] Ulaby, F.T., and Long, D.G. 2014. *Microwave Radar and Radiometric Remote Sensing*. University of Michigan Press.

[39] San Francisco State University Fault Creep Monitoring. What is Creep?. http://funnel.sfsu.edu/creep/WhatsCreepPage.html ((accessed on February 20th, 2023 at 10.00).

[40] Hasil Long Form Sensus Penduduk 2020. Badan Pusat Statistik Indonesia.

https://www.bps.go.id/pressrelease/2023/01/30/2039/hasil-long-form-sensus-penduduk-2020.html (accessed on January 20th, 2023 at 13.00).

[41] Martinelli, G., and Judd, A. 2004. Mud volcanoes of Italy. Geological Journal, 39, 49-61.

[42] Bonini, M. 2009. Mud volcano eruptions and earthquakes in the Northern Apennines and Sicily, Italy. Tectonophysics, 474, 723-735.

[43] Anderlini, L., Serpelloni, E., Belardinelli, M. 2016. Creep and locking of a low-angle normal fault: Insights from the Altotiberina fault in the Northern Apennines (Italy). Geophysical Research Letters, 43, 4321 - 4329.

[44] Polcari, M., Anderlini, L., Bignami, C., Pepe, A., Solaro, G., Serpelloni, E., Moro, M., Albano, M., Chiaraluce, L., Stramondo, S., 2014. Advanced InSAR and GPS measurements for the detection of surface movements along the Alto Tiberina (Italy) normal fault system: data modeling and future perspectives. AGU Fall Meeting Poster. T13C-4678.

[45] Xu, X., Ward, L., Jiang, J., Smith-Konter, B.,Tymofyeyeva, E., Lindsey, E.O., Sylvester, A.G., and Sandwell, D.T. 2018. Surface creep rate of the southern San Andreas fault modulated by stress perturbations from nearby large events. Geophysical Research Letters, 45, 10259–10268.

[46] Rolandone, F., Ryder, I., Agram, P., Burgmann, R., and Nadeau, R. 2010. Investigating the Creeping Segment of the San Andreas Fault using InSAR time series analysis. Geophysical Research Abstracts, 12, EGU2010-6732.

[47] Xu, X., Sandwell, D., Klein, E., and Bock, Y. 2021. Integrated Sentinel-1 InSAR and GNSS Time-Series Along the San Andreas Fault System. Journal of Geophysical Research: Solid Earth, 126, e2021JB022579. https://doi.org/10.1029/2021JB022579.

[48] Champenois, J., Fruneau, B., Pathier, E., Deffontaines, B., Lin, K.C., Hu, J.C. 2012. Monitoring of active tectonic deformations in the Longitudinal Valley (Eastern Taiwan) using Persistent Scatterer InSAR method with ALOS PALSAR data. Earth and Planetary Science Letters, 337-338, 144-155.

[49] Liu, F., Lu, P., Griffith, C., Hedges, S.W., Soong, Y., Hellevang, H., Zhu, C. 2012. CO_2–brine–caprock interaction: Reactivity experiments on Eau Claire shale and a review of relevant literature. International Journal of Greenhouse Gas Control, 7, 153-167.

[50] Soong, Y., Howard, B.H., Hedges, S.W., Haljasmaa, I., Warzinski, R.P., Irdi, G., McLendon, T.R. 2014. CO_2 Sequestration in Saline Formation. Aerosol and Air Quality Research, 14, 522–532.

[51] Sun, C.H., Chang, S.C., Kuo, C.L., Wu, J.C., Shao, P.H., Oung, J.N. 2010. Origins of Taiwan's mud volcanoes: Evidence from geochemistry. Journal of Asian Earth Sciences, 37 (2), 105-116.

IOP Conf. Series: Earth and Environmental Science 1245 (2023) 012020 doi:10.1088/1755-1315/1245/1/012020

The risk of ignorance of DEM Height Reference System on coastal inundation modeling (the case of Jakarta)

Heri Andreas, Firman Juliandri, Dhota Pradipta, Dina A. Sarsito, Hasanuddin Z. Abidin,

Geodesy Research Group, Faculty of Earth Science and Technology, Institute of Technology Bandung, Indonesia, LABTEX IXC, Jl. Ganesha 10 Bandung 40132 West Java Indonesia

Heriandreas49@gmail.com

Abstract. As a result of continuing land subsidence which is also followed by sea level rise, some lowland coastal areas will experience coastal inundation. The lowlands will gradually become lower than sea level at high tide and even during normal conditions. Some areas where even it is not high tide would experience coastal inundation permanently in certain time. Coastal inundation modeling is important to create, so that we can map potential hazards and disaster risks. With coastal conditions that are usually inhabited by many people, become economic centers and others, then when coastal inundation occurs, material losses will certainly occur. To create a coastal inundation model, parameters are needed including the DEM. The coastal lowlands are represented from the DEM. From the corrected DEM for land subsidence, sea level rise and tidal hydraulic models, with some calculation we will get a coastal inundation model. The most important thing related to this model is the accuracy of the DEM and the height of the reference system. If we do not use an accurate DEM, and do not use the height reference system correctly then we will not get a reliable coastal inundation model. The impact is that we will not be get a correct map the hazards and disaster risks due to the coastal inundation. This paper will demonstrate the risk when we ignore the height reference system, so that we get a miss leading in the dangers and risks of coastal inundation in Jakarta as a case study area. As a result of investigation we can see with incorrect height reference system we found underestimate and or overestimate model and disputed the risk.

1. Introduction

The impact of land subsidence is seen clearly in cities or urban areas in the coastal region. These impacts can be in the form of coastal inundation (see figure 1), infrastructure damage, expansion of flooding, and sea water intrusion to a decrease in environmental quality [1, 2, and 3]. The significant impact of land subsidence is categorized as an ecological disaster. Million of dollar is expected need to be spent due to that ecological disaster in one urban area. In this regard, mitigation and adaptation are very important in reducing disaster risk. In the future if actions taken are less, there will be more significant disaster. One important step on mitigation and adaptation is understanding the coastal inundation through modelling (see also figure 1). Coastal inundation modelling is important to do so that we get potential hazards and disaster risks from the occurrence of it. With coastal conditions that are usually inhabited by many people, become economic centres and others, then when coastal inundation occurs, material losses will certainly occur as well.

By definition, Land Subsidence is defined as a decrease in the surface of the ground with respect to a reference system of height such as the sea surface, geoid or ellipsoid surface. Geodetic measurements (e.g., Leveling, GNSS and InSAR) can identify rates per year and magnitude even is small value [4, 5, 6, 7, 8, and 9] in certain periods of times in some part on the surface of the ground. In the term of rate, generally 1-20 centimeter per year and a total 1-4 meter are being observed.

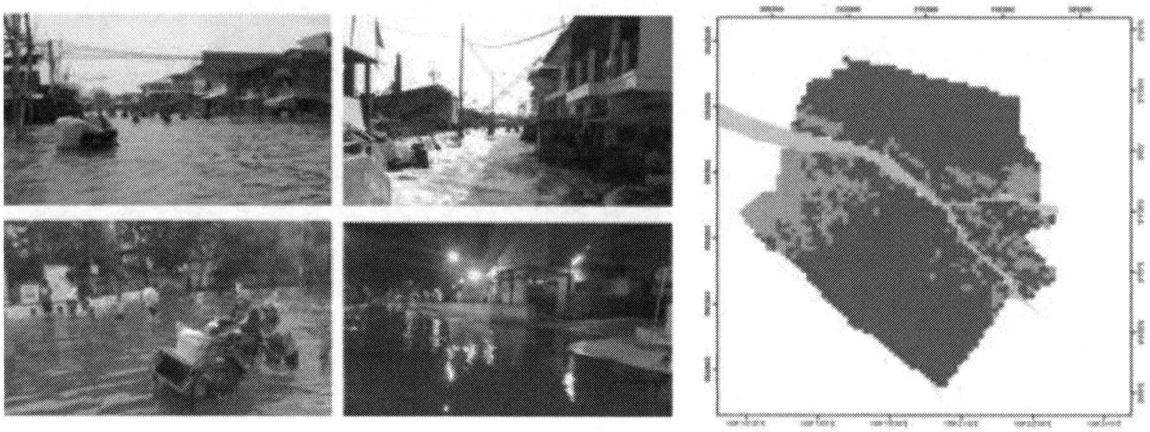

Figure 1. Documentation of coastal inundation in the field (case of Jakarta) along with example of coastal inundation modeling

To create a coastal inundation model, parameters are needed including the Digital Elevation System (DEM), land subsidence information, sea level rise information and tidal wave to define height reference system. The coastal lowlands are represented from the DEM. From the corrected DEM for land subsidence, sea level rise and tidal hydraulic models including definition of height reference system, we will get a coastal inundation model (see figure 2 for illustration). To get accurate data from DEM, leveling surveys are usually used. However, this survey will take a very long time for a large area. Currently the use of LiDAR is widely used to obtain an accurate DEM, although it cannot replace survey leveling. To obtain good land subsidence information, GNSS and InSAR surveys are usually carried out. Meanwhile to get good sea level rise data obtained from satellite altimetry.

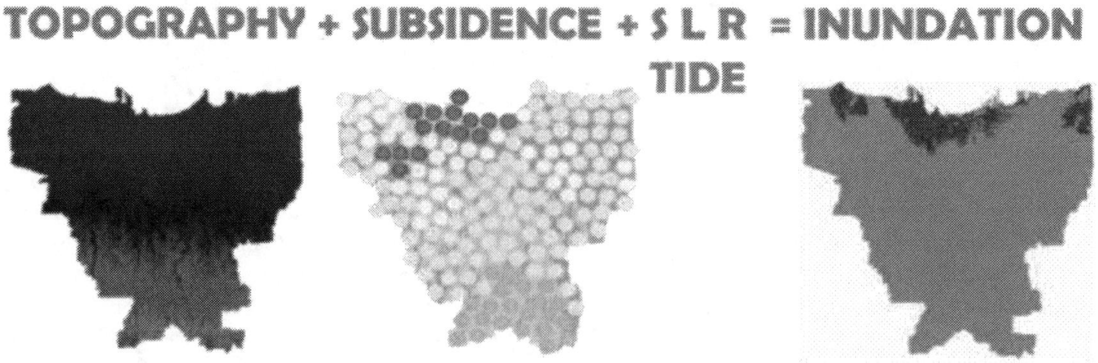

Figure 2. The figure shows how to create coastal inundation model. From the corrected DEM for land subsidence, sea level rise and tidal hydraulic models including definition of height reference system, we will get a coastal inundation model

The most important thing related to coastal inundation modeling are the accuracy of the DEM and the correct choose of height of the reference system. If we do not use an accurate DEM, and do not use the height reference system correctly (in this case it should be High Tide or Mean Sea Level), then we will not get a good or reliable coastal inundation model. The impact is that we will not be correct map the hazards and disaster risk due to the coastal inundation. This paper will demonstrate the risk when we ignore the height reference system, so that we get a miss leading in the dangers and risks of coastal inundation in Jakarta as a case study area.

2. Data and methods

As mentioned above, when we ignore the height reference system, so that we get a miss leading in the dangers and risks of coastal inundation in Jakarta as a case study area. The ignorance can come from ignorance of the various high reference systems or mistakes in choosing the high reference system. Conceptually there are in fact many high reference systems. Some of them are the High Tide System, Mean Sea Level (MSL), Geoid, Low Tide, Chart Datum, Ellipsoid Surface, and many other systems. For illustrations of various height reference systems, see Figure 3 below. As mentioned above, for modeling coastal inundation the high system should have reference to High Tide or at least Mean Sea Level. In this way we will not find cases of overestimate or under estimate.

Figure 3. Illustration of difference type of height reference system

Mean Sea Level is obtained from the average tide over some periods of observation (e.g. Ideally for about 18 years), while for high tide and low tide is obtained from the tidal range and the average maximum and minimum tide. The geoid is obtained from gravity measurements, while the Ellipsoid Surface is determined from the Earth Ellipsoid and Geodetic Datum modelers.

This paper will show the use of several different reference systems for the formation of coastal inundation modeling in Jakarta. The high systems used include Ellipsoid Surface, Geoid EGM2008, Mean Sea Level and High Tide. From the use of some of these systems, we will both understand that when we ignore the height reference system, so that we get a miss leading in the dangers and risks of coastal inundation in Jakarta as a case study area.

The DEM data used comes from modified LiDAR data. This DEM data is quite accurate in modeling coastal inundation. DEM has an accuracy of about 20-40 centimeters. The selection of an accurate DEM is important so that a coastal inundation simulation using several difference height reference systems will give clear results, including when it is correlated with the occurrence of coastal inundation in the field, the results will also be appropriate.

3. Result and discussion

The results of the study show that when we use the EGM2008, Ellipsoid Surface and Geoid, we will get an underestimated model for the case in Jakarta. Seems with this parameter, the coastal inundation seems doesn't exist. Conversely, when we use high tide and Mean Sea Level, the coastal inundation model will be reasonable. When validated with inundation events in the field, the most suitable is when we use the High Tide model. The complete research results can be seen in the following paragraphs, complete with illustrations in Figures 4, 5 and 6. It should be noted that apart from using the right reference system, we still must pay attention to other accuracy, such as the optimum accuracy of DEM and the right value of land subsidence information.

Figure 4a shows when we use the ellipsoid surface height reference system in coastal inundation modeling in Jakarta. In the model results there is no potential for coastal inundation in Jakarta even though there is indeed in the field. The use of an ellipsoid surface shows an underestimate result. Figure 4b shows when we use the EGM2008 Geoid height reference system in coastal inundation modeling in Jakarta. In the same way as the model using the ellipsoid surface, it can be seen that in the model results there is no potential for coastal inundation in Jakarta even though there is facts in the field. In this case the use of the 2008 EGM Geoid shows underestimated results. Figures 4c and 4d use the Mean Sea Level and High Tide as Height reference systems, respectively. From the modeling results, there is a potential for coastal inundation in Jakarta. If it is correlated with events in the field, this model is generally suitable.

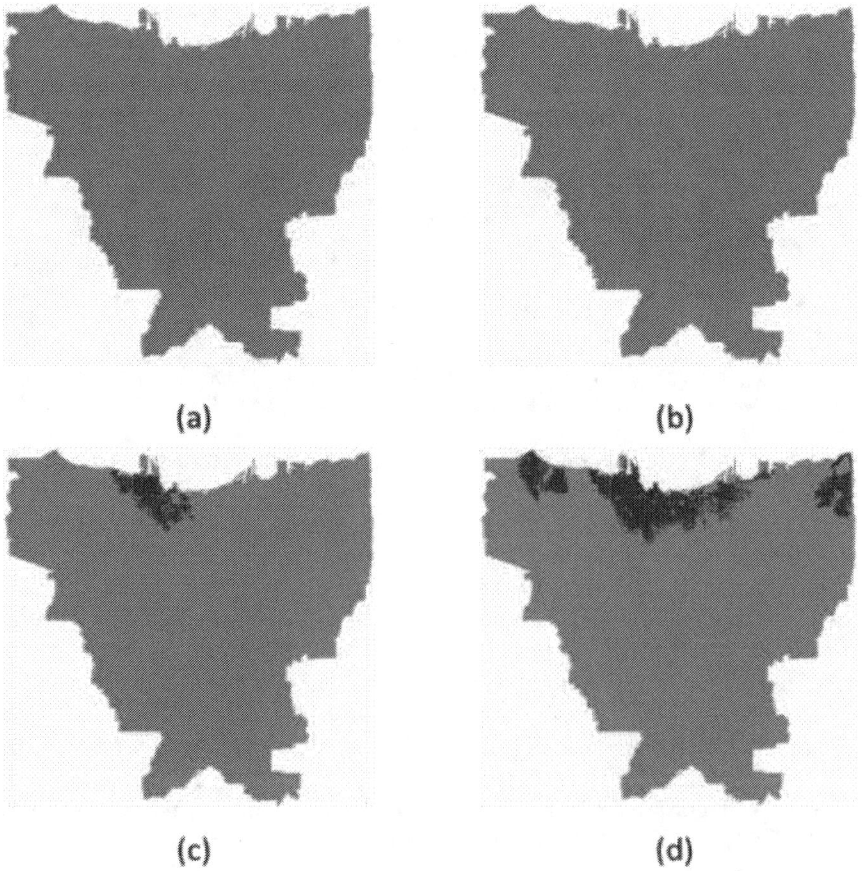

Figure 4. Visualization of coastal inundation each by using height reference system of ellipsoid surface (a), Geoid EGM 2008 (b), Mean Sea Level (c) and High Tide (d)

IOP Conf. Series: Earth and Environmental Science 1245 (2023) 012020 doi:10.1088/1755-1315/1245/1/012020

The most appropriate height reference system for modeling coastal inundation is the High Tide System Model d on figure 4. By using logic, when high tides occur in the field, there will be potential for flooding in several places whose positions are already lower than sea level. It should be noted here that the High Tide system is very local in nature, it will be different in each region. For this reason, good tide monitoring will be the key.

Figures 5a and 5b respectively show when we use the ellipsoid surface height reference system and the 2008 Geoid EGM in coastal inundation modeling in Jakarta overlaid with plotting of inundation events in the field. It can be seen that in the model results there is no potential for coastal inundation in Jakarta even though there is clear fact in the field. The use of the ellipsoid surface and Geoid EGM 2008 once again shows underestimated results. Figures 5c and 5d respectively use the High Tide and Mean Sea Level high reference system in the modeling of coastal inundation in Jakarta overlaid with plotting of inundation events in the field. From the modeling results, there is a potential for coastal inundation in Jakarta. If it is correlated with events in the field, this model is suitable especially when we used High Tide.

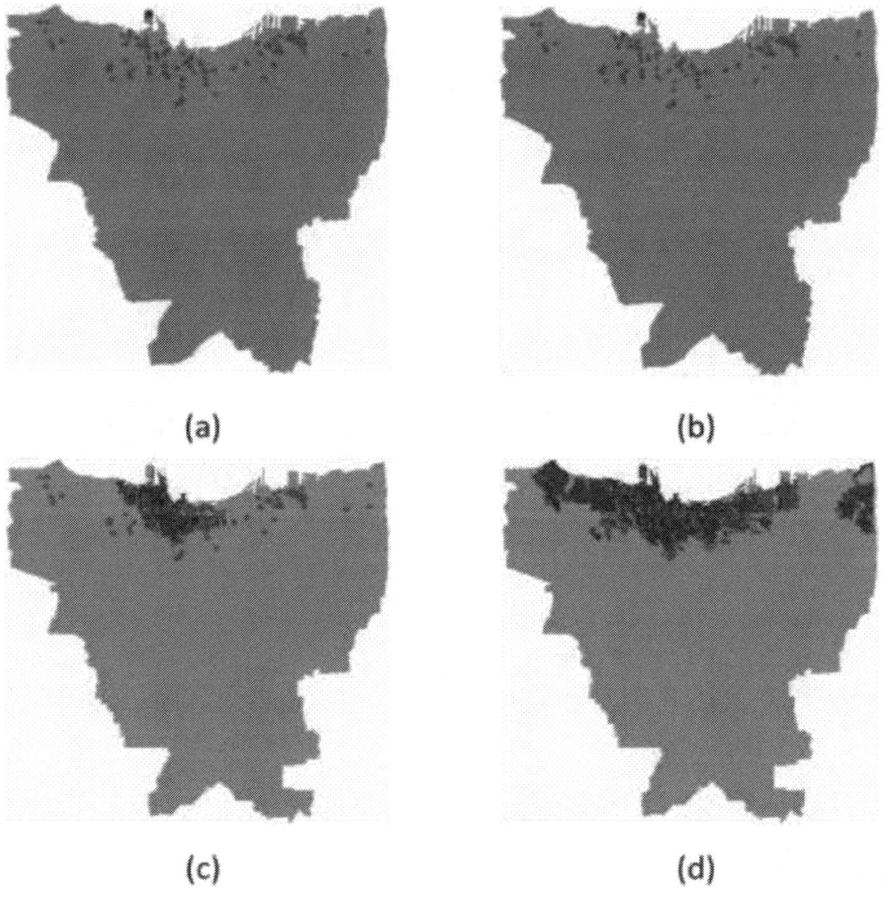

Figure 5. Visualization of coastal inundation along with plotting coastal inundation in the field each by using heigh reference system ellipsoid surface (a), Geoid EGM 2008 (b), Mean Sea Level (c) and High Tide (d)

To get a clearer picture that the most reliable way of modeling is to use the High Tide system, you can see the image below showing the modeling results along with the documentation of coastal inundation in the field. By using logic, when high tides occur in the field, there will be potential for flooding in several places whose positions are already lower than sea level. It should be noted here that the High Tide system is very local in nature, it will be different in each region. For this reason, good tide monitoring will be the key.

Figure 6. Visualization of coastal inundation along with plotting coastal inundation in the field and photo field documentations

So from our investigation, it is very clear that the model generated from the High Tide system provides results that are suitable for conditions in the field, so that we can use the results of the model for disaster risk mapping, including making decisions on how to reduce disaster risk. Once again, we can imagine that if the model is underestimated or overestimated, disaster risk reduction efforts can be misguided.

4. Conclusions

One important step on mitigation and adaptation against land subsidence and coastal inundation is understanding the coastal inundation through modelling. Coastal inundation modelling is important to do so that we get potential hazards and disaster risks from the occurrence of it. With coastal conditions that are usually inhabited by many people, become economic centres and others, then when coastal inundation occurs, material losses will certainly occur.

The most important thing related to coastal inundation modeling are the accuracy of the DEM and the correct choose of height of the reference system. If we do not use an accurate DEM, and do not use the height reference system correctly, then we will not get a good coastal inundation model. The impact is that we will not be correct map the hazards and disaster risk due to the coastal inundation. The prove we can see from research above in the case of Jakarta. The most appropriate height reference system for modeling coastal inundation is the High Tide System.

So, do not ignore in using the heigh reference system when we modelled coastal inundation. For some cases we still see publication with ignorance in the use heigh reference system, in this case model produced seem overestimate or under estimate. Latter on there will be miss calculation of real risk of disaster.

5. Acknowledgment

Thanks to all research associate in Institute of Technology Bandung. Also, thanks to student in Institute of Technology Bandung who help research in the field. Last but not least is very much thanks to the local people for sharing experience about where and when they suffered coastal inundation. That is data validation in the field

References

[1] Andreas, H., Zainal Abidin, H., Pradipta, D., Anggreni Sarsito, D., Gumilar, I. (2018) Insight look the subsidence impact to infrastructures in Jakarta and Semarang area; Key for adaptation and mitigation. MATEC Web of Conference 147, 08001 (2018)

[2] Sutanta, H., Rachman. A., Sumaryo and Diyono (2005) Predicting Land Use Affected by Land Subsidence in Semarang Based on Topographic Map of Scale 1:5.000 and Leveling Data. Proceedings of the Map Asia 2005 Conference (in CDRom), Jakarta, 22-25 August 2005.

[3] Mazzotti, S., Lambert, A., Van der Kooij, M., & Mainville, A. (2009) Impact of anthropogenic subsidence on relative sea-level rise in the Fraser River delta. Geology, 37(9), 771–774, http://dx.doi.org/10.1130/G25640A.1.

[4] Abidin, H.Z R. Djaja, D. Darmawan, S. Hadi, A. Akbar, H. Rajiyowiryono, Y. Sudibyo, I. Meilano, M. A. Kusuma, J. Kahar, C. Subarya (2001) Land subsidence of Jakarta (Indonesia) and its geodetic-based monitoring system. *Nat. Hazards J.* Int. Soc. Prevention Mitigation Nat. Hazards 23 (2/3), 365–387

[5] Bell, J., Amelung, F., Ramelli, A., & Blewitt, G. (2002) Land subsidence in Las Vegas, Nevada, 1935–2000: New geodetic data show evolution, revised spatial patterns, and reduced rates. *Environmental and Engineering Geoscience*, 8(3), 155–174.

[6] Cabral-Cano, E., Dixon, T. H., Miralles-Wilhelm, F., Diaz-Molina, O., Sanchez-Zamora, O., & Carande, R. E. (2008) Space geodetic imaging of rapid ground subsidence in Mexico City. *Geological Society of America Bulletin*, 120(11–12), 1556–1566, http://dx.doi.org/10.1130/B26001.1.

[7] Chaussard Estelle, Amelung F, Abidin HZ, San Hoon Hong (2013) Sinking cities in Indonesia: ALOS PALSAR detects rapid subsidence due to groundwater and gas extraction. *Remote Sensing of Environment* 128 (2013) 150 -161 http://dx.doi.org/10.1016/j.rse.2012.10.015

[8] Dixon, T. H., Amelung, F., Ferretti, A., Novali, F., Rocca, F., Dokka, R., et al. (2006) Space geodesy: Subsidence and flooding in New Orleans. *Nature*, 441(7093), 587–588, http://dx.doi.org/10.1038/441587a.

[9] Koudogbo, F. N., Duro, J., Arnaud, A., Bally, P., Abidin, H. Z., and Andreas, H. (2012) Combined X- and L-band PSI analyses for assessment of land subsidence in Jakarta, *Proc. SPIE, Remote Sensing for Agriculture, Ecosystems, and Hydrology* XIV, 8531, 853107, doi:10.1117/12.974821, 2012.

Fragmentation Evaluation of Some Selected Mudrock Indiced by Physical and Mineralogical Properties

Misbahudin[1], P R D Perdana[1], D Y Fatimah[1], F F Amanda[1], D P Kusumastuti[2], B Wicaksono[2], A S N Chairat[3]

[1]Geology Engineering Program, Faculty of Exploration and Production Technology, Universitas Pertamina, Teuku Nyak Arief Simprug Street, Kebayoran Lama, South Jakarta 12220, Indonesia
[2]Civil Engineering Program, Institut Teknologi Perusahaan Listrik Negara (IT-PLN), Menara PLN, Lingkar Luar Barat Street, Duri Kosambi, West Jakarta 11750, Indonesia
[3]Industrial Engineering Program, Institut Teknologi Perusahaan Listrik Negara (IT-PLN), Menara PLN, Lingkar Luar Barat Street, Duri Kosambi, West Jakarta 11750, Indonesia

misbahudin@universitaspertamina.ac.id

Abstract. Not many mudrock in Indonesia have been evaluated for slaking characteristics. Mudrock disintegrates rapidly when interacting with the atmosphere and hydrosphere. This research investigates the fragmentation of slaked claystone from the Kerek Formation. These formations are known for causing road, slope, and land-strengthening engineering problems along Jalan Semarang-Bawen, Middle Java, Indonesia. In this research, the method includes quantitative analysis of rock slaking level, which collides with the physical and mineralogical properties of the rocks. The slaking test with natural exposure for four cycles was carried out on as many as five samples. It is found the durability parameter value for all samples generally shows a slight difference. The durability index and Fragment Size Distribution Ratio in the 4-month cycle may be considered to evaluate the degree of fragmentation of material mudrock. RGL.0708.10 have an average DIR of 11.81%, and RGI.0908.1 is 2.18%. The value of FSD_{Ratio} for RGI.0908.2, RGL.0908.1, RGI.0908.1, and RGI.0708.3 is 1.00, whereas RGI.0708.10 is 0.9997. The indices in this research can be projected and considered to be used in another formation sample with engineering problems. The mineralogical composition of smectite minerals does not show a significant relationship to affect the fragmentation of mudrock with R^2 .0.268 in month-4 for the D_{IR} and 0.1322 for the FSD_{Ratio}. On the microscopic texture of the sample, there is the presence of calcite minerals, and a modest relationship exists between calcite content and its effect on the fragmentation of mudrock with R^2 for D_{IR} and FSD_{Ratio}, respectively 0.6797 and 0.4732. In addition, physical properties such as natural water content highly influence mudrock's durability with R^2 for D_{IR} and FSD_{Ratio}, respectively 0.8321 and 0.9023.

Keywords: fragmentation, mudrock, physical properties, slaking, smectite.

1. Introduction

Mudrock is a sedimentary rock exposed to atmospheric interactions; the mudrock becomes smaller fragments with fragment size variations—the mudrock classification in this research is based on [14]. The disintegration of the rock resulted in numerous construction problems, such as slope stability, foundations, and underground excavation problems [1]. For example, [4] found that the Surco flysch in Spain has many slopes problem and vast outcroppings with material variety due to the deterioration of clay-bearing rock. The degree of disintegration can vary depending on the type of mudrock. Mudrocks disintegrate rapidly due to interaction between the atmosphere and the water environment [5].

This research uses the method from [6] to assess mudrock durability. Durability is a measure to evaluate the degree of mudrock resistance to weathering or disintegration phenomena affected by water content changes. This assessment includes the wetting-drying process. Wetting and drying cycle tests usually assess the disintegration behavior of mudrocks in slake-durability tests [7]. Therefore, wetting and drying tests without mechanical interference on slake durability were performed among the disintegration index methods [3]. After the sample was wet-dried, the mudrocks slaked into smaller fragments. According to [8], the slake durability test is unsuitable for concluding the long durability period. However, according to [9], it conducted the slake-durability test with modification and proposed a new decay index that considers the duration time of the test and disintegrated fragment size. The slake-durability test is overestimated to the force and stress that work on samples, whereas the weathering condition in the field is more complex than in a laboratory. According to [5], the high durability of mudrock correlates to the induration factor indicated by dry density and void ratio. Contrarily, absorption and specific gravity in high value will affect the durability in a lower value.

Based on [10], it conducted a study and field survey that some clay shales in West Java, Indonesia, push some soldier piles and cause a wall collapse. This phenomenon triggers the wall at the toe part and landslide the road, and the culvert breaks, severed by seepage and indicating that the water influences durability. According to [11], it found some foundation failures in the Hambalang area, West Java, due to weathering rock of the Jatiluhur Formation. Based on [12], it tested the samples from this area and shows Jatiluhur Formation has more durable characteristics than the Kerek Formation from Semarang-Bawen, Middle Java.

Not many mud rocks in Indonesia have been evaluated for slaking characteristics so are Kerek Formation. These formations are known for causing road, slope, and land-strengthening engineering problems along the Semarang-Bawen toll road. The climate condition of this area has high rainfall and intense weathering levels that allow the natural agitation of rocks to continue rapidly. However, when it takes a field survey, many disintegrations that took place intensively led to severe slope issues and foundations for toll roads. The problem is also present in slope stability, such as surface erosion and landslide on the many slopes. Based on [13], this formation consists of an alternation of claystone, marl tuffaceous, sandstone, conglomerate, volcanic breccia, and limestone. Description for claystone is light grey-dark, calcareous, partly interlayered with siltstone or sandstone; locally contains forams, mollusca, and coral colonies. The conglomerate layer occurs in claystone at Kali Kripik; and in sandstone. Limestone is commonly bedded, crystalline, and sandy, with a total thickness of more than 400 m. This formation formed in the Middle Miocene age. In delivering the research problem, this research investigates the mudstones from the Kerek Formation and then evaluates their fragmentation resulting from slaking. The sampling is taken from 5 outcrops, as shown in figure 1. Some new parameter is proposed in this research to evaluate the durability of mudrock.

2. Materials and Method

2.1. Rock sampling

The material for laboratory examination is collected from the fresh condition rock. Rock's description and name classification are based on [14]. This field rock shows the layering; when interacting with water, it is easy to form the smaller fragment. The procedure for a treat the sample in the field is based on [15]. The sample is in the condition of undisturbed material to prevent the release of stress that makes

IOP Conf. Series: Earth and Environmental Science 1245 (2023) 012021 doi:10.1088/1755-1315/1245/1/012021

the sample not natural again. If the release stress is allowed, the sample will cause rapid disintegration. The dimension of samples is determined by cubic shape with 18 cm x 18 cm x 18 cm. Next, the sample is appropriately protected by wrapping plastic and aluminum foil—this protects to prevent the change of water content and temperature of the sample when it takes to the laboratory. Finally, the box-shaped samples are placed into multiplex slabs with bubble wrap filling to prevent mechanical shocks during transportation.

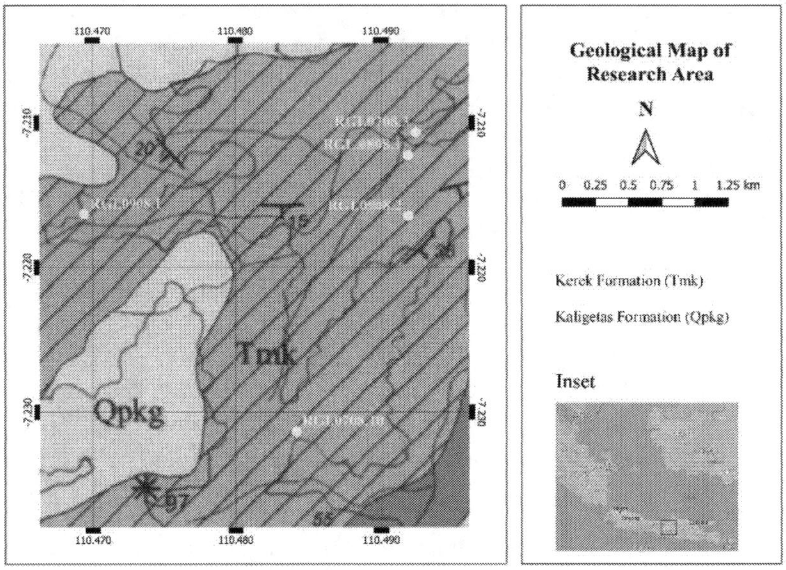

Figure 1. Sampling location of the research area overlying the geological map that source from [13].

2.2. *Natural Expose Slaking Test*

Sample preparation in 10 fragments with each mass of 45-60 g. The sample is prepared by drying under sun exposure until constant weight before continuing to be naturally exposed on the mesh size 2 mm for a 4-month cycle of natural agitation (figure 2). The sample will experience wetting and drying from the natural environment in natural exposure. After one month and in dry condition, the sample is sieved by a series of mesh sizes of 25.4 mm, 19 mm, 12.7 mm, 6.3 mm, 4 mm, and the last 2 mm. After natural exposure slaking, the slaked material will be fragmented into distributed sizes. Each sample retained in the sieve is weighed. Each weight is then plotted on the fragment size distribution curve. Some parameters are considered to evaluate the material's durability to disintegration. Dry weight after each cycle (W_n) and initial condition (W_0) is calculated for the Durability Index of retained material (D_{IR}) following equation (1).

$$D_{IR} = \frac{W_n}{W_0} \times 100\% \quad (1)$$

Fragment Size Distribution Ratio (FSD_{Ratio}) is proposed to measure the material's durability degree in more detail, considering the retained material in each mesh size.

2.3. *Physical-Mineralogical Properties*

The physical properties test is based on [16] procedure of water content, absorption, density, and porosity. The pore-related parameter is immersed by ethylene glycol against the effect of water on the disintegration of mudrock. The density parameter is measured by dividing the samples' dry mass and volume, whereas the porosity is calculated by dividing pore volume by sample volume. Water content is related to the mass of water in natural conditions compared to the dry mass of rock, whereas absorption uses the mass of water in every pore in rock. X-Ray Diffraction (XRD) is used to obtain the material's composition and identify the presence of smectite group minerals based on semiquantitative analysis in Match software. The semiquantitative analysis considers the mineral's peak intensity value and the difference between minimum and maximum Bragg 2θ in each crystal phase. The database of minerals used in XRD is from JCPDS International Centre for Diffraction Data. The smectite group in the samples can be identified from this card database as vermiculite and montmorillonite. The observation found that the sample contains calcareous; the calcite content is determined to identify the effect on durability. For the microscopic observation, the Scanning Electron Microscope (SEM) is conducted to overview the textural condition of samples.

Figure 2. Natural exposure slaking test on the claystone samples.

3. Results and Discussion

3.1. Rock Description and Mineralogical Examination
The outcrop condition in the field has a great variety of weathering degrees. The lithology that was found is equivalent to Kerek Formation. The fresh condition of mudrock samples lay on road slopes and river slopes (figure 3). Some outcrops show slight weathering with some intense cracks formed by natural processes. The megascopic description of mudrock consists of determining the lithology of claystone. Claystone has brownish grey with clay grain size. The lithology has calcareous, low compactness, and conchoidal structure. When the rock is dripped with chloride acid, the reaction occurs, shown by a bubble reaction on the surface. To obtain more details of the mineralogical composition, it conducts X-Ray Diffraction with the result shown in table 1. Smectite group minerals consist of vermiculite and montmorillonite; calcite content represents the calcareous condition in samples.

3.2. Physical Properties Test Results
Table 1 summarizes the physical properties of absorption, dry density, water content, and void ratio. Natural water content ranges from 6.27% to 15.76% for all samples. RGI.0908.2 has the highest water content value, and RGI.0708.10 has the lowest. From the dry density result, RGI.0908.1 has the highest value of 1.96 g/cm3, and RGI.0908.2 and RGI.0708.10 has the lowest value with a similar result of 1.80 g/cm^3. The void ratio ranges between 0.40 and 1.82 for all samples. Therefore, the lowest and the highest, respectively, are RGL-0908.1 and RGI.0708.3. Those parameters indicate the compactness and induration level. The absorption value ranges from 15.79 to 33.38%. Therefore, the absorption value of RGI.0708.3 has the highest value, and the lowest is RGL.0908.1. When the void ratio is high, the absorption is also a high value. The value variation for the smectite group of minerals is shown for all samples except RGL-0908.1, which has no smectite minerals. RGI.0708.3 has the highest percentage of

smectite group minerals as the absorption and void ratio are highest. Based on calcite content, RGI.0908.2 has the highest content, whereas the lowest is in RGI.0908.1.

Figure 3. The typical outcrop of Kerek claystone in the field lie on the river slope showing the bedding planes and conchoidal structure.

Table 1. Physical properties of mudrock samples.

Sample	Water content (%)	Absorption (%)	Void ratio	Dry density (g/cm^3)	Smectite Group Minerals (%)	Calcite Content (%)
RGI.0908.2	15.76	16.40	0.42	1.80	4.70	28.90
RGL.0908.1	12.41	15.79	0.40	1.82	0.00	11.50
RGI.0908.1	9.14	21.91	0.76	1.96	3.90	6.80
RGI.0708.3	9.18	33.38	1.82	1.93	7.80	13.50
RGI.0708.10	6.27	29.15	1.11	1.80	6.10	20.00

3.3. Natural Expose Observation
After going through the sun drying to obtain initial dry weight, samples start with a hairline crack. The samples were then exposed to natural conditions during the first month. The cycle is repeated until the fourth month. After one month, the sample is sieved with various mesh sizes. The conditioned sample per month is shown in figure 4. Based on the figure, after the first month, all samples had experienced complete disintegration, with fragmented material still retained in notable appearance. In month 2, all samples begin to be retained less fragmented material and continue in month 3. Finally, in the last month, all samples have no retained material; even so, the sample remains fragments with almost no amount.

3.4. Durability Index of Retained Mudrock
After the sample is exposed to natural slaking, the sample is sieved by various mesh sizes. The various fragmented material sizes are plotted in each month's fragment size distribution curve. Each sample has its distribution curve, as shown in figure 5. Each sample represents the various characteristics of the percentage of material retained in a series of sieve sizes. The mesh has a size of 2 mm, 4 mm, 6.3 mm,

12.7 mm, 19 mm, and 25.4 mm. The quantitative analysis is applied to this curve as represented by the Durability Index of Retained Fragment (D_{IR}). This index refers to the amount of retained material in a cumulative 2 mm sieve that is dried after a specific cycle and then divided by the initial dry condition weight. All samples break into smaller fragments at the end of the cycle. In month-1, all samples break into smaller fragments, as samples experience complete disintegration with various mesh sizes. In month-2, only RGI.0708.10 had fragments retained in 2 mm mesh. No fragments were retained in another mesh size. In month-3 and month-4, all samples have no retained material in all mesh. This condition shows that the claystone sample in the research area has low durability.

Sample | Initial Condition | After Month-1 | After Month-2

RGI.0908.2

RGL.0908.1

RGI.0908.1

RGI.0708.3

RGI.0708.10 

Figure 4. Condition of samples on 2 mm mesh size from the fragmented sample after the cycle of disintegration for four months.

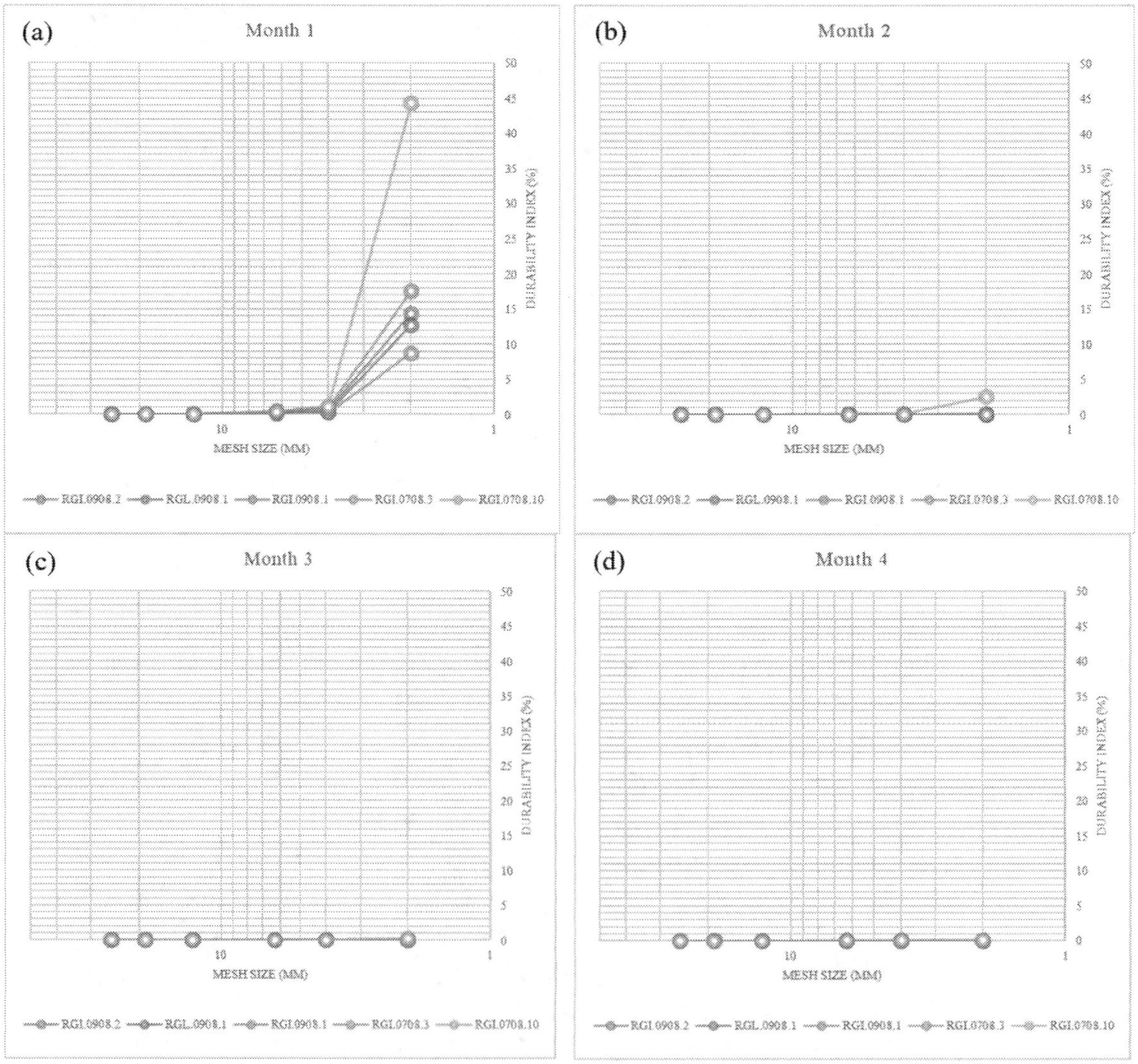

Figure 5. Fragment size distribution curve for (a) Month-1, (b) Month-2, (c) Month-3, (d) Month-4.

Based on the average value of D_{IR} during the entire cycle for all rock samples, RGI.0708.10 are more durable than other rock samples. These samples have an average D_{IR} of 11.81%, while RGI.0908.1 has a lowest D_{IR} value of 2.18%. It should be noted that lower the D_{IR} value than lower the durability of the sample to disintegration. When the durability index is used for evaluating the material's degree of resistance to weathering in only 2 mm size, it is proposed to obtain the different indices that reveal how fragment size represents a characteristic of durability material. The proposed parameters, the Fragment Size Distribution Ratio (FSD_{Ratio}), are modifications taken from Erguler and Shakoor [17]. In this modification, the area above the curve divided by the total area encompassing the distribution curve is used as equation (2).

$$FSD_{Ratio} = \frac{Area\ above\ curve}{Area\ total} \tag{2}$$

For example, as shown in Figure 6, the RGI.0708.10 have an area above curve 2291.57, and the total area coverage is 2340, the FSD_{Ratio} value in month 1 is 0.9793. The value of the FSD_{Ratio} ranges from 0 to 1. Value approached 1 shows a high level of disintegration; in contrast with close to 0, the samples experience low disintegration.

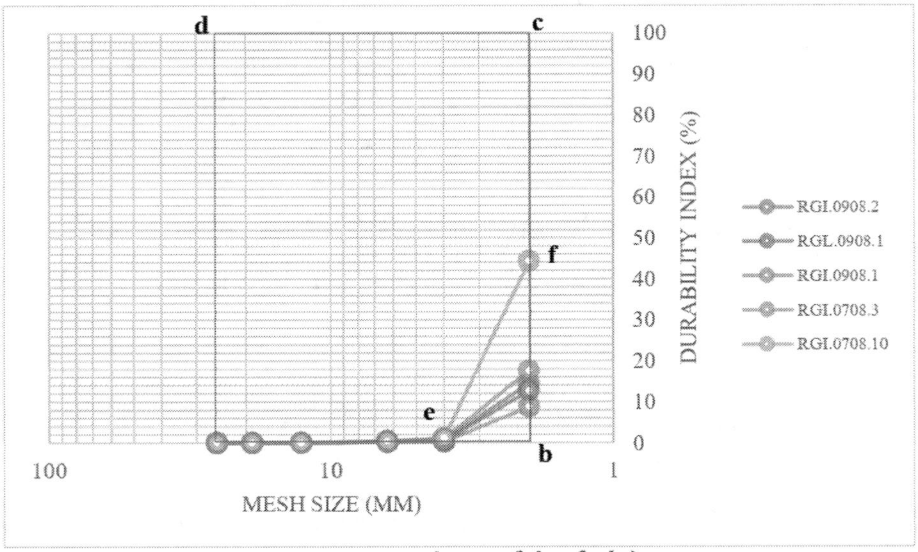

$$FSD_{Ratio} = \frac{Area\ of\ (aefcda)}{Area\ of\ abcda}$$

Figure 6. The proposed method to calculate the FSD_{Ratio} using an example Month 1 graph.

Table 2 shows the recapitulation of the value FSD_{Ratio}. The tabulation is made for month-1 until month-4. Most samples have the FSD_{Ratio} range of 0.9793 - 1 in the first month and reach the maximum value of 1 in month-4. All samples show low durability to disintegration, indicated by the high value of the FSD_{Ratio}.

Table 2. FSD_{Ratio} value for each month.

Sample	FSD_{Ratio} Month-1	FSD_{Ratio} Month-2	FSD_{Ratio} Month-3	FSD_{Ratio} Month-4
RGI.0908.2	0.9927	1.0000	1.0000	1.0000

RGL.0908.1	0.9939	1.0000	1.0000	1.0000
RGI.0908.1	0.9957	1.0000	1.0000	1.0000
RGI.0708.3	0.9907	1.0000	1.0000	1.0000
RGI.0708.10	0.9793	0.9986	0.9996	0.9997

For the rate of disintegration, it is plotted the cycle month and FSD_{Ratio} (figure 7). All samples reached a maximum of 1 of the FSD_{Ratio} in month 4. The rate disintegration value ranges from 0.0013 to 0.0062, with the unit $FSD_{Ratio}/$ month. This rate is obtained from the gradient of the line equation.

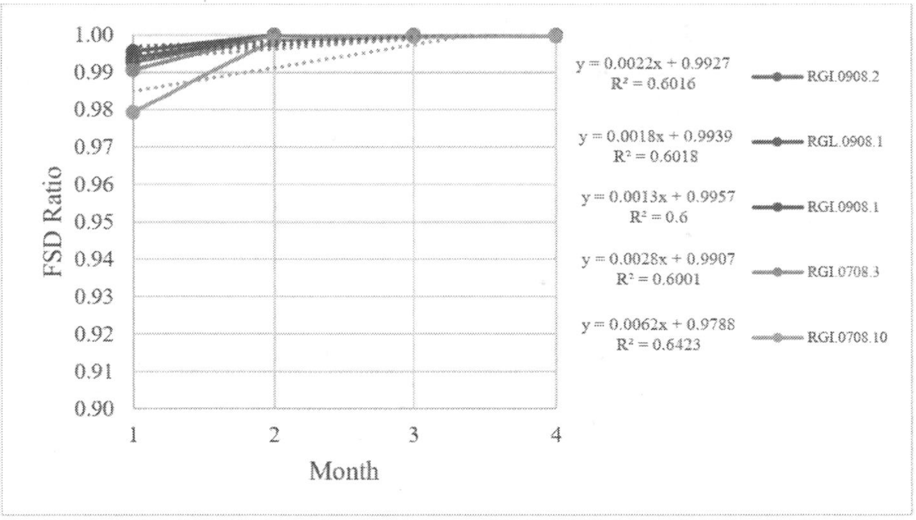

Figure 7. The rate of disintegration is calculated from the line equation of the graph between the FSD_{Ratio} and cycle.

The regression analysis method is used to show the trend of the two variables [18, [19]. How strongly the independent variable on the dependent variable is indicated by deterministic coefficients (R^2). The x-axis in the correlation graph is the independent variable, whereas the y-axis is the dependent variable.

For the correlation between D_{IR} and FSD_{Ratio}, the independent variables are the durability index, while the dependent variable is the Fragment Size Distribution Ratio. In figure 8, there is a significant correlation between the D_{IR} value and FSD_{Ratio} following satisfies the deterministic coefficient (R^2) with a linear trend. This significant influence occurred in each of the cycle months.

3.5. Correlation of Physical-Mineralogical Properties to the Durability Parameter

The various geological and engineering parameters express the disintegration variety of mudrock [5], [20]. In this section, the correlation is obtained by finding the influence parameter on the durability of mudrock. The x-axis in this correlation is a void ratio, absorption, dry density, water content, smectite group minerals content, and calcite content, while the D_{IR} and FSD_{Ratio} is plotted as the y-axis. Table 3 and 4 summarizes the correlation between the physical and mineralogical properties and durability indices. The durability parameter in month 4 shows a high value of R^2. Therefore, the D_{IR} and FSD_{Ratio} in month 4 are considered the durability evaluator to slaking.

The R^2 value between D_{IR} to void ratio, water content, absorption, dry density, smectite content, and calcite content are 0.5748, 0.8321, 0.5209, 0.0765, 0.2680, and 0.6797. From this value, the regression analysis between the D_{IR} and the physical properties of mudrock indicate a significant influence on the water content and modest for the calcite content, void ratio, and absorption. The identic correlation occurred in bivariate analysis between the FSD_{Ratio} and physical properties. The R^2 value between FSD_{Ratio} to void ratio, water content, absorption, dry density, smectite content, and calcite content are 0.5725, 0.9023, 0.3839, 0.1694, 0.1322, and 0.4732. This value shows a significant influence of water content on durability and a modest influence on the calcite content and void ratio to fragmentation. On the other hand, Smectite group minerals that consist of montmorillonite and vermiculite do not significantly affect slaking material. Same with the dry density that does not significantly influence durability. The result is that all correlations show a high correlation in the parameter natural water content; for the void ratio and calcite content, the level is modest.

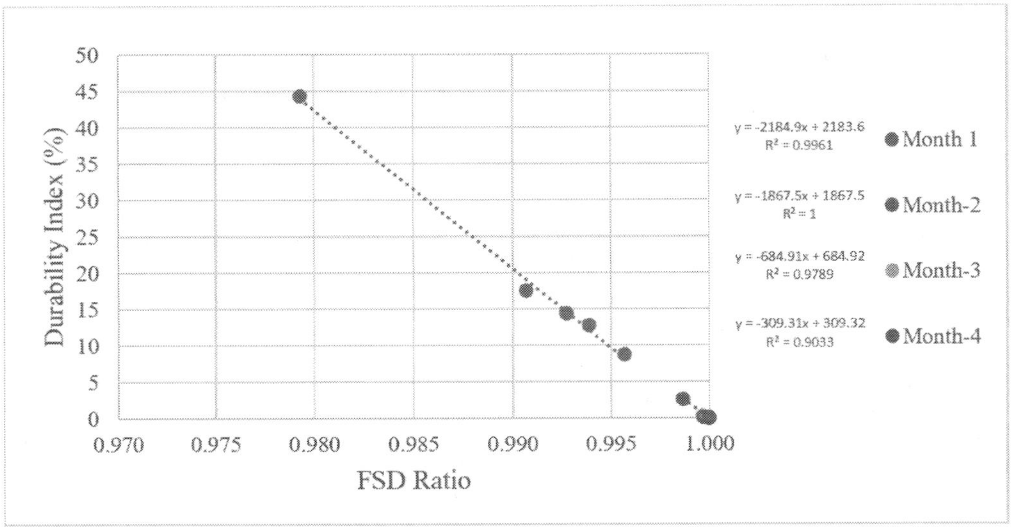

Figure 8. The significant correlation between D_{IR} and FSD_{Ratio}.

According to [5], the high claystone absorption value correlates to larger pore space. The pore water pressure in the rock makes the grain lose cohesion and produces a crack that disintegrates the mudrock. Natural water content in this research has a high value of R^2 that significantly affect the disintegration. According to [9], the percent of water absorption is the most critical parameter controlling mudrocks' durability. However, the properties that may significantly affect the disintegration of mudrock are not always the parameters that will significantly affect the disintegration. Refer to [22]; it revealed no significant relationship between slake durability in the second cycle with chemical composition and mineralogy. This research found that calcite content moderately influences slaking, and the absorption is a low correlation, although the void ratio is moderate. To investigate this, the Scanning Electron Microscope is conducted to evaluate the microstructure of samples.

Table 3. Correlation between D_{IR} and physical and mineralogical properties.

Cycle	Void Ratio	Water Content (%)	Absorption (%)	Dry density (g/cm^3)	Smectite Content (%)	Calcite Content (%)
Month 1	$y = -34.542x^2 + 83.626x - 18.157$ $R^2 = 0.4431$	$y = 0.92x^2 - 22.948x + 149.56$ $R^2 = 0.8588$	$y = -0.1003x^2 + 5.8058x - 56.449$ $R^2 = 0.3176$	$y = 1911x - 7.634$ $R^2 = 0.2357$	$y = 2.0305x + 10.405$ $R^2 = 0.1739$	$y = -0.1781x^2 + 7.0508x - 37.407$ $R^2 = 0.6175$
Month 2	$y = -3.4165x^2 + 7.9072x - 2.8586$	$y = 0.0718x^2 - 1.8088x + 10.872$	$y = -0.0141x^2 + 0.7447x - 8.4987$	$y = -12.07\ln(x) + 8.0496$	$y = 0.122x - 0.007$	$y = -0.0122x^2 + 0.4774x - 3.2726$

	$R^2 = 0.5676$	$R^2 = 0.8909$	$R^2 = 0.3513$	$R^2 = 0.202$	$R^2 = 0.0986$	$R^2 = 0.4332$
Month 3	$y = -0.2935x^2 + 0.7129x - 0.2554$	$y = 0.0061x^2 - 0.1584x + 0.9926$	$y = -0.001x^2 + 0.0551x - 0.6427$	$y = -0.993\ln(x) + 0.6849$	$y = 0.0149x - 1E-04$ $R^2 = 0.1562$	$y = -0.0014x^2 + 0.0545x - 0.357$
	$R^2 = 0.5414$	$R^2 = 0.869$	$R^2 = 0.3987$	$R^2 = 0.1448$		$R^2 = 0.5729$
Month 4	$y = -0.0933x^2 + 0.2431x - 0.0882$	$y = 0.002x^2 - 0.0534x + 0.3485$	$y = -0.0002x^2 + 0.0142x - 0.1747$	$y = -0.274\ln(x) + 0.1997$	$y = 0.0074x - 0.0042$ $R^2 = 0.268$	$y = -0.0006x^2 + 0.0226x - 0.1445$
	$R^2 = 0.5748$	$R^2 = 0.8321$	$R^2 = 0.5209$	$R^2 = 0.0765$		$R^2 = 0.6797$

Table 4. Correlation between FSD_{Ratio} and physical and mineralogical properties.

Cycle	Void Ratio	Water Content (%)	Absorption (%)	Dry density (g/cm³)	Smectite Content (%)	Calcite Content (%)
Month 1	$y = 0.0149x^2 - 0.0368x + 1.0074$	$y = -0.0004x^2 + 0.0104x + 0.9314$	$y = 4E-05x^2 - 0.0023x + 1.0219$	$y = 0.0712\ln(x) + 0.9462$	$y = -0.001x + 0.9951$ $R^2 = 0.215$	$y = 8E-05x^2 - 0.0033x + 1.0172$
	$R^2 = 0.439$	$R^2 = 0.8526$	$R^2 = 0.3432$	$R^2 = 0.2143$		$R^2 = 0.6514$
Month 2	$y = 0.0018x^2 - 0.0042x + 1.0015$	$y = -4E-05x^2 + 0.001x + 0.9942$	$y = 8E-06x^2 - 0.0004x + 1.0046$	$y = 0.0065\ln(x) + 0.9957$	$y = -6E-05x + 1$ $R^2 = 0.0977$	$y = 2E-06x^2 - 7E-05x + 1.0005$
	$R^2 = 0.5705$	$R^2 = 0.8914$	$R^2 = 0.3526$	$R^2 = 0.2014$		$R^2 = 0.4596$
Month 3	$y = 0.0005x^2 - 0.0011x + 1.0004$	$y = -1E-05x^2 + 0.0002x + 0.9985$	$y = 2E-06x^2 - 0.0001x + 1.0012$	$y = 0.0016\ln(x) + 0.9989$	$y = -2E-05x + 1$ $R^2 = 0.1116$	$y = 2E-06x^2 - 7E-05x + 1.0005$
	$R^2 = 0.567$	$R^2 = 0.8916$	$R^2 = 0.3654$	$R^2 = 0.1837$		$R^2 = 0.4596$
Month 4	$y = 0.0004x^2 - 0.0009x + 1.0003$	$y = -8E-06x^2 + 0.0002x + 0.9988$	$y = 1E-06x^2 - 8E-05x + 1.0009$	$y = 0.0013\ln(x) + 0.9992$	y = 1e-2E-05x $R^2 = 0.1322$	$y = 1E-06x^2 - 6E-05x + 1.0004$
	$R^2 = 0.5725$	$R^2 = 0.9023$	$R^2 = 0.3839$	$R^2 = 0.1694$		$R^2 = 0.4732$

3.6. Microstructure texture

For identifying the fabric of mudrock, there are some representative figures about the result of the Scanning Electron Microscope from RGI.0708.3 (figure 9). From figure 9, based on the crenulate morphology, element content, and their concentration ratio such as Si, Al, Fe, Mg, Ca, and O, mudrock samples show the presence of montmorillonite with the formula of this mineral $(Ca)_{0.33}(Al, Mg)_2(Si_4O_{10})(OH)_2 \cdot nH_2O$. In this photograph, it can also be seen that samples have some pores like inter or intragranular porosity generated from the irregular arrangement of the fossil shells made of calcium carbonate. Furthermore, it shows the abundant presence of calcareous nannofossils such as coccolith, discoaster, and numerous fossil fragments. This calcium carbonate from the samples is also confirmed by the XRD result, which reflects the presence of calcite in considerable amounts. This observation leads to whether fossil/ calcite content affects the durability of mudrock. From the XRD analysis (figure 10), RGI.0708.3 has the montmorillonite and calcite. Along the other samples both minerals are also present in other samples (figure 11).

IOP Conf. Series: Earth and Environmental Science 1245 (2023) 012021 doi:10.1088/1755-1315/1245/1/012021

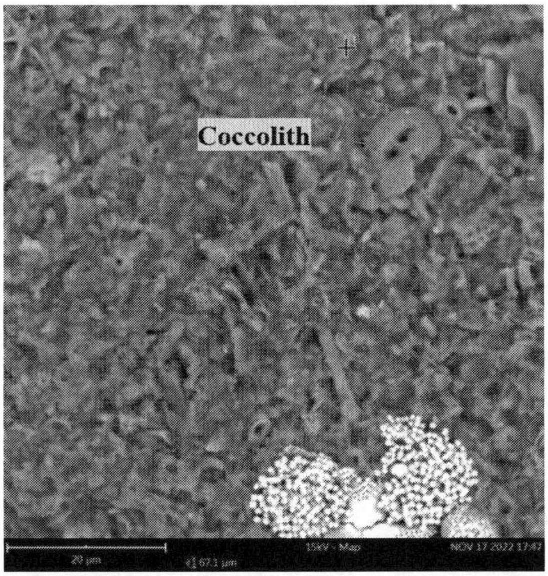

Element Number	Element Symbol	Element Name	Atomic Conc.	Weight Conc.
8	O	Oxygen	65.55	52.71
14	Si	Silicon	8.19	11.56
20	Ca	Calcium	4.39	8.85
26	Fe	Iron	3.01	8.44
6	C	Carbon	12.74	7.69
13	Al	Aluminium	5.05	6.85
49	In	Indium	0.39	2.27
51	Sb	Antimony	0.17	1.01
12	Mg	Magnesium	0.51	0.62

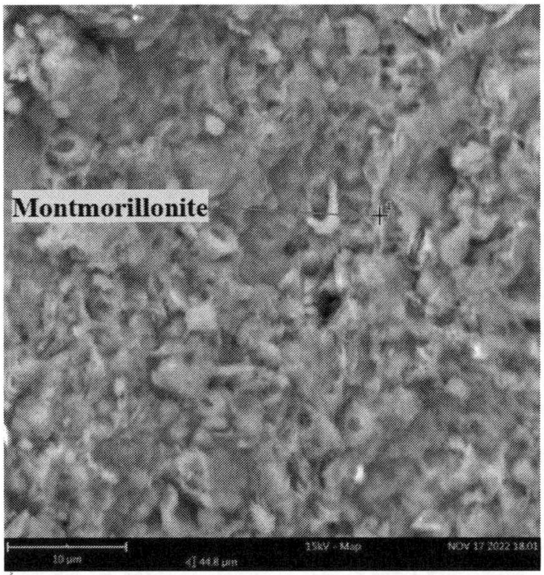

Element Number	Element Symbol	Element Name	Atomic Conc.	Weight Conc.
8	O	Oxygen	66.51	55.63
14	Si	Silicon	8.85	12.99
20	Ca	Calcium	3.94	8.26
6	C	Carbon	12.91	8.10
13	Al	Aluminium	4.90	6.92
26	Fe	Iron	1.43	4.19
49	In	Indium	0.41	2.48
12	Mg	Magnesium	0.96	1.22

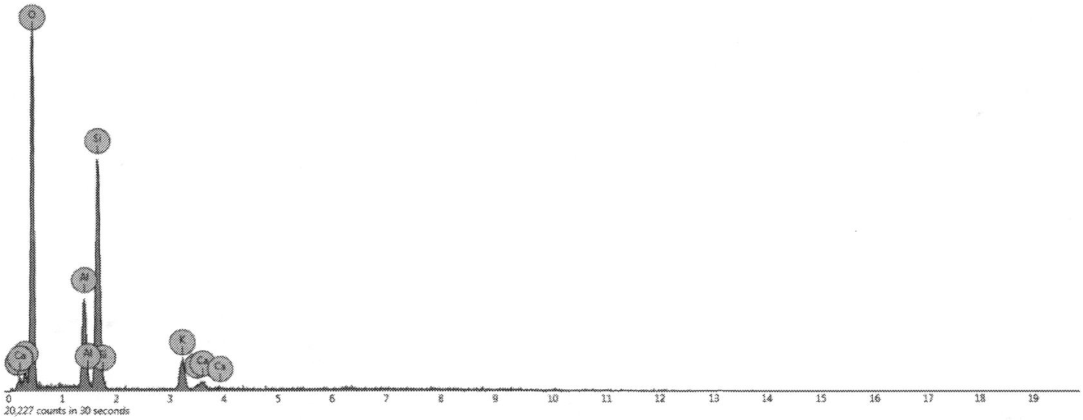

Figure 9. The Scanning Electron Microscope image with EDS results for RGI.0708.3.

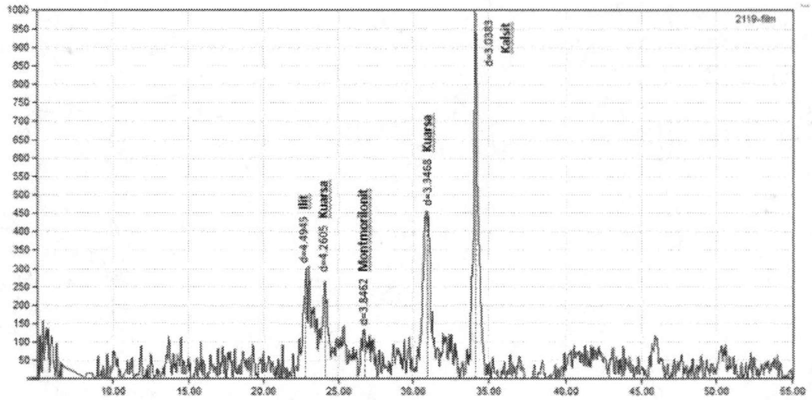

No	Mineral	%
1	Palygorskite	26.8
2	Illite	25.5
3	Kaolinite	17.3
4	Calcite	13.5
5	Quartz	9.2
6	Montmorillo nite	4.2
7	Vermiculite	3.5

Figure 10. The XRD results from Sample RGI.0708.3.

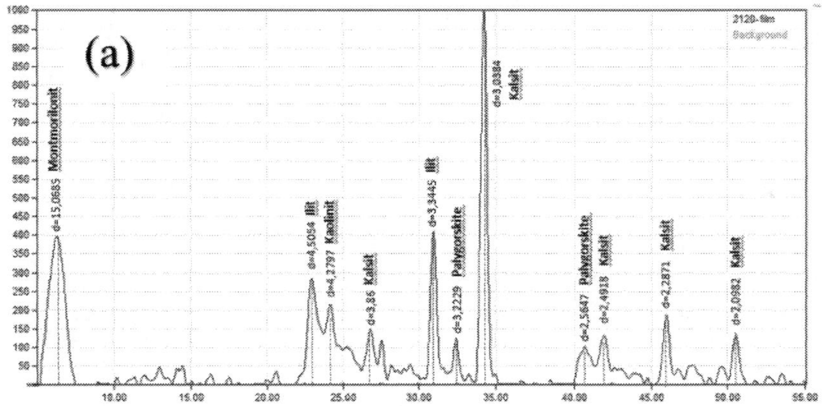

No	Mineral	%
1	Calcite	42.4
2	Kaolinite	22.0
3	Quartz	15.1
4	Illite	12.3
5	Montmorillonite	3.7
6	Palygorskite	3.5
7	Vermiculite	1.0

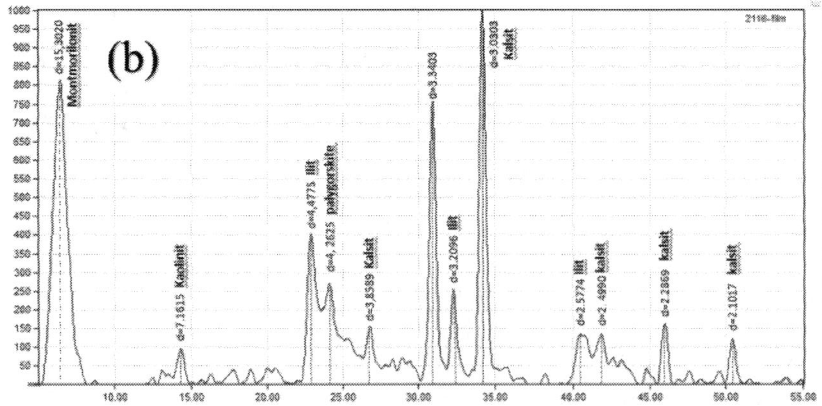

No	Mineral	%
1	Illite	36.9
2	Palygorskite	22.6
3	Kaolinite	18.8
4	Calcite	11.5
5	Quartz	10.2
6	**Montmorillonite**	**0**
7	**Vermiculite**	**0**

IOP Conf. Series: Earth and Environmental Science 1245 (2023) 012021 doi:10.1088/1755-1315/1245/1/012021

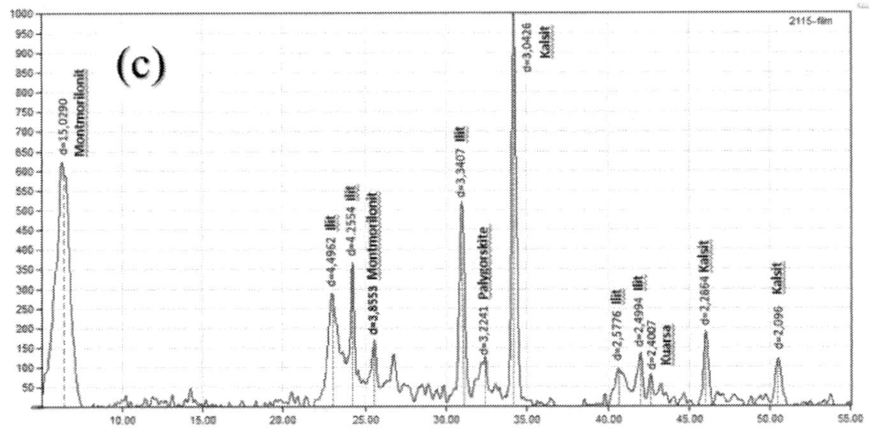

No	Mineral	%
1	Illite	79.2
2	Calcite	6.8
3	Kaolinite	4.7
4	Quartz	4.2
5	Montmorillonite	3.7
6	Palygorskite	1.2
7	Vermiculite	0.2

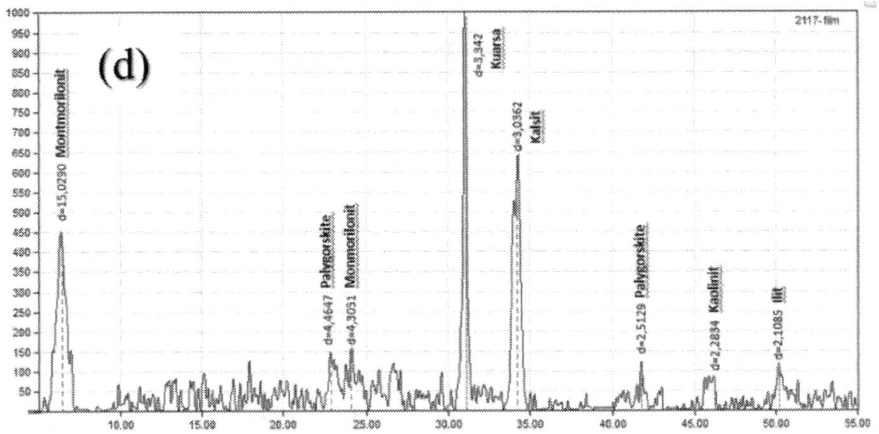

No	Mineral	%
1	Palygorskite	36.5
2	Calcite	20.0
3	Quartz	18.4
4	Illite	12.2
5	Kaolinite	6.8
6	Vermiculite	4.9
7	Montmorillonite	1.2

Figure 11. XRD diffractogram shows calcite and smectite group mineral present for (a) RGI.0908.2, (b) RGL.0908.1, (c) RGI.0908.1, and (d) RGI.0708.10.

From the table, the correlation between calcite content and FSD_{Ratio}, and D_{IR} shows in month-4 a modest correlation. It has a trend of higher calcite content, the durability higher to disintegration. From the graph, calcite content from fossils fills the claystone pore and may prevent water from entering the pore and make pore water pressure cause the slaking.

4. Conclusion

The mudrock of the Kerek Formation may indicate various degrees of disintegration based on quantitative parameters. The natural exposure slaking test simulates the weathering condition where the rock is exposed. Most samples disintegrate into smaller fragments even during natural exposure's first month. The D_{IR} and FSD_{Ratio} show each other that all samples in the research area show low durability to disintegration in the range of value 2.18-11.81%. After month-4, all samples have a value of FSD_{Ratio} 1.00. This value expresses complete disintegration and no retained fragment on the mesh. Evaluating the mudrock for an extended period of exposure is essential. The samples have no retained material for the case of claystone in this research area in month-4. The proposed method to calculate the durability parameter can be applied to other mudrock samples in Indonesia. Mudrock with low durability and rapid disintegration can affect engineering activities, such as causing slope failure or foundations in the road where highway construction in the research area takes place. Some parameters, such as calcite and natural content, have considerable influence on the durability of mudrock with R^2 respectively 0.6797 and 0.8321 for the D_{IR} whereas for the FSD_{Ratio} is respectively 0.4732 and 0.9023. Further investigation through microfabric observation is considered to identify the characteristics in more detail of the internal structure of the samples.

Acknowledgments
Thank Universitas Pertamina for the collaboration research scheme with Institut Teknologi Perusahaan Listrik Negara (IT-PLN) for supporting the funding of this research.

References
[1] Moon, V.G., Beattie, A.G., 1995. Textural and microstructural influence on the durability of Waikato coal measures mudrocks. Quarterly Journal of Engineering Geology 28, 303-312.
[2] Sadisun, I.A., Shimada, H., Ichinose, M., Matsui, K., 2005. Study on the physical disintegration characteristics of Subang claystone subjected to a modified slaking index test. Geotech. Geol. Eng. 23, 199-218. doi:10.1007/s10706-003-6112-6.
[3] Misbahudin, Sadisun, I.A., 2019. The use of disintegration ratio in evaluating rock durability in selected mudrock samples in Indonesia. Journal of Geoscience, Engineering, Environment, and Technology 4 (3), 158-170. doi: 10.25299/jgeet.2019.4.3.2331.
[4] Cano, M., Tomas, R., Riquelme, A., 2017. Relationship between monitored natural slaking behavior, field degradation behavior and slake durability test of marly flysch rocks: preliminary results. Procedia Engineering 191: 609-617.
[5] Dick, J.C., Shakoor, A., Wells, N., 1994. A geological approach toward developing a mudrock-durability classification system. Can. Geotech. J. 31, 17-27. doi:10.1139/t94-003.
[6] ASTM, 1996. Annual Book of ASTM Standards, Soil and Rock, Construction: v. 8, Section 4, West Conshohocken, PA. 950 p.
[7] Franklin, J.A., Chandra, R., 1972. The slake-durability test. Int. J. Rock Mech. Min. Sci. 9, 325–328. doi:10.1016/0148-9062(72)90001-0.
[8] Gautam, T.P., Shakoor, A., 2016. Comparing the slaking of clay-bearing rocks under laboratory conditions to slaking under natural climatic conditions. Rock Mechanics and Rock Engineering 49: 19-31.
[9] Heidari, M., Momeni, A., Mohebbi, Y., 2018. Durability assessment of clay-bearing soft rocks by using new decay index. Periodica Polytechnica Civil Engineering. doi: 10.3311/PPci.11284.
[10] Oktaviani, R., Rahardjo, P.P., Sadisun, I.A., 2018. Landslides induced by slaking of geomaterial. MATEC Web of Conferences 229: 1-6.
[11] Sani, R.A., Muslim, D., Zakaria, Z., Misbahudin, 2017. Daya dukung tanah pondasi dangkal pada Bukit Hambalang Selatan, Kecamatan Citeureup, Kabupaten Bogor. Jurnal Lingkungan dan Bencana Geologi 8 (3), 143-152.
[12] Alatas, I.M., Kamaruddin, S.A., Nazir, R., Irsyam, M., 2016. Effect of weathering on disintegration and shear strength reduction of clay shale. Jurnal Teknologi (Sciences & Engineering) 78: 93-99.
[13] Thanden RE, Sumadirdja H, Richards PW, Sutisna K and Amin TC 1996. Geological map of the Magelang and Semarang sheets, Jawa, Pusat Survei Geologi, Bandung.
[14] Potter, P.E., Maynard, J.B., and Pryor, W.A., 1980. Sedimentology of Shale. Springer-Verlag, New York.
[15] Clayton, C. R. I., Matthews, M. C., dan Simons, N. E., 1987. Site Investigation, Department of Civil Engineering, University of Surrey, 15p.
[16] ISRM, 2007. The complete ISRM suggested methods for rock characterization, testing, and monitoring: 1974-2006 in Ulusay, R., Hudson, J.A. (Eds.). Suggested methods prepared by the commission on testing methods, International Society for Rock Mechanics (ISRM), Ankara, Turkey.
[17] Erguler ZA, Shakoor A. 2009. Quantification of the fragment size distribution of clay-bearing rocks after slake-durability testing. Environ Eng Geosci. 15, 81-89.

[18] Davis, J.C., 1973. Statistics and data analysis in geology. John Willey & Sons, New York.
[19] Czajka, R., 1994. Determination of variability of physical properties in a selected layer (from the Baltic cliff) using statistical methods. Bull. Int. Assoc. Eng. Geologists 49, 33–39. doi: 10.1007/BF02594998
[20] Gautam, T.P. and Shakoor, A 2013. Slaking behavior of clay-bearing rocks during a one-year exposure to natural climatic conditions, Engineering Geology, 166: 17-25.
[21] Czerewko, M.A. Cripps, J.C., 2001. Assessing the durability of mudrocks using modified jar slake index test. Quarterly Journal of Engineering Geology and Hydrogeology 34, 153–163. doi: 10.1144/qjegh.34.2.153
[22] Koralegedara, N.H., Maynard, J.B., 2017. Chemical, mineralogical and textural properties of the Kope Formation mudstones: how they affect its durability. Engineering Geology 228: 312-322. doi: 10.1016/j.enggeo.2017.08.025.

IOP Conf. Series: Earth and Environmental Science 1245 (2023) 012022 doi:10.1088/1755-1315/1245/1/012022

Velocity Changes Associated to the 2019 Ambon 6.5 Mw Earthquake Using Ambient Seismic Noise

A Amiruddin[1*], N T Puspito[2], D P Sahara[2], A D Nugraha[2], Zulfakriza[2]

[1] Earth Science Master's Program, Faculty of Earth Science and Technology, Institut Teknologi Bandung, Indonesia
[2] Global Geophysics Research Group, Faculty of Mining and Petroleum Engineering, Institut Teknologi Bandung, Indonesia

corresponding author's e-mail : ahmadamir2626@gmail.com

Abstract. Seismic velocity changes which occurred before, during and after large magnitude earthquake carry information about damage fault surrounding the area. Research focused on Ambon 6.5 Mw earthquake on September 26, 2019. We use ambient noise cross-correlation technique to compute empirical Green's Function between station pairs from January to December 2019. We use doublet method to determine perturbation calculated in frequency domain by using Moving Windows Cross Spectrum in frequency between 0.1 to 1.0 Hz. Our results show no obvious significant velocity change before the mainshock. Velocity fluctuation average about ±0.1% caused by non-isotropic of noise source around the area. Sudden velocity decrease occurred and reached average value of -0.35% in 7 days after mainshock. After reaching its lowest value, velocity gradually increased about +0.4% in 14 days. Velocity fluctuations about ±0.2% also occurred for three months after the earthquake which may caused by the aftershocks around the earthquake epicentre. Following the large magnitude mainshock, velocity reduction may be induced by cracks opening and fractures due to deformation. While the velocity increase may caused by post-seismic relaxation due to elastic behaviour and crack healing of the fault throughout the surrounding region.

1. Introduction

Monitoring of crustal properties in earthquake-prone region is very useful to obtain the information of properties and seismic activity. Seismic surface wave velocities always change based on elastic behavior of earth's crust and stress activity around the area. Ambient seismic noise can be used to evaluate the crustal velocity changes which associated to stress change underneath. Velocity changes which happened after mainshock can be governed by the post seismic stress relaxation in crustal area around the fault zone [3].

Ambon 6.5 Mw earthquake on September 2019 cause deformation and strong shaking in crustal and surface area around the fault zone. These strong shaking and deformation in seconds cause changes in seismic velocity because the changes in elastic properties. In this research, we investigate the temporal changes of seismic velocity in the epicentral region of mainshock by using vertical component from 4 seismometers around the area (Figure 1a). To determine the velocity changes, we use ambient noise cross correlation method to obtain the empirical Green's Function that mimics the impulse response between two stations [6]. To calculate the velocity changes, we use Moving Window Cross Spectrum (MWCS) analysis, using earthquake multiplets [10].

IOP Conf. Series: Earth and Environmental Science 1245 (2023) 012022 doi:10.1088/1755-1315/1245/1/012022

Ambon earthquake on 6.5 Mw located between Seram island and Ambon Island detected by BMKG station network. Ambon earthquake triggered by north-south fault plane orientation [11] and cause the displacement around ~4 cm horizontal and `10 in vertical [8]. In this study, the station pair spacing are around 45 km for the nearest station AAI – KRAI while the furthest station pair is MSAI – NLAI which around 210 km. We need to note that all stations are located on different island except for KRAI and

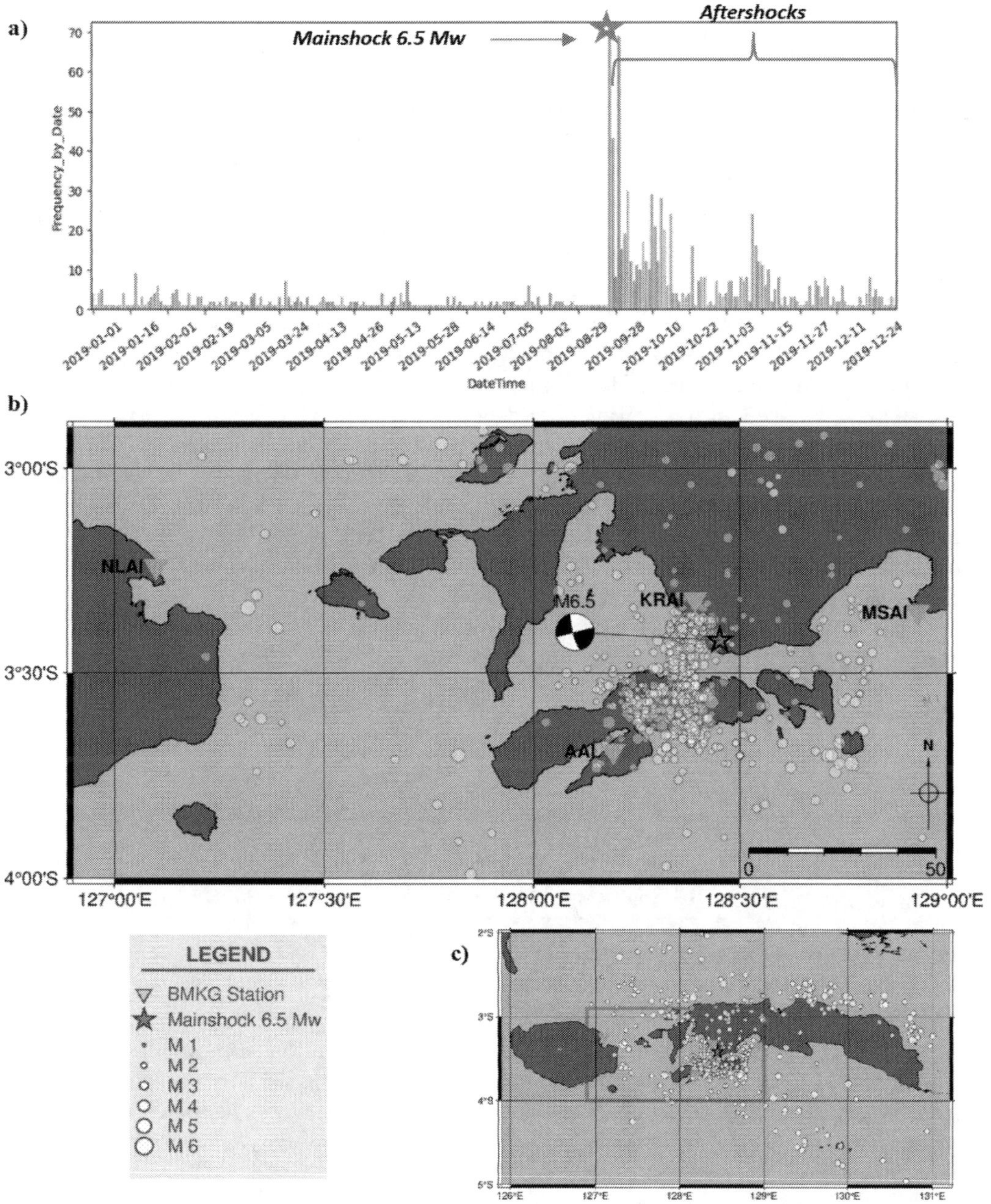

Figure 1. a) Daily seismicity frequency detected around the area of Ambon in map c). Red star denotes the mainshock. b) Map of epicenter of earthquake 6.5 Mw inside the red rectangle area c). White circles show the earthquake events January – Desember 2019. Earthquake events catalog, location and focal mechanism obtained from BMKG catalog.

MSAI because the source of ambient noise mainly from ocean microseism which may cause non isotropic source distribution [5]. This study aims to obtain seismic velocity changes in pre-seismic, co-seismic and post- seismic phase which associated to the 26 September Ambon 6.5 Mw mainshock.

2. Data and Method

This study uses waveform data acquired from continuous measurement on vertical component from 4 seismometers around Ambon: AAI KRAI, MSAI and NLAI recorded by BMKG from January to December 2019. The missing and incomplete daily waveform are detected in manual checking especially after the mainshock on NLAI station. To reduce the effect of large phases caused by earthquakes, we used filter and adjustments applied to waveform each station. The filter and analysis of waveform generally follows the common procedure [2] to obtain reliable Green's Function using the MSNoise package in Python [7].

First, we demeaned the waveform to extinguish the trend effect of waveform reading. Then we applied bandpass filter to between 0.1 – 1.0 Hz to obtain the ambient noise frequency [1]. For temporal normalization, we used one bit normalization to as it is most robust to remove amplitude caused by earthquake aftershocks [2]. Meanwhile for spectral whitening which aimed to suppress frequency, we used several experimental frequency ranges adjusted in the bandpass and whitening filter to evaluate process in MWCS (Table 1) and analyze which frequency shows the clearest velocity change. Then after the signal pre-processed, the daily waveform data cross correlated between two stations (Figure 2a). For daily pair of stations, signals were cross correlated in 30 mins windows resulting in positive and negative time lags and then stacked in linear method. 9 months of daily GF processed for reference stacking from January 1 to September 1, 2019, or before the mainshock occurred to obtain stable Green's Function. Meanwhile for current stacking, we used 10 days to 20 days of moving windows. The result of cross correlation and current stacking also can be used to analyze the direct and scattered arrival wave to determine the window length for MWCS process.

Table 1. Filter used in filtering and MWCS processing.

Bandpass	Whitening	MWCS
0.1 - 1.0 Hz	0.1 - 1.0 Hz	0.1 - 1.0 Hz
0.1 - 1.0 Hz	0.1 - 0.8 Hz	0.1 - 0.8 Hz
0.1 - 1.0 Hz	0.1 - 0.6 Hz	0.1 - 0.6 Hz
0.1 - 1.0 Hz	0.1 - 0.4 Hz	0.1 - 0.4 Hz
0.1 - 1.0 Hz	0.1 - 0.2 Hz	0.1 - 0.2 Hz

We use the MWCS method to obtain the relative velocity variation measurement. MWCS applied to time-series which computed by cross correlating noise between two stations. The reference stacking which has longer stacking duration as representative of background value when the medium is relatively stable (Figure 2b). Meanwhile, current stacking has shorter duration than reference stacking represents the situation at a given time period of crust (Figure 2c). The time delay between two Green's Function (reference and stacking) determined in the unwrapped phase of cross spectrum, proportional to frequency [7]. The delay for each window calculated by using weighted least square within the frequency needed.

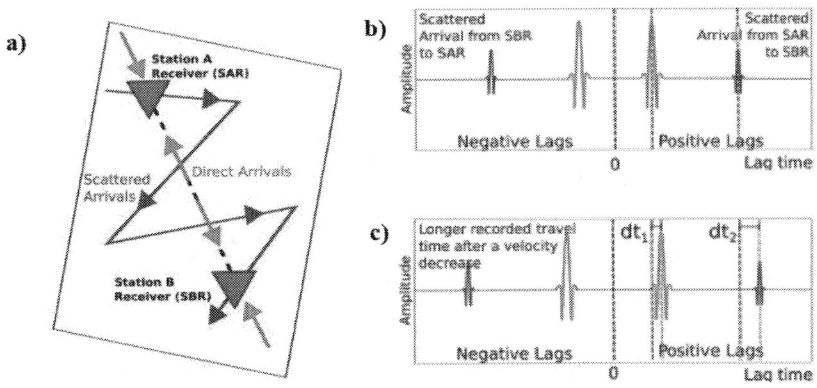

Figure 2 Adapted from [12], shows illustration of how velocity changes measured from cross correlation when a) direct and scattered arrival recorded on two stations. Cross correlation in stable medium produces cross correlation in figure b) shows in positive and negative lags have no delay. Meanwhile in c) when in unstable condition, the cross correlation will produce time lags.

3. Result and Discussion

3.1. Cross Correlation Result

After the cross correlation and stacking process, we need to analyse the amplitude or energy recovered in positive and negative time lags to determine the window length used for MWCS process. The cross correlated energy in Green's Function criteria must be enough to process in MWCS and obtain the recovered energy in coda wave. To enhance the reliability, we also do the SNR analysis to obtain the coherent signal energy in time domain for different frequencies (Figure 4g). Clear direct arrivals are observed in all station pairs in zero-time lag in red arrow (Figure 4a – f). On the other hand, the scattered arrival which shown in yellow arrows cannot be observed in station pair AAI-NLAI, KRAI-NLAI and MSAI-NLAI, while observed clearly in other station pairs. From the cross-correlation and SNR analysis, we decided to choose the positive and negative time lag between 40 s and 90 s to avoid the direct and scattered arrival (Figure 4 a – f).

3.2. Velocity Variation

To obtain the velocity change around the epicentral region which caused by nonlinear strong ground motion, we observed the highest velocity reduction and correlated with the event timeline. While the high fluctuation after the mainshock may prompted by the aftershock around the area. Several research about temporal velocity variation using ambient noise cross correlation have shown that temporal changes were mainly constrained around hypocentre and ruptured area [4]. Seismic velocity in cracked medium also dependent on stress. When crack density decrease, the velocity increased caused by an increase of pressure without any fluid participation [9].

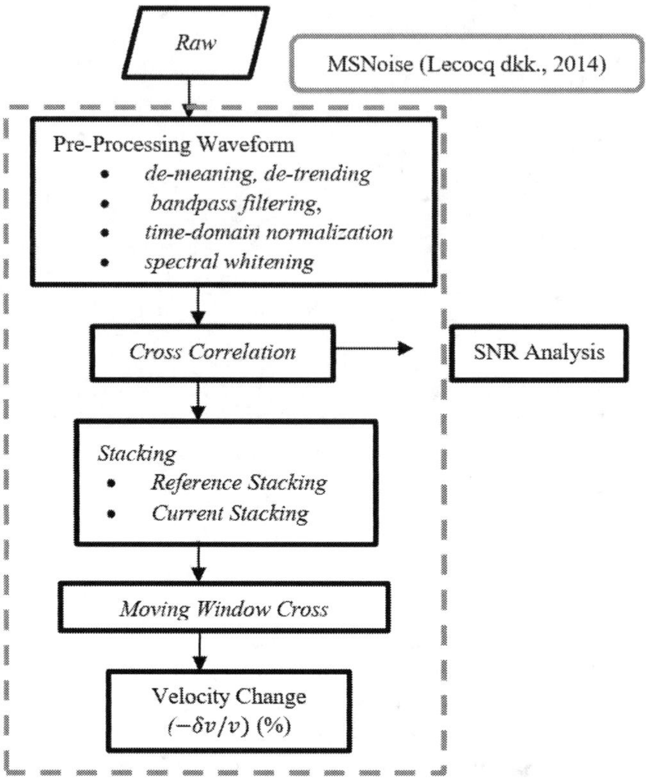

Figure 3 Workflow data processing

The velocity variation result presented in Figure 5 in different frequency ranges a) 0.1 – 0.2 Hz, b) 0.1 – 0.4 Hz, c) 0.1 – 0.6 Hz, d) 0.1 – 0.8 Hz and e) 0.1 – 1.0 Hz and different current stack days 2, 5, 8, 10, 12, 15, 18 and 20 days. We focused on velocity variation of co-seismic phase. In overall velocity variation, the fluctuations before the mainshock vary between ± 0.25% but shows very high fluctuation in filter range 0.1 – 0.2 Hz, 0.1 – 0.4 Hz and 0.1 – 1.0 Hz which may provoked by high non-isotropic ambient noise source around the area, so we decided to exclude them for analysis. While frequency range 0.1 – 0.6 Hz and 0.1 – 0.8 Hz show more clear fluctuations and velocity changes when mainshock occurred. Then the mainshock 6.5 Mw followed by hundreds of aftershocks around the epicentral region.

The velocity variation in co-seismic phase should only associated with nonlinear strong ground motion and deformation around the epicentral area. In this case, we focused on investigating the velocity variation with small obscuration of noise in frequency range between 0.1 – 0.6 and 0.1 – 0.8 Hz on 10 days and 12 days moving windows current stacking which have relatively low fluctuation. Before the mainshock occurred, the average temporal velocity relatively fluctuates between ±0.1% (Figure 6a and 6b). Ambient noise source mostly interfered by ocean microseism [1] while in this research also interfered by hollow waveform data few days after the earthquake because the velocity calculation composed by stacked correlation.

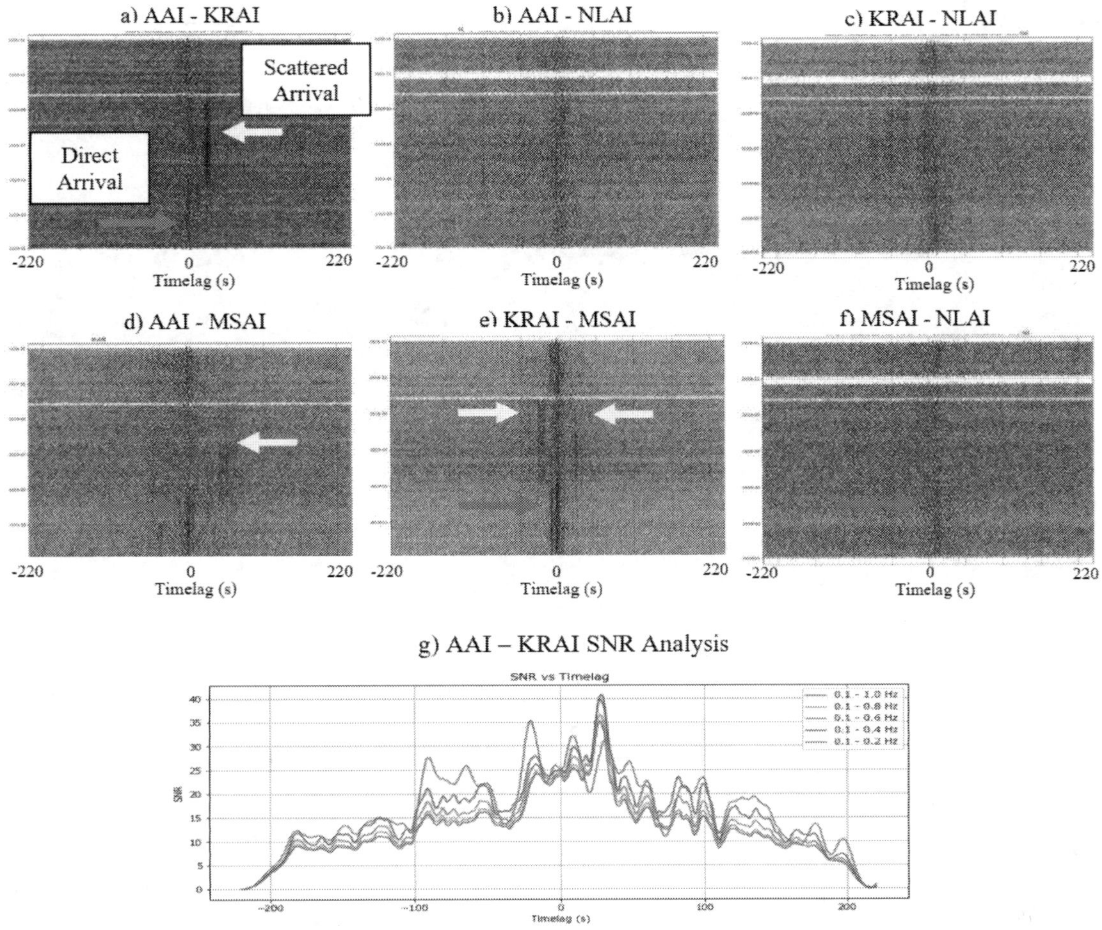

Figure 4. (a – f) Daily cross correlation result from each station pairs to analyse the direct and scattered arrival wave in positive and negative time lags. Red arrows denote the direct arrival wave and yellow arrows shows the clear scattered arrival around the time lags. And SNR analysis also applied to all cross correlation in different frequencies, as shown in figure g) for AAI – KRAI pair.

Significant temporal changes of velocity occurred when mainshock occurred with average of velocity reduction -0.25% between all station pairs. Seismic velocity dramatically decreased in all frequency bands and observed clearly in 10 days and 12 days current stack resolution (Figure 6a and 6b). The velocity decreased after the mainshock and keep declining gradually for 7 days. This declining pattern observed on all frequency range. The lowest value clearly observed -0.35% in frequency range 0.1 – 0.6 Hz in current stack 10 days (Figure 6c). After reached the lowest value, the velocity recovered in average of +0.4% in 14 days and bounced back to 100% to the value before the mainshock occurred. This pattern observed clearly in all frequency range. After the significant decreased and increased velocity change, the pattern of velocity become highly fluctuated around average of ±0.3%. This pattern also observed in all frequency range.

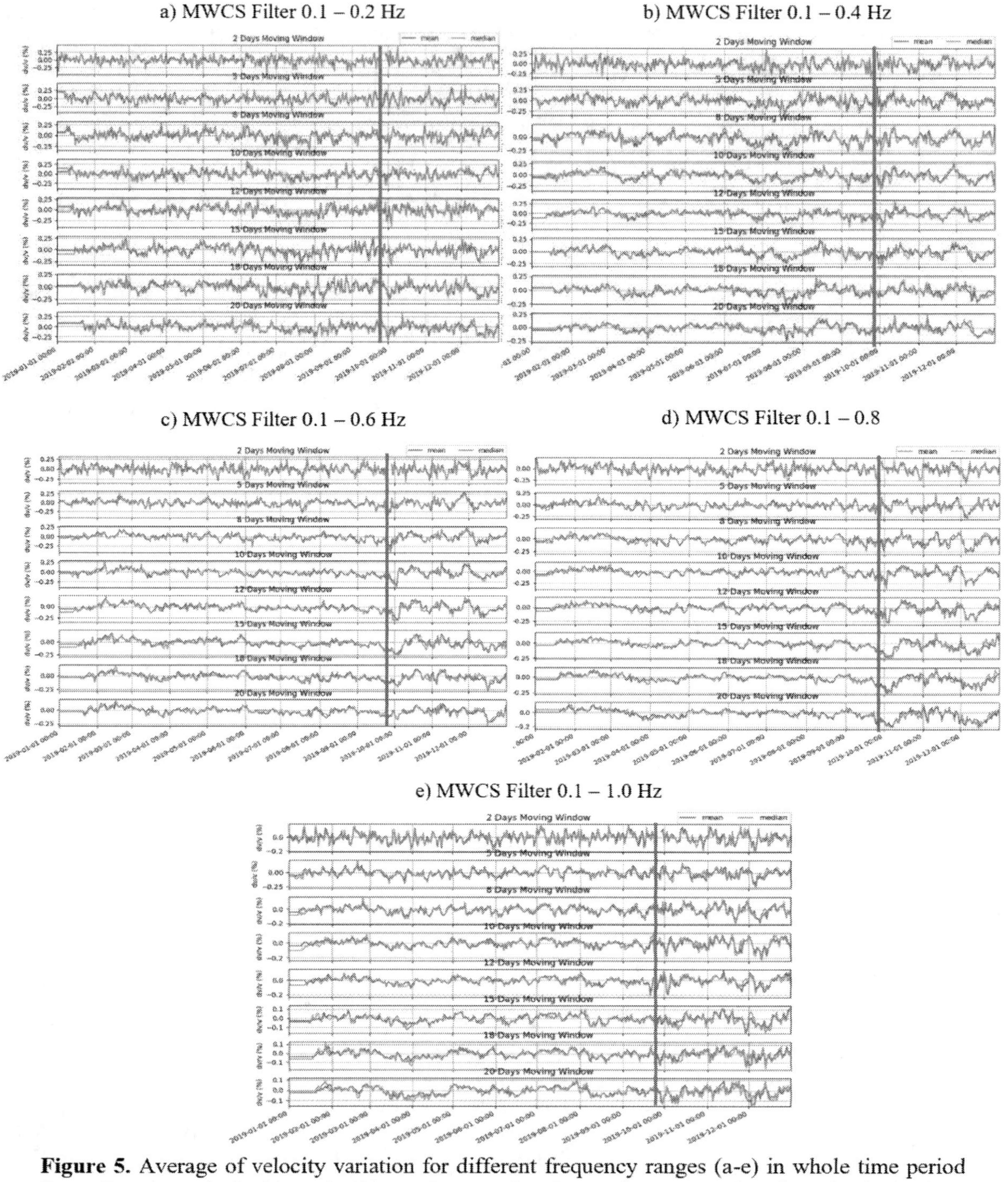

Figure 5. Average of velocity variation for different frequency ranges (a-e) in whole time period from all station pairs in this study. Blue and orange line denote the mean and median of velocity from all station pairs. Red line denotes when the mainshock on September 26, 2019.

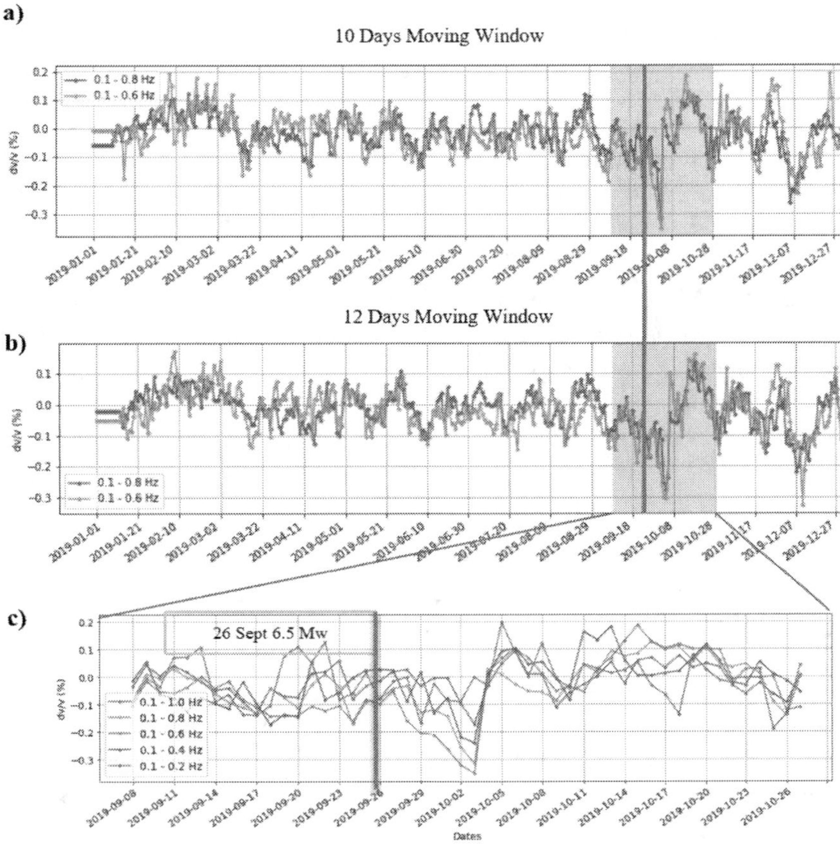

Figure 6. Velocity change focused on 0.1 –0.6 Hz filter of 10 and 12 days current stacking days in whole period study a) and b). Analysis focused on co-seismic phase with 0.1 – 0.6 Hz of 10 days current stack in c). Vertical red line denotes the mainshock time occurred.

The strong ground motion and deformation are the main cause of velocity decrease of Ambon. The velocity decreases mainly dominated in the crust area around the hypocentre depth and rupture area. The deformation process causes significant decrease and gradually decreased because of the deformation process. While the velocity increase after the mainshock shows the process of post-seismic relaxation. The recovery need 14 days to recover to 100% as before the mainshock occurred. And following the mainshock, the hundreds of aftershock events are responsible for the high fluctuation after the recovery (±0.2%). These mechanisms are often located deep around the hypocentre depth and are caused by cracks opening and closing.

4. Conclusion

The main objection of this study is to analyse the velocity change which associated to Ambon 6.5 Mw earthquake. We choose frequency band and current stack parameter which produced minimum fluctuation to make sure minimize the non-isotropic noise interference. We decided to use frequency band 0.1 – 0.6 Hz and current stack 10 days as the clearest velocity variation graph. We investigate there are no significant velocity change before the mainshock (January – August). The average temporal velocity relatively fluctuates around ±0.1% before the mainshock. When the mainshock occurred on September 26, the velocity decreased significantly in 7 days to -0.35%. This process mainly caused by deformation and rupture process around the epicentral region. After reach the lowest value, the velocity increased +0.4% or recovered 100% to the value before the mainshock occurred in 14 days. We interpret this shows post seismic relaxation when the rocks are in healing process after deformed. Following the mainshock, hundreds of aftershock events are responsible for the high fluctuation of velocity variation (±0.2%) until the end of observation period.

References

[1] Asten, M. W. dan Henstridge, J. D. (1984): Array estimators and the use of microseisms for reconnaissance of sedimentary basins. Geophysics, 49(11):1828–1837, ISSN 0016-8033.

[2] Bensen, G. D., Ritzwoller, M. H., Barmin, M. P., Levshin, A. L., Lin, F., Moschetti, M. P., Shapiro, N. M., and Yang, Y. (2007): Processing seismic ambient noise data to obtain reliable broad-band surface wave dispersion measurements. Geophysical Journal International, 169(3):1239–1260

[3] Brenguier, F., Shapiro, N. M., Campillo, M., Ferrazzini, V., Duputel, Z., Coutant, O., and Nercessian, A. (2008): Towards forecasting volcanic eruptions using seismic noise. Nature Geoscience, 1(2), 126–130.

[4] Brenguier, F., M. Campillo, T. Takeda, Y. Aoki, N. M. Shapiro, X. Briand, K. Emoto, and H. Miyake (2014), Mapping pressurized volcanic fluids from induced crustal seismic velocity drops, Science, 345(6192), 80–82, doi:10.1126/science.1254073.

[5] Froment, B., Campillo, M., Roux, P., Gouedard, P., Verdel, A., and Weaver, R. (2010): Estimation of the effect of nonisotropically distributed energy on the apparent arrival time in correlations. Geophysics, 75(5):SA85

[6] Lobkis, O. I., and R. L. Weaver (2001), On the emergence of the Green's function in the correlations of a diffuse field, J. Acoust. Soc. Am., 110(6) 3011–3017

[7] Lecocq, T., Caudron, C. and Brenguier, F. (2014): MSNoise, a Python Package for Monitoring Seismic Velocity Changes Using Ambient Seismic Noise, *Seismological Research Letters*, 85(3), 715-726.

[8] Meilano, I., Salman, R., Rahmadani, S., Shi, Q., Susilo, Lindsey, E., Supendi, P., and Daryono (2021): Source Characteristics of the 2019 Mw 6.5 Ambon, Eastern Indonesia, Earthquake Inferred from Seismic and Geodetic Data. Seismological Research Letters. 92.

[9] O'Connell, R. J. dan Budiansky, B., (1974): Seismic velocities in dry and saturated cracked solids, J. Geophys. Res.,79, 5412-5426.

[10] Ratdomopurbo, A. and Poupinet, G. (1995): Monitoring a temporal change of seismic velocity in a volcano: Application to the 1992 eruption of Mt. Merapi (Indonesia). Geophysical Research Letters, 22(7):775–778.

[11] Sahara, D. P., Nugraha, A. D., Muhari, A., Rusdin, A. A., Rosalia, S., Priyono, A., Zulfakriza, Widiyantoro, S., Puspito N.T., Rietbrock, A., Lesmana, A., Kusumawati, D., Ardianto, Baskara, A. W., Halauwet, Y., Ash Shiddiqi, H., Rafie, M. T., Pradisti, R., Mozef, P. W., Tuakia, M. Z., and Elly, E., (2021): Source mechanism and triggered large aftershocks of the Mw 6.5 Ambon, Indonesia earthquake, Tectonophysics, Volume 799, 2021, 228709, ISSN 0040-1951.

[12] Yates, A. S., Savage, M. K., Jolly, A. D., Caudron, C., and Hamling, I. J. (2019): Volcanic, coseismic, and seasonal changes detected at White Island (Whakaari) volcano, New Zealand, using seismic ambient noise. Geophysical Research Letters, 46, 99–108.

Geochemical and geological properties of tin related to ore deposits: A Review

A D Handoko[1,2], B Sapiie[3], Syafrizal[4], A Rudyawan[3]

[1] Doctoral Program of Geological Engineering, Faculty of Earth Sciences and Technology, Institut Teknologi Bandung, Jl. Ganesha No.10, Indonesia, Bandung 40132, 32020004@mahasiswa.itb.ac.id, https://orcid.org/0000-0002-1136-4243
[2] Research Center for Geological Resources, National Research and Innovation Agency, Indonesia, aryo002@brin.go.id, https://orcid.org/0000-0002-1136-4243
[3] Department of Geological Engineering, Faculty of Earth Sciences and Technology, Institut Teknologi Bandung, Jl. Ganesha No.10, Indonesia, Bandung 40132
[4] Department of Mining Engineering, Faculty of Mining and Petroleum Engineering, Institut Teknologi Bandung, Jl. Ganesha No.10, Indonesia, Bandung 40132

Abstract Tin have atomic number 50 and atomic mass 118,71, commonly found as cassiterite in nature. It is one of important elements in low carbon technologies. Nowadays, tin utilized in broad application range from tin can to electronics. Despite many uses, tin is only produced from limited areas. Because of broad application, understanding of tin in nature is important. Tin usually can be found together with wolfram, although it often occurs at different locations. Tin may occur in various ore deposit types such as greisen, vein, skarn, and placer. This article will describe the geochemical and geological properties of tin in nature, especially that related to ore deposits.

Keywords: tin, geology, geochemistry, ore deposits

1. Introduction

Tin is one of the important elements in green technology. Besides used in green technology, tin also used in tinplate, alloy, solder, and bar tin [1]. Researcher found and studied tin deposits around Atlantic Ocean namely Andean Belt, East Brazilian Belt, Rondônia-Guyana Belt, Rocky Mountain Belt, Appalachian Belt, Central African Belt, South West Africa-Nigeria Belt, Iberian Belt, Armorican Belt, and Erzgebirge Province [2]. Besides that, tin also can be found in Southeast Asia Tin Belt in which most of tin (more than 50%) originated from [3]. Figure 1 shows the global tin deposits location with historical production.

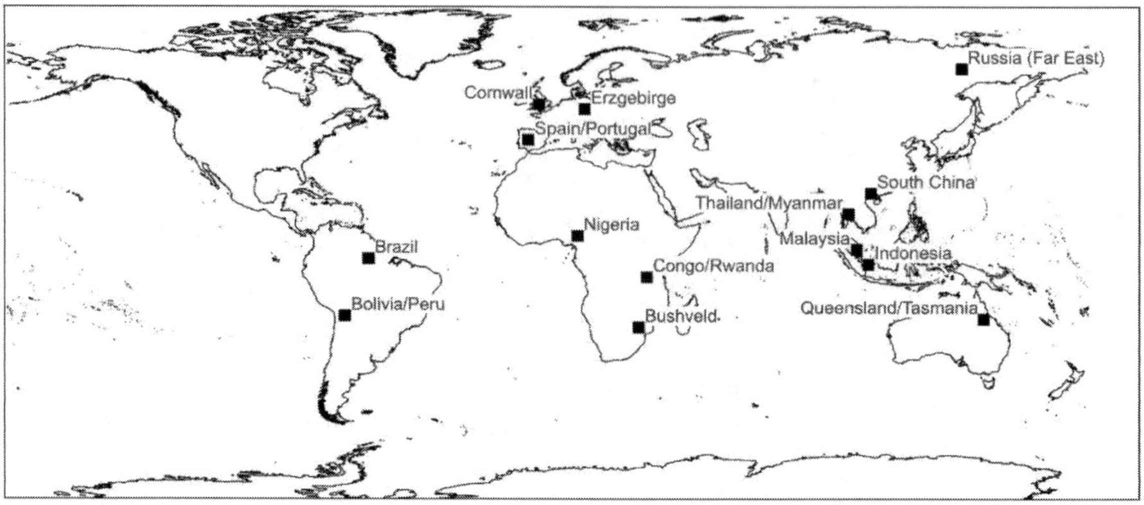

Figure 1 Distribution of tin deposits around the world with black rectangles showing areas with historical tin production [4]. Coastline adapted from North American Cartographic Information Society [5]

The global tin consumption increased from around 250 Gg in 1999 to 381 Gg in 2017 with the biggest producers being China, Indonesia, and Myanmar. Currently, tin produced by China and Myanmar are used by China, that has become the largest tin consumer in the world replacing the US and European Union. Tin consumption increase in China is the biggest contributor to global tin consumption raise [6].

Concentration of tin in bulk crust is around 1.7 ppm with concentration in upper crust around 2.1 ppm and in lower crust around 1.1 ppm [4]. Meanwhile the lowest percentage of mineable tin deposits are around 0.1 percent with most of the deposits ranging from 0.25 0.75 percent [7]. Additionally, each deposit type shows a different tonnage and Sn grade with highest grade contained in lode while greisen shows low grade-high tonnage characteristics (Figure 2).

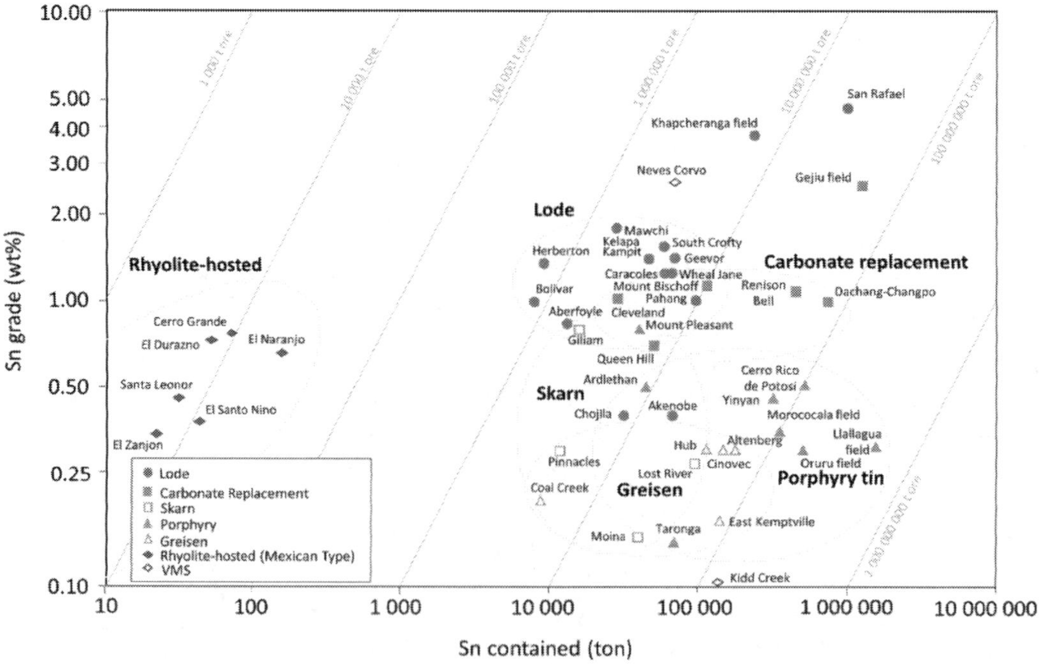

Figure 2 Global ore grade-contained Sn diagram of tin deposits [7]

2. Geochemical properties of tin

Tin atomic number is 50 with electron configuration $1s^2 2s^2 2p^6 3s^2 3p^6 4s^2 3d^{10} 4p^6 5s^2 4d^{10} 5p^2$ or [Kr] $5s^2 4d^{10} 5p^2$ [8]. Tin in nature can be found as Sn^{2+} (ionic radii 1.12 angstrom) or Sn^{4+} (ionic radii 0.71 angstrom) [9]. Sn^{4+} ionic radii are similar to Fe^{3+} (0.64 angstrom), Ti^{4+} (0.68 angstrom), Nb^{5+} (0.68 angstrom), Ta^{5+} (0.69 angstrom), and Sc^{3+} (0.81 angstrom) [10]. Thus, tin ion can be substituted by those ions, and vice versa. The substitution mechanism can be divided into trivalent, tetravalent, pentavalent, and hexavalent [11]. Table 1 shows the substitution mechanism of tin in cassiterite that divided into several classes.

Table 1 Substitution mechanism of tin in cassiterite [11]

Class	Coupling	Equation	ionic radii (VI) Å
Trivalent	Uncoupled	$Fe^{3+} + 1/3\, Fe_i^{3+} \rightleftharpoons Sn^{4+}$	0.645
		$Fe^{3+} + OH^- \rightleftharpoons Sn^{4+} + O^{2-}$	0.645
Tetravalent	Uncoupled	$Si^{4+} \rightleftharpoons Sn^{4+}$	0.400
		$Mn^{4+} \rightleftharpoons Sn^{4+}$	0.530
		$Ti^{4+} \rightleftharpoons Sn^{4+}$	0.605
		$W^{4+} \rightleftharpoons Sn^{4+}$	0.660
		$Nb^{4+} \rightleftharpoons Sn^{4+}$	0.680
		$Ta^{4+} \rightleftharpoons Sn^{4+}$	0.680
			0.690
		$Zr^{4+} \rightleftharpoons Sn^{4+}$	0.720
Pentavalent	1:1	$Nb^{5+} + Fe^{3+} \rightleftharpoons 2Sn^{4+}$	0.643
		$Ta^{5+} + Fe^{3+} \rightleftharpoons 2Sn^{4+}$	0.643
		$Nb^{5+} + Mn^{3+} \rightleftharpoons 2Sn^{4+}$	0.643
		$Ta^{5+} + Mn^{3+} \rightleftharpoons 2Sn^{4+}$	0.643
	2:1	$2Nb^{5+} + Fe^{2+} \rightleftharpoons 3Sn^{4+}$	0.687
		$2Ta^{5+} + Fe^{2+} \rightleftharpoons 3Sn^{4+}$	0.687
		$2Nb^{5+} + Mn^{2+} \rightleftharpoons 3Sn^{4+}$	0.703
		$2Ta^{5+} + Mn^{2+} \rightleftharpoons 3Sn^{4+}$	0.703
		$2W^{5+} + Fe^{2+} \rightleftharpoons 3Sn^{4+}$	0.673
		$2W^{5+} + Mn^{2+} \rightleftharpoons 3Sn^{4+}$	0.690
Hexavalent	1:1	$W^{6+} + Fe^{2+} \rightleftharpoons 2Sn^{4+}$	0.690
		$W^{6+} + Mn^{2+} \rightleftharpoons 2Sn^{4+}$	0.635
	1:2	$W^{6+} + 2Fe^{3+} \rightleftharpoons 3Sn^{4+}$	0.630
		$W^{6+} + 2Mn^{3+} \rightleftharpoons 3Sn^{4+}$	0.587

Sn isotope varies from ^{112}Sn to ^{124}Sn except ^{123}Sn with most abundant tin isotopes in decreasing order are ^{120}Sn (32.58%), ^{118}Sn (24.22%), and ^{116}Sn (14.54%) (Table 2). Most of tin isotope are stable, only ^{113}Sn and ^{121}Sn not stable with half-life 113 days and 27 hours, respectively [12]. Besides as parent element, Sn also product of other element radioactive decay (daughter element). It is commonly identified as a daughter element of indium and antimony. But most half-life of tin parent elements short, ranging from minutes to several seconds (Table 3). Thus, we can assume that all of parent elements already decayed to tin.

Table 2 Tin isotope abundance in nature [13]

Stable Isotope	Atomic Mass	Abundance (%)
^{112}Sn	111.904825	0.97%
^{114}Sn	113.9027801	0.66%
^{115}Sn	114.9033447	0.34%
^{116}Sn	115.9017428	14.54%
^{117}Sn	116.902954	7.68%
^{118}Sn	117.901607	24.22%
^{119}Sn	118.903311	8.59%
^{120}Sn	119.902202	32.58%
^{122}Sn	121.90344	4.63%
^{124}Sn	123.905277	5.79%

Table 3 Sources of tin isotope with its half life

Isotope	Decay	Half life	Isotope	Decay	Half life	References
$^{120}_{49}In$	β^-	3.08 s	$^{120}_{50}Sn$	Stable	-	[14]
$^{120}_{51}Sb$	β^+	15.89 min	$^{120}_{50}Sn$	Stable	-	[14]
$^{118}_{49}In$	β^-	5 s	$^{118}_{50}Sn$	Stable	-	[15]
$^{118}_{51}Sb$	β^+	3.6 min	$^{118}_{50}Sn$	Stable	-	[15]
$^{116}_{49}In$	β^-	14 s	$^{116}_{50}Sn$	Stable	-	[16]
$^{116}_{51}Sb$	β^+	15.8 min	$^{116}_{50}Sn$	Stable	-	[16]

Variation of tin isotope in cassiterite observed in several observation, and was caused by fluctuations in mineralizing fluid (Figure 3). Core-rim cassiterite mineral compositions study by laser ablation shows that fluid evolution can occur in fluid episode that result in lighter tin isotope as well as lower Ta concentration [17]. Besides that, the results also shows that tin mineral can be formed by more than one fluid episodes.

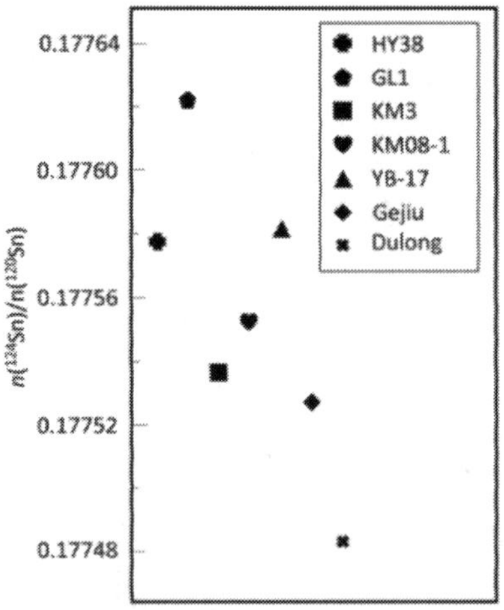

Figure 3 Sn isotope variation in cassiterite from China [18]

Most important tin mineral is cassiterite that contain 78.8 percent tin and chemical formula SnO_2. Because cassiterite is very resistant to weathering and can be classified as heavy mineral (7.15 g/cm^3), many cassiterite accumulates in heavy mineral sands, and profitable to be mined [4]. Besides cassiterite, other important tin minerals are stannite, herzenbergite, nordenskioldine, and malayaite in which listed in Table 4. Study shows that in system with low acid volatile like skarn, malayaite and Sn bearing calc-silicate mineral stable, while in system that rich in B_2O_3, CO_2 and F_2O like greisen, cassiterite stable except in system with low SiO_2 and rich B_2O_3, nordenskioldine can develop [19].

Table 4 Important tin mineral [10]

No	Mineral	Chemical Formula	Tin Percentage (wt%)
1	Cassiterite	SnO2	78.8
2	Stannite	Cu_2FeSnS_4	27.61
3	Herzenbergite	SnS	78.73
4	Nordenskioldine	$CaSn^{4+}(BO_3)_2$	42.95
5	Malayaite	$CaSnSiO_5$	44.48

Cassiterite is the most important tin mineral, becoming a subject in various studies to understand the characteristics of it. Geochemistry of cassiterite as well as trace elements in cassiterite research shows that cassiterite mineral shows banded geochemical substitution of Sn^{4+} with Zr, Ti, Fe, Mn, Nb, Ta, and W, with W, Fe, Nb, and Ta shows sector zonation [11].

Besides in tin ore mineral as mentioned in Table 4, tin may also present in other minerals as impurities in ore deposits. Tin concentrations in various minerals in Furong tin deposits study discovered that tin concentrations in minerals related to skarn process relatively high up to around 44000 ppm (Table 5) [20]. It is also noted that tin impurities concentrations in those minerals related to substitution of Al^{vi} in magnetite and silicate as well as replace Fe^{3+} in epidote and low Al^{vi} silicates.

Table 5 Tin concentration from various mineral in Furong tin deposit [20]

Mineral name	Host rock	Analysis number	Fe^{3+} (apfu)		Al^{vi} (apfu)		Sn (ppm)	
			Range	Average	Range	Average	Range	Average
Plagioclase	Main-phase granite	8	0-0.01	0.01	0	0	0-4	2
k-feldspar		7	0-0.01	0	0	0	1-4	2
Hornblende		8	0.48-0.70	0.6	0-0.07	0.02	87-204	144
Biotite-1		7	0.11-0.20	0.17	0	0	44-62	55
Biotite-2		7	0.20-0.24	0.21	0.04-0.10	0.07	276-321	293
Chlorite	Altered main-phase granite	12	0-0.14	0.07	0.95-1.24	1.04	2-21	9
Muscovite		6	0.21-0.41	0.28	1.53-1.87	1.75	299-583	492
Garnet	Stage I-Skarn	7	0.84-0.98	0.91	0.98-1.16	1.08	4285-10511	6171
Diopside		5	0.03-0.05	0.04	0.01-0.02	0.02	47-232	110
Vesuvianite	Stage II-Skarn	8	1.82-2.46	2.21	8.09-8.83	8.52	398-1962	1070
Pargasite		5	0.20-0.46	0.34	0.49-0.74	0.64	9973-13771	11785
Phlogopite		4	0.03-0.04	0.04	0.01-0.04	0.02	4-7	6
Epidote		6	0.68-0.89	0.78	2.08-2.39	2.23	19830-44241	32675
Scapolite		5	0-0.01	0	0	0	1-4	2
Magnetite		5	1.97-1.98	1.97	0.02-0.04	0.02	1668-4318	2704

3. Geological properties of tin

Primary tin deposits usually occur in syn-subduction tectonic setting. It usually related to highly fractionated low oxidation granitic intrusions [21] (Figure 4) that melt enriched tin sedimentary sources (S-type) [22]. Sedimentary tin deposits developed by weathering rocks that contain tin minerals, e.g., cassiterite. Because of its resistance to weathering and characteristics, cassiterite will be enriched in heavy mineral sands.

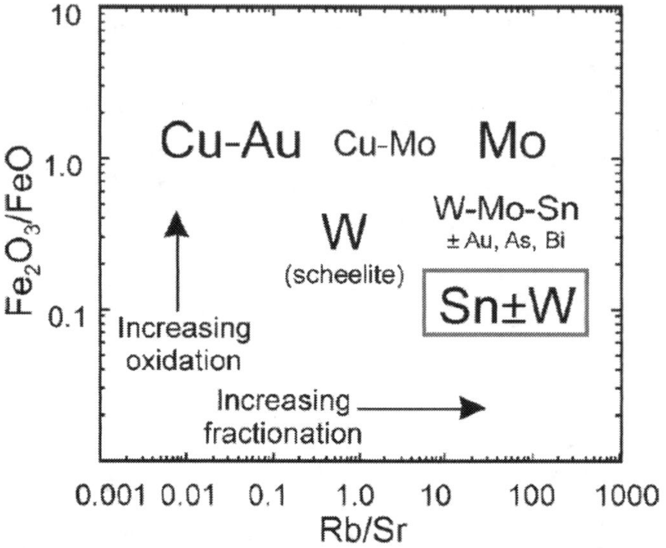

Figure 4 Oxidation-fractionation graph of various metal deposits [21]. Tin (red box) marked with high fractionated low oxidation.

Tin deposits can be classified as several different types, depending on the classification scheme. Dill [10] divided tin deposits into two large groups namely magmatic and sedimentary tin deposits. Magmatic tin deposits consist of pegmatite, granitic skarn, greisen-vein and breccia pipe, porphyry, and volcanic massive sulfide (VMS). Additionally, sedimentary tin deposits consist of SEDEX, laterite (residual placer) and placer deposits related to alluvial-fluvial and marine (Table 6).

Table 6 Tin deposits classification [10]

Classification scheme for tin deposits
(1) Magmatic tin deposits
(1) Sn-Ta-Nb pegmatites
(2) Granitic skarn deposits
(1) Proximal Sn-(W) skarn deposits (Erzebirge-Cornwall-type)
(2) Distal skarn/replacement deposits (Manto-type)
(3) Post-granitic endo- and exogranitic greisen-vein in S- and A-type granites and breccia pipe deposits
(4) Porphyry-type and subvolcanic deposits
(1) Sn-(Ag-Pb-Zn-Cu) vein-type deposits
(5) Stratiform Sn-W deposits
(6) Volcanic massive sulfide Pb-Zn-Cu deposits (VMS)
(1) Iberian-type
(2) Sedimentary tin-tungsten deposits
(1) SEDEX/SMS Pb-Zn-Ag-FeS-(Sn) deposits
(2) Sn laterite (residual placer) deposits
(3) Sn placer deposits
(1) Alluvial-fluvial
(2) Marine

Geologically, tin enrichment in granitic rocks has taken place because melt produced from Sn rich source, fractional crystallization continued by hydrothermal process, or restite enrichment from melt extraction [23]. Tin in nature is contained in common rock-forming minerals such as muscovite, biotite, titanite, and rutile so tin will not be enriched if those minerals are not melted [23]. This phenomenon is well recorded in Nanling Region, China where granites products from lower temperature melting process did not host Sn, but higher temperature product from higher temperature biotite-dehydration process contains Sn [24]. Other observation from metamorphism process shows that tin concentration in biotite and muscovite decreases as metamorphism occur, suggesting that the tin released and enriched in melt (Figure 5) [25]. Figure 6 shows tin concentration in biotite from granite that can reach up to 1000 ppm. Petrological experiment shows that Cl and F are important as solubility control of SnO_2 in peraluminous melt with oxygen fugacity near Ni-NiO with strong effect from temperature, oxygen fugacity as well as Al concentration [26].

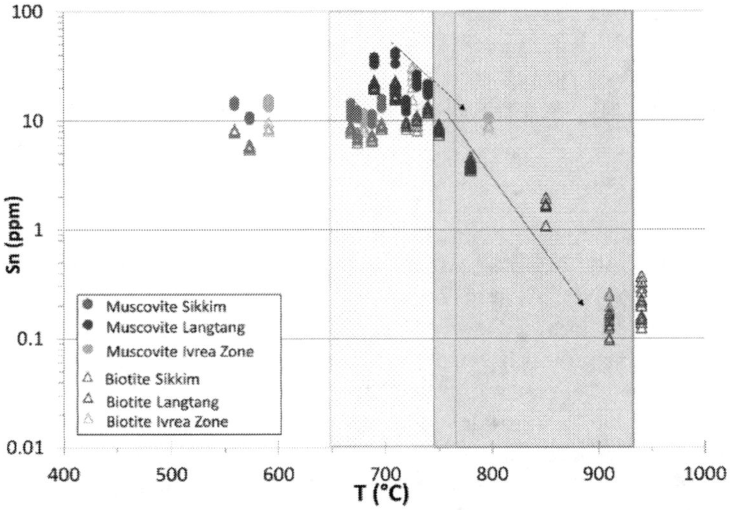

Figure 5 Tin concentrations from biotite and muscovite minerals based on LA-ICP-MS analysis related to metamorphism [25]. The yellow boxes and green boxes, respectively, are areas where muscovite melting (650-760°C) and biotite melting (740-940°C) occur. Dotted line represents muscovite mineral data trend, while solid line represents biotite mineral data trend.

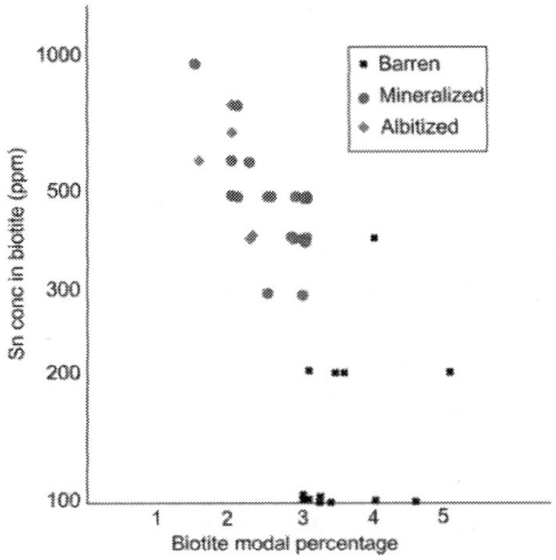

Figure 6 Tin content in biotite from granites [27]

Many researchers show correlation between tin deposits and specific regions globally. Study shows that tin deposits can be developed because of accumulation of tin-enriched sedimentary rocks subjected to heat that led to melting and release of tin to fluid [22]. The heat itself is not restricted to specific tectonic settings. Link between enriched protolith and tin deposits also observed by other researchers from examination of tin mineralization around Atlantic. Tin mineralization around Atlantic is believed to develop because they linked with tin rich areas [2]. Besides that, observation in tin fertile granite, shows low concentrations of tin that indicates scavenging of tin from granitic rocks by later hydrothermal fluids [28].

Tin is usually transported as chlorine complex in hydrothermal fluids, either as $SnCl_2$ [29] or $SnCl_4$ [30]. HCl molality is the important parameter for tin transport and precipitation, thus tin can be precipitated because of HCl molality change caused by fluid-wall reaction, dilution, as well as boiling [31]. Fluid-wall reaction caused HCl neutralization and tin precipitation. There are different mechanisms in HCl neutralization in carbonate (dissolution) and schist (conversion feldspar and biotite to muscovite) [32]. Dilution is another method to precipitate tin. This method decreases HCl concentration by mixing mineralizing fluid with low HCl concentration fluid (e.g. meteoric fluid), thus encouraging tin precipitation [31]. Additionally, tin precipitation may be encouraged by boiling process. The process may be induced by fault activity that led to vapor loss and precipitation of tin [30].

4. Conclusions

Tin can be found as ten different isotopes in the nature, with three largest tin isotope abundance in nature are ^{120}Sn (32.58%), ^{118}Sn (24.22%), and ^{116}Sn (14.54%). Tin may be replaced by other elements, such as Fe, Mn, Si, Ti, W, Nb, Ta and Zr by trivalent, tetravalent, pentavalent, and hexavalent substitutions. Variations in tin isotopes may reflect mineralization history of ore minerals. Many tin deposits developed in tin-rich protolith suggest the importance of protolith enrichment in tin deposit genesis. Cassiterite (SnO_2) is the most important tin mineral with other tin minerals such as stannite, herzenbergite, nordenskioldine, and malayaite may be found. Because cassiterite is resistant to weathering and high-density characteristics, many tin deposits are formed in heavy mineral sands. Tin transport and precipitation are mainly controlled by concentration of HCl, and mechanisms that can affect the HCl in fluid such as fluid-rock interactions, fluid mixing and vapor loss due to degassing.

5. Acknowledgement

We thank ISEST that organize conference in which this article presented. ADH thank Indonesian National Research and Innovation Agency for doctoral study support in ITB.

References

[1] U.S. Geological Survey 2022 Mineral commodity summaries 2022. In: *Mineral Commodity Summaries,* (Reston, VA p 202
[2] Schuiling R D 1967 Tin belts on the continents around the Atlantic Ocean *Economic Geology* **62** 540-50
[3] Schwartz M, Rajah S, Askury A, Putthapiban P and Djaswadi S 1995 The southeast Asian tin belt *Earth-Science Reviews* **38** 95-293
[4] Lehmann B 2020 Formation of tin ore deposits: A reassessment *Lithos* 105756
[5] North American Cartographic Information Society n.d. Natural Earth quick start kit.
[6] Li H, Qin W, Li J, Tian Z, Jiao F and Yang C 2021 Tracing the global tin flow network: highly concentrated production and consumption *Resources, Conservation and Recycling* **169** 105495

[7] Mlynarczyk M 2005 Constraints on the genesis of lode-style tin mineralization: Evidence from the San Rafael tin-copper deposit, Peru. In: *Department of Earth & Planetary Sciences,* (Montreal: McGill University) p 360
[8] Lide D R 2004 *CRC handbook of chemistry and physics* vol 85: CRC press)
[9] Railsback L B 2003 An earth scientist's periodic table of the elements and their ions *Geology* **31** 737-40
[10] Dill H G 2010 The “chessboard” classification scheme of mineral deposits: Mineralogy and geology from aluminum to zirconium *Earth Science Reviews* **100** 1-420
[11] Bennett J 2021 On the Geochemistry of Cassiterite. In: *School of Earth Sciences*: The University of Western Australia)
[12] n.d. Tin (Sn) Element Properties https://www.webmineral.com/chem/Chem-Sn.shtml
[13] Holden N E, Coplen T B, Böhlke J K, Tarbox L V, Benefield J, Laeter J R d, Mahaffy P G, O’Connor G, Roth E, Tepper D H, Walczyk T, Wieser M E and Yoneda S 2018 IUPAC Periodic Table of the Elements and Isotopes (IPTEI) for the Education Community (IUPAC Technical Report) *Pure and Applied Chemistry* **90** 1833-2092
[14] Kitao K, Tendow Y and Hashizume A 2002 Nuclear Data Sheets fo A = 120 *Nuclear Data Sheets* **96** 241-390
[15] Kitao K 1995 Nuclear Data Sheets Update for A = 118 *Nuclear Data Sheets* **75** 99-198
[16] Blachot J 2010 Nuclear Data Sheets for A = 116 *Nuclear Data Sheets* **111** 717-895
[17] Liu P, Mao J, Lehmann B, Weyer S, Horn I, Mathur R, Wang F and Zhou Z 2021 Tin isotopes via fs-LA-MC-ICP-MS analysis record complex fluid evolution in single cassiterite crystals *American Mineralogist* **106** 1980-6
[18] Yamazaki E, Nakai S, Yokoyama T, Ishihara S and Tang H 2013 Tin isotope analysis of cassiterites from Southeastern and Eastern Asia *GEOCHEMICAL JOURNAL* **47** 21-35
[19] Burt D M 1978 Tin silicate-borate-oxide equilibria in skarns and greisens; the system cao-sno2-sio2-H2O-B2O3-CO2-F2O- 1 *Economic Geology* **73** 269-82
[20] Chen S-C, Yu J-J, Bi M-F and Lehmann B 2022 Tin-bearing minerals at the Furong tin deposit, South China: Implications for tin mineralization *Geochemistry* **82** 125856
[21] Cerny P, Blevin P L, Cuney M and London D 2005 *One Hundredth Anniversary Volume,* ed J W Hedenquist, *et al.* (Littleton - Colorado: Society of Economic Geology) p 1133
[22] Romer R L and Kroner U 2016 Phanerozoic tin and tungsten mineralization—Tectonic controls on the distribution of enriched protoliths and heat sources for crustal melting *Gondwana Research* **31** 60-95
[23] Wolf M, Romer R L, Franz L and López-Moro F J 2018 Tin in granitic melts: The role of melting temperature and protolith composition *Lithos* **310-311** 20-30
[24] Zhao P, Yuan S, Williams-Jones A E, Romer R L, Yan C, Song S and Mao J 2022 Temporal Separation of W and Sn Mineralization by Temperature-Controlled Incongruent Melting of a Single Protolith: Evidence from the Wangxianling Area, Nanling Region, South China *Economic Geology* **117** 667-82
[25] Kunz B E, Warren C J, Jenner F E, Harris N B W and Argles T W 2022 Critical metal enrichment in crustal melts: The role of metamorphic mica *Geology* **50** 1219-23
[26] Bhalla P, Holtz F, Linnen R L and Behrens H 2005 Solubility of cassiterite in evolved granitic melts: effect of T, fO2, and additional volatiles *Lithos* **80** 387-400
[27] Imeokparia E 1982 Tin content of biotites from the Afu younger granite complex, Central Nigeria *Economic Geology* **77** 1710-24
[28] Lehmann B and Harmanto 1990 Large-scale tin depletion in the Tanjungpandan tin granite, Belitung Island, Indonesia *Economic Geology* **85** 99-111
[29] Heinrich C A 1990 The chemistry of hydrothermal tin(-tungsten) ore deposition *Economic Geology* **85** 457-81
[30] Korges M, Weis P, Lüders V and Laurent O 2017 Depressurization and boiling of a single magmatic fluid as a mechanism for tin-tungsten deposit formation *Geology* **46** 75-8

[31] Schmidt C 2018 Formation of hydrothermal tin deposits: Raman spectroscopic evidence for an important role of aqueous Sn(IV) species *Geochimica et Cosmochimica Acta* **220** 499-511
[32] Eugster H P 1985 Granites and hydrothermal ore deposits: a geochemical framework *Mineralogical Magazine* **49** 7-23

Design and Preliminary Results of Debris Flow Analog Model in Laboratory Scale, Case Study of Bentarsari Sub-watershed

Imam A. Sadisun[1,2], Indra A. Dinata[1]*, and Lambok M. Hutasoit[1]
[1]Applied Geology Research Group, Faculty of Earth Sciences and Technology, Institut Teknologi Bandung, Indonesia
[2]Research Centre for Disaster Mitigation, Institut Teknologi Bandung, Indonesia
*Corresponding author: indraandradinata@gmail.com

Abstract. Bentarsari sub-watershed is located at Brebes Regency administrative area, Central Java. This sub-watershed is surrounded by hills composed by volcanic breccia of Kumbang Formation (Tpk) and consists of sedimentary rocks at the lower part of the watershed. Volcanic breccia is the oldest rock in the watershed and appears to have a more intensive weathering process, making it more susceptible to landslides which are mostly followed by debris flow. Accordingly, the aim of this research is to conduct a laboratory scale analog modeling of debris flow event in this sub-watershed. The main materials used for modeling are moderate to completely weathered volcanic breccias to residual soils. The rainfall scenario is 50 mm/day as lower bound of heavy rain, 100 mm/day as lower bound of very heavy rain and 150 mm/day as lower bound of extreme rain. As a first step, the computer simulation is used for analyzing material movement of the model, using spherical materials with a density of 1.84 g/cm^3 taken from average value of laboratory test results. The simplification process is carried out on spherical material with 4 cm and 2 cm diameter to depict moving material. The solid material volume used in the simulation is 0.07 m^3 while the liquid is 0.03 m^3. Then, a laboratory scale analog model has been made with a tub having dimensions of 0.5 m high, 2 m long and 1 m wide. When this analog model is tilted 50 degrees, the resulting slip plane then extends to the backward of the slope and the flowing materials move downward colliding with the previously deposited material, forming a flow which bends slightly to the left at the foot slope. The results of this research are expected to lead to a new method for modelling debris flows using analog models at laboratory scale.

Keywords: debris flow, analog model, deterministic modelling

1. Introduction

Indonesia is a country which is very prone to landslide disasters. Many landslides occur in the areas with steep to very steep slopes, intensive weathering processes, and high rainfall intensity. Landslide which has a mixture of solid and liquid materials can turn into a debris flow. Several landslides were followed by debris flow, including the Cililin landslide on April 21, 2004, the Jember landslide on January 2, 2006, the Ciwidey landslide on February 23, 2010, the Karangkobar landslide on December 12, 2014, and the Hululais landslide on April 28, 2016, have claimed many lives and caused extensive damage. For example, the debris flow event in Jemblung, Sampang Village, Karangkobar District, Banjarnegara Regency, Central Java, or what is often known as the Karangkobar Landslide, has claimed more than 100 lives, dozens of residents were injured, buried around 40 houses and cut village roads reaching 500 m [1]. Another example is Pasir Panjang debris flow event at Bentarsari sub-watershed, Salem District,

Brebes Regency, Central Java on February 22, 2018 which has claimed 18 lives, cut off the road connecting the districts and damaged hectares of rice fields. The aim of this research is to conduct a laboratory scale analog modeling of debris flow event in this sub-watershed, especially at the northern slope of where Pasir Panjang debris flow occurred.

2. Debris Flow

According to Cruden and Varnes (1978) [2] landslide can be divided into six categories based on the type of movement, which are falls, topples, slides, lateral spread, flows, and combined (complex) type. According to Stiny (1910) [3], the debris flow begins with flooding on steep mountain slopes and transports large amounts of suspended load and bed load. When the amount of sediment carried by the flow increases to a certain extent, the flows change into a viscous mass of water, soil, sand, gravel, stones and logs mixed together through the valley channel. According to Varnes (1978) [4] debris flow is a mass containing solid material, water, and air flowing as a liquid stream. The debris flow has three parts, namely the source area, the flow track and the run-out lobe/depositional area.

The source area is a place for accumulation of debris flows initial material which found on the upper slopes. The accumulation of debris flow initial material can be in the form of slope-forming material or forming natural dams. Natural dams are formed when an initial landslide occurs in a source area which blocks an existing river. The blockage which occurs then produces a puddle that saturates the clogging materials. The saturated materials will reach a critical point for a landslide to occur followed by debris flow. Apart from the presence of natural dams, the debris flow can also occur through sudden loading and liquefaction. The occurrence of landslides above the river slopes causes a sudden loading on saturated river deposits. This condition will initiate liquefaction in river deposits. The flowing liquefaction process developed into a debris flow with material originating from landslide and river sediments (Figure 1).

The flow track is the place where the debris flow flows toward the valley associated with the river flow. The deposition area is a plain as a place for the deposition of debris flow materials. At the boundary between the flow track and the deposition area, the debris flow will form a fan formation. The distribution of debris flows in the deposition area depends on the topography of the deposition area. If in the depositional area still has terrace geometry, then a second flow path can be formed which will form a second depositional area (multiple depositional area).

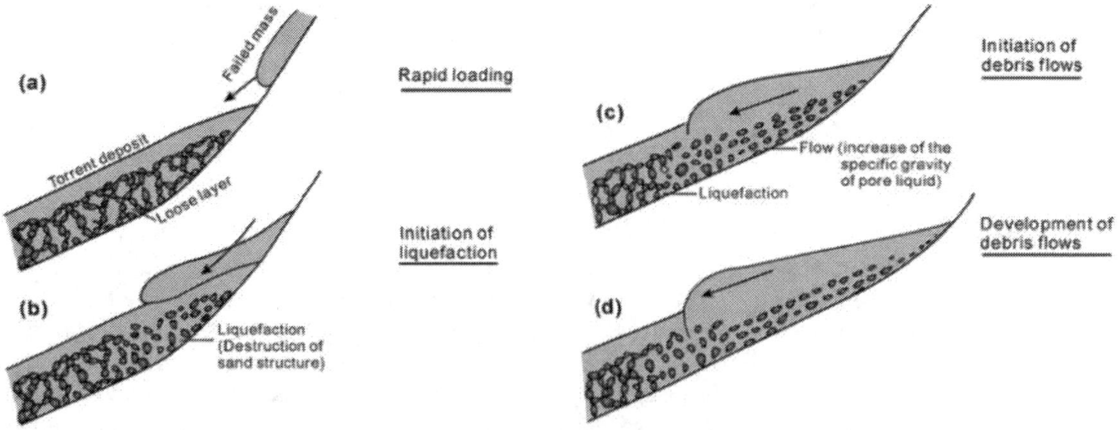

Figure 1. Illustration of the initiation of debris flow due to sudden loading and liquefaction [3].

Debris flow can be divided into two types, they are hillslope and channelised debris flows [5]. Both types of debris flow are formed based on the topography and geological conditions of debris flow location. The hillslope debris flow moves directly down the valley and the source area is relatively visible from the deposition area. The channelised debris flow moves along existing channel features (rivers, valleys, basins) and the source area is relatively invisible from the deposition area (Figure 2).

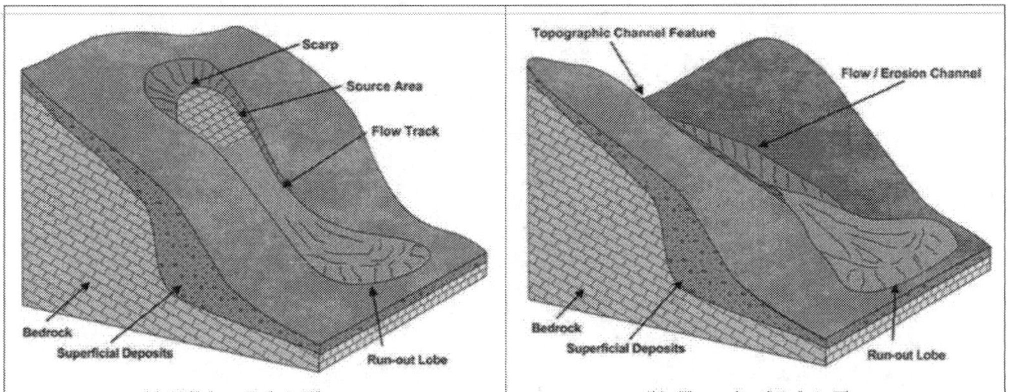

(a) *Hillslope Debris Flow* (b) *Channelised Debris Flow*

Figure 2. The debris flow type: (a) hillslope and (b) channelised [5].

Debris flow often occurs in areas which have hilly morphology with high rainfall. The slope angle and the river channel are determining factor in the process of debris flow velocity. Debris flow will occur if there is an abundant supply of debris material with a large supply of water as flowing medium. Debris flow most frequently occurs after or during heavy rains. The debris flow has high specific gravity so that large chunks can be carried by the flow to the end of deposition area.

Debris flow generally follows pre-existing surface flows but can move down slopes and across unobstructed alluvial fan surfaces. This is because the debris flow tends to build its own channel due to the natural embankments which form at the lateral flow boundaries. The debris flow moves down the slope in a series of waves with periods ranging from a few seconds to several hours. The debris flow front is higher than the other sections and contains large moving boulders. The frontier flow is followed by more fluid and turbulent mud with high concentrations of suspended sediment. This more liquid phase continues until the next wave arrives or until the debris flow activity ceases [6]. The speed of the debris flow varies because it is influenced by the size of debris material, concentration, sorting of debris material, and channel geometry which includes shape, slope, width, and curve.

The debris flow can be highly erosive as it moves in the flow track [7]. During movement in the flow track, the density of the debris flow can be twice as large as the density of the accompanying fluid. The debris flow can also increase the shear stress up to 6 times at the bottom of the river flow. There is an increase in shear stress for 20 seconds after the front of the debris material enters the flow track [8]. The increase in shear stress is caused by a sudden loading at the location of landslide in the source area. An additional attractive characteristic of debris flows is the ability to move boulder over great distances. The mobility of the debris flow is highly dependent on the presence of clay-sized material (even only 1% - 2%). Clay-sized materials reduce permeability and increase pore pressures, thereby increasing the mobility of the larger materials.

3. Analog Model

The analog model has been made with dimensions of 1.75 m height, 2 m length, and 1 m width (Figure 3). A tub with dimensions of 0.5 m high, 2 m long and 1 m wide and with one side open is placed on a support. This tub will be used to place the debris materials for modeling analogously on a laboratory scale. The tub will be supported by an electric hydraulic pump capable of tilting the table up to 75° (Figure 4). The hydraulic pump is expected to be able to tilt the tub which has been filled with 2 tons of debris flow materials.

IOP Conf. Series: Earth and Environmental Science 1245 (2023) 012024 doi:10.1088/1755-1315/1245/1/012024

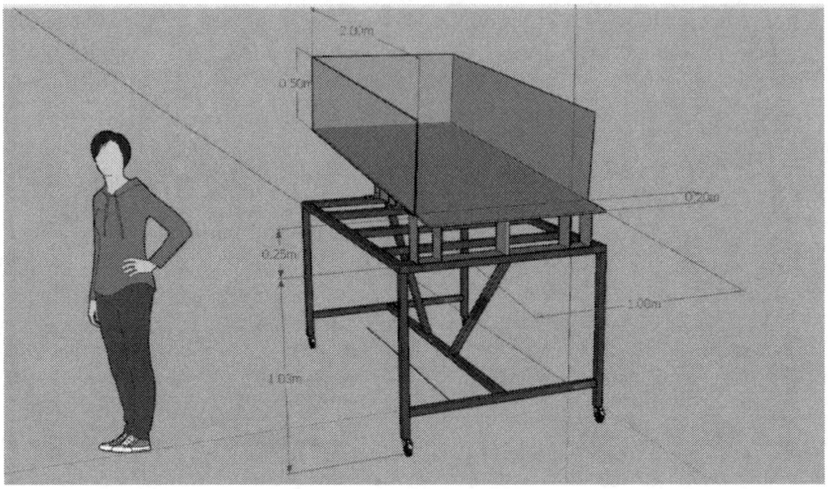

Figure 3. Debris flow laboratory scale analog model design.

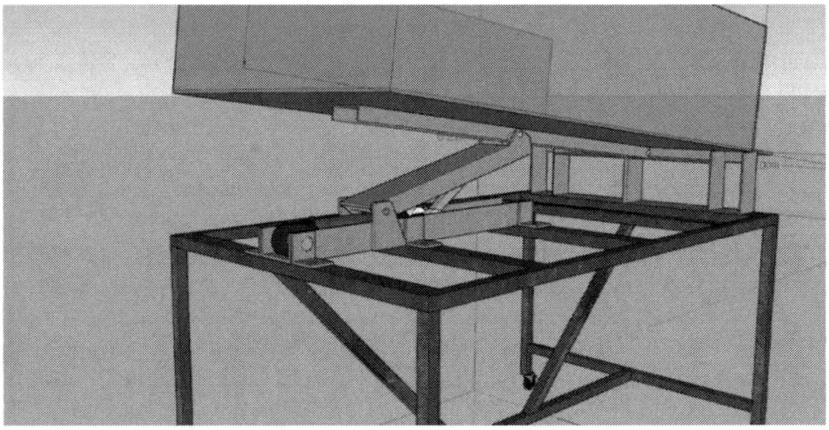

Figure 4. Electric hydraulic pump which is under the tempered glass.

The Bentangsari sub-watershed is surrounded by hills composed by volcanic breccia of Kumbang Formation (Tpk) and consists of sedimentary rocks at the lower part of the watershed. Volcanic breccia is the oldest rock in the watershed and appears to have a more intensive weathering process, making it more susceptible to landslides which are mostly followed by debris flow. The slope to be modeled in this research is the northern slope of the Bentangsari sub-watershed which is composed of the Kumbang Formation (Tpk). This formation is then in detailed classified into five rock and soil units, which are Fresh - Slightly Weathered Volcanic Breccia Unit, Fresh - Slightly Weathered Sandstone Unit, Moderate - Highly Weathered Volcanic Breccia Unit, Medium - Highly Weathered Sandstone Unit, and Completely Weathered – Residual Soil of Volcanic Breccia unit. The main materials for modeling debris flow are Moderately - Strong Weathered Volcanic Breccia Unit and Completely Weathered Volcanic Breccias - Residual Soil Unit.

Hydrograph supply (water) will affect the flow viscosity so that the rheological conditions of the research area can be ascertained. The rheological property of the debris flow is the relationship between deformation and flow due to shear stress changes. Rheology can be approximated by the parameter of viscosity (concentration). The rheology of the flow will describe the general bulk density, flow strength, fluid type, and the resulting landscape and deposits.

The nearest rainfall station to the research area is PK.5c Dukuhjeruk (± 25 km). The PK.5c Dukuhjeruk monthly rainfall value is used as the basis for determining the rainfall scenario which will be used in the analog model. PK.5c Dukuhjeruk has a maximum rainfall of 493 mm/month which occurred in March 2010, has an average rainfall range of 24 mm/month – 369 mm/month in 2009-2015, and there are no rain events in July – September (Table 1 and Figure 5). If it is assumed that the rain that occurs in one month accumulates on certain days, the rainfall scenario which will be used is 50 mm/day as lower limit for heavy rain class, 100 mm/day as lower limit for very heavy rain class and 150 mm /day as lower limit of extreme rain.

Table 1. PK.5c Dukuhjeruk monthly rainfall data for 2009-2015.

Year	Jan	Feb	Mar	Apr	May	Jun	Jul	Aug	Sep	Oct	Nov	Des
2009	266	287	195	190	207	109	19	0	0	22	262	22
2010	391	295	493	299	302	181	35	100	155	90	130	283
2011	81	459	481	171	208	37	43	0	0	70	245	475
2012	223	434	269	23	131	15	0	0	0	34	169	291
2013	415	198	193	197	148	289	91	0	0	34	131	202
2014	233	286	256	170	149	83	65	7	0	7	125	289
2015	227	143	177	152	119	9	8	0	0	0	49	197
Max	391	459	493	299	302	181	43	100	155	90	262	475
Ave	240	369	360	171	212	86	24	25	39	54	202	268
Min	81	287	195	23	131	15	0	0	0	22	130	22

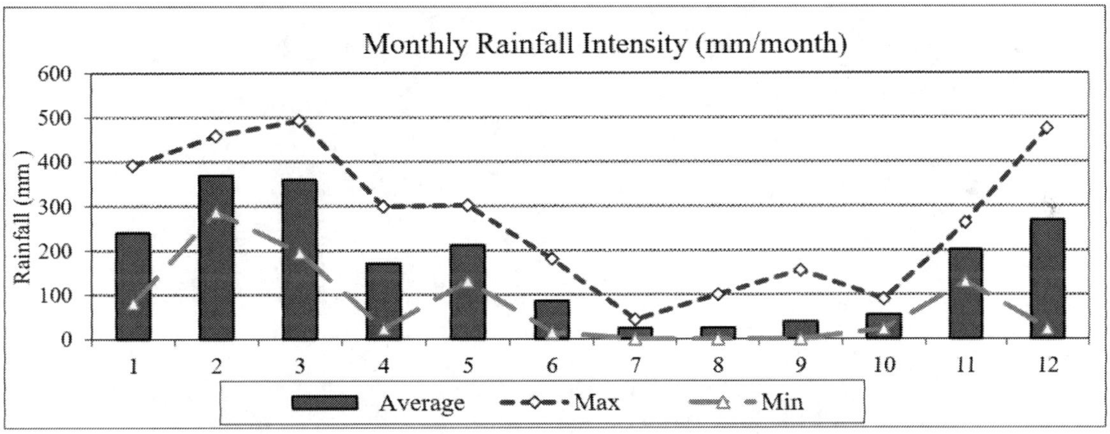

Figure 5. Graph of maximum, average and minimum rainfall of PK.5c Dukuhjeruk in 2009-2015.

4. Preliminary Result

The first analog modeling experiment used materials taken from the Pasir Panjang debris flow. As a first step, the computer simulation is used for analyzing material movement of the model, using spherical materials with a density of 1.84 g/cm^3 taken from average value of laboratory test results. The simplification process is carried out on spherical material with 4 cm and 2 cm diameter to depict moving material. A laboratory scale analog model was then carried out using material with a volume of 0.07 m^3 and supply hydrograph with 0.03 m^3 water volume which added slowly (Figure 6). The comparison between the solid and liquid volumes of 7:3 is based on previous research result [9]. The tub slope angle slowly increases until 20°, a failure occurs which causes the saturated materials to move downwards. This movement only occurs in saturated materials, so the slip planes which occur are on the boundary of the saturated and unsaturated zones (Figure 7). When the tub is tilted 50°, the slip plane then extends to the backward of the slope and the flowing materials move downward colliding with the previously deposited material, forming a flow which bends slightly to the left at the foot slope (Figure 8).

Figure 6. The initial conditions of the analog model.

Figure 7. Material conditions with 20° slope scenario.

Figure 8. Material conditions with 50° slope scenario.

5. Concluding Remarks

A laboratory scale analog model using a tub with dimensions of 0.5 m high, 2 m long and 1 m wide, with a volume of 0.07 m^3 material and supply hydrograph with 0.03 m^3 water volume shows that when the tub with 20° slope angle, a failure occurs which causes the saturated material to move downward with slip planes on the boundary of saturated and unsaturated zones. When the tub is tilted 50°, the slip plane then extends to the backward of the slope and the flowing materials move downward colliding with the previously deposited material, forming a flow which bends slightly to the left at the foot slope. The results of this research are expected to lead to a new method for modelling debris flows using analog models at laboratory scale. For further improve the reliability of the analog model results, subsequent modeling is further suggested to be carried out with the same material by adding several parameters, such as artificial rain tools, artificial slope surface profiles, and static cameras at four locations to record the movement of the modeled materials.

6. Reference

[1] Sadisun I A Kartiko R D and Dinata I A 2018 Numerical Simulation of Some Debris Flow Events in Central Java for Predicting Run-out Distributions *Proceedings of SEAGS-AGSSEA Conference 2018* (Jakarta) pp 357-360

[2] Cornforth D H 2005 *Landslides in Practice: Investigation, Analysis, Remedial/Preventive Options in Soil*, (Hoboken: John Wiley & Sons)

[3] Jakob M and Hungr O 2005 *Debris-flow Hazard and Related Phenomena* (Chichester: Praxis Publishing Ltd)

[4] Blijenberg H M 2007 Application of physical modelling of debris flow triggering to field conditions: limitations posed by boundary conditions *Engineering Geology* **91** pp 25-33

[5] Nettleton I M Martin S Hencher S and Moore R 2005 *Debris flow types and mechanisms (Scottish Road Network Landslides Study)* ed Winter M G MacGregor F and Shackman L (Edinburgh) pp 45-67

[6] Pierson T C 1980 Erosion and deposition by debris flows at Mount Thomas, North Canterbury, New Zealand *Earth Surface Process* **5** pp 227-247

[7] Janda R J Scott K M Nolan K M and Martinson H A 1981 *Lahar movement, effects, and deposits (The 1980 eruptions of Mount St. Helens)* ed Lipman P W and Mullineaux D R (Washington) pp 461-478

[8] Berger C Schlunnegger F and McArdell B W 2011 Direct measurement of channel erosion by debris flows, Illgraben, Switzerland *Journal of Geophysical Research* **116** pp 1-18

[9] Sadisun I A Dinata I A and Kartiko R D 2019 *Run-out Distribution of Pasir Panjang Landslides Followed by the Debris Flow, Central Java, Indonesia* (Jeju: IAEG ARC12 2019)

First occurrence of planktonic foraminiferal species *Boliella adamsii* as a marker for the Pleistocene–Holocene Boundary in the sea around Sumba Island

[1,2]R D W Ardi;[1]Aswan;[1]K A Maryunani;[3]E Yulianto;[1,3]P S Putra;[1,3]S H Nugroho
[1]Geological Engineering Study Program, Faculty of Earth Sciences and Technology, Institut Teknologi Bandung, Bandung, Jawa Barat 40132, Indonesia
[2]Agrotechnology Study Program, Faculty of Science and Technology, Universitas Nahdlatul Ulama Purwokerto, Banyumas, Jawa Tengah 53144, Indonesia
[3]Research Center for Geological Disaster, National Research and Innovation Agency, Kompleks LIPI, Bandung, Jawa Barat 40135, Indonesia
Corresponding author: 32019301@mahasiswa.itb.ac.id (RDWA)

Abstract. The expensive cost of Accelerator Mass Spectrometry radiocarbon dating (AMS ^{14}C), the main tool for geochronology in Late Pleistocene–Holocene paleoclimate and paleoceanography study using marine sediment cores, hinders Indonesian researchers to produce competing research outputs. The first occurrences of planktonic foraminiferal species which have been applied as Pleistocene–Holocene Boundary markers in previous Quarternary marine sediment studies in the Indonesian region need to be confirmed as these species might actually occur since the Late Pleistocene. Quantitative foraminifera determination and oxygen isotope ($\delta^{18}O$) analysis of planktonic foraminiferal species *Globigerinoides (Gs.) ruber* have been carried out in marine sediment cores ST10 (off south Sumba) and ST14 (Sumba strait). $\delta^{18}O$ of *Gs. ruber* was correlated to North Greenland Ice Core Project (NGRIP) Greenland Ice Core Chronology 2005 (GICC05) $\delta^{18}O$ to determine the depth of the Pleistocene–Holocene Boundary (11653 BP) in core ST10 and ST14. Pleistocene–Holocene Boundary depth in core ST10 (31 cm) and ST14 (28 cm) are nearly coeval with the First occurrence (FO) of planktonic foraminifera *Boliella adamsii,* which at 34–35 cm depth interval in core ST10 and at 26–27 cm depth interval in core ST14. This indicated the practicality of *Boliella adamsii* FO as a Pleistocene–Holocene Boundary marker in the sea around Sumba Island, in case radiocarbon ages and proper planktonic foraminiferal $\delta^{18}O$ data are not available. Future studies using a similar method in other Indonesian regions are needed to confirm the practicality of *Boliella adamsii* FO as a Pleistocene–Holocene Boundary marker in the Indonesian region.

Keywords: biostratigraphy; marine sediment; micropaleontology; Late Quaternary.

1. Introduction

Geochronology, the study of time as it relates to Earth's history, is an important aspect of geological studies including paleoclimate and paleoceanography [1]. Accelerator Mass Spectrometry radiocarbon dating (AMS ^{14}C) is highly utilized as the main tool to generate absolute age for geochronology in paleoclimate and paleoceanography studies using marine sediment cores, especially for Late Pleistocene–Holocene time scale. This method is able to produce a high resolution geochronology for geological studies covering the last 55,000 years using relatively small amounts of materials (less than 1 mg) [2]. AMS ^{14}C instruments, however, are not available in most developing countries' research facilities, incuding Indonesia. Researchers from these countries need to analyse their samples at overseas laboratories, usually in developed countries (i.e. United States, Germany, Japan). The price for AMS ^{14}C analysis in developed countries' laboratories is considered expensive by Indonesian standards. The analysis fee for each sample in those laboratories could cost around 500–750 USD or around 7,600,000–11,400,000 IDR (1 USD = 15,200 IDR) [3,4]. The expensive price for AMS ^{14}C analysis makes it not affordable for Indonesian researchers, especially those with a limited budget such as university students and small university researchers. This limitation hinders them from producing competing research outputs. Due to the limited research budget, those researchers usually rely on planktonic foraminiferal biostratigraphy to know the geochronology of their research location even though it is less reliable compared to the AMS ^{14}C.

In Late Pleistocene–Holocene paleoclimate and paleoceanography studies, a geological event that can be easily recognized from biostratigraphy is the Pleistocene–Holocene Boundary. In the case when planktonic foraminiferal biostratigraphy is applied, the Pleistocene–Holocene Boundary can be indicated by the First Occurrence (FO) of *Globorotalia fimbriata*, *Boliella adamsii*, or *Globigerinella calida* [5,6]. Previous Quarternary marine sediment studies in the Indonesian region applied either the FO of *Globorotalia fimbriata* [7,8], *Boliella adamsii* [9,10], or *Globigerinella calida* [11,12]. However, there are arguments that those species might actually occur earlier, since the Late Pleistocene [5,13–15]. These contradictive arguments need to be confirmed in order to increase the applicability of planktonic foraminiferal biostratigraphy in inferring the Pleistocene–Holocene Boundary. This research utilize quantitative planktonic foraminiferal distribution and foraminiferal oxygen isotope ($\delta^{18}O$) records of marine sediment cores taken from off south Sumba and Sumba strait (**Figure 1**) to confirm the most practical biostratigraphic event as the Pleistocene–Holocene Boundary marker, especially in the sea around Sumba Island.

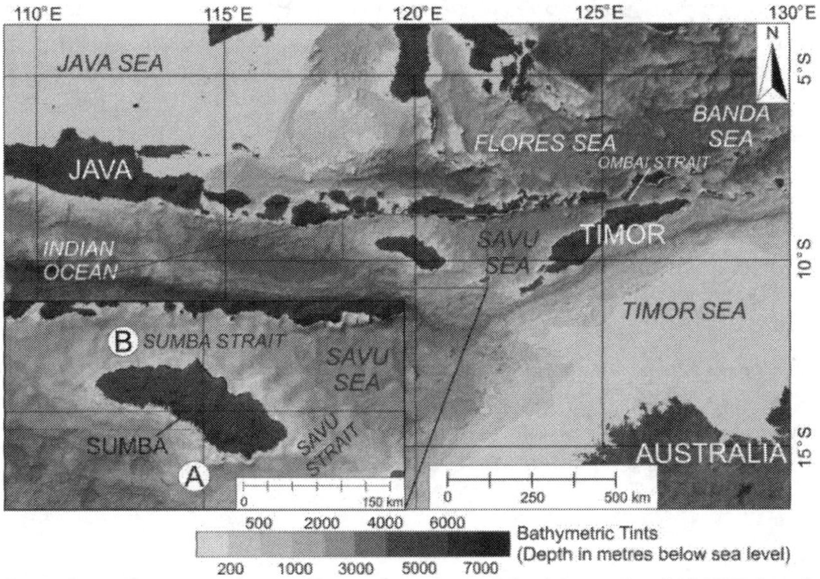

Figure 1. Location of two marine sediment cores used in this study: (A) ST10 and (B) ST14.

Quarternary marine sediments in the sea around Sumba Island consist of calcareous pelagic and hemipelagic sediment [12,16]. These sediments are rich in foraminiferal specimens [7,8,17,18] and contain transported pollen grains from nearby landmasses [19]. Terrestrial materials are also present and their amount indicates the variability of Australian-Indonesian monsoon rainfall [19,20].

2. Data and Method

Two gravity sediment cores were collected from off south Sumba (ST10) and Sumba strait (ST14) as a part of the 2016 Widya Nusantara Expedition (E-WIN) using the Baruna Jaya VIII research vessel operated by the Research Center for Oceanography, Indonesian Institute of Sciences (LIPI) (now Research Center for Oceanography, National Research and Innovation Agency/BRIN) (**Table 1**). To produce planktonic foraminiferal distribution, foraminiferal identification in the 1 cm-thick sub-samples from the sediment cores is needed. Those sub-samples were taken at 2 cm intervals, resulting in a total of 191 sub-samples, 69 sub-samples from core ST10 and 122 sub-samples from core ST14. Sub-samples were weighed to ensure their weight consistency (~5 g each) and dried in a drying oven for ~6–8 hours at 40° C afterward. Then, the sub-samples were washed on the top of a 230 mesh (63 μm) sieve to clean the foraminiferal specimens from mud. The mud-free residue was then dried in a drying oven for ~6 hours at 60 °C.

Table 1. Information of sediment cores used in this study.

Core	Location	Coordinate		Depth (meters below sea level)
		Latitude (°E)	Longitude (°S)	
ST10	Off south Sumba	119.885	10.491	1004
ST14	Sumba strait	119.245	9.079	1283

For quantitative foraminiferal identification, about 300 specimens of foraminifera were counted from each sub-sample as counting 300 specimens is considered enough to represent all species that might occur within [21]. In the case of a sub-sample with very abundant foraminiferal specimens, each was split several times until each partition contains only around 300 foraminiferal specimens. Sub-samples were then observed under a stereo microscope in 50X magnification and foraminiferal specimens were determined based on previous highly-cited studies [5,15,22]. All specimens in one partition were determined and counted while new species identified in other partitions were counted as one specimen. The normalizations were done against the number of splits and the weight of samples (5 g). Samples preparation and determination were done at the Sedimentary Laboratory of the Research Center for Geological Disaster of the National Research and Innovation Agency (BRIN) and the Micropaleontology Laboratory of Institut Teknologi Bandung (ITB).

For planktonic foraminiferal $\delta^{18}O$ analysis, ~6 specimens of *Globigerinoides* (*Gs.*) *ruber* sized 250–350 μm (measured with a digital microscope) from 4–8 cm intervals were picked. Afterward, those specimens were washed with an ultrasonic cleaner for 15–60 seconds to clean the mud infillings and then dried in a drying oven for ~3–6 hours at 60 °C. The $\delta^{18}O$ analysis was conducted in the Benedum Stable Isotope Laboratory of Brown University, USA. In order to infer the depth of the Pleistocene–Holocene Boundary in core ST10 and ST14, *Gs. ruber* $\delta^{18}O$ records of core ST10 and ST14 were correlated to North Greenland Ice Core Project (NGRIP) Greenland Ice Core Chronology 2005 (GICC05) $\delta^{18}O$ records (reference present time = 1950 AD) [23]. In NGRIP GICC05 $\delta^{18}O$ records, the age of the Pleistocene–Holocene Boundary is 11653 BP [24].

3. Result and Discussion

Based on the correlation of *Gs. ruber* $\delta^{18}O$ records of core ST10 to NGRIP GICC05 $\delta^{18}O$ records (**Figure 2**), the Pleistocene –Holocene Boundary was inferred at a depth of 31 cm. The FO of *Boliella adamsii* was detected at the depth interval of 34–35 cm which is relatively near to the depth of the Pleistocene–Holocene Boundary. *Globigerinella calida* was present throughout the entire core depth intervals. *Globorotalia fimbriata* first occurred at the depth interval of 90–91 cm, however, its occurrence was relatively inconsistent at the depth interval of 131 cm upward.

In core ST14, the correlation between *Gs. ruber* $\delta^{18}O$ records and NGRIP GICC05 $\delta^{18}O$ records (**Figure 3**) indicate the depth of 28 cm as the Pleistocene–Holocene Boundary. The FO of *Boliella adamsii* was indicated relatively near to the depth of the Pleistocene–Holocene Boundary, at the depth interval of

26–27 cm. *Globigerinella calida* occurred in the entire core depth intervals while *Globorotalia fimbriata* first occurred at the depth interval of 96–97 cm. Similar to core ST10, the occurrence of *Globorotalia fimbriata* was also relatively inconsistent at the depth interval of 97 cm upward.

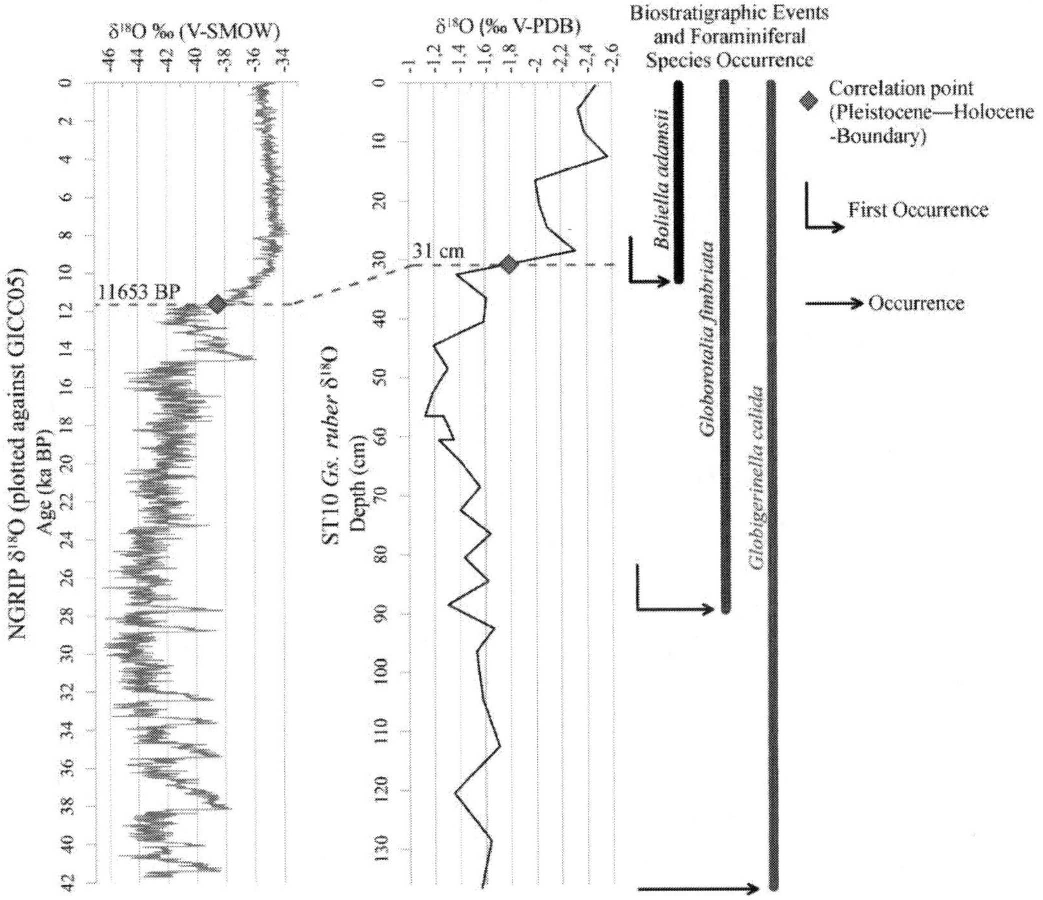

Figure 2. Comparison of biostratigraphic events in core ST10 to the Pleistocene–Holocene Boundary inferred from the correlation between ST10 *Gs. ruber* $\delta^{18}O$ records and NGRIP GICC05 $\delta^{18}O$ records.

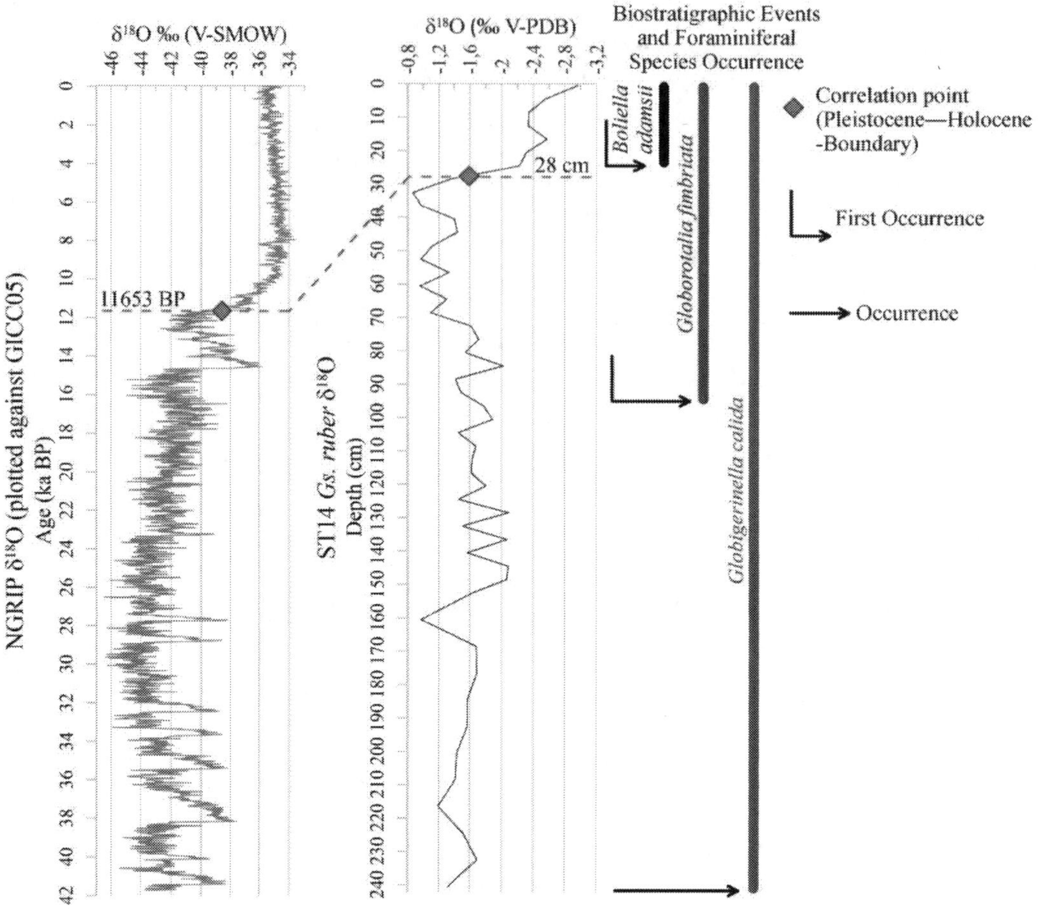

Figure 3. Comparison of biostratigraphic events in core ST14 to the Pleistocene–Holocene Boundary inferred from the correlation between ST14 *Gs. ruber* $\delta^{18}O$ records and NGRIP GICC05 $\delta^{18}O$ records.

In both core ST10 and ST14, the FO of *Boliella adamsii* was detected at the adjacent depth or nearly coeval to the Pleistocene–Holocene Boundary. This depth adjacency indicates that the FO of *Boliella adamsii* is applicable as the marker of the Pleistocene–Holocene boundary at least in the sea around Sumba Island. On the other hand, the use of *Globorotalia fimbriata* and *Globigerinella calida* FO as a Pleistocene–Holocene boundary markers is not recommended. These species were most likely occurred at least since the Late Pleistocene which was also suggested by previous studies [5,13,15].

Similar studies which combine the planktonic foraminiferal biostratigraphic event, particularly *Boliella adamsii* FO with oxygen isotope ($\delta^{18}O$) chronology in other Indonesian regions are needed in the future. These future studies will confirm the applicability of *Boliella adamsii* FO as a Pleistocene–Holocene Boundary marker in the Indonesian region. For now, new evidence of *Boliella adamsii* FO applicability as a Pleistocene–Holocene Boundary marker from this study hopefully will assist Indonesian researchers, especially the low-budgeted students and small university researchers to study the Late Pleistocene–Holocene paleoclimate and paleoceanography using marine sediment cores.

4. Conclusion

Based on its depth adjacency (near coeval) to the depth of Pleistocene–Holocene Boundary in both core ST10 and ST14, the FO of *Boliella adamsii* is applicable as the marker for the Pleistocene–Holocene Boundary in the sea around Sumba Island if radiocarbon ages and proper planktonic foraminiferal $\delta^{18}O$

data are not available. The use of FO of other previously applied species (i.e. *Globorotalia fimbriata* and *Globigerinella* calida) as a Pleistocene–Holocene Boundary marker is not recommended as those species were most likely occurred at least since the Late Pleistocene. Future studies with a similar method in other Indonesian regions are also needed to confirm the applicability of *Boliella adamsii* FO as a Pleistocene–Holocene Boundary marker in the Indonesian region.

5. References

[1] Harmon R S 2021 Introduction and Overview Encycl. Geol. 1–12
[2] Hajdas I, Ascough P, Garnett M H, Fallon S J, Pearson C L, Quarta G, Spalding K L, Yamaguchi H and Yoneda M 2021 Radiocarbon dating Nat. Rev. Methods Prim. 1 1–26
[3] Woods Hole Oceanographic Institution 2023 Fees for Radiocarbon Analyses https://www2.whoi.edu/site/nosams/radiocarbon-services/fees/, accessed February 20 2023
[4] Beta Analytic 2023 How much does carbon dating cost? https://www.radiocarbon.com/how-much-does-carbon-dating-cost.htm, accessed February 16 2023
[5] Bolli H M and Saunders J B 1985 Oligocene to Holocene low latitude planktic foraminifera Plankton Stratigraphy ed H M Bolli, J B Saunders and K Perch-Nielsen (New York: Cambridge University Press) pp 155–262
[6] Blow W H 1969 Late Middle Eocene to Recent planktonic foraminiferal Biostratigraphy Proceedings of the First International Conference on Planktonic Microfossils 1 ed R Bronniman and H H Renz (Geneva) pp 199–421
[7] Ardi R D W, Maryunani K A, Yulianto E, Putra P S and Nugroho S H 2019 Biostratigrafi dan Analisis Perubahan Kedalaman Termoklin di Lepas Pantai Barat Daya Sumba sejak pleistosen Akhir berdasarkan Kumpulan Foraminifera Planktonik Bul. Geol. 3 355–62
[8] Ardi R D W, Aswan A, Maryunani K A, Yulianto E, Putra P S and Nugroho S H 2021 Changes of thermocline depth at the Sumba Island offshore based on planktonic foraminiferal assemblages and its implication to eutrophication since the Last Deglaciation (~18 ka BP): a preliminary study Rud. Zb. 36 31–45
[9] Nugroho S H, Putra P S and Pamungkas A D G 2020 Karakteristik Sedimen Holosen – Pleistosen Berdasarkan Ukuran Butir pada Sedimen Inti EW17-08, Tepi Barat Laut Paparan Sunda J. Geosaintek 6 107–116
[10] Silalahi I R, Adisaputra M K, Saputro E and Ali A 2018 Keberadaan Fosil Penunjuk Foraminifera Plankton Kuarter di Perairan Bengkulu, Pantai Barat Sumatera J. Geol. Kelaut. 16 25–36
[11] Yosephin Y, Nugroho S H, Putra P S, Suedy S W A and Izzati M 2019 Palinologi Laut di Selat Sumba, Nusa Tenggara Timur Ris. Geol. dan Pertamb. 29 43–52
[12] Putra P S and Nugroho S H 2020 Holocene Climate Dynamics in Sumba Strait, Indonesia: A Preliminary Evidence From High Resolution Geochemical Records and Planktonic Foraminifera Stud. Quartenaria 32 91–99
[13] Knappertsbusch M 2016 Evolutionary prospection in the Neogene planktic foraminifer Globorotalia menardii and related forms from ODP Hole 925B (Ceara Rise, western tropical Atlantic): evidence for gradual evolution superimposed by long distance dispersal? Swiss J. Palaeontol. 135 205–48
[14] Coxall H K, Wilson P A, Pearson P N and Sexton P F 2007 Iterative evolution of digitate planktonic foraminifera Paleobiology 33 495–516
[15] Adisaputra M K, Hendrizan M and Kholiq A 2010 Katalog Foraminifera Perairan Indonesia (Bandung: Puslitbang Geologi Kelautan)
[16] Putra P S and Nugroho S H 2017 Distribusi Sedimen Permukaan Dasar Laut Perairan Sumba, Nusa Tenggara Timur Oseanologi dan Limnol. di Indones. 2 49–63
[17] Damanik A, Putra P S, Nugroho S H and Kapid R 2019 The vertical relationship between foraminiferal abundance and characterisitics of core sediment in Sumba Strait, East Nusa Tenggara J. Geol. Kelaut. 17 19–32

[18] Ardi R D W, Aswan A, Maryunani K A, Yulianto E, Putra P S and Nugroho S H 2023 Sea Surface Paleoproductivity Reconstruction Based on Foraminiferal Accumulation Rate in the Western Savu Strait Since the Last Glacial Maximum (~23 Ka Bp) *Rud. Zb.* 38 167–78

[19] Ardi R D W, Aswan, Maryunani K A, Yulianto E, Putra P S, Nugroho S H and Istiana 2020 Last Deglaciation—Holocene Australian-Indonesian Monsoon Rainfall Changes Off Southwest Sumba, Indonesia Atmosphere (Basel). 11 932

[20] Ardi R D W, Aswan, Maryunani K A, Yulianto E, Putra P S and Nugroho S H 2022 Australian–Indonesian monsoon rainfall responses to the northern hemisphere climatic changes prior to the Last Glacial Maximum: an early indication IOP Conf. Ser. Earth Environ. Sci. 1047 012031

[21] Dennison J M and Hay W W 1967 Estimating the Needed Sampling Area for Subaquatic Ecologic Studies J. Paleontol. 41 706–8

[22] Kenneth J P and Srinivasan M S 1983 Neogene planktonic foraminifera: A phylogenetic atlas (Stroudburg, Pennsylvania: Hutchinson and Ross)

[23] Andersen, Katrine K; Bigler, Matthias; Buchardt, Susanne L; Clausen, Henrik Brink; Dahl-Jensen, Dorthe; Davies, Siwan M; Fischer, Hubertus; Goto-Azuma, Kumiko; Hansson, Margareta E; Heinemeier, Jan; Johnsen, Sigfús Jóhann; Larsen, L B; Muscheler, Raimund; B M Greenland Ice Core Chronology 2005 (GICC05) and 20 year means of oxygen isotope data from ice core NGRIP

[23] Rasmussen S O, Andersen K K, Svensson A M, Steffensen J P, Vinther B M, Clausen H B, Siggaard-Andersen M-L, Johnsen S J, Larsen L B, Dahl-Jensen D, Bigler M, Röthlisberger R, Fischer H, Goto-Azuma K, Hansson M E and Ruth U 2006 A new Greenland ice core chronology for the last glacial termination J. Geophys. Res. 111 D06102

6. Appendices

Appendix A

Images of the Pleistocene–Holocene Boundary marker species: *Boliella adamsii* (1 and 2), *Globorotalia fimbriata* (3 and 4), and *Globigerinella calida* (5 and 6).

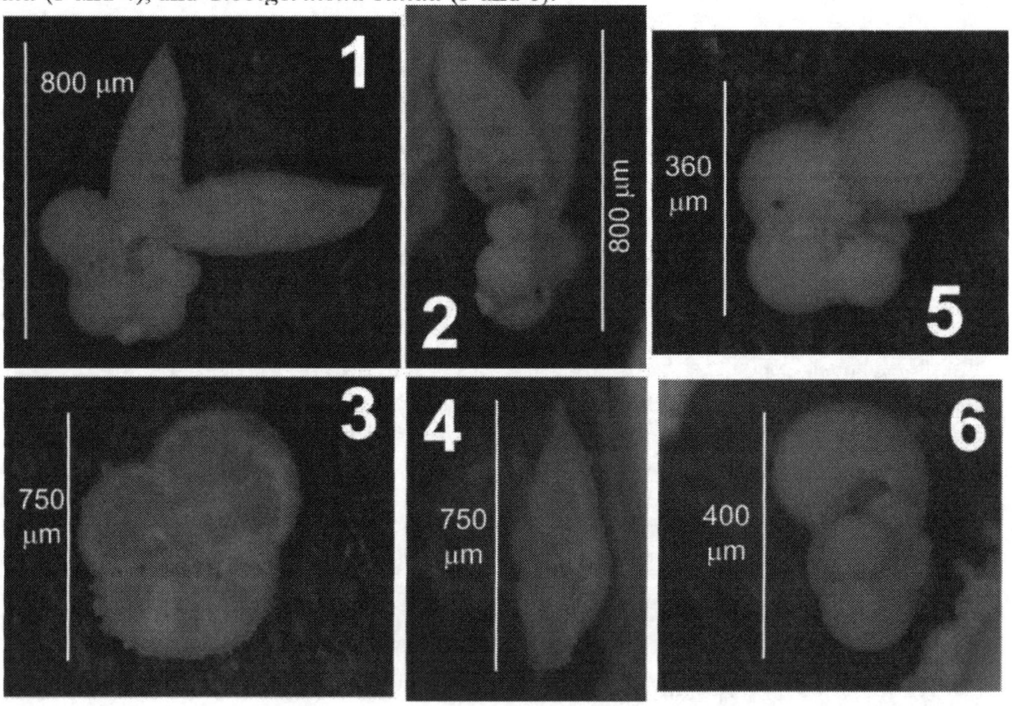

IOP Conf. Series: Earth and Environmental Science 1245 (2023) 012025 doi:10.1088/1755-1315/1245/1/012025

Appendix B
Detailed distribution of the Pleistocene–Holocene Boundary marker species: *Boliella adamsii*, *Globorotalia fimbriata*, and *Globigerinella calida* in core ST10.

Core	Epoch (based on oxygen isotope correlation)	Interval (cm)		Biostratigraphic Event	Species		
		Top	Bottom		*Boliella adamsi*	*Globorotalia fimbriata*	*Globgerinella calida*
ST10	Holocene	0	1			1	739
ST10	Holocene	2	3		2		6217
ST10	Holocene	4	5		1	2	4375
ST10	Holocene	6	7		4		7957
ST10	Holocene	8	9		128	128	2659
ST10	Holocene	10	11		2	512	5596
ST10	Holocene	12	13		10		3013
ST10	Holocene	14	15		4	256	10694
ST10	Holocene	16	17		128	5	2412
ST10	Holocene	18	19		431		3440
ST10	Holocene	20	21		512		1252
ST10	Holocene	22	23		512	512	7491
ST10	Holocene	24	25		3		2202
ST10	Holocene	26	27		4	64	5202
ST10	Holocene	28	29		256		1661
ST10	Holocene	30	31		1	1	4151
ST10	Pleistocene	32	33	↳ (First Occurrence)	2	128	1937
ST10	Pleistocene	34	35	*Boliella adamsii*	181		1998
ST10	Pleistocene	36	37				969
ST10	Pleistocene	38	39				467
ST10	Pleistocene	40	41			215	1118
ST10	Pleistocene	42	43				1941
ST10	Pleistocene	44	45				2779
ST10	Pleistocene	46	47				2726
ST10	Pleistocene	48	49			1	838
ST10	Pleistocene	50	51				3144
ST10	Pleistocene	52	53				724
ST10	Pleistocene	54	55				1616
ST10	Pleistocene	56	57			2	1433
ST10	Pleistocene	58	59				2092
ST10	Pleistocene	60	61				946
ST10	Pleistocene	62	63			37	2379
ST10	Pleistocene	64	65				867
ST10	Pleistocene	66	67				1174
ST10	Pleistocene	68	69			459	3216
ST10	Pleistocene	70	71			1	3135
ST10	Pleistocene	72	73				647

↳ First Occurrence

→ Occurrence

IOP Conf. Series: Earth and Environmental Science 1245 (2023) 012025 doi:10.1088/1755-1315/1245/1/012025

ST10		74	75				479
ST10		76	77			1	1177
ST10		78	79				1222
ST10		80	81				561
ST10		82	83				869
ST10		84	85			1	2151
ST10		86	87				664
ST10		88	89				3245
ST10		90	91	*Globorotalia fimbriata* ↳		128	1955
ST10		92	93				4075
ST10		94	95				2229
ST10		96	97				3164
ST10		98	99				1856
ST10		100	101				1916
ST10		102	103				1724
ST10		104	105				2458
ST10		106	107				3002
ST10		108	109				1629
ST10		110	111				4353
ST10		112	113				1624
ST10		114	115				4664
ST10		116	117				5088
ST10		118	119				1866
ST10		120	121				1539
ST10		122	123				3592
ST10		124	125				615
ST10		126	127				3225
ST10		128	129				2446
ST10		130	131				1751
ST10		132	133				919
ST10		134	135				2950
ST10		136	137	*Globigerinella calida* →			1324

Appendix C

Detailed distribution of the Pleistocene–Holocene Boundary marker species: *Boliella adamsii*, *Globorotalia fimbriata*, and *Globigerinella calida* in core ST14.

Core	Epoch (based on oxygen isotope correlation)	Interval (cm)		Biostratigraphic Event	Species		
		Top	Bottom		*Boliella adamsi*	*Globorotalia fimbriata*	*Globgerinella calida*
ST14	Holocene	0	1			131	1047
ST14		2	3		1		1396
ST14		4	5				1782

↳ First Occurrence

→ Occurrence

ST14		6	7				1177
ST14		8	9		3		1292
ST14		10	11		1		890
ST14		12	13		1		1767
ST14		14	15				1119
ST14		16	17			1	453
ST14		18	19		74		1785
ST14		20	21		1		953
ST14		22	23		74		3268
ST14		24	25		1		968
ST14		26	27	*Boliella adamsii* →	37		1479
ST14		28	29				1786
ST14		30	31			1	2660
ST14		32	33			167	336
ST14		34	35				557
ST14		36	37			1	332
ST14		38	39				2075
ST14		40	41			1	159
ST14		42	43			288	863
ST14		44	45				660
ST14		46	47			1	374
ST14		48	49				155
ST14		50	51			1	2187
ST14		52	53				303
ST14		54	55				3098
ST14		56	57			1	629
ST14		58	59				3495
ST14		60	61				324
ST14		62	63			1	1335
ST14		64	65			1	167
ST14	Pleistocene	66	67			144	577
ST14		68	69				320
ST14		70	71				553
ST14		72	73				478
ST14		74	75				1493
ST14		76	77				934
ST14		78	79				1098
ST14		80	81				475
ST14		82	83			37	888
ST14		84	85			1	636
ST14		86	87				1344
ST14		88	89				319
ST14		90	91				710
ST14		92	93				0
ST14		94	95			1	1161
ST14		96	97	*Globorotalia fimbriata* →		2	154
ST14		98	99				584
ST14		100	101				944
ST14		102	103				669

IOP Conf. Series: Earth and Environmental Science 1245 (2023) 012025 doi:10.1088/1755-1315/1245/1/012025

ST14		104	105			440
ST14		106	107			717
ST14		108	109			244
ST14		110	111			1002
ST14		112	113			573
ST14		114	115			928
ST14		116	117			437
ST14		118	119			465
ST14		120	121			122
ST14		122	123			1022
ST14		124	125			653
ST14		126	127			808
ST14		128	129			140
ST14		130	131			853
ST14		132	133			920
ST14		134	135			484
ST14		136	137			74
ST14		138	139			1156
ST14		140	141			583
ST14		142	143			1217
ST14		144	145			339
ST14		146	147			371
ST14		148	149			1051
ST14		150	151			874
ST14		152	153			151
ST14		154	155			897
ST14		156	157			460
ST14		158	159			891
ST14		160	161			168
ST14		162	163			463
ST14		164	165			687
ST14		166	167			1089
ST14		168	169			342
ST14		170	171			927
ST14		172	173			237
ST14		174	175			861
ST14		176	177			381
ST14		178	179			791
ST14		180	181			620
ST14		182	183			548
ST14		184	185			292
ST14		186	187			842
ST14		188	189			449
ST14		190	191			747
ST14		192	193			487
ST14		194	195			848
ST14		196	197			292
ST14		198	199			789
ST14		200	201			676

ST14		202	203				833
ST14		204	205				440
ST14		206	207				1008
ST14		208	209				309
ST14		210	211				1039
ST14		212	213				1239
ST14		214	215				1378
ST14		216	217				282
ST14		218	219				1349
ST14		220	221				477
ST14		222	223				1368
ST14		224	225				146
ST14		226	227				1231
ST14		228	229				626
ST14		230	231				368
ST14		232	233				24
ST14		234	235				792
ST14		236	237				560
ST14		238	239				824
ST14		240	241				798
ST14		242	243	*Globigerinella calida* →			1331

7. Acknowledgments

This research is part of the Doctoral Dissertation Research (contract number: 007/E5/PG.02.00.PL/2023) funded by Directorate General of Higher Education, Ministry of Education and Culture, Republic of Indonesia, granted to Aswan, Khoiril Anwar Maryunani, and Ryan Dwi Wahyu Ardi. We also thank the Research Center for Oceanography of Indonesian Institute of Science (now BRIN), especially Udhi Hermawan, PhD as the chief scientist of E-WIN 2016 for the permission of data usage and administrative assistance and all crews of the Baruna Jaya VIII R.V., especially Singgih Adi Wibowo for technical supports and data collecting. The Research Center Geological Disaster of BRIN and Geological Engineering Department of ITB are thanked for the laboratory facilities. We also express our gratitude to Indonesian Endowment Fund for Education (LPDP) (contract number: 202001110215954) for funding Ryan Dwi Wahyu Ardi's Doctoral study.

IOP Conf. Series: Earth and Environmental Science 1245 (2023) 012026 doi:10.1088/1755-1315/1245/1/012026

Cianjur M5.6 Earthquake Aftershock Survey (CEAS) Using ITB Temporary Seismographs

M F Azhari[1], A D Nugraha[2], Z Zulfakriza[2], RV Ry[2], A Lesmana[3], A S Putra[3], F Muttaqy[4], N Heryandoko[35], A Saputra[5], A Mawadah[1], D A Putri[6], H S Saputra[6]

[1] Master Program of Earth Science, Faculty of Earth Science and Technology, Institut Teknologi Bandung, Indonesia
[2] Global Geophysics Research Group, Faculty of Mining and Petroleum Engineering, Institut Teknologi Bandung, Indonesia
[3] Doctor Program of Geophysical Engineering, Faculty of Mining and Petroleum Engineering, Institut Teknologi Bandung, Indonesia
[4] Research Center for Geological Disaster, National Research and Innovation Agency (BRIN), Indonesia
[5] Meteorological, Climatological and Geophysical Agency (BMKG), Indonesia
[6] Bachelor Program of Geophysical Engineering, Faculty of Mining and Petroleum Engineering, Institut Teknologi Bandung, Indonesia

Email : fikriazharimuhammad@gmail.com

Abstract. Western Java, part of the Sunda Arc, is an active tectonic region with high seismicity. The tectonic activity has produced active faults which are located near densely populated areas. On November 21, 2022, a significant earthquake rocked the Cianjur region, West Java, causing severe damage and loss of life. In this study we acquired aftershock data which was generated from the Cianjur M5.6 earthquake. Twenty stations consisting of 19 Smartsolo IGU-16HR 3C 5 Hz Seismometer (short-period type) and 1 Guralp 6TD 30s to 100 Hz Seismometer (broadband type) were deployed to monitor aftershocks from 22 November to 23 December 2022. The seismograph stations were deployed covering the earthquake source area and tectonic features in the Cianjur region. The results show that the aftershock events is clearly recorded at 20 stations with an average difference between the arrival times of the S waves and P waves (Ts-Tp) of 2 seconds. A total of ±500 earthquake events were obtained consisting of ±4000 P wave phases and ±3000 S wave phases with a Vp/Vs ratio value of 1.72 and RMS error less than 0.1 second. The Cianjur M5.6 earthquake was classified as a shallow crustal seismicity with a strike slip fault mechanism, hypocenters were generally distributed at a depth of 3-10 km.

1. Introduction

West Java is a region with complex tectonic characteristics. Located north of the megathrust subduction zone, where the Indo-Australian oceanic plate subducts beneath the surface of Sundaland, on the southern boundary of the Eurasian Plate [1]. Tectonic activity from the subduction of the Indo-Australian Plate under the Eurasian Plate has resulted in several active faults in West Java including the Cimandiri Fault, Lembang Fault and Baribis Fault, which are located close to densely populated areas [2]. The existence of these local faults has an impact on high seismic activity in West Java and its surrounding area.

On November 21, 2022, a significant earthquake rocked the Cianjur, West Java and surrounding area with the 5.6 magnitude and 10 km focus depth [3]. The Meteorological, Climatological and Geophysical Agency (BMKG) stated that the mainshock event had generated quite strong shocks (MMI III-VIII) around the earthquake source area in Cianjur, West Java. The BMKG further stated that this earthquake was included in the mainshock-aftershock type of shallow crustal earthquake, namely the main earthquake which was then followed by a series of aftershocks [4]. The National Disaster Management Agency (BNPB) reported that the incident caused damage to buildings and public facilities, especially in the Cianjur area, West Java. Based on information from BNPB on 27 December 2022, the Cianjur M5.6 earthquake damaged more than 56,000 buildings and killed 602 people. There were more than 114,000 people, originating from 16 districts in Cianjur, who were evacuated due to the high intensity of aftershocks in the area.

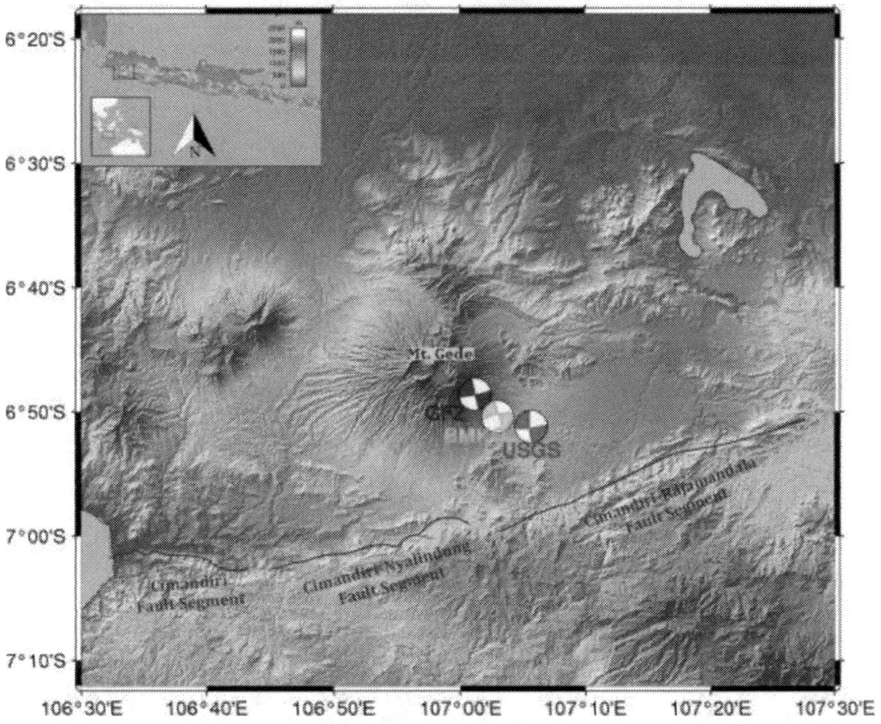

Figure 1. Epicenter of The Cianjur M5.6 Earthquake Based on Data from BMKG (Green), GFZ (Blue) and USGS (Red)

The interesting things about Cianjur earthquake, the structure which caused this earthquake is still being debated considering that the epicenter of the earthquake was quite far from the Cimandiri fault and also Gede volcano as shown in Figure I.1. Where the epicenter is estimated to be at a distance of 20 km north of the Cimandiri Fault Rajamandala Segment, this fault has a left-lateral slip mechanism [5]. There was possibility the Cianjur earthquake was produced by unidentified fault. Based on historical records of seismicity, this area experienced devastating earthquakes in 1844, 1879, 1900 [6] and also 1962, M5.5 1982 and M5.4 2000, Cianjur earthquakes [7]. Along with the series of earthquake events that occurred in Cianjur, West Java, it certainly has implications for changes in tectonic conditions in the region. Therefore, to provide further information of the recent tectonic dynamics, studies are needed to identify the characteristics of the earthquake mechanism, so further investigation is needed regarding the complexity of the structure and the source characteristic of the Cianjur earthquake.

2. Methods

2.1 Data Acquisiton

The data used in this study is aftershock recording data using the ITB temporary seismograph network deployed after the Cianjur M5.6 earthquake. In this research, 20 stations were used with a data recording period of 1 month from November 22 to December 23, 2022. There are 2 types of seismometers used in this study, including 1 unit of broadband seismometer Guralp 6TD 30s to 100 Hz 2 x 1200 V/m/s and 19 units of Smartsolo seismometer (short-period) IGU-16HR 3C 5 Hz.

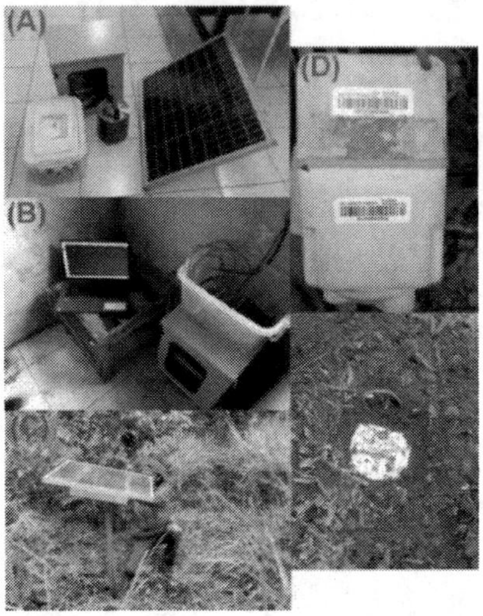

Figure 2. The Aftershock Data Acquisition. Guralp seismograph component (broadband) (A), Guralp installation and waveform monitoring (B), Solar panel installation for seismograph battery (C), Smartsolo seismometer component (short-period) (D), Smartsolo installation (E)

The seismograph stations was installed to cover the earthquake source area covering several areas, including Cianjur, Sukabumi, West Bandung and Bandung Regencies. Station distribution can be seen in the following figure.

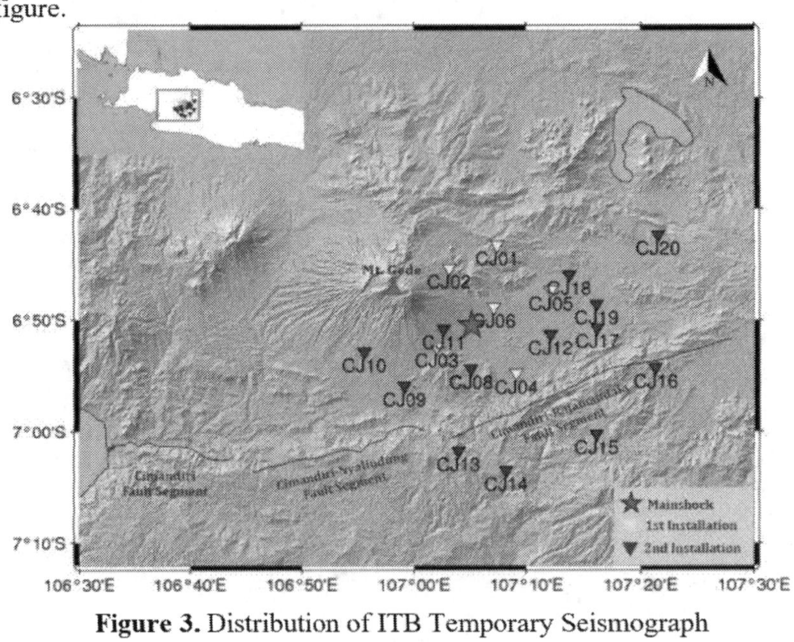

Figure 3. Distribution of ITB Temporary Seismograph

IOP Conf. Series: Earth and Environmental Science 1245 (2023) 012026 doi:10.1088/1755-1315/1245/1/012026

The seismograph was installed in two stages. The first stage was carried out on November 22, 2022, with 6 seismometer stations installed. While the second stage will be carried out on November 26, 2022, 14 seismometer stations will be installed. Data recording time ranges from 26-31 days. The following is a data record table.

Table 1. Aftershock Data Recording

No	Station Name	Coordinat			Recording Duration (Day)
		Lat	Lon	Elev	
1	CJ01	-6.72120	107.12368	824	31
2	CJ02	-6.75547	107.05235	1097	31
3	CJ03	-6.87365	107.03391	846	31
4	CJ04	-6.91330	107.15194	431	32
5	CJ05	-6.78862	107.20229	314	32
6	CJ06	-6.81413	107.11842	540	32
7	CJ07	-6.81413	107.11842	540	28
8	CJ08	-6.90607	107.08447	651	28
9	CJ09	-6.93205	106.98528	615	27
10	CJ10	-6.88073	106.92621	691	27
11	CJ11	-6.84683	107.04405	880	27
12	CJ12	-6.85487	107.20183	343	28
13	CJ13	-7.02960	107.06640	1068	27
14	CJ14	-7.05844	107.13723	1034	27
15	CJ15	-7.00411	107.26916	966	27
16	CJ16	-6.90479	107.35485	678	27
17	CJ17	-6.84509	107.27035	291	27
18	CJ18	-6.76571	107.22900	273	27
19	CJ19	-6.80978	107.26922	271	27
20	CJ20	-6.70583	107.35854	354	26

2.2 Data Processing

The stages of processing the earthquake aftershock data include several stages.

2.2.1 Waveform Identification and Picking P & S Wave

We have identified the aftershock events which were recorded at a minimum of four stations and pick the arrival time of the P and S waves at the event. This process is done manually using Seisgram2K Software.

2.2.2 Velocity Model Parameter Setting

Second, the 1D seismic velocity model is determined according to the characteristics of the study area. In this case, the ANT velocity model from Rosalia (2022) has been used, which has modeled the seismic velocity structure in each subsurface layer in the western part of Java [8].

Table 2. 1D Velocity Model

Depth (km)	P Velocity (km/sec)	S Velocity (km/sec)
0	3.77	2.14
1	4.43	2.61
2	4.54	2.69
3	4.64	2.75
4	4.74	2.81
5	4.85	2.88
6	5.04	3.00
7	5.22	3.10
8	5.39	3.19
9	5.50	3.25
10	5.49	3.25
11	5.43	3.22
12	5.38	3.19
13	5.34	3.17
14	5.39	3.19
15	5.42	3.21
16	5.47	3.24
17	5.55	3.28
18	5.64	3.33
19	5.75	3.39
20	5.83	3.44

2.2.3 Hipocenter Determination

Third, we determine the location of the hypocenter using the non-linear method of the NLLoc program [9]. This program implements a probabilistic inversion approach, a direct search method that the algorithm employed by Tarantola and Valette [10]. In addition, the hypocenter mapping is determined using GMT program.

3. Result and Discussion

Events recorded by temporary seismograph stations have been identified. Followed by determining the arrival time of the P and S waves by picking the first motion for each wave. P and S wave picking is carried out at a minimum of four stations that record earthquake events. From the picking results, more than 500 events were recorded and the arrival times of the P and S waves were successfully determined. The process is carried out using the Seisgram2K software. An example of a successfully recorded seismogram is shown in the following figure.

IOP Conf. Series: Earth and Environmental Science 1245 (2023) 012026 doi:10.1088/1755-1315/1245/1/012026

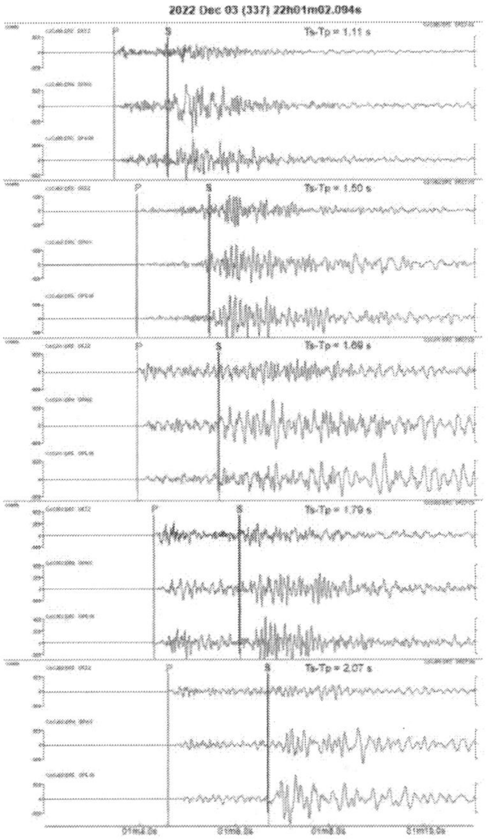

Figure 4. Determination of the P and S wave arrival time for one of the events on Desember 3, 2022. The green line shows the arrival time of P waves and the blue line shows the arrival time of the S waves.

In determining the arrival time of P and S waves, P waves can be easily observed on the vertical component (Z) seismogram, while S waves can be observed on the horizontal component (N and E) seismograms. The quality of picking results can be controlled by using Wadati Diagrams. The diagram depicts the arrival time of the P wave (Tp) as the X-axis and the difference between the arrival time of the S wave and the arrival time (Ts-Tp) as the Y-axis to determine the Vp/Vs ratio. The following is an example of the results of the Wadati Diagram plotting in one of the picked events.

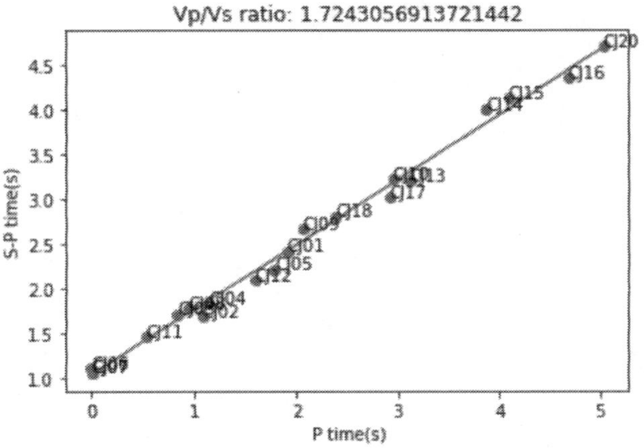

Figure 5. Wadati Diagram of P and S wave picking results in one of the earthquake events

Furthermore, the arrival time observation data of the entire event is plotted into a Wadati Diagram to see the linear relationship between the phases. From the plotting results, it can be seen that the gradient of the arrival time distribution trend is 0.7, so that the value of the Vp/Vs ratio is 1.72. It shows the P-wave and S-wave arrival times that we observed fit the linear relationship, and the quality is good enough for earthquake hypocenter determination.

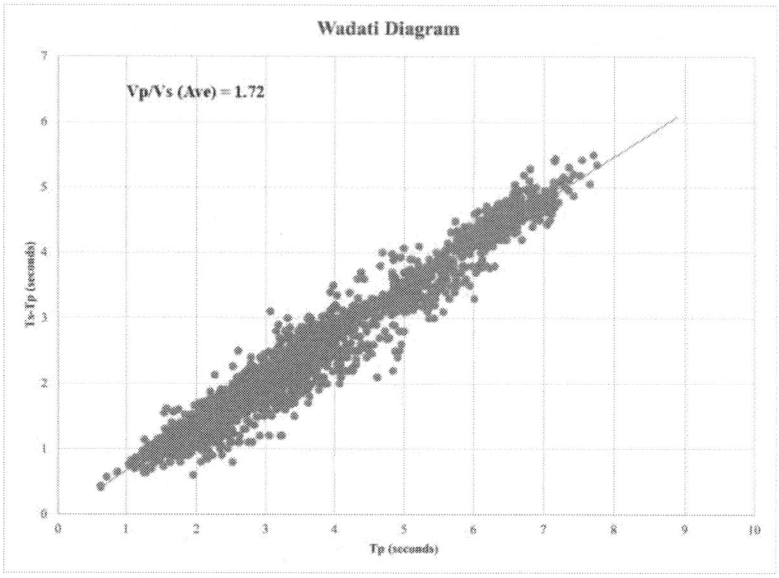

Figure 6. Diagram of the overall earthquake events. The Vp/Vs ratio value is 1.72.

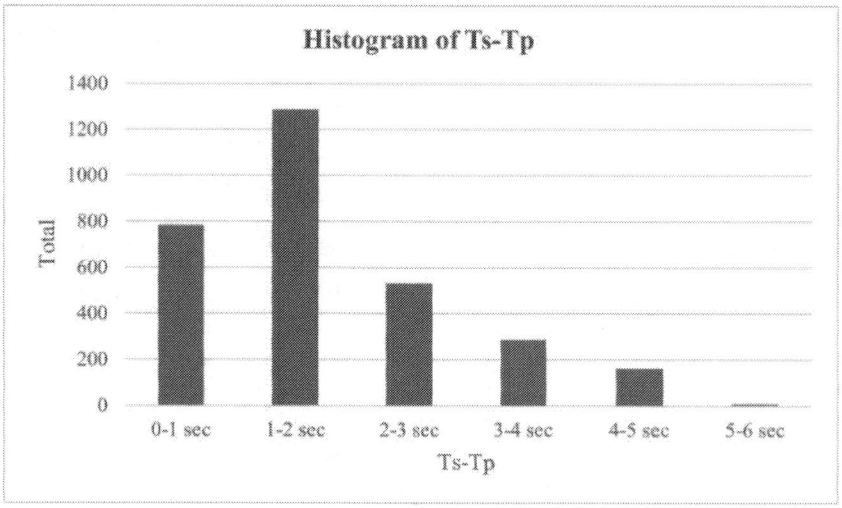

Figure 7. Distribution of the difference value between the arrival times of the P and S waves on average for all events

The distribution of the difference in the arrival time of the P and S waves on average for all events is dominated by events with a difference in arrival time of less than 2 seconds. This shows that the recorded event location is very close to the seismometer station. Beside that, this can minimize the error value obtained from determining the hypocenters.

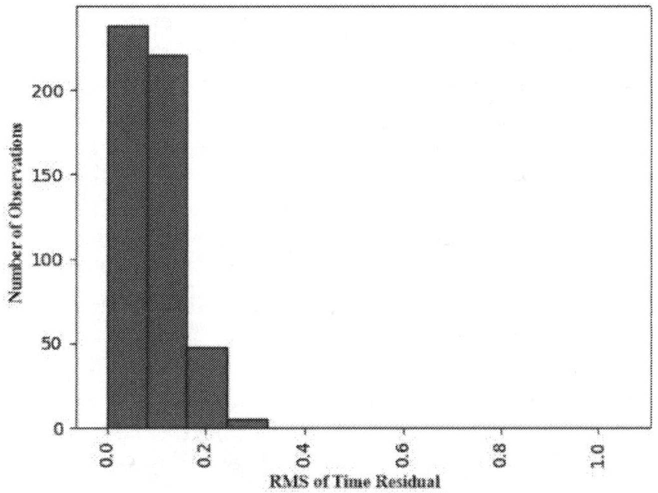

Figure 8. Distribution of the RMS residual error for all events

The results of determining the location of the hypocenter aftershock for the Cianjur M5.6 earthquake 2022 show that the non-linear method (NonLinLoc) has good statistical abilities, seen from the distribution of residual RMS values for each specified event. In general, the residual RMS value is under 1 second with the highest RMS value being under 0.1 second.

From the results of hypocenter determination, each parameter was obtained, such as origin time, azimuth gap, RMS error and location of the epicenter. The location of the earthquake hypocenter was plotted using the GMT program. The distribution of hypocenters is shown in figure below.

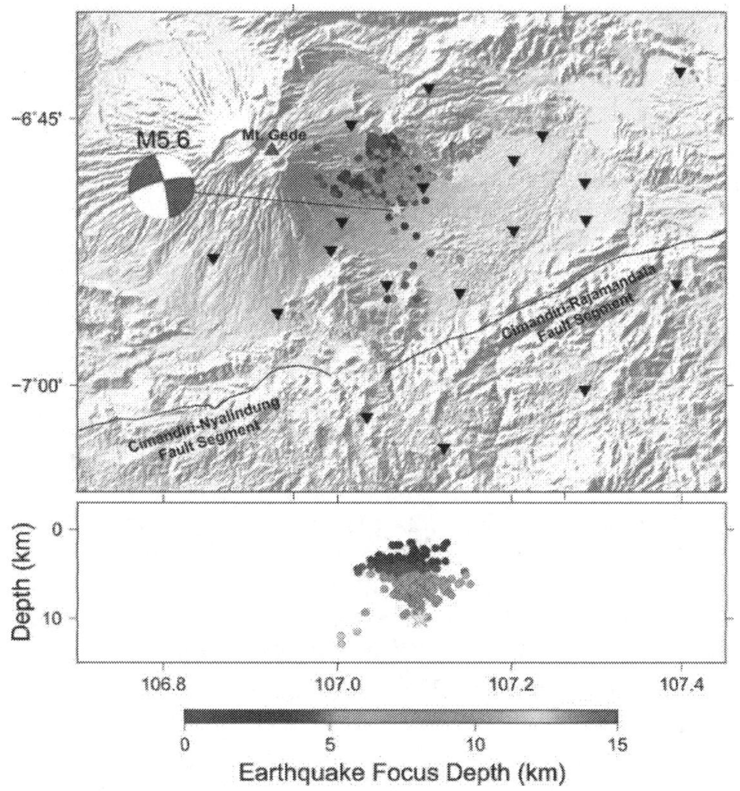

Figure 9. Cianjur Aftershock Map. Focal mechanism reference was obtained by BMKG data.

Based on hypocenter determination using the NonLinLoc program, more than 500 aftershock events were successfully recorded by the ITB temporary seismograph stations. Total phases obtained are ±4000 P-arrival time phases and ±3000 S-arrival time phases. The hypocenters of Cianjur aftershock were also distributed at a depth of between 3-10 km. Aftershock cluster was located at the north of the M5.6 mainshock (shown by a yellow star). It can be seen that the seismicity pattern of the aftershocks for the Cianjur M5.6 earthquake has two distinct segments, including the North-South segment and the West-East segment which are consistent with the focal mechanism parameters with a strike slip fault mechanism. In addition, the North West-South East (NW-SW) segment looks more dominant based on the event distribution. In the previous study, moderate earthquake could affect aftershock evolution, for example the seismogenic fault of Alxa M5.8 earthquake associated with a NEE-striking blind fault that becomes steeply dipping toward the south [11].

The Cianjur aftershock map shows that this earthquake is an earthquake originating from a fault that has not been identified before (blind fault), not related to the Cimandiri fault segment. There is a possibility that this earthquake caused fault conjugation, but this finding needs to be confirmed by several advanced method to gain comprehensive understanding. Futhermore, the findings show that this earthquake did not cause volcanic activity from Gede volcano. Gede volcano is monitored visually and instrumentally from the Volcano Observation Center in Cianjur. The latest data shows that Gede volcano erupted in 1957 with the characteristics of an explosive eruption in the form of thick gray to black ash eruptions, with an eruption column height of 3 km above the peak, but the activity level of Gede volcano at this time is Level I (Normal). The latest developments in Gede volcano activity following the Cianjur earthquake from 21 November 2022 to 27 November 2022 show that there has been no increase in the number of volcanic earthquakes since the Cianjur earthquake occurred on 21 November 2022 to 27 November 2022. Only 2 deep volcanic earthquakes were recorded with an amplitude of 21-45 mm and an earthquake lasting 5-8 seconds [12]. The location of Gede volcano, which is relatively close to the center of the Cianjur earthquake, is feared to affect the magmatic activity of Gede volcano. Linde and Sacks (1998) state that volcanoes can be triggered by eruptions up to a distance of 100 km from the epicenter of an earthquake with a magnitude < M 8.0 and up to a distance of 50 km for earthquakes with a magnitude ≥M8.0 [13]. Manga and Brodsky (2006) estimate that only 0.4% of explosive volcanic eruptions occur a few days after a major earthquake [14]. Whereas Sawi and Manga (2018) define that there are volcanoes that are sensitive (ever triggered by an eruption) and insensitive (never been triggered by an eruption). Sensitive volcanoes tend to be more active than those that are not sensitive and erupt more frequently, only occur in subduction zones, and are dominated by andesite and basaltic andesite eruptions. Eruptions triggered by earthquakes do not show a relationship between magnitude and distance. Meanwhile, there is a 5-12% increase in eruptive activity in the 2 months to 2 years after the main earthquake [15].

4. Conclusion

Based on data acquisition, it shows good results with clear waveforms recorded at 20 stations. The difference between the arrival time of the P and S waves (Tp-Ts) is ≤ 2 seconds. A total of ±500 earthquake events were obtained consisting of ±4000 P wave phases and ±3000 S wave phases with a Vp/Vs ratio value of 1.72 and RMS error less than 0.1 second. From hypocenter determination shows that the aftershock of Cianjur M5.6 earthquake was dominated by shallow crustal seismicity with a depth of 3-10 km, it was produced by a strike slip mechanism. For the case of Gede volcano, there was no significant increase in volcanic seismic activity following the Cianjur M5.6 earthquake. However, the future research is necessary to consider the possibility of volcanic and tectonic interaction especially in West Java region, Indonesia.

Acknowledgments

We are grateful to the Global Geophysics Research Group, Bandung Institute of Technology (ITB) for allowing their instruments to be used in this study, and to the Institute of Research and Community Service (LPPM) ITB for partial support of the fieldwork. We also thanks to the LPDP scholarship granted to main author from The Indonesia Endowment Fund for Education, The Ministry of Finance for partially funding this study.

References

[1] Hall R, Cottam M.A., and Wilson E.J. (2011): The SE Asia Gateway: history and tectonics of the Australia-Asia Collision. pp. 75–109 (London Geological Society 355, Special Publication).
[2] Irsyam M, Widiyantoro, S., Natawidjaja, D.H., Meilano, I., Rudyanto, A., Hidayati, S., Triyoso, W., Hanifa, N.R., Djarwadi, D., Faizal. L., dan Sunarjito. (2017): Peta sumber dan bahaya gempa Indonesia tahun 2017, Cetakan pertama. Pusat Penelitian dan Pengembangan Perumahan dan Permukiman, Badan Penelitian dan Pengembangan, Kementerian Pekerjaan Umum, Bandung.
[3] BMKG. (2022): Analisis Gempabumi Cianjur (Jawa Barat) Mw 5.6 Tanggal 21 November 2022. https://bmkg.go.id.
[4] Mogi K. (1963): Some Discusions on Aftershocks, Foreshocks and Earthquake Swarms: The Fracture of a Semi-Infinite Body Caused by Inner Stress Origin and Its Relation to the Earthquake Phenomena, Bulletin of the Earthquake Research Institute 41, 615–658.
[5] Supendi P, Nugraha AD, Puspito NT, Widiyantoro S, and Daryono D. (2018): Identification of active faults in West Java, Indonesia, based on earthquake hypocenter determination, relocation and focal mechanism analysis. *Geoscience Letters*, 5(1), 1-10.
[6] Visser S.W. (1922): Inland and Submarine Epicentra of Sumatra and Java Earthquakes, Javasche Boekhandel En Drukkerij.
[7] Supartoyo, dan Surono. (2009): Kegempaan di Wilayah Jawa Barat dan Kejadian Gempabumi Jawa Barat Selatan Tanggal 2 September 2009, Jurnal Gunungapi dan Mitigasi Bencana Geologi, 1, No.2 July 2009.
[8] Rosalia S, Widiyantoro S, Cummins PR, Yudistira T, Nugraha AD, Zulfakriza Z, and Setiawan A. (2022): Upper crustal shear-wave velocity structure Beneath Western Java, Indonesia from seismic ambient noise tomography. Geoscience Letters, 9 (1), 1-14.
[9] Lomax A. (2005): A reanalysis of the hypocentral location and related observations for the great 1906 California earthquake. Bulletin of the Seismological Society of America, 95(3), 861-877.
[10] Tarantola A, and Valette B. (1982): Inverse Problems = Quest for Information, J. Geophys. 50 159– 70.
[11] Zeng, X., Li, X., Li, M., and Li, W. (2023): Investigation of the 2015 Ms5. 8 Alxa Left Banner earthquake sequence: Aftershock evolution and seismogenic structure. Frontiers in Earth Science, 11.
[12] PVMBG. (2022): Press Release Aktivitas Vulkanik Gunungapi Gede Pasca Gempa Cianjur 21 November 2022. https://geologi.esdm.go.id/.
[13] Linde, A. T., and Sacks, I. S. (1998): Triggering of volcanic eruptions. Nature, 395(6705), 888-890.
[14] Manga, M., and Brodsky, E. (2006): Seismic triggering of eruptions in the far field: Volcanoes and geysers. Annu. Rev. Earth Planet. Sci, 34, 263-291.
[15] Sawi, T. M., and Manga, M. (2018): Revisiting short-term earthquake triggered volcanism. Bulletin of Volcanology, 80, 1-9.

IOP Conf. Series: Earth and Environmental Science 1245 (2023) 012027 doi:10.1088/1755-1315/1245/1/012027

Volcanic Geohazard Analysis - Case Study from the Ujung Lemah Abang Nuclear Power Plant Area, Central West Java

P Sumintadiredja[1], A Rudyawan[1*], M Surmayadi[2], IGBE Sucipta[1]

[1]Geological Engineering Study Program, Faculty of Earth Sciences and Technology, Institut Teknologi Bandung, Jalan Ganesa No. 10 Bandung 40132
[2] Pusat Volkanologi dan Mitigasi Bencana Geologi, Jl. Diponegoro 57, Bandung 40122

Email:*alfend@geodin.net; alfend@itb.ac.id

Abstract. The paper discusses the proposed development plan of a nuclear power plant in the Ujung Lemah Abang (ULA) region, located in the northern part of Mount Muria in Central Java, Indonesia. The study evaluates the potential volcanic disasters at the proposed site of the Ujung Lemah Abang nuclear power plant (NPP-ULA) near Mount Muria using the Bayesian method, which is based on a probabilistic model of the spatial distribution of volcanic events in the past and the geological processes model.

Mount Mulia is a polygenetic volcano that undergoes periodic eruptions alternating between lateral and central eruption-dominated phases. Semivariogram analysis shows a range of 150,000 years. This corresponds to the average duration of volcanic activity and the time between eruptions. Mount Muria volcanic complex can be classified as an extinct volcano with limited future eruption potential based on its last activity 320,000 years ago. The probability of eruption occurrences in Muria within the next 100 years is 1.21856×10^{-5}. With this probability of eruptions, it appears that the Muria Mtn is not going to be erupted anytime soon. However, when a new magma system emerges, resulting in epicenter concentrations beneath Mount Muria, micro-earthquake monitoring is required to detect magmatic activities beneath the Muria Volcanic Complex. To estimate the likelihood of phreatomagmatic eruptions, geothermometer monitoring in deep wells is required as an active monitoring of hydrothermal activity in the Muria Volcanic Complex.

1. Introduction

The proposed nuclear power plant site is north of the Muria Peninsula on the Muria Volcanic Complex (Figure 1), which includes Mount Muria in the center, Mount Genuk to the north, and Mount Patiayam to the south. To mitigate volcanic disasters, mitigation engineering solutions are required, particularly for energy generation facilities located in volcanic areas. According to IAEA recommendations, when determining the location of power plants, volcanic activity must be considered [1]. The study assesses the potential volcanic disasters caused by Mount Muria's central eruptions, which can result in pyroclastic material fall, ballistic projectile, lava flow, pyroclastic flow, and volcanic material lahar. Maar formation has the potential for pyroclastic material fall and ballistic fragment disasters, accompanied by the material throwing process, lava flow, and release of toxic gases into the air.

The simulation of the potential disaster caused by the maar eruption is carried out in the northwest part of Mount Muria's slope, where the farthest combination of maar occurs. Mount Muria has a lava dome in the crater and at least four lava pyroclastic domes on its slope. The parasitic lava dome on the Mount Muria slope is the source of the potential disaster of dome extrusion and its association. Except for lava flow, there is no geological or historical evidence that the dome extrusion event produces pyroclastic material. Regionally, the area surrounding the Muria Volcanic Complex has been affected by recent earthquake activity caused by some of the East Java Basin's fault systems, specifically the E-W Sakala - Sepanjang Faults and the NE - SW faults system, which are related to the older basement structural grain. However, there are significantly fewer earthquake activities in the Muria area.

Depending on the type of disaster assessment, evaluating volcanic and seismic disasters that affect nuclear facilities differs in several ways [2,3]. The most significant distinction is that the primary focus is on disasters at the specific location of the nuclear facility, in this case, the Tapak PLTN location, rather than disasters in the larger area surrounding the volcano or fault. On the left end of Figure 2, the level of risk acceptance for the nuclear facility is shown, which means that volcanic or earthquake eruptions could result in a loss of control over nuclear power generation or nuclear material leakage.

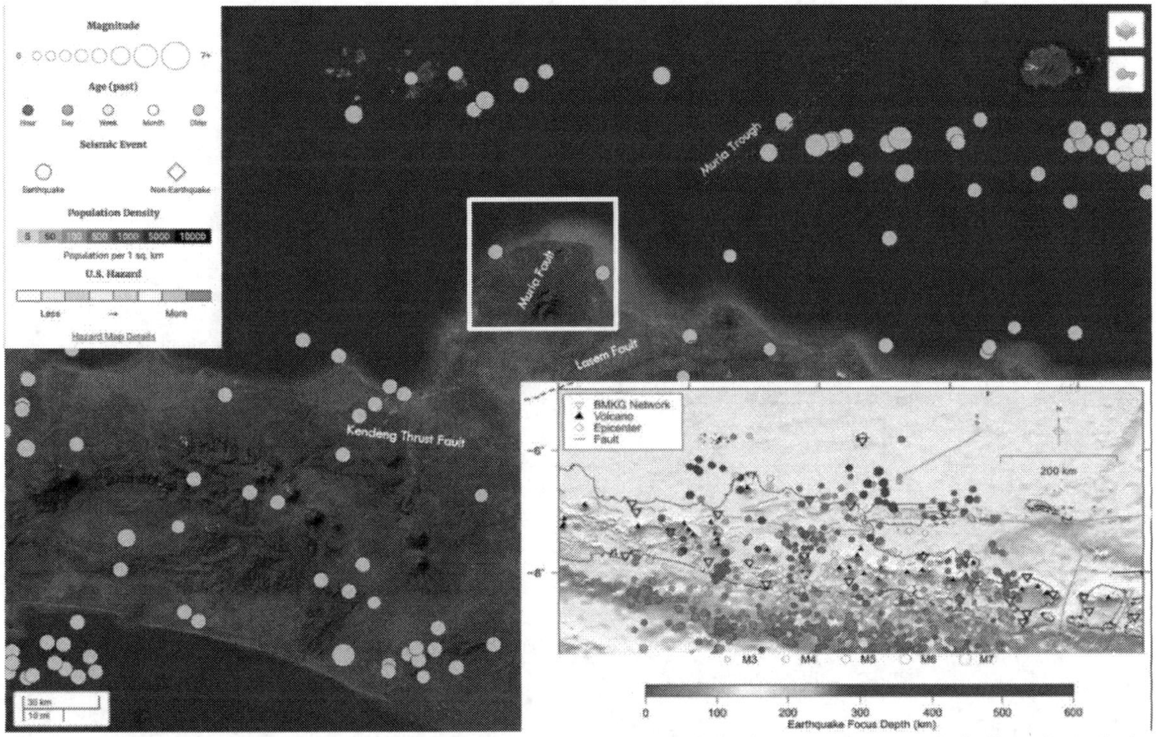

Figure 1. Location of study area for the power plant. Earthquake map on top left (inzet) and the interpreted faults around Muria Volcanic Complex [4] and the earthquake data were acquired from USGS.

The estimation of disasters and risks based on a small number of volcanic events is challenging either because they occur infrequently or are unidentifiable [6]. Because geological records are incomplete, a small number of volcanic and seismic events preserved in geological records can determine risk analysis. One approach to improving disaster assessment based on multiple models is to modify probabilistic analysis by incorporating additional data via Bayesian inference [6,7,8].

During the first stage of the investigation, all data is gathered in order to identify potentially hazardous volcanic phenomena at the proposed site. Volcanic disasters are evaluated using deterministic and probabilistic methods to determine the acceptability of proposed area.

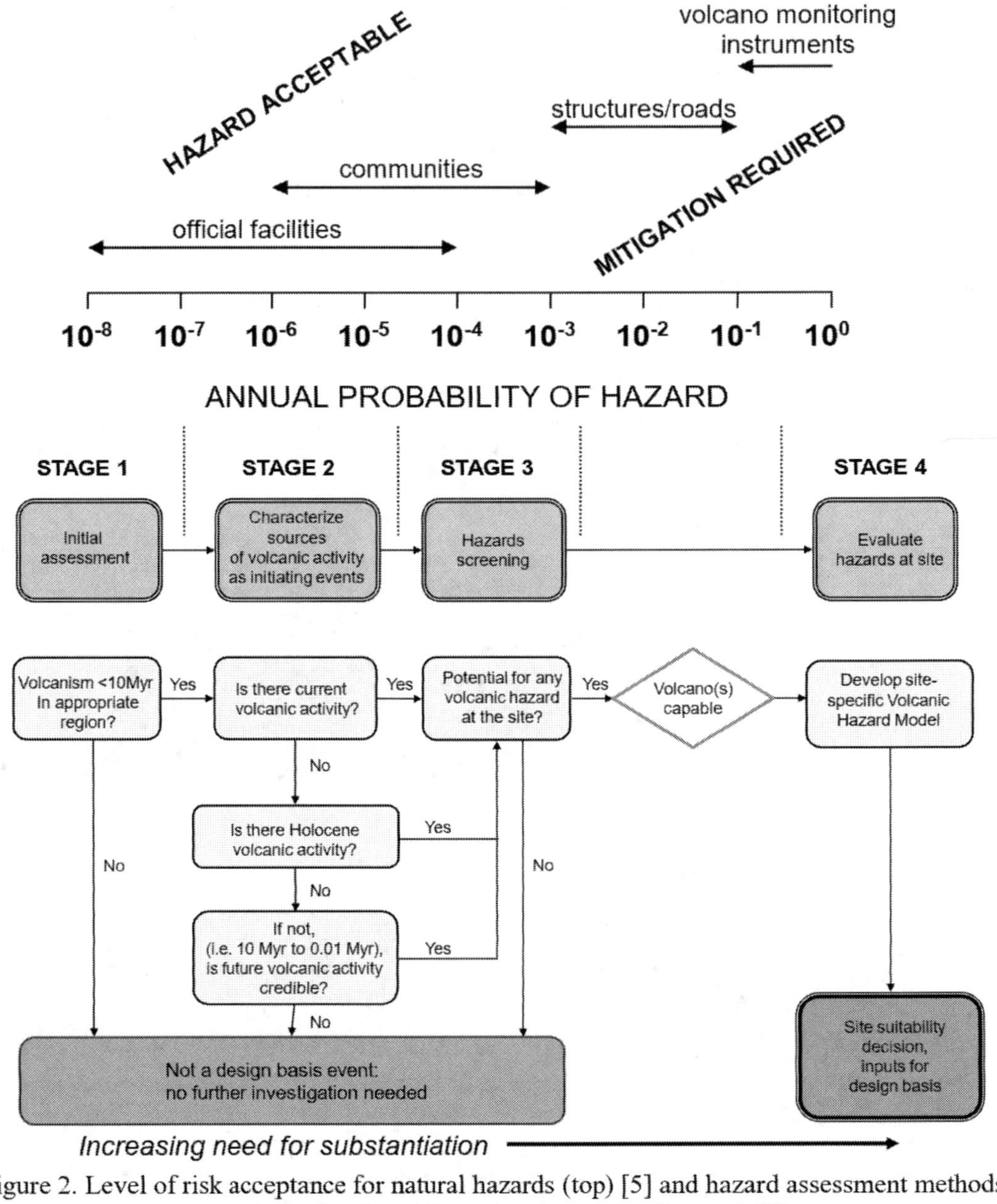

Figure 2. Level of risk acceptance for natural hazards (top) [5] and hazard assessment methods especially developed for volcanic activity (bottom) [9]

2. Evaluation of Volcanic Disasters

Deterministic conservative evaluations for any geological phenomenon need to be conducted in order to determine the SDV (distance value). The evaluation includes the following assumptions: (1) Every eruption occurs at every volcanic center or in the vicinity of a nuclear power plant. (2) The maximum potential for eruptions for that volcano is postulated by the size and duration of eruptions. There are geological phenomena that can cause disasters at nuclear power plant sites, such as pyroclastic fall, where wind sources blow directly towards the site at maximum possible speeds and at heights with high eruption column density during the eruption duration. The sources of avalanches, lahars, debris flows, and pyroclastic flows are assumed to be the most observations of the form and activity of comparable volcanoes that can be compared. As a result of this assessment, an area surrounding the volcano has a specific SDV-related geological phenomenon. The presence of volcanic deposits around the location is direct evidence that the location is within the SDV for that phenomenon, which needs to be investigated further. This evaluation is part of determining whether technical solutions to protect the site can be implemented.

Further evaluation decisions are made based on whether the phenomenon is associated with active volcanoes during the Holocene, which necessitates deterministic evaluation, or with pre-Holocene or historical events, which necessitate statistical/stochastic analysis to determine the recurrence rate or probability that the event will occur [9]. Statistical/stochastic evaluations of source activity or similar volcanoes can be performed to determine patterns and trends.

Petrological Analyses
The HK volcanic rock series or Muria Muda volcanic rocks formed at lower degrees of partial mantle melting than the K volcanic rock series or Muria Tua volcanic rocks [10] Since the Holocene, strike-slip faults have dominated the structural pattern in the Java Sea surrounding the Muria Volcanic Complex (10,000 years). This indicates that the primary tectonic force at the time was a compression regime.

Plotting SiO2 (Figure 3) volcanic rock versus historical age using available data reveals that the most recent eruption products resulted from the differentiation of previous magma sets. If no new magma is added, the Muria Volcanic Complex's volcanic activity will be caused by magma that cools from the previous magma set. When heat from the cooling magma contacts groundwater, it usually causes phreatic eruptions. Monitoring volcanic earthquakes beneath the Muria Volcanic Complex can detect new magma input. However, there is no indication of a concentration of epicenters beneath the Muria Volcanic Complex [10]. Mount Muria is thus a non-capable volcano for magmatic eruptions soon.

3. Spatial and Temporal Probabilistic Analysis

To investigate the spatial and temporal correlation of volcanic eruptions, stochastic models such as the Poisson process and the Kernel are used. The development of stochastic models is used to determine the probability of new volcanic eruptions using two spatial variables and one temporal variable that can depict volcanic eruptions, namely the location of the eruption center, the area affected by eruption products, and the frequency of eruption occurrences.

To identify the characteristics of volcanic hazard threats, descriptive methods and deterministic models are used. In this case, information on the frequency of areas affected by volcanic hazards is required. Hazards are a random function [11] corresponds to a "moving average" of randomly generated location functions.

The number of eruptions within a given time frame is the temporal component. According to observations, eruption events are frequently clustered in time, indicating a potential temporal correlation. The stochastic process must take into account the timing of volcanic activity. In geological space and time, the Poisson process model can identify and quantify volcanic phenomena (spatial and temporal analysis).

The first step in understanding the nature of volcanoes is determining eruption data and constructing eruption chronologies or time series of past eruptions. A table with a specific time interval covering the eruption period is one simple way to illustrate eruption data. The chronology of Muria volcano eruptions based on rock age dating is shown in Figure 3 that also shows data of the number of eruptions with a 25,000-year time interval based on confirmed ages [10].

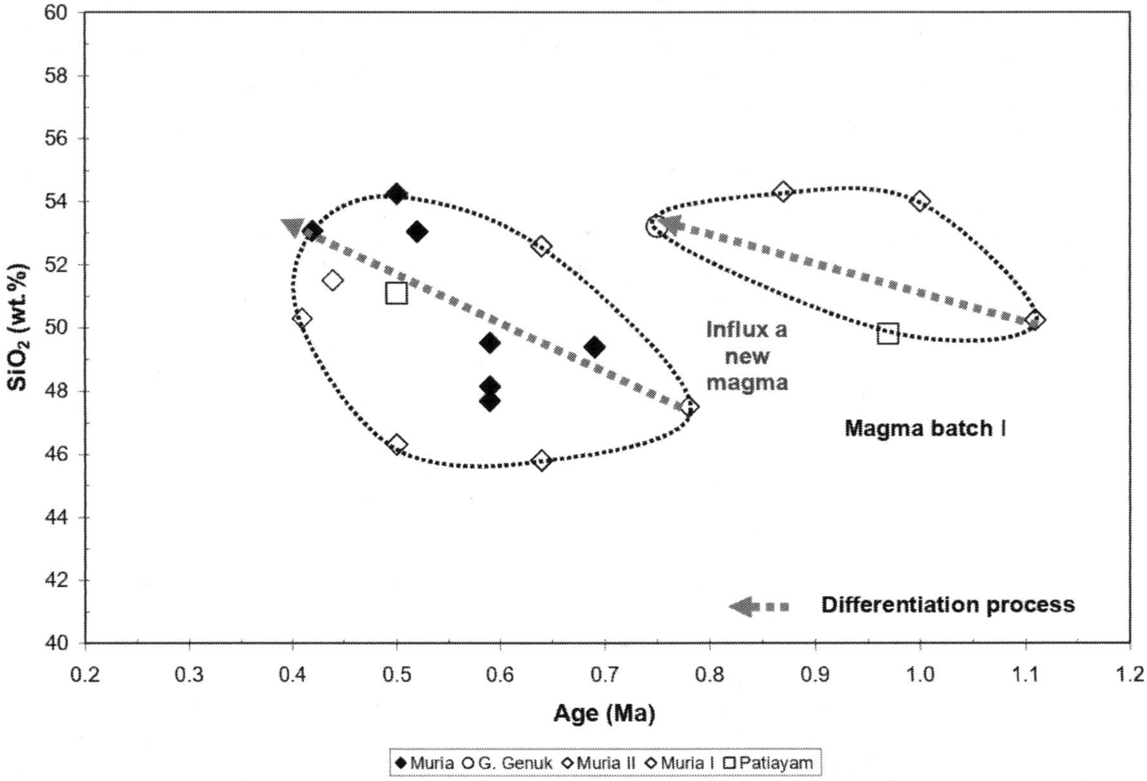

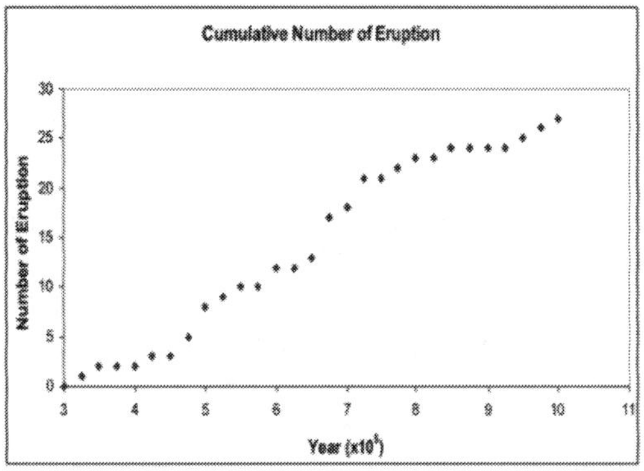

Umur (Ma)	Interval Waktu (juta tahun)
1	
	0.03
0.97	
	0.04
0.93	
	0.09
0.84	
	0.13
0.77	
	0.02
0.75	
	0.03
0.72	
	0
0.72	
	0.01
0.71	
	0.01
0.7	

Umur (Ma)	Interval Waktu (juta tahun)
	0.01
0.69	
	0
0.69	
	0
0.69	
	0
0.69	
	0.05
0.64	
	0.04
0.6	
	0.01
0.59	
	0.06
0.53	
	0.01
0.52	
	0.02

Umur (Ma)	Interval Waktu (juta tahun)
0.5	
	0
0.5	
	0
0.5	
	0.03
0.47	
	0.01
0.46	
	0.04
0.42	
	0.08
0.34	
	0.02
0.32	

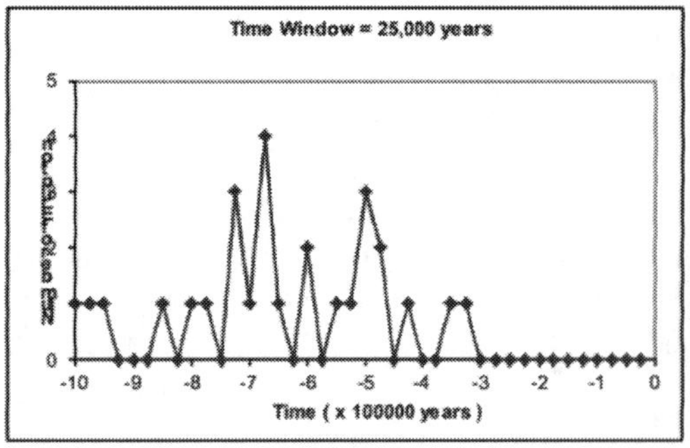

Time interval x 10^5 (years)	Number of Eruptions
0.25	0
0.5	0
0.75	0
1	0
1.25	0
1.5	0
1.75	0
2	0
2.25	0
2.5	0
2.75	0
3	0
3.25	1
3.5	1

Time interval x 10^5 (years)	Number of Eruptions
3.75	0
4	0
4.25	1
4.5	0
4.75	2
5	3
5.25	1
5.5	1
5.75	0
6	2
6.25	0
6.5	1
6.75	4
7	1

Time interval x 10^5 (years)	Number of Eruptions
7.25	3
7.5	0
7.75	1
8	1
8.25	0
8.5	1
8.75	0
9	0
9.25	0
9.5	1
9.75	1
10	1

Figure 3. Plots of different geochemical composition of Gunung Muria (top) and its cumulative number of eruptions.

A cumulative plot depicting the number of eruptions that occur during or before time "t" is used to conduct time series analysis of Muria volcano complex eruptions. This type of plot clearly shows the fluctuating pattern and eruption rate that occurred over a specific time period. Correlating the plotting points and measuring the correlation slope yields the average number of eruptions in each time unit over a given period. The cumulative plot of the number of eruptions in the Muria volcano complex (Figure 3) shows no regime differences and indicates a stationary pattern or a non-time-dependent process.

Additional information comes from the time intervals between eruptions. The distribution of time intervals between eruptions is defined as a time function (t):

$$\Psi_N(t) = \text{the proportion of repose intervals longer than } t. \quad (1)$$

where N is the number of intervals [12]

The number of repose periods N(T) with a duration greater than T (25,000 years) for Muria volcano is presented in Figure 4 which also shows the survivor function with a very good proportion for short and long repose periods. This diagram depicts an exponential distribution as an indication of the Poisson process that is identical to a stationary process.

T (25,000 years)	N(T) Repose Period
0	17
1	7
2	2
3	1

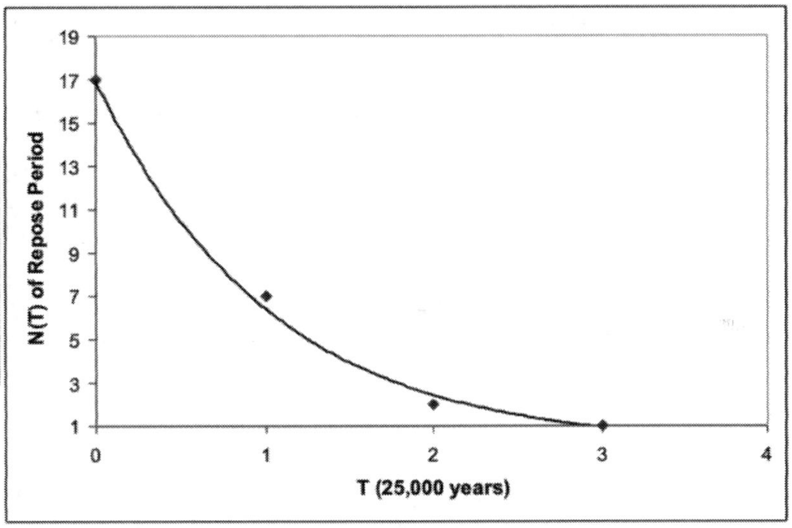

Figure 4. Distribution of observed and calculated repose period with a duration window of 25000 years in Muria Volcanic Complex.

The Poisson distribution provides a suitable model for understanding the nature and hypotheses of random data. The Poisson process is used to explain stochastic phenomena that are highly variable and sporadically occur over time, generally under random conditions. The number of events, X, per unit time has a probability of:

$$P(X = x) = e^{-\lambda}\lambda^{x}/x! \quad (2)$$

where λ is the average annual occurrence of volcanic eruptions, and x = 0, 1, 2,....

The estimated average occurrence, λ, is the average number of eruptions during the observation period, defined as:

λ = Total number of eruptions during the observation period / observation Period

Assessing the potential occurrence or recurrence of volcanic eruptions in a sequence or interval of time is a fundamental problem in statistical calculations. As in natural processes, there are two possible events that can occur, either it occurs or does not occur. Similarly, in volcanic processes, there is a possibility of eruption or non-eruption. In the world of statistics, this type of process is known as a Bernoulli sequence, which is based on three main assumptions, which are: (1) there are only two possible events, occurrence or non-occurrence (eruption in this case) and (2) the probability of the event is constant; and (3) statistical experiments are independent.

The probability of success p (and the probability of non-occurrence is 1-p), the number of successes in a number of ***n*** Bernoulli trials is depicted as a Bernoulli distribution. The probability of the occurrence, ***x***, is expressed as:

$$B(x) = {}_{n}C_{x}\, p^{x}\, (1 - p)^{n-x} \tag{3}$$

where ${}_{n}C_{x}$ is the number of combinations of n and x at a time.

Ignoring the time dependency, the initial estimate of the average probability of eruption occurrence in the Muria Volcano Complex during activity is m = 0.675 per 25,000 years during the observation period, or 2.7 x 10-5 per year. Considering that the youngest eruption of Muria occurred approximately 320,000 years ago, the probability of an eruption in Muria in the next 100 years, equivalent to the following 13 periods (320,000 + 100/25,000 years), is 1.21856 x 10-5. Meanwhile, the probability of one eruption event occurring within the same time interval is B(x = 0) + B(x = 1) = 1.80528 x 10-5.

In reality, volcanic eruptions can occur during a certain period of time. In this case, the occurrence of eruptions can be depicted as a Poisson process. Volcanic eruption events are often grouped by time, and they do not occur randomly but tend to be correlated with time. Detection and quantification of such events can be done using variogram analysis. Although this statistical analysis is often applied in spatial variables, variogram analysis can also be applied to the study of stochastic processes in calculating time properties within large class intervals. With the stationary hypothesis, the variogram $\gamma(\tau)$ is defined as:

$$\gamma(\tau) = \Sigma\, (z_i - z_{i+h})^2 / 2n$$

where:

$\gamma(\tau)$	: the value of variogram with a time interval (t) to others
z_i	: the number of eruptions at time interval i
z_{i+h}	: the number of eruptions to other eruptions at the time interval
$\Sigma(z_i - z_{i+h})2$	: the sum of the square of differences between the number of eruptions of all pairs of data within a certain time interval
n	: the number of data pairs separated by a certain time interval

During a period of 1 million years, eruptions at Muria Complex are grouped into time intervals of 25,000 years (Figure 5). The figure shows the "measurement scale" as the grouping of eruption events into time intervals. This grouping is done randomly to obtain maximum resolution and stability of statistical parameter estimates. The temporal correlation of data variables is analyzed by the semi-variogram. This correlation shows a "range" with a value of 150,000 years, which indicates the stabilization of the first-level variogram of a number of eruptions in different time intervals (Figure 5). The range corresponds to a memory effect that can be interpreted as the average duration of volcanic activities and repose periods.

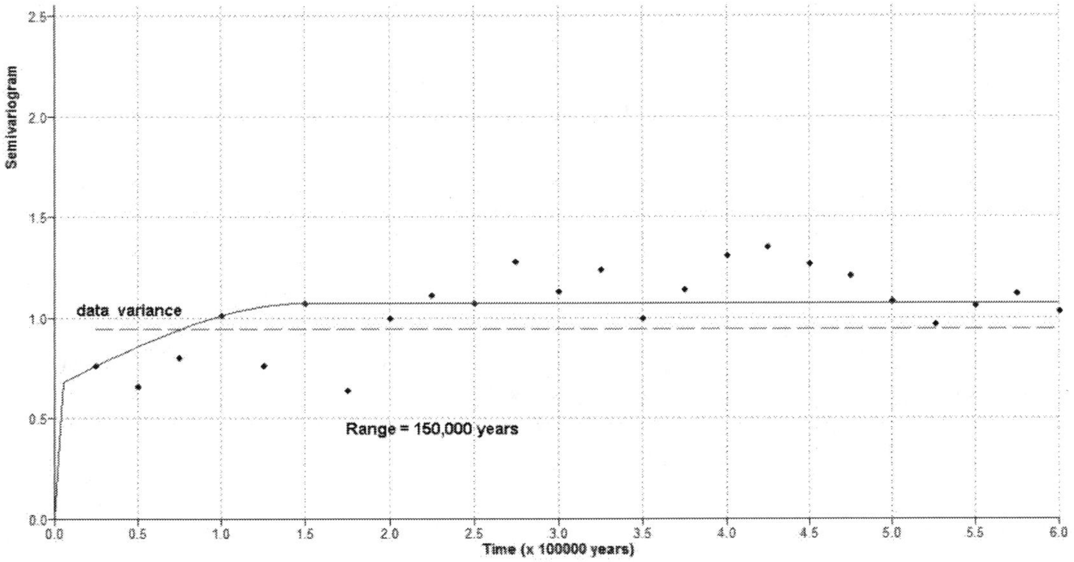

Figure 5. *Memory effect* determination with variogram analyses showing the different number of intervals on eruption period.

Volcano geological data can be used as the foundation for quantitative hazard forecasting. Volcanic hazard forecasting frequently requires evaluating extremely rare events with severe consequences. The acceptable probability of hazards for facilities with high consequences, such as nuclear facilities, ranges from 10-4 to 10-8 events per year [5].

The first step in developing this probability model is to specify the events that will be examined [8, 13]. In the case of the Muria Complex, the events are limited to the formation of new eruption points associated with the Patiayam, Genuk, and Muria volcanic systems. This study's hazard forecasting model excludes hazards associated with eruptions such as pyroclastic flows, lava, and lahars. The volcanic hazard forecast in this study focuses on the possibility of new eruption points forming in the Muria Peninsula, which could have an impact on the Ujung Lemah Abang nuclear power plant.

The volcanic data used in this model come from monogenetic and polygenetic volcanic models, and include craters, lava domes, and maars. Each eruption source is treated equally and is considered independent of the others, regardless of the type of eruption. The possibility of magma injection without a new eruption is ignored. An alternative model for determining events can be used by weighting factors such as age, eruption type, distribution, and activity episodes.

The next stage in this model is to develop the mathematical probability model. The probability of an event occurring in a specific location is expressed as:

$$P[N \geq 1] = 1 - e^{-\lambda_t \lambda_s A t} \tag{4}$$

Where A is the study area, the Muria Volcanic Complex that may impact the Ujung Lemah Abang nuclear power plant. The time t used in this formula is assumed to be the 100 years of the nuclear power plant's operation. The next modeling stage is to determine the spatial density, λs (the number of events that occur in each km), and the temporal recurrence rate, λt (the events per year). For the Muria Volcanic Complex, the estimated temporal recurrence rate is 2.7 x 10-5 events per year according to the Poisson distribution. This means that one volcanic eruption event is likely to occur every 37,037 years.

Although the distribution of volcanic eruptions in the Muria Complex exhibits an exponential distribution as a Poisson process or stationary process, the spatial distribution is assumed as a separate discrete event in space and time. Therefore, the Gaussian kernel function is chosen for modeling, expressed as:

$$\lambda_s(x,y)=\frac{1}{2\pi Nh^2}\sum e^{\frac{1}{2}\left(\frac{d_i}{h}\right)^2} \tag{5}$$

The model calculates the probability density function based on the distance (d_i) between a point (x,y) and the nearest volcano, using a smoothing parameter (h) and the number of volcanoes (N) in the area and number of events (λs) . The spatial density is then integrated across the entire spatial map to obtain a bivariate probability density function. The probability calculations depend on the value of h, with smaller h values resulting in higher probabilities for nearby volcanoes and larger h values producing more uniform probability distributions. The Gaussian kernel factor is equivalent to the standard deviation of the bivariate Gaussian distribution, and the choice of kernel depends on geological and statistical assumptions.

Various smoothing parameter values ranging from 12 km to 35 km were used in this study (Figure 6). The kernel model was chosen because, unlike the homogeneous approach, it allows for the comparison of geological information and does not require the determination of volcanic activity zones. The resulting spatial density map depicts the likelihood of new volcanic eruptions in the Muria Peninsula, ranging from 1.42 x 10-4 to 1.50 x 10-5. It is important to note that the probability values are extremely sensitive to the smoothing parameter chosen, implying that there is no definitive method for determining the optimal h value.

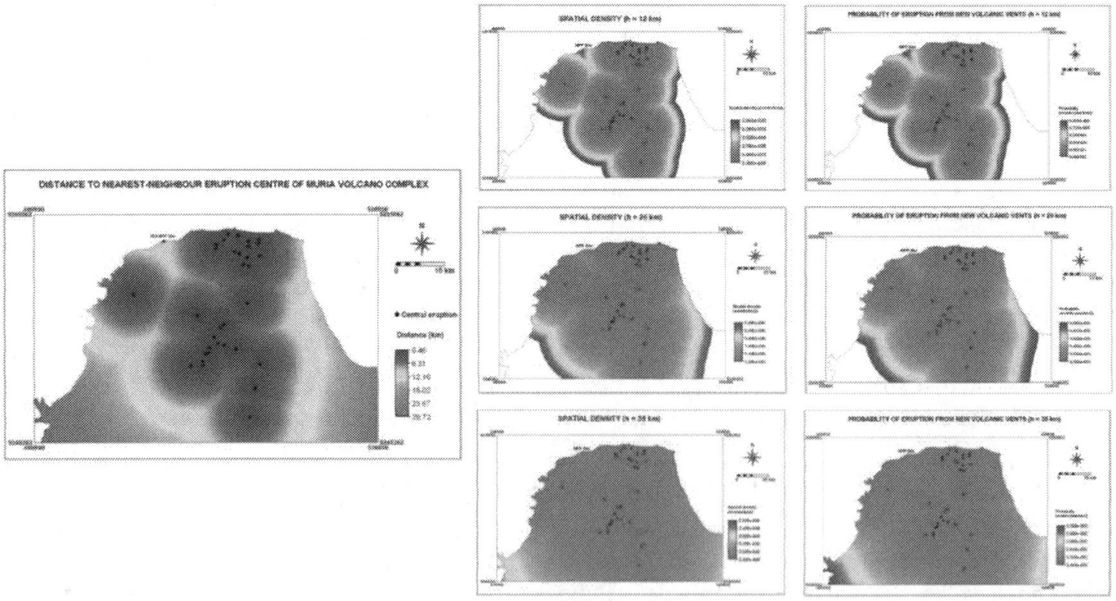

Figure 6. Spatial density map shows the probability of new volcanic eruptions in the Muria Peninsula, with probabilities ranging from 1.42×10^{-4} to 1.50×10^{-5}.

4. Conclusions

Volcano geological data is used to create a quantitative hazard forecasting model. The focus is on evaluating rare events with severe consequences, particularly for high-consequence facilities like nuclear power plants. The acceptable probability range for such hazards is between 10-4 and 10-8 events per year. In the specific case of the Muria Complex, the model examines the formation of new eruption points associated with the Patiayam, Genuk, and Muria volcanic systems. Hazards like pyroclastic flows, lava, and lahars are not considered. The goal is to forecast the possibility of new eruption points in the Muria Peninsula that could affect the Ujung Lemah Abang nuclear power plant.

The model incorporates volcanic data from various sources and treats each eruption source equally and independently. The possibility of magma injection without a new eruption is disregarded. Alternative models could consider factors such as age, eruption type, distribution, and activity episodes.

The Muria Volcanic Complex's stratovolcano system and caldera are non-capable volcanoes, according to vulcanological data. The magma system beneath the Muria Volcanic Complex has cooled, leaving only magma-derived volcanic gases, altered rocks, and anomalous temperature gradients in some areas. Future magmatic eruptions are impossible unless new magma influxes into the Muria Volcanic Complex's magma chamber. Seismic studies using micro-earthquake monitoring show no concentration of epicenters beneath Mount Muria, implying that there is currently no new magma influx into the Muria Volcanic Complex's magma chamber.

While volcanic eruptions in the Muria Complex exhibit an exponential distribution, the spatial distribution is treated as a discrete event. The model utilizes a Gaussian kernel function, considering the distance between a point and the nearest volcano, along with a smoothing parameter (h) and the number of volcanoes and events in the area. The choice of kernel depends on geological and statistical assumptions. The Gaussian kernel model is preferred because it allows for the comparison of geological information and does not require the determination of volcanic activity zones. It is important to note that the probability values are highly sensitive to the chosen smoothing parameter, implying that there is no definitive method for determining the optimal value.

The Poisson distribution shows an initial estimation of the average probability of eruption occurrences in the Muria Volcanic Complex during activity, which is m = 0.675 per 25,000 years during the observation period, or 2.7×10^{-5} per year. Considering the youngest Muria eruption occurred approximately 320,000 years ago, the probability of eruption occurrences in Muria within the next 100 years is 1.21856×10^{-5}. With this probability of eruptions, it appear that the Muria Mtn is not going to be erupted anytime soon. However, when a new magma system emerges, resulting in epicenter concentrations beneath Mount Muria, micro-earthquake monitoring is required to detect magmatic activities beneath the Muria Volcanic Complex. To estimate the likelihood of phreatomagmatic eruptions, geothermometer monitoring in deep wells is required as an active monitoring of hydrothermal activity in the Muria Volcanic Complex.

5. References

1. P. Sumintadiredja, M.L. Tobing, B. Wibowo, I G.B.E. Sucipta, M. Nurhasanudin, H. Wibowo, 2007, Volcanic Hazard Analysis for Proposed Nuclear Power Plant Siting in Central Java, Indonesia, Proceeding IAMG 2007, Beijing, pp.171-175.
2. IAEA, 1997a, Volcanoes and Associated Topics in Relation to Nuclear Power Plant Siting. Provisional Safety Standards Series No.1, International Atomic Energy Agency (IAEA), IAEA-PSS-01, Vienna.

3. IAEA, 1997b, Site Safety Review Mission: Final Review of the Siting Studies at Muria Peninsula, Indonesia (External Events Topics). TC project INS/9/021, IAEA-RU-6847.
4. Muttaqy, F., Nugraha, A.D., Puspito, N.T. *et al.* Double-difference earthquake relocation using waveform cross-correlation in Central and East Java, Indonesia. *Geosci. Lett.* **10**, 5 (2023). https://doi.org/10.1186/s40562-022-00259-2
5. Mc. Birney, A. R., Serva, L., Guerra, M., and Connor, C.B., 2003, Volcanic and Seismic Hazards at A Proposed Nuclear Power Site in Central Java. Journal of Volcanology and Geothermal Research, 126: pp. 11-30.
6. Von Mises, R., 1957, Probability, Statistics, and Truth, 2nd. Rev. Ed. Dover, New York, 244 pp.
7. Connor, C.B., Stamatakos, J.A., Ferrill, D.A., Hill, B.E., Ofoegbu, G.I., Conway, M.F., Sagar, B., and Trapp, J., 2000, Geologic Factors Controlling Patterns of Small-volume Basaltic Volcanism: Application to A Volcanic Hazards Assessment at Yucca Mountain, Nevada. Journal of Geophysical Research, 105(1): pp. 417-432.
8. Martin, A. J., Umeda, K., Connor, C.B., Weller, J.N., Zhao, D., and Takashi, M., 2004, Modeling Long-term Volcanic Hazard through Bayesian Inference: An Example from the Tohoku Arc, Japan. Journal of Geophysical Research, 109(B10208): pp. 1-20.
9. IAEA, 2012, Volcanic Hazards In Site Evaluation For Nuclear Installations, Specific Safety Guide 21,Vienna.
10. National Technical Team (NTT), 2000, Feasibility Study of Nuclear Power Plant at Muria Peninsula, Central Java, Indonesia. Final Report on Volcanology, BATAN, unpublished.
11. Lantuéjoul, C. 1994. Non conditional simulation of stationary isotropic multigaussian random functions. In Geostatistical simulations. Edited by M. Armstrong and P.A. Dowd. Kluwer Academic Publishers
12. D. R. Cox, P. A. W. Lewis, The Statistical Analysis of Series of Events . The Annals of Mathematical Statistics 37 (1966), no. 6, 1852--1853
13. Connor CB, Hill BE, 1995, Three nonhomogeneous Poisson models for the probability of basaltic volcanism: Application to the Yucca Mountain region. Journal of Geophysical Research 100:12 107-10 125

Acknowledgments
Authors wishing to acknowledge assistance from Geological Engineering Study Program and the Faculty of Earth Sciences and Technology at ITB, and the Centre of Volcanology and Geological Hazard Mitigation Institute.

Review of morphological changes of Anak Krakatau before and after 2018 eruption using Aerial photogrammetry

Wilfridus F S Banggur[1,2*], Aditya Pratama[1,2], Taufik Ismail[2], Dini Nurfiani[1], Ardy Setya Prayoga[3], M N Kartadinata[3], Devy Kamil Syahbana[3], Deni Mardiono[3], Philipson Bani[4]

[1]Research Center for Geological Disaster, National Research and Innovation Agency (BRIN), Bandung, Indonesia
[2]Faculty of Earth Sciences and Technology, Bandung Institute of Technology, Indonesia
[3]Centre for Volcanology and Geological Hazard Mitigation, Bandung, Indonesia
[4]Universite de la Nouvelle Caledonie, Noumea, New Caledonia, France
*Corresponding author: wilf001@brin.go.id

Abstract. We carried out drones photogrammetry on Anak Krakatau in 2013, 2019, 2020 and 2021. Changes in the morphology of Anak Krakatau occurred so quickly as its volcanic activity never stopped. Its pyroclastic cone in 2013 reached a height of 334 meters with a coastline circumference of 6.8 km. The circumference of the pyroclastic cone calculated from the traces of the old crater wall has a volume of 31971 m^3. During the 2018 eruption, a large part of the cone was pulverised, resulting in a large crater of 400 m in diameter. Paradoxically, while the top of the volcano has dropped to 156 m, the island-volcano has grown relatively large, with an additional circumference of 7.2 km. The body of Anak Karakatu has experienced an addition to the coastline, especially on the South, East and North sides. This happened due to the eruption and tsunami accompanied by avalanches during the December 2018 eruption. During the 2020 and 2021 eruption periods, the coastline has not changed. This period was marked by the growth of a pyroclastic cone in the crater that had formed previously during the 2019 eruption. The pyroclastic cone was opened to the southwest by lava flows that flowed westward as far as 600 meters.

1. Introduction

The use of aerial photogrammetry methods has been widely used to examine volcanic activity. Some of which have been successfully carried out such as research on monitoring of lava dome growth in Redoubt volcano (Diefenbach et al., 2013), geomorphological analysis and growth of lava dome at Molodoy Sviveluch Volcano (Shevchenko et al., 2015), analysis of precise volume of volcanic ash-plume to understand the dynamic of gas emission at Yasur Volcano in Vanuatu (Gomez & Kennedy, 2018), assesment of changes in morphology and structur at lava dome in Merapi Volcano (Darmawan et al, 2018), assesment of emplaced lava with gas composition and emision rate in Agung volcano (Syahbana et al., 2019), lava hazard forecasting in Kilauea (Diefenbach et al., 2018), and lava dome changes at Agung volcano (Andaru & Rau, 2019).

Some important points from the results of the above research studies include preparation that's understanding the characteristics of the volcano to be studied, including knowing the morphology, peak height, and orientation of the crater opening. Knowing the object or volcanic activity that will be carried out to collect aerial photography data, then planning a flight path, and it can be done starting from the slope of the mountain, to collect data on the slopes, then to the top to collect situation data inside the crater. And for high mountains the flight path can be made to circle vertically from starting point to the level with the height of the top mountain, and then flown horizontally to the crater peak. This can prevent the uav from falling due to wind and battery drain. After knowing morphological aspects, peak height, made a flight path, then determine the type of drone can be used.

Anak Krakatau volcano located is Sunda Strait (Figure 1a), has a low slope and topography, we can fly from the southern slopes (Figure 1b), on this side the coastline and morphology are good for a starting point for flying drones. To deal with strong winds and crater smoke, we waited for the wind to calm down, then took aerial photography inside the crater. Morphological changes in the Anak Krakatau Complex are strongly influenced by volcanic activity and rock resistance, as well as seawater abrasion (Ismail et al., 2020), in 2019 a crater lake was formed (Figure 1c), and until now its eruption activity has formed a pyroclastic cone (cinder cone) and lava flows stretching westward (Figure 1d).

Research on the history of the eruption of Karakatau has been carried out. According to (Self, 1992) the 1883 eruption of Krakatau was affected by overpressure in the magma chamber by saturation of volatiles which then triggered the eruption, and this catastrophic eruption spewed around 12.5 km^3 of magma with composition in the magma chamber composed of rhyodacit, the zoning underneath is Dacit magma, then in the lowest zone in the magma chamber is an andesitic magma (Mandeville et al., 1996). The death toll in this eruption is estimated at 36,417 people (Auker et al., 2013).

(a) (b)

(c) (d)

Figure 1. The research area is located on Mount Anak Krakatau, Sunda Strait (a). Photos (b) taken with hellicopter in 2013, Photos (c), and (d) were taken using a drone from the south. Situation of the cone of Mount Anak Krakatau in 2013 (b). Crater Lake with a diameter of 400 m in 2019 (c), and Crater Lake filled with pyroclastic cone and lava flow in 2020 (d).

Anak Krakatau is the third evolution of the Krakatau volcano, which has grown since 1929, with a rest period between 1-8 years, with the intensity of eruptions occuring 1-6 times a year (Sutawidjaja, 2006). According to Abdurrachman et al. (2018) Anak Krakatau growth was affected by the partial melting of the subduction system between Java and Sumatra, and mantle upwelling that infiltrate beneath Mount Anak Krakatau.

Meanwhile, on 22 December 2018 an eruption accompanied by an avalanche triggered a tsunami in the Sunda Strait causing 430 fatalities along the densely populated coasts of Banten and Lampung (Borrero et al., 2020). According to (Grilli et al., 2021) with numerical method, estimates of the volume of landslides ranging from 0.175 to 0.313 km^3 and the geometry of the body of Mount Anak Krakatau which collapsed is the peak part of up to 50% of the body volume below sea level, and after the 2018 eruption activity, the potential danger of the eruption is vulcanian and strombolian eruptions which can be accompanied by ejection of incandescent rocks and heavy ash rain and lava flows within a radius of 2 km from the center of the crater (Kristianto et al., 2021).

Several studies using aerial photography and remote sensing on Anak Krakatau have been carried out. Volcanoclastic deposits in early 2020 (Darmawan et al., 2021), study about flank collapse in southwest part of Anak Krakatau (Esteban et al., 2021), growth of Anak Krakatau after 2018 using Sentinel 1 (Achmad et al., 2019), the changes of Anak Krakatau landscape after 2018 December eruption (Ginting et al., 2021), and eruption dinamic of Anak Krakatau (Hochfeld et al., 2022).

In this study, we will present data from morphological observations from field observations and also drone data to estimate the volume of pyroclastic cones and lava flows for the 2013, 2019, 2020 and 2021 eruption periods.

2. Methods

Aerial photography data collection at Anak Krakatau can be done in a day, with 2 to 4 flights. When flying the Phantom 4 drone at Anak Krakatau, we used manual flight mode with the DJI Go application, and the drone's flying situation was monitored, both through the application and the drone's situation at altitude, and the morphology situation. Monitoring is observing the crater smoke and wind direction. The first to second flights were focused on the south and east sides (Fig. 1c, d). The 3rd and 4th flights from the summit of Anak Krakatau, to take the situation of the crater and also the morphology of the western and northern parts (Fig. 5a,b), (Fig. 7c).

Specifically for 2013 data, aerial photography data was carried out using a helicopter with a Canon EOS 5D Mark III camera that has been integrated with GPS as a coordinate reference. Flying altitude of 200 m from the top of Anak Krakatau. Field photo data as many as 105 photos.

During the 2019-2021 aerial photography data collection period, we used the Phantom 4 drone, which is a quadcopter drone that is easy to carry and use on every flight path. It has a stable camera gimbal for taking aerial photos perpendicular with 20 Megapixel camera specifications, and has been integrated with the GPS and GLONASS satellite systems as coordinate references. Data collection, vehicles used, and altitude can be seen in Table 1.

Table 1. Period aerial photographs of Anak Krakatau

Date	Vehicle	Altitude (m)	Number of flights	Photos	Time each Flight (Minute)
1/04/ 2013	Helicopter	200	2	105	15
19/03/2019	Drone Phantom 4	200	3	120	15
06/12/2020	Drone Phantom 4	250	4	130	15
06/06/2021	Drone Phantom 4	250	4	145	15

Field photo data for each collected and then selected based on the quality of the photos. Oblique and blurred photos are not used for processing. The selected photos are then processed using the Agisoft Metashape Professional application to generate 3D spatial data (DEM). After generating the DEM, we use Arcgis 10.8 to process and analyze DEM to calculated the volume of the avalanche by comparison between the avalanches on the southwest side of Anak Krakatau based on DEM 2013 and DEM 2019 data, and calculate the volume of pyroclastic cones and lava flows based on 2019, 2020, and 2021 DEM.

IOP Conf. Series: Earth and Environmental Science 1245 (2023) 012028 doi:10.1088/1755-1315/1245/1/012028

3. Results

Four DEMs (Figure 2) were generated in this work that depicts changes in the shape of the topographical surface of Anak Krakatau representing the 2013, 2019-2021 eruption periods, and significant increase in volume and circumference occurred in 2019-2021 period (Table 2). We made the 2013 DEM and orthopoto a reference as the morphology and body volume of the cone of Mount Anak Krakatau before the the 22 December 2018 tsunami. Body volume landslides were calculated by the difference in volume between DEM 2013 and DEM 2019. Pyroclastic cones and lava flows grew inside the Anak Krakatau crater during the 2020 and 2021 eruption periods. growth rate by the difference between DEM 2019, 2020 and DEM 2021 volumes.

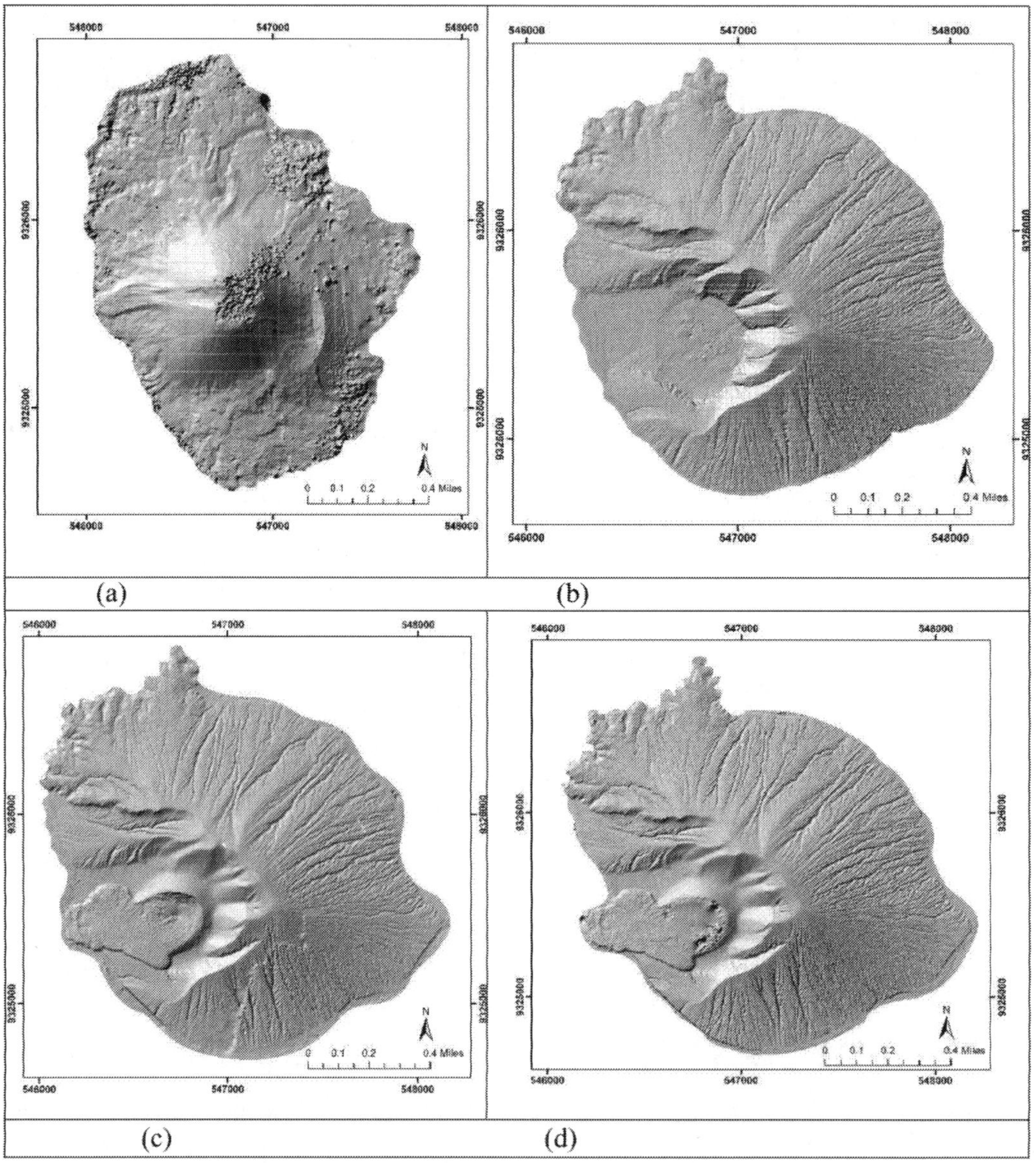

Figure 2. A relief morphology Anak Krakatau periode 2013 (a), periode 2019 (b), periode 2020 (c), and periode 2021 (d).

Table 2. Morphological changes of Anak Krakatau evidenced by DEMs

Period	Area (km^2)	Circumference (km)	Volume (km^3)	Volcano peak (m)
2013	2.49	6.58	90.41	334
2019	3.04	6.91	143.08	156
2020	3.22	7.21	183.90	160
2021	3.16	7.30	278.99	162

A comparison of the volume of the 2013 DEM of Mount Anak Krakatau with the 2019 DEM (Figure 3) shows that the lost body mass of Anak Krakatau reached 0.245 km^3, and the additional volcanic mass on the east side due to eruptions and avalanches reached 0.124 km^3. The east side of the body also experienced an increase in volume of 0.084 km^3. This mass is thought to be the result of an eruption followed by an avalanche and then caused a tsunami on 22 December 2018.

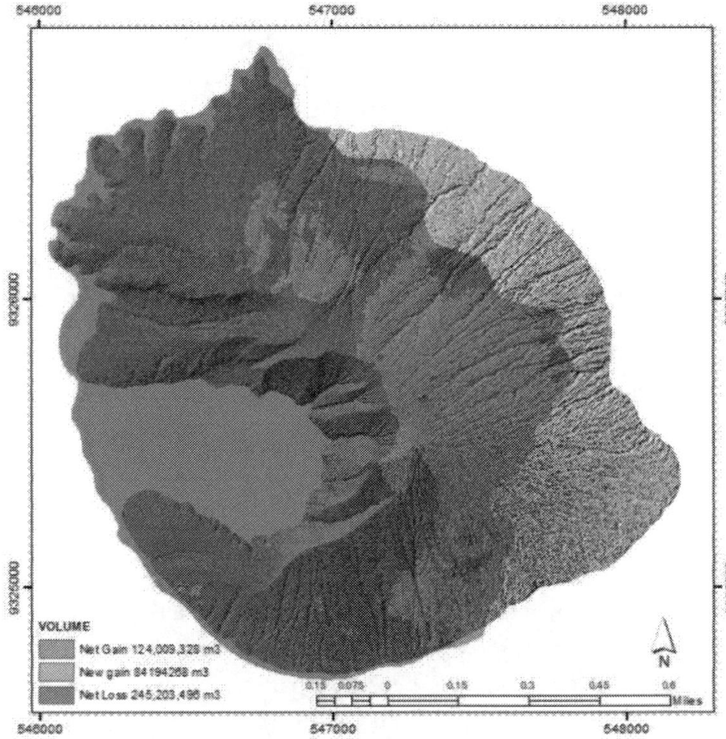

Figure 3. Comparison between body shapes that experience the addition and reduction of Mount Anak Krakatau. The colored part is DEM 2013 data, and the relief is DEM 2019.

From the results of a comparison of the 2013 DEM and the 2019 DEM as well as the delineation on the west side which experienced avalanches (Figure 4), the estimated volume about 0.107 km^3. The addition of the coastline to the south and east of Mount Anak Krakatau was more affected by the landslide and tsunami of December 22 2018. The mass slide of Mount Anak Krakatau entered the sea water, then was blown back by sea waves and deposited on the coast line on the south and east side of Anak Krakatau. A comparison of the 2019, 2020 and 2021 DEMs provides an estimated volume of a pyroclastic cone (cinder cone) of 0.001 km^3, and a volume of lava flows of 0.017 km^3.

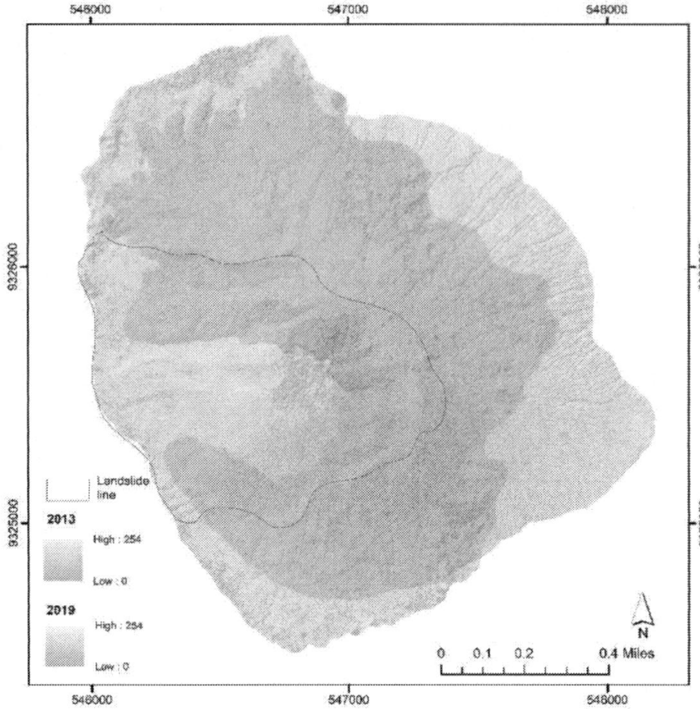

Figure 4. Delineation of the landslide area of Mount Anak Krakatau based on the 2013 DEM and 2019 DEM.

Reconstruction of the collapse of the southwest side slope using SAR satellite imagery for the period 19 November 2018 to 2 February 2019 by (Williams et al., 2019) shows that the peak of Anak Krakatau on 22 December 2018 is still intact but collapsed in 28 December 2018 , so that the volume of Anak Krakatau's body that avalanche does not include its peak. This means that the volume of the southwestern slope which failed was relatively small. In construction he estimated about ~0.1 km^3, and this volume was much less than the volume estimated by (Giachetti et al., 2012) of 0.28 km^3.

The delineation that we do (Figure 4), apart from using the DEM comparison, we also use the results of our field observations and we found a crack in the slope which is currently the highest peak of Anak Krakatau (Figure 5a), and we predict that this fracture zone will become a path of west slope instability, which will then trigger avalanches, if Anak Krakatau's cone continues to grow. The addition of the coastline has occurred significantly (Figure 6), and this is directly propotional to the addition of Anak Krakatau's body volume since the 2019 period, and the coastline has not changed in the 2020-2021 period.

IOP Conf. Series: Earth and Environmental Science 1245 (2023) 012028 doi:10.1088/1755-1315/1245/1/012028

Figure 5. the crack path at the top of Anak Krakatau (a), Crack path at the top of Anak Krakatau with visuals using drones (b), crater front situation (c), and (d). Photo taken in March 2019.

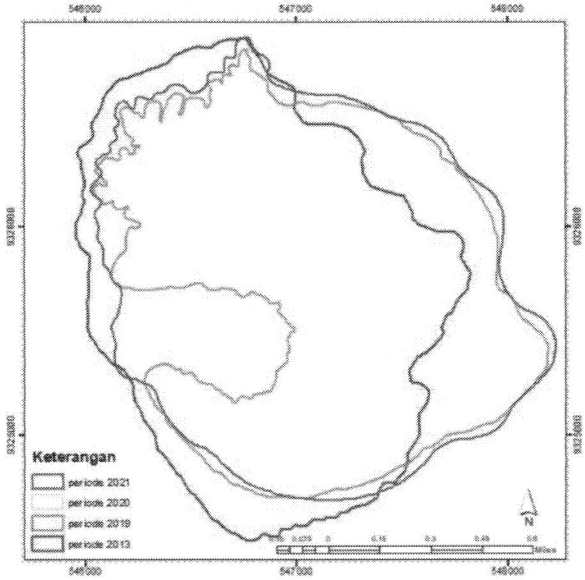

Figure 6. Changes in the coastline of Mount Anak Krakatau, showing an increase in area and circumference in the east.

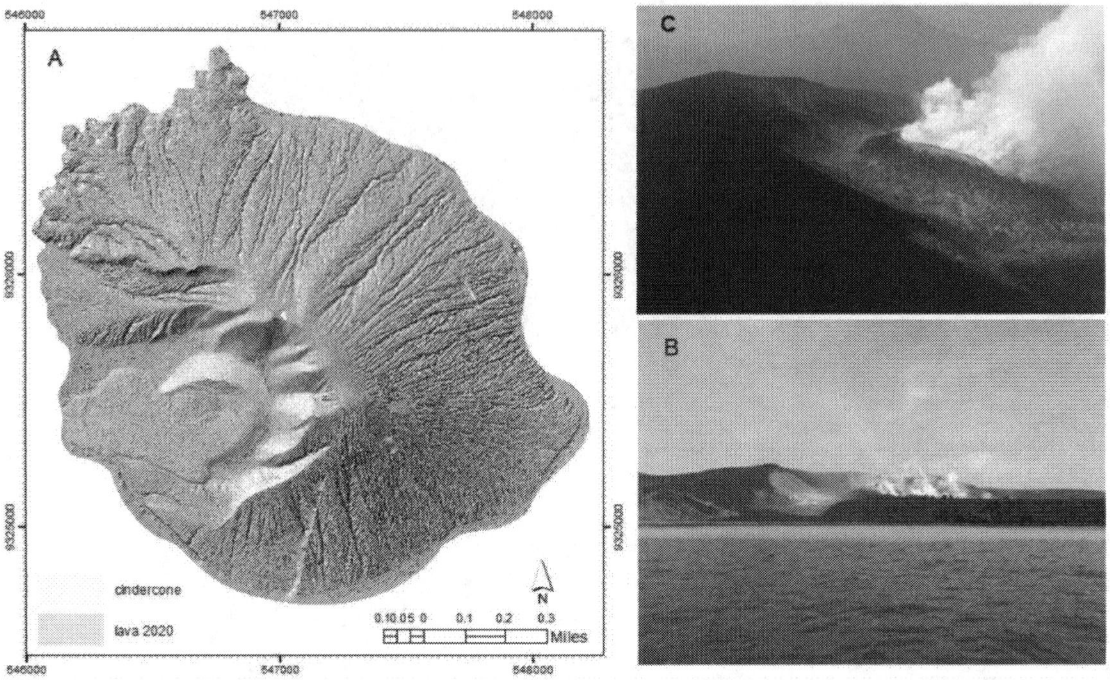

Figure 7. Delineation of volcanic products from the eruption of Mount Anak Krakatau in the form of pyroclastic cones and lava flows (A). The 2020 eruption lava flows to the west (B), and the pyroclastic cone of Anak Krakatau which has started to grow since the 2020 eruption, photo in 2021 (C).

4. Discussion

The use of drones to support monitoring of volcanism activity and changes in volcanic morphology is a technological breakthrough, especially its use in eruption crises. The selection of the drone that will be used needs to be adjusted to the location of the volcano that will be used as the object of data collection. Preparation for flight, and flight execution to retrieve visual data requires accuracy so as to produce good visual data. The quality of the resulting DEM is influenced by the type of camera used, flying altitude, and data representation.

Based on drone visual data and field observations, in the 2019 period the body of Mount Anak Krakatau experienced an increase in area in the eastern part, which was caused by an avalanche followed by a tsunami. Then throughout the 2019 period, sea water on the west and south sides tends to be orange brown, this is thought to be related to magma activity which indicates a supply from depth, near surface activity, and surface activity in the form of emissions. The energy supply from this depth reaches its energy release in the 2020 period with an eruption in April 2020 in the form of a volcanic and Strombolian eruption, and also accompanied by an explosion. The eruption products of this period produce pyroclastic cones and lava flows. During the 2019-2021 eruption period, it contributed greatly to increasing the height of the peak and also the distribution of tephra that scattered on the peak, north and east (Figure 7c).

5. Conclusion

The use of drones for monitoring volcanic activity, especially for Mount Anak Krakatau, needs to be improved in technology. With Krakatau's high level of volcanic activity, drones are needed that can be flown remotely, especially from the Volcano Observation Post in Pasauran. Gas detection equipment and infrared thermal can also be additional components to support monitoring activities.

Taking into account the rapid changes in the morphology of Anak Krakatau, it shows that Anak Krakatau is still growing. The periodic use of the photogrammetric method using drones to monitor Anak Krakatau's volcanism is absolutely necessary as a mitigation effort, especially in monitoring the accelerated growth of its volcanic cone which can lead to a repeat of the December 2018 tsunami.

6. Acknowledgment

We would like to thank the PVMBG for providing the opportunity to participate in monitoring the activities of Mount Anak Krakatau. This study was supported financially by the Earth Science and Maritime Research Organization, National Research and Innovation Agency of the Republic of Indonesia through Rumah Program Kebencanaan (No. SP DIPA-124.01.1.690501/2023) grant. Additional funding was also received from Faculty of Earth Sciences and Technology, Institut Teknologi Bandung through Post-Doctoral Research Grants to AP (No. 1607A/IT1.C01/SPK-KP.01/2023) and was also supported by the Post-Doctoral Research Grant from National Research and Innovation Agency Republic of Indonesia (BRIN) to DN (No. 64/II/HK/2022).

References

Abdurrachman, M., Widiyantoro, S., Priadi, B., & Ismail, T. (2018). Geochemistry and Structure of Krakatoa Volcano in the Sunda Strait, Indonesia. *Geosciences*, *8*(4), 111. https://doi.org/10.3390/geosciences8040111

Achmad, A. R., Syifa, M., Park, S. J., & Lee, C.-W. (2019). Geomorphological Transition Research for Affecting the Coastal Environment due to the Volcanic Eruption of Anak Krakatau by Satellite Imagery. *Journal of Coastal Research*, *90*(sp1), 214. https://doi.org/10.2112/SI90-026.1

Andaru, R., & Rau, J.-Y. (2019). LAVA DOME CHANGES DETECTION AT AGUNG MOUNTAIN DURING HIGH LEVEL OF VOLCANIC ACTIVITY USING UAV PHOTOGRAMMETRY. *The International Archives of the Photogrammetry, Remote Sensing and Spatial Information Sciences*, *XLII-2/W13*, 173–179. https://doi.org/10.5194/isprs-archives-XLII-2-W13-173-2019

Auker, M. R., Sparks, R. S. J., Siebert, L., Crosweller, H. S., & Ewert, J. (2013). A statistical analysis of the global historical volcanic fatalities record. *Journal of Applied Volcanology*, *2*(1), 2. https://doi.org/10.1186/2191-5040-2-2

Borrero, J. C., Solihuddin, T., Fritz, H. M., Lynett, P. J., Prasetya, G. S., Skanavis, V., Husrin, S., Kushendratno, Kongko, W., Istiyanto, D. C., Daulat, A., Purbani, D., Salim, H. L., Hidayat, R., Asvaliantina, V., Usman, M., Kodijat, A., Son, S., & Synolakis, C. E. (2020). Field Survey and Numerical Modelling of the December 22, 2018 Anak Krakatau Tsunami. *Pure and Applied Geophysics*, *177*(6), 2457–2475. https://doi.org/10.1007/s00024-020-02515-y

Darmawan et al. - 2018—Morphological and structural changes at the Merapi.pdf. (n.d.).

Darmawan, H., Worosuprojo, S., Wahyudi, W., Harijoko, A., Wibowo, H. E., Mutaqin, B. W., Haerani, N., Syarifuddin, S., Albab, M. A. U., Jati, R., & Asriningrum, W. (2021). *Morphological Changes of Anak Krakatau after the 22 December 2018 Flank Collapsed* [Preprint]. In Review. https://doi.org/10.21203/rs.3.rs-230137/v1

Diefenbach, A. K., Adams, J., Burton, T., Koeckeritz, B., Sloan, J., & Stroud, S. (2018). *The 2018 U.S. Geological Survey-Department of Interior UAS Kīlauea Eruption Response. 2018*, V23D-0107.

Diefenbach, A. K., Bull, K. F., Wessels, R. L., & McGimsey, R. G. (2013). Photogrammetric monitoring of lava dome growth during the 2009 eruption of Redoubt Volcano. *Journal of Volcanology and Geothermal Research*, *259*, 308–316. https://doi.org/10.1016/j.jvolgeores.2011.12.009

Esteban, M., Takabatake, T., Achiari, H., Mikami, T., Nakamura, R., Gelfi, M., Panalaran, S., Nishida, Y., Inagaki, N., Chadwick, C., Oizumi, K., & Shibayama, T. (2021). Field Survey of Flank Collapse and Run-up Heights due to 2018 Anak Krakatau Tsunami. *Journal of Coastal and Hydraulic Structures*, 1 Pages. https://doi.org/10.48438/JCHS.2021.0001

Giachetti, T., Paris, R., Kelfoun, K., & Ontowirjo, B. (2012). Tsunami hazard related to a flank collapse of Anak Krakatau Volcano, Sunda Strait, Indonesia. *Geological Society, London, Special Publications*, *361*(1), 79–90. https://doi.org/10.1144/SP361.7

Ginting, F. I., Gusnidar, Nelson, M., Rudiyanto, Minasny, B., & Fiantis, D. (2021). Changes in Anak Krakatau landscape after December 2018 eruption. *IOP Conference Series: Earth and Environmental Science*, *708*(1), 012088. https://doi.org/10.1088/1755-1315/708/1/012088

Gomez, C., & Kennedy, B. (2018). Capturing volcanic plumes in 3D with UAV-based photogrammetry at Yasur Volcano – Vanuatu. *Journal of Volcanology and Geothermal Research*, *350*, 84–88. https://doi.org/10.1016/j.jvolgeores.2017.12.007

Grilli, S. T., Zhang, C., Kirby, J. T., Grilli, A. R., Tappin, D. R., Watt, S. F. L., Hunt, J. E., Novellino, A., Engwell, S., Nurshal, M. E. M., Abdurrachman, M., Cassidy, M., Madden-Nadeau, A. L., & Day, S. (2021). Modeling of the Dec. 22nd 2018 Anak Krakatau volcano lateral collapse and tsunami based on recent field surveys: Comparison with observed tsunami impact. *Marine Geology*, *440*, 106566. https://doi.org/10.1016/j.margeo.2021.106566

Hochfeld, I., Hort, M., Schwalbe, E., & Dürig, T. (2022). Eruption dynamics of Anak Krakatau volcano (Indonesia) estimated using photogrammetric methods. *Bulletin of Volcanology*, *84*(8), 73. https://doi.org/10.1007/s00445-022-01579-z

Kristianto, Indrastuti, N., Basuki, A., Purnamasari, H. D., Adi, S., Patria, C., & Haerani, N. (2021). Potential Eruption and Current Activity of Anak Krakatau Volcano, Indonesia. *IOP Conference Series: Earth and Environmental Science*, *873*(1), 012021. https://doi.org/10.1088/1755-1315/873/1/012021

Mandeville, C. W., Carey, S., & Sigurdsson, H. (1996). Magma mixing, fractional crystallization and volatile degassing during the 1883 eruption of Krakatau volcano, Indonesia. *Journal of Volcanology and Geothermal Research*, *74*(3–4), 243–274. https://doi.org/10.1016/S0377-0273(96)00060-1

Self, S. (1992). Krakatau revisited: The course of events and interpretation of the 1883 eruption. *GeoJournal*, *28*(2), 109–121. https://doi.org/10.1007/BF00177223

Shevchenko, A. V., Dvigalo, V. N., & Svirid, I. Yu. (2015). Airborne photogrammetry and geomorphological analysis of the 2001–2012 exogenous dome growth at Molodoy Shiveluch Volcano, Kamchatka. *Journal of Volcanology and Geothermal Research*, *304*, 94–107. https://doi.org/10.1016/j.jvolgeores.2015.08.013

Sutawidjaja, I. (2006). Pertumbuhan Gunung Api Anak Krakatau setelah letusan katastrofi s 1883. *Indonesian Journal on Geoscience*. https://doi.org/10.17014/ijog.vol1no3.20063

Syahbana, D. K., Kasbani, K., Suantika, G., Prambada, O., Andreas, A. S., Saing, U. B., Kunrat, S. L., Andreastuti, S., Martanto, M., Kriswati, E., Suparman, Y., Humaida, H., Ogburn, S., Kelly, P. J., Wellik, J., Wright, H. M. N., Pesicek, J. D., Wessels, R., Kern, C., … Lowenstern, J. B. (2019). The 2017–19 activity at Mount Agung in Bali (Indonesia): Intense unrest, monitoring, crisis response, evacuation, and eruption. *Scientific Reports*, *9*(1), 8848. https://doi.org/10.1038/s41598-019-45295-9

Williams, R., Rowley, P., & Garthwaite, M. C. (2019). Reconstructing the Anak Krakatau flank collapse that caused the December 2018 Indonesian tsunami. *Geology*, *47*(10), 973–976. https://doi.org/10.1130/G46517.1

Petrogenesis of Adang and Talaya volcanic rocks, in the Mamuju area and its surrounding, West Sulawesi

M A Zaidani[1], B Priadi[2], I G B E Sucipta[2], V Susanto[2] and A Saepuloh[2]

[1] Geological Engineering, Bandung Institute of Technology, Indonesia
[2]Petrology, Volcanology, and Geochemistry Research Group, Geological Engineering, Faculty of Earth Science and Technology, Bandung Institute of Technology, Indonesia

E-mail : arafatzaidani@gmail.com

Abstract. The Adang Volcanic and Talaya Volcanic complex located in the Mamuju area are ancient volcanic complexes of the Neogene period. These two complexes consist of pyroclastic, volcaniclastic, and basic-intermediate lava containing the main modal minerals of leucite and clinopyroxene, making these rocks unique and rarely found in Indonesia. This study aims to determine the petrological and geochemical characteristics as well as the tectonic setting of rock formation. The rocks are generally porphyritic and vitrophyric in texture with various rock types as a result of magma differentiation. The Adang Volcanic Lava is composed of trachyandesite, tephriphonolite, trachyte, phonolite, and leucitite. Talaya Volcanic Lava is composed of basalt and trachybasalt rocks. Based on geochemical content, rocks have high alkalinity with wide range concentrations of K_2O (0.27 – 10.12%), Na_2O (0.54 – 11.06%), and total alkali (4.6 – 12.6%), as well as evidenced by the presence of feldspathoids. Rocks are divided into three groups, sodic, potassic, and ultrapotassic. The ultrapotassic rocks have similarities with Group III (Active Orogenic Zone) and Group I (Lamproite). The intensive enrichment of incompatible elements such as Rb and Th as well as the negative Nb and Ti anomalies reflect that magma derived from mantle metasomatism and is associated with subduction events. The genesis of the rock occurred in the tectonic setting of alkaline arc magmatism or continental extensional zone which was formed after the collision of the Banggai-Sula microcontinent with West Sulawesi. This process is also indirectly influenced by pre-collision subduction of the oceanic lithosphere.

Keywords: Petrogenesis, Mamuju, Alkali, Geochemistry, Potassic, Ultrapotassic, Alkaline Arc.

1. Introduction

Sulawesi Island is an area that is at the junction of the Eurasian, Pacific, and Indo-Australian plates. The interaction of those major plates produces regions with various unique and complex geological characteristics. Sulawesi Island is divided into three geological provinces [1,2,3], namely West Sulawesi (West and North Sulawesi Plutonic-Volcanic Arc), Central Sulawesi (Sulawesi Metamorphic Belt), and East Sulawesi (East Sulawesi Ophiolite Belt). Sulawesi is also composed of the Banggai-Sula Micro-continent in the eastern and southeasternmost part of the island. Rocks formed up the West Sulawesi are more continental in nature consisting of magmatic rock, sedimentary rock (Mesozoic – Quaternary), and metamorphic rock (Cretaceous) [4]. Adang and Talaya Volcanic is the product of West Sulawesi magmatism in Neogene Period, located in Mamuju area where potassic and ultrapotassic are found.

There are many implications regarding the tectonic settings and the mechanism of formation of the Adang and Talaya volcanic rocks. Therefore, this study aims to determine the petrological and geochemical characteristics as well as the tectonic setting of rock formation.

2. Regional Geology and Sample Location

In the western part of the regional geological map of Mamuju area [5], there were several rocks units the oldest to the youngest, namely the Latimojong Formation, Talaya Volcanic, Adang Volcanic, Intrusive Rocks, Mamuju Formation, and Member of Tapalang Mamuju Formation. Lithological formations which are associated with the study area are the Talaya Volcanic and Adang Volcanic which is at the Middle-Late Miocene Epoch. The Talaya Volcanic comprises lithologies of andesitic basaltic volcanic breccia, tuff, and lava, with intercalations of sandstone and marl, locally coal. The Adang Volcanic comprises lithologies of leucite-basaltic tuff, lava, and volcanic breccia, mainly micaceous. Based on volcanostratigraphy, Adang Volcanic encompasses several hummocks: Tapalang, Malunda, Adang, Ampalas, Karampuang, Sumare, and Labuang Rano, while Talaya Volcanic comprises of Kaluku Hummock.

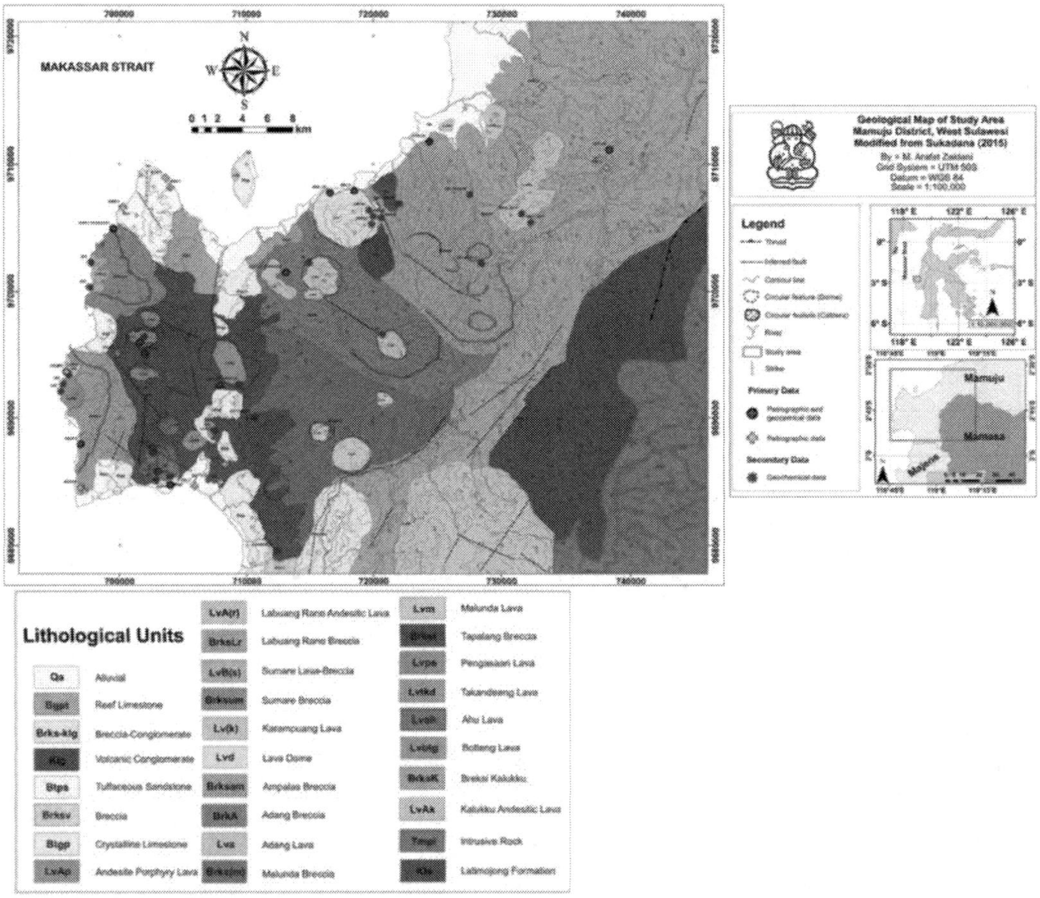

Figure 1. Geological map of the study area, Mamuju, West Sulawesi (modified after Sukadana, 2015) [9] with petrographic and geochemical data distributions.

IOP Conf. Series: Earth and Environmental Science 1245 (2023) 012029 doi:10.1088/1755-1315/1245/1/012029

3. Data and methods

The study area is located in some parts of Mamuju and Mamasa District. The methods used for this study are petrographic and geochemical analyses of primary and secondary data. The primary data has been collected directly from field observation and sampling of relatively fresh to slightly weathered rocks at several points in the study area (figure 1). The samples were processed into 39 thin sections for petrographic observation and 11 samples were carried out in Intertek Laboratories for XRF and ICP-MS/OES analyses to get the major oxides, trace elements, and REE data. However, primary data did not cover some areas well, so secondary data was needed for more representative and precise interpretation. There are 28 additional geochemical data acquired from various authors of previous studies [6,7,8,9], some samples do not have complete REE and trace elements data. For this study, all data were grouped into several clusters based on the location and volcanostratigraphy. The Adang Volcanic encompasses Tapalang, Ampalas, Adang, Sumare, and Labuang Rano groups, while Talaya Volcanic is represented by the Kalukku group. These data were plotted into several diagrams to determine geochemical characteristics such as rock types, magma series, magma evolution, and tectonic setting of volcanic rocks in the study area. The sampling locations of petrographic and geochemical data and their correlation with age, lithology, formation, and stratigraphy are summarized in figure 1 and figure 2.

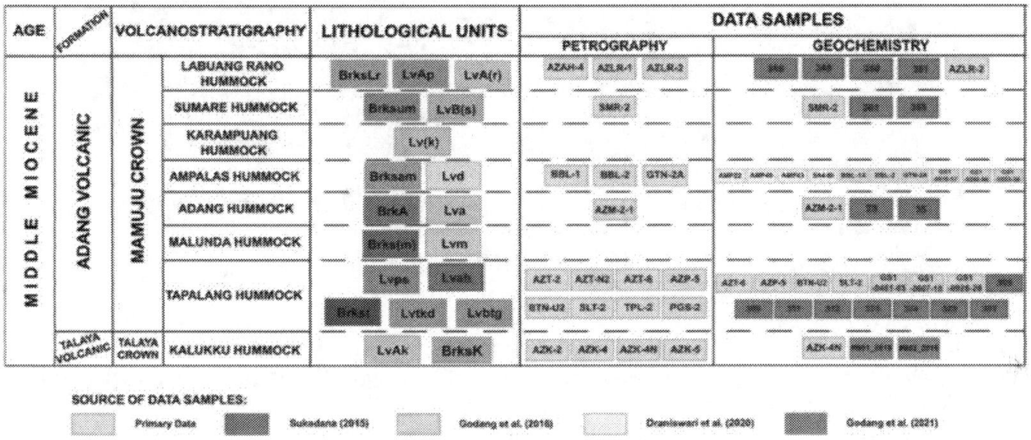

Figure 2. The correlation of petrographic and geochemical data with age, formation, stratigraphy, and lithology. The color differences of data in data samples columns shows different data sources.

4. Results

4.1. Petrography

The outcrops found in the study area are lava, pyroclastic rocks, and boulders. The igneous rock classification used to determine the name of the rocks is the classification by Williams et al. (1982) [10]. Based on the observation of thin sections, rocks from Adang and Talaya Volcanics generally have a porphyritic texture. The mineralogy of rocks is dominated by the presence of leucite, clinopyroxene, and a small amount of biotite as phenocrysts with volcanic glass and opaque minerals as groundmasses. Those rocks are classified as leucitite. Some of the phenocryst minerals also occur as a groundmass. Some of the rock samples show vitrophyric and trachytic textures, composed of volcanic glass and plagioclase or alkali feldspar as groundmass. Phonolite and trachyte also occurred.

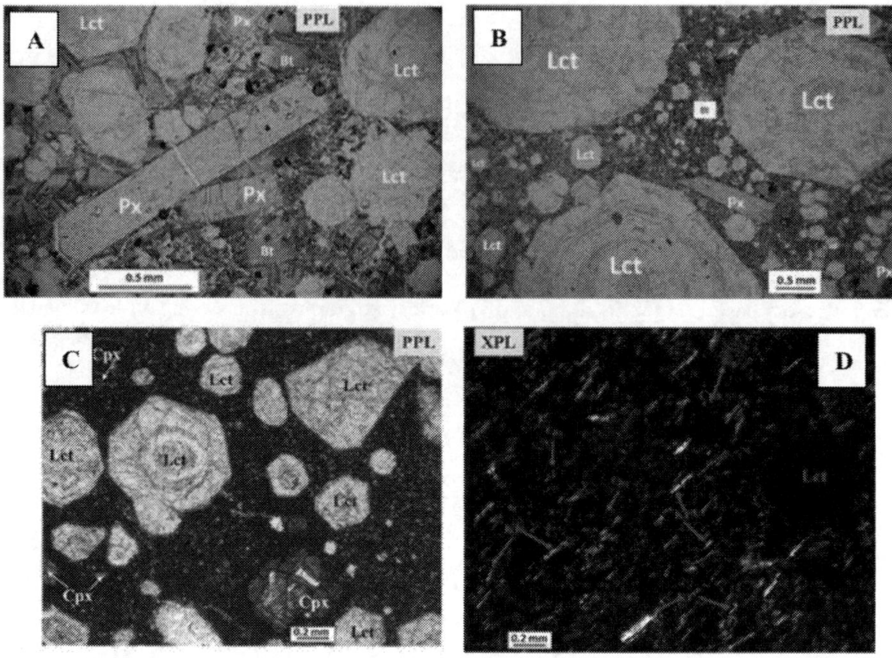

Figure 3. Igneous rocks thin sections of leucitite that show porphyritic texture with leucite and pyroxene as phenocrysts (A-C) and leucite phonolite with trachytic texture (D). Exp: Lct = Leucite; Px = Pyroxene; Bt = Biotite; Cpx = Clinopyroxene; Kf = Alkali feldspar.

4.2. Geochemical analysis

Based on the major oxides data, the silica content of the Adang and Talaya Volcanics lava is (47.06 – 65.7%) so the magma is categorized as basaltic-intermediate magma. Magma formed high alkalinity rocks with wide range concentrations of K_2O (0.27 – 10.12%), Na_2O (0.54 – 11.06%), and total alkali (4.6 – 12.6%). The rocks found in the lava, fragments, and blocks of the Adang Volcanic are classified as trachyandesite, basaltic trachyandesite, trachyte, tephriphonolite, phonotephrite, and andesite. Lava from Talaya Volcanic was plotted within the basalt and trachybasalt fields, with fragments from the volcanic breccia plotted within the dacite field. Almost all of the samples are categorized as alkaline rocks based on their total alkali value (figure 4A). Based on the diagram of Na_2O vs K_2O (figure 4C) [11] there are three groups of alkaline rocks in the study area: sodic, potassic, and ultrapotassic. Magma series also shows a wide variety from tholeiite to shoshonite based on the diagram of K_2O against SiO_2 by Peccerillo and Taylor [12] (figure 4B). However, samples might fall in the tholeiite series field due to the significant number of LOI content. Samples of the shoshonite series are the most common.

The trace elements in rocks could be used to determine the process of magma evolution such as crystal fractionation and crustal contamination. By using a variation diagram of Y and Nb elements, the level of crustal contamination could be determined. Rocks formed with high levels of crustal contamination condition have low Y and Nb content. Based on Y vs Nb amd Ba/Th vs Th/Nb diagrams (figure 4E and figure 4F), Adang and Talaya volcanic rocks formed as the result of a wide range level of crustal contamination. For example, samples of Adang Volcanics with high levels of crustal contamination are from the Ampalas group, while samples from the Tapalang group have a high Y and Nb ratio that shows the crustal contamination was not significant. Samples from Talaya in Kalukku also have a high level of crustal contamination. Based on the ratio of Ba/Th and Th/Nb the contamination process dominantly occurred in the lower crust (figure 4F).

Table 1. Major oxides abundances of primary data sample

	Adang Volcanic										Talaya Volcanic
	Lava							Fragment or boulder			
	GTN-2A	BBL-1A	BBL-2	AZP-5	AZT-6	BTN-U2	SLT-2	SMR-2 (Boulder)	AZLR-2 (Boulder)	AZM-2-1 (Boulder)	AZK-4N (Fragment)
SiO_2	48.93	47.06	55.18	53.79	51.96	49.39	49.38	50.51	49.25	56.63	65.34
TiO_2	1.48	1.33	0.89	1.53	1.45	1.51	1.66	1.01	1.07	0.90	0.53
$Al2O_3$	12.36	11.17	13.34	13.75	14.98	11.70	11.15	9.79	13.07	15.95	13.70
Fe_2O_3	10.34	9.46	7.96	9.80	10.09	11.07	12.17	8.16	8.90	7.08	3.82
MnO	0.18	0.16	0.13	0.15	0.20	0.16	0.25	0.13	0.15	0.09	0.07
MgO	5.12	5.00	4.68	2.11	2.20	5.13	5.12	7.57	6.04	1.71	2.96
CaO	6.81	7.53	7.08	2.79	4.00	5.96	4.18	11.50	8.18	3.19	4.10
Na_2O	0.59	4.68	2.74	8.61	6.32	5.27	3.20	0.62	4.51	2.79	2.55
K_2O	10.12	0.71	6.15	0.55	0.27	2.62	4.53	5.52	1.55	8.13	4.26
P_2O_5	0.736	0.806	0.558	0.178	0.197	0.872	0.632	0.788	0.896	0.276	0.214
LOI	2.18	8.83	1.14	6.41	6.75	5.47	6.08	4.05	5.95	2.83	1.37
Total	98.85	96.74	99.85	99.67	98.42	99.15	98.35	99.64	99.56	99.57	98.92

Tectonic setting determination using Al_2O_3 vs TiO_2 diagram (after Muller and Groves, 2003) [13], shows the rocks of the study area are plotted into both arc-related and within-plate fields (figure 5C). The samples that have more arc-related characteristics are from Ampalas (Adang Volcanic) and Kalukku (Talaya Volcanic) groups, while most samples that show within-plate characteristics are from Tapalang. This might be a clue that the tectonic setting of the study area was transitioning from subduction-related to within-plate magmatism. This can also be seen from the La/Yb vs Th/Nb and La/Yb vs Nb/La diagrams (after Hollocher et al., 2009) [14] (figure 5A and figure 5B) which shows the tectonic setting of rock formation is an alkaline arc. However, there is one sample from the Ampalas group plotted in the continental arc field and two other samples (Ampalas and Kalukku) plotted near the continental arc borderline. Based on Ta/Yb vs Th/Yb diagram (figure 5E) the samples were plotted in the shoshonite series with an alkaline arc tectonic setting. However, one sample from the Ampalas group falls in the calc-alkali series with a volcanic arc tectonic setting. The diagram also shows samples from Ampalas and Kalukku categorized as group III potassic rocks. The variation diagram of Zr vs Nb (figure 5D) indicates potassic rocks formed in subduction and post-collision phase and zone with the characteristic of low content of Zr and Nb and then turned into more evolved due to the enrichment of both elements.

Only seven samples have the criteria for ultrapotassic rocks following the definition of ultrapotassic rocks according to Foley et al. (1987) [15] so those seven samples each from Tapalang, Adang, Ampalas, and Sumare areas were used to determine the ultrapotassic rocks classifications. There are three groups of ultrapotassic rock: Group I (lamproites), Group II (kamafugites), and Group III (active orogenic zone products) [15]. Based on the diagram (figure 6) the rocks from the study area are categorized as a combination of group I and group III ultrapotassic rocks. The diagram shows the samples were plotted close with the group borderline. This might indicate those rocks were formed in a transition zone of active orogenic into more within-plate tectonic setting.

The primitive mantle normalized [17] spider diagram for rocks in the study area shows strong enrichment of the incompatible elements like Rb, Ba, Th, and K relative to OIB and MORB patterns. The negative anomaly of the trace elements is shown by Nb, Sr, and Ti while the positive anomaly of trace elements is shown by Hf and Zr (figure 7A). LREE pattern (La, Ce, Pr, Nd, Sm) also shows strong enrichment relative to HREE elements (Eu, Gd, Tb, Dy, Y, Ho, Er, Tm, Yb, Lu) in the REE diagrams (figure 7B). The samples from the Tapalang group were the most enriched, while the samples from Kalukku group (Talaya Volcanics) show the least enrichment. Some of the samples have a negative anomaly of Eu. Most of the sample show a similar pattern to the group I (Lamproite) and group III ultrapotassic rocks. This shows that Adang and Talaya Volcanics might have the same tectonic setting as both groups I and III.

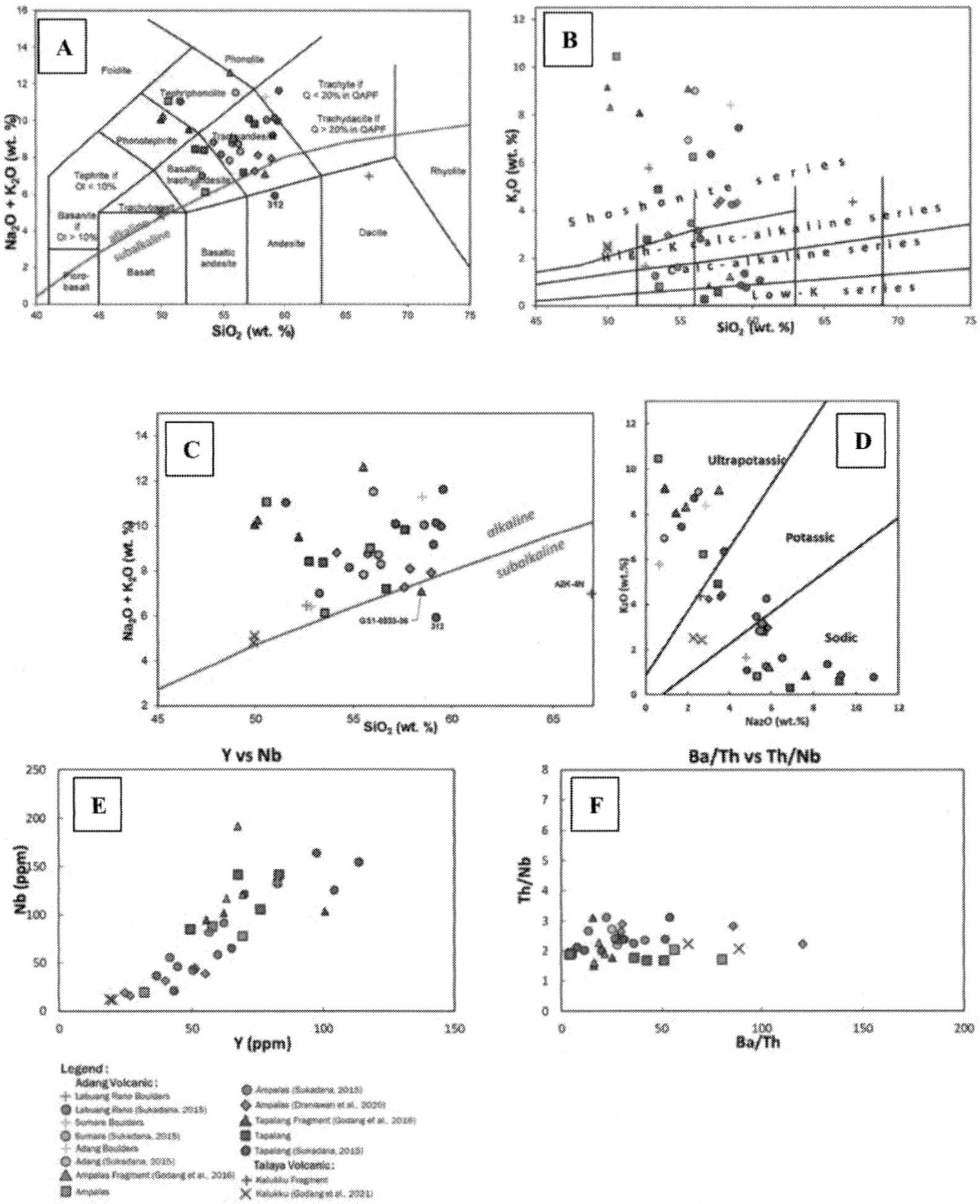

Figure 4. Volcanic Rocks Classification discriminant diagrams A-D: A). SiO_2 vs $Na_2O + K_2O$ (Le Bas et al., 1986) [21], B). SiO_2 vs K_2O (Pecerillo and Taylor,1976) [12], C). SiO_2 vs $Na_2O + K_2O$ (Irvine and Baragar, 1971) [22], D). Na_2O vs K_2O (Middlemost, 1975) [11], and E). Y vs Nb, F). Ba/Th vs Th/Nb

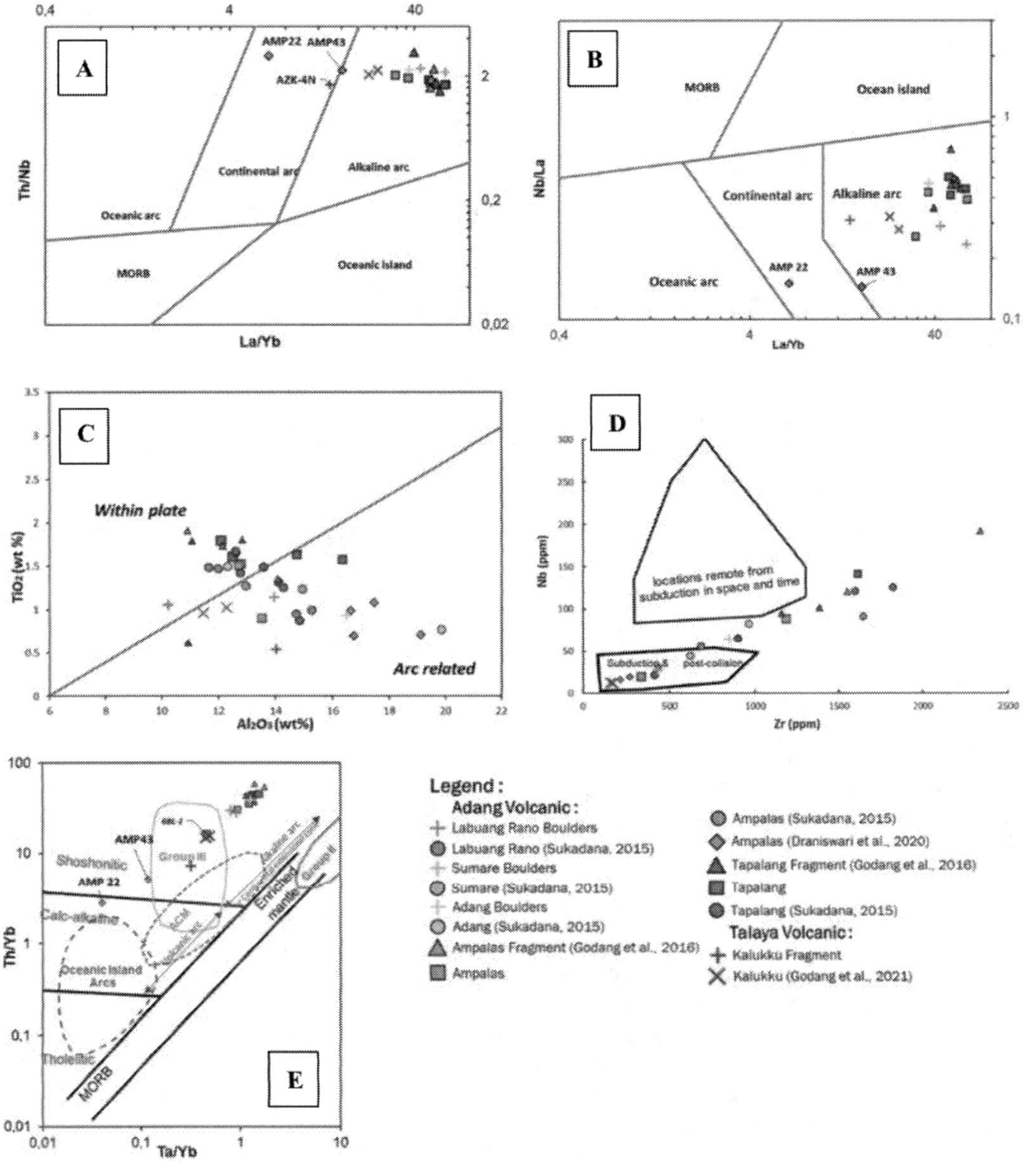

Figure 5. Discriminant diagrams: A). La/Yb vs Th/Nb and B). La/Yb vs Nb/La (Hollocher et al., 2012) [12], C). TiO_2 vs Al_2O3 (Muller and Groves, 2000) [13], D). Zr vs Nb (Thompson and Fowler, 1986) [23], E). Ta/Yb vs Th/Yb (Pearce,1983) [24].

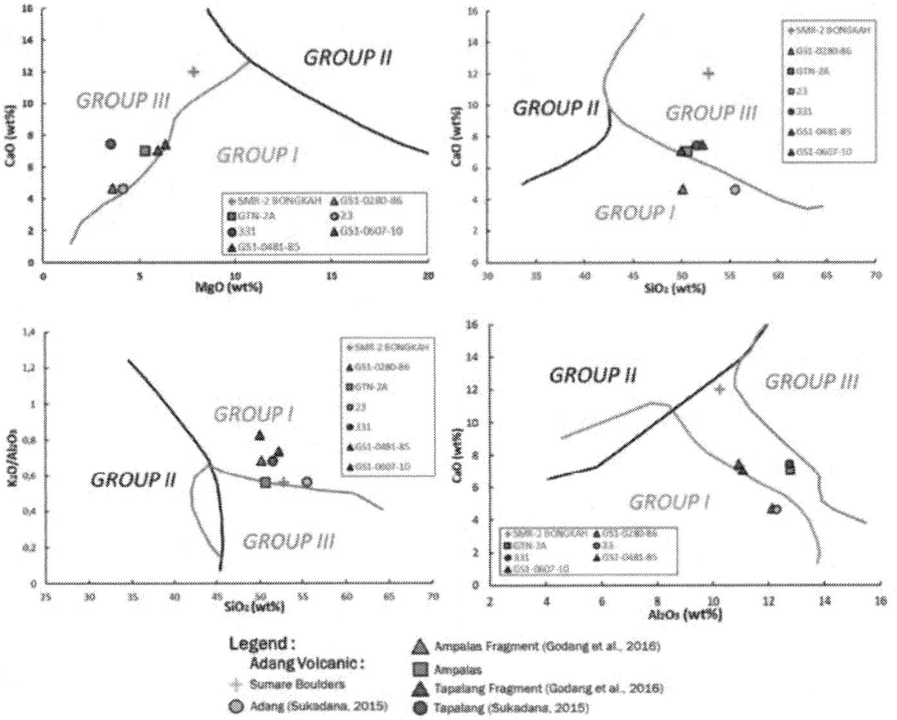

Figure 6. Diagram discriminant of ultrapotassic rocks classification using major oxides (Foley et al., 1987) [15]

5. Discussion

Based on the petrographic analysis, Adang and Talaya volcanic rocks generally show a porphyritic texture, indicating the minerals are formed in two different cooling stages. First, when magma was far from the surface in the magma chamber by crystallizing coarse-medium-sized minerals to form phenocrysts such as leucite and clinopyroxene, then when it was near the surface so that fine-grained minerals and volcanic glass were crystallized as groundmass. Trachytic texture also shows that magma flows as it rises toward the surface. Crystallization of flow materials causes a tendency for the crystals to position themselves in the direction of the flow. Leucite is a feldspathoid mineral that characterizes potassic and ultrapotassic rocks. This rock is classified as an alkaline rock with a very high K_2O content. Usually appears when the magma has low silica saturation. If the magma was saturated with silica, the minerals would appear as feldspar.

Based on the major oxides data, it was found that magma is basaltic-intermediate in composition. This indicates that the magma has partially differentiated. Based on the total alkali (K_2O + Na_2O) vs silica (SiO_2) content (figure 4A), the rocks that appear the most are trachyandesite rocks. Rocks are generally categorized as alkaline rocks. The alkalinity of Adang and Talaya volcanic rocks are also grouped into three types, sodic (Na_2O is more dominant than K_2O), potassic (K_2O is slightly dominant), and ultrapotassic (K_2O is extremely dominant). The Magma series also vary from tholeiite to shoshonite, with most samples are shoshonite series. The high LOI content might be the cause of the samples plotted as tholeiite series.

IOP Conf. Series: Earth and Environmental Science 1245 (2023) 012029 doi:10.1088/1755-1315/1245/1/012029

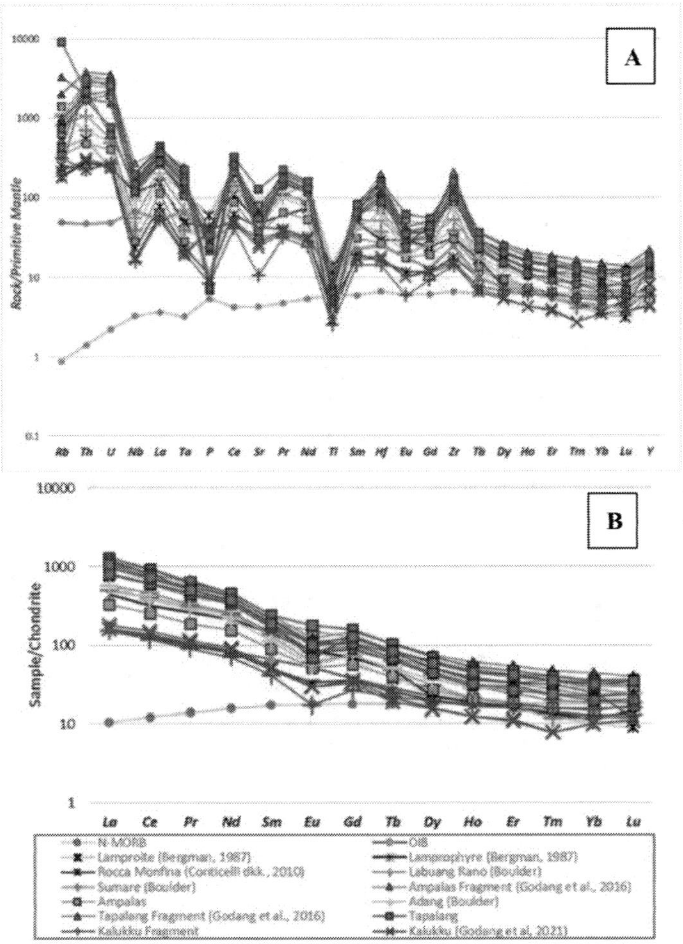

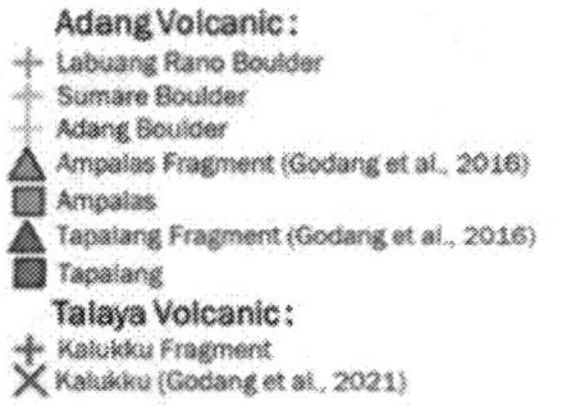

Figure 7. A). Spider diagram of trace element with primitive mantle normalization [17] B). Chondrite normalized of REE spider diagram [17].

Crystal fractionation and crustal contamination occurred during the magma evolution process from calc-alkali to the shoshonite series. Crustal contamination plays a role in the enrichment of incompatible elements such as Ba and Th. The high ratio of Ba/Th shows crustal contamination dominantly occurred in the lower crust.

The Ultrapotassic rocks which have high K_2O concentration from the study area were classified as group I and group III ultrapotassic rocks (figure 6). Group I (lamproites) ultrapotassic rocks are typically formed as the product of post-orogenetic magmatic phenomena in regions that have experienced a continental collision with underlying fossil of Beniof zones and are commonly associated with lineaments or fault trends. Group III ultrapotassic rocks are rocks of active orogenic zone tectonic setting with the characteristics of high Al_2O_3 [18]

IOP Conf. Series: Earth and Environmental Science 1245 (2023) 012029 doi:10.1088/1755-1315/1245/1/012029

The tectonic setting of the Adang and Talaya Volcanics is interpreted as an alkaline arc based on La/Yb vs Th/Nb, La/Yb vs Nb/La, and Ta/Yb vs Th/Yb diagrams (figure 4A, figure 4B, and figure 4E). The tectonic setting of the study area is an alkaline arc with a combination of subduction-related and within-plate magmatism. The transitional environment from arc-related to within-plate magmatism is shown by the TiO_2 vs Al_2O_3 diagram (figure 5C) where samples from Kaluku (Talaya Volcanic) and Ampalas group have higher Al_2O_3 content as typical arc-related characteristics and samples with high TiO_2 characterizes the within plate tectonic setting.

The petrogenesis of volcanic rocks in the study area started in Miocene. In Early Miocene, the collision of the Banggai-Sula microcontinent with West Sulawesi occurred [19]. Collision causes thickening of the crust, subduction stops, and slab detachment [20]. Crustal contamination from previous subduction causes enrichment of REE and incompatible elements such as K, Rb, and Ba so that for a long time it will produce high K calc-alkali magma. In Middle Miocene, magma that previously formed in the continental arc tectonic setting began to evolve into a potassic-ultrapotassic alkaline magma forming porphyritic rocks such as leucitite and phonolite with the main mineral of leucite and pyroxene. The initial stage was initiated by the volcanism at Kalukku (Talaya Volcanic) and Ampalas (Adang Volcanic) areas, characterized by relatively lower Th/Yb, Ta/Yb, Nb/La, and La/Yb ratios. This is supported by the lower position of the spider diagram pattern (figure 7A and figure 7B) relative to other samples from another area such as the Tapalang area. The characteristic feature of this stage is the high intensity of crustal contamination and low Ti, Nb, and Zr content. Furthermore, magma develops along with the collision process. Then in the Late Miocene, there was an uplift and decompression of the lithosphere, the mantle became unstable and triggered an asthenosphere rise and delamination of the lower lithosphere [20]. The lower litosphere and asthenosphere interaction form a new type of shoshonite/potassic-ultrapotasic magma such as in the Tapalang area. This magma is also found in Ampalas and other locations. This was indicated by increased Ti and Nb contents, higher Th/Yb, Ta/Yb, La/Yb, and Nb/La ratios, more enriched concentrations of REE and incompatible elements, and higher Zr and Hf. This indicates that in this process the within-plate began to develop but the influence of the previous subduction is still visible, based on the pattern of the spider diagram shown with the negative anomaly of Ti and Nb.

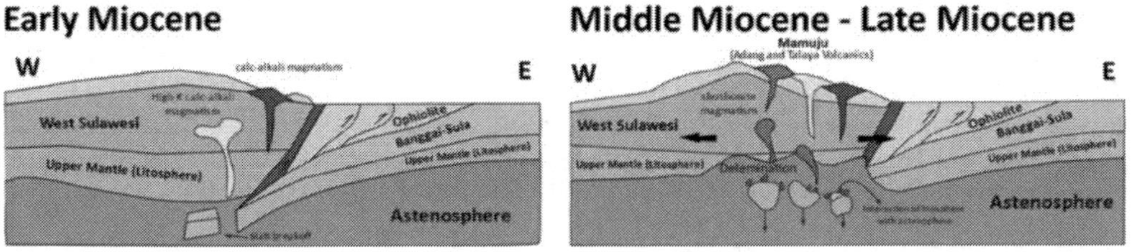

Figure 8. Tectonic model for the Miocene evolution of West Sulawesi microcontinent (modified after Liu et al., 2021) [20].

6. Conclusion

The Adang and Talaya Volcanics have petrological characteristics of leucitite, phonolite, and trachyte rocks with porphyritic textures with the main mineral of leucite and clinopyroxene as phenocrysts in volcanic glass groundmass. Vitrophiric and trachytic textures are also present. The geochemical characteristics of the Adang and Talaya Volcanics are basaltic-intermediate rocks with concentration of TiO_2 (0.71 – 1.68%), K_2O (0.27 – 10.12%), and Na_2O (0.54 – 11.06%). Adang Volcanic magma has a wide range of alkalinity from sodic to ultrapotassic with varying magma series of tholeiitic, calc-alkali, high-K calc-alkali, and shoshonite magmas. The type of rocks based on geochemical data are basaltic trachyandesite, trachyandesite, andesite, tephriphonolite, and trachyte. Talaya Volcanic magma is potassic alkaline and has a shoshonite series with rock types of basalt and trachybasalt. Magma is very enriched by rare earth elements and incompatible elements such as Rb, Ba, Th, U, K, and Zr. The Adang and Talaya Volcanics formed in the alkaline arc tectonic settings at the continental margin by a combined mechanism of subduction, collision, and within-plate magmatism.

References

[1] Sukamto 1975 *Geologi Sulawesi* (Bandung: Pusat Penelitian dan Pengembangan Geologi)
[2] Hamilton W 1979 *Tectonics of the Indonesian region* (Denver: United States Geological Survey) Professional Paper No. 1078
[3] Smith R B, 1987. *Sedimentology and tectonics of a Miocene collision complex and overlying late orogenic clastic strata, Buton Island, Eastern Indonesia.* University of California, Santa Cruz (1983), p. 255
[4] van Leeuwen T, Allen C M, Elburg M, Massonne H, Palin J M and Hennig J 2016 *J. Asian Earth Sci.* The Palu metamorphic complex, NW Sulawesi, Indonesia: origin and evolution of a young metamorphic terrane with links to Gondwana and Sundaland **115** 133–152
[5] Ratman N and Atmawinata S 1993 *Peta Geologi Lembar Mamuju, Sulawesi Barat skala 1:100.000* (Bandung: Pusat Penelitian dan Pengembangan Geologi)
[6] Draniswari W A, Kusuma S, Adimedha T and Sukadana I G 2020 *EKSPLORIUM* peran kontaminasi kerak pada diferensiasi magma pembentuk batuan vulkanik Sungai Ampalas, Mamuju, Sulawesi Barat **73** 41(2)
[7] Godang S, Fadlin, Priadi B, Idrus A and Sukadana I G 2021 *Indo. J. of Geosci.* komatiitic lamprophyre in West Sulawesi: first evidence for >1350°c and 3.5 - 3.8 gpa mantle melts **18(1)** 39-58
[8] Godang S, Fadlin and Priadi B 2016 *Indo. J. of Geosci.* geochemical signatures of potassic to sodic Adang Volcanics, Western Sulawesi: Implications for their tectonic setting and origin **3(3)** 195-216
[9] Sukadana I G 2015 *Petrogenesis Batuan Vulkanik Adang dan Kaitannya Dengan Keterdapatan Mineral Radioaktif di Kabupaten Mamuju, Sulawesi Barat* (Yogyakarta: Universitas Gadjah Mada)
[10] Williams H, Turner F J and Gilbert C M 1982 *Petrography: An Introduction to the Study of Rocks in Thin Sections (Second Edition)* (San Francisco: W. H. Freeman) 626p
[11] Middlemost E A K 1975 *Earth Science Reviews* The basalt clan **11** 337-364
[12] Peccerillo A and Taylor S R 1976 *Contributions to Mineralogy and Petrology* geochemistry of Eocene calc-alkaline volcanic rocks from the Kastamonu area, Northern Turkey **58** 63-81
[13] Muller D and Groves D I 2000 *Potassic igneous rocks and associated gold–copper mineralization 3rd ed.* Berlin Heidelberg (New York: Springer)
[14] Hollocher K, Robinson P, Walsh E and Roberts D 2012 *Am. J. Sci.* geochemistry of amphibolite-facies volcanics and gabbros of The Støren Nappe in extensions west and southwest of Trondheim, western gneiss region, Norway: a key to correlations and paleotectonic settings **312** 357-416
[15] Foley S F, Venturelli G, Green D H and Toscani L 1987 *Earth Sci. Rev.* the ultrapotassic rocks: characteristics, classification and constraints for petrogenetic models **24** 81–134
[16] Bergman S C 1987 *Geo. Soc. Lon. Spl. Pub.* lamproites and other potassium-rich igneous rocks : a review of their occurrence, mineralogy, and geochemistry **30** 103-109
[17] Sun S and McDonough W F 1989 *Geo. Soc. Lon. Spl. Pub* chemical and isotopic systematics of oceanic basalts: implications for mantle composition and processes **42** 313-345
[18] Wilson M 1989 *Igneous Petrogenesis: A Global Tectonic Approach* (London: Unwin Hyman Inc)
[19] Hall R 1996 *Geo. Soc. Lon. Spl. Pub* reconstructing Cenozoic SE Asia **106** 153–184
[20] Liu J, Zhang J, Hsia J, Xian W, Yin C, Djoko F, Cheng C, Zhao C, Liu X, Chen Y and Wang X 2020 *J. Asian Earth Sci.* Late Miocene to Pliocene crustal extension and lithospheric delamination revealed from the 5 ma Palopo granodioritic intrusion in Western Sulawesi,

Indonesia **201**
[21] Le Bas M J, Le Maitre R W, Streckeisen A and Zanettin B 1986 *J. of Petrology* a chemical classification of volcanic rocks based on the total alkali-silica diagram **27** 745-750
[22] Irvine T N and Baragar W R A 1971 *Can. J. of Earth Sci.* a guide to the chemical classification of the common volcanic rocks **8** 523-548
[23] Thompson R N and M B Fowler 1986 *Contrib. Mineral Petrol.* subduction-related shoshonitic and ultrapotassic magmatism: a study of a Siluro-Ordovician syenites from the Scottish Caledonides **94** 507-522
[24] Pearce J A 1983 role of the sub-continental lithosphere in magma genesis at active continental margins. In: Hawkesworth C J and Norry M J (eds) *Continental Basalts and Mantle Xenoliths* (Nantwith: Shiva), 230-24

IOP Conf. Series: Earth and Environmental Science 1245 (2023) 012030 doi:10.1088/1755-1315/1245/1/012030

Peak Ground Velocity (PGV) estimation of Mw 5.6 Cianjur earthquake November 11, 2022 using Global Positioning System (GPS)

Irwan Meilano[1*], Susilo Susilo[2], Sidik Tri Wibowo[3]

[1]Faculty of Earth Science and Technology, Institute of Technology Bandung (ITB), Indonesia

[2]Research Center for Geospatial, National Research and Innovation Agency, Indonesia

[3]Geospatial Information Agency, Indonesia

E-mail: irwanm@itb.ac.id

Abstract. The primary purpose of GPS is to be used for positioning, navigation, and timing (PNT). In the last decade, research on GPS for earthquake hazards has increased rapidly, such as for tectonic deformation, GPS seismology, modeling earthquake source parameters, estimation of magnitude, etc. In this study, following the Mw 5.6 Cianjur earthquake on November 11, 2022, we estimate the peak ground velocity from the CJUR GPS station, which is located about 15 km from the epicenter. We compute GPS velocity from the high-rate GPS receiver independent exchange (RINEX) data and broadcast ephemeris. Using a narrow line combination of L1 and L2 observations, we invert velocity epoch by epoch. The result shows the CJUR GPS station has a peak ground velocity of about 11.5 cm/s and equal to VI MMI scale. The peak ground velocity from the GPS station could be important for deriving the earthquake shakemap. By increasing the number of GPS stations in Indonesia, it is very important to support earthquake hazard monitoring.

1. Introduction

On November 21st, 2022 at 13:21:10 Western Indonesia Time (06:21:10 UTC) the *Mw* 5.6 Cianjur earthquake occurred northwest of Cimandiri Fault (Figure 1). The epicenter is located at latitude = 6.87^o S, longitude = 107.01^o E, and depth = 11 km (https://inatews.bmkg.go.id). The mainshock of this event felt in Cianjur city up to V-VI MMI (Modified Mercalli Intensity), IV-V MMI in Garut and Sukabumi, III MMI in Cimahi, Lembang, Bandung city, Cikalong Wetan, Rangkasbitung, Bogor and Bayah, II-III MMI in Tangerang Selatan, Jakarta and Depok (Figure 2) [1]. The Meteorology, Climatology and Geophysics Agency (BMKG) reported 424 aftershocks recorded until December 17th, 2023. Based on the National Disaster Management Authority (BNPB) on December 12, 2023, this event caused 334 people dead, and 35.601 damaged buildings.

The utilization of the high-rate GNSS data are widely used to estimate displacement due to the earthquake and potential used to support earthquake early warning system (EEWS) [2]. In the recent studies show the availability of high-rate GNSS data to determined earthquake magnitude rapidly. However, research on the ground velocity waveform using high-rate GNSS data to earthquake studies or monitoring is still lacking.. The recent study [3], show the consistency betwwen velocity from GNSS observations with the ground motion models and macroseismic intensity observations. The residual standard deviation of ground motion model for Peak Ground Velocity (PGV) from GNSS is about 0.58 [3].

In this research, we applied the "variometric approach" proposed by [4, 5] to capture velocity waveform from high-rate GNSS due to the *Mw* 5.6 Cianjur earthquake. Due to the lack of GNSS stations, we only use one GNSS station that located close to the epicenter. The velocity waveform can be used as complementary constraints to ShakeMaps and to determine the earthquake magnitude using PGV scaling law [6].

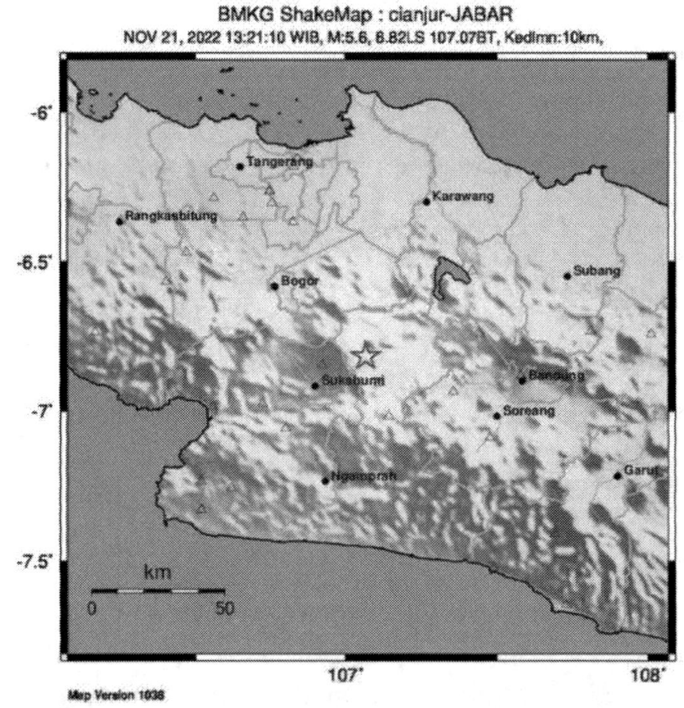

PERCEIVED SHAKING	Not felt	Weak	Light	Moderate	Strong	Very strong	Severe	Violent	Extreme
POTENTIAL DAMAGE	none	none	none	Very light	Light	Moderate	Mod./Heavy	Heavy	Very Heavy
PEAK ACC.(%g)	<0.05	0.3	2.8	6.2	12	22	40	75	>139
PEAK VEL.(cm/s)	<0.02	0.1	1.4	4.7	9.6	20	41	86	>178
INSTRUMENTAL INTENSITY	I	II-III	IV	V	VI	VII	VIII	IX	X+

Scale based upon Worden et al. (2011)

Figure 1. Shakemap of M5.6 Cianjur earthquake on 21 November 2023

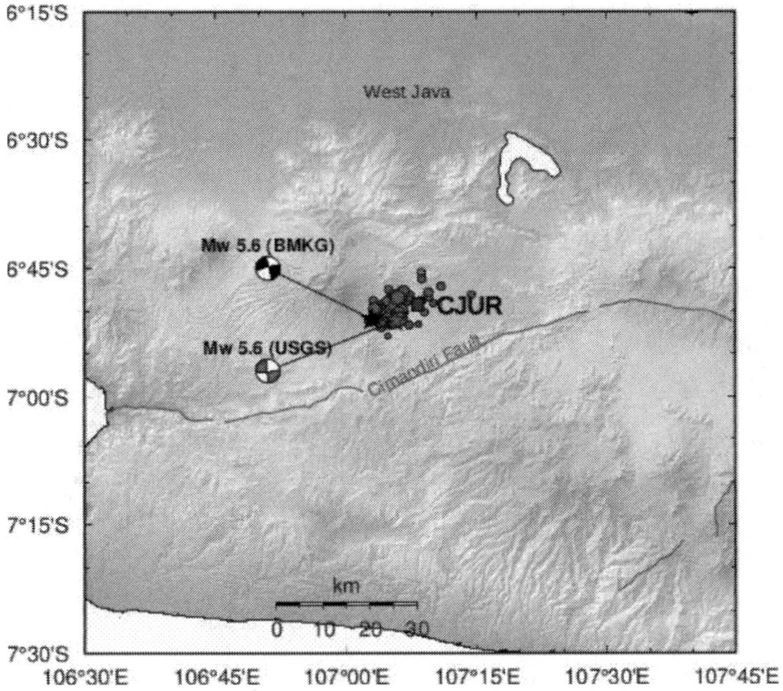

Figure 2. Location of CJUR GNSS site and the epicenter of Cianjur earthquake (BMKG and USGS)

2. Methods

In this research, we use the continuous GPS high-rate data provided by BIG. The nearest cGPS site is CJUR station, which is located about 5 km to the southeast of the epicenter. CJUR station was built at the top of the building and equipped with Septentrio *SPT POLARX5* GNSS receiver with the *LEIAR25* GNSS antenna. The station recorded GNSS data at 1 Hz intervals and send the data in real-time to the BIG processing center using Virtual Private Network. Figure 3 shows the monumentation of CJUR GNSS site (srgi.big.go.id).

Figure 3. Monumentation of CJUR GNSS site.

The "variometric approach" is performed a single time difference on the GNSS phase observation and orbital position[3]. In this methods, we taking a difference between the current time observation and the prior time observation depend on the interval of observation[3]. This method was implemented on Satellite Navigation-Derived Instantaneous Velocities (SNIVEL) software package [7]. Using SNIVEL we derived the velocity waveform high-rate GNSS data with 1 Hz intervals. We used a broadcast ephemeris and high-rate Receiver Independent Exchange (RINEX) GPS data to derived the velocity waveform. The SNIVEL software needs the L1 and L2 phase observables GPS signal. With the narrow lane combination from the L1 and L2 signal, velocity waveform was derived on an epoch by epoch observation. We processed 5 minutes data starting from 06:21:00 UTC.

3. Result and discussion

Figure 4 shows the velocity waveform in the north-south, east-west, and vertical component starting from 06:21:00 - 06:26:00 and the Ground Velocity (GV) waveform. Within 5 minutes length of the data, the high-rate GPS data captured the ground velocity due to the Cianjur earthquake. The maximum velocity which recorded by CJUR GNSS site is about 11.5 cm/s (see Figure 4). The maximum value reached about 4s from the origin time of the earthquake. Regarding the ShakeMap documentatiom ([8]), the ground velocity 11.5 cm/s equal to VI MMI scale. This MMI value is quite different with the color scale in the Figure 1. Our suggestion, this discrepancy due to the calculation of the shakemap in the Figure 1 used model and update with accelerometer data and not included in situ observation. From this result, we show that the high-rate GPS data could be complementary constraints to ShakeMap.

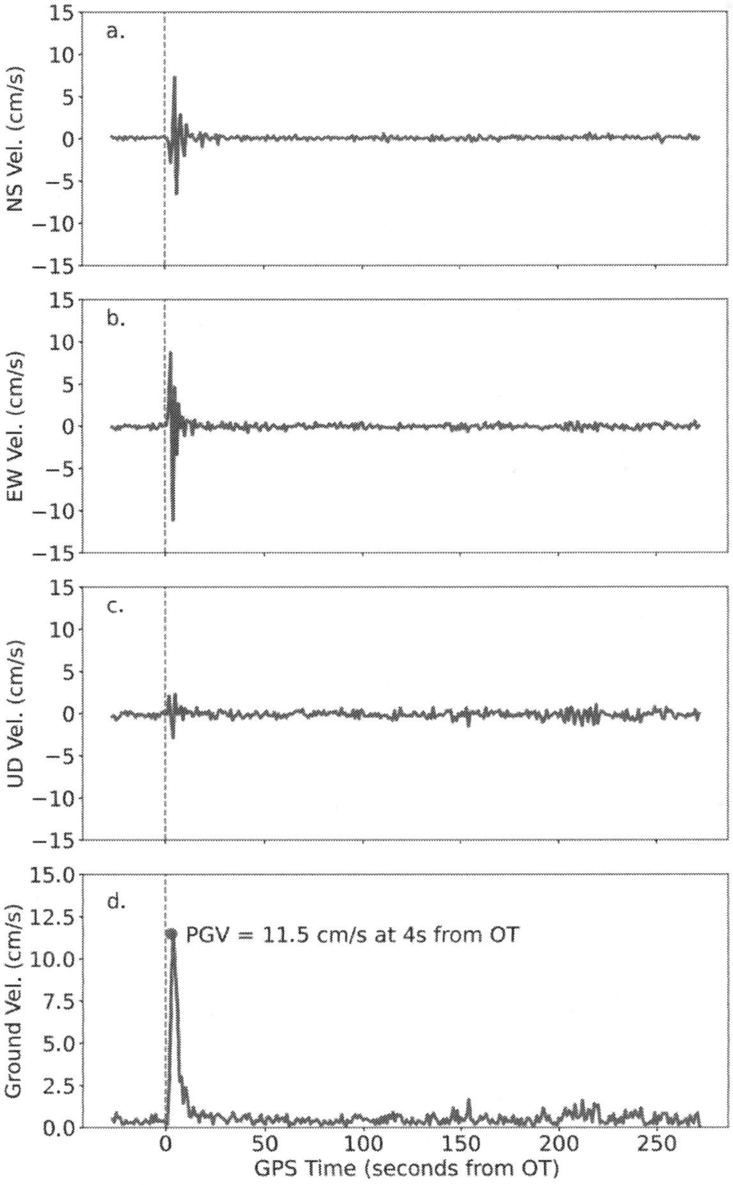

Figure 4. Velocity waveform of CJUR GNSS site. a) is the velocity waveform in the nourth-south component. b) is velocity waveform in the east-west component. c) is is velocity waveform in the vertical component. d) is Ground velocity (GV). Red line indicate the earthquake origin time in GPS time.

4. Conclusion

This research demonstrates the ability to use a high-rate GPS data to determine the ground velocity waveform by post processing method. The result shows the velocity in the horizontal component of CJUR GNSS site are significant, while the vertical velocity is not clearly significant. the maximum ground velocity reached up to 11.5 cm/s at 4 s from earthquake origin time. It value is equal to VI MMI scale. This result show the availability of high-rate GNSS data in the earthquake study or monitoring. The most important is distribution of the GNSS site, more dense the GNSS network is more accurate of the result and vice versa.

Acknowledgments

We thank the Geospatial Information Agency for providing GNSS data. SS is supported by by the Disaster Research seed grant of the National Research and Innovation Agency from the DIPA Research Organization for Earth Sciences and Maritime, National Research and Innovation Agency Nomor: SP DIPA-124.01.1.690501/2023. Most figures were generated by Generic Mapping Tools [9].

References

[1] Supendi P, Priyobudi, Jatnika J, Sanipar D, Ali Y H, Heryandoko N, Daryono, Adi S P, Karnawati D, Anugerah S D, Fatchurochman I and Sudrajat A 2022 *Report https://www.bmkg.go.id/berita/?p=42632lang=IDtag=cianjur*

[2] Allen R M and Ziv A 2011 *Geophysical Research Letters* **38**

[3] Crowell B, DeGrande J, Dittmann T and Ghent J 2023 *Seismica* **2**

[4] Colosimo G, Crespi M and Mazzoni A 2011 *Journal of Geophysical Research: Solid Earth* **116**

[5] Grapenthin R, West M, Tape C, Gardine M and Freymueller J 2018 *Seismological Research Letters* **89** 1040–1048

[6] Fang R, Zheng J, Geng J, Shu Y, Shi C and Liu J 2021 *Seismological Society of America* **92** 227–237

[7] Crowell B W 2021 *Seismological Research Letters* **92** 840–848

[8] Worden C and Wald D 2016 *URL http://usgs. github. io/shakemap. http://dx. doi. org/10.1234* **12345678**

[9] Wessel P, Smith W H, Scharroo R, Luis J and Wobbe F 2013 *Eos, Transactions American Geophysical Union* **94** 409–410

Rain Scanner Radar and Optical Flow Combination For Early Identification of The Incoming Bow Echo Heavy Precipitation

G A Nugroho[1], Syahrul[2], E Maryadi[2], A Rahmawati[3], T Sinatra[1], A Indrawati[1], N Cholianawati[1], A Awaludin[1]

[1]Research Center for Climate and Atmosphere, National Research and Innovation Agency (BRIN), Bandung, Indonesia
[2]Research Center for Artificial Intelligence and Cyber Security, National Research and Innovation Agency (BRIN), Bandung, Indonesia
[3]Research Center for Smart Mechatronics, National Research and Innovation Agency (BRIN), Bandung, Indonesia

ginaldi.ari.nugroho@brin.go.id

Abstract. In this study, heavy precipitation in the form of bow echo is observed, identified, and tracked using a rain scanner and optical flow method. Two case studies of bow echoes with a range of properties from 15 to 33 km were observed in the afternoon on 1st February and 4th February 2021. The rain scanner developed based on X-band radar could obtain the first detection of the bow echo shape. The difference between these two bow echoes events is from their initiation. The first bow echo cases are initiated from weakly organized cells. Meanwhile, the second case of bow echo is initiated from a squall line event. Bow echo is initially observed in the western part of Bandung city. Approximately twenty minutes after the first detection, the bow echo passed the center of Bandung city. Image processing using edge detection could track the bow echo from the radar data. Applied Lucas-Kanade optical flow was then used to retrieve the bow echo movement. Vector flow from optical flow showed the dominant movement toward the east direction in both case studies. A simple method to predict the incoming time of the bow echo using the estimated speed from vector flow and the image translation is also presented.

Keywords: Bow echo, rain scanner, radar, optical flow

1. Introduction

Heavy precipitation in the form of a bow echo is associated not only with high rainfall intensity but also with strong wind occurrence on the surface [1, 2, 3]. Information on the early detection of the bow echo could become valuable information for the incoming heavy precipitation.

The initiation of bow echo comes from sources that evolve and transform into a bow echo shape. A previous study by Klimowski et al [1] classified the initiation phase of the bow echo that comes from three sources: weakly organized cells, squall lines, and supercells. The first two sources (weakly organized cell and squall line) are more common to occur compared with the last source (supercell). The weakly organized cell comes from the merger of several cells into a bow-echo shape. The bow echo movement tends to follow the cell that initiates the merger process [1]. A squall line is also another possibility of the source that could transform into a bow echo shape.

To detect the initiation phase, it is essential to be able to detect the bow echo pattern. Several studies have proposed methods using various techniques. Kamani et al [4] use the skeleton matching method to detect the bow echo as an indicator of severe weather. The bow echo feature from the radar data is extracted using the skeletonization method and skeleton context to match the shape pattern of the bow echo. A different approach is done by Fanlin and Jinyi [5] which measured the contour to find the curve of the bow echo. Another method used pseudo reflectivity combined with the Convolutional Neural Network to detect the bow echo [6].

In this study, we present an alternative method to detect the bow echo using image processing. Two case studies of bow echo observed by RSR are used as an example of the applied alternative method. The objective of this study is to detect the early bow echo shape that capture by radar and then predicted the movement using optical flow and image translation. The structure of this paper is described as follows. Bow echo detection is discussed in section 2. Meanwhile, section 3 presents the bow echo nowcast. Two case studies to test the proposed method are described in section 4. The conclusion is given in section 5.

2. Bow Echo

2.1. Overview of the Bow Echo from Rain Scanner Radar

The two main data used in this study are Rain Scanner Radar (RSR) and Automatic Weather Station (AWS). RSR is a modified marine X-band radar dedicated to rain observation using single polarization (horizontal). The maximum coverage area is 44 km with a spatial resolution of 120 x 120 meters. The RSR is located in Bandung city (107.59° E, 6.89° S) which has been tested concerning the accuracy and reliability of rain detection [7,8,9]. The RSR observation dataset is collected from 2019 to 2021 to obtain the bow echo event. However, in this study, we present only two case studies of the bow echo event (February 1st and February 4th, 2021). The confirmation of the heavy precipitation carried out by the bow echo event is obtained from AWS measurement data.

2.2. Bow Echo Detection

Two methods are used to determine the bow echo, a visual method and a proposed method of shape recognition. The visual method is applied by analyzing the image sequence visually. Meanwhile, shape recognition using image processing is used to find the bow-like shape from the image sequence which resembles the bow echo.

Shape recognition consists of three steps, a) identifying the object, b) measuring the properties of the object, and c) determining the bow shape. The first step of identifying the object consists of three parts, convert the image into grayscale, image filter by removing the small object, and labeled the object. The second step is to measure the properties of the object. This is done by utilizing the regionprops function [10]. Four parameters from the regionprops output (perimeters, filled area, circularities, eccentricity, orientation, and equivalent diameter) are used as a threshold to determine whether the object suits the criteria of a bow echo shape or not.

Perimeter is the distance between each adjacent pair of pixels surrounding the region's border by measuring the distance between them. The filled area specifies the number of pixels within a region where all holes are filled in. Circularities are the measure of the roundness of an object, with a value of 1 representing a circled object and more than 1 showing a less circular object. Circularities are calculated by using two parameters (Filled area and Perimeter) using Equation 1. Eccentricity is defined as the ratio of the lengths of the focal point and major axis of an ellipse, with a range value between 0 and 1. Eccentricity with the value of 0, is interpreted as a circle object, meanwhile, if the eccentricity with the value of 1 represents the line segment object. Orientation is defined as the angle between the major axis of the object with the x-axis.

$$\text{Circularities}=\frac{\text{perimeters}^2}{4 \times \pi \times \text{Filled Area}} \quad (1)$$

Several tests using the combination of circularities, eccentricity, and orientation are carried out towards several bow echo shape objects. Thirty images with a possible of bow shape object are tested using the combination of these three parameters. From this test, we proposed a new method to detect the bow echo by using these three parameters with certain thresholds described in Table 1.

Table 1. Proposed threshold for bow echo detection using shape recognition

Parameter	Threshold
Circularities	6.8 < circularities < 18
Eccentricity	eccentricity >= 0.85
Orientation	-61 <= orientation <= 61

The proposed method of shape recognition using threshold from three parameters (circularities, eccentricity, and orientation) is then applied to two case studies of bow echo occurrence observed by the RSR.

3. Bow Echo Prediction

3.1. Lucas Kanade Optical Flow

The advantage of the RSR is its high temporal resolution [11]. The high interval of the image sequence generated by the radar could give valuable information on the movement of the precipitation echo. One method to calculate the precipitation movement from the image sequence is the optical flow. Optical flow is a method to identify the movement of an object from a sequence of images [12]. Optical flow is defined as the apparent motion of individual pixels on the image plane. It often serves as a good approximation of the true physical motion projected on the image plane. The more rapid the interval of the image sequence, the more information about the object movement could be retrieved.

Several approaches could be applied to the optical flow such as the local differential, global variational, and phase-based [13]. Two common optical flow categories often used are global variation and local differential. Global variation utilizes the entire image pixel in the calculation of the vector field. Meanwhile, the local differential only uses a small neighborhood area in the calculation. Lucas Kanade is the optical flow type that uses the local differential principle [14]. Lucas Kanade intention is to calculate the vector of optical flow using an assumption that the small area near the neighborhood will be similar to the vector.

3.2. Prediction

In general, the nowcast process consists of three elements, namely identification, tracking, and prediction. Identification and tracking of the bow echo are already been done by shape recognition. Meanwhile, prediction is the process of guessing or predicting the next bow echo movement, in this case on a short-term scale, namely the next 20 minutes (prediction interval of 2 minutes). 20 minutes is considered sufficient to provide information for users in making decisions. Since the movement of the bow echo in both case studies is linear, a simple image translation is carried out. The input for the pixel movement is taken from the average motion flow value calculated by Lucas Kanade optical flow.

Image translation is a simple tool to shift the image towards a certain pixel location. In this study, we use the assumption that the bow echo is commonly has a bigger, uniform shape and only affected by the environmental wind condition. The properties of how many translations of the object are based on the average motion flow. The prediction result was tested using RMSE (Root Mean Square Error) and IoA (Index of Agreement).

4. Result

4.1. Case study bow echo 1st February 2021

4.1.1. Observation from RSR and AWS

Visual analysis is conducted toward the image of the RSR to locate the bow echo sequence. The result shows that the bow echo sequence occurred from 17:16 -17:38 LT (Local Time) on February 1st, 2021. The source of the bow echo was also captured by the radar at 17:04 – 17:14 LT. Several convective cells that merge into one big cell are moving in the east direction depicted at 17:04 LT. The merging cell is established at 17:16 LT in the form of a bow shape alike. The bow echo shape is well preserved until 17:38 LT and started to unravel after that.

AWS instrument captured the incoming heavy precipitation that was carried out by the bow echo. The graphical data of rainfall from AWS started to increase after 17:30 LT which the nearest location of the bow echo towards the AWS location is at that time. The bow echo started to unravel after 17:38 LT is probably because the rainfall occur at the radar location which is the same location as the AWS. This condition creates high attenuation around the radar location. However, the wind speed parameter from AWS data is not capturing any high fluctuation of wind speed. The dominant movement of the bow echo towards the east direction is strengthened by the AWS measurement which shows a wind direction value of 270° at 17:40 LT.

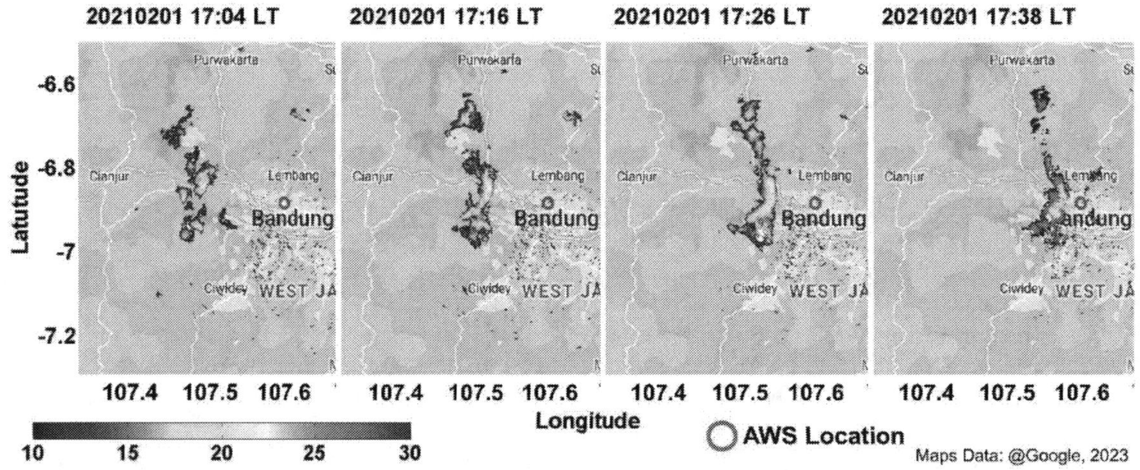

(a)

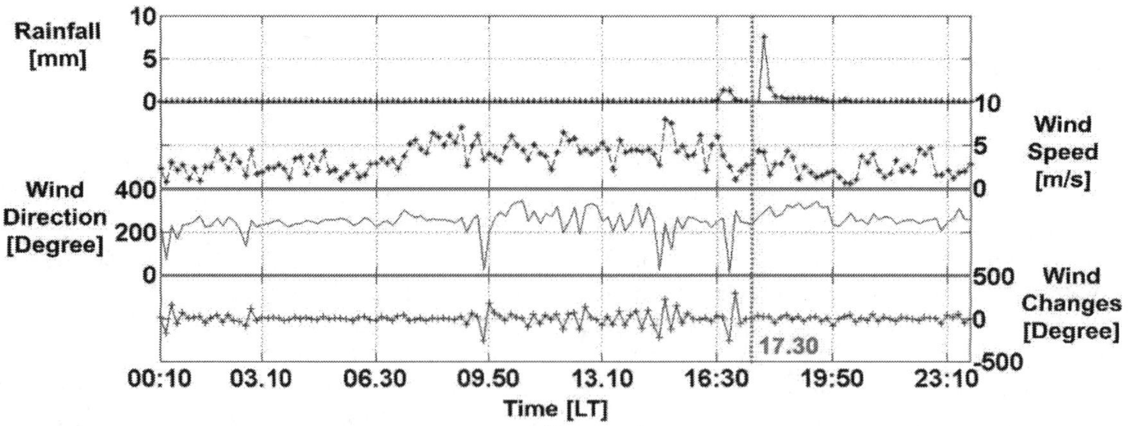

(b)

Figure 1. Heavy precipitation on Case Study 1, a) RSR observation, b) AWS observation

4.1.2. Bow echo detection and prediction

In order to quantify the exact detection of the bow echo, the proposed method is applied to the RSR data. The image from RSR is then processed using image processing to obtain the object feature. Figure 2 is the bow echo identification process in case study no 1. The figure compared the bow echo identification method within the image at 17:14 LT and 17:16 LT. There are four image panels (A until D) on each figure. The first image panel A is the image from the RSR. The second panel B shows the grayscale-filtered image. The third panel C is the object detection, and the fourth panel D is the image of the bow echo detection. If bow echo detection exists, then the object will be marked with a red border.

In the third panel C on both Fig 2a and 2b, the number of objects detected is different. Edge detection acknowledges that there is five object and three objects at 17:14 LT and 17:16 LT, respectively. Notice that the bow echo did not consider to occur at 17:14 LT since the cell was still in a fragment of object number two, three, and four (Fig 2a panel C). After 17:16 LT, those cell fragment is well combined into one single cell as depicted in Fig 2b panel C (object number 2). This object is suited to the criteria of the proposed method of bow echo detection in section 2.2. Therefore, this object is marked by a red border with additional information on the region parameter is also displayed in Fig 2b panel D.

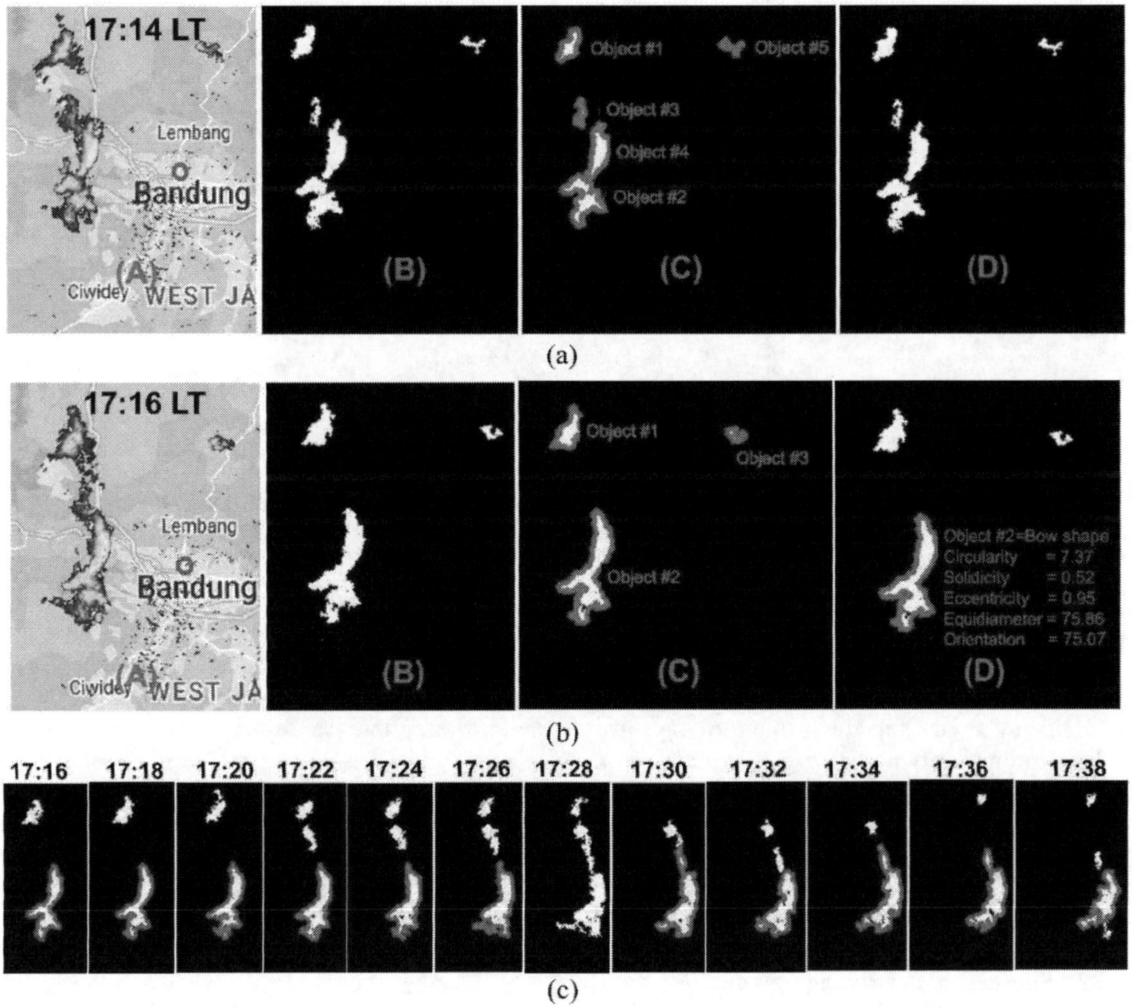

Figure 2. Bow echo detection on case study 1, a) Bow echo at 17:14 LT, b) Bow echo at 17:16 LT, c) Bow echo track

By using this procedure, we detect and track the bow echo shape from 17:16 LT to 17:38 LT. With an interval of two minutes, the proposed method could track eleven from a possible twelve-bow echo shape (Fig 2c). The reason why the object at 17:28 LT is that the circularity value is more than the criteria on the proposed method in section 2.2.

The early information of the bow echo at 17:16 LT could be utilized to predict the incoming heavy precipitation. Optical flow using Lucas Kanade is utilized to obtain the average motion flow value (Fig 3 panel A). The image used is from 17:10 LT to 17:18 LT to gather the information before, during, and after the early detected bow echo. The average motion flow is 7.36 (rounded to 7) pixels per 2 minutes. Panel B showed the nowcast result with the evaluation comparing the actual and prediction result depicted in panel C in Fig 3. Since the prediction method only use a simple translation method (without considering the growth decay of the precipitation echo), the RMSE and the IoA did not show a good result.

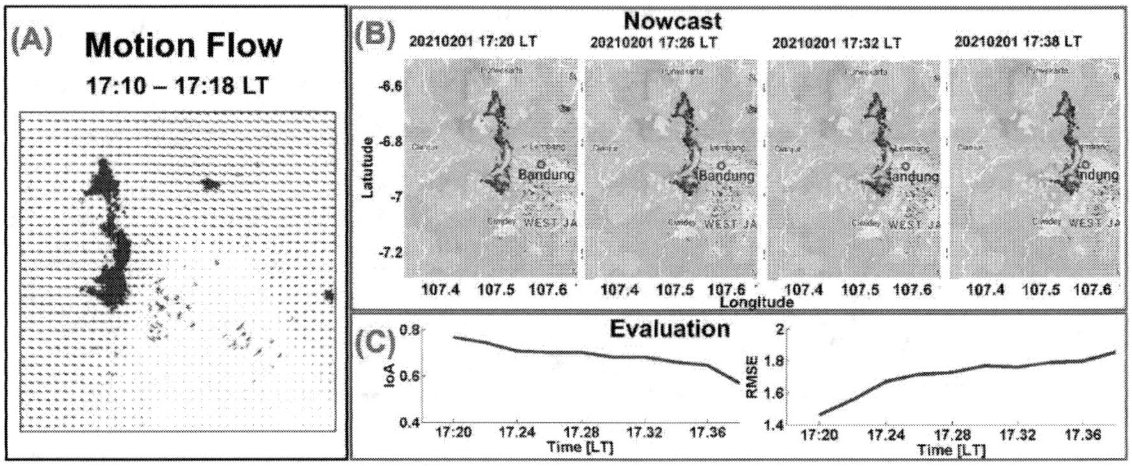

Figure 3. Nowcast bow echo case study 1, a) Motion flow from Lucas Kanade Optical Flow, b) Nowcast result, c) Evaluation between prediction and actual result.

4.2. Case study bow echo 4th February 2021

4.2.1. Observation from RSR and AWS

The same procedure is also carried out to the image sequence in case study number two. The bow echo is observed on February 4th, 2021 from 15:56 to 16:24 LT based on the visual image sequence of the RSR (Figure 4). The source of this bow echo is well observed from 15:42 until 15:52 LT. Long stretched precipitation echo or a squall line moving towards the east direction are observed during this period. Several minutes after that, the bow echo shape is formed with the movement is still similar to the previous squall line.

The bow echo approached the AWS location at 16:24 LT and was followed by the increased reading of AWS wind speed value (9.71 m/s) at 16:30 LT. Later, rainfall is detected at 16:40 LT (3.6 mm) with the maximum reached at 16:50 LT (4.3 mm). Wind direction showed small changes during the period from 16:20 (294°) until 16:30 LT (225°).

4.2.2. Bow echo detection and prediction

Investigating the first bow echo in case study 2, we compared the image radar data at 15:54 LT and 15:56 LT. From a visual point of view, these two images from panel A (Fig.5a and Fig.5b) look like a bow-type shape. After we processed it using the same procedure as case study 1 (converting to grayscale, image filtering, edge detection, calculating the object, applied bow echo detection method), Fig 5a does not fulfill the criteria of a bow echo shape, because the circularities value (circularity = 6.05) is below the criteria on the proposed method.

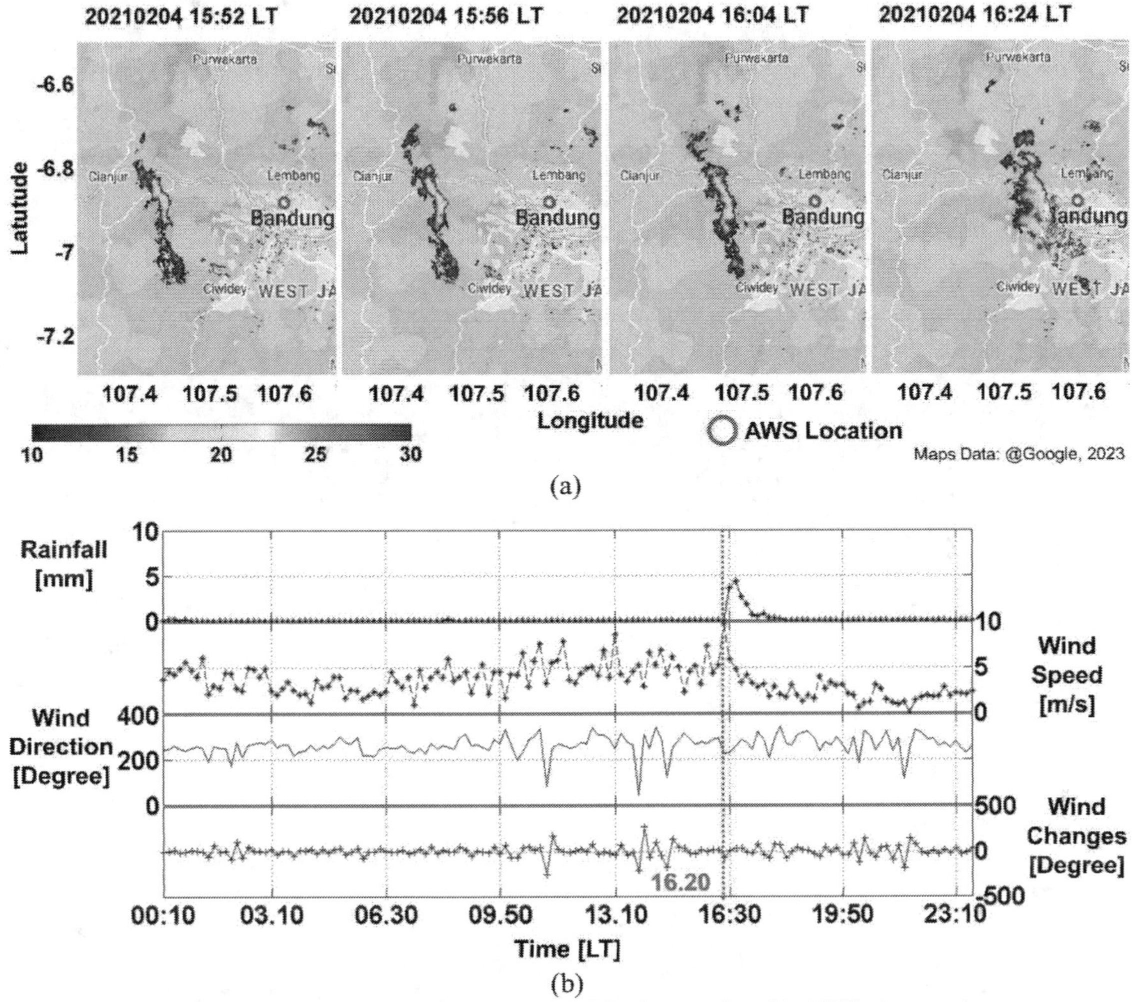

Figure 4. Heavy precipitation on Case Study 2, a) RSR observation, b) AWS observation

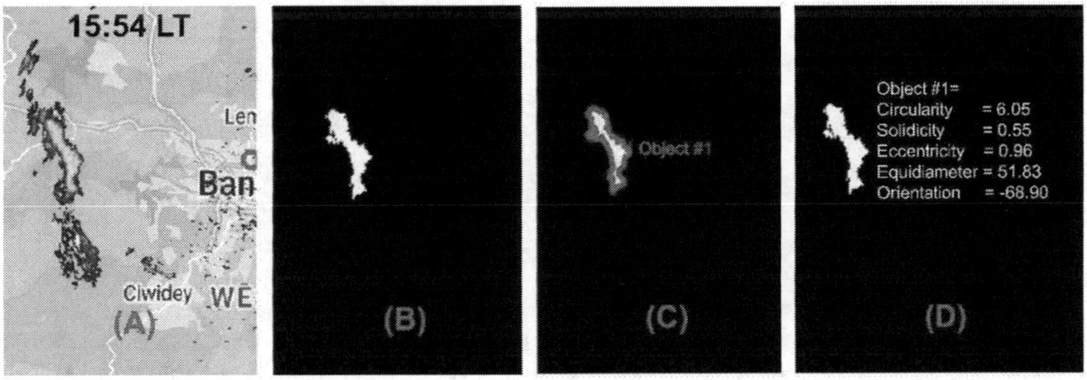

(a)

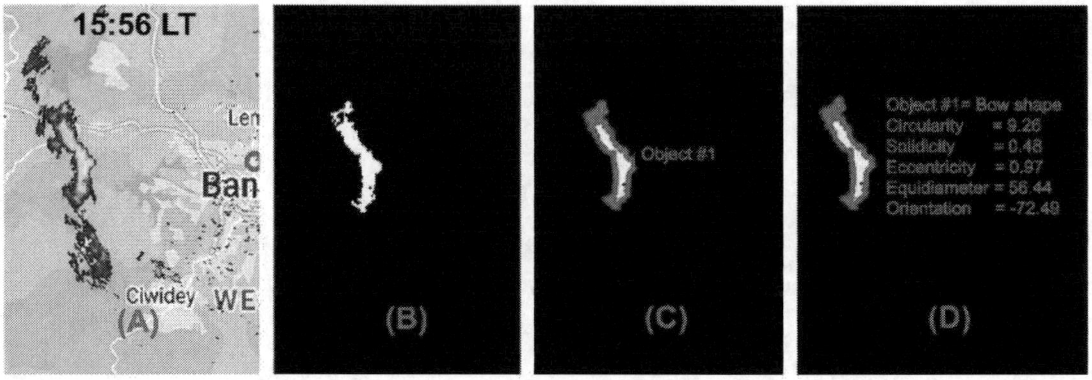

(b)

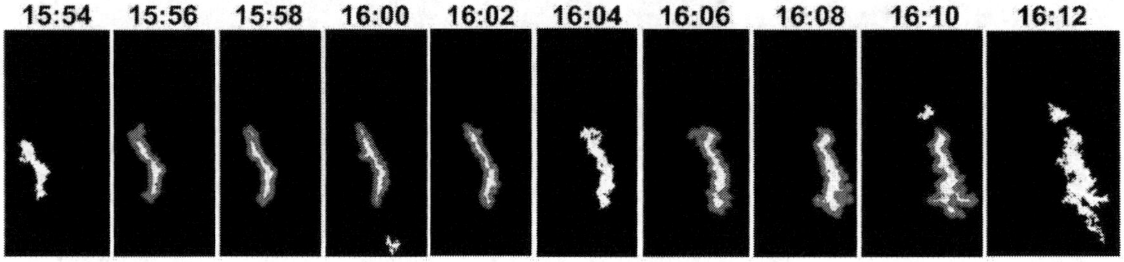

(c)

Figure 5. Bow echo detection on case study 2, a) Bow echo at 15:54 LT, b) Bow echo at 15:56 LT, c) Bow echo track. The different between Fig 5a and Fig 5b is that the parameter label value have a different color which shows that Fig 5a is not in a bow echo criteria (circularity = 6.05 which is below the threshold in Table 1).

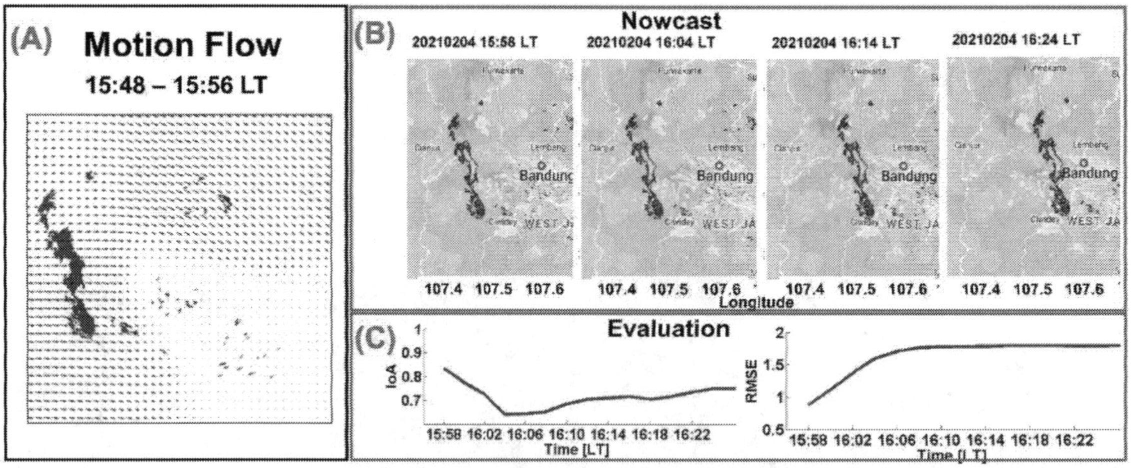

Figure 6. Nowcast bow echo case study 2, a) Motion flow from Lucas Kanade Optical Flow, b) Nowcast result, c) Evaluation between prediction and actual result.

We then checked the bow echo sequence and track it from 15:54 to 16:24 LT. However, the bow echo only detects until 16:12 LT. The proposed method did not consider the image from 16:14 to 16:24 as a bow echo shape type. The proposed method also did not detect the bow echo pattern at the 16:04 LT image. The reason is similar to the previous case study number one, where the circularities value is lower than the threshold of the proposed method. Again, we utilized the information of the early detection of bow echo at 15:56 LT. The dataset used to calculate the average motion flow are from 15:48 to 15:56 LT. Panel A in Figure 6 showed the average motion flow with the value of 5.15 (rounded to 5) pixels per 2 minutes. The translated image for the next twenty minutes starting from 15:58 LT are depicted on panel B. Meanwhile, the evaluation for the RMSE and IoA are shown in panel C Figure 6.

5. Discussion

The proposed method using a combination of three parameters (circularities, eccentricity, and orientation) is presented in this study. This method is different from other previous research [4,5,6] since it uses a combination of three parameters from the shape measurement of an object image to detect bow echo. In this section, the limitation of this study, as well as future works, are discussed.

5.1. Limitation of this study

Two case studies demonstrate the ability of the proposed method to detect heavy precipitation in the form of bow echo coming from the west towards the east side of the Bandung basin. The incoming heavy precipitation is thought to be influenced by the west monsoon condition since occurs in the February period [15]. However, other patterns of precipitation distribution occurred in the Bandung basin such as the precipitation distribution between the north-south area [16]. It will be very interesting to see whether the proposed method could observe the bow echo in different distribution patterns. Additional case studies are needed to ensure the capability and accuracy of the proposed method could detect the bow echo with different patterns.

5.2. Future Works

The early identification of bow echo has the potential to be utilized for early detection and prediction of extreme rainfall. Two case studies demonstrated a lead time of more than 20 min that could be used not only for early mitigation but also for nowcast application. This study presented the preliminary nowcast using optical flow and image translation. Evaluation of the nowcast result with the actual RSR observation using RMSE and IoA showed low accuracy and needs more improvement on the prediction method. Improvement should be done by considering the factor of growth and decay. Machine learning could also be considered to be applied to the nowcast radar image [17].

6. Conclusion

In this paper, we proposed an alternative method to detect the bow echo pattern using the combination of the measured circularities, eccentricity, and orientation of the object. The threshold value of this combination is then tested in two case studies of a bow echo. The bow echo event has a similar characteristic where is detected from the west direction moving east towards the AWS location in Bandung city. The difference is the initiation of the bow echo where the first case study comes from several weak convective cells that combine into a cell with a bow echo shape. Meanwhile, the bow echo is generated from the squall line in the second case study. The incoming heavy precipitation carried out by the bow echo is confirmed by the AWS reading on both case studies. Furthermore, an early increase in wind speed is also observed in the second case study. The proposed method could determine and quantify the first bow echo several minutes before it reaches the AWS location. Simple predictions for the next 20 minutes are carried out starting from the first detection of the bow echo.

References

[1] B A Klimowski, M R Hjelmfelt, and M J Bunkers 2004 Radar Observations of The Early Evolution of Bow Echoes *Weather and Forecasting* **19(4)** 727-734

[2] N T Atkins and M St. Laurent 2009 Bow Echo Mesovortices. Part I: Processes That Influence Their Damaging Potential *Monthly Weather Review* **137(5)** 1497-1513

[3] E LMauri and W A Gallus Jr. 2021 Differences Between Severe And Nonsevere Warm-Season, Nocturnal Bow Echo Environments *Weather and Forecasting* **36(1)** 53-74

[4] M M Kamani, F Farhat, S Wistar, and J Z Wang 2018 Skeleton Matching with Applications in Severe Weather Detection *Applied Soft Computing* **70** 1154-1166

[5] Z Fanlin and H Jinyi 2018 Automatic Detection of Bow Echoes from Weather Radar Images *2018 13th World Congress on Intelligent Control and Automation (WCICA)* IEEE

[6] A Mounier, L Raynaud, L Rottner, M Plu, P Arbogast, M Kreitz, and B Touzé 2022 Detection of bow echoes in kilometer-scale forecasts using a convolutional neural network *Artificial Intelligence for the Earth Systems* **1(2)** e210010

[7] T Sinatra, A Awaludin, F Nauval, and Purnomo C 2021 Calibration of Spatial Rain Scanner using Rainfall Depth of Rain Gauges *IOP Conference Series: Earth and Environmental Science* **893(1)** p. 012064 IOP Publishing

[8] A Awaludin, T Sinatra, G A Nugroho, and F Nauval 2021 Clutter Removal Improvement of Marine Radar for Weather Observation *AIP Conference Proceedings* **Vol. 2366 No. 1** p. 060023 AIP Publishing LLC

[9] P Hidayatulloh, B B S Wicaksana, S Darmawan, A Rahmawati, T Sinatra, and A Awaludin 2022 An Improved Rainfall Detection of Rain Radar Using High-Precision Circuit and Speckle Filtering *2022 International Conference on Radar, Antenna, Microwave, Electronics, and Telecommunications (ICRAMET)* 123-127 IEEE

[10] J Ananthanarasimhan, P Leelesh, M S Anand, and R Lakshminarayana 2020 Validation of Projected Length of The Rotating Gliding Arc Plasma Using 'Regionprops' Function *Plasma Research Express* **2(3)** 035008

[11] G A Nugroho and T Sinatra 2019 Comparison and Analysis of The First Signal Detection of Two X-Band Radar with RGB Himawari Satellite Data before Localized Heavy Rain March 17 2017 Bandung Basin Indonesia *IOP Conference Series: Earth and Environmental Science* **303(1)** p. 012056 IOP Publishing

[12] D Fleet and Y Weiss 2006 Optical Flow Estimation *Handbook of mathematical models in computer vision* 237-257

[13] S S Beauchemin and J L Barron 1995 The Computation of Optical Flow *ACM computing surveys (CSUR)* **27(3)** 433-466

[14] B D Lucas and T Kanade 1981 An Iterative Image Registration Technique with An Application to Stereo Vision *IJCAI'81: 7th international joint conference on Artificial intelligence* **2** 674-679

[15] H K B Tjasyono, R Gernowo, B H S Woro, and I Juaeni 2008 The Character of Rainfall in the Indonesian Monsoon *The International Symposium on Equatorial Monsoon System, Yogyakarta*

[16] R Fitriani, N J Trilaksono, and T W Hadi, 2019 Investigation of the Regeneration of Precipitating Convective Cloud in Basin Topography Area *IOP Conference Series: Earth and Environmental Science* **303(1)** p. 012015) IOP Publishing

[17] L Foresti, I V Sideris, D Nerini, L Beusch, and U Germann 2019 Using a 10-Year Radar Archive for Nowcasting Precipitation Growth and Decay: A Probabilistic Machine Learning Approach. *Weather and Forecasting* **34(5)** 1547-1569

SIGNIFICANT PRECIPITATION ANOMALIES OVER INDONESIA IN THE AFTERMATH OF IOD EVENTS

Chairul Nazli Nasution[1], Faiz Rohman Fajary[2], Muhammad Rais Abdillah[2] and Aditya Rakhmat Kartadikaria[3]

[1]Graduate Program in Earth Science, Faculty of Earth Sciences and Technology, Institut Teknologi Bandung, Bandung, Indonesia
[2]Atmospheric Science Research Group, Faculty of Earth Sciences and Technology, Institut Teknologi Bandung, Bandung, Indonesia
[3]Oceanography Research Group, Faculty of Earth Sciences and Technology, Institut Teknologi Bandung, Bandung, Indonesia

E-mail: chairulnazli1@gmail.com

Abstract. Indian Ocean Dipole (IOD) is one of the prominent climate phenomena affecting Indonesia. The well-documented impacts of IOD are dry and wet anomalies during positive and negative IOD events, respectively, which are clearly detected during the development and mature phases of IOD in the boreal summer to autumn (June-November). IOD impacts on other seasons were questionable because of unclear associated signals. Here, we find significant precipitation anomalies in the aftermath of IOD events, which reach a peak in the following spring (March-May). The delayed pattern is distinctive with the typical impact of IOD. During positive IOD, the quick recovery and warming of IOD-induced sea surface temperature (SST) in the eastern Indian Ocean due to increasing shortwave radiation is suggested to play a role. When the influence of El Niño-Southern Oscillation (ENSO) is linearly removed, the aftermath impacts of IOD in the western regions appear to be suppressed because of weaker positive (SST) anomalies. While the eastern regions show anomalous increase on precipitation due to anomalous increase of convergence. This finding reveals the complex impacts of IOD and ENSO and may benefit to developing a better seasonal forecast over Indonesia.

1. Introduction

Precipitation in Indonesia is modulated by many global factors, such as monsoon and other climate variabilities (Aldrian & Susanto, 2003). One of the climate variabilities that affects Indonesia's precipitation is Indian Ocean Dipole (IOD). IOD is an atmosphere-ocean phenomenon characterized by strong sea surface temperature (SST) dipole on Tropical Indian Ocean accompanied by changes in the equatorial wind direction (Saji, et al., 1999; Webster et al., 1999). IOD events can be identified by the change in Dipole Mode Index (DMI), calculated from the difference of SST anomalies on western (50°E-70°E, 10°S–10°N) and south-eastern (90°-110°E, 10°S–EQ) Tropical Indian Ocean (Saji, et al., 1999). IOD has two phases, the positive and negative phases. On positive phase, eastern Tropical Indian Ocean has lower SST anomaly than the western Tropical Indian Ocean. Easterly wind anomaly appear on Tropical Indian Ocean, and eastern Africa has increased convective activities, while western Indonesia has reduced convection (Webster et al., 1999). On the negative phase, Eastern SST anomaly is higher, and westerly wind anomaly appears with lower magnitude than positive phase's easterly wind anomaly (Vinayachandran et al., 2002).

Previous studies have been conducted on the effects of IOD on June-November to the precipitation over Indonesian region during its phases. On positive IOD, precipitation on Indonesia is lower, while it increases on negative IOD (Ashok et al., 2003; Saji et al., 1999; Vinayachandran et al., 2002). Regional view of the IOD effects show bigger impacts over the central and southern part of Sumatra, Java Island, and smaller region of influence on Borneo, Sulawesi, and Papua. When the effect of El Niño-Southern Oscillation (ENSO) is removed, affected regions reduced to only Sumatra, Java, and Borneo (Kurniadi et al., 2021). However, on other seasons the impact of IOD is still unclear. Fig. 1 shows increased precipitation after positive IOD events.

This study investigates the pattern of precipitation on Indonesian region related to IOD events, focused on other seasons after the events have occurred (December to March). We employed total correlation and partial correlation method to describe IOD and precipitation relationships using Dipole Mode Index and other variables explained on Section 2. In Section 3, the results presented and explained and then summarized in section 4.

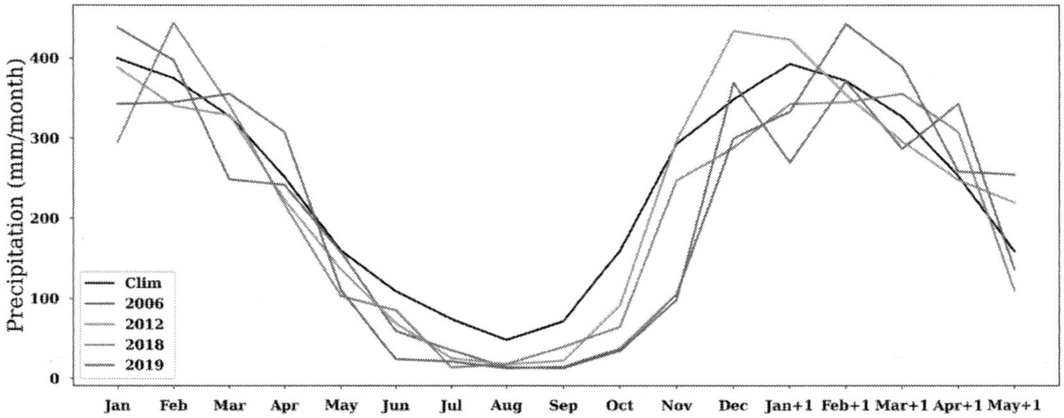

Fig. 1 Precipitation pattern during positive IOD years. Notation of +1 shows the year after IOD. Black line depicts precipitation climatology over 30 years (1981-2010). Coloured lines show precipitation during IOD occurred in 2006, 2012, 2018, and 2019.

2. Data and Method

2.1. Data

In this study, we analyzed precipitation, SST, velocity potential on 850mb, and outgoing longwave radiation (OLR) data. DMI and Nino 3.4 indices were used to determine IOD and ENSO years. We obtained SST data from ECMWF Reanalysis v5 (ERA5) Reanalysis dataset, precipitation data from Global Precipitation Climate Project (GPCP) monthly dataset, velocity potential data from NCEP/NCAR Reanalysis Monthly Means dataset, and OLR data from NOAA Interpolated Outgoing Longwave Radiation dataset. The time period for the DMI and Nino 3.4 ranges from 1979 to 2021 and for other variables we use data from 1979-2020.

The intensity of IOD is measured from SST anomaly differences between the Western Tropical Indian Ocean (50°E-70°E and 10°S-10°N) and South-Eastern Tropical Indian Ocean (90°E-110°E and 10°S-0°N). These values are used as indices for IOD (DMI) (Saji, et al., 1999). We could then identify IOD years and its phases using these indices. DMI is accessible from NOAA ESRL Physical Sciences Laboratory website (https://psl.noaa.gov/gcos_wgsp/Timeseries/DMI/).

Nino 3.4 indices are calculated from the mean of SST on Nino 3.4 region (5°N-5°S, 150°W-90°W). Five-month running mean calculated from these indices can be used to identify ENSO years and phases (Trenberth, 1997). This study uses Nino 3.4 indices calculated from Extended Reconstructed

SST version 5 dataset. These indices are accessible from NOAA Physical Science Laboratory website https://psl.noaa.gov/gcos_wgsp/Timeseries/Nino34/.

SST data were obtained from ERA5 Reanalysis Dataset with spatial resolution of 0.25° and temporal resolution of 1-month. Dataset is accessible from the website https://ecmwf.int/en/forecasts/datasets/reanalysis-datasets/era5

The GPCP Monthly Dataset provides precipitation data merged from satellite and in-situ observation with the spatial resolution of 2.5°. Dataset is accessible from the website https://psl.noaa.gov/data/gridded/data.gpcp.html.

The NCEP/NCAR Reanalysis 1 project provides atmospheric data on a spatial resolution of 2.5°. This study used this dataset to obtain monthly velocity potential data on 850 mb pressure height. Dataset is accessible from the website https://psl.noaa.gov/data/gridded/data.ncep.reanalysis.html.

OLR data were obtained from NOAA Interpolated Outgoing Longwave Radiation dataset and has the spatial resolution of 2.5° and monthly temporal resolution. OLR values in this dataset came from satellite observations with gaps filled by spatial and temporal interpolation. Data is accessible from the website https://psl.noaa.gov/data/gridded/data.olrcdr.interp.html.

2.2. Method

At first, we calculated climate variable anomalies constructed from removing its monthly climatological mean and then smoothed with three-month running average to remove intraseasonal signals from these values. We then quantified the relations between IOD and precipitation anomaly by calculating their correlation coefficients. Since precipitation anomaly is influenced by IOD and ENSO phases, partial correlation can be applied to remove the influence of ENSO. Partial correlation shows the relations between two variables (in this case, IOD and precipitation) after the relation of other variables (ENSO) is removed, or supposed to be constant (Cramer, 1946). For this study, the equation for partial correlation is given by the formula

$$\rho_{PR,DMI.N34} = \frac{\rho_{PR,DMI} - \rho_{PR,N34}\rho_{DMI,N34}}{\sqrt{(1-\rho_{PR,N34}{}^2) - (1-\rho_{DMI,N34}{}^2)}} \quad (1)$$

where $\rho_{PR,DMI.N34}$ is the partial correlation of precipitation anomaly (PR) and DMI after removing Nino 3.4 influence, $\rho_{PR,DMI}$ is the correlation between PR and DMI, $\rho_{DMI,N34}$ is the correlation between DMI and Nino 3.4, and $\rho_{PR,N34}$ is the correlation between PR and Nino 3.4. Correlation between two variables are calculated using Pearson's correlation coefficient formula:

$$\rho = \frac{n\sum PR \cdot DMI - (\sum PR)(\sum DMI)}{\sqrt{n\sum PR^2 - (\sum PR)^2} - \sqrt{n\sum DMI^2 - (\sum DMI)^2}} \quad (2)$$

where ρ is the Pearson correlation coefficient. We also applied Student's t test to determine area with over 95% confidence level. The test statistic for H_0: $\rho_{PR,DMI.N34} = 0$:

$$t = \rho\sqrt{\frac{n-2-k}{1-\rho^2}} \quad (3)$$

where k is the number of conditioning variables (Nino 3.4). Since in this study we need to determine the relationships of IOD to precipitation anomalies not only when dipole mode events occured, but also after the events have transpired, we applied both total and partial correlation of DMI to precipitation anomalies and other climate variables on the periods of September-November (lag +0 month) and two seasons after dipole mode events, December-February (lag +3 month) and March-May (lag +6).

3. Results

Partial correlation coefficients of IOD (DMI) with variables precipitation, SST, OLR and velocity potential anomalies were calculated using ENSO (Nino 3.4) as controlled variable to see relationship of IOD after ENSO influence is removed. The results were also compared with total (bivariate) correlation coefficients between DMI and other variables. Using correlation method, September-November averaged values of DMI and Nino 3.4 were used to test their relationship with September to November (lag +0 months), December to following year's February (lag +3 months), and following year's March to May (lag +6 months) averaged values of other mentioned variables.

3.1. ENSO and IOD indices

Fig. 2 shows the 3-monthly time series of DMI and Nino 3.4 index over 43 years. Both the DMI and Nino 3.4 tends to fluctuate at timescales larger than 1 year period, indicating interannual phenomena.

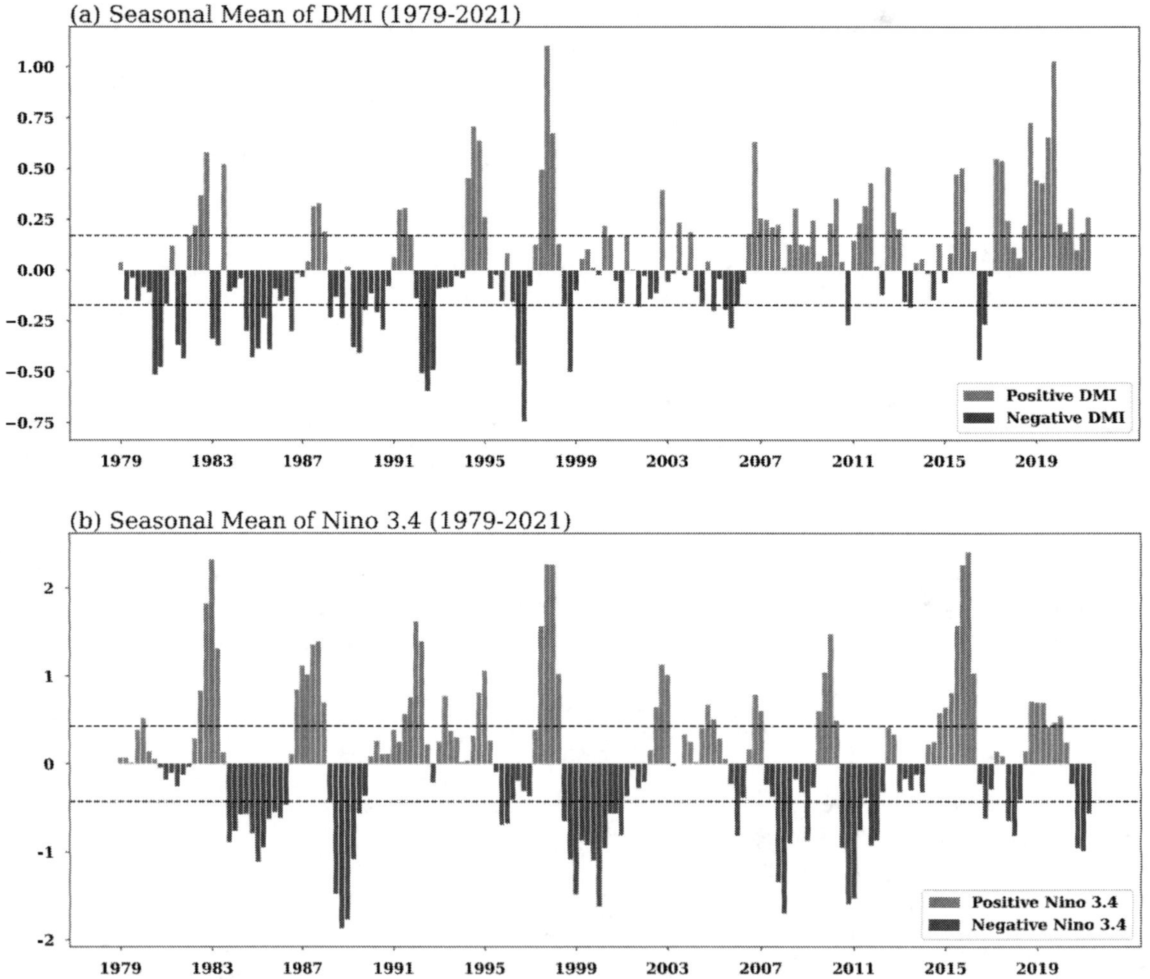

Fig. 2 Time series of seasonal mean (3 monthly) of (a) DMI and (b) Nino 3.4 indices during the period of 1979-2021. Dotted lines are the threshold for identifying years of each respective climate variabilities based on the definition of Verdon and Franks (2005) for IOD events (a) and the definition of Trenberth (1997) for ENSO events (b).

3.2. Precipitation

Total correlation of DMI and PR (Fig. 3) shows negative correlations within the majority of the Indonesian region on September-November (lag +0 month) indicating decreased precipitation anomalies on positive IOD events, while precipitation anomalies will be increased on negative IOD events. This shows similar results with previous study (Kurniadi et al., 2021). Partial correlation shows smaller regions of influence and magnitudes.

On lag +3 months (from December to February of the following year), partial correlation in the region also shows smaller region influenced by IOD compared to total correlation. The change is quite large in comparison, region with correlation magnitude ≥ 0.2 reduced by more than half of region shown from total correlation. There are small regions with positive correlations (Sumatra and Papua) that indicates positive precipitation anomalies occured after positive IOD events and negative precipitation anomalies occured after negative IOD events on these regions.

On lag +6 months (following year's March to May), total correlation shows majority of Indonesian regions are positively correlated with DMI, especially around the islands of Java and Sumatra, indicating increased precipitation anomalies occured after positive IODs, and reduced precipitation anomalies after negative IODs. Partial correlation shows smaller correlation magnitudes around Java and Sumatra, while eastern Indonesian region has stronger relationship in terms of correlation magnitude compared to the results shown by total correlation. Java and Sumatra region has the magnitude around 0.2-0.4 while eastern Indonesian region has magnitudes above 0.4.

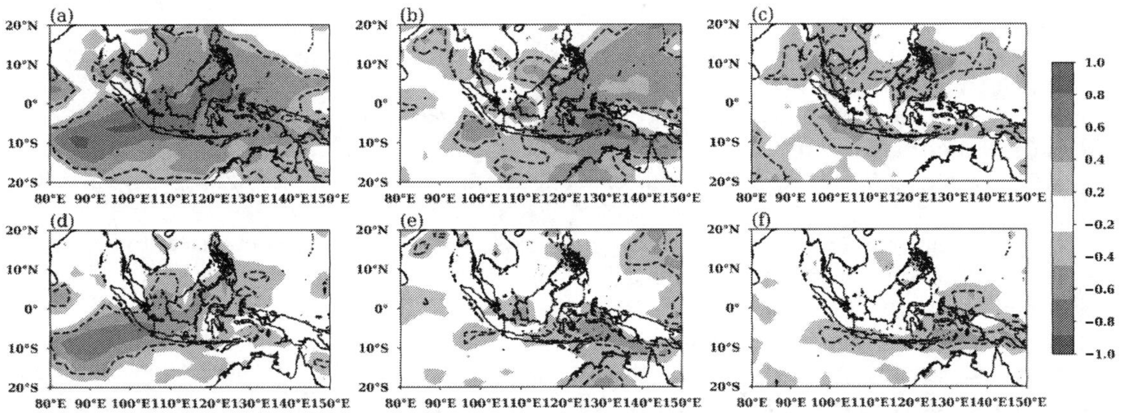

Fig. 3 Spatial distribution of time-lagged total correlation coefficients (a-c) and partial correlation coefficients (d-f) of DMI and PR. From left column to right, average of September-November DMI values are correlated with average precipitation anomalies on September-November (a,d), December-February (b,e), and March-May (c,f). Dotted lines show area with 95% confidence level.

3.3. Sea Surface Temperature

Fig. 4 Shows the total and partial correlation of DMI and SST anomalies. On lag +0 month, total correlation shows most of the sea region has negative correlation with magnitude ≥ 0.6, which indicates the heating of sea surface around indonesia is highly influenced by the IOD. Partial correlation shows the region affected by IOD is reduced in size, and the magnitude of the correlation coefficients are reduced.

On lag +3 and +6 months of total and partial correlation shows reduced area of influence with positive correlation coefficients after the effect of ENSO is removed. This indicates stronger heatings occured after IOD events and the heatings are influenced both by IOD and ENSO events. Stronger heatings affects evaporations and cloud formations on these affected regions.

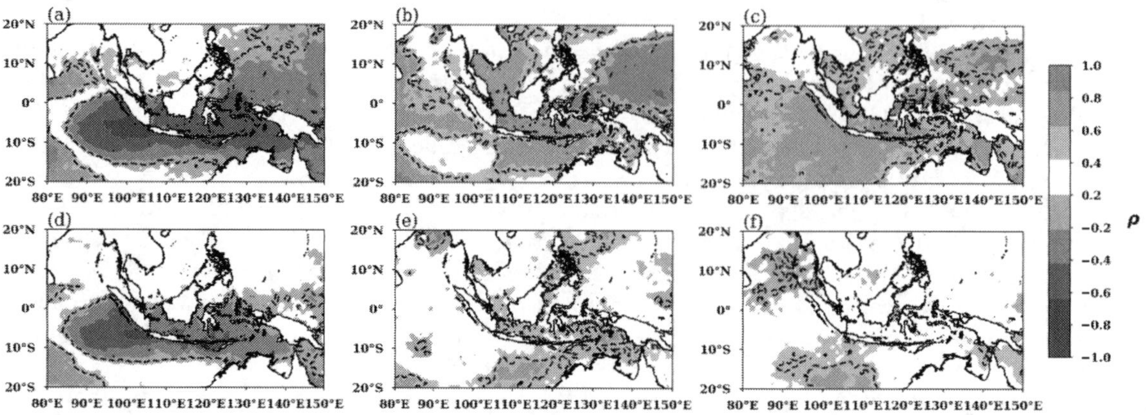

Fig. 4 Same as in Fig. 3 but with SST anomalies.

3.4. Outgoing Longwave Radiation

Total correlation of OLR (Fig. 5) on lag +0 months shows positive correlation coefficients between DMI and OLR anomalies on Indonesian region. On positive IOD events, the formations of clouds are reduced while on negative IOD events cloud formations are increased. Partial correlation result has smaller magnitudes of coefficient compared to total correlations, but still yields positive signs around the region.

On lag +3 months partial correlation result yields smaller regions of influence compared to total correlation. This means there is significant influence of ENSO on the OLR anomalies. From December to February after Positive IOD events more clouds appear on the region, smaller in magnitude than during the IOD events which will reduce the heating of the surface. The opposite pattern will occur after negative IODs, with smaller magnitude compared to during negative IOD events.

On lag +6 months total correlation yields large region with positive correlation around Java Island, and after ENSO effect is removed, the affected region around Java Island is reduced. Previous test with SST shows that the area around Java Island has lower SST heating after the effect of ENSO is removed which explains smaller magnitude on the region. Reduced heating will lead to smaller evaporation, and resulted in smaller precipitation anomaly around March-May. However, the pattern in the eastern region of Indonesia is different. After the effect of ENSO is removed, the region has negative correlation coefficients compared to total correlation result.

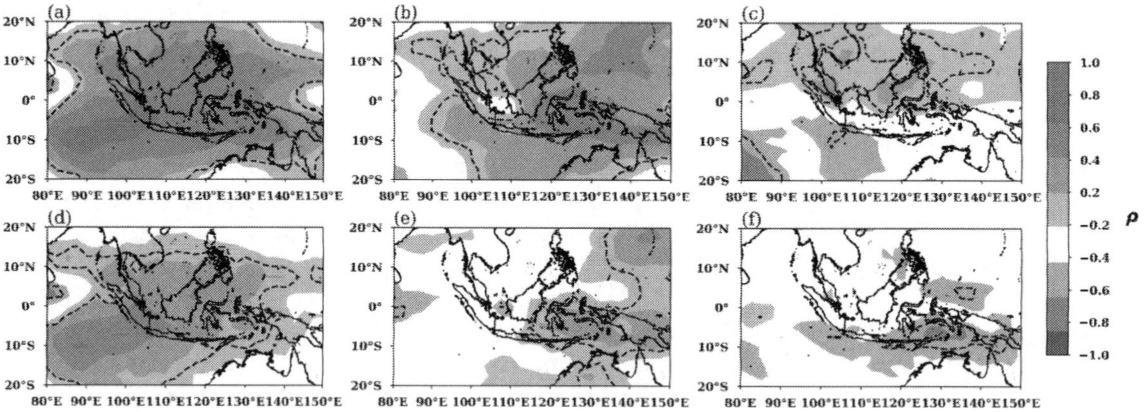

Fig. 5 Same as Fig. 3 but with OLR anomalies.

3.5. *Velocity Potential*

Fig. 6 shows total and partial correlation of DMI and velocity potential anomaly. During positive (negative) IOD events, Indonesian region has decreased (increased) convection. Partial correlation shows smaller magnitude which indicates the decrease (increase) of convection activities are also influenced by ENSO.

On lag +3 months partial correlation resulted in increased region with positive correlation coefficients on Sumatra, Java and part of Borneo. And on lag +6 months,there are regions with positive correlation coefficients on eastern Indonesia. This means after positive IOD events convection around these areas are increased, causing more clouds to appear around the eastern region.

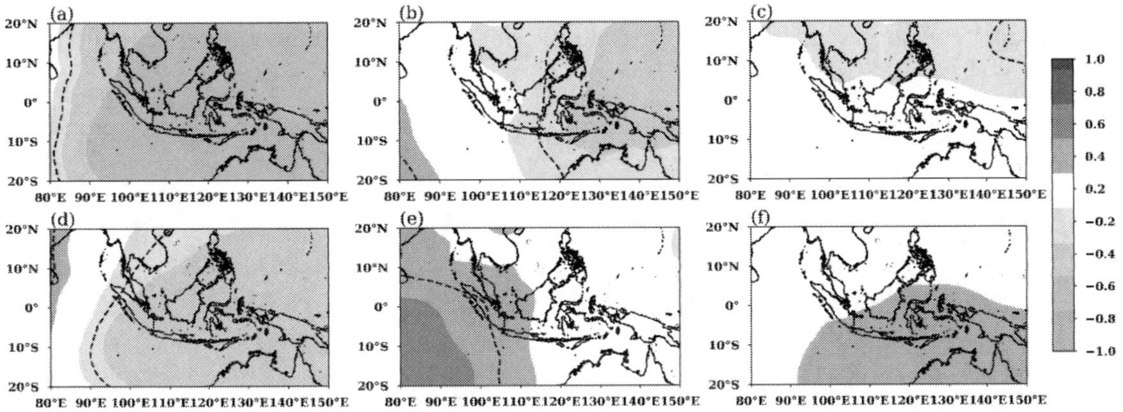

Fig. 6 Same as in Fig. 3 but for velocity potential anomalies.

4. Summary

The result of our analysis shows that after positive IOD events, there are significant precipitation anomaly on Indonesian region. After the effect of ENSO is removed, the magnitude and the affected region of the anomaly around Java Island is reduced, while around eastern Indonesia precipitation anomaly increased in both magnitude and affected region.

Analysis of variables SST, OLR, and velocity potential shows the two regions mentioned above has different mechanisms that affects their precipitation anomalies. Around Java Island, after the events of IOD changes on the SST heatings occured. After the effect of ENSO is removed, the magnitude is reduced and the decreases the rate of the cloud formation. Eastern Indonesian region shows different mechanism, after the effect of ENSO is removed, there are increased convective activities which resulted in more coulds to form above the region.

References

[1] Saji N H, Vinayachandran P N and Yamagata T 1999 A dipole mode in the tropical Indian Ocean *Natura* **401** 360–3

[2] Chang C-P, Wang Z, McBride J and Liu C-H 2005 Annual Cycle of Southeast Asia — Maritime Continent Rainfall and the Asymmetric Monsoon Transition *Journal of Climate* **18** 287–301

[3] Webster P J, Moore A M, Loschnigg J P and Leben R R 1999 Coupled ocean–atmosphere dynamics in the Indian Ocean during 1997–98 *Nature* **401** 356–60

[4] Aldrian E and Susanto R D 2003 Identification of three dominant rainfall regions within Indonesia and their relationship to sea surface temperature *International*

Journal of Climatology **23** 1435–52

[5] Kurniadi A, Weller E, Min S-K and Seong M-G 2021 Independent ENSO and IOD impacts on rainfall extremes over Indonesia *International Journal of Climatology* **41** 3640–56

[6] Vinayachandran P N, Iizuka S and Yamagata T 2002 Indian Ocean dipole mode events in an ocean general circulation model *Deep Sea Research Part II: Topical Studies in Oceanography* **49** 1573–96

[7] Verdon D C and Franks S W 2005 Indian Ocean sea surface temperature variability and winter rainfall: Eastern Australia *Water Resources Research* **41**

[8] Ashok K, Guan Z and Yamagata T 2003 Influence of the Indian Ocean Dipole on the Australian winter rainfall *Geophysical Research Letters* **30**

[9] Cramer H 1946 *Mathematical Methods of Statistics* (Princeton University Press)

[10] As-syakur Abd R, Adnyana I W S, Mahendra M S, Arthana I W, Merit I N, Kasa I W, Ekayanti N W, Nuarsa I W and Sunarta I N 2014 Observation of spatial patterns on the rainfall response to ENSO and IOD over Indonesia using TRMM Multisatellite Precipitation Analysis (TMPA) *Int. J. Climatol* **34**

[11] BLAIR T A 1918 PARTIAL CORRELATION APPLIED TO DAKOTA DATA ON WEATHER AND WHEAT YIELD *Monthly Weather Review* **46** 71–3

[12] Saji N H and Yamagata T 2003 Possible impacts of Indian Ocean Dipole mode events on global climate *Clim Res* **25** 151–69

[13] Sobel A H, Burleyson C D and Yuter S E 2011 Rain on small tropical islands *Journal of Geophysical Research* **116**

[14] Klein, et al. 1999 Remote sea surface temperature variations during ENSO: Evidence for a tropical atmospheric bridge *Journal of climate* **12**

[15] Zhang T, Yang S, Jiang X and Zhao P 2016 Seasonal–Interannual Variation and Prediction of Wet and Dry Season Rainfall over the Maritime Continent: Roles of ENSO and Monsoon Circulation *Journal of Climate* **29** 3675–95

[16] Forbush S E, Pomerantz M A, Duggal S P and Tsao C H 1983 Statistical considerations in the analysis of solar oscillation data by the superposed epoch method *Solar Physics* **82** 113–22

[17] Wilks D S 2006 *Statistical Methods in the Atmospheric Sciences. 2nd Edition* (London: Academic Press)

[18] Saji N H and Yamagata T 2003 Structure of SST and Surface Wind Variability during Indian Ocean Dipole Mode Events: COADS Observations *Journal of Climate* **16** 2735–51

[19] Alexander M A, Bladé I, Newman M, Lanzante J R, Lau N-C and Scott J D 2002 The Atmospheric Bridge: The Influence of ENSO Teleconnections on Air-Sea Interaction over the Global Oceans *Journal of Climate* **15** 2205–31

[20] Trenberth K E 1997 The definition of El Nino *Bulletin of the American Meteorological Society* **78** 2771–8

THE EFFECT OF RAINFALL ANOMALIES DUE TO ENSO ON TEA PRODUCTIVITY IN WEST JAVA INDONESIA (CASE STUDY: BANDUNG REGENCY)

Rahmatika[1], Joko Wiratmo[1]

[1]Meteorology Study Program, Faculty of Earth Sciences and Technology, Bandung Institute of Technology

Email: joko.wiratmo@meteo.itb.ac.id

Abstract. This study aims to determine the effect of rainfall anomalies due to ENSO on tea productivity in Bandung Regency West Java Indonesia. This study used data on monthly rainfall, tea productivity, and the Nino 3.4 index from 2003-2020. The methods used are Fast Fourier Transform (FFT), Butterworth Bandpass Filter, and Inverse Fast Fourier Transform (IFFT) for the processing of these three data. The results of this study indicate that the rainfall pattern in the Bandung Regency area in 4 districts (Rancabali, Pasirjambu, Pangalengan, and Kertasari) is monsoon. ENSO correlation to rainfall anomalies and rainfall anomalies correlation to tea productivity in each district is low. The highest correlation between ENSO to rainfall anomalies is found in Pangalengan District (-0.309) and the highest correlation between rainfall anomalies to tea productivity is found in Kert asari District for lag 0 and 1 which are -0.304 and -0.285 respectively. The low correlation (-0.148 to - 0.309) showed that the effect of rainfall anomalies due to ENSO on tea productivity of each district in Bandung Regency has a small effect but with a significance value of less than 5% (0.000 to 0.030) which means it has a significant relationship.

Keywords: ENSO, Precipitation, Tea productivity

1. Introduction

The territory of Indonesia is an area traversed by the equator and is located between two continents and oceans, as well as the place of convergence of two major circulations in the world, namely Walker and Hadley circulation which greatly affects the level of rainfall variability in Indonesia. This condition causes Indonesia as a tropical region to have high rainfall variability both in space and time scales. Moreover, both the spatial and temporal rainfall variability is also amplified: higher intensity and more frequent rainfall have occurred during the wet season and longer drought during the dry season (Held & Soden 2006;Zhang & Fueglistaler 2019) An extensive study using a global rainfall data series had suggested an increase in global annual maximum daily precipitation (Westra, Alexander & Zwiers 2013). Positive trends were also indicated through regional studies, such as in northern Australia (Roderick, Wasko & Sharma 2019), India (Goswami et al 2006), and many other regions. (As-syakur et al 2013) observed an increasing rainfall over Indonesia, especially in Kalimantan, Java, Sumatra, and Papua, from 1998 to 2010, based on the TRMM multisatellite dataset.

According to Aldrian and Susanto (2003), the rainfall spectrumin Indonesia shows the dominance of annual and semi-annual climate signals. Rainfall in Indonesia is influenced by various weather or climate factors, both globally and locally (Yulihastin et al., 2010). Globally, rainfall conditions in Indonesia are influenced by the Asian- Australian monsoon, ENSO, and Indian Ocean Dipole (IOD) phenomena (Zubaidah, 2012). ENSO and Indian Ocean Dipole (IOD) phenomena are related to inter-annual rainfall variations in Indonesia. The main factor affecting climate variability in Indonesia is ENSO. ENSO is the existence of Walker circulation disorder in the

Pacific Ocean (Aldrian and Susanto, 2003). The El Nino phenomenon is caused by the warming of the sea surface temperature in the equatorial Pacific Ocean which causes a decrease in rainfall so that the dry season is longer than the rainy season Oceanwhich causes a decrease in rainfall so that the dry season is longer than the rainy season Ocean which causes a decrease in rainfall so that the dry season is longer than the rainy season while La Nina is caused by the cooling of the sea surface temperature in the equatorial Pacific Ocean which causes an increase in rainfall so that the rainy season is longer than the dry season (As-syakur, 2010). In addition, the high variability of rainfall is also influenced by local conditions. The amount of rainfall variability can have an impact on the agricultural sector which indirectly affects food, welfare, life, and social economy in the community (Baidu et al., 2017).

West Java province is a major tea producer that produces 79% of tea in Indonesia. West Java province has a plantation area of 484.234 ha. A totalof 84.70 thousand ha are tea plantations. PT. Perkebunan Nusantara VIII exports 40% of the tea that has been processed abroad. According to the West Java provincial plantation office in 2020, WestJava produced 101.441 tons out of 128.016 tons, equivalent to 79% of national tea production. Bandung Regency produced 35.633 tons, equivalent to 35% of West Java province's tea production in 2020. This indicates that Bandung Regency is one of the largest tea producers in Indonesia. However,tea production in West Java over the past 10 years (2011-2020), especially in Bandung Regency varies so that there is a decrease and increase that causes instability of tea productivity shown in Figure 1.1.

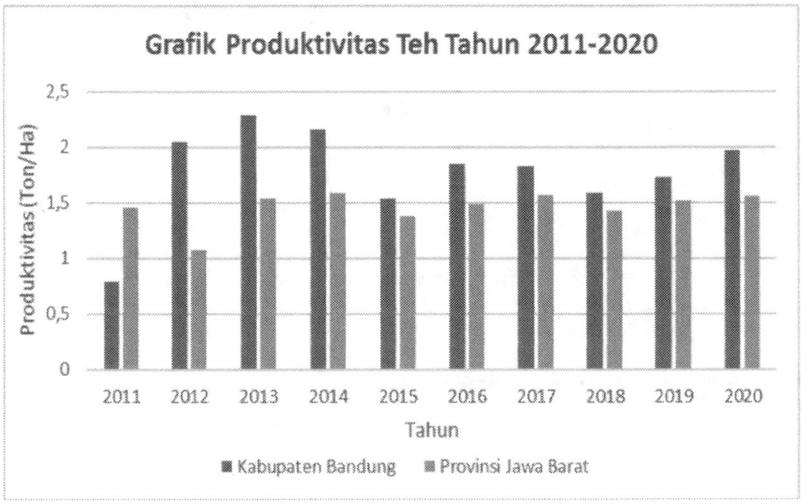

Figure 1.1 Graph of tea productivity in Bandung Regency and West Java Province in 2011- 2020.

The 1987 drought reported by Sukasman (1987) showed that almost all tea plants in West Java experienced severe drought resulting in decreased production and high mortality in young plants, especially in the lowlands. Then, when the drought occurred in 1997 the tea plant experienced a drought that resulted in a decrease in production to reach 53% (Kartawijaya, 1992). During the period 2005- 2014 El-Nino phenomenon occurred with the strongest intensity in late 2009 and early 2010 (Dalimoenthe et al., 2016). According to Dalimoenthe (2016), there was a decrease in the amount of rainfall in the highlands (Bandung) in 2010-2014 (after El-Nino) compared to 2005-2009 (before El Nino). Also, in 2015 there were El-Nino symptoms with very strong intensity, with the Nino index reaching its peak at 2.483 (BMKG, 2015). Rainfall intensity in West Java and sea surface temperatures on the north and south coasts of WestJava decreased during the 2015 El Nino.

Tea productivity in Bandung Regency is influenced by climate variability, one of which is ENSO. In 2012 there was an event of La Nina effect on the length of the rainy season and high rainfall that caused the productivity of tea fell in Bandung Regency. Meanwhile, in 2015 there was an El Ninoeffect on the length of the dry season and low rainfall which caused annual rainfall in Bandung Regency of 2311 mm and had an impact on slowing the rate of decline in tea productivity (Rahman, 2017). It also proves that El Nino and La Nina play

a role in reducing and increasing rainfall in Bandung Regency. Therefore, there is a need for research onthe effect of rainfall anomalies due to (only) ENSO on tea productivity in West Java with case study of Bandung Regency). This means that the research consider only ENSO not the others and also be done for very limited area.

2. Data and Methods

2.1 Data

In this study, Nino 3.4 index data from 2003-2020 from KNMI Climate Explorer is the result of data processing taken from NOAA's SST Optimum Interpolation (OI) version 2 (v2) with monthly temporal resolution. Then, rainfall data from ERA5-Land reanalysis data with a spatial resolution of $0.1° \times 0.1°$ and monthly temporal resolution from 2003-2020. In addition, monthly tea productivity data from the Sinumbra plantation in Rancabali District, Rancabolang plantation in Pasirjambu District, Pasirmalang plantation in Pangalengan District, and Sedep plantation in Kertasari District in Bandung Regency in 2003-2020 were obtained from PT Perkebunan Nusantara VIII, Bandung, West Java. The study area was in Bandung Regency, West Java, located at 6° 49 '– 7° 18' LS and 107° 14' – 107°56 ' BT with an area of 176.238,67 ha.

Figure 2.1 Administrative map of Bandung Regency (Source: Bandung Regency RPJPD 2005-2025)

(Source: Bandung Regency RPJPD 2005-2025)

2.2 Methods

This study used methods to see the impact of ENSO on rainfall anomalies and rainfall anomalies on tea productivity by spectral analysis method (FFT), bandpass filter, IFFT, and correlation. First of all, there is a process called "calculating rainfall anomalies". The rainfall anomaly is the difference between the actual value of precipitation and the average value of precipitation for the long term (climatology). Rainfall anomalies are calculated using Equation 2.1. (Source: https:// cds.climate.copernicus.eu).

Rainfall anomaly $= Xi - (\bar{X})$ .... (2.1)

where:

Xi = amount of rainfall in month i (mm)

$\bar{X}$ = Long-term monthly average rainfall (January 2003-December 2020)

The methods used in this study are as follows.

2.2.1 Spectral Analysis Method

The spectral analysis method is a method to transform from the time domain to the frequency domain, so it looks at periodicity to then determine the type of weather patterns involved as a peak spectral energy at a particular frequency. The spectral analysis method used is Fast Fourier Transform (FFT). This method is used in the data Nino 3.4 and rainfall anomalies to see the periodicity of ENSO events.

2.2.2 Butterworth Bandpass Filter

The butterworth Filter is one of the most widely used classical filter methods (Paarman, 2003). Butterworth

Bandpass Filter method is used to separate the signal, weaken, eliminate certain frequency signals that are not wanted to escape the desired signal, then determined the value of the low cutoff, high cutoff, and the order of the bandpass. Determination of cutoff value in this study is reviewed from the results of Fast Fourier Transform (FFT). This method is used in Nino 3.4 index data, rainfall anomalies, and tea productivity to see the signal of ENSO events.

2.2.3 Invers Fast Fourier Transform (IFFT)

The inverse Fast Fourier Transform (IFFT) methodis the opposite of the Fast Fourier Transform (FFT) used in digital signal processing applications. IFFT is a method of converting the frequency domain into the time domain. In other words, in the IFFT stage, there will be a process of returning the signal from the frequency domain to return to the time domain after the FFT-bandpass filter process.

2.2.4 Correlation

Correlation is one of the statistical analysis techniques used to find the relationship between two variables (independent variables with dependent variables) that are quantitative. Interpretation the value of correlation can be seen in the box below.

Table 2.1 Interpretation of the strength ofcorrelation values (Sugiyono, 2017).

Coefficient Interval Correlation	Relationship Strength Levels
0.00 – 0.199	Very low
0.20 – 0.399	Low
0.40 – 0.599	Enough
0.60 – 0.799	Strong
0.80 – 1.000	Very strong

The hypotheses used in a correlation analysis are as follows:
1. The significance value of $t < \alpha = 0{,}05$ (5%), then H_0 is rejected and H_1 is accepted means that there is a significant influence between the independent variable to dependent variable.
2. The significance value of $t > \alpha = 0{,}05$ (5%), then H_0 is accepted and H_1 is rejectedmeans that there is no significant influence between the independent variable to dependent variable.

3. Results

3.1 Rainfall Pattern in Bandung Regency

Based on the results of the calculation of the average monthly rainfall for 18 years (2003-2020) in Figure 3.1 shows that the rainfall pattern of four districts in Bandung Regency, West Java representing each estate (Rancabali, Pasirjambu, Pangalengan, and Kertasari) has a monsoon rainfall pattern that is unimodial because of the characteristics that are strongly influenced by the Asian-Australian monsoon system, these results November- December (Rancabali, Pangalengan, Pasirjambu) and the peak of the dry season occurs in June-August. Then, in Kertasari District, the peak of the rainy season occurs in December-February because it coincides with the west monsoon while the peak of the dry season occurs in June-August to coincidewith the east monsoon.

The Western Monsoon occurs around October to April. The Western Monsoon occurs due to the position of the sun in the southern hemisphere. As a result, the Australian Continent experiences summer so that it has low pressure, while the Asian Continent is colder so that it has high pressure. Due to the difference in pressure between the two continents, the wind moves from the Asian Continent to the Australian continent. The wind thatblows from the Asian Continent through the vast ocean that is the Pacific Ocean and the South China Sea then the wind brings a lot of water vapor that causes rain in Indonesia and its surroundings

The Eastern Monsoon occurs around April to October. Eastern Monsoon occurs due to the position of the sun in the northern hemisphere. As aresult, the Asian Continent experiences summer so that it has low pressure, while the Australian Continent is colder so that it has high pressure. Due to the difference in pressure between the two continents, the wind moves from the Australian Continent to the Asian Continent. The wind blows from the Australian Continent through the desert in northern Australia and only passes through a little ocean, the wind is dry and only passes a little water vapor that causes drought in Indonesia and surrounding areas.

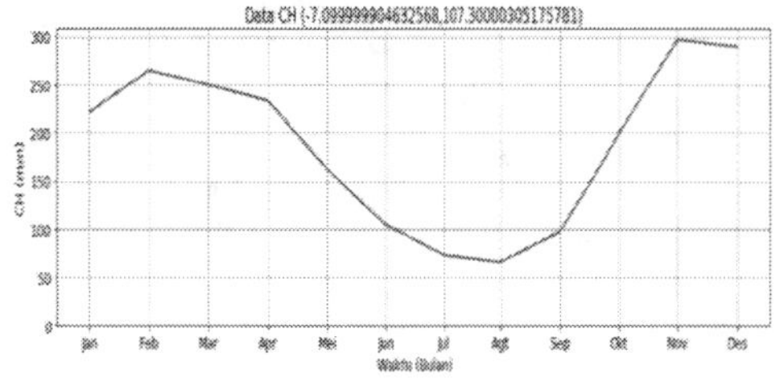

(a)

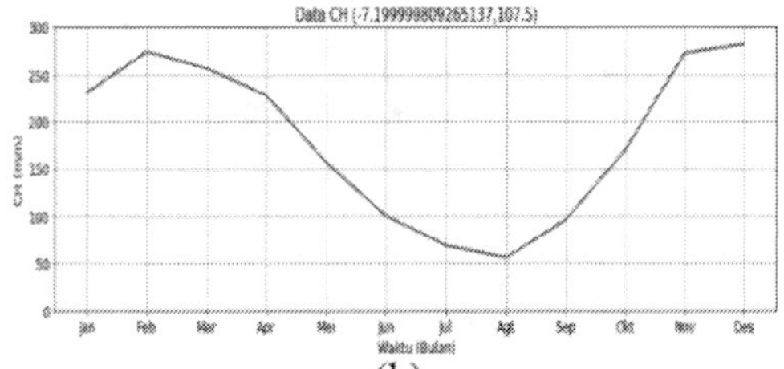

(b)

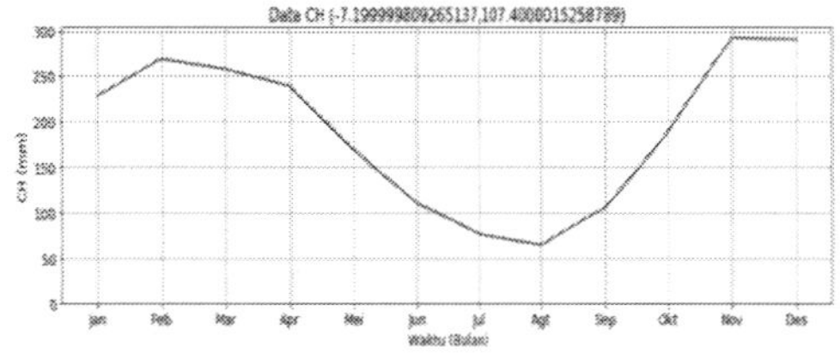

(c)

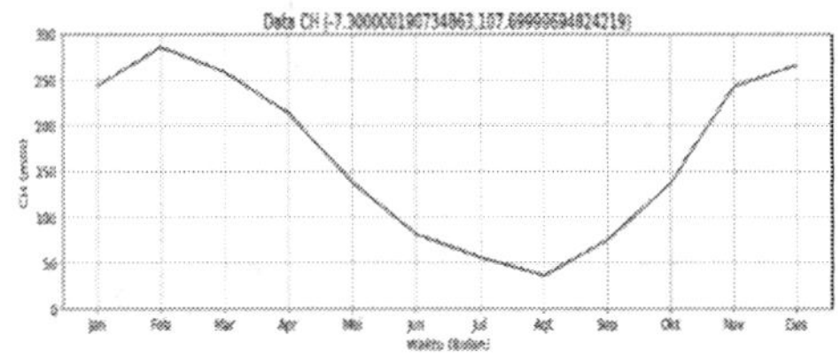

(d)

Figure 3.1 The results of the graph plot the averagemonthly rainfall in four districts in Bandung Regency ((a) Rancabali. (b) Pasirjambu, (c) Pangalengan, and (d) Kertasari).

3.2 *The Effect of Interannual Phenomena onRainfall Anomalies*

To determine the effect of interannual phenomena, namely El Nino and Southern Oscillation (ENSO) on rainfall anomalies in four sub-districts in Bandung Regency, West Java is done with steps as below.

3.2.1 FFT-Bandpass Nino 3.4

The results of FFT-bandpass and the original FFT data from Nino 3.4 are shown in Figure 3.3, which shows that the results of FFT-bandpass and FFT areappropriate

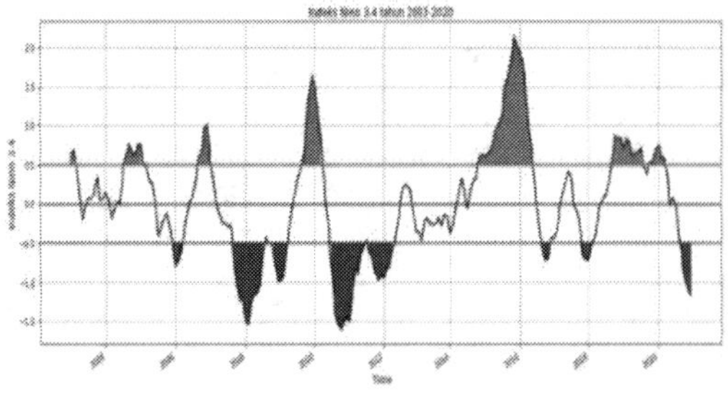

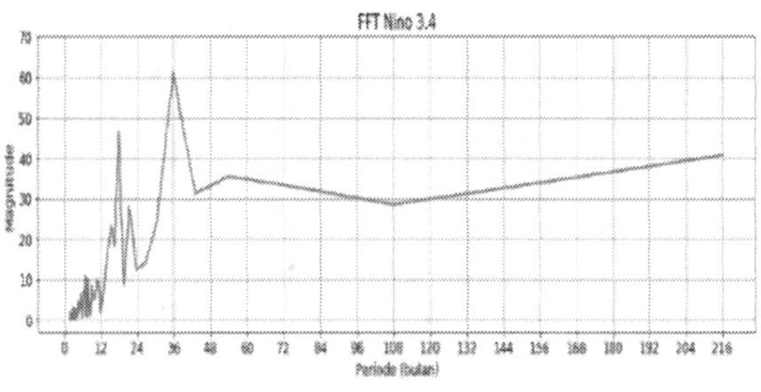

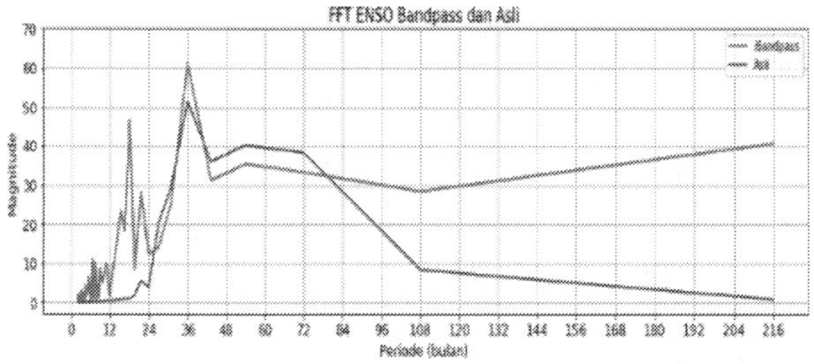

Figure 3.3 Result of the Nino 3.4 index plot for 2003-2020 (28 years), (b) comparisonofFFT-bandpass and FFT original Nino 3.4 data.

Figure 3.3 (a) and (b) show that the red area is an El Nino event, and the blue area is a La Nina event that shows the pattern produced by FFT-bandpass already represents the original data (made overlay plots index Nino 3.4 and FFT-bandpass). It is because the peaks and valleys of the bandpass filter are in the places that correspond to the original data.

El Nino and La Nina events according to the Ocean Nino Index (ONI) in the region of 3.4 in 2003-2020. There were 7 El Nino events and 8 La Nina events from 2003-2020 shown in Table 3.1.

Tabel 3.1 El Nino and La Nina events in 2003-2020 (Source: https:// ggweather.com/ enso/ oni.htm).

El Nino and La Nina Events 2003-2020					
El Nino			**La Nina**		
Weak	**Moderate**	**Strong**	**Weak**	**Moderate**	**Strong**
2004-2005	2002-2003	2015-2016	2005-2006	2011-2012	2007-2008
2006-2007	2009-2010		2008-2009	2020-2021	2010-2011
2014-2015			2016-2017		
2018-2019			2017-2018		

The results of overlay FFT-bandpass and FFT original data in Figure 3.3 (b), selected 1/24 as high cutoff and 1/96 as low cutoff, the value is taken based on interannual phenomena that occur around 2-8 years. From the figure shows the results of spectral analysis (FFT) Nino 3.4 there is the strongest period (peak spectral energy) which is 36, 54 and 72 means that there is interannual variation corresponding to the period of the ENSO phenomenon cycle (El Nino and La Nina) which is 3-6 years where the period or phenomenon is quite influential at the point of study.

3.2.2 FFT-Bandpass Rainfall Anomaly

The results of FFT-bandpass and FFT rainfall anomalies in four districts in Bandung Regency showed the suitability of the pattern as shown in Figure 3.4 below.

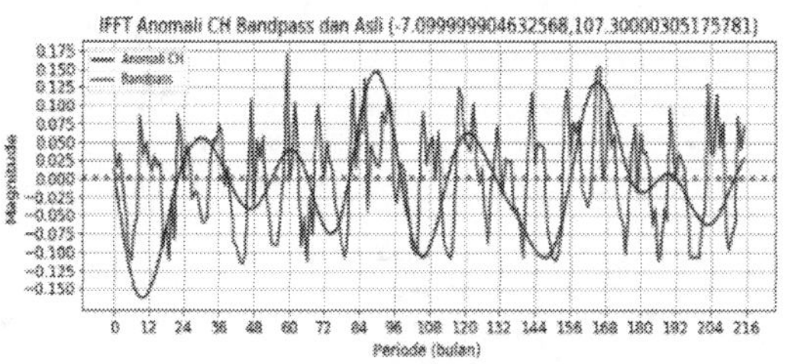

(a) Rancabali District

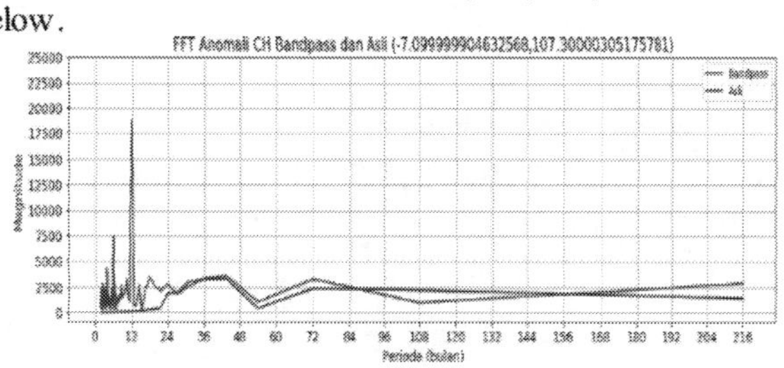

(b) Rancabali District

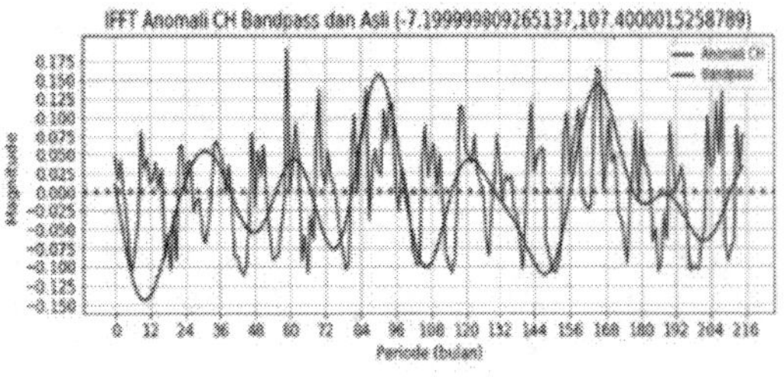

(c) Pasirjambu District

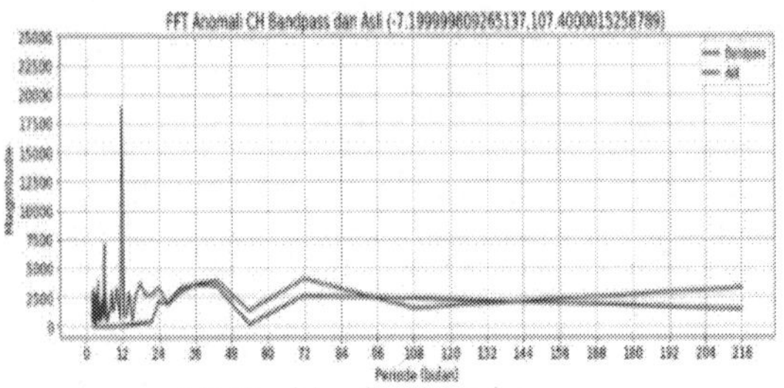

(d) Pasirjambu District

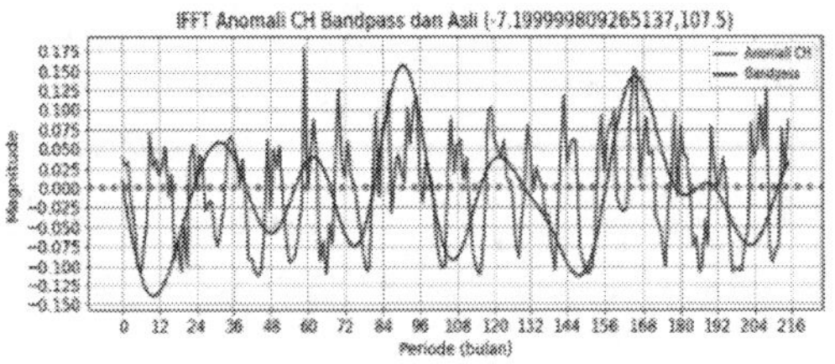

(e) Pangalengan District

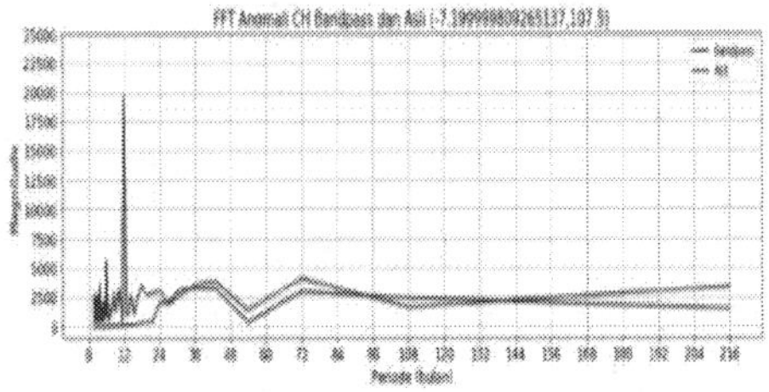

(f) Pangalengan District

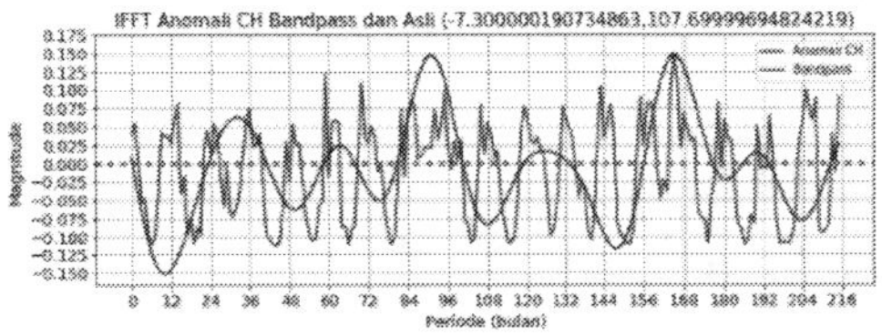

(g) Kertasari District

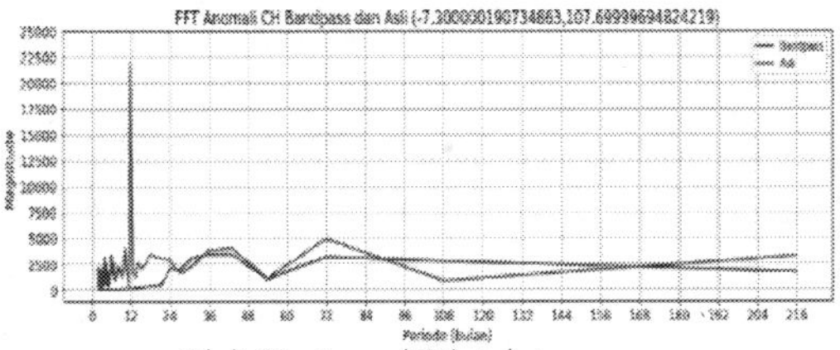

(h) Kertasari District

Gambar 3.4 The results of time series plots, (a), (c),(e), (g) comparison of IFFT-bandpass rainfall anomalies and original rainfall anomaly data of each sub-district in Bandung Regency, (b), (d), (f), (h) FFT of bandpass data and original rainfall anomaly data of each sub-district in Bandung Regency.

3.2.3 The effect of ENSO Phenomenon on RainfallAnomaly

The results of the IFFT-bandpass Nino 3.4 and the rainfall anomaly in each sub-district are presented in the plot of Figure 3.5. The black and blue horizontal solid wave lines are Nino 3.4 and rainfall anomalies, respectively. The red (greater than 0.5) and blue (smaller than -0.5) dotted lines show the thresholds for El Nino and La Nina events, respectively. The red and blue vertical lines indicate El Nino and La Nina occurrences, respectively.

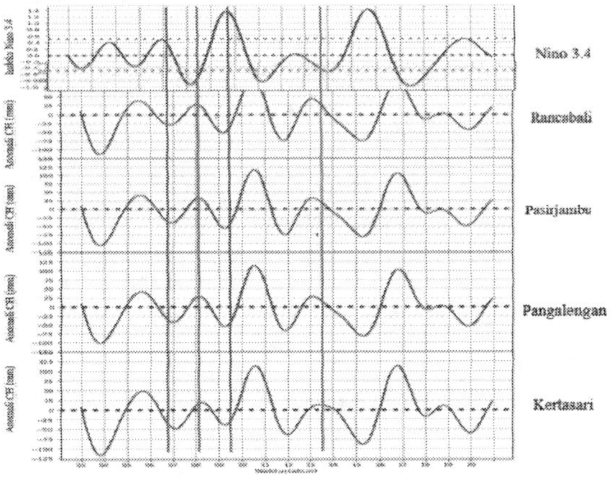

Figure 3.5 The result of the IFFT-bandpass Nino 3.4 time series plot and rainfall anomalies in each sub-

district in study are in 2003-2020.

Based on Figure 3.5 at the time of the El Nino was in 2009 and strong El Nino in 2015 in general in the four districts at the point of the study there was a negative rainfall anomaly means a slight decrease in rainfall. Meanwhile, in the strong La Nina events in 2008 and 2011 in general, positive rainfall anomalies mean an increase in rainfall.

3.3 The effect of Rainfall Anomalies due to ENSO on Tea Productivity

Analysis of the effect of rainfall anomalies due to ENSO on tea productivity was conducted in the districts of Rancabali, Pasirjambu, Pangalengan, and Kertasari from the results of IFFT-bandpass Nino 3.4, rainfall anomalies, and tea productivity.

3.3.1 Rancabali District

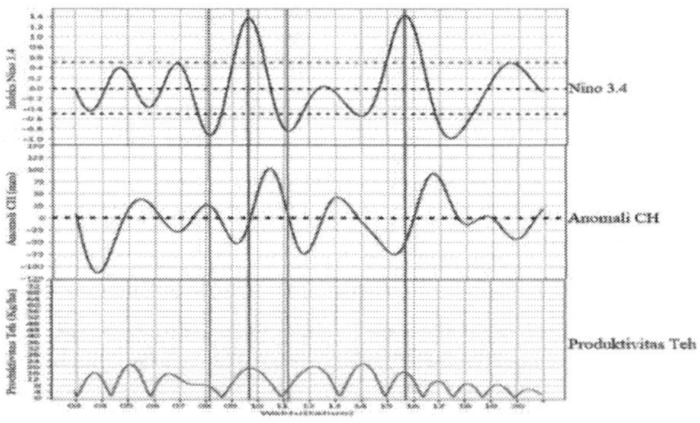

Figure 3.6 The results of the IFFT-bandpass Nino 3.4 times series plot, rainfall anomalies, and tea productivity in Rancabali District in 2003-2020.

Based on Figure 3.6, during the moderate El Nino event in 2009 there was a decrease in rainfall which was marked by a negative rainfall anomaly. This was accompanied by an increase in tea productivity. During the strong El Nino event in 2015 there was a decrease in rainfall which was marked by a negative rainfall anomaly. This was accompanied by an increase in tea productivity.

During the strong La Nina event in 2008 there was an increase in rainfall which was indicated by a positive rainfall anomaly. This was accompanied by a decrease in tea productivity. During the strong La Nina event in 2011 there was an increase in rainfall which was marked by a positive rainfall anomaly. This was accompanied by a decrease in tea productivity.

3.3.2 Pasirjambu District

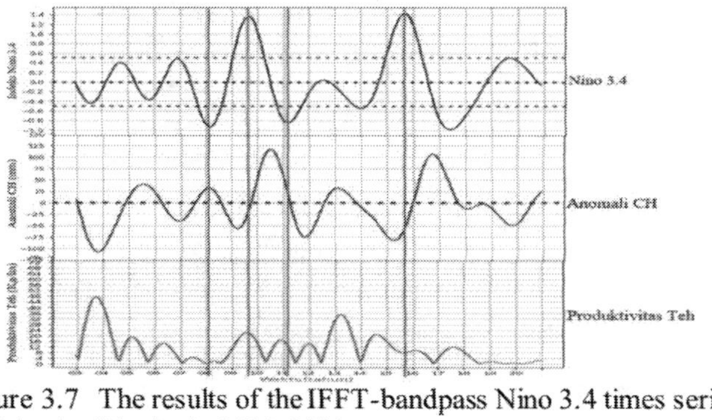

Figure 3.7 The results of the IFFT-bandpass Nino 3.4 times series plot, rainfall anomalies, and tea productivity in Pasirjambu District in 2003-2020.

Based on Figure 3.7, during the moderate El Nino event in 2009 there was a decrease in rainfall which was marked by a negative rainfall anomaly. This was accompanied by an increase in tea productivity. During a strong El Nino event in 2015 there was a slight decrease in rainfall which was marked by a negative rainfall anomaly. This was accompanied by a decrease in tea productivity.

During the strong La Nina event in 2008 there was an increase in rainfall which was indicated by a positive rainfall anomaly. This was accompanied by a decrease in tea productivity. During the strong La Nina event in 2011 there was a slight increase in rainfall which was indicated by a positive rainfall anomaly. This was accompanied by a decrease in tea productivity.

3.3.3 Pangalengan District

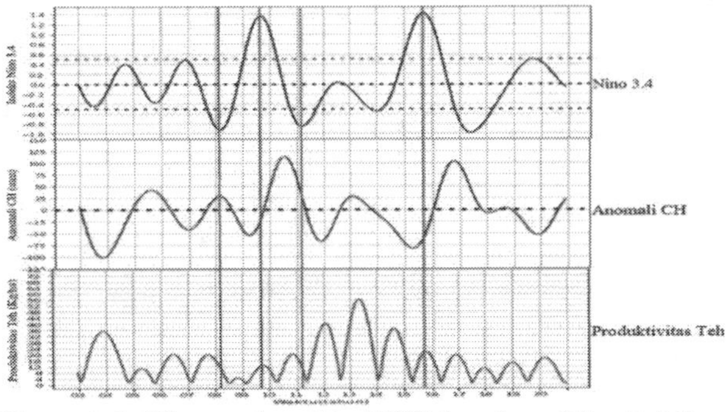

Figure 3.8 The results of the IFFT-bandpass Nino 3.4 times series plot, rainfall anomalies, and tea productivity in Pangalengan District in 2003-2020.

Based on Figure 3.8, during the moderate El Nino event in 2009 there was a slight decrease in rainfall which was marked by a negative rainfall anomaly. This was accompanied by an increase in tea productivity. During the strong El Nino event in 2015 there was a decrease in rainfall which was marked by a negative rainfall anomaly. This was accompanied by an increase in tea productivity.

During the strong La Nina event in 2008 there was an increase in rainfall which was indicated by a positive rainfall anomaly. This was accompanied by a decrease in tea productivity. During the strong La Nina event in 2011 there was a slight increase in rainfall which was indicated by a positive rainfall anomaly. This was accompanied by a decrease in tea productivity.

3.3.4 Kertasari District

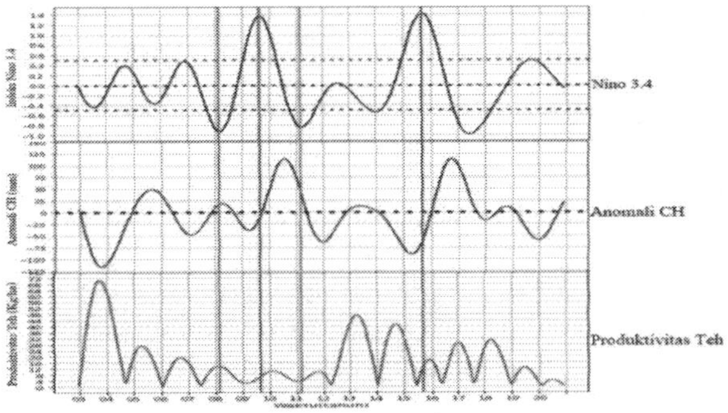

Figure 3.9 The results of the IFFT-bandpass Nino 3.4 times series plot, rainfall anomalies, and tea productivity in Kertasari District in 2003-2020.

Based on Figure 3.9, during the moderate El Nino event in 2009 there was a slight decrease in rainfall which was marked by a negative rainfall anomaly. This was accompanied by an increase in tea productivity. During the strong El Nino event in 2015 there was a decrease in rainfall which was marked by a negative rainfall anomaly. This was accompanied by an increase in tea productivity.

During the strong La Nina event in 2008 there was an increase in rainfall which was indicated by a positive rainfall anomaly. This was accompanied by an increase in tea productivity. During the strong La Nina event in 2011 there was a slight increase in rainfall which was indicated by a positive rainfall anomaly. This was accompanied by a decrease in tea productivity.

3.4 *Tea Productivity Pattern in Bandung Regency*

The monthly average tea productivity in the study area is shown in Figure 3.10. In May generally shows an increase in productivity while September is the month with the lowest tea productivity. Although other factors such as management also have an effect, in this study this was ruled out.

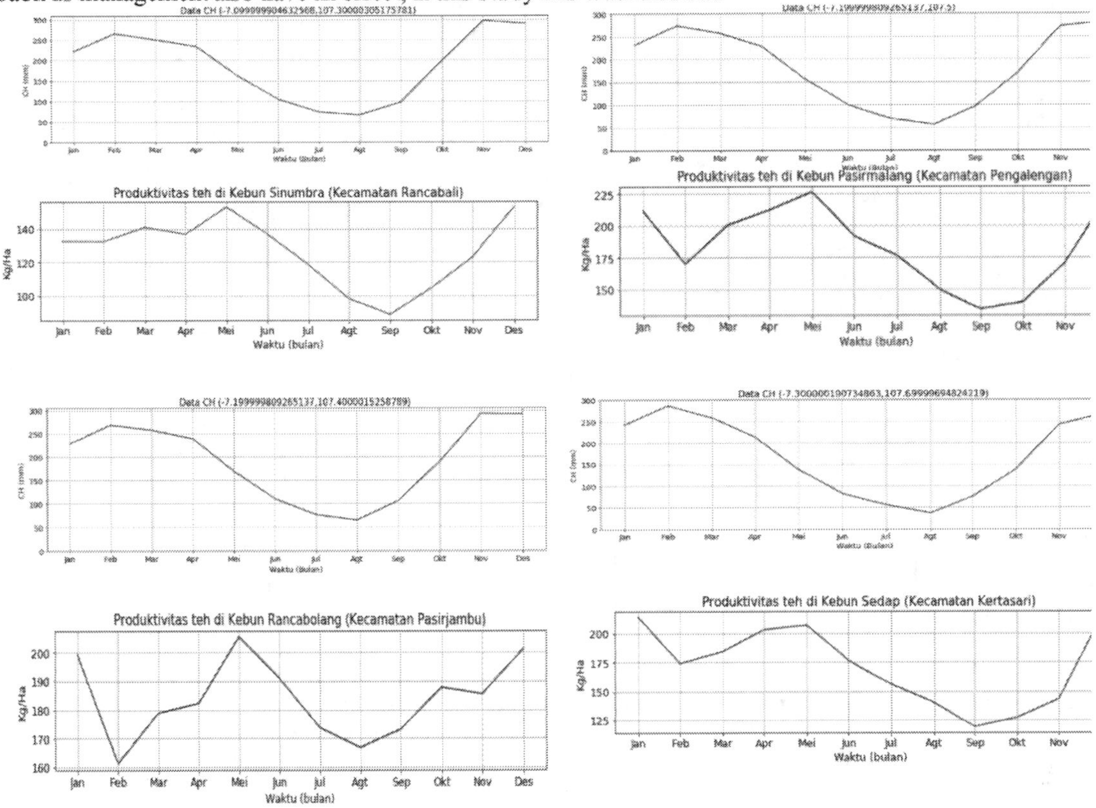

Figure 3.10 The results plot the average monthly rainfall and tea productivity in four districts in Bandung Regency ((a) Rancabali, (b) Pasirjambu, (c) Pangalengan, and (d) Kertasari).

When compared to the rainfall pattern chart with thetea productivity pattern chart in Figure 3.10 shows there is a one-month time lag between rainfall patterns and tea productivity in the four districts. For example, when rainfall begins in May, productivity begins to decline 1 month later (lag 1).

3.5 *Correlation of ENSO with Rainfall Anomalies*

The results of IFFT-bandpass index Nino 3.4 correlated to IFFT-bandpass rainfall anomalies.

Table 3.2 Correlation and significance level between Nino index 3.4 and rainfall anomalies

District	Correlation Value	Significant Value
Rancabali	-0.204	0.003
Pasirjambu	-0.264	0.000
Pangalengan	-0.309	0.000
Kertasari	-0.269	0.000

Table 3.2 shows the results of the correlation of theNino 3.4 index with the rainfall anomaly in 2003- 2020. The correlation values are low in the four districts with a significance value of less than 0.05 (5%) which means that there is a significant influence between the Nino 3.4 index (ENSO) on rainfall anomalies.

All study points have a negative correlation showing an inverse relationship between rainfall anomalies and the Nino 3.4 index. This means that if the Nino 3.4 index increases (higher positive) above 0.5 then the rainfall anomaly will decrease (more negative). Rising sea surface temperatures above normal conditions in the central and eastern Pacific Ocean (El Nino) which will weaken the trade winds blowing from east to west (increasing temperatures in the eastern Pacific) causing a decrease in rainfall in Indonesia, including in four districts of Bandung Regency. Meanwhile, if the Nino 3.4 index decreases (more negative) below -
0.5 then the rainfall anomaly will increase (more positive). The decrease in sea surface temperature below normal conditions in the central and eastern Pacific Ocean is known as the La Nina phenomenon. The trade winds blowing from east to west are strengthening (cold temperatures in the eastern Pacific) causing increased rainfall in Indonesia including four sub-districts, Bandung Regency.

3.6 Correlation of Rainfall Anomalies with TeaProductivity

The results of IFFT-bandpass rainfall anomalies correlated to IFFT-bandpass tea productivity.
Table 3.3 Correlation and significance level between rainfall anomalies and tea productivity

District	Correlation Value		Significant Value	
	Lag 0	Lag 1	Lag 0	Lag 1
Rancabali	-0.148	-0.201	0.030	0.003
Pasirjambu	-0.148	-0.201	0.030	0.000
Pangalengan	-0.208	-0.203	0.002	0.003
Kertasari	-0.304	-0.285	0.000	0.000

Table 3.3 shows the results of the correlation of rainfall anomalies with tea productivity in 2003- 2020. The correlation values at lags 0 and 1 are low in the all districts with a significance value of less than 0.05 (5%), meaning that the rainfall anomaly has a significant effect on tea productivity in the four districts. The highest correlation value of all districts is Kertasari District at lag 0 and lag 1 which are -0.304 and -0.285 with a significance value of 0.000 respectively.

The results of this correlation and significance levelshow that the effect of rainfall anomaly on tea productivity is relatively weak so it can be said that rainfall has a small effect on tea productivity in the region. This is also following the results of research conducted by Hendro Prawiro (2009) and Lou et al(2020), who said that a decrease in rainfall is not always followed by a decrease in tea productivity. Climate conditions with high rainfall can cause fungi activity resulting in a decrease in production which results in a decrease in tea productivity. Meanwhile, when rainfall decreases, sunlight becomes more increase. Thus, the process of photosynthesis runs faster to produce more energy and growth goes fast.

4. Summary

Based on the results of this study, several conclusions were obtained, which is the correlation of ENSO to rainfall anomalies in each district with the highest value of Pangalengan District (-0.309) with a significance value of 0.000. The correlation of rainfall anomalies to tea productivity in each district is low with the highest value of Kertasari District for lag 0 and lag 1 are -0.304 and -0.285 with a significant value of 0.000 respectively. Rainfall anomalies due to ENSO on tea productivityin each district of Bandung Regency can be said to have small effect, this is evidenced by the low correlation value of -0.148 to -0.309 and has a significance value of less than 5% of 0.000 to 0.030 so thatthere is a significant relationship. So although rainfall anomalies impact on tea productivity is small but this value is significant. We suggest to investigate more closely this relationship for larger scale and for the other phenomenons such as Monsoon, Dipole Mode, Madden Julian Oscillation, Semi Annual Oscillation, etc.

References

[1] Abd. Rahman As-syakur, Tasuku Tanaka, Takahiro Osawa & Made Sudiana Mahendra (2013) Indonesian rainfall variability observation using TRMM multi-satellite data, International Journal of Remote Sensing, 34:21, 7723-7738, DOI: 10.1080/01431161.2013.826837

[2] B. N. Goswami et al., (2006). Increasing Trend of Extreme Rain Events Over India in a Warming Environment Science314,1442-1445. DOI:10.1126/science.1132027

[3] Aldrian, E. and Dwi Susanto, R. (2003), Identification of three dominant rainfall regions within Indonesia and their relationship to sea surface temperature. Int. J. Climatol., 23: 1435-1452. https://doi.org/10.1002/joc.950

[4] As-syakur, A.R., Adnyana, I.W.S., Mahendra, M.S., Arthana, I.W., Merit, I.N., Kasa, I.W., Ekayanti, N.W., Nuarsa, I.W. and Sunarta, I.N. (2014), Observation of spatial patterns on the rainfall response to ENSO and IOD over Indonesia using TRMM Multisatellite Precipitation Analysis (TMPA). Int. J. Climatol., 34: 3825- 3839. https://doi.org/10.1002/joc.3939

[5] Baidu M, Amekudzi LK, Aryee JNA, Annor T. Assessment of Long-Term Spatio-Temporal Rainfall Variability over Ghana using Wavelet Analysis. *Climate.* 2017; 5(2):30 https://doi.org/10.3390/cli5020030

[6] Dalimoenthe, S. L., Wulansari, R., and Rezamela,E. (2016). *Impact of Climate Changes on leaves Productivity in Various Elevation Levels.* Gambung Tea and Quinine Research Center, Bandung.

[7] Kartawijaya, W. S. (1992*). Evaluation of the Effectof the 1991 Long Drought on Production in Several Tea Plantations.* Warta Tea and Quinine3 (3/4): 55-70.

[8] Lou, W., Sun, K., Zhao, Y., Deng, S., Zhou, Z. (2020). *Impact of climate change on inter- annual variation in tea plant output in Zhejiang, China.* Xinchang Weather Bureau and XinchangTea Station, Xinchang County, China.

[9] Plantation Office. (2018). *LKPJ Plantation Officeof West Java Province in 2017.* Retrieved on 10 February2022, from http://disbun.jabarprov.go.id/cassets/libs/uploads/dokumen/LKPJ2017.pdf.

[10] Rahaden, B. H., Radjawane, I., Arifatus, H. R. (2019). *The dynamics of negative Indian Ocean Dipole (nIOD) and its relation to the anomalous high rainfall in West Jawa Province, Indonesia.* Earth and Enviromental Science.

[11] Rahman, A. (2017). *Climate and Land Suitability Analysis of Tea Productivity for Projection in 2030 (Case Study: Bandung Regenncy).* Bandung: Final Project, Bandung Institute of Technology.

[12] Sirait, R. (2020*). The Effect of Rainfall Variation onPalm Oil Productivity in North Sumatera.* Bandung: Final Project, Bandung Institute of Technology.West Java Province Plantation Office. (2021). Plantation Statisics Book 2020. Bandung.

[13] Yulihastin E., Febrianti, N., Trismidianto. (2010). *Impact of El Nino and IOD on the Indonesian Climate. National institute of Aeronautics and Space (LAPAN).* Indonesia

[14] Y., Zhang, & Fueglistaler, S. (2019). Mechanism for increasing tropical rainfall unevenness with global warming. *Geophysical Research Letters*, 46, 14836–14843. https://doi.org/10.1029/2019GL086058

[15] Zubaidah, A. (2012*). Analysis of Change in RainfallSatellite Tropical Measuring Mission (TRMM)in 2009 and 2010.* Research in Field of Environment and Disaster Mitigation-Pusfatja LAPAN

[16] Held, I. M., & Soden, B. J. (2006). Robust Responses of the Hydrological Cycle to Global Warming, Journal of Climate, 19(21), 5686-5699. doi: https://doi.org/10.1175/JCLI3990.1

[17] Westra, S., Alexander, L. V., & Zwiers, F. W.(2013). Global Increasing Trends in Annual Maximum Daily Precipitation, Journal of Climate, 26(11), 3904-3918. doi: https://doi.org/10.1175/JCLI-D-12-00502.1

[18] Roderick, T. P., Wasko, C., & Sharma, A. (2019). Atmospheric moisture measurements explain increases in tropical rainfall extremes. Geophysical Research Letters, 46, 1375– 1382. https://doi.org/10.1029/2018GL080833

Prediction of Convective Available Potential Energy and Equivalent Potential Temperature using a Coupled WRF and Deep Learning for Typhoon Identification

Mamad Tamamadin[1,3], Changkye Lee[2], Seong-Hoon Kee[1,2], Jurng-Jae Yee[1,2]

[1] Department of ICT Integrated Ocean Smart Cities Engineering, Dong-A University, Busan, Korea
[2] University Core Research Center for Disaster-free and Safe Ocean City Construction, Dong-A University, Busan, Korea
[3] Atmospheric Science Research Group, Institut Teknologi Bandung, Bandung, Indonesia

E-mail: jjyee@dau.ac.kr

Abstract. To predict typhoons in the western North Pacific Ocean, it is required to predict the determinants of typhoon activities. The formation of the typhoon can be controlled by Convective Available Potential Energy (CAPE) and Equivalent Potential Temperature (theta-e). To predict the variables, a mesoscale numerical model of Weather Research and Forecasting (WRF) can be used. However, the output of WRF needs to improve to obtain a more accurate CAPE and theta-e prediction. This study uses a coupled WRF model and Deep Learning (DL) Multilayer Perceptron Regressor approach to increase CAPE and theta-e prediction skills. Simulation with dataset scenarios with WRF outputs as predictors and sounding data as predictors are developed and tested to obtain the most appropriate package of deep learning simulation. The study found that coupled models provide increased mean accuracy of theta-e and CAPE, namely 16.6% and 32.0% higher than using original WRF, respectively. This study also shows the difference of skill scores in the spatial distribution of CAPE and theta-e of WRF result and its coupled model.

Keywords: accuracy improvement, CAPE, deep learning, theta-e, typhoon, WRF

1. Introduction

Typhoon is one of the hydrometeorological hazards [1] threatening life and disrupting economic activity. The impact of typhoons is very detrimental, especially to inhabitants in coastal areas. Especially in the western North Pacific region, there are more than 5 typhoons that occur in a year. Therefore, this area is known as the most active typhoon occurrence area in the world [2], [3]. To reduce the risk of this impact, accurate early identification of typhoons as early warning [4] is needed. Typhoon prediction accuracy will greatly determine the strategy and priority of anticipatory steps in potential areas.

Several studies have shown that typhoons can be identified using two variables, namely the equivalent potential temperature (theta-e) and convective available potential energy (CAPE) [5]–[10] as stated in [11]. Theta-e is the temperature that an air parcel would reach to diagnose atmospheric instability. CAPE measures the fuel potential for a thunderstorm to form by estimating the strength of an upward current within a thunderstorm and describes the instability of the atmosphere. For early identification of typhoon, both variables can be predicted using a numerical prediction. However, due to several limitations of the numerical model, such as extreme conditions which occasionally cannot be represented by physical equations, low resolution, and inaccuracy in determining the initial and boundary values, the accuracy of these two variables still needs to be improved. One possible solution is the use of deep learning (DL) to improve the numerical model. However, the hyperparameter of the deep learning should be determined to get the appropriate model configuration for each variable.

The increasing number of deep learning models are being considered for weather forecasting due to their well capability to address with important and complex nonlinear problems [12]. Previous studies have conducted experiments using a combination of DL and numerical models, particularly weather research and forecasting (WRF), to predict the weather. DL-based parameterizations were successfully conducted to WRF model [13]. Hybrid downscaling by WRF and Convolutional Neural Network (CNN) model to get the precipitation in higher resolution [14]. Wind signal prediction was produced by using a hybrid of WRF and deep learning neural network [15]. The use of machine learning has also been carried out to improve the output of the WRF for wind speed values [16]. To the best author's knowledge, there is a lack of study in discussing the combination of machine learning and WRF to get the theta-e and CAPE prediction.

This study aims to develop a prediction framework that combines Weather Research and Forecasting (WRF) and Deep Learning (DL) as a coupled WRF+DL model to produce theta-e and CAPE predictions with better accuracy than using the original WRF. Several steps have been taken to achieve this goal, including WRF simulation of past typhoons, calculation of theta-e and CAPE values, deep learning optimization for the coupled WRF+DL model, and evaluation of model performance.

The discussion of the paper is divided as follows. The data and method is described in Section 2. Section 3 discusses the results and discussion. The conclusion is given in Section 4.

2. Data and Method

2.1 Data

The coarser meteorological forecast of the Global Forecast System (GFS) has been used for input of WRF model [17]. Typhoon track from Japan Meteorological Agency (JMA) is used to see the position of typhoons and test the identification of typhoons through the position of CAPE and Theta-e values. The typhoon track data can be accessed at Digital Typhoon: Typhoon Images and Information managed by National Institute of Informatics (NII) [18]. Sounding data from Wyoming university [19] which is used as a target in deep learning. There are 14 sounding stations spread across in the western North Pacific region.

2.2 Method

This study aims to produce theta-e and CAPE predictions that have better accuracy by using coupled WRF+DL in identifying typhoons. Several steps were carried out to achieve this goal, including WRF simulation, calculating theta-e and CAPE values, deep learning optimization for the coupled WRF+DL model, and testing the prediction results.

2.2.1. Configuration and simulation of WRF for CAPE dan Theta-e calculation.

In this study, the regional Weather Research and Forecasting (WRF) model version 4.2 [20] is run for periods of 15 typhoon occurrences with double-nested modeling system. The WRF model is used to downscale GFS data to get the higher resolution based on the double-nested domain. The mother domain with 27 km spatial resolution is centered at 28.686°N, 127.753°E, and finer resolution domain (i.e., 9 km) is nested with the same center location (see Figure 1).

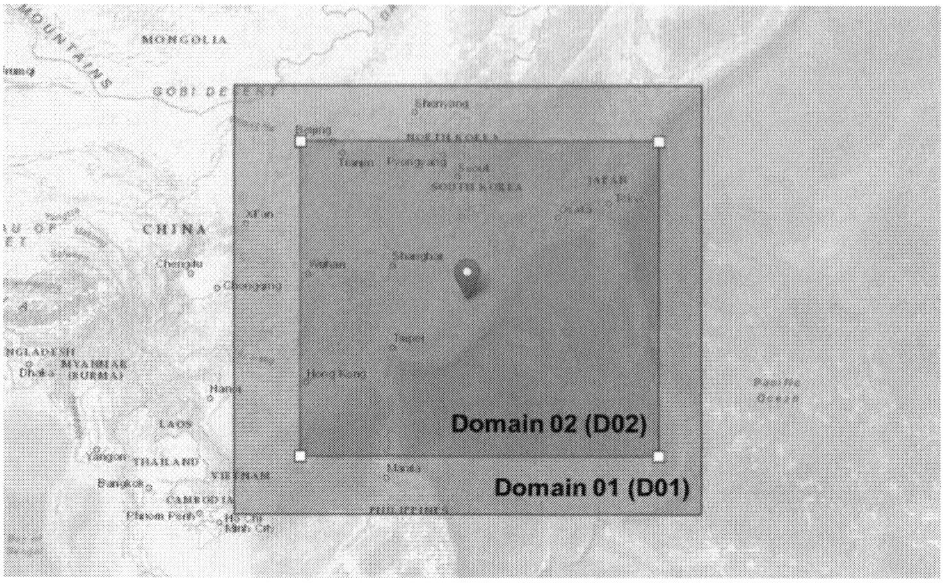

Figure 1. Domain for WRF simulation

The physical parameterization in the model input is adjusted with previous studies performed in the western North Pacific region [21]. The scheme comprises of Kain–Fritsch convective parameterization scheme [22], WRF single-moment 3-class microphysics (WSM3) scheme [23], the Revised MM5 surface layer scheme [24], NCAR Community Atmosphere Model (CAM 3.0) radiation scheme, Yonsei University (YSU) planetary boundary layer scheme, and Noah land surface model. The same configuration is applied for the mother and the nested domain.

To accommodate data that includes the presence or absence of typhoons, the dataset is compiled from the results of WRF simulations and sounding on 15 past typhoon events that pass through the western North Pacific region. The WRF output is used to calculate the theta-e and CAPE values using equation 1 and equation 2 [11]. The two variable values from the WRF are as predictors and sounding observations are as predictand. Therefore, the WRF grid was chosen at a location adjacent to the sounding location. There are several sounding stations that have been passed/approached by the typhoon path and some other areas have not. To calculate CAPE, equation (1) is used.

$$\mathrm{CAPE} = \int_{Z_f}^{Z_n} \mathrm{g}\left(\frac{T_{\mathrm{v,parcel}} - T_{\mathrm{v,env}}}{T_{\mathrm{v,env}}}\right) \mathrm{Rdz} \tag{1}$$

where Z_f and Z_n are the height of the level of free convection and equilibrium level (neutral buoyancy), $T_{v,parcel}$ and $T_{v,env}$ is the virtual temperature of the specific parcel and the virtual temperature of the environment, respectively, R is the gas constant and g is the acceleration due to gravity.

Theta-e is calculated using equation (2).

$$\theta_e = T_e\left(\frac{p_0}{p}\right)^{K_d} \approx \left(T + \frac{L_v}{C_{pd}} r\right)\left(\frac{p_0}{p}\right)^{\frac{\mathrm{R_d}}{\mathrm{C_{pd}}}} \tag{2}$$

where θ_e is the equivalent potential temperature, T_e is equivalent temperature, p_0 is a reference pressure (usually 100 kPa), p is the partial pressure of dry air, T is the temperature, L_v is the latent heat of vaporization, C_{pd} is the heat capacity at constant pressure of dry air, r is the total water mixing ratio, c is the heat capacity of liquid water and R_d is the gas constant for dry air. For theta-e calculations, several main variables that can be taken from the WRF output include air temperature, relative humidity, air pressure.

2.2.2. Deep learning.

The deep learning algorithm used in this study is the Multilayer Perceptron (MLP) by using scikit-learn as python library [25]. MLP is a supervised learning algorithm that can learn a number of input dimensions to produce a number of output dimensions through a feature set $X=x_1,x_2,\ldots,x_m$ and target y which is processed by one or more hidden layers [26]. In this study, MLP will generate the improved accuracy of CAPE and theta-e prediction based on predictors generated by WRF. Because the supervised algorithm is used, the target is needed by taking data from the sounding. The target is required since the supervised algorithm uses data from the sounding. Correction weight values are determined automatically during MLP simulation iterations. To get the optimal configuration of the MLP algorithm, hyperparameter tuning is needed as shown in Figure 2. However, since the number of data is large, a python script is needed that can automatically be used by an iterative process to find the optimal hyperparameter configuration for predicting theta-e and CAPE values that have a better accuracy. To prevent indefinite looping, upper and lower threshold values for each parameter have been established.

There are several parameters required to obtain the optimal prediction, namely activation, solver, alpha, hidden layer size, and learning rate. The output of a node is determined by its activation function given an input or set of inputs. In MLP, the available activation function consists of identity (linear), logistic, heaviside (tanh), and relu (rectified linear unit).

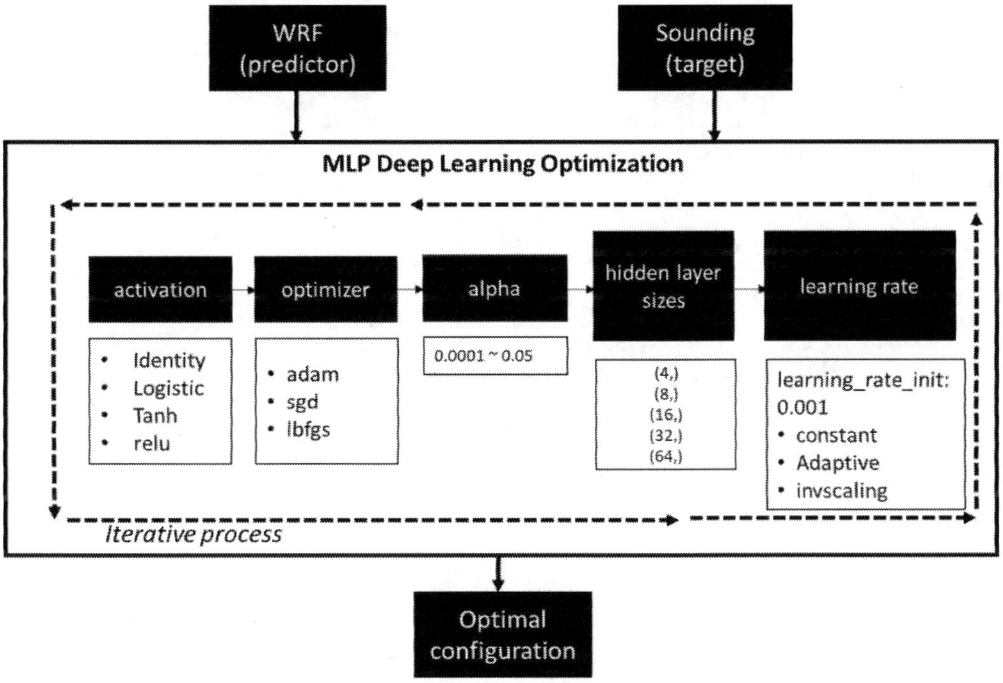

Figure 2. MLP Deep Learning optimization

In addition, deep learning requires an optimizer or solver in order to reduce the error function or increase output effectiveness. Adam (Adaptive Moment Estimation), SGD (Stochastic Gradient Descent), and Limited-memory BFGS (LBFGS) are among the optimizer algorithms offered by MLP. By storing the decaying average of the previous gradients and the decaying average of the previous squared gradients from the estimated values, the Adam optimizer calculates adaptive learning rates for each parameter. The SGD method uses the proper smoothness technique to iteratively reduce the value of the loss function. LBFGS, a member of the family of quasi-Newton algorithms, uses a small amount of computer memory to approximate the Broyden-Fletcher-Goldfarb-Shanno algorithm (BFGS). On unconstrained input values, LBFGS minimizes the loss function, which is a differentiable scalar function. The regularization term uses the alpha parameter to choose weights with large magnitudes for preventing overfitting.

The hidden layer size is determined to identify weighted dimensions that will be used in the output value computation procedure. The size of the hidden layer demonstrates the algorithm's backpropagation's time complexity. The step-size in the space search parameter scheduled using the chosen criteria, such as constant, adaptive, or invscaling, is controlled by the learning rate.

3. Evaluation of CAPE dan Theta-e prediction

The prediction of CAPE and theta-e is evaluated using Pearson's correlation for skill scores. Comparing the visualization of the vertical profile for theta-e among deep learning results, observation, and WRF is another method of evaluation. Additionally, both theta-e and CAPE's temporal variation are used in the comparison.

3. Result and Discussion

This study examines the ability of coupled WRF+DL to predict theta-e and CAPE which are used to identify typhoons in the future. There are 3 important discussions including testing the prediction results, vertical profile analysis of theta-e values, and temporal analysis for theta-e and CAPE values using the Bavi 2020 typhoon case. After going through an optimization process in using the WRF+DL to predict theta-e and CAPE, the most appropriate configuration is obtained for use, as shown in Table 1. To get the optimal result, the learning rate, solver, and activation parameters for the two variables (theta-e and CAPE) should be the same, namely logistic, lbfgs, and constant, respectively. However, there are also different hidden layer and alpha sizes, including (1) 0.22 and (64,) for theta-e variables, and (2) 0.2 and (32,) for CAPE.

Table 1. Optimal hyperparameter tuning of MLP.

Variable	Parameter	Value
Theta-e	activation	logistic
	alpha	0.22
	solver	lbfgs
	batch_size	392
	max_iter	5000
	hidden_layer_sizes	(64,)
CAPE	activation	logistic
	alpha	0.2
	solver	lbfgs
	batch_size	392
	max_iter	5000
	hidden_layer_sizes	(32,)

Each chosen parameter is examined. Logistics is the proper activation value for both parameters. This indicates that there is a chance that the two variables (theta-e and CAPE) will have data patterns that resemble a logarithmic curve. For weight optimization for both atmospheric variables, lbfgs is the proper solver. The prediction accuracy of CAPE and Theta-e can be increased by fitting the dataset with the quasi-Newton method employed in lbfgs. The "lbfgs" solution does not require the learning rate parameter, hence it was not chosen.

The difference in alpha and batch size values between the two variables is not large. Small modifications, on the other hand, can have considerably different results. Because the alpha value prevents overfitting in the model, also known as L2 regularization, the alpha value of each data series of a variable will vary depending on the nature of the data.

The batch size parameter specifies the amount of training samples gathered from the first to the last sample number. Processing a smaller quantity of data compared to the overall data becomes valuable when memory is limited. A batch size that is excessively small, on the other hand, will result in less reliable predicted results. Because the entire amount of data trained in this study is 3920, the effective batch size for 10 iterations of training is 392.

There are no standard criteria for choosing hidden layer size because it is determined by the complexity of the problem, the amount of data, and the model architecture rather than the number of features. As a result, numerous possibilities were considered in order to reach the best values, namely 64 for theta-e and 32 for CAPE. Overfitting can occur when hidden layer sizes are excessively large. As a result, in addition to calculating the proper hidden layer size, alpha determination is also considered at this stage.

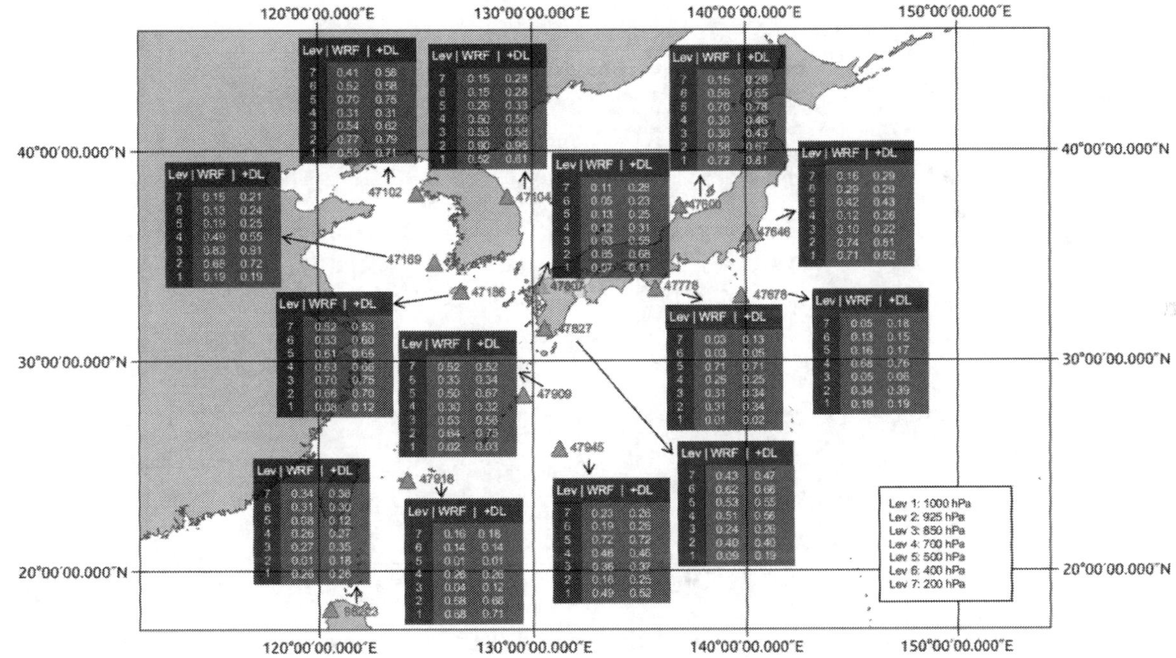

Figure 3. Skill score for theta-e prediction using WRF and coupled WRF-Deep Learning

The testing results of the theta-e prediction are shown on a sounding location map in which each sounding location has a correlation result for 17 altitude levels as shown in Figure 3. In the figure, there is comparison of the capabilities of using only WRF and coupled WRF+DL evaluated using correlation called as skill score in this study. There are 14 sounding stations used to test the capabilities of the proposed model. The highest score for coupled WRF + DL is 0.95 which is the result of increased accuracy after the value of 0.90 was achieved by using only WRF, reached at station 47104. The lowest score of WRF+DL is 0.02 after improving from 0.01 reached at station 47778. The mean accuracy of WRF+DL use is 0.43, while mean accuracy of WRF use is 0.36.

For CAPE, each station has one score as seen Figure 4. The highest value is shown at station 47104 with a score of 0.84 after using the coupled model. Previously, WRF could only predict CAPE with a score of 0.70 at the same station. The prediction accuracy score of CAPE is generally higher than the prediction accuracy of theta-e. The lowest score of WRF+DL is 0.23 after improving from 0.14 reached at station 47909. The mean accuracy of WRF+DL use is 0.54, while mean accuracy of WRF use is 0.41.

The stations inside the red box are located on the ocean offshore with lower score compared to the other stations that are located near coastal regions (see Figure 4). The coastal regions benefit from various land surface features that can influence weather patterns, such as mountains, forests, and urban areas. These features can create local variations in temperature, moisture, and wind patterns, which contribute to convective initiation (CI). In contrast, ocean offshore areas lack such significant surface features, leading to a relatively homogeneous environment that may not promote the same level of convective initiation. The ocean offshore contributes to the convection propagation (CP) after the triggering of coastal CI [27]. The pattern of convection propagation tends to be more complex than CI activity since CP refers to the movement, growth, and evolution of existing convective systems. Once convective storms have initiated, their behavior becomes more complex and dynamic. This condition can generally cause the model to be difficult to predict the CAPE.

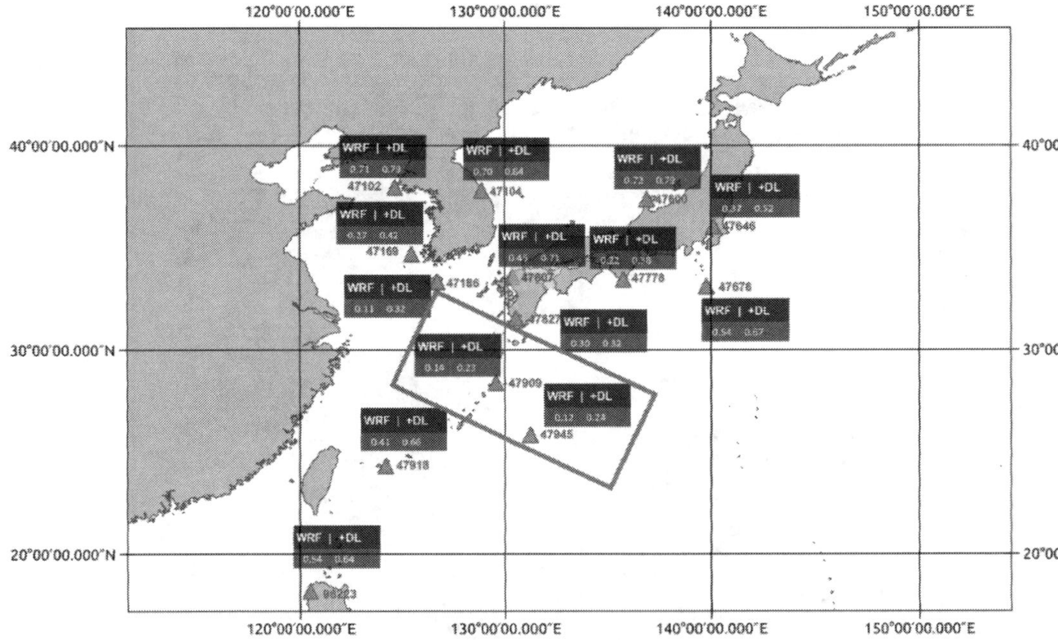

Figure 4. Skill score for CAPE prediction using WRF and coupled WRF-Deep Learning

Based on data from all stations and a variety of altitude levels, we conclude that the mean accuracy of theta-e prediction using WRF+DL is 16.6% higher than using original WRF, and the mean accuracy of CAPE prediction using WRF+DL is 32.0% higher than using original WRF. For both variables, the mean accuracy improvement is 24.3%. This percentage demonstrates the considerable improvement in accuracy with the use of paired WRF+DL. However, certain accuracy scores at some locations are still extremely low, and it is difficult to enhance using this methodology considerably. This issue arises since we generalize the DL hyperparameter configuration to all sounding locations, which is unable to resolve certain particular cases.

In this study, the typhoon "Bavi 2020" was used to see the results of deep learning in improving WRF prediction results for theta-e and CAPE. This typhoon was developed on 22 August 2020 at 00 UTC and disappeared on 27 August 2020 at 06 UTC. The lifetime of this typhoon is around 126 hours. Minimum pressure, maximum wind, average speed, travel length of this typhoon is 950 hPa, 157.4 km/h, 18.6 km/h, and 2346 km, respectively.

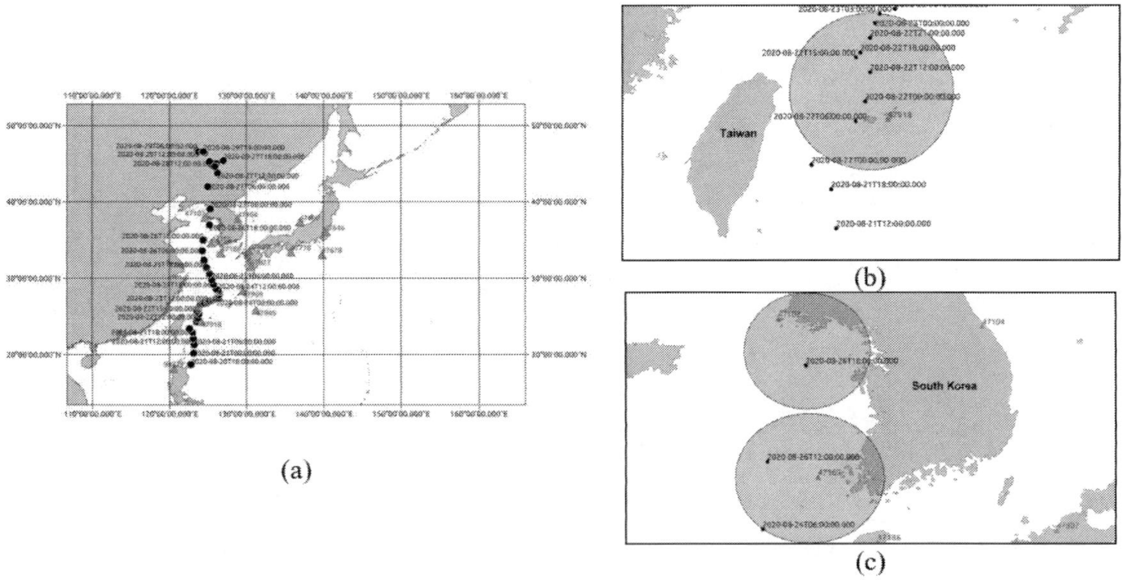

Figure 5. Sounding station and typhoon track of Bavi 2020. Track and sounding distribution in (a), and soundings impacted by typhoon in (b) and (c)

Based on the map shown in Figure 5b and 5c, there are 3 sounding stations affected by the Bavi 2020 typhoon, namely 47918 (on 22 August 2020, 06:00 UTC), station 47169 (on 26 August 2020, 12:00 UTC), 47102 (on August 26, 2020, at 18:00 UTC). Furthermore, station 47918 is used to analyze the predicted results of Theta-e and CAPE during the Bavi 2020 typhoon event since it is the closest station to the observed typhoon track.

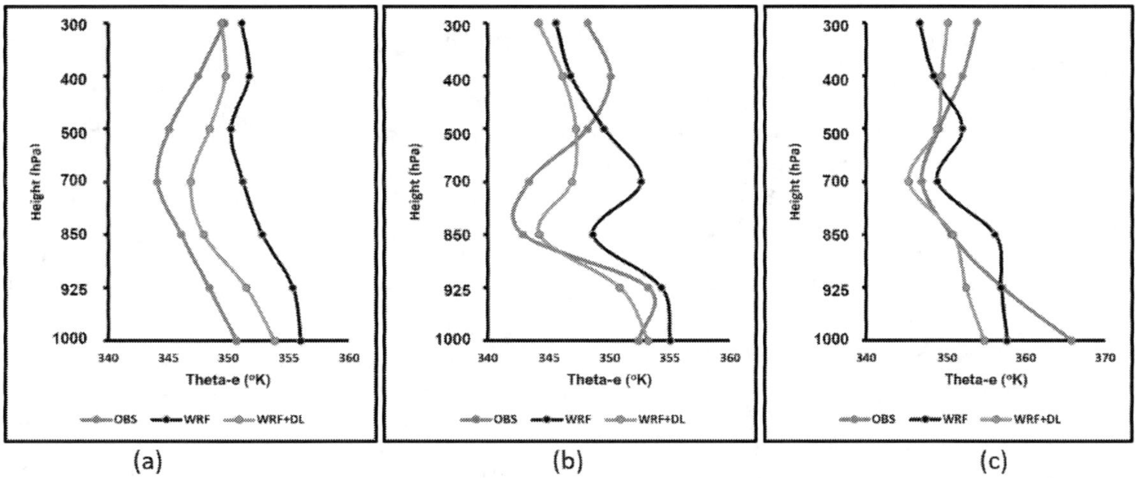

Figure 6. Vertical Profile of theta-e before (a), during (b), and after (c) typhoon passing through the sounding station of 47918

The vertical profile of the equivalent potential temperature (theta-e) before, during and after the typhoon passed the sounding station area of 47918 is shown in Figure 6 by comparing the predicted results of WRF, coupled WRF+Deep Learning, and observations. It can be seen from the three charts that the vertical pattern of the coupled WRF+DL results is generally more similar to the observations compared to only using the WRF model. If seen in Figure 6a, the vertical profile of theta-e is undisturbed. However, the vertical pattern changed with the many variations of values as the typhoon "Bavi 2020" passing through the area near the observation station. The vertical pattern of theta-e began to return to normal condition as the typhoon move away from the station area.

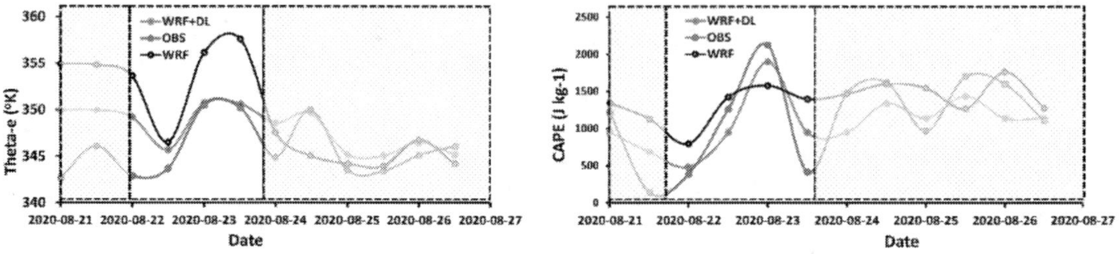

(a) (b)

Figure 7. Temporal variation of Theta-e (a) and CAPE (b) near sounding station of 47918

Based on the comparison shown in Figure 7a, the time series pattern of the predicted theta-e value generated by the coupled WRF+DL is closer to the observation data than the WRF models. The figure shows that as typhoon Bavi begins to approach, the theta-e value shows an increase. Likewise, the comparison for the prediction of the CAPE value (see Figure 7b) shows the same conditions where there is an increase in CAPE when a typhoon approaches the sounding area.

The results of this study will then be used to continue the development of a typhoon prediction model using the ensemble k-NN algorithm to analyze the atmospheric similarity of current and past typhoons. The initial development of the typhoon track prediction model has been carried out previously based on the closeness to the track from past typhoons [28].

4. Conclusion

The purpose of this study is to compare the prediction performance of coupled WRF+DL to original WRF for theta-e and CAPE values. The two variables are required for predicting typhoon events, particularly those that occur often in the western North Pacific region each year. This study has three primary components that show the results of the proposed model, examination of the vertical profile of theta-e values, and temporal analysis for theta-e and CAPE using the typhoon Bavi 2020 case. The study results reveal that coupled WRF+DL predictions of theta-e and CAPE improve the original WRF predictions. The mean accuracy of WRF+DL is 16.6% higher than using original WRF, and the mean accuracy of CAPE prediction using WRF+DL is 32.0% higher than using original WRF. The vertical profile of theta-e values and temporal patterns of the two variables (theta-e and CAPE) were also evaluated, with the coupled model producing better results. The study is conducted before, during, and after the typhoon passing through the station area.

The simulations are limited to the use of deep learning only. Therefore, there is an opportunity to improve the accuracy of WRF using other machine learning algorithms. The training dataset also only inlvolves typhoon occurrences affecting the Korea region. As a result, if there is a possibility of different hyperparameter tuning that should be optimized.

Acknowledgement

The authors would like to thank the Basic Science Research Program through the National Research Foundation of Korea (NRF), funded by the Ministry of Education (2016R1A6A1A03012812).

References

[1] A. Tounsi and M. Temimi, "A systematic review of natural language processing applications for hydrometeorological hazards assessment," *Nat. Hazards*, Feb. 2023, doi: 10.1007/s11069-023-05842-0.

[2] R. Zhan, Y. Wang, and M. Ying, "Seasonal Forecasts of Tropical Cyclone Activity Over the Western North Pacific: A Review," *Trop. Cyclone Res. Rev.*, vol. 1, no. 3, pp. 307–324, Sep. 2012, doi: 10.6057/2012TCRR03.07.

[3] A. D. Magee, A. S. Kiem, and J. C. L. Chan, "A new approach for location-specific seasonal outlooks of typhoon and super typhoon frequency across the Western North Pacific region," *Sci. Rep.*, vol. 11, no. 1, Art. no. 1, Sep. 2021, doi: 10.1038/s41598-021-98329-6.

[4] A. Susandi *et al.*, "Development of Hydro-Meteorological Hazard Early Warning System in Indonesia," *J. Eng. Technol. Sci.*, vol. 50, no. 4, pp. 461–478, Oct. 2018, doi: 10.5614/j.eng.technol.sci.2018.50.4.2.

[5] S. J. Camargo, K. A. Emanuel, and A. H. Sobel, "Use of a Genesis Potential Index to Diagnose ENSO Effects on Tropical Cyclone Genesis," *J. Clim.*, vol. 20, no. 19, pp. 4819–4834, Oct. 2007, doi: 10.1175/JCLI4282.1.

[6] M. Lee and T. Frisius, "On the role of convective available potential energy (CAPE) in tropical cyclone intensification," *Tellus Dyn. Meteorol. Oceanogr.*, vol. 70, no. 1, p. 1433433, Jan. 2018, doi: 10.1080/16000870.2018.1433433.

[7] D. S. Nolan and M. G. McGauley, "Tropical cyclogenesis in wind shear: climatological relationships and physical processes," in *Cyclones: formation, triggers, and control.*, K. Oouchi and H. Fudeyasu, Eds., Nova Science Publishers, Happauge, New York, p. 34.

[8] M. K. Tippett, S. J. Camargo, and A. H. Sobel, "A Poisson Regression Index for Tropical Cyclone Genesis and the Role of Large-Scale Vorticity in Genesis," *J. Clim.*, vol. 24, no. 9, pp. 2335–2357, May 2011, doi: 10.1175/2010JCLI3811.1.

[9] I. G. Watterson, J. L. Evans, and B. F. Ryan, "Seasonal and Interannual Variability of Tropical Cyclogenesis: Diagnostics from Large-Scale Fields," *J. Clim.*, vol. 8, no. 12, pp. 3052–3066, Dec. 1995, doi: 10.1175/1520-0442(1995)008<3052:SAIVOT>2.0.CO;2.

[10] Y. Zhang, H. Wang, J. Sun, and H. Drange, "Changes in the tropical cyclone genesis potential index over the western north pacific in the SRES A2 scenario," *Adv. Atmospheric Sci.*, vol. 27, no. 6, pp. 1246–1258, Nov. 2010, doi: 10.1007/s00376-010-9096-1.

[11] M. Wahiduzzaman and A. Yeasmin, "A kernel density estimation approach of North Indian Ocean tropical cyclone formation and the association with convective available potential energy and equivalent potential temperature," *Meteorol. Atmospheric Phys.*, vol. 132, no. 5, pp. 603–612, Oct. 2020, doi: 10.1007/s00703-019-00711-7.

[12] Y. Wang, R. Zou, F. Liu, L. Zhang, and Q. Liu, "A review of wind speed and wind power forecasting with deep neural networks," *Appl. Energy*, vol. 304, p. 117766, Dec. 2021, doi: 10.1016/j.apenergy.2021.117766.

[13] X. Zhong, Z. Ma, Y. Yao, L. Xu, Y. Wu, and Z. Wang, "WRF–ML v1.0: a bridge between WRF v4.3 and machine learning parameterizations and its application to atmospheric radiative transfer," *Geosci. Model Dev.*, vol. 16, no. 1, pp. 199–209, Jan. 2023, doi: 10.5194/gmd-16-199-2023.

[14] T. Tu, K. Ishida, A. Ercan, M. Kiyama, M. Amagasaki, and T. Zhao, "Hybrid precipitation downscaling over coastal watersheds in Japan using WRF and CNN," *J. Hydrol. Reg. Stud.*, vol. 37, p. 100921, Oct. 2021, doi: 10.1016/j.ejrh.2021.100921.

[15] Y. Qin *et al.*, "Hybrid forecasting model based on long short term memory network and deep learning neural network for wind signal," *Appl. Energy*, vol. 236, pp. 262–272, Feb. 2019, doi: 10.1016/j.apenergy.2018.11.063.

[16] M. Tamamadin, C. Lee, S.-H. Kee, and J.-J. Yee, "Improved Wind Speed Prediction at Two Height Levels Using Different Machine Learning Algorithms for Typhoon Track Identification," *Archit. Inst. Korea*, vol. 42, no. 2, pp. 430–431, 2022.

[17] National Oceanic and Atmospheric Administration (NOAA), "GFS Data." Mar. 15, 2022. Accessed: Mar. 15, 2022. [Online]. Available: https://nomads.ncep.noaa.gov/pub/data/nccf/com/gfs/

[18] "Digital Typhoon: Typhoon Images and Information - National Institute of Informatics." http://agora.ex.nii.ac.jp/digital-typhoon/ (accessed Mar. 04, 2023).

[19] University of Wyoming, "Sounding data." Apr. 20, 2022. Accessed: Apr. 20, 2022. [Online]. Available: http://www.weather.uwyo.edu/upperair/sounding.html

[20] W. C. Skamarock *et al.*, "A Description of the Advanced Research WRF Model Version 4," *Natl. Cent. Atmospheric Res.*, vol. 145, no. 145, p. 550, 2019.

[21] D.-H. Cha, C.-S. Jin, D.-K. Lee, and Y.-H. Kuo, "Impact of intermittent spectral nudging on regional climate simulation using Weather Research and Forecasting model," *J. Geophys. Res. Atmospheres*, vol. 116, no. D10, 2011, doi: 10.1029/2010JD015069.

[22] J. S. Kain and J. M. Fritsch, "A One-Dimensional Entraining/Detraining Plume Model and Its Application in Convective Parameterization," *J. Atmospheric Sci.*, vol. 47, no. 23, pp. 2784–2802, Dec. 1990, doi: 10.1175/1520-0469(1990)047<2784:AODEPM>2.0.CO;2.

[23] S.-Y. Hong, J. Dudhia, and S.-H. Chen, "A Revised Approach to Ice Microphysical Processes for the Bulk Parameterization of Clouds and Precipitation," *Mon. Weather Rev.*, vol. 132, no. 1, pp. 103–120, Jan. 2004, doi: 10.1175/1520-0493(2004)132<0103:ARATIM>2.0.CO;2.

[24] P. A. Jiménez, J. Dudhia, J. F. González-Rouco, J. Navarro, J. P. Montávez, and E. García-Bustamante, "A Revised Scheme for the WRF Surface Layer Formulation," *Mon. Weather Rev.*, vol. 140, no. 3, pp. 898–918, Mar. 2012, doi: 10.1175/MWR-D-11-00056.1.

[25] O. Kramer, "Scikit-Learn," in *Machine Learning for Evolution Strategies*, O. Kramer, Ed., in Studies in Big Data. Cham: Springer International Publishing, 2016, pp. 45–53. doi: 10.1007/978-3-319-33383-0_5.

[26] "Neural network models (supervised)," *scikit-learn*. https://scikit-learn/stable/modules/neural_networks_supervised.html (accessed Apr. 05, 2023).

[27] L. Bai, Z. Meng, K. Sueki, G. Chen, and R. Zhou, "Climatology of tropical cyclone tornadoes in China from 2006 to 2018," *Sci. China Earth Sci.*, vol. 63, no. 1, pp. 37–51, Jan. 2020, doi: 10.1007/s11430-019-9391-1.

[28] M. Tamamadin, C. Lee, S.-H. Kee, and J.-J. Yee, "Regional Typhoon Track Prediction Using Ensemble k-Nearest Neighbor Machine Learning in the GIS Environment," *Remote Sens.*, vol. 14, no. 21, Art. no. 21, Jan. 2022, doi: 10.3390/rs14215292.

Correlation Analysis of GNSS-Derived Precipitable Water Vapor (PWV) with Rainfall Data in Indonesia

Azhari Barbarosa[1], Dudy D Wijaya[2], Zamzam A J Tanuwijaya[2], Nabila S E Putri[3], Wedyanto Kuntjoro[2]

[1]Geodesy and Geomatics Engineering Study Program, Bandung Institute of Technology, Ganesha 10 St., Bandung 40132, Indonesia
[2] Geodesy Research Group, Bandung Institute of Technology, Ganesha 10 St., Bandung 40132, Indonesia
[3] Surveying and Cadastre Research Group, Bandung Institute of Technology, Ganesha 10 St., Bandung 40132, Indonesia

E-mail Address: AzhariBarbarosa@students.itb.ac.id

Abstract. GNSS (Global Navigation Satellite Systems) have become an important tool for various activities related to positioning, such as navigation, construction projects, and deformation measurement. Additionally, GNSS can also estimate Precipitable Vapor Vapor (PWV) for meteorological purposes. PWV is a measure of atmospheric water vapor, which can eventually precipitate as rain. Since Indonesia's climate is mainly characterized by the change in rainfall, monitoring precipitation is a crucial step in understanding its pattern. In this study, we aim to analyze the correlation between PWV and rainfall in Indonesia. We used GNSS observations from the InaCORS network across Indonesia, as well as rainfall data from GSMaP at the InaCORS stations in 2019 from 1st January to 31st December. Both of these data are normalized every five days (pentad days) and compared to each other to obtain the Pearson's correlation coefficients. From these results, we generated a heatmap of the Pearson's correlation coefficients between pentad PWV and rainfall at InaCORS locations in Indonesia, with the highest correlation of 0.753819 in Garut Regency and the lowest correlation of 0.090146 in Fakfak Regency. Moreover, we also analyzed the timeseries of PWV and rainfall comparison at several sampling stations. From the results we found, PWV and rainfall correlate each other in positive way. Southern regions of Indonesia have higher correlation compared to northern regions of Indonesia.

1. Introduction

In modern times, Global Navigation Satellite System (GNSS) has become an integral part of daily activities in the public. GNSS is a terminology that refers to a set of satellite systems and technologies for positioning and navigation [1]. In addition, GNSS can be used in any weather without experiencing significant obstacles. GNSS is included in the extraterrestrial measurement method because it utilizes objects or objects in space. GNSS has been widely used in various countries for activities and activities that especially require information about position, navigation and time [2]. Currently, there are various GNSS satellites such as GPS (Global Positioning System) from the United States, GLONASS

(Global'naya Navigatsionnaya Sputnikovaya Sistema) from Russia, Galileo from Europe, Compass or BeiDou from China, QZSS (Quasi-Zenith Satellite System) from Japan, and Indian Regional Navigation Satellite System (IRNSS) / Navigation Indian Constellation (NavIC) from India.

GNSS works by sending data from satellites in the form of longitude, latitude, altitude, and time signals to GNSS signal receivers on the surface, commonly called GNSS receivers. At first, the data was generally used in daily needs such as navigation for travel, as well as for research purposes such as plate tectonics, deformation, and earthquake monitoring. However, with time, GNSS data can be used for more than just positioning. One implementation of GNSS development is the estimation of Precipitable Water Vapor (PWV).

Precipitable Water Vapor (PWV) is a variable that indicates the water vapor present in the atmosphere if the water vapor becomes dew [3]. In technical terms, PWV can also be defined as integrated water vapor in the vertical direction, expressed as the equivalent height of the water column in liquid form [4]. According to Junbo Shi [5] and Prashant Kishor [6], PWV has an important role in atmospheric water transport, energy conversion, and climate change, so it is important to monitor and understand its situation and change mechanism. PWV also plays a role in the rainfall that occurs on this earth.

Rainfall is the height of rainwater that collects in a flat place and does not experience evaporation, infiltration into the soil, and drainage to other places. The value of rainfall is expressed in millimeters (mm). Rainfall of 1 (one) millimeter means that on a flat place covering an area of one meter, there is water with a height of one millimeter or a volume of one liter [7]. Indonesia has diverse rainfall in each region. This is because each region of Indonesia has a different altitude. Rainfall will tend to increase in mountainous areas [8,9]. In general, Indonesia has high annual rainfall due to Indonesia as a maritime country located in the tropics. Tropical regions have high rainfall because of the convection process and the formation of convective clouds [9,10].

This research aims to find a correlation between PWV and rainfall in Indonesia in 2019. In this year, Indonesia experienced a weak El Nino from the beginning of the year to June, as well as positive IOD (Indian Ocean Dipole) from May to the end of the year with index value of 0.4-2.5 °C [11]. Researchers have a hypothesis that PWV derived from GNSS observations has a relationship with the amount of rainfall that occurs. The results will be obtained as a heatmap of the Pearson's correlation coefficients between pentad PWV and rainfall at InaCORS locations in Indonesia and comparison models between PWV and rainfall from InaCORS samples. The results is expected to help in various other purposes, such as rainfall prediction.

2. Study Area

The Republic of Indonesia is a country lies between two continents, Asia and Australia, and between two oceans, the Pacific Ocean and the Indian Ocean [12]. Indonesia has approximately 17.000 islands, which is located between 6 degrees north and 11 degrees south latitude, and from 95 to 141 degrees east longitude. There are five largest islands in Indonesia, which are Sumatra, Java, Kalimantan (Borneo), Sulawesi, and Papua (New Guinea) [13]. Indonesia has a tropical climate because it is located in an area passed by the equator [14].

For the locations of station samples, they are spread across the major islands of Indonesia, as shown in Figure 1. These islands are Sumatra, Java, Kalimantan, Sulawesi, and Papua. For Sumatra, Java, and Papua, each island used two InaCORS stations to compare lowland and highland InaCORS stations. For Kalimantan and Sulawesi, each island uses only one InaCORS station because these islands do not have an InaCORS station in the highlands.

The latitude and longitude coordinates and altitude above the ellipsoid for each InaCORS station studied can be seen in Table 1. The coordinates and altitude were obtained from the link http://nrtk.big.go.id/ managed by Badan Informasi Geospatial (Geospatial Information Agency) [15]. Based on the table, the InaCORS stations in highland areas are InaCORS stations CBKT, CLBG, and CWMN. InaCORS stations, other than these three stations, are located in lowland areas.

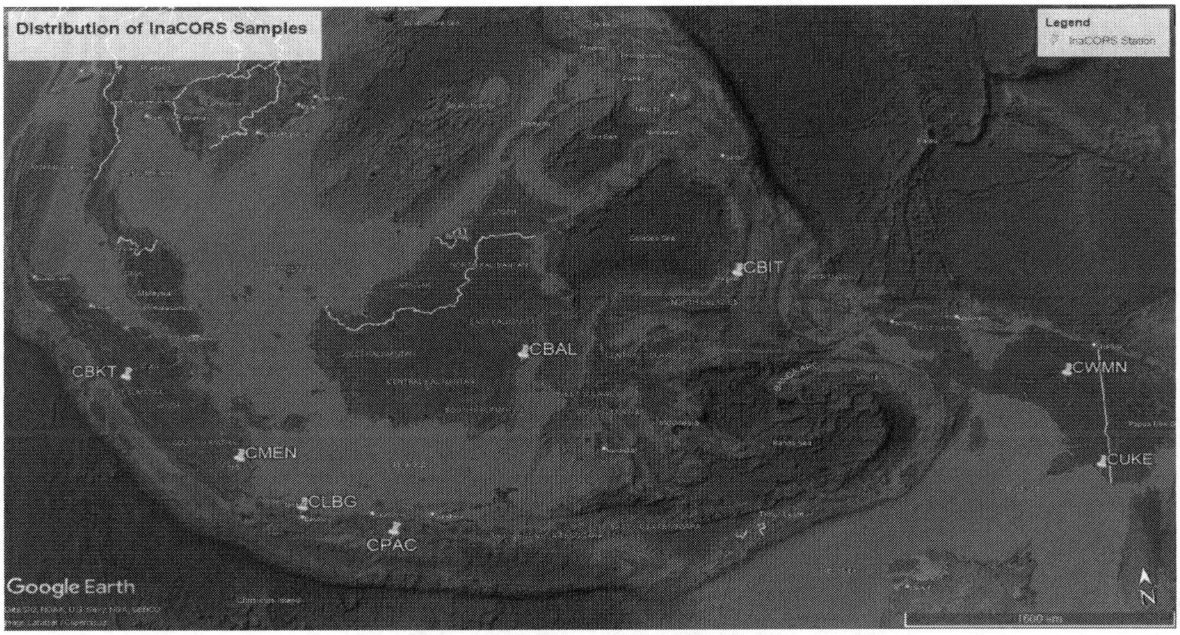

Figure 1. InaCORS samples distribution

Table 1. City/regency, coordinates, and altitude of InaCORS samples.

InaCORS station name	City/Regency of InaCORS Station	Latitude	Longitude	Height above the ellipsoid (m)
CBKT	Bukit Tinggi City	0° 18' 32" S	100° 22' 16" E	929.3
CMEN	Tulang Bawang Regency	4° 28' 27" S	105° 15' 10" E	40
CLBG	West Bandung Regency	6° 49' 28" S	107° 36' 56" E	1329.7
CPAC	Pacitan Regency	8° 11' 46" S	111° 05' 52" E	37.6
CBAL	Balikpapan City	1° 15' 22" S	116° 50' 23" E	75
CBIT	Bitung City	1° 26' 35" N	125° 11' 12" E	78.3
CWMN	Jayawijaya Regency	4° 05' 44" S	138° 56' 33" E	1744.4
CUKE	Merauke Regency	8° 29' 23" S	140° 23' 39" E	80.2

3. Data Used

3.1. Precipitable Water Vapor (PWV) data

These PWV data were originally GNSS observation data from the InaCORS station. GPS data obtained from InaCORS station observations can be processed to obtain tropospheric delay parameters. This data processing uses software called GAMIT (GNSS at MIT). The party that does the processing with the software is the Geospatial Information Agency (BIG).

GAMIT is software that has been specially designed by MIT (Massachusetts Institute of Technology), Harvard-Smithsonian Center for Astrophysics (CfA), Scripps Institution of Oceanography (SIO), and Australian National University [16]. The software can estimate coordinates and deformation velocities of observing stations, stochastic or functional representation of post-seismic deformation, atmospheric deceleration, satellite orbits, and earth orientation parameters. To run the software, Fortran and C are the programming languages used. GAMIT can be ran on various operating systems, such as Windows, LINUX, and MacOS [16].

The InaCORS data come from 237 different observation stations spread across Indonesia. For InaCORS samples, 18 stations were used. These data are then further processed to obtain values related to the disturbance parameters in GNSS positioning. The values in the processed data are GNSS observation time, Zenith Total Delay (ZTD), Zenith Hydrostatic Delay (ZHD), Zenith Wet Delay (ZWD), pressure, temperature, and PWV. The time span of the data obtained is more than 16 years from 1st January 2004 to 10th October 2021, with an interval between data every 1 hour.

3.2. Rainfall data

The rainfall data used belong to the secondary data type, just like the PWV data discussed earlier. These data include rainfall data from March 1, 2014, to September 30, 2021. The data are presented as a text document (.txt). Rainfall data can be accessed through JAXA Global Rainfall Watch (GSMaP). GSMaP has a global rainfall database that can be downloaded for free, although user registration is required. GSMaP rainfall data can be accessed at https://sharaku.eorc.jaxa.jp/GSMaP/.

GSMaP estimates rainfall for high resolution global rainfall maps using satellite data [17,18]. GSMaP implements algorithms provided by the microwave radiometer rainfall retrieval algorithm [19] microwave sounder rainfall retrieval algorithm [20], microwave imager/sounder rainfall retrieval algorithm [21], and the microwave-infrared (IR) fusion algorithm [22]. GSMaP has spatial resolution of 0.1 degree (~11 km) and temporal resolution of one hour [23,24].

As previously stated, the rainfall data used in this study are from 237 InaCORS locations. In the data, there are several values listed. The first value is the time of the rainfall observation. Rainfall observation time is divided into year, month, day, and hour. The next value is the amount of rainfall that coincides with the observation time. The amount of rainfall intensity in an hour is expressed in millimeter [23,24].

4. Methods

4.1. PWV determination from GPS data

In determining the value of PWV mathematically, PWV is closely related to the disturbing parameters in GNSS positioning caused by atmospheric deceleration. These parameters, namely ZTD, ZHD, and ZWD, can be calculated with several formulas. Based on Nilsson et al. [25], ZTD can be determined in relation to the total tropospheric refractivity in the following formulation:

$$ZTD=10^{-6}\int_{h_l}^{\infty} N_t(h)\,\mathrm{dh} \tag{1}$$

From the equation, h_l is the height of the lowest level of the troposphere and $N_t(h)$ is the total refractivity of the troposphere. The total tropospheric refractivity can be attributed to the effects of dry gas and water vapor.

Based on equation (1), *ZTD* is affected by dry gas and water vapor. Dry gas is related to *ZHD* and water vapor has a correlation with *ZWD*. In simple terms, *ZTD* can be expressed by the following equation [1]:

$$ZTD=ZHD+ZWD \tag{2}$$

To obtain PWV value, the first value to be found is the ZWD value. With equation (3), the ZWD value can be determined.

$$ZWD=ZTD\text{-}ZHD \tag{3}$$

After the ZWD value is obtained from equation (3), the PWV value can be found with the following equations:

$$PWV=\pi\times ZWD \tag{4}$$

$$\pi=\frac{10^6}{\rho_W R_V\left[\left(\frac{K_3}{T_m}\right)+K_2'\right]} \tag{5}$$

The conversion factor of ZWD to PWV is given by equation (4). In equation (5), ρ_w is the density constant of water (10^3 kg/m^3). R_V is the gas constant for PWV (461 J.kg^{-1}.K^{-1}) [26]. K_2' and K_3 are the atmospheric refractive constants, where $K_2' = 16.48$ K.hPa^{-1} and $K_3 = (3.776 \pm 0.014) * 10^5$ K^2.hPa^{-1} [27]. T_m is the mean temperature of the troposphere [4].

4.2. Pentad days normalization

In this study, the normalization method of every five days (pentad days) was carried out for two reasons. The first reason for this method is to smoothen the appearance of the data when plotted onto a graph. The raw rainfall and PWV data used had a one-hour interval during 2019. If the raw data is plotted into a graph for one year, the plot of the graph will make it difficult to draw analysis and conclusions from it. Instead of using hourly rainfall and PWV data, the data were converted into pentad days to make it easier to highlight the characteristics of rainfall and PWV at the stations under study [28]. Therefore, 1 year with 8760 hours can be divided into 73 pentad days.

To see how much difference there is between the graphical plots before and after normalization every five days, Figure 2 and Figure 3 can be observed. Both figures are graphical plots of the comparison of PWV and rainfall at CUKE Station before and after normalization every five days. The increase and decrease of PWV and rainfall can be more easily seen in the Figure 3 compared to the Figure 2.

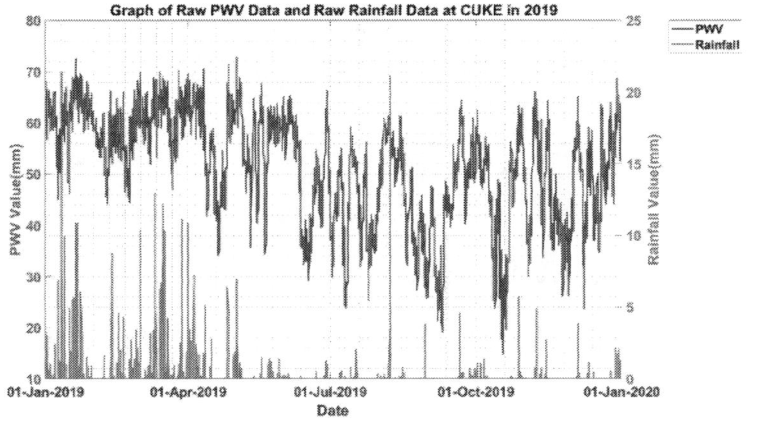

Figure 2. Graph of PWV and rainfall data before pentad days normalization.

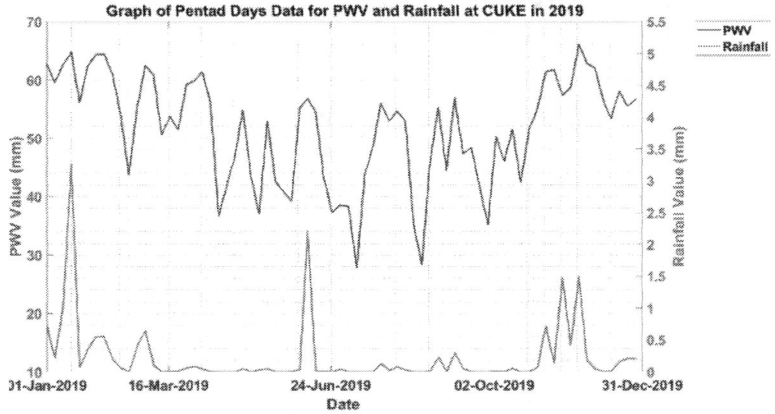

Figure 3. Graph of PWV and rainfall data after pentad days normalization.

The second reason the pentad days data normalization method is used is to reduce the effects of random high-frequency temporal variability of the weather system [29]. This can provide an opportunity to evaluate data for monitoring extreme weather and climate events in the short term [30]. By doing so, the data used can illustrate the presence of drought and rainfall extremes that occur over relatively short periods of time [31].

5. Results and Analysis

5.1. PWV and rainfall spatial distribution in Indonesia, 2019

Before delving deeper into the correlation coefficient value, the spatial distribution values of each PWV and rainfall must be considered. Throughout 2019, northern Indonesia, except for parts of northern Sumatra Island, had a higher spatial distribution of PWV compared to southern Indonesia, especially Java Island and its surrounding islands. This behavior can be seen in Figure 4.

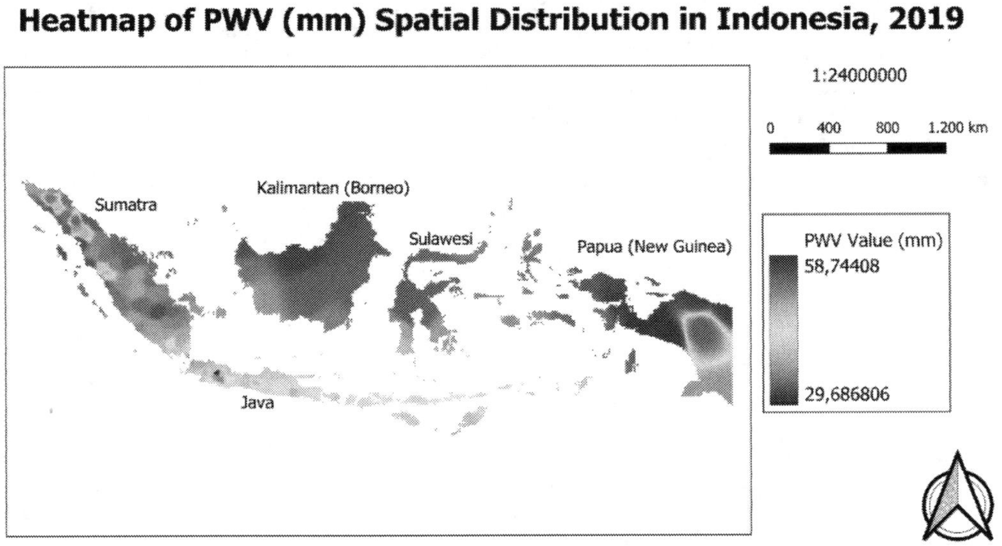

Figure 4. Heatmap of PWV (mm) spatial distribution in Indonesia, 2019.

On the other hand, the spatial distribution pattern for rainfall in Indonesia is more varied, as shown in Figure 5. The majority of Indonesia, especially Sumatra, Java, northern Sulawesi, and the Maluku Archipelago, have low spatial distribution values of rainfall.

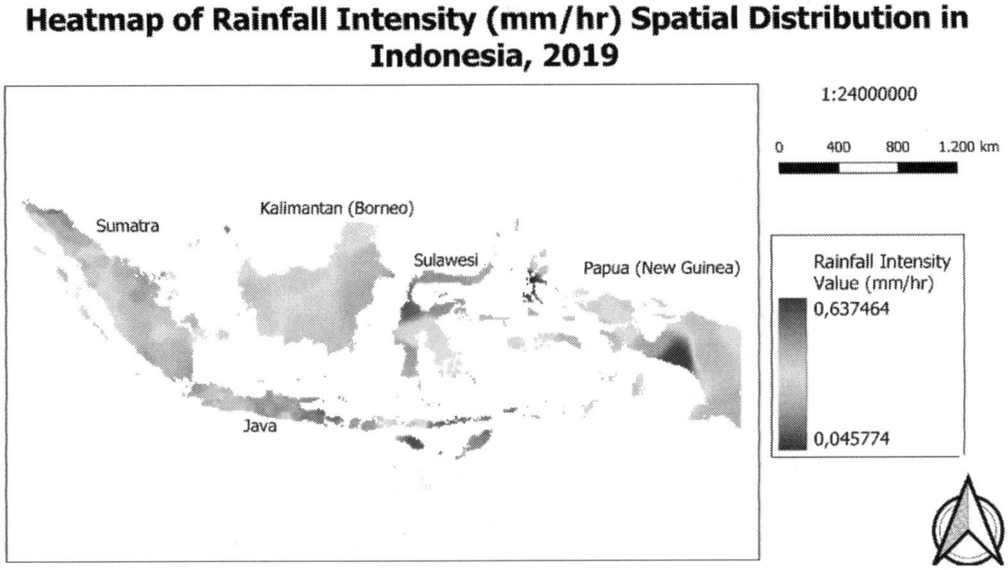

Figure 5. Heatmap of rainfall (mm/hr) spatial distribution in Indonesia, 2019.

IOP Conf. Series: Earth and Environmental Science 1245 (2023) 012035 doi:10.1088/1755-1315/1245/1/012035

5.2. PWV and rainfall correlation in Indonesia, 2019

Based on Figure 6, Indonesia has a positive correlation coefficient. The lowest correlation coefficient value is at the CFAK station, which is located in Fakfak Regency, West Papua, and has a value of 0.090146. Meanwhile, the highest value is at the CRUT station, with a value of 0.753819, which is located in Garut Regency, West Java. From the figure, southern Indonesia, especially Java Island, tends to have a higher correlation than northern Indonesia.

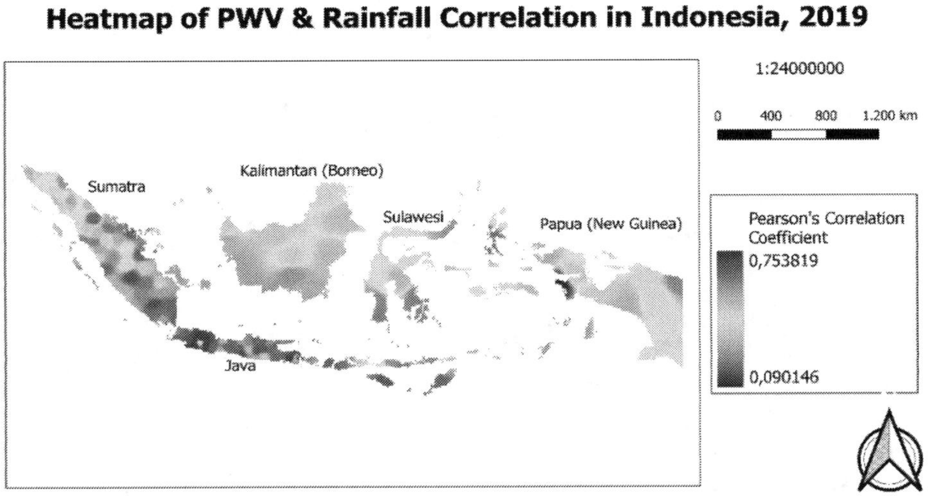

Figure 6. Heatmap of PWV and rainfall correlation in Indonesia, 2019.

5.3. Timeseries of PWV and rainfall comparison at InaCORS samples

5.3.1. InaCORS samples on Sumatra. On the island of Sumatra, The first InaCORS station on Sumatra Island is CMEN station in Tulang Bawang Regency, Lampung Province. Figure 7 shows that the PWV values obtained normalized every five days during 2019 are in the 28-65 mm range. In this area, rainfall is relatively low from mid-March to mid-October. The rainfall trend line increases as the PWV trend line does the same. Furthermore, based on the distribution plot of pentad days of rainfall compared to PWV at CMEN station, as shown in Figure 8, rainfall has a range of values of 0-1.5 mm. The scatter plot at CMEN station shows that the intensity of rainfall increases with increasing PWV value, particularly when PWV exceeds 50 mm. This is also shown by the trend line of the scatter plot. The resulting correlation coefficient between PWV and rainfall data after normalizing every five days is 0.64985.

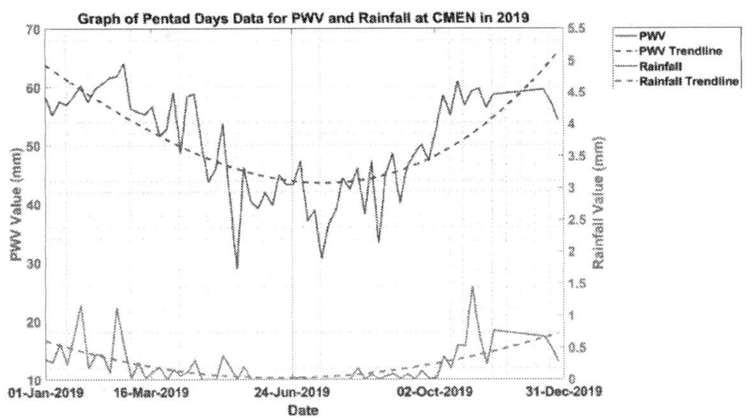

Figure 7. Comparison graph of pentad days PWV with rainfall at CMEN station in 2019.

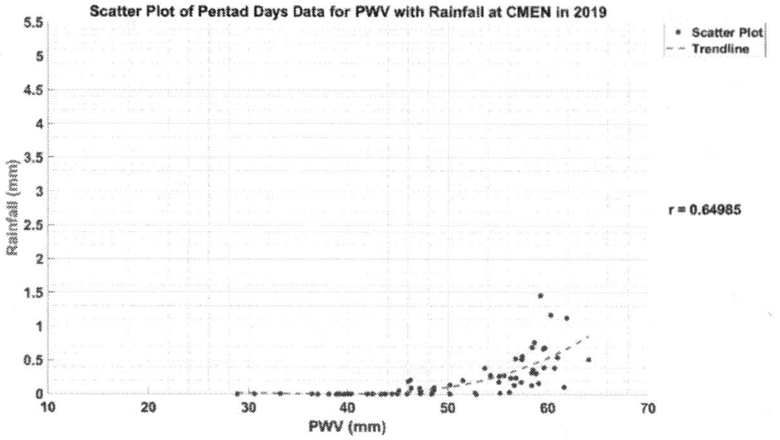

Figure 8. Scatter plot of pentad days PWV with rainfall at CMEN station in 2019.

The next InaCORS station on Sumatra Island is the CBKT station in Bukit Tinggi, West Sumatra Province. Based on the plot graph in Figure 9, the PWV value after being normalized every five days is found to be in the range of 25-48 mm. The graph also shows the value of PWV and rainfall, which decreased from the end of April to mid-September. The rainfall trend line generally increases when the PWV trend line does the same. In addition, based on the distribution plot of pentad days of rainfall compared with PWV, as shown in Figure 10, the most rainfall occurs in the range of 0-1.3 mm, although there are rainfall values around 2.4 mm. The distribution plot at CBKT station shows that the intensity of rainfall increases with increasing PWV value, particularly when the PWV value is greater than 35 mm. This is also indicated by the trend line of the scatter plot. The correlation coefficient between PWV and rainfall got after normalizing every five days at CBKT station is 0.53156.

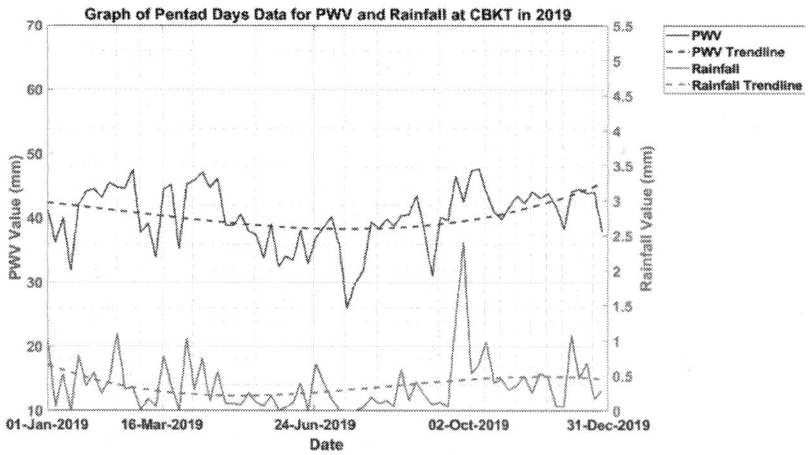

Figure 9. Comparison graph of pentad days PWV with rainfall at CBKT station in 2019.

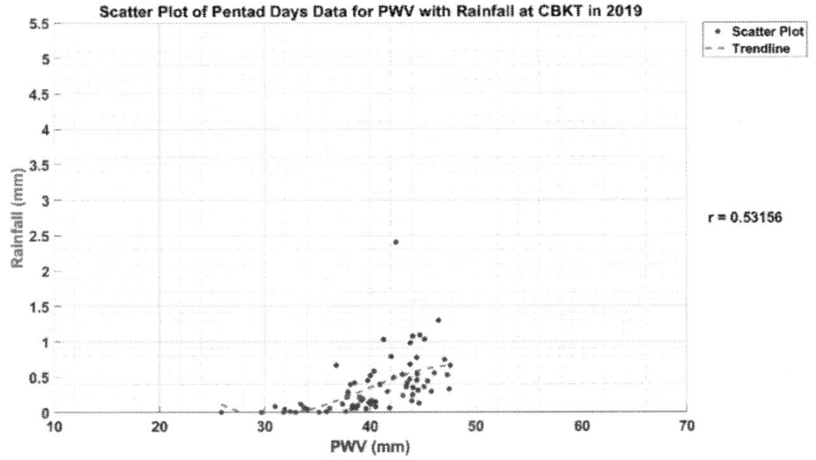

Figure 10. Scatter plot of pentad days PWV with rainfall at CBKT station in 2019.

5.3.2. InaCORS samples on Java. On the island of Java, the first InaCORS sample studied was the CLBG station located in West Bandung Regency, West Java Province. In Figure 11, the PWV value obtained after normalizing every five days is between about 14-43 mm during 2019. From the figure, rainfall decreased in intensity from mid-March to mid-September. The rainfall trend line generally increases when the PWV trend line does the same. Then, the distribution pattern of pentad days PWV against rainfall for CLBG station can then be seen in Figure 12. In the figure, rainfall has a value between 0-1.2 mm. The scatter plot shows that the intensity of rainfall events at CLBG station increases with increasing PWV values, especially when PWV has values above 20 mm. This is also indicated by the trend line of the scatter plot. The resulting correlation coefficient between PWV and rainfall data after normalizing every five days is 0.75047.

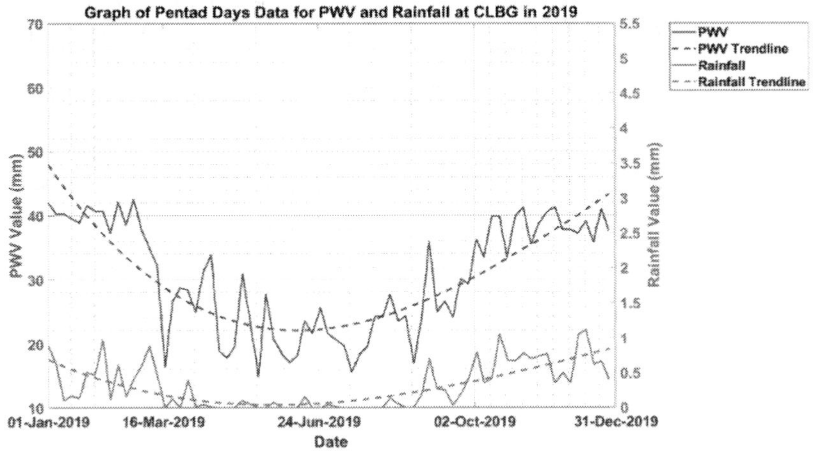

Figure 11. Comparison graph of pentad days PWV with rainfall at CLBG station in 2019.

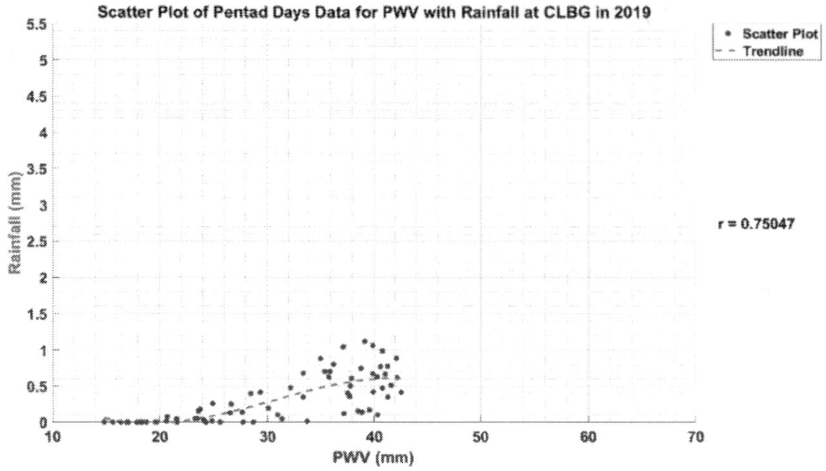

Figure 12. Scatter plot of pentad days PWV with rainfall at CLBG station in 2019.

The second InaCORS station studied on Java Island is the CPAC station located in Pacitan Regency, East Java Province. In Figure 13, the PWV value obtained after normalizing every five days is between about 27-63 mm during 2019. From the figure, rainfall almost does not occur during mid-March to early October. The rainfall trend line increases when the PWV trend line experiences the same thing. Then, in Figure 14, the distribution pattern of PWV pentad days against rainfall for CPAC station can be seen. In the figure, most rainfall occurs between 0-0.1 mm with PWV of around 27-50 mm. The scatter plot at CPAC station shows that rainfall with intensity around 0-0.1 mm still occurs a lot even though the PWV value has increased. However, the scatter plots also show that rainfall events at CPAC stations with intensities ranging above 0.1 mm increase with increasing PWV values. The trend line of the scatter plot also shows this. The resulting correlation coefficient between PWV and rainfall data after normalizing every five days is 0.6456.

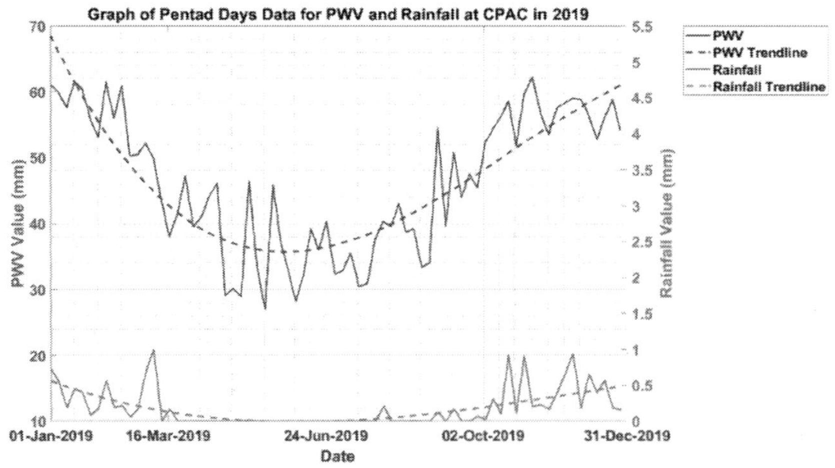

Figure 13. Comparison graph of pentad days PWV with rainfall at CPAC station in 2019.

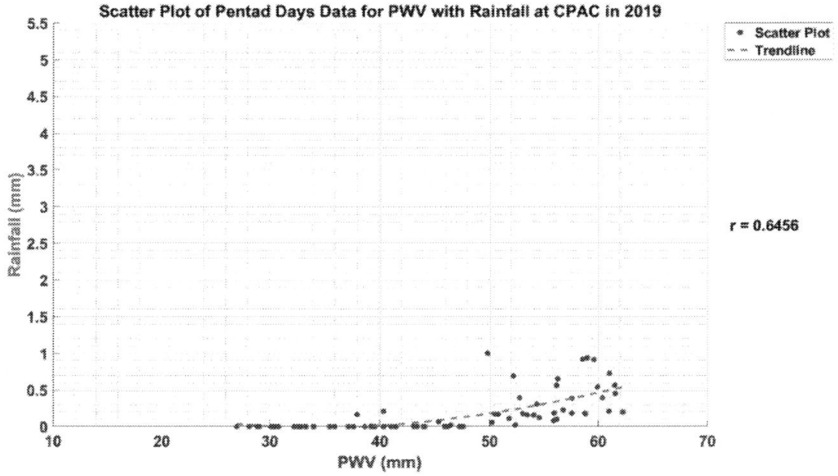

Figure 14. Scatter plot of pentad days PWV with rainfall at CPAC station in 2019.

5.3.3. InaCORS samples on Kalimantan. The InaCORS station on Kalimantan Island studied is CBAL Station, in Balikpapan City, East Kalimantan Province. In Figure 15, the PWV value obtained after normalizing every five days has a range between 37-67 mm during 2019. The rainfall trend line generally increases when the PWV trend line does the same, although there are some times when the rainfall trend line is inversely proportional to the PWV trend line. Figure 16 shows the distribution plot of pentad days of rainfall compared to pentad days of PWV data at CBAL station. According to the graph, rainfall is generally distributed between 0 and 0.9 mm. However, there is one rainfall value of 1.875 mm. Rainfall with intensity in the 0-0.1 mm range occurs mostly when PWV is in the range of 37-61 mm. The scatter plot also shows that rainfall events at CBAL station with intensities above 0.1 mm increase with increasing PWV values. The increase in rainfall intensity above 1 mm starts at a PWV of 50 mm and goes up to over 60 mm. This is also indicated by the trend line of the scatter plot. The resulting correlation coefficient between PWV and rainfall data after normalizing every five days is 0.51522.

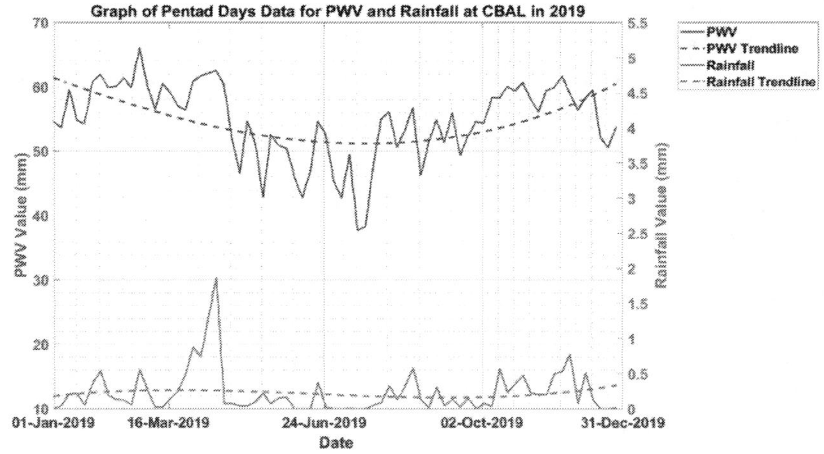

Figure 15. Comparison graph of pentad days PWV with rainfall at CBAL station in 2019.

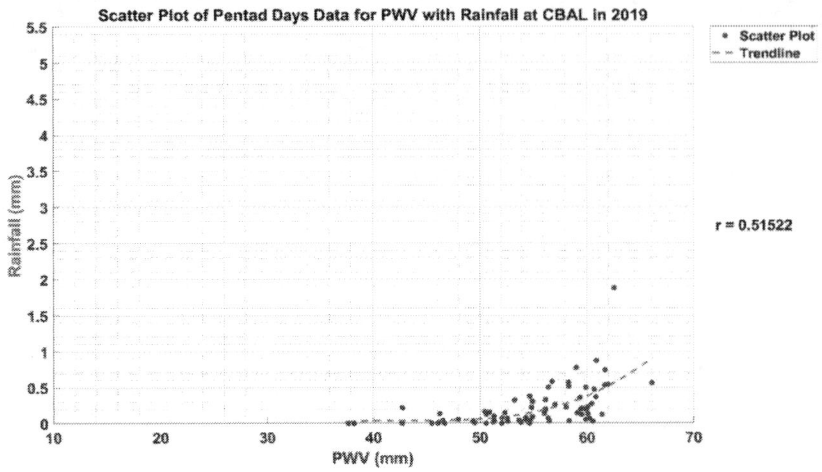

Figure 16. Scatter plot of pentad days PWV with rainfall at CBAL station in 2019.

5.3.4. InaCORS samples on Sulawesi. The InaCORS station on Sulawesi Island studied was CBIT Station, located in Bitung City, North Sulawesi Province. In Figure 17, the PWV value obtained after normalizing every five days has a range of 38-67 mm during 2019. The rainfall trend line generally increases when the PWV trend line experiences the same thing. Figure 18 shows the distribution plot of pentad days of rainfall data compared to pentad days of PWV data at CBIT station. Based on the figure, rainfall has a value between 0-1.3 mm. Rainfall in the range of 0-0.1 mm occurs most often with PWV in the range of 38-58 mm. The distribution plot also shows that rainfall events at CBIT station with intensities above 0.1 mm increase with increasing PWV values. The increase in rainfall intensity above 1 mm starts when the PWV is 47 mm and goes up to over 65 mm. This is also indicated by the trend line of the scatter plot. The resulting correlation coefficient between PWV and rainfall data is 0.59358.

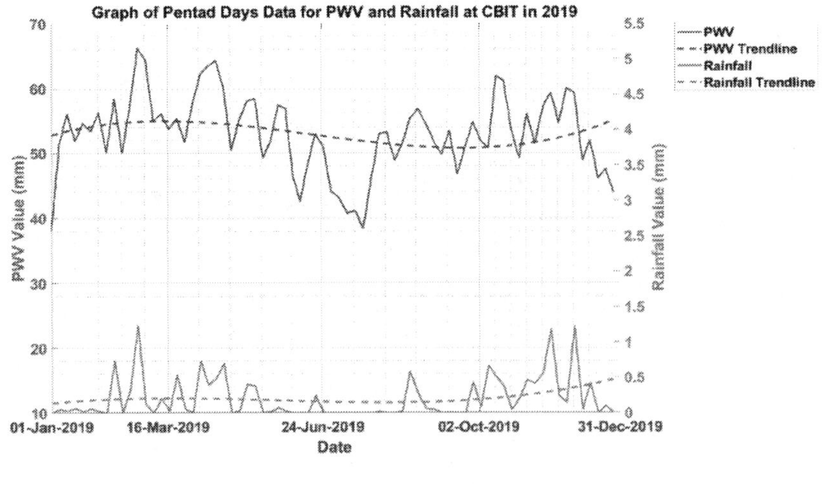

Figure 17. Comparison graph of pentad days PWV with rainfall at CBIT station in 2019.

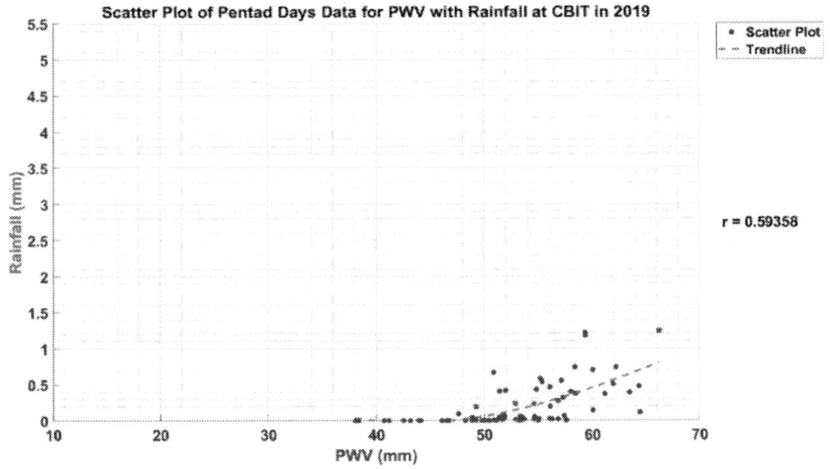

Figure 18. Scatter plot of pentad days PWV with rainfall at CBIT station in 2019.

5.3.5. InaCORS samples on Papua. The first InaCORS station in Papua studied is the CWMN station located in Jayawijaya Regency, Papua Province. The pentad days PWV value obtained in Figure 19 ranges between 24 and 40 mm in 2019. Rainfall in this area occurs mostly throughout the year. The rainfall trend line generally increases when the PWV trend line does the same. Figure 20 shows the distribution plot of pentad days of rainfall compared to pentad days of PWV at station CWMN. Based on the figure, most rainfall occurs in the range of 0 mm to 1.0159 mm. The scatter plot at station CWMN shows that rainfall intensity tends to increase with increasing PWV values. This is also indicated by the trend line of the scatter plot. The resulting correlation coefficient between PWV and rainfall data is 0.46827.

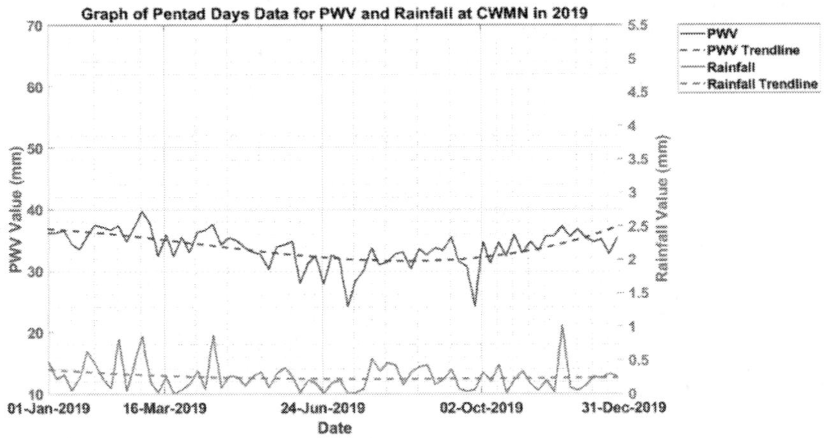

Figure 19. Comparison graph of pentad days PWV with rainfall at CWMN station in 2019.

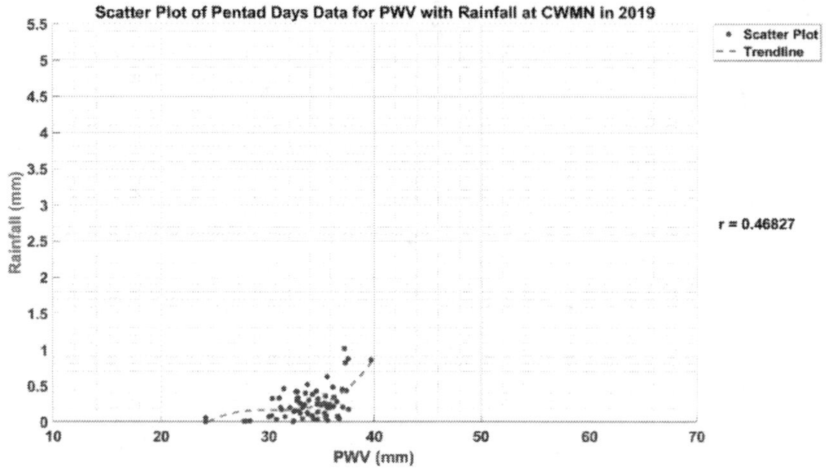

Figure 20. Scatter plot of pentad days PWV with rainfall at CWMN station in 2019.

The last InaCORS station in Papua studied is the CUKE station located in Merauke Regency, Papua Province. In Figure 21, the pentad days PWV values obtained have a range between 27-67 mm during 2019. Rainfall in this area is relatively low from March until the end of October. The rainfall trend line generally increases when the PWV trend line experiences the same thing. Figure 22 shows the distribution plot of pentad days of rainfall compared to pentad days of PWV at station CUKE. Based on the figure, rainfall has values in the range of 0-3.2616 mm. Most rainfall occurs in the range of 0-0.1 mm, with PWV in the range of 27-62 mm. The scatter plot at CUKE station shows that rainfall intensity in the range of 0-0.1 mm still occurs a lot even though the PWV value has increased. However, the scatter plot also shows that rainfall events at CUKE station with intensities above 0.1 mm increase with increasing PWV values. This is also indicated by the trend line of the scatter plot. The increase in rainfall intensity above 1 mm starts when the PWV is 55 mm. The resulting correlation coefficient between pentad days PWV data and pentad days rainfall data is 0.44127.

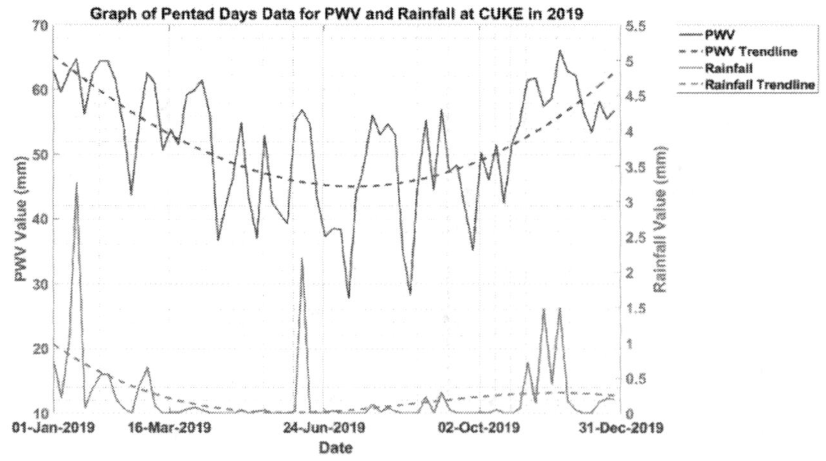

Figure 21. Comparison graph of pentad days PWV with rainfall at CUKE station in 2019.

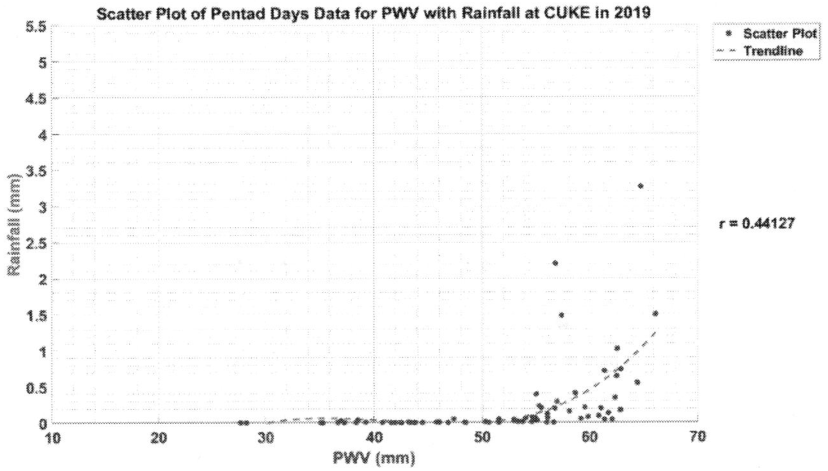

Figure 22. Scatter plot of pentad days PWV with rainfall at CUKE station in 2019.

5.4. Analysis of PWV and rainfall correlation and comparison

Based on the results obtained, the correlation coefficient values obtained have a positive value throughout 2019. The correlation coefficient values of the sampling stations located in the highland and lowland areas can be expressed in Table 2 and Table 3.

Table 2. Correlation coefficient of PWV with rainfall at highland InaCORS samples

InaCORS Station Name	Height above the ellipsoid (m)	Pearson's Correlation Coefficient
CBKT	929.3	0.53156
CLBG	1329.7	0.75047
CWMN	1744.4	0.46827

Table 3. Correlation coefficient of PWV with rainfall at lowland InaCORS samples.

InaCORS Station Name	Height above the ellipsoid (m)	Pearson's Correlation Coefficient
CMEN	40	0.64985
CPAC	37.6	0.6456
CBAL	75	0.51522
CBIT	78.3	0.59358
CUKE	80.2	0.44127

Based on the results obtained and the two tables above, the highest correlation coefficient from sampling stations is shown by the CLBG station on Java Island with a value of 0.75047 at 1329.7 m above the ellipsoid. In addition, the InaCORS samples on Java Island have a larger correlation coefficient value compared to other islands, although the PWV values are more volatile. This shows that the relationship between PWV and rainfall on Java Island tends to be stronger. The tables above also show that stations in lowland areas do not always have a smaller correlation coefficient value compared to stations in highland areas, although the range of the highest PWV values in the lowlands tends to be greater than the range of the highest PWV values in the highlands.

This positive correlation indicates that PWV and rainfall have a directly proportional relationship. This unidirectional relationship is also visible in the graph comparing pentad days PWV with rainfall at each sampling station. On the graph, rainfall tends to occur when PWV has increased before. The opposite is true when rainfall decreases and PWV also decreases.

IOP Conf. Series: Earth and Environmental Science 1245 (2023) 012035 doi:10.1088/1755-1315/1245/1/012035

A positive correlation between rainfall and PWV in the Eulerian atmospheric moisture budget equation. The equation is expressed as follows [32]:

$$E\text{-}P=\frac{\partial PWV}{\partial t}+\nabla.\frac{1}{g}\int_{p_t}^{p_s} qv\,dp \tag{6}$$

In equation (6), E and P are variables for total evaporation and precipitation, respectively. Time is represented by the variable t.g is the gravitational constant, which has a value of 9.8 m/s^2. The variables p_s and p_t are the pressures at the Earth's surface and top of atmosphere, respectively. The pressure at the top of the atmosphere can be considered zero [33]. The variable q is a value that states the specific humidity in a particular area. v is a variable that states the wind vector.

Based on equation (6), if an area has $E\text{-}P>0$, it is experiencing water evaporation. Conversely, if an area has $E\text{-}P<0$, it is experiencing precipitation. The value of E is greater when the gradient of PWV increases, and the value of P is greater when the gradient of PWV decreases. This is in accordance with the results obtained, where rainfall occurs when the PWV increases before the rain occurs. In contrast, during the rain, PWV decreased.

On the other hand, there are some rainfall intensities in the range of 0-0.1 mm based on the distribution plot of rainfall with PWV at each station, and the intensity increases with increasing PWV values. The phenomenon of a lot of rainfall occurring when it has a value between 0-0.1 mm can also be explained by equation (6). A lot of rainfall between 0-0.1 mm can be caused by the influence of wind that makes the specific humidity in an area change.

Then, if the altitude of each InaCORS station is considered, InaCORS stations in lowland areas have lower PWV values than InaCORS stations in highland areas. The event where PWV is different in the highlands and lowlands can be caused by significant pressure and temperature differences in the area. These pressure and temperature differences also cause differences in the humidity present in the two areas. To understand why differences in humidity at low and high altitudes can affect PWV, the equation below expresses its correlation with specific humidity [34].

$$PWV=\frac{1}{g}\int_{p_t}^{p_s} q\,dp \tag{7}$$

The specific humidity used is proportional to the mixing ratio of water vapor in the atmosphere. The mixing ratio of water vapor is closely related to air pressure. The relationship is expressed as follows:

$$r=r_s\left(\frac{p}{p_s}\right)^{\lambda} \tag{8}$$

From equation (8), r is the variable for the mixing ratio at pressure p at various altitudes of interest. r_s is the variable for the mixing ratio at ground level or at the reference altitude of interest. The power λ is a variable that depends on altitude, season, and atmospheric conditions [33].

Equation (7) and equation (8) show that PWV has a relationship with air pressure. The air pressure in the lowlands is high. Conversely, air pressure in the highlands is low. In addition, based on Gay-Lussac's Law, air pressure is directly proportional to air temperature. This is under the actual environmental conditions where lowlands tend to have higher ambient temperatures than highlands. Therefore, lowlands have high PWV values and highlands have low PWV values.

In addition, the results obtained also show that the PWV in areas near the equator is higher compared to areas far from the equator when excluding areas in the highlands. This is because areas close to the equator have higher temperatures. This leads to higher evaporation rates and increases the specific humidity.

6. Conclusion

In this study, researchers used PWV data derived from GPS data and rainfall data to determine the correlation that occurred in Indonesian regions during 2019. There are some interesting findings that can be highlighted in this study. The southern part of Indonesia, especially Java Island, has a stronger relationship between PWV and rainfall than other islands, although the PWV value is more volatile. This is indicated by the relatively higher correlation coefficient of PWV with rainfall on Java Island. For lowland areas, the correlation coefficient values for CMEN station in Sumatra, CPAC station in Java, CBAL station in Kalimantan, CBIT station Sulawesi, CUKE station in Papua are 0.64985, 0.6456, 0.51522, 0.59358, and 0.44127, respectively. For highland areas, the correlation coefficient values for CBKT station in Sumatra Island, CLBG station in Java Island, and CWMN station in Papua Island are 0.53156, 0.75047, and 0.46827, respectively.

The correlation coefficient between pentad days of PWV data and pentad days of rainfall data obtained at all InaCORS stations studied is positive. The positive correlation coefficient value indicates that the PWV derived from GNSS data is directly proportional to the rainfall that occurs in Indonesia. The positive correlation between pentad days of rainfall data with PWV and rainfall that mostly occurs at 0-0.1 mm intensity can be explained by the Eulerian atmospheric moisture budget equation.

If the altitude of the InaCORS station location is considered, the highlands have a lower range of PWV values compared to the lowlands. In addition, the PWV will be higher and more stable the closer to the equator. These are due to the difference in specific humidity. This difference is caused by differences in pressure and temperature in these areas which affect the level of evaporation that occurs. The difference in specific humidity at low and high altitudes and in areas close and far to the equator can be explained by the PWV equation with specific humidity and the mixing ratio equation.

Acknowledgments. I would like to express my deepest gratitude to Dr. Techn. Dhota Pradipta, S.T., M.T. who has provided advices and inputs so that the writing of this research article can be maximized. In addition, I would also like to thank anonymous Reviewer 1 and Reviewer 2 for providing reviews to the writing of this paper.

References

[1] Wijaya DD. GNSS Meteorologi. Bandung: ITB Press; 2019.
[2] Abidin HZ. Penentuan Posisi dengan GPS dan Aplikasinya. Jakarta: PT Pradnya Paramita 2007.
[3] Syaifullah MD. Potensi atmosfer dalam pembentukan awan konvektif pada pelaksanaan teknologi modifikasi cuaca di DAS Kotopanjang dan DAS Singkarak 2010. Jurnal Sains & Teknologi Modifikasi Cuaca 2011;12:9–16. https://doi.org/10.29122/jstmc.v12i1.2185.
[4] Bevis M, Chiswell S, Businger S, Herring TA, Bock Y. Estimating wet delays using numerical weather analyses and predictions. Radio Sci 1996;31:477–87. https://doi.org/10.1029/96RS00008.
[5] Shi J, Xu C, Guo J, Gao Y. Real-Time GPS precise point positioning-based precipitable water vapor estimation for rainfall monitoring and forecasting. IEEE Transactions on Geoscience and Remote Sensing 2015;53:3452–9. https://doi.org/10.1109/TGRS.2014.2377041.
[6] Kishore P, Venkat Ratnam M, Namboothiri SP, Velicogna I, Basha G, Jiang JH, et al. Global (50°S–50°N) distribution of water vapor observed by COSMIC GPS RO: Comparison with GPS radiosonde, NCEP, ERA-Interim, and JRA-25 reanalysis data sets. J Atmos Sol Terr Phys 2011;73:1849–60. https://doi.org/10.1016/J.JASTP.2011.04.017.
[7] Triatmodjo B. Hidrologi Terapan. Yogyakarta: Beta Offset; 2008.
[8] Supari S, Sudibyakto, Ettema J, Aldrian E. Spatiotemporal Characteristics of Extreme Rainfall Events over Java Island, Indonesia. Indones J Geogr 2012;44:62–86. https://doi.org/10.22146/indo.j.geog.2391.
[9] Tjasyono HKB, Gernowo R, Woro BHS, Ina J. The character of rainfall in the Indonesian monsoon. The International Symposium on Equatorial Monsoon System, Yogyakarta, 2008.
[10] Mulyono D. Analisis karakteristik curah hujan di wilayah Kabupaten Garut Selatan. Jurnal Konstruksi 2016;12. https://doi.org/10.33364/konstruksi/v.12-1.274.
[11] BMKG. Catatan Kondisi Iklim, Kualitas Udara, dan Gas Rumah Kaca Tahun 2019. Jakarta: 2020.

[12] Cribb R, Ford M. Indonesia as an Archipelago: Managing Islands, Managing the Seas. In: Cribb R, Ford M, editors. Managing an Archipelagic State, ISEAS Publishing; 2009, p. 1–27. https://doi.org/doi:10.1355/9789812309815-005.
[13] CBS/Indonesia CB of S-, NFPCB/Indonesia SM of PFPCB-, MOH/Indonesia M of H-, International M. Indonesia Demographic and Health Survey 1997. Calverton, Maryland, USA: CBS and Macro International; 1998.
[14] Curry JA, Webster PJ. Thermodynamics of Atmospheres and Oceans. Elsevier; 1998.
[15] Badan Informasi Geospasial. Site Status 2018. http://nrtk.big.go.id/sbc/User/SiteMap/SiteMap (accessed February 14, 2023).
[16] Herring TA, King RW, McClusky SC. Introduction to gamit/globk. Massachusetts Institute of Technology, Cambridge, Massachusetts 2010.
[17] Kubota T, Shige S, Hashizume H, Aonashi K, Takahashi N, Seto S, et al. Global precipitation map using satellite-borne microwave radiometers by the GSMaP Project: Production and validation. IEEE Transactions on Geoscience and Remote Sensing 2007;45:2259–75. https://doi.org/10.1109/TGRS.2007.895337.
[18] Okamoto K i., Ushio T, Iguchi T, Takahashi N, Iwanami K. The Global Satellite Mapping of Precipitation (GSMaP) Project. Proceedings. 2005 IEEE International Geoscience and Remote Sensing Symposium, 2005. IGARSS '05., vol. 5, 2005, p. 3414–6. https://doi.org/10.1109/IGARSS.2005.1526575.
[19] Aonashi K, Awaka J, Hirose M, Kozu T, Kubota T, Liu G, et al. GSMaP passive microwave precipitation retrieval algorithm : Algorithm description and validation. Journal of the Meteorological Society of Japan Ser II 2009;87A:119–36. https://doi.org/10.2151/jmsj.87A.119.
[20] Shige S, Yamamoto T, Tsukiyama T, Kida S, Ashiwake H, Kubota T, et al. The GSMaP precipitation retrieval algorithm for microwave sounders—Part I: Over-Ocean Algorithm. IEEE Transactions on Geoscience and Remote Sensing 2009;47:3084–97. https://doi.org/10.1109/TGRS.2009.2019954.
[21] Kubota T, Shige S, Kachi M, Aonashi K. Development of SSMIS rain retrieval algorithm in the GSMaP project. Proc 28th ISTS, 2011.
[22] Ushio T, Sasashige K, Kubota T, Shige S, Okamoto K, Aonashi K, et al. A kalman filter approach to the Global Satellite Mapping of Precipitation (GSMaP) from combined passive microwave and infrared radiometric data. Journal of the Meteorological Society of Japan Ser II 2009;87A:137–51. https://doi.org/10.2151/jmsj.87A.137.
[23] Kubota T, Aonashi K, Ushio T, Shige S, Takayabu YN, Kachi M, et al. Global Satellite Mapping of Precipitation (GSMaP) Products in the GPM Era. In: Levizzani V, Kidd C, Kirschbaum DB, Kummerow CD, Nakamura K, Turk FJ, editors. Satellite Precipitation Measurement: Volume 1, Cham: Springer International Publishing; 2020, p. 355–73. https://doi.org/10.1007/978-3-030-24568-9_20.
[24] GPM Global Rainfall Map Algorithm Development Team. Global Satellite Mapping of Precipitation (GSMaP) for GPM: Algorithm Theoretical Basis Document (ATBD) (Algorithm Ver.6 Prepared). 2014.
[25] Nilsson T, Böhm J, Wijaya DD, Tresch A, Nafisi V, Schuh H. Path Delays in the Neutral Atmosphere. In: Böhm J, Schuh H, editors. Atmospheric Effects in Space Geodesy, Berlin, Heidelberg: Springer Berlin Heidelberg; 2013, p. 73–136. https://doi.org/10.1007/978-3-642-36932-2_3.
[26] Wallace JM, Hobbs P V. Atmospheric science: an introductory survey. 2nd ed. Elsevier; 2006.
[27] Davis JL, Herring TA, Shapiro II, Rogers AEE, Elgered G. Geodesy by radio interferometry: Effects of atmospheric modeling errors on estimates of baseline length. Radio Sci 1985;20:1593–607. https://doi.org/10.1029/RS020i006p01593.
[28] Ji Y, Sun X, Xu Y, Yao J, Yang X-Q. Summer Regional Pentad Extreme Precipitation in Eastern China and Their Possible Causes. Front Earth Sci (Lausanne) 2021;8.

[29] World Meteorological Organization. Guide to Climatological Practices. 2018th ed. Geneva: 2018.
[30] World Meteorological Organization. Space-Based Weather and Climate Extremes Monitoring (SWCEM) 2019. https://public.wmo.int/en/programmes/wmo-space-programme/swcem (accessed June 11, 2022).
[31] Paterson L, Berod D, de Coning E, Harding J, Holmlund K, Honda Y, et al. Harnessing Earth System Science, Technology and Services to reduce Disaster Risk–WMO contributions. WMO Bulletin 2022;71:3–11.
[32] Shi Y, Jiang Z, Liu Z, Li L. A Lagrangian Analysis of Water Vapor Sources and Pathways for Precipitation in East China in Different Stages of the East Asian Summer Monsoon. J Clim 2020;33:977–92. https://doi.org/10.1175/JCLI-D-19-0089.1.
[33] Viswanadham Y. The Relationship between Total Precipitable Water and Surface Dew Point. J Appl Meteorol Climatol 1981;20:3–8. https://doi.org/10.1175/1520-0450(1981)020<0003:TRBTPW>2.0.CO;2.
[34] Hagos SM, Leung LR, Garuba OA, Demott C, Harrop B, Lu J, et al. The Relationship between Precipitation and Precipitable Water in CMIP6 Simulations and Implications for Tropical Climatology and Change. J Clim 2021;34:1587–600. https://doi.org/10.1175/JCLI-D-20-0211.1.

IOP Conf. Series: Earth and Environmental Science 1245 (2023) 012036 doi:10.1088/1755-1315/1245/1/012036

Utilization of Jason-3 Satellite Altimetry Data for Observation of TC Seroja

Hananda D. Mahetran[1], Dudy D. Wijaya[1,2], Zamzam A. J. Tanuwijaya[1,2], Wedyanto Kuntjoro[1,2], Agustinus B. Setyadji[1,2], Fathin Nurzman[2], Ivonne M. Radjawane[3], Hansan Park[4]

[1]Department of Geodesy and Geomatics Engineering, Institut Teknologi Bandung, Indonesia
[2]Geodesy Research Group, Institut Teknologi Bandung, Indonesia
[3]Oceanography Research Group, Institut Teknologi Bandung , Indonesia
[4]Korea-Indonesia Marine Technology Cooperation Research Center (MTCRC), Cirebon, Indonesia

hanandadmahetran@gmail.com

Abstract. The Jason-3 Satellite Altimetry has two main instruments: the Poseidon-3B Altimeter and the Advanced Microwave Radiometer. The two sensors produce many output data. Satellite altimetry data commonly used to observe TC are Sea Level Anomaly (SLA) and Significant Wave Height (SWH). This research tries something new by adding three other data, namely Precipitable Water Vapor (PWV), Liquid Water Content (LWC), and Wind Speed (WS). The object of this research is TC Seroja which occurred from 2 to 12 April 2021. This research was conducted to obtain all variable's spatial and temporal profiles. The results of this study indicate that TC Seroja causes an increase in SLA, PWV, SWH, LWC, and WS. The highest SLA was 0.495 m, SWH was 4.649 m, PWV was 0.10760 m, LWC was 2.08 kg/m^2, and WS was 18.42 m/s. In addition, the spatial model for the five variables in April 2021 obtained using IDW (Inverse Distance Weighting) can also provide information on the influence of TC in various places.

1. Introduction

A tropical cyclone (TC) is a form of extreme weather disturbance that begins with a tropical depression or an intensive center of low pressure over the ocean, thus triggering intensive convection processes and cloud formation [1]. TC can occur in warm waters with temperatures over 26.5°C [2]. Most TC form at latitude 10°-20° from the equator [3]. TC that form below 10° from the equator are few due to the small Coriolis force, even though the temperature in that area is high. Conversely, TCs that form above latitude 20° from the equator are few due to the low temperature in that area, even though the Coriolis force in that area is more significant.

Indonesia is located at latitude 6°N to 11°S and longitude 95°E to 141°E. Even though they are located at low latitudes, TCs rarely occur in Indonesia [4], [5]. The characteristics of TC in Indonesia are divided by region, namely TC in southern Indonesia and TC in northern Indonesia. Most TC in southern Indonesia form from December to April. Meanwhile, most TCs in northern Indonesia are formed from

July to October [6]. Several TCs that have occurred in Indonesia are TC Vamei in 2001, Rosie in 2008, Cempaka in 2017, Dahlia in 2017, Flamboyan in 2018, Kenanga in 2018, and Seroja in 2021 [7].

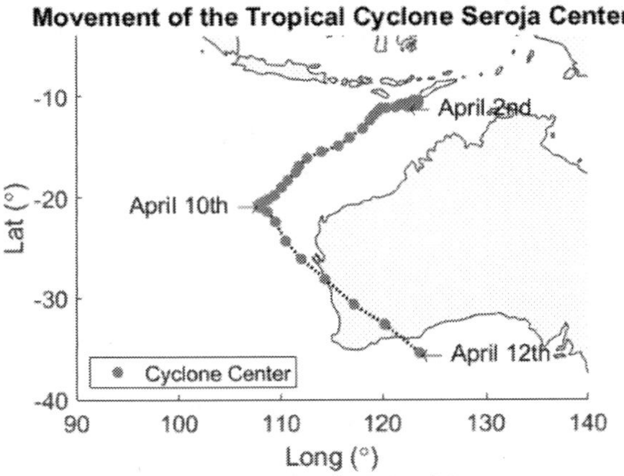

Figure 1 Movement of TC Seroja's center

TC can be observed using data from satellite altimetry [8]. The satellite altimetry data commonly used in TC observations is Sea Level Anomaly (SLA) and Significant Wave Height (SWH). Several studies use this data to observe TC. One study showed that the TC Cempaka and Dahlia caused an increase in SWH in the south, north, and strait areas from Lampung to Central Java with the highest value was 3.75 m and the largest percentage of SWH increase was 1028.31% [9]. Another study used SLA data from satellite altimetry to observe sea level height in southern Java when TC Cempaka occurred. This study's results indicate a decrease in SLA reaching 0.2 m at the center of the cyclone and an increase in SLA reaching 0.4 m in the southern coastal areas of Java [10]. In addition, there is also research to examine SLA changes on the Lampung Coast when TC Dahlia occurs. This study uses SLA data from satellite altimetry from 30 November to 2 December 2017. As a result, TC Dahlia increased SLA in Kota Agung and Krui waters [11].

One of the huge TCs that recently happened was TC Seroja which occurred from 2-12 April 2021. On 31 March 2021, wind speeds over the island of Timor began to increase, followed by the formation of a depression zone in the Savu Sea on 1 April 2021. The depression zone continued to strengthen to form a low-pressure system. On 2 April 2021, a low-pressure system formed a TC [12]. Madden-Julian Oscillation (MJO) at that time which was at the turn of phase 4 and phase 5 formed a supportive environment for cyclogenesis [13]. The TC moves southwards until it ends in southern Australia on 12 April 2021. The movement of TC Seroja's center from its beginning to its end can be seen in Figure 1. Other research found that TC Seroja influenced the intensity of extreme rainfall and wind speed in Nusa Tenggara Timur and western Australia [13], [14]. This extreme rainfall triggered flooding in several areas [15], [16].

Not only to analyze TC Seroja's impact, but this research also aims to analyze the potential of satellite altimetry data to observe TC. Other research said that TC Seroja caused extreme rainfall and wind speed based on rainfall and wind speed data from other sources. This study tries to observe the impact of TC on rainfall and wind speed using satellite altimetry data. Satellite altimetry has two main instruments, namely radar altimeter and radiometer. To study wind speed, wind speed data from altimeter radar sensor measurements are used. While rainfall was studied indirectly using Liquid Water Content (LWC) and Wet Tropospheric Correction (WTC) data from radiometer measurements. In addition, this study also adds SLA and SWH data from radar altimeter measurements to observe the impact of TC Seroja on these two ocean parameters.

This study describes the feasibility of satellite altimetry data to observe TC. If proven usable, it is hoped that satellite altimetry data can be combined with other data to conduct more comprehensive research.

2. Methodology

2.1. Data

The data used in this study are SLA, SWH, WS, WTC, and LWC from the Jason-3 satellite altimetry. Data were obtained using the Radar Altimetry Database System (RADS) in the ITB Geodesy Research Group room. The Jason-3 satellite altimetry has 9,9156 days repeat cycle period and 254 passes within a cycle, ascending (asc) passes are odd-numbered and descending (dsc) passes are even-numbered [17].

SLA is the height of the sea level relative to the Mean Sea Surface (MSS). SLA can be obtained by reducing satellite altitude data by satellite distance to sea level, propagation correction, bias, and MSS models [18]. The measurement of the distance between the satellite and the sea surface is carried out by an active radar altimeter sensor. The signal emitted by the altimeter radar will be reflected by the sea surface and again captured by the altimeter sensor. The time difference between signal transmission and signal capture is again recorded and then processed into data measuring the distance between the satellite and the sea surface [19].

SWH is the average height of 1/3 of the highest ocean waves [20]. The presence of waves on the surface of the sea changes the shape of the return wave which is measured by the radar altimeter. The leading edge of the returned waveform is stretched as a result of the previous return from the crest of the wave and then the return from the trough of the wave. This stretching increases with increasing wave height. Therefore, wave conditions can be estimated from the slope of the leading edge of the returned waveform [19]. The Jason-3 satellite mission measures SWH using two frequencies, namely Ku-Band and C-Band [18]. TC cause rain in the affected areas. Signals at the Ku-Band frequency get more significant interference from rain when compared to signals at the C-Band frequency [21]. Therefore, in this study, SWH measurements were used at the C-band frequency to observe the influence of TC Seroja on the characteristics of sea waves.

WTC is a correction to signal propagation disturbances by water vapor and cloud grains in the tropospheric medium [19]. And LWC is the amount of water content in the atmosphere in liquid form. WTC is obtained using brightness temperature (BT) measurements measured at a multi-frequency radiometer. Channels with a frequency of 23.8 GHz are very sensitive to water vapor which captures most of the WTC information. Higher frequency channels are also used to capture LWC contributions. This second channel has a frequency of 34 GHz. The radiometer used by the Jason mission also uses a third channel at 18.7 GHz, in addition to good sensitivity to water vapor, providing additional information about the sea surface roughness [23].

The last data used is WS. In addition to measuring the waveform of the received signal to estimate SWH, the altimeter also measures the energy of the received signal (backscatter coefficient) which can be used to estimate WS at sea level [19], [22]. The wind will reduce the energy of the signal as it travels through the atmosphere. The smaller the energy of the received signal, the greater the wind speed in the area traversed [18]. Two kinds of WS data can be obtained from the Jason-3 satellite altimetry, namely altimeter wind speed and water vapor radiometer wind speed. In rainy conditions, the wind speed measurement by the water vapor radiometer sensor has worse accuracy than the wind speed by the radar altimeter sensor [23], [24]. Therefore, the wind speed measurement results from the altimeter sensor were used in this study.

2.2. Convert WTC to PWV (Precipitable Water Vapor)

Altimetry satellites have an on-board microwave radiometer whose original function is to measure the WTC for their measurement using radar altimeter sensor. The WTC, as it reflects the amount of water vapor in the atmosphere, could be used to derive the PWV [25]. This section will describe an algorithm for obtaining the PWV from the WTC generated by satellite altimetry.

The WTC obtained by the satellite altimetry is negative of the Zenith Wet Delay (ZWD) (Equation 1). To get PWV from ZWD can use equation 2, with an approximate value for Π being 0.16 [26].

$$WTC = -ZWD \quad (1)$$

$$PWV = \Pi . ZWD \quad (2)$$

2.3. Inverse Distance Weighting (IDW) Interpolation

The first law of geography states, "Objects that are close together in space are always more similar than those that are far apart" [27]. IDW is an interpolation method that fulfills the first law of geography and is one of the most frequently used interpolation methods in geographic information systems [28]. IDW performs interpolation by assigning a weight to the sample data based on the distance. Closer sample points will have greater weight [29].

3. Result and Discussion

SLA, SWH, WTC, LWC, and WS data were obtained from RADS in areas ranging from latitude 4°S to 40°S and longitude 90°E to 135°E. This area was taken to determine the extent of the influence of TC Seroja on SLA, SWH, PWV, LWC, and WS. In this area, there are 44 satellite altimetry passes consisting of 22 ascending and 22 descending passes, as seen in Figure 2. It should be noted that satellite altimetry does not pass through all of these passes at the same time as the TC Seroja. Therefore, it is necessary to select the pass that the satellite passes at the same time as TC Seroja occurs. Data processing is divided into 3, namely spatial profiles, temporal profiles, and spatial models.

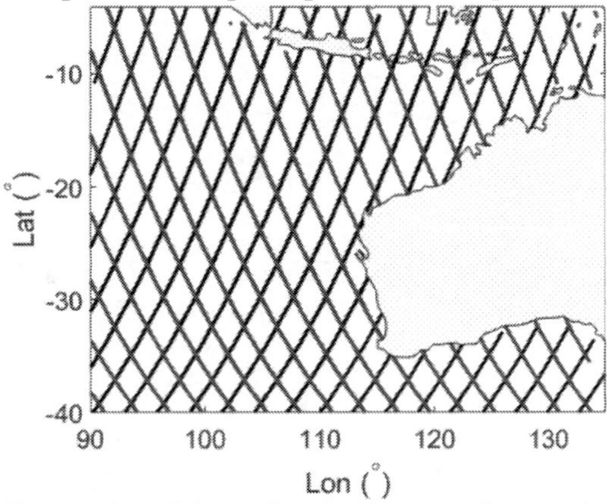

Figure 2 Ascending (black) and descending (blue) passes of Jason-3 in the research area

3.1. Spatial Profile

The Jason-3 satellite altimetry orbits at a speed of 7.2 km/s in outer space or the equivalent of 5.8 km/s on the earth's surface. During operation, satellite altimetry retrieves data every 1.0187 seconds. Thus, the distance between the nearest observation points on the same pass is 5.8636 km. This sampling distance is close enough to observe TC, which generally has a radius of up to hundreds of kilometers [3].

In this area, there are 44 satellite altimetry passes. But only pass 64 (dsc) that close to TC Seroja's center (< 150 km). This pass was passed by satellite altimetry on April 7, 23:26 UTC. This pass is the closest to the center of TC Seroja compared to the other 43 passes. The distance between pass 64 and the center of TC Seroja is 23,231 km.

The profiles of the five variables in pass 64 can be seen in Figure 3. The figure shows data gaps in the SLA profile at -14° to -16° latitude. This deletion can occur because the SLA value in that range is above the threshold set by RADS, so it was considered as outlier and deleted automatically. However, the SLA profile still shows the influence of TC. The figure shows that the value of SLA decreases near the center of the TC and increases in value when it is away from the cyclone's center. The largest SLA value occurs near Java Island, with a value of 0.495 m at 943.610 m from the TC's center. This indicates constructive interference between storm surge waves and daily tides, causing higher tides.

The PWV and LWC profiles show a significant increase at -15° latitude. This shows a huge increase in water vapor and liquid water content. The most significant water evaporation occurs at the cyclone walls [30]. The PWV value reaches 10.760 cm with 143.165 m from the TC's center. At the same time, the largest LWC value is 2.08 kg/m2. The figure also shows that the increase and decrease in PWV are slower and gentler than LWC.

The SWH and WS profiles have the same characteristics: they generally increase as they approach the center of a TC. This shows the magnitude of the influence of the wind on the formation of sea waves when a TC occurs. The wind that blows on the water's surface will form waves in the water [31]. When the wind speed increases, the waves will grow [32]. The highest SWH value reached 4.649 m at a distance of 129.532 km from the TC's center. At the same time, the highest WS value occurs at a distance of 63.104 km at 18.42 m/s or 66.321 km/hour.

The WS profile shows fluctuations in value when approaching -15° latitude. The WS of the altimeter is obtained using backscatter coefficient (sigma 0) data. Fluctuations in WS values around -15° latitude indicate disturbances from the amount of rainfall in the area. Measurements on the Ku-band are greatly affected by rain, while measurements on the C-band are not so affected [33].

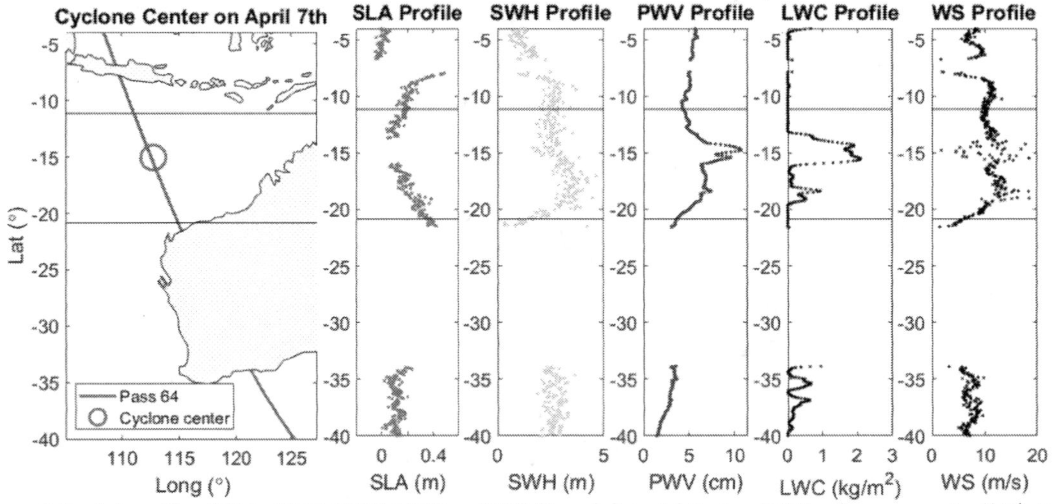

Figure 3 Spatial profile of SLA, SWH, PWV, LWC, and WS (from left to right) when TC Seroja occur on April 7th, 2021

The effect of rain on altimeter measurements can be explained in more detail by comparing the backscatter coefficient values of the Ku-band and C-band measurements, as seen in Figure 4. The backscatter coefficient on Ku-band will experience a significant decrease in value when measuring in cloudy areas and rainy conditions. This can also be seen from the Ku-band's backscatter coefficient, which generally has a smaller value than the backscatter coefficient of the C-band. The figure also shows a significant difference between the Ku-band's backscatter coefficient and the C-band's backscatter coefficient at latitude 14S° to 19S°. From this, it can be concluded that TC Seroja resulted in high rainfall in the area.

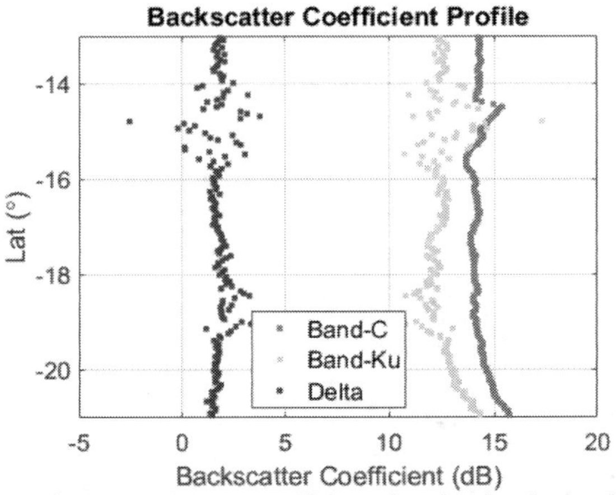

Figure 4 Backscatter coefficient of Ku-band and C-band

3.2. Temporal Profile

Processing of the temporal profile for the five variables was carried out at one sample point (-16.117°, 112.970°). This processing was carried out to compare the values of each variable from March to May 2021. Since the period for satellite altimetry to pass through the same point for the second time is 9.9156 days (10 days less 2 hours), there are 9 times to be compared. The temporal profile can be seen in Figure 5. This figure shows that the highest values of SWH, LWC, PWV, and WS occurred during TC Seroja.

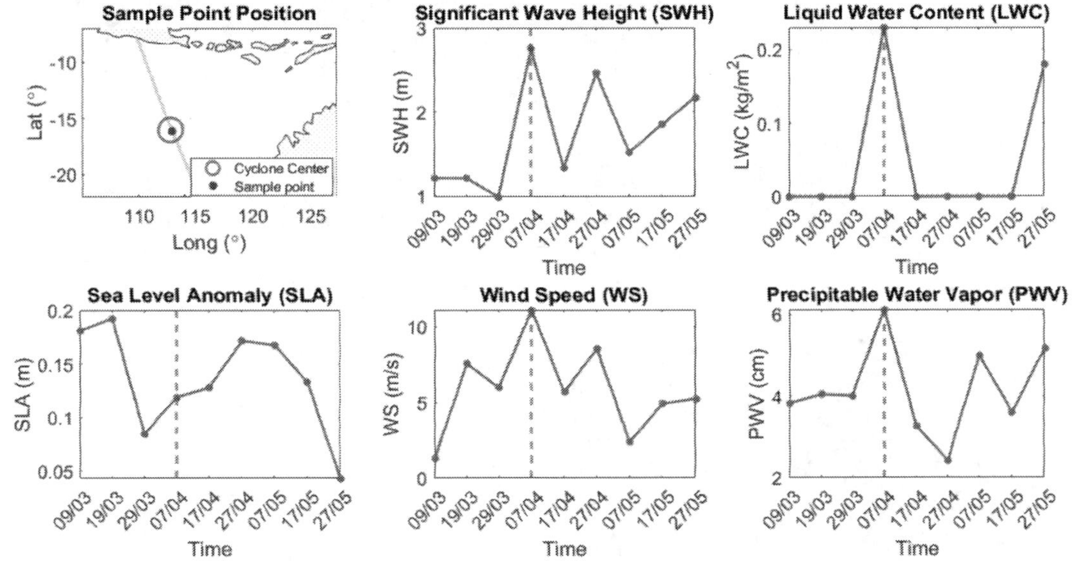

Figure 5 Temporal profile

3.3. Spatial Model

The spatial modeling of SLA, SWH, PWV, and WS during April 2021 was carried out by visualizing the five variables, which were divided into five timeframes, namely week-0, week-1, week-2, week-3, and week-4. Week-0 is used as a visual reference for week-1 to week-4. The data used in week-0 is data from 22 to 31 March 2021. The data used in week-1 is data from 1 to 7 April 2021. The data used in week-2 is data from 8 to 14 April 2021. The data used in week 3 is data from 15 to 21 April 2021. The data used in week 4 is data from 22 to 30 April 2021. The difference in the period results in a difference in the distribution of pass used in data processing each week, so the reduction process by week-0 data cannot be done directly. Therefore, an interpolation process is carried out to change the data distribution into a grid measuring 15' x 15'. The interpolation process is carried out using cubic interpolation or third-degree polynomials. After the data has been spread evenly on a 15' x 15' grid, the process of reducing week-1 to week-4 data by week-0 can be carried out.

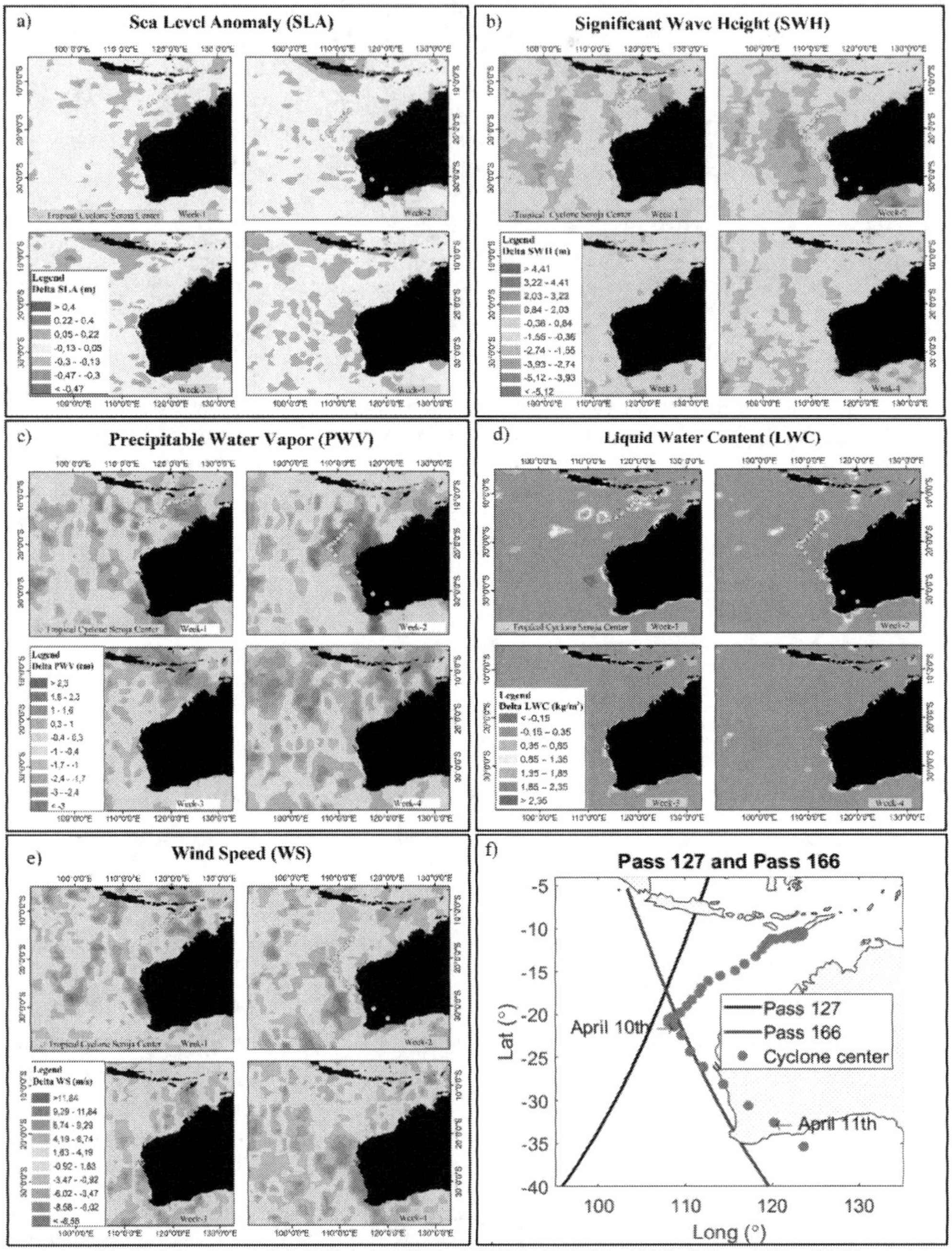

Figure 6 Spatial model a) SLA, b) SWH, c) PWV, d) LWC, e) WS, f) Pass 127 and 166

Data processing is continued by performing IDW interpolation to create a spatial model for each variable. The SLA model every week can be seen in Figure 5.a. The figure shows that TC Seroja only occurs in week-1 and week-2. The figure shows a visualization of the difference in SLA from week 1 to 4 to week 0. From this figure, the influence of TC Seroja on the SLA is unclear. The area around the center of the TC Seroja is dominated by class 4 with an SLA difference ranging from -0.13 m to 0.05 m. This shows a tiny change in SLA value in that area.

Figure 5.b. shows the weekly model of SWH in April. The figure shows an increase in SWH values around the cyclone's center during week-1 and week-2. The rise in SWH in this area was not seen in week-3 and week-4. Thus, it can be concluded that the increase in SWH in the region during week-1 and week-2 is the result of TC Seroja. SWH values at week-1 and week-2 ranged from 1 to 3 meters higher than SWH at week-0. In week 2, it can be seen that the increase in SWH is not right at the center of TC Seroja but is to the west of the center of TC Seroja. This can happen because the satellite altimetry data obtained in the area is from pass 127 (asc) observed on 10 April 2021 and pass 166 (dsc) observed on 11 April 2021. Both passes are located to the west of the center of TC Seroja. The high SWH in this area is also caused by another TC, namely TC Odette.

The weekly PWV model is shown in Figure 5.c. Basically, the PWV value will be greater as it gets closer to the equator. This is due to the higher water temperature causing greater evaporation. The PWV value at the equator during week-0 ranged from 4.7 cm to 7.5 cm. Like SWH, the PWV value around the center of TC Seroja also increased during week-1 and week-2. In week-1, the PWV value increased to 2.3 cm higher than week-0. While the increase in PWV at week-2 has a value of more than 2.3 cm higher than week-0. This is indicated by areas with the highest class (red).

From the distribution of PWV in week-1 and week-2, it can be seen that the PWV value decreased significantly in the waters south of NTT. This can be seen from the color of the area. During week-1, that aresa has an orange color which indicates the PWV value of week-1 is higher than the PWV value of week-0. However, during week-2, that area is blue, which indicates the PWV value during week-2 in that region becomes smaller than the PWV value during week-0. This shows that the amount of water vapor content has decreased quite a lot after the occurrence of a TC. The decrease in water vapor can occur due to moisture divergence or because there has been a huge amount of precipitation before.

Similar to SWH, the magnitude of the PWV value in week 2 was also influenced by the presence of TC Odette. However, the effect of using data from pass 127 and pass 166 is not as big as that found in the SWH week-2.

The LWC model for each week in April 2021 can be seen in Figure 5.d. The image shows rain when TC Seroja occurs. However, the phenomenon of rain when a TC occurs cannot be observed as a whole. This is because the spatial and temporal resolution of the Jason-3 satellite is not good for rain observations. The phenomenon of rain only occurs in a small area.

The weekly WS model in April 2021 can be seen in Figure 5.e. Just like SWH and PWV, the figure shows an increase in WS values around the center of TC Seroja during week-1 and week-2. The decrease in WS values from week-1 to week-2 was also seen in the waters south of NTT. In addition, a decrease in the value of WS in the waters southwest of Australia was also seen from week 2 to week 3. This decrease in the value of WS again confirms the findings on the temporal profile. The WS distribution in the 2nd week shows an inappropriate increase in the center of TC Seroja. This is the same as the SWH distribution in week-2. One reason is the use of data from pass 127 and pass 166 which are west of the center of TC Seroja week-2.

The spatial model created can be used to determine the distribution of areas affected by TC Seroja. Even so, there are still things lacking in using satellite altimetry data in modeling phenomena with a short period and narrow coverage. For example, as seen in the SWH and WS week-2 models. The increase in the values of these two variables did not occur exactly at the center of TC Seroja, but to the west. This can happen due to limited data availability. In week 2, sample data when the TC Seroja occurred was obtained from pass 127, passed by the satellite on 10 April 2021 at 10.14 UTC and pass 166, passed by the satellite on 11 April 2021 at 23.05 UTC. The positions of the two passes can be seen in Figure 5.f.

The SWH and WS values in the two passes have a higher value than the other passes used in the interpolation. Because the interpolation method makes it impossible to make the maximum value at a point other than the sample point, the point with the highest value remains in the two passes. This is why the increase in SWH and WS in the 2nd week is not right at the center of TC Seroja. One effort that can be made to overcome this is to use satellite altimetry data with different missions. Different satellite missions have different orbits.

4. Conclusion

From the research, satellite altimetry can be used to observe TC. The spatial profiles of SLA, SWH, PWV, LWC, and WS that have been made can show that TC Seroja increases SWH, PWV, LWC, and WS in areas with latitudes close to the cyclone's center. Meanwhile, SLA decreased when it was close to the cyclone's center and increased when it was away from the cyclone's center up to a certain distance. The highest SLA was 0.495 m, SWH was 4.649 m, PWV was 0.10760 m, LWC was 2.08 kg/m2, and WS was 18.42 m/s.

The temporal profile from March to May showed that the SWH, PWV, LWC, and WS values when TC Seroja occurred were the highest. In addition, PWV, LWC, and WS also show a significant decrease in value up to 10 days after TC Seroja. This shows that the amount of water vapor content has decreased quite a lot after the occurrence of a TC. The decrease in water vapor can occur due to moisture divergence or because there has been a huge amount of precipitation before.

The obtained SLA spatial model does not show the influence of TC Seroja. The SWH, PWV, LWC, and WS spatial models show the influence of TC Seroja on these variables. This effect can be seen by increasing the values of the four variables near the center of TC Seroja in week-1 and week-2. However, the spatial and temporal resolution of the Jason-3 satellite in carrying out data collection causes several deficiencies in the resulting spatial model so that it does not represent the actual conditions. In order for TC observations to provide better results, a combination of satellite altimetry data can be performed with other data that have a higher spatial and temporal resolution.

Acknowledgement

This paper can be completed with the help of various parties. Therefore, on this occasion, the authors would like to thank Nabila S.E. Putri who has helped the author in understanding the impact of TC on PWV; and Muhammad Syahrullah Fathulhuda who has provided basic knowledge about satellite altimetry and RADS. This research is also part of projects titled "Korea-Indonesia Marine Technology Cooperation Research Center (20220512)" and "Ocean and Coastal Basic Survey and Capacity Enhancement in Cirebon, Indonesia (PG53340)" which are funded by the Ministry of Oceans and Fisheries, Korea.

References

[1] N. S. Haryani dan A. Zubaidah, "Dinamika Siklon Tropis di Asia Tenggara Menggunakan Data Penginderaan Jauh," Teknologi, vol. 29, no. 324, 2012.
[2] A. D. Nugroho dan N. H. Muzaki, "Study of Surface and Vertical Sea Temperatures during the Process of Tropical Cyclone Formation in the Territory of Indonesia (Case Study 2019-2021)," Sci. IOP Conf. Ser. Earth Environ., hal. 0–9, 2022, doi: 10.1088/1755-1315/989/1/012006.
[3] J. P. Terry, Tropical Cyclone Climatology and Impacts in the South Pacific. Springer, 2007.
[4] E. E. S. Makmur dkk., "Strengthening the Early Detection and Tracking of Tropical Cyclones near Indonesian Waters," IOP Conf. Ser. Earth Environ. Sci., vol. 925, no. 1, 2021, doi: 10.1088/1755-1315/925/1/012010.
[5] N. S. Ningsih, F. Hanifah, T. S. Tanjung, L. F. Yani, dan M. Al Azhar, "The Effect of Tropical Cyclone Nicholas (11–20 February 2008) on Sea Level Anomalies in Indonesian Waters," J. Mar. Sci. Eng., vol. 8, no. 11, hal. 1–17, 2020, doi: 10.3390/jmse8110948.
[6] M. D. Syaifullah, "Siklon Tropis, Karasteristik dan Pengaruhnya di Wilayah Indonesia pada Tahun 2012," J. Sains Teknol. Modif. Cuaca, vol. 16, no. 2, hal. 61–71, 2015, doi: 10.29122/jstmc.v16i2.1048.
[7] Avrionesti, F. Khadami, dan D. W. Purnaningtyas, "Ocean Response to Tropical Cyclone Seroja at East Nusa Tenggara Waters," IOP Conf. Ser. Earth Environ. Sci., vol. 925, no. 1, hal. 0–8, 2021, doi: 10.1088/1755-1315/925/1/012045.

[8] International Altimetry Team, "Altimetry for the Future: Building on 25 Years of Progress," Adv. Sp. Res., vol. 68, no. 2, hal. 319–363, 2021, doi: 10.1016/j.asr.2021.01.022.
[9] W. Windupranata, C. A.D.S. Nusantara, D. D. Wijaya, dan K. Prijatna, "Impact Analysis of Tropical Cyclone Cempaka-Dahlia on Wave Heights in Indonesian Waters from Numerical Model and Altimetry Satellite," KnE Eng., vol. 2019, hal. 203–214, 2019, doi: 10.18502/keg.v4i3.5851.
[10] H. Burrahman, N. K. Ginting, dan A. Rahmat, "Study of the Influence of Cempaka Tropical Cyclone on the Height of Sea Waves in the South Java Sea Using the Delft 3D Application," IOP Conf. Ser. Earth Environ. Sci., vol. 739, no. 1, 2021, doi: 10.1088/1755-1315/739/1/012043.
[11] F. S. Ariwibowo, A. Fadlan, dan D. Triatmoko, "Kajian Perubahan Tinggi Muka Laut pada saat Siklon Tropis Dahlia di Pesisir Lampung," J. Widya Climago, vol. 2, no. 2, hal. 41–52, 2021.
[12] R. Y. Setiawan dkk., "Impacts of Tropical Cyclone Seroja on the Phytoplankton Chlorophyll-a and Sea Surface Temperature in the Savu Sea, Indonesia," IEEE Access, vol. 9, hal. 152938–152944, 2021, doi: 10.1109/ACCESS.2021.3125605.
[13] B. Latos dkk., "The Role of Tropical Waves in the Genesis of Tropical Cyclone Seroja in the Maritime Continent," Nat. Commun., vol. 14, no. 856, hal. 1–12, 2023, doi: 10.1038/s41467-023-36498-w.
[14] D. S. Krisnayanti, P. de Rozari, V. C. Garu, A. C. Damayanti, D. Legono, dan H. Nurdin, "Analysis of Flood Discharge due to Impact of Tropical Cyclone," Civ. Eng. J., vol. 8, no. 9, hal. 1752–1763, 2022, doi: 10.28991/CEJ-2022-08-09-01.
[15] R. Kurniawan dkk., "The Impact of Tropical Cyclone Seroja to the Rainfall and Sea Wave Height in East Nusa Tenggara," IOP Conf. Ser. Earth Environ. Sci., vol. 925, no. 1, hal. 0–12, 2021, doi: 10.1088/1755-1315/925/1/012049.
[16] A. B. Sekaranom, N. H. Putri, dan F. C. Puspaningrani, "The impacts of Seroja Tropical Cyclone towards extreme weather in East Nusa Tenggara," E3S Web Conf., vol. 325, hal. 01020, 2021, doi: 10.1051/e3sconf/202132501020.
[17] CNES, EUMETSAT, JPL, dan NOAA/NESDIS, Jason-3 Products Handbook, no. 1.5. 2018.
[18] R. Scharroo, "RADS Data Manual," 2021.
[19] L. L. Fu dan A. Cazenave, Satellite Altimetry and Earth Sciences. 2001.
[20] L. Q. Avia, "A Comparative Analysis of the Wind and Significant Wave Height on the Extreme Weather Events (TC Cempaka and TC Dahlia) in the Southern Sea of Java, Indonesia," IOP Conf. Ser. Earth Environ. Sci., vol. 572, no. 1, 2020, doi: 10.1088/1755-1315/572/1/012033.
[21] Y. Quilfen, J. Tournadre, dan B. Chapron, "Altimeter Dual-Frequency Observations of Surface Winds, Waves, and Rain Rate in Tropical Cyclone Isabel," J. Geophys. Res., vol. 111, hal. 1–13, 2006, doi: 10.1029/2005JC003068.
[22] D. Stammer dan A. Cazenave, Satellite Altimetry over Oceans and Land Surfaces. 2018.
[23] A. Ribal dan I. R. Young, "Global Calibration and Error Estimation of Altimeter, Scatterometer, and Radiometer Wind Speed Using Triple Collocation," Remote Sens., vol. 12, no. 12, 2020, doi: 10.3390/rs12121997.
[24] T. Meissner dan F. J. Wentz, "Wind-Vector Retrievals Under Rain with Passive Satellite Microwave Radiometers," IEEE Trans. Geosci. Remote Sens., vol. 47, no. 9, hal. 3065–3083, 2009, doi: 10.1109/TGRS.2009.2027012.
[25] F. Nurzaman, D. D. Wijaya, Z. A. J. Tanuwijaya, dan W. Kuntjoro, "Utilization of Altimetry Satellites Microwave Radiometer for Atmospheric Observation over Indonesia," IOP Conf. Ser. Earth Environ. Sci., vol. 1047, no. 1, hal. 0–12, 2022, doi: 10.1088/1755-1315/1047/1/012002.
[26] A. R. Reis, J. Catalao, G. Vieira, dan G. NIco, "Mitigation of Atmospheric Phase Delay in Insar Time Series using Era-Interim Model, GPS, and Modis Data: Application to the Permafrost Deformation in Hurd Peninsula, Antarctica," no. 1, hal. 3454–3457, 2015.
[27] W. R. Tobler, "A Computer Movie Simulating Urban Growth in the Detroit Region," in Economic Geography, 1970, vol. 46, hal. 234–240, doi: 10.2307/143141.
[28] W. L. Cerón, R. V. Andreoli, M. T. Kayano, T. Canchala, Y. Carvajal-Escobar, dan R. A. F.

Souza, "Comparison of Spatial Interpolation Methods for Annual and Seasonal Rainfall in Two Hotspots of Biodiversity in South America," An. Acad. Bras. Cienc., vol. 93, no. 1, hal. 1–22, 2021, doi: 10.1590/0001-3765202120190674.

[29] P. Wei, S. Xie, L. Huang, dan L. Liu, "Ingestion of GNSS-Derived ZTD and PWV for Spatial Interpolation of PM2.5 Concentration in Central and Southern China," Int. J. Environ. Res. Public Health, vol. 18, no. 15, hal. 1–26, 2021, doi: 10.3390/ijerph18157931.

[30] R. A. Anthes, Tropical Cyclones: Their Evolution, Structure, and Effects. American Meteorological Society, 1983.

[31] T. Li dan L. Shen, "The Principal Stage in Wind-wave Generation," J. Fluid Mech., vol. 934, hal. 1–50, 2022, doi: 10.1017/jfm.2021.1153.

[32] J. Proudman, Ocean Surface Waves: Their Physics and Prediction, vol. 11. 2005.

[33] S. M. Bhandari dan A. K. Varma, "Potential of Simultaneous Dual-Frequency Radar Altimeter Measurements from TOPEX/Poseidon for Rainfall Estimation over Oceans," Remote Sens. Environ., vol. 58, no. 1, hal. 13–20, 1996, doi: 10.1016/0034-4257(96)00050-8.

Data Reconstruction of Sea Surface Temperature in Indonesia's Fish Management Area 713 (IFMA-713) Using Machine Learning

S Nurdjaman[1,2], R A Wicaksono[2], A R Kartadikaria[1,2] and M R Abdullah[3]

[1]Oceanography Group, Faculty of Earth Science and Technology, Bandung Institute of Technology, Bandung, Indonesia
[2]Oceanography Study Program, Bandung Institute of Technology, Bandung, Indonesia
[3]Atmospheric Science Research Group, Faculty of Earth Science and Technology, Bandung Institute of Technology, Bandung, Indonesia

E-mail: rheno.armand07@gmail.com

Abstract. The IFMA-713 in Indonesia is water that has dynamic of temperature changes due to interactions with the Pacific Ocean and the surrounding. Sea surface temperature data can be obtained by measuring with satellite imagery. However, satellite imagery measurements of sea surface temperature can be incomplete due to cloud cover. In this study, a machine learning method was used to reconstruct sea surface temperature data using a backpropagation neural network algorithm. The data used in this research is data captured with MODIS Satellite. Then, the reconstruction of sea surface temperature data is carried with four scenarios with missing data percentages: empty data, zero values, average values at the point of data collection, and Indonesia's average sea surface temperature. Accurate results were obtained in reconstructing sea surface temperature where the scenarios had a positive correlation. The most accurate scenarios for reconstructing sea surface temperature data with missing data were those in which the empty data was filled with average values at the point of data collection or Indonesia's average sea surface temperature.

Keywords: Backpropagation Neural Network, Data Reconstruction, IFMA-713, Machine Learning, Sea Surface Temperature (SST).

1. Introduction

Indonesia as an archipelagic country has a sea area of 5.8 million km^2 or around 62% of the entire Indonesian state. As an archipelagic country with such a large sea area it has an impact on the climate around its territory. Indonesia's geographical location between the Pacific Ocean and the Indian Ocean, flanked by two large bodies of water, makes its climate maritime rather than continental [1]. This maritime climate is affected by many oceanographic phenomena, such as El Niño–Southern Oscillation (ENSO), Indian Ocean Dipole (IOD), and monsoon wind cycles. These phenomena can cause variations in sea surface temperature in Indonesia.

The temperature in the sea has a tendency to vary both horizontally and vertically. With variations in temperature with depth, the surface layer tends to have a homogeneous temperature. However, as the depth increases, the water under the homogeneous layer does not have time to experience rapid temperature changes, causing a considerable difference with the water layer above it. The layer where there is a sudden vertical change in temperature is known as the thermocline layer. Under the thermocline layer, temperature changes will occur slowly so that the layer can be said to be almost homogeneous. In each waters, the thickness of the homogeneous layer and/or the depth of the thermocline layer will vary. Sea Surface Temperature (SST) is the temperature of the waters around sea level. The temperature of the water, especially on the surface, is highly dependent on the amount of heat from the sun. Sea surface temperature is very important to know because the distribution of sea surface temperature can provide information about fronts, upwelling, currents, weather/climate and fishing grounds.

Sea surface temperature (SST) data can be obtained using MODIS Aqua satellite imagery. However, the MODIS Aqua satellite has some limitations when collecting sea surface temperature (SST) data. Clouds can block the satellite's sensor, preventing it from collecting data from areas that are covered by clouds. This can lead to missing or inaccurate data [2]. Atmospheric disturbances due to clouds not only interfere with satellite image post-processing, but also interfere with the process of image recognition and classification [3]. Several methods can be used to fill in the gaps in satellite image data due to cloud cover. One method to replace cloud-covered regions with satellite images of the same area that are not covered by clouds. This method is not always practical due to the difficulty of finding replacement data and the high costs involved.

The rapid development of technology and the use of good data archives allows SST data processing to be integrated with Artificial Intelligence (AI) to facilitate processing of sea surface temperature data and to reconstruct blank data due to cloud cover. One of the Artificial Intelligence (AI) approaches used is Machine Learning (ML) technology or machine learning which has advantages especially in automation. Machine learning is a powerful tool for spatial reconstruction. It has the potential to improve the quality of spatial reconstructions in a variety of applications. As machine learning algorithms continue to improve, spatial reconstruction will become more accurate, more efficient, and more widely used.

2. Study Area

This study focuses on the Republic of Indonesia State Fisheries Management Area 713 (IFMA 713), which is located at coordinates 1 degree north and 8.5 degree south latitude and from 114.4 to 122.7 degree east longitude. IFMA 713 covers several water areas, including the Makassar Strait, Bone Bay, Flores Sea, and Bali Sea. It is bordered by nine provincial governments: East Kalimantan, South Kalimantan, East Java, Bali, West Nusa Tenggara, East Nusa Tenggara, South Sulawesi, Central Sulawesi, North Sulawesi, and West Sulawesi.

The waters of WPPNRI 713 are influenced by the Pacific Ocean and are also part of the western Equatorial Pacific System which plays an important role in the phenomenon of sea and atmosphere interaction, namely ENSO (El Niño–Southern Oscillation). In addition, the territorial waters of WPPNRI 713 are part of the Trans-World Belt Flow (The Great Conveyor Belt) that crosses Indonesian territory, better known as Arlindo (Indonesian Cross Flow). Arlindo carries a mass of warm water from the Pacific Ocean to the Indian Ocean via the Makassar Strait, Lombok Strait, Timor Sea, Ombai Strait and Lifamatola. Arlindo was also identified as gathering in the warm pools of northern Papua [4]. The study area can be seen in Figure 1

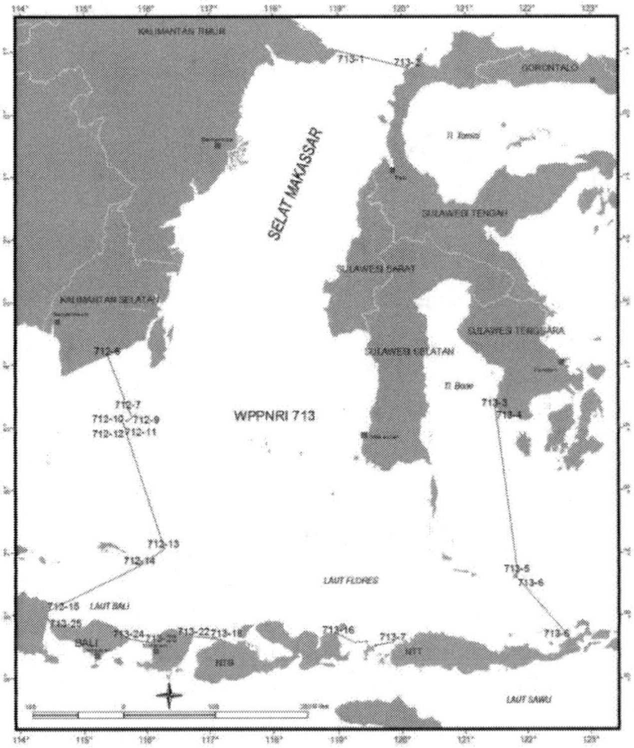

Figure 1. Indonesia's Fish Management Area 713

3. Data Used

The data to be used is Sea Surface Temperature (SST) data for the period from 1 January 2003 to 31 December 2021 (19 years). The SST data is obtained from the MODIS (Moderate-Resolution Imaging Spectroradiometer) satellite which can be accessed on the OPeNDAP website. The data has a spatial resolution of 4.63 km and a temporal resolution of 1 day. The SST data obtained from the MODIS satellite uses mid-infrared waves captured in bands 22 and 23 with a wavelength of 3.5 to 4.2 micrometer to capture SST data. The data used in this band is very suitable because it has less variation than bands 31 and 32.

In addition to using data from the MODIS satellite, daily sea surface temperature data was used for January 2003 to December 2021, which was sourced from Optimum Interpolated Sea Surface Temperature (OISSTv2) which can be accessed at NOAA. This data has a daily average temporal resolution. This data will be used to test machine learning performance to determine hyperparameters.

4. Methods

4.1. Application of backpropagation method

Backpropagation is a supervised learning method that trains multilayer perceptrons to recognize patterns and respond correctly to similar patterns, even if they are not exactly the same as the patterns used during training[5].

The backpropagation algorithm accepts the input pattern and performs a computational process based on the randomly obtained initial weights. If the output from the network is different from the expected target, the network will make adjustments to the existing weights. The process will continue until the output from the network and the expected target are the same. The learning process takes a long time to reach that value. Therefore, the learning process is limited and will stop if the difference between the output and the target has reached a value that is smaller than the tolerance value (error rate). The amount of weight adjustment in each learning cycle is determined by a parameter called learning rate

[6]. There are 3 process in backpropagation method which is forward pass, backpropagation and weight adjustment

4.1.1. Forward pass

During the input phase, the network receives n inputs (x_i) to the and produces an output (y_k) The output is then compared with the target to be achieved (t_k), a hidden layer consisting of p inputs, and m output units. the error of the model can be calculated from the difference between $t_k$and y_k. If there is an error smaller than the tolerance limit, the model iteration will be stopped. However, if the error is still greater than the tolerance limit, then the weight of each line in the network will be modified to reduce the errors that occured. Forward pass value can be determined using the following formulation:

$$Z_{net_j} = v_{j0} + \sum_{i=1}^{n} x_i v_{ji} \quad (1)$$

$$Z_j = f(z_{netj}) = \frac{1}{1+e^{-z_{net_j}}} \quad (2)$$

where $Z_{net_j}$is the j hidden unit, v_{j0} is the weight of the line connecting the bias from the input unit to the Z_j hidden layer unit, v_{ji} is the weight of the i line bias, f is the activation function and Z_j is the output in the hidden unit. Next, calculate the network output in y_k units with the following formulation:

$$y_{\text{net}_k} = W_{k0} + \sum_{j=1}^{p} Z_j W_{kj} \quad (3)$$

$$y_k = f(y_{netk}) = \frac{1}{1+e^{-y_{net_k}}} \quad (4)$$

Where y_{net_k} is the k hidden unit, W_{k0} is the line bias weight form hidden layer unit to the Z_j output unit to the y_k, W_{kj} is the weight of the j unit and y_k is output unit.

4.1.2. Backpropagation

After calculating the output layer in the first phase, the error from the output layer (δ_k) will be calculated to distribute the error in unit y_k to all hidden units connected to the value y_k. δ_k is also used to make changes to the line weights that are directly related to units. In the same way, the error calculation is carried out for each unit in the hidden layer (δ_j) as the basis for changing the weight of the previous hidden layer unit. This calculation is performed for all δ factors in the hidden layer unit associated with the input unit. The δ factors can be calculated using the following formulation:

$$\delta_k = (t_k - y_k) f'(y_{net_k}) = (t_k - y_k) y_k (1 - y_k) \quad (5)$$

where t_k is the target to be reached, y_k is the output, and f is the activation function. Then, the calculation of the W_{kj} weight change rate (ΔW_{kj}) is used to change the W_{kj} weight with the learning rate (a) using the following equation:

$$\Delta W_{kj} = a \delta_k Z_j \qquad ; k = 1, 2, \dots, m \ ; j = 0, 1, \dots, p \quad (6)$$

Next, calculate the δ factor based on the error in each hidden layer unit Z_j and the weight change rate v_{ji} (Δv_{ji})

$$\delta_{net_j} = \sum_{k=1}^{m} \delta_k w_{kj} \quad (7)$$

$$\delta_j = \delta_{net_j} f'(z_{net_j}) = \delta_{net_j} z_j (1 - z_j) \quad (8)$$

$$\Delta v_{ji} = a\, \delta_j x_i \qquad ; j = 1, 2, \dots, p \ ; i = 0, 1, \dots, n \quad (9)$$

Where δ_{net_j} is the error value in the j hidden layer unit, δ_j is the error value in the hidden layer unit and a is the learning rate.

4.1.3. Weight adjustment

After all the δ factors are calculated, weight adjustments will be carried out on all lines. Changes in the weight of a line will be adjusted based on factors in the neurons in the previous layer using the following equation

$$w_{kj}(new) = w_{kj}(old) + \Delta w_{kj} \quad (10)$$

$$v_{ji}(new) = v_{ji}(old) + \Delta v_{ji} \quad (11)$$

4.2. Pearson's Correlation Coefficient

Correlation analysis is an analysis that is often used to determine the relationship between two different variables. The correlation coefficient is often denoted by r where the correlation value r requires magnitude and direction between negative and positive. The interval values used are between negative to positive with values that are absolute and nondimensional without a unit of measure. When the value of the coefficient r is zero, there is no measurable association of the two variables, whereas when the coefficient of r is positive, then there will be a strong association between the two variables. A positive correlation value shows the relationship between the two variables is positive or directly proportional to each other. Meanwhile, the negative correlation value shows the relationship between the variables is negative or inversely proportional. The value of the Pearson correlation coefficient can be calculated using the following equation.

$$r = \frac{(\sum(x_i-\bar{x})(y_i-\bar{y}))}{\sqrt{\sum(x_i-\bar{x})^2 (y_i-\bar{y})^2}} \quad (12)$$

Where r is the correlation coefficient, x_i is the sample value of the x-variable, $\bar{x}$ is the average value of the x-variable, y_i is the sample value of the y-variable, $\bar{y_i}$ is the average value of the y-variable.

4.3. Root Mean Square Error

Performance evaluation needs to be done to test the accuracy of the model being run so that in the future the model can be applied to other problems. Root Mean Square Error (RMSE) is one method to evaluate the performance of the model. RMSE calculations are used to determine the magnitude of the error rate of prediction results from a dataset where the smaller the value of the RMSE (close to zero), The more accurate the prediction results, the closer they will be to the actual value. RMSE is calculated using the following equation:

$$RMSE = \sqrt{\frac{\sum(y_i-y_n)^2}{n}} \quad (13)$$

Where y_i is actual data value, y_n is the predicted value, and n is the amount of data being tested.

4.4. Model scenario

The downloaded SST data is divided into training data and test data. Training data is used as input or to train the backpropagation model. Meanwhile, test data is used to test model performance when making predictions on actual data [7]. After dividing the training data and test data, the SST data from MODIS satellite imagery is averaged monthly to fill in the gaps in the training data and test data. The SST data is also used as input for determining machine learning parameters.

The machine learning program scenarios carried out in this study are divided based on several treatments for training data and test data. In training data, treatment is carried out by providing variations in the length of training data for 15 years, 10 years and 5 years and test data in the period of 2018-2021 period to test program performance with variations length of training data. Then, after performance testing, treatment is carried out on the available test data by removing the datum by 20%, to simulate empty data on satellite imagery and test program performance on test data. For details of each scenario, see table 1

Table 1. Scenario model

Scenario	Test data period	Missing data percentage	Treatment for missing data
1	2018-2021	20%	Empty data is not filled
2	2018-2021	20%	Empty data is filled with zero value
3	2018-2021	20%	Empty data is filled with the average SST value at data collection point
		30%	
		40%	
		50%	
4	2018-2021	20%	Empty data is filled with the average SST value at Indonesia
		30%	
		40%	
		50%	

5. Result and Analysis

5.1. Timeseries plot

The performance test of the machine learning program was carried out using daily SST data to determine the hyperparameter to increase accuracy for performance as visualize in Figure 2.

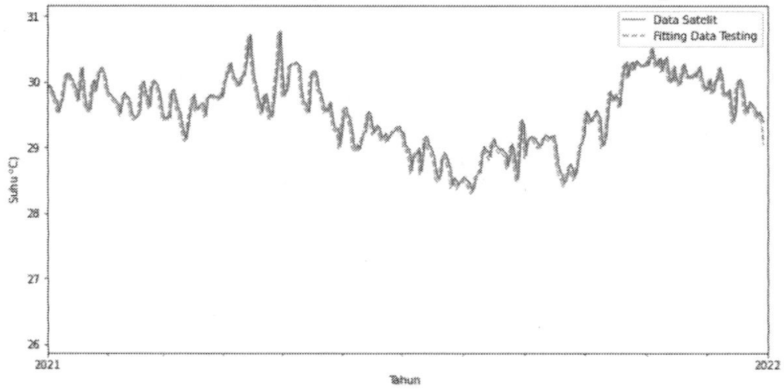

Figure 2. Timeseries plot with machine learning on daily SST data

The results of the machine learning performance test show a good correlation with a value of 0.99 with an RMSE of 0.38 °C. Furthermore, using SST data from MODIS, cloud cover simulation is carried out on satellite sensors by removing some data at times with the greatest cloud cover, namely during the rainy season or October to April [8]. After the data is emptied, a data reconstruction trial is carried out using a machine learning program with the same hyperparameter that has been run previously with daily SST data.

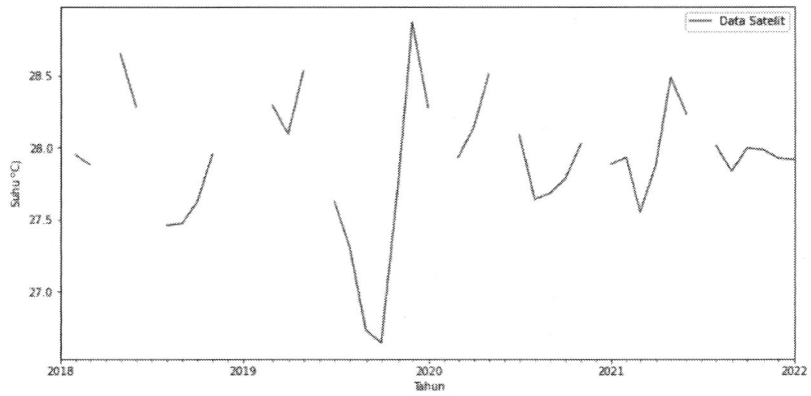

Figure 3. Simulation of data loss satellite imagery in MODIS

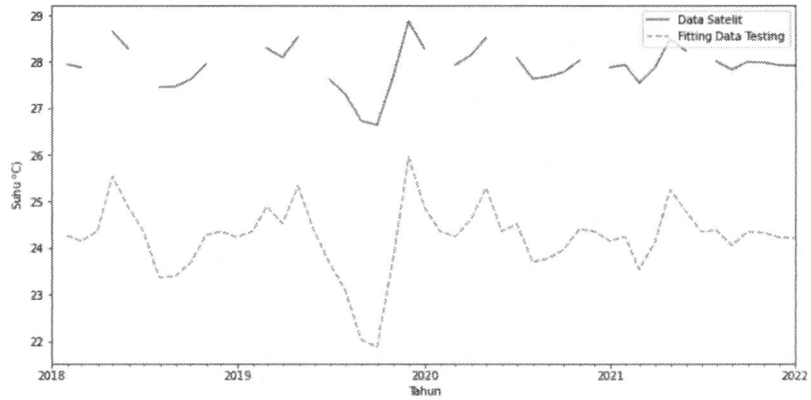

Figure 4 Timeseries plot with machine learning using the same parameters

As seen in Figure 4, after the program was run with the same parameters on the SST daily data, the reconstruction results showed a fairly good correlation of 0.99. However, the RMSE results obtained from the program are 4.56 °C so the model becomes less accurate. Therefore, it is necessary to repair the machine learning program before running it again for data reconstruction by performing hyperparameter tuning.

According to [9], hyperparameter tuning is the process of finding the optimal value of the hyperparameter of a machine learning model to improve the performance of the machine learning model. This is done by trying various hyperparameter values and comparing the results with performance metrics such as accuracy. After making adjustments to the hyperparameter. The following is the result of the reconstruction of empty sea surface temperature data on satellite data shown in Figures 5.

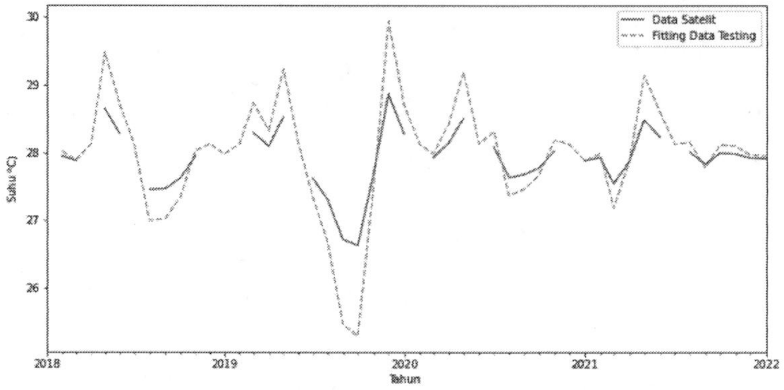

Figure 5. Timeseries plot after adjusting hyperparameter

Table 2. Evaluation performance on variation of training data length

Training Data (year)	**RMSE (°C)**		Correlation
	Training	Test	
15	0,7	0,75	0,99
10	0,69	0,76	0,99
5	0,78	0,79	0,99

Based on Table 2 above, the best performance results are shown by the length of training data for 15 years with the smallest RMSE value of 0.75 °C, followed by the model with the length of training data for 10 years with an RMSE value of 0.76 °C and the worst performance is shown by program with a length of training data for 5 years. However, the three programs have been able to determine the pattern of increase/decrease in sea surface temperature as evidenced by having a correlation value of 0.99 even though they have different RMSE values. The RMSE value varies depending on the length of the training data. If the training data is too short, the model will not be able to learn the patterns in the data and the accuracy will decrease. If the training data is too long, the model will overfit the data and the accuracy will decrease due to an increase in bias and a decrease in variation.

5.2. Reconstructed value with scenario

The results of the reconstruction model show good performance in reconstructing sea surface temperature data. However, it is necessary to carry out further program testing to see the model's performance in reconstructing data with certain scenarios. The results of testing the model against predetermined scenarios are shown in Figure 6 to Figure 9.

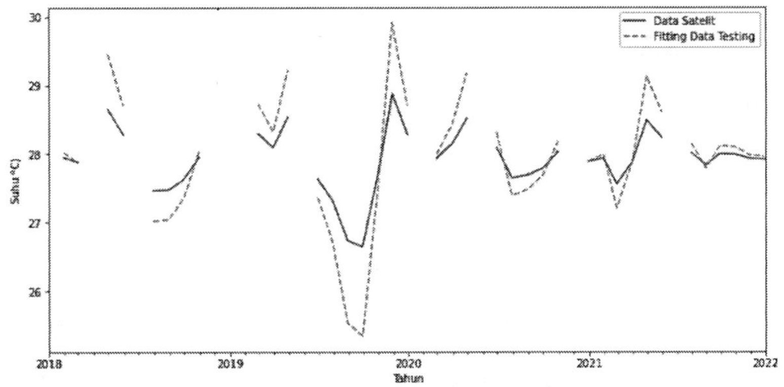

Figure 6 Data reconstruction scenario 1

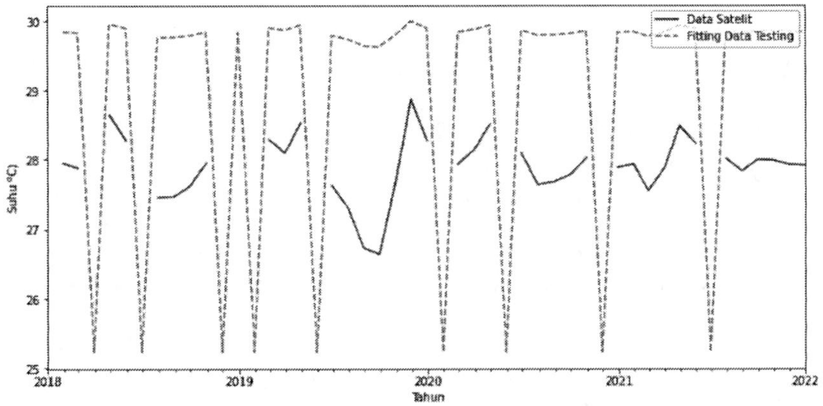

Figure 7. Data reconstruction scenario 2

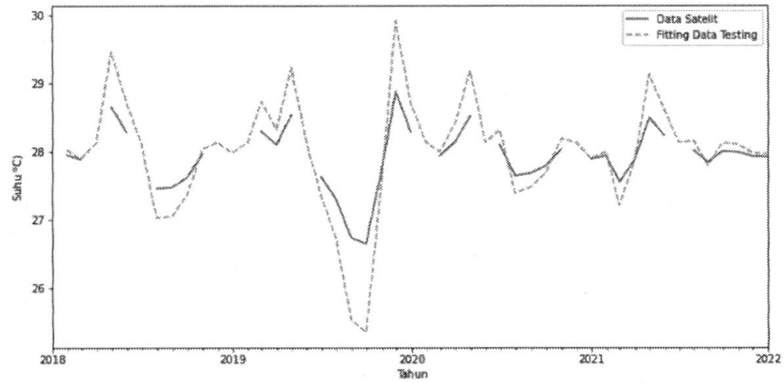

Figure 8. Data reconstruction scenario 3

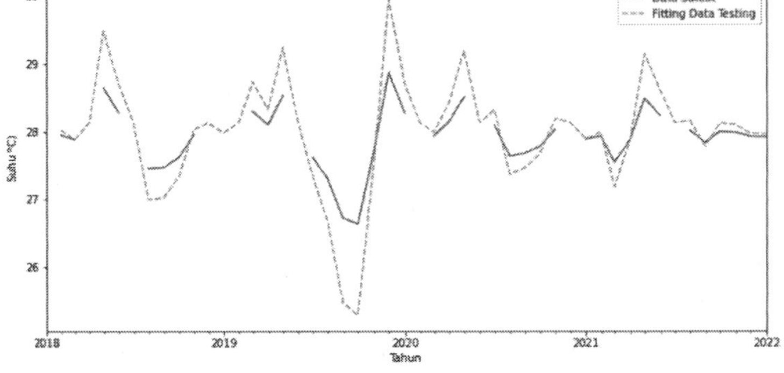

Figure 9 Data reconstruction scenario 4

The machine learning program that runs with 4 scenarios in this study aims to see differences in program performance in reconstructing sea surface temperature data in WPPNRI-713. The scenarios that are run use variations in the treatment of empty data in the test data and variations in the length of the training data. Variations in the treatment of blank data in test data are used to review the performance of machine learning programs in reconstructing data. While variations in the length of the training data are used to see whether the program is able to capture patterns of increase/decrease from the sea surface temperature data in IFMA-713. The results from the graphs in Figure 6 to Figure 9 show that scenario 1 and scenario 2 show poor performance when compared to scenario 3 and scenario 4. Evaluation of the performance of each scenario can be seen in Table 3

Table 3. Evaluation performance

Evaluation	Reconstruction	**Scenario 1**	**Scenario 2**	**Scenario 3**	**Scenario 4**
RMSE (°C)	0,75	154,3	12,76	0,51	0,51
r	0,99	-0,201	-0,175	0,96	0,96

The results from scenario 3 and scenario 4 show that the reconstruction performance is quite close to the initial data reconstruction. This is because the blank data values for sea surface temperature reconstructed in scenario 3 and scenario 4 are filled with the average value and 28 °C of all data so that the machine learning program can perform the reconstruction well when compared to scenario 1, namely without filling in blank data and scenario 2 with empty data filled with zero values. The results of scenario 3 and scenario 4 reconstruction can also capture the pattern of increase/decrease in sea surface temperature in IFMA-713.

The performance produced by scenario 1 and scenario 2 performed poorly that it cannot be used as a program to reconstruct data. From Table 2 it can be seen that the performance of scenario 3 and scenario 4 has the same performance and is close to the initial reconstruction. so that scenario 3 and scenario 4 are tested again by reconstructing the missing data by 30%, 40% and 50%. The following graphs are generated for the reconstruction of missing data by 30%, 40% and 50% in Figures 10 to Figure 15.

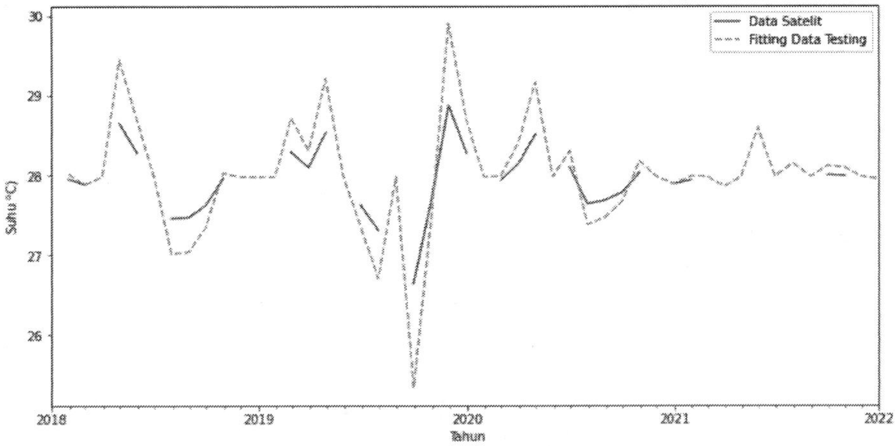

Figure 10. Data reconstruction scenario 3 with 30% missing data

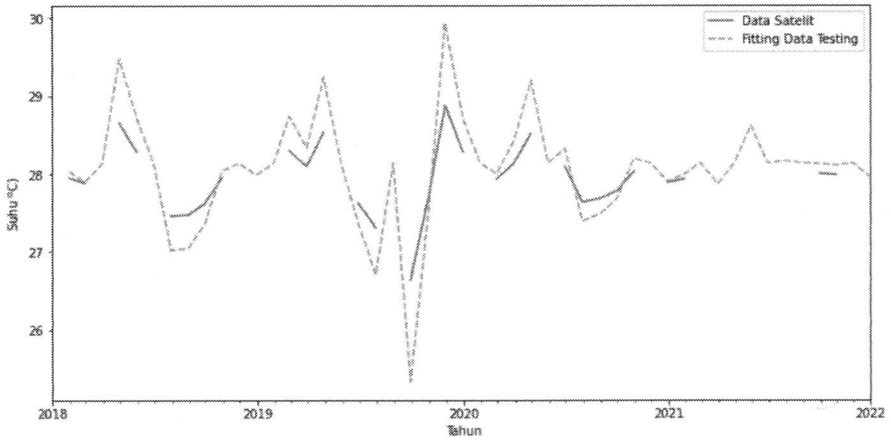

Figure 11. Data reconstruction scenario 4 with 30% missing data

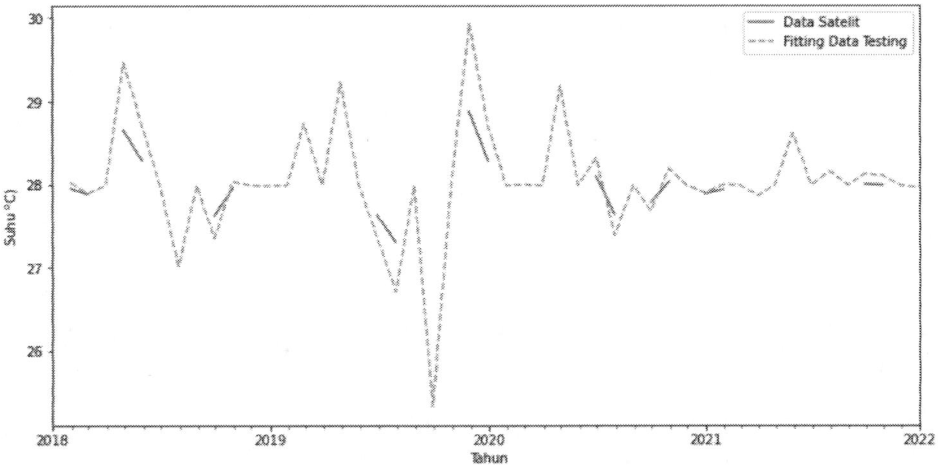

Figure 12. Data reconstruction scenario 3 with 40% missing data

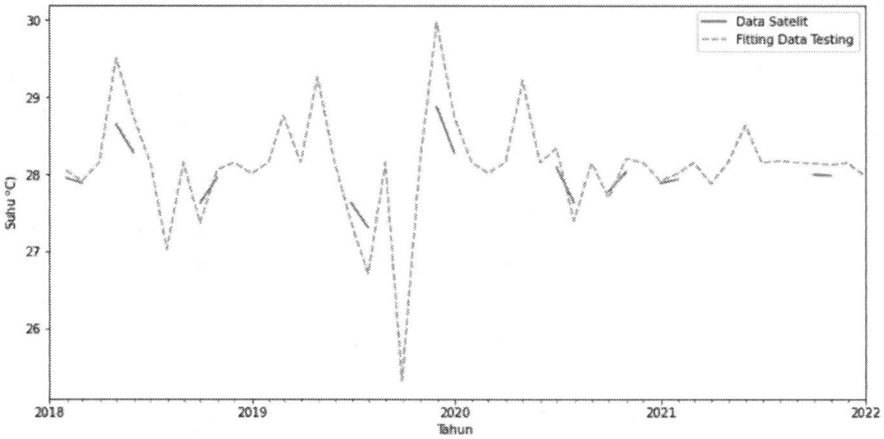

Figure 13. Data reconstruction scenario 4 with 40% missing data

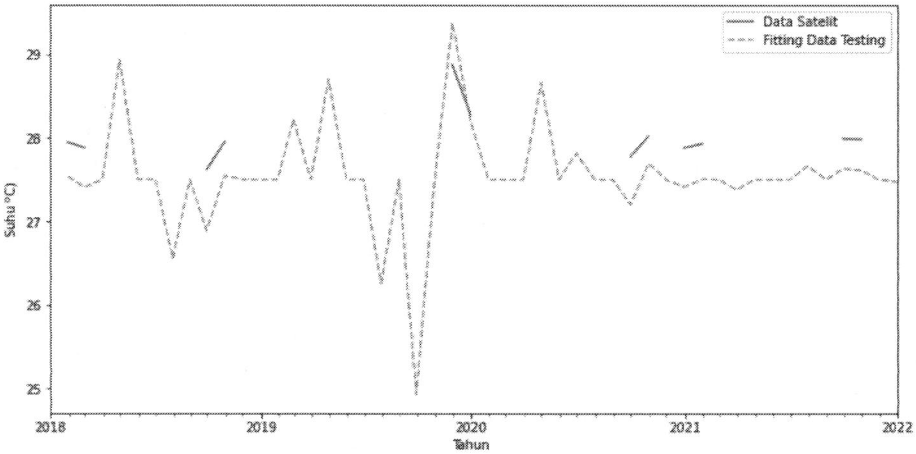

Figure 14. Data reconstruction scenario 3 with 50% missing data

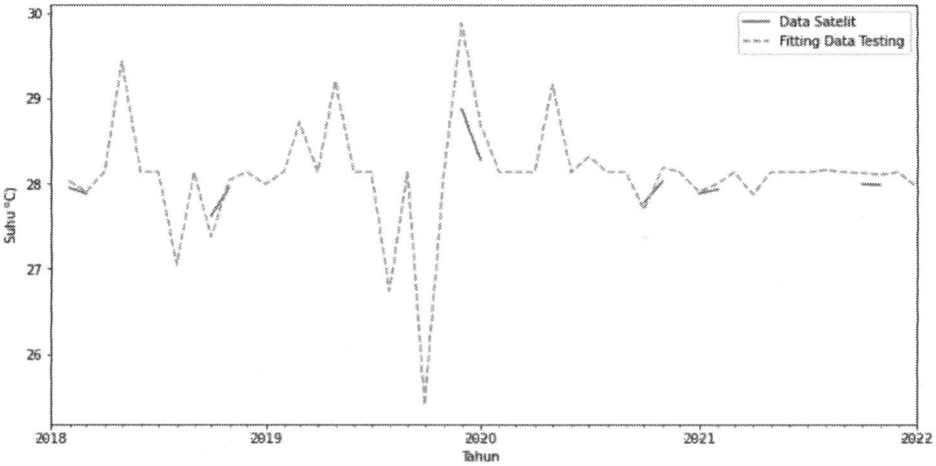

Figure 15. Data reconstruction scenario 4 with 50% missing data

The performance results of scenario 3 and scenario 4 have a difference that is not significant enough. This can be seen from the RMSE value which has a value that is not much different. However, the two scenarios have similarities, namely as the missing data increases, the resulting correlation value will decrease when compared to sea surface temperature data without missing data with the resulting correlation values of 0.82, 0.79 and 0.76 with a decrease a correlation value of 0.3 for each addition of 10% missing data.

This is because when the missing data values increase, it will affect the performance of the model in reconstructing the data due to the tendency when the reconstruction is carried out, missing values can shift the input pattern, and the backpropagation network must complete a complex solution to fulfill all the reconstructed patterns [10].

5.3. Spatial plot

Spatial visualization of the reconstructed data in WPPNRI 713 with the aim of knowing the spatial distribution of sea surface temperature and program performance in all study areas. The reconstruction is carried by applying backpropagation to all data points in the study area in one period of time. The results of the spatial plot can be seen in Figure 16

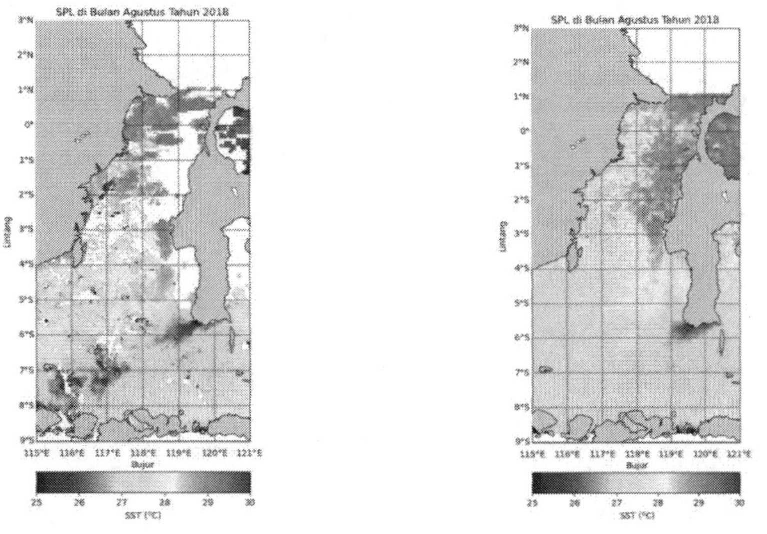

Figure 16. Spatial reconstruction at IFMA-713 in August 2018

In Figure 16, spatial plots with empty data that is not filled with a value (NaN) will show blank spaces marked in white, which makes the information displayed by the data less informative. When data reconstruction was carried out for missing values, it was seen that the spatial plot became more informative with more complete data compared to the previous spatial plot. Even though the spatial plots of the data reconstruction results are good, it can be seen in the reconstructed spatial plots. The program adjusts certain values to make the resulting data more similar to the seasonal patterns of the training data, this is due to the insufficient training data length so that the program has limited ability to reconstruct data with very high accuracy values. As can be seen in the spatial plot image resulting from data reconstruction, the program has been able to capture a significant difference in sea surface temperature with sea surface temperatures in the north of the Makassar Strait being higher when compared to the Bali Sea.

Machine learning programs show better performance when compared to statistical methods in reconstructing data. However, that does not mean that statistical methods cannot be used at all to reconstruct data. There are several limitations and advantages possessed by both methods based on needs.

According to [11], the biggest difference from using statistical methods and machine learning is in the development of the model. In machine learning methods, models are built by algorithms based on data availability, while model development in statistical methods is done manually by users. The advantage of machine learning programs is that they can develop models with complex relationships between data variables. another advantage of machine learning is also found in the amount of data processed. The more data that is processed, the resulting model will be better. while statistical methods are more often used on data that has medium and short length sizes to be processed.

6. Conclusion

The machine learning model can reconstruct accurately with a value of r = 0.99 for all variations in length of training data in the WPPNRI-713 region. The results of the machine learning program show that the longer the training data used, the program results will have better performance in reconstructing the data as evidenced by the smallest RMSE value is a program with 15 years of training data with an RMSE value of 0.75 °C. The machine learning program cannot perform data reconstruction if there are blank values in the test data as shown by scenario 1 and does not have good performance when the blank values are filled with zero values or scenario 2. However, the program can perform reconstruction if the missing values are fill with the average value (scenario 3) or fill in the average value of sea surface temperature in Indonesia (scenario 4) and both scenarios (scenario 3 and scenario 4) show performance that is close to the initial reconstruction. Scenario 3 and scenario 4 are able to reconstruct missing data by 30%, 40% and 50% with good performance as 20 % missing data scenario. The results of the spatial reconstruction have been able to reconstruct the sea surface temperature of the IFMA-713 area by filling in the blank values with data from the machine learning program.

Reference

[1] Habibie M N dan Nuraini T A 2014 Karakteristik dan tren perubahan suhu permukaan laut di indonesia periode 1982-2009 *J. Meteorol. dan Geofis.* **15**

[2] Strabala K 2004 MODIS Cloud Mask User's Guide *Reading* 32

[3] Nahari R V, Ulum M dan Alfita R 2017 Cloud Removal Pada Citra Satelit Aqua Modis *Semin. Nas. Inov. Teknol.* 165–70

[4] Koeshendrajana I S dan Rusastra I I W 2019 POTENSI SUMBER DAYA KELAUTAN DAN PERIKANAN WPPNRI 572

[5] Siang J J 2005 Jaringan Syaraf Tiruan & Pemrogramannya Menggunakan Matlab

[6] Santika D D, Susanti B, Anderson W dan Wongso K 2007 Implementasi Metode Backpropagation Dalam Klasterisasi Objek *CommIT (Communication Inf. Technol. J.* **1** 30–37

[7] Kartini D 2017 Penerapan Data Mining dengan Algoritma Neural Network (Backpropagation) Untuk Prediksi Lama Studi Mahasiswa *Pros. SISFOTEK* **1** 235–41

[8] Giarno G, Dupe Z L dan Mustofa M A 2012 Kajian Awal Musim Hujan Dan Awal

Musim Kemarau Di Indonesia *J. Meteorol. dan Geofis.* **13** 1–8

[9] Yu T dan Zhu H 2020 Hyper-parameter optimization: A review of algorithms and applications *arXiv Prepr. arXiv2003.05689*

[10] Gupta A dan Lam S M 1998 Weight decay backpropagation for noisy data *Neural networks* **11** 1127–1138

[11] Ley C, Martin R K, Pareek A, Groll A, Seil R dan Tischer T 2022 Machine learning and conventional statistics: making sense of the differences *Knee Surgery, Sport. Traumatol. Arthrosc.* **30** 753–7

IOP Conf. Series: Earth and Environmental Science 1245 (2023) 012038 doi:10.1088/1755-1315/1245/1/012038

Coastline Change using Time Series Satellite Images in the North Coast of Java: Study Case Indramayu Regency, West Java, Indonesia

Erva Aulia Fauzunora[1], Asep Saepuloh[1*], Ayi Tarya[2]

[1] Geological Engineering Study Program, Faculty of Earth Sciences and Technology, Bandung Institute of Technology, Indonesia
[2] Earth Science Study Program, Faculty of Earth Sciences and Technology, Bandung Institute of Technology, Indonesia

Email: ervaaulia89@gmail.com
*Corresponding author e-mail: saepuloh@itb.ac.id

Abstract. Coastal areas have continuous change due to the interaction between sea and land, which generally causes abrasion and accretion. Indramayu Regency has a coastline of around 147 km. Indramayu Coastal has the characteristics of a sandy beach with a dominant abrasion process triggered by digging beach sand and changes in the function of mangrove land into ponds. We applied remote sensing observation to obtain coastline data on a regular basis. Currently, satellite images support the availability of Landsat data in a relatively short time and cover a wide area. Landsat imagery has been used in the last ten years to detect coastline changes in study areas based on the BILKO and Digital Shoreline Analysis System (DSAS) methods. Abrasion in 2012 – 2022 covers an area of 805.71 ha and accretion covers an area of 941.63 ha. The widest abrasion is in Pasekan with an area of 266,451 ha. The abrasion process in Pasekan is very high due to the influence of high currents and waves from the sea, and the accretion process in Cantigi is very high due to the high supply from the Cimanuk River and the many mangroves that collect sediment that enters the sea.

1. Introduction

The coastal area is the area that connects land with water. Coastal regions experience continuous changes due to the interaction between sea and land. The waves and wind along the coast cause constant abrasion of the rocks. This causes environmental pollution, coastal intrusion, and changes in coastline patterns. The northern coastal area of West Java has a lowland with several bays and headlands, has a curved coastline with a high accretion rate [1]. The coastline is the meeting line between land and sea water which is not fixed and can change or move according to the tides and coastal erosion that occurs [2].

Indramayu Regency is located on the north coast of Java Island, consisting of 9 sub-districts that are bordered by the sea with a coastline length of 147 km [3]. Indramayu Regency is also one of the districts that depend on the regional economy by utilizing the coastal area. According to data reported by Bapeda-JABAR in 2007, abrasion in the Indramayu coastal area was around 47.87% of the total length of the

North coast of West Java, while the stable rate was around 5.75%. This greatly affects the extent of coastal areas and communities [1].

Data is needed to see changes in the coastline and beach materials, which can be obtained periodically with the rapid development of remote sensing technology, especially satellite imagery. Satellite imagery can support the availability of data on the characteristics of an area in a relatively short time and cover a wide area so that it can provide the information needed for monitoring. The use of remote sensing data is moderately more effective than field surveys which require a long time and are difficult to monitor [4]. The aim of this research to analyze changes in the coastline that occurred in Indramayu Regency from 2012 – 2022. This can help determine which sub-districts experience abrasion and accretion based on the period.

Coastline changes can be described using spatial data in the form of satellite imagery. Satellite images with high resolution can illustrate coastline change in more detail and with the temporal resolution of satellite it can detect these changes easily. Usage of remote sensing data is very helpful to determine coastline changes, we can calculate the change by temporal resolution mean that we will get the information from other years before and we can study large area using this method. The use of satellite images with medium spatial resolution such as Landsat is suitable for spatial dynamics monitoring of coastline [5].

Based on previous study [6], the coastline in Indramayu Regency, between the 1990s and 2019, experienced changes in both abrasion and accretion (sedimentation), which could harm people living on the coast. Also, sedimentation causes river estuaries silting and land conflicts such as the struggle for land ownership. Furthermore, abrasion is very detrimental to the community because it damages ponds and buildings. The high level of abrasion in this Regency, especially between the 1989–2009 period, destroyed and damaged agricultural lands, pond areas, and buildings. Therefore, the government and society are making efforts to minimize abrasion. In fact, the government is building a wave barrier, both large river stones and concrete. This is considered quite effective, because between the 2009–2019 period, the level of abrasion decreased. These efforts not only reduce and minimize abrasion but also cause accretion. The accretion that appears in the coastal area is utilized by the community in various fields, including aquaculture land such as shrimp and milkfish ponds, as well as tourism. Meanwhile, accretions that appear in areas where accessibility is difficult are used for fishponds, while areas with easy access are used as tourist attractions.

The research location is in the Indramayu Regency area, it comprises 9 sub-districts stretching from East to West, including Krangkeng, Karangampel, Balongan, Indramayu, Pasekan, Cantigi, Kandanghaur, Patrol and Sukra. The aim of this research is to determine changes in the coastline in the study area in 2012 – 2022 using satellite imagery data. Figure 1 showed research location in Indramayu Regency.

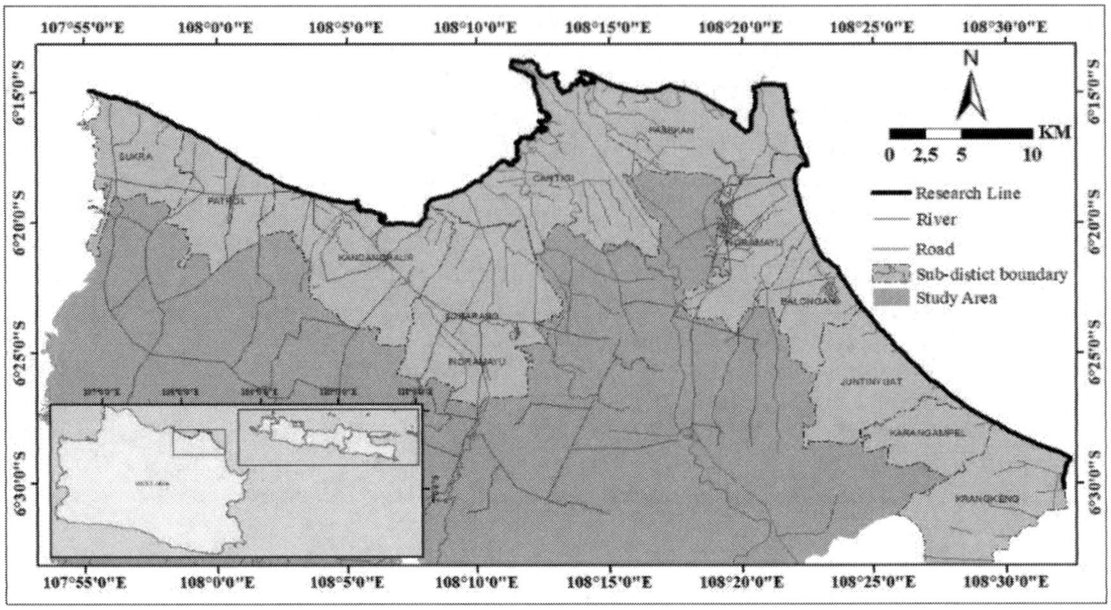

Figure 1. The Indramayu coastal is in Indramayu Regency and the black line is the coastline that is the focus of this research.

2. Data Collection

The data used in the research process is Landsat 7 ETM+ and Landsat 8 OLI/TIRS data from 2012 to 2022, downloaded from the usgs.gov website. Landsat 7 has a radiometric resolution of 8 bits or 256 digital values (0-255), the spatial resolution of ETM imagery is 30 meters (channels 1-5 and 7), thermal data has a spatial resolution of 60 meters (channel 6), and panchromatic data (channel 8) with a spatial resolution of 15 meters. Image utilization is carried out for various fields, such as land and water management, pollution monitoring, detection of land surface changes, and mapping of coastline changes. Landsat 8 has an On-Board Operational Land Imager (OLI) and Thermal Infrared Sensor (TIRS) sensor with a total of 11 channels, 9 channels (bands 1-9) on OLI and 2 channels (bands 10 and 11) on TIRS. The OLI sensor has 1 near-infrared channel and 7 visible channels, which include wavelengths reflected by objects on the earth's surface, with a spatial resolution of 30 meters. The TIRS sensor has 2 thermal bands, providing a more accurate surface temperature. Image data selection is because both Landsat 7 and Landsat 8 have Near Infrared (NIR) and Shortwave Infrared (SWIR) bands which can be used to distinguish the content of groundwater and vegetation, making it easier to draw coastlines. Image data is taken from the time of tide from the study area [7]. The data uses 11 satellite image data with 10 years to detect coastline changes at the study site. It is sufficiently detailed to see coastline changes and the rate of abrasion and accretion in the study area. This study uses GIS software to describe the coastline at the study area. Table 1 showed Landsat satellite sensor type and acquisition date.

Table 1. Description of Landsat Satellite image data.

No	Scene ID	Date	Recording Time	Cloud Cover	Elevation	Tidal Condition
1	LE07_L1TP_121064_20120620_20200908_02_T1	20-06-2012	2:48:21	9	-0,225	towards receding
2	LC08_L1TP_121064_20130615_20200912_02_T1	15-06-2013	2:55:52	7,52	-0,171	towards receding
3	LC08_L1TP_121064_20140922_20200910_02_T1	22-09-2014	2:54:00	9,77	0,067	towards tide
4	LC08_L1TP_121064_20150520_20200909_02_T1	20-05-2015	2:52:57	6,85	-0,273	towards receding
5	LC08_L1TP_121064_20161029_20200905_02_T1	29-10-2016	2:26:15	27,86	0,185	towards tide
6	LC08_L1TP_121064_20170525_20200903_02_T1	25-05-2017	2:53:26	3,33	-0,082	towards tide
7	LC08_L1TP_121064_20180816_20200831_02_T1	16-08-2018	2:53:20	3,08	-0,072	towards tide
8	LC08_L1TP_121064_20190819_20200827_02_T1	19-08-2019	2:54:03	2,05	-0,065	towards tide
9	LE07_L1TP_121064_20200525_20200820_02_T1	25-05-2020	2:26:47	3,00	-0,331	towards receding
10	LC08_L1TP_121064_20210605_20210614_02_T1	05-06-2021	2:54:00	5,44	-0,05	towards receding
11	LC08_L1TP_121064_20220726_20220802_02_T1	26-07-2022	2:56:16	7,01	-0,085	towards tide

3. Temporal Coastline Change Detection

Before analysis or identifying the coastline at the research location, the downloaded image is first prepared for optimal results. The corrections made are radiometric corrections that are processed in GIS software. Radiometric corrections function to correct pixel values so that they match what they should be by considering the factor of atmospheric disturbances as the primary source of error [8]. Radiometric correction improves image quality due to reflection errors on the earth's surface and other factors, such as the direction of sunlight, weather conditions, and atmospheric conditions, so that the resulting information will be more accurate. Radiometric correction is useful for analysis multitemporal and multi-sensor data used for continuous interpretation and detection of changes. The resulting image will be sharper than the image before it was corrected. Next is an atmospheric correction which functions to correct atmospheric disturbances such as fog, dust, or smoke, which can cause bias and reflection effects on objects. Atmospheric correction was carried out using the Fast Line of Sight Atmospheric Analysis of Spectral Hypercubes (FLAASH) method. The FLAASH method is the reflectance received by the sensor and then remodelled by adding factors that affect the reflectance received by the sensor, such as the reflection of surrounding objects (adjacency effect) and the reflectance scattered by the atmosphere [6]. Atmospheric correction can remove noise from the atmosphere, optimizing the image in the analysis process.

In delineating the coastline, this study uses the BILKO method to determine the boundary between land and sea by utilizing the Brightness Value (BV). The required band is the infrared band [9]; on Landsat 7 required band 4, and on Landsat 8 required band 5. Infrared waves have a high reflectance to land and low reflectance on water. The BV value is required in the separation of land and sea. Meanwhile, the BILKO formula used is as follows:

$$BILKO = \left(\left(\frac{Ld}{((N \times 2)+1)} \times (-1) \right) \right) + 1) \quad (1)$$

Where, N is the minimum BV value for Landsat imagery, Landsat 7 (30) and Landsat 8 (7000), L_d is the NIR or SWIR band (Band 4 for Landsat 7 and Band 5 for Landsat 8)

The following are the before and after results of the BILKO method in the study area to obtain the coastline every year. Visually, after BILKO method, the resulting image will be brighter, and the boundary between land and sea will look clearer (See fig 2).

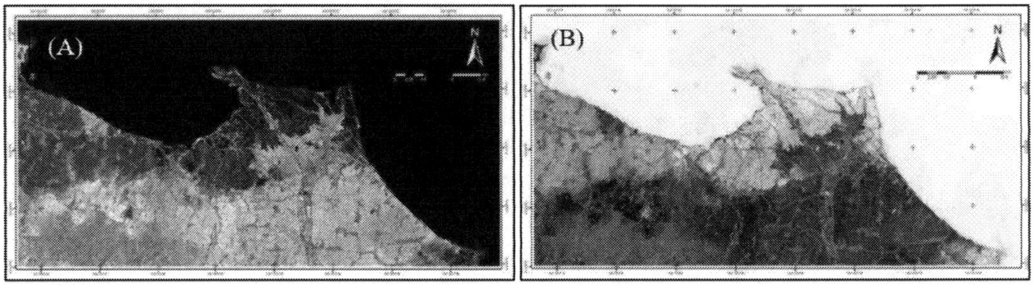

Figure 2. (A) Landsat 8 of 2022 before BILKO method, (B) Landsat 8 of 2022 after BILKO method.

The following analysis is the Digital Shoreline Analysis System (DSAS) which function to calculate the change in coastline position based on time statistically and geospatially based. This method uses a point as a measurement reference, resulting from the intersection of the transects made by the researcher with the coastline based on time. The DSAS method consists of several analysis, such as [10]; (1) Shoreline Change Envelope (SCE) is a total measurement of coastline changes considering all existing coastline positions, (2) Net Shoreline Movement (NSM) is a measurement of the distance between coastline changes. Results that are positive (+) mean the coastline is advancing, and if it is negative (-), it means the coastline is retreating, (3) End Point Rate (EPR) calculates the rate of change of coastline by dividing the distance between one coastline and another. Data with positive values (+) experience accretion, and data with negative values (-) experience abrasion, and (4) Linear Regression Rate (LRR), is a statistical analysis of the rate of change using linear regression, which can be determined using the least-square regression line for all points of intersection of the coastline with the transect.

4. Result

The data used in the study used Landsat 7 ETM+ and Landsat 8 OLI/TIRS from 2012 to 2022. The figure below shows the BILKO method used in drawing coastlines. The results of this analysis show the difference between land and sea. Visually, the sea will appear brighter than the land. Whereas before the analysis using sea and land, BILKO looks dim. The following shows the before and after images using the BILKO method in with drawing the coastline every year. From the results of the Digital Shoreline Analysis System (DSAS), the abrasion process dominates compared to the accretion process, with an abrasion area of 805,71 ha and an accretion area of 941,63 ha (See fig 3).

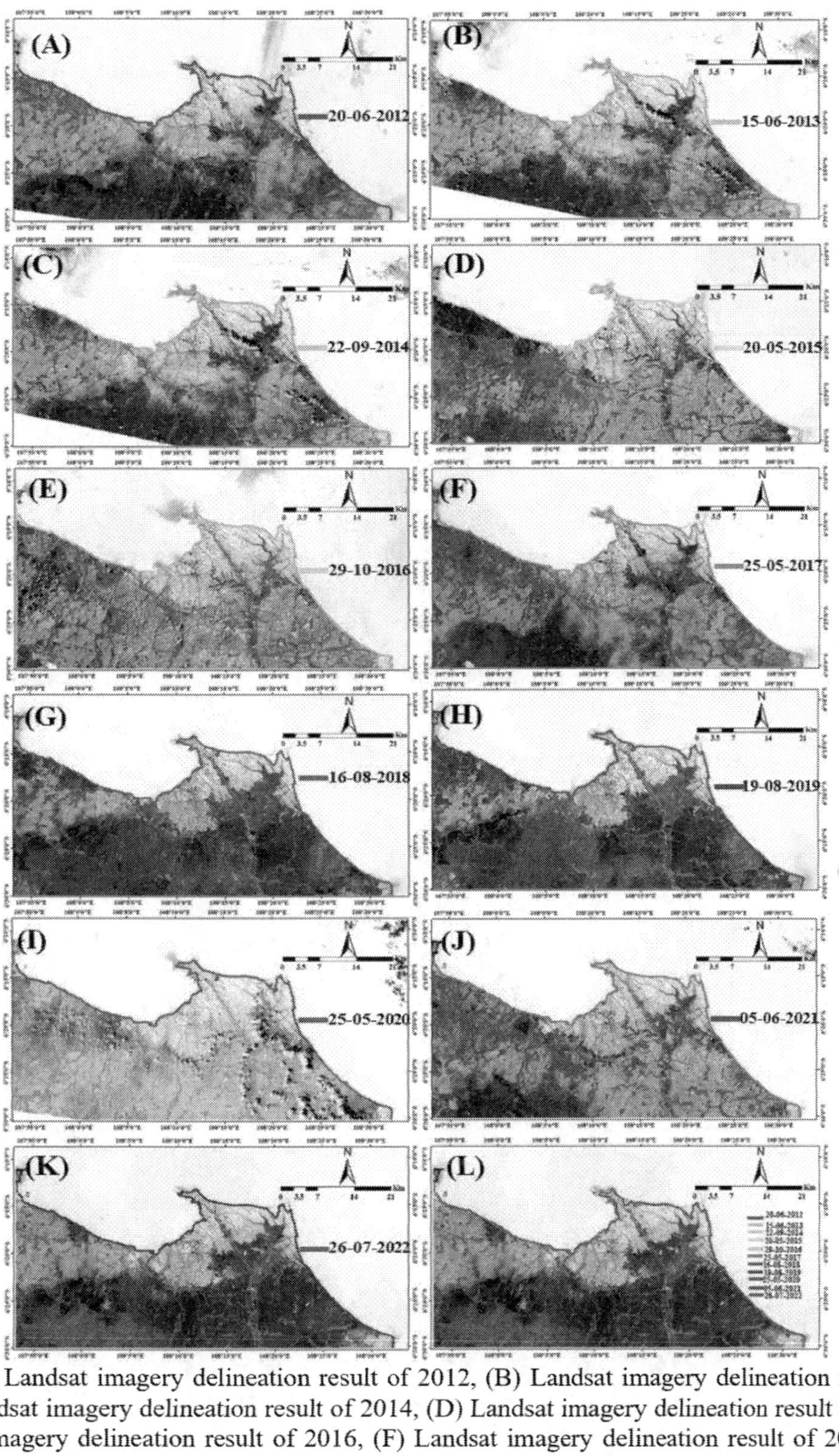

Figure 3. (A) Landsat imagery delineation result of 2012, (B) Landsat imagery delineation result of 2013, (C) Landsat imagery delineation result of 2014, (D) Landsat imagery delineation result of 2015, (E) Landsat imagery delineation result of 2016, (F) Landsat imagery delineation result of 2017, (G) Landsat imagery delineation result of 2018, (H) Landsat imagery delineation result of 2019, (I) Landsat imagery delineation result of 2020, (J) Landsat imagery delineation result of 2021, (K) Landsat imagery delineation result of 2020, (L) Landsat imagery delineation result of 2012 – 2022.

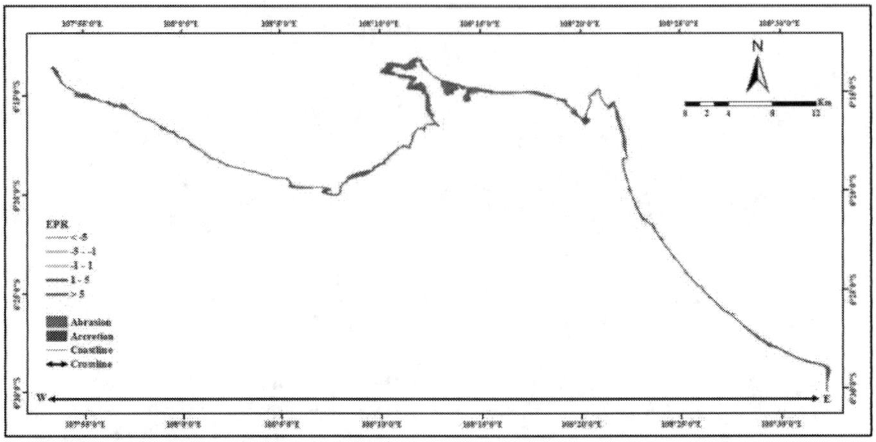

Figure 4. Map of abrasion and accretion of Indramayu Regency. The red line indicates the abrasion area, and the blue line indicates the accretion area.

The DSAS method is used to analyze coastline changes to see which areas have high to low intensities of abrasion and accretion can be seen by classification [9] in Table 2. The transects used have intervals of 100 m with a transect length of 35 km along the line W – E. The highest accretion rate is in the Cantigi sub-district with a value of 29,223 m/year with a movement of 290,154 meters. the highest abrasion rate is in Karangampel sub-district with a value of -12,105 m/year with a movement of -120,192 meters (table 3).

Table 2. Classification for coastline change (EPR and LRR values).

Values (m/year)	Class
< -5	High Abrasion
-5 to -1	Moderate Abrasion
-1 to 1	Stable
1 – 5	Moderate Accretion
>5	High Accretion

Table 3. Rate and change of coastline for each sub-district along the Indramayu coastal.

Sub-district	2012-2022	
	EPR (m/y)	NSM (m)
Sukra	-5,767	-57,267
Patrol	-6,035	-59,923
Kandanghaur	-3,6	-36,35
Cantigi	29,223	290,154
Pasekan	-6,976	-69,269
Indramayu	-1,909	-18,954
Balongan	-0,115	-1,143
Juntinyuat	-2,185	-21,703
Karangampel	-12,105	-120,192
Krangkeng	-0,614	-6,1

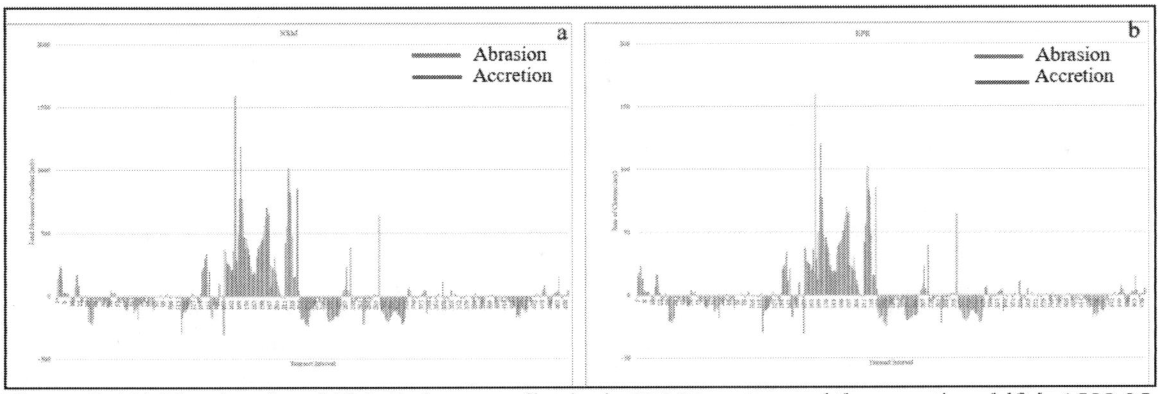

Figure 5. (A) The abrasion shift in Indramayu district is 306,91 meters and the accretion shift is 1588,25 meters. (B) The abrasion rate in Indramayu district is 30,91 m/year and the accretion rate is 159,96 m/year.

5. Discussion

Coastline delineation using the BILKO method produces numbers for each coastline yearly. This can occur due to factors caused by nature, for example, tides, wave size and sedimentation the elements generated by humans, for example, land conversion. Coastline processing using the Digital Shoreline Analysis System (DSAS), we can find out that the processes that occur are divided into two, namely abrasion and accretion. We can also find out the value of the rate of change and movement of the coastline and calculate the area of abrasion and accretion in each sub-district along the Coastal Indramayu. From DSAS process resulted in 473 transects stretching along the coastline, divided into 293 abrasion transects and 180 accretion transects. Abrasion in 2012 – 2022 covered an area of 805,71 ha and accretion covered an area of 941,63 ha. The widest abrasion is in Pasekan covering an area of 266,451 ha. The widest accretion is in Cantigi, with an area of 595,541 ha. The difference between abrasion and accretion in the study area is minimal, which can be caused by the range of years taken in this study. The use of Landsat imagery per year for a period of 10 years is due to see changes in the coastline in more detail. The accretion process at Cantigi is very high because the high sediment supply from the Cimanuk River and the many mangroves that hold the sediment that enters the sea. The abrasion process in Pasekan is very high because the high influence of currents and waves from the sea.

Based on previous study [5], coastline changes were analysed from 1989-2019, using Landsat 5 and Landsat 8 satellite imagery data, using the BILKO method and the overlay method. The results were for the value of the abrasion rate in 1989-2019 covering 2,127 ha and accretion covering an area of 746 ha. Whereas in the current study with 1 year, from 2012-2022, using the BILKO and DSAS methods, the value of the abrasion area is 805,71 ha and the accretion area 941,63 ha. There are differences in values, and this is due to the time distance used to analyse, the method used, as well as the level of accuracy of researchers in analysis.

6. Conclusion

The research conducted is research that focuses on changes in the coastline in Indramayu Regency in 2012-2020 using satellite imagery data. The methods used include radiometric correction, which functions to sharpen the image before analysis, atmospheric correction, which functions to remove noise from the atmosphere, which can optimize the image in the analysis process. The BILKO method is used to delineate coastlines that utilize the NIR and SWIR bands. The DSAS method is used to analyse the coastline and the rate of abrasion and accretion in the study area. From the results of the Digital Shoreline Analysis System (DSAS), the abrasion process dominates compared to the accretion process, with an abrasion area of 805,71 ha and an accretion area of 941,63 ha. The widest abrasion is in Pasekan covering an area of 266,451 ha.

The widest accretion is in Cantigi, with an area of 595,541 ha. The difference between abrasion and accretion in the study area is minimal, which can be caused by the range of years taken in this study. The use of Landsat imagery per year for a period of 10 years is due to see changes in the coastline in more detail. The accretion process at Cantigi is very high because the high sediment supply from the Cimanuk River and the many mangroves that hold the sediment that enters the sea. The abrasion process in Pasekan is very high because the high influence of currents and waves from the sea.

Acknowledgments
This research is part of projects titled "Korea-Indonesia Marine Technology Cooperation Research Center (20220512)" and "Ocean and Coastal Basic Survey and Capacity Enhancement in Cirebon, Indonesia (PG53340)" which are funded by the Ministry of Oceans and Fisheries, Korea.

Credit author statement
Erva Aulia Fauzunora: Making first draft and revising the paper. **Asep Saepuloh**: Revising substantial context of the paper, interpreting the results, and designing of the work. **Ayi Tarya**: Analysing the data and revising the important intellectual content.

Reference
[1] Solihuddin T and Jayawiguna M H 2020 Pantura Jawa Kementrian Kelautan dan Perikanan.
[2] Triatmojo B 2008 Teknik Pantai, Cetakan kelima (Beta Offset, Provinsi Daerah Istimewa Yogyakarta).
[3] Badan Pusat Statistik 2022 Kabupaten Indramayu dalam Angka 2022 (Badan Pusat Statistik Kabupaten Indramayu).
[4] Prost G L 2013 Remote sensing for geoscientists (CRC Press New York).
[5] Natih N M N, Pasaribu R A, Sangadji M S and Kusumaningrum E E 2020 Study on shoreline changes using Landsat imagery in Sangsit Region, Bali Province IOP Conf. Ser.: Earth Environ. Sci. 429 012059
[6] Setiawan Y and Nurjaya I W 2022 Coastline Changes in Indramayu Regency Between 1989-2019 Jurnal Pengelolaan Sumberdaya Alam dan Lingkungan (Journal of Natural Resources and Environmental Management) **12** 543–54.
[7] Module F 2009 Atmospheric correction module: Quac and flaash user's guide Version **4** 44.
[8] Schott J R 2007 Remote sensing: the image chain approach (Oxford University Press on Demand).
[9] Hanifa N R, Djunarsjah E and Wikantika K 2004 Reconstruction of Maritime Boundary between Indonesia and Singapore Using Landsat-ETM Satellite Image. TS9 Marine Cadastre and Coastal Zone Management 3rd FIG Regional Conference.
[10] Himmelstoss E A, Henderson R E, Kratzmann M G and Farris A S 2018 Digital coastlineanalysis system (DSAS) version 5.0 user guide (US Geological Survey).

Characterizing Coastline and Sediment Interaction in The North Coast of Java using Satellite Images

Efrat N. Nainggolan[1] and Asep Saepuloh[2*], Susanna Nurdjaman[3]

[1,2]Geological Engineering Study Program, Faculty of Earth Sciences and Technology, Bandung Institute of Technology, Indonesia
[3]Earth Science Study Program, Faculty of Earth Science and Technology, Bandung Institute of Technology, Indonesia

Email: efratngln@gmail.com
*Corresponding author e-mail: saepuloh@itb.ac.id

Abstract. The large amount of sedimentation process caused rapid changes in the coastline presented by capes form in the coastal area. To characterize coastline and sediment interaction used coastline change analysis, hydrological and vegetation indices using Landsat imagery. Accordingly, we successfully characterized the Cirebon coastline into four categories: abrasion, accretion, abrasion to accretion, accretion to abrasion. The NSM range values -1.123,67 m – 1.026,19 m. The highest abrasion (EPR, LRR, WLR) is 33,18 m/y, 31,16 m/y and 31,16 m/y. The highest accretion (EPR, LRR, WLR) 30,3 m/y, 29,16 m/y, 29,16 m/y. The results of the hydrological index parameters show that humidity levels vary from drought (low humidity) to high humidity, but in the eastern part it includes the water surface. The results of the vegetation index parameter show that the vegetation level was dense, moderate to sparsely grown with vegetation, canopy cover varying from low to medium. Clay material (mud) has impermeable properties, so the study area is dominated by accretion because there is large supply sediment from rivers that enter the sea. Based on these data, it can be concluded that the characteristics of the beach are divided into muddy beaches, and coastal structures such as harbor.

1. Introduction

Sediments are the main material forming the morphology (topography and bathymetry) of the coastal area. Sediment originated from rock fragmentation due to weathering which can take place physically, chemically or biologically. Changes in the morphology of coastal areas can occur through erosion, transportation, and deposition mechanisms. Sedimentation in the coastal environment originates from the presence of sediments originating from the mainland and is basically the main factor in forming the coastal area. The displaced sediments are sediment that lies on the surface of the bottom of the waters[1].

Sedimentation in the coastal environment originates from the presence of sediments originating from the mainland and basically is the main factor in forming the coastal. Sedimentation is closely related to the geomorphic agents working in it. The main geomorphic agents that cause or influence the processes and dynamics of coastal waters are waves, currents and wind [2]. In the coastal environment, sediments are dynamic which will have erosion, transportation and deposition on a spatial and temporal scale. The strong sedimentation process causes rapid changes in the coastline and causes capes and land to form in coastal areas.

Research about coastal characteristics is needed to be carried out in Cirebon Coastal Area to obtain information regarding the physical condition of coastline including the ongoing processes in coastal area. The study of the Cirebon Coastal sedimentation process was carried out in the context of collecting geological data and information on coastal and offshore waters using satellite imagery to determine coastline changes, accretion or abrasion processes and determine the characteristics of beach types.

Cirebon Coastal Area is defined as an area consisting of the Cirebon District and Cirebon City. This area is located on the north coast of Java Island, Indonesia. This area covers a shoreline of approximately 50 km long[3]. The research area is located along the coastal area of, which is bordered by Indramayu District in the northwest (Figure 1).

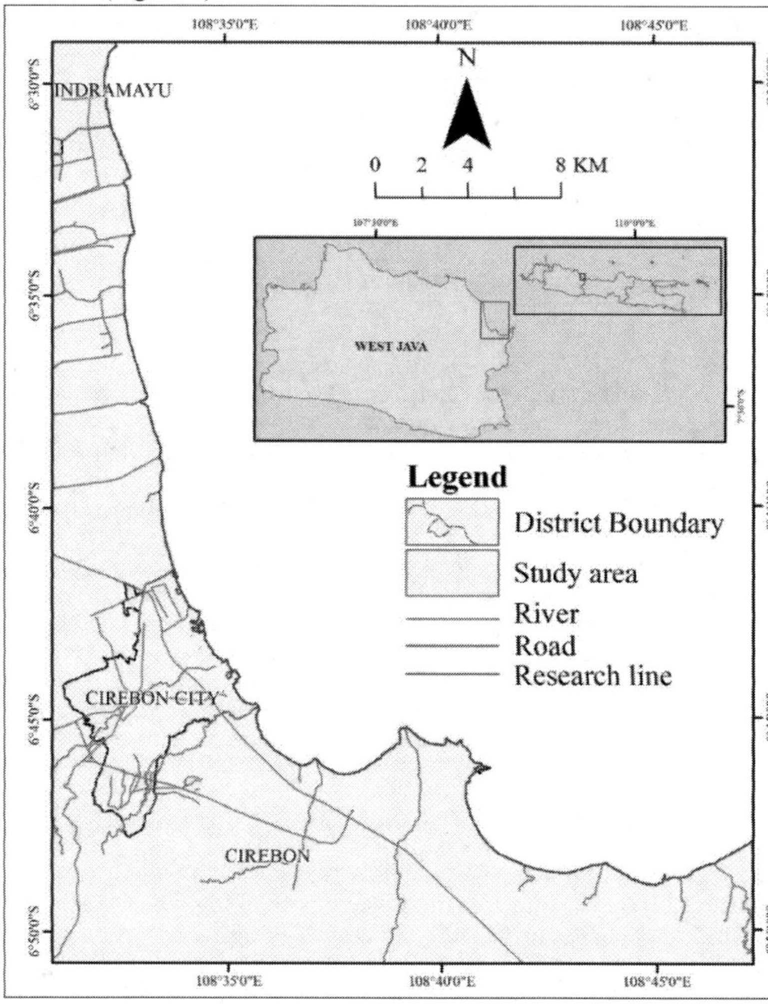

Figure 1. The Cirebon coastal is in Cirebon Regency and Cirebon City which are bordered by Indramayu Regency in the northwest and the purple line is the coastline that is the focus of this research

The surface geology of the Cirebon coastal area and its surroundings consists of alluvium (Qa) deposits. These sedimentary facies are increasingly towards the coast covered by coastal deposits (Qac) [4]. Coastal deposits consist of silt from swamp deposits, silt, and gray clay containing mollusk shells which are deposited around the coast with a thickness of up to several meters. While alluvium deposits can be distinguished into gravel, sand and gray clay, which are deposited along the river floodplain with a thickness of approximately 5 m.

In the geological map of the West Java Muara-Cirebon Quaternary Sheet, Scale 1:50,000, which is based on the results of shallow drilling with a scale of 1:50,000, it is said that Qac belongs to the BM

IOP Conf. Series: Earth and Environmental Science 1245 (2023) 012039 doi:10.1088/1755-1315/1245/1/012039

(beach on marine) facies, namely coastal bund deposits above nearshore/shallow sea deposits. It was also stated that Qa is included in the FM (floodplain on marine) facies, namely floodplain deposits above sediments near the coast [5]. The repetition of sedimentation processes between linear clastic (sea and coast), fluviatile, and swamp deposits in the study area, has inspired that these deposits are related to basin filling and environmental changes.

2. Image data collection

The data used is Landsat satellite imagery which is considered to have the availability of long enough time series image data, free of charge and has good resolution (spatial, temporal, radiometric) (intermediate level). The satellite images used were taken in 1988, 2003 and 2022 using the Landsat 5 TM, Landsat 7 ETM+ and Landsat 8 OLI/TIRS satellites which can be downloaded from the website http://earthexplorer.usgs.gov/ [6] (Table 1, Figure 2). In addition to using Landsat satellite imagery, geological maps and Quaternary geological maps of the study area are also used to support research results and discussion.

Table 1. Details of Landsat data used in this study

No	Satellite Image	Product-ID	Acquisition Date	Scene Center Time	Cloud Cover
1	Landsat-5 TM	LT05_L1TP_121065_19880914_20200917_02_T1	14/09/1988	02:25:03.3750880Z	28
2	Landsat-7 ETM+	LE07_L1TP_121065_20030119_20200916_02_T1	19/01/2003	02:42:46.5070176Z	6
3	Landsat-8 OLI/TIRS	LC08_L1TP_121065_20220726_20220802_02_T1	26/07/2022	02:54:40.0405670Z	5,26

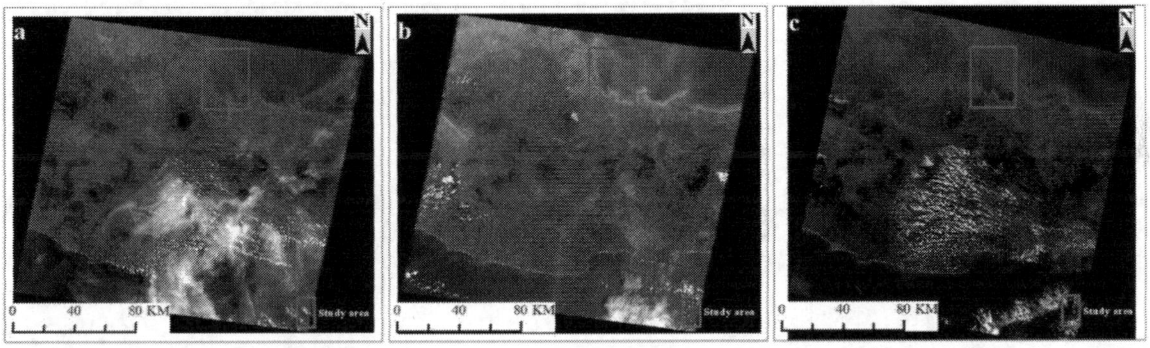

Figure 2. Geotiff results with metadata on Landsat images: Landsat-5 with acquisition time of 14 September 1988 (a); Landsat-7 with acquisition time 19 January 2003 (b) and Landsat-8 with acquisition time 26 July 2022 (c)

3. Research methodology

Coastal characteristics and sediment material in this study used 3 parameters, namely coastline changes, hydrological indices and vegetation indices. The following describes the method used in this study.

3.1. Image processing

Reflectance, radiance, and atmospheric conditions all affect the digital number (DN) values of Landsat data. Radiometric calibration employs algorithms and processes that improve Landsat data. This is done by converting the DN values of the data to spectral radiance (at the sensor), and then to reflectance (also at the sensor). This is followed by the removal of atmospheric effects, which are due to absorption and scattering, to perform atmospheric correction (reflectance at the surface) [7].

FLAASH (Fast Line-of-sight Atmospheric Analysis of Hypercubes) is a first-principles atmospheric correction tool that corrects wavelengths in the visible through near-infrared and shortwave infrared regions, up to 3 μm. FLAASH works with most hyperspectral and multispectral sensors [8]. Water vapor and aerosol retrieval are only possible when the image contains bands in appropriate wavelength positions. FLAASH can correct images collected in either vertical (nadir) or slant-viewing geometries. Atmospheric correction results in this study (Figure 3).

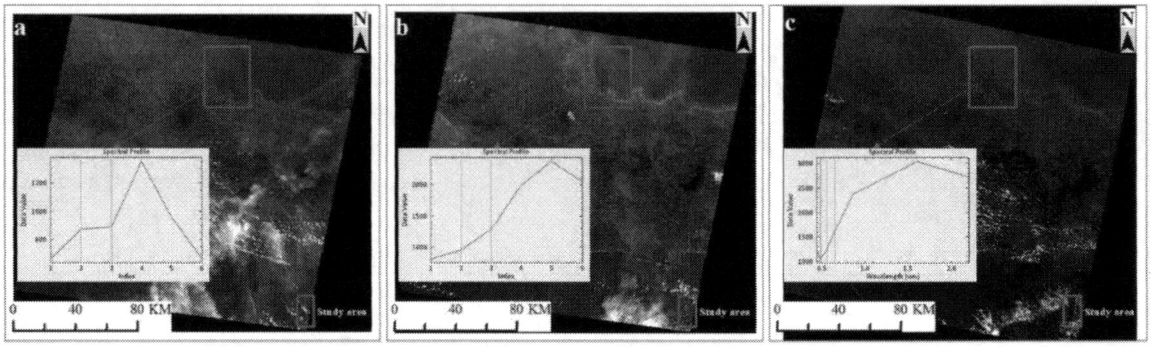

Figure 3. Atmospheric correction results using the FLAASH method are shown by the spectral profile of the blue band values lower than the green band on Landsat imagery: Landsat-5 with acquisition time of 14 September 1988 (a); Landsat-7 with acquisition time 19 January 2003 (b) and Landsat-8 with acquisition time 26 July 2022 (c)

3.2. Identifying land boundaries using BILKO method

Determination of the boundary between land and sea is done by utilizing the brightness value or Brightness Value (BV) of land and sea. For Landsat 5 and 7 required band 4 while Landsat 8 uses band 5 which is the infrared band [9]. Infrared waves have a reflectance low to water and high reflectance against the mainland (Figure 4). This method requires a BV value for the lowest land and the highest oceanic BV values. The BV value is required in the separation of land and ocean [9], [10]. The use of the BILKO formula is as follows (1)

$$BILKO = \left(\frac{\emptyset}{(N \times 2)+1)}\right)(\times -1))+1) \quad (1)$$

Where N is minimum BV value of Landsat 5 and 7 (30) and Landsat 8 (7000) imagery and Ø is Band 4 (Landsat 5 dan 7) or Band 5 (Landsat 8)

The results of applying the BILKO formula can be seen in Figure 4. The boundaries of land and sea are clearly visible before digitizing the coastline.

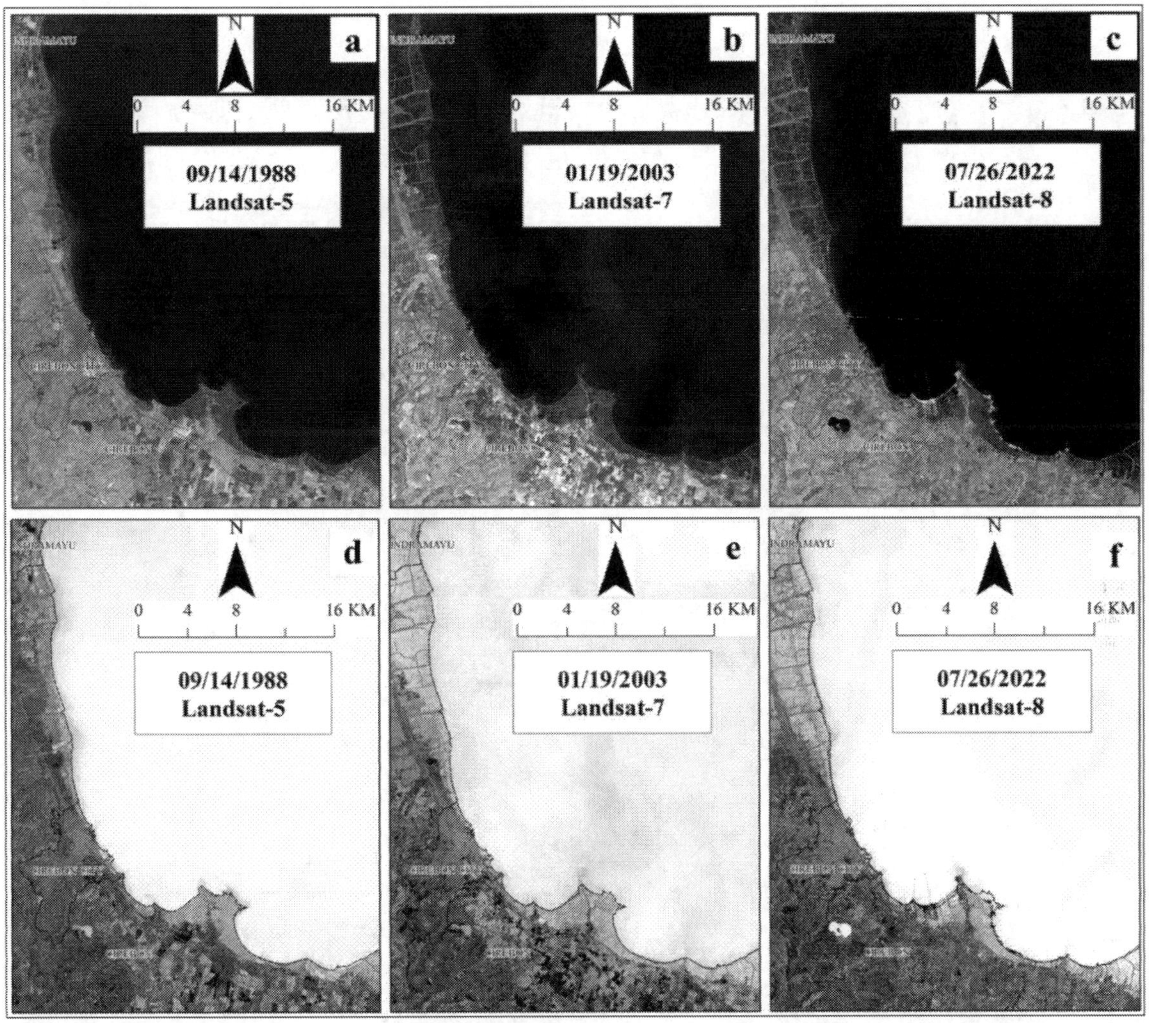

Figure 4. The results of applying the BILKO formula to separate land and sea parts on Landsat to make it easier to digitize coastlines: Landsat-5 before applying the BILKO formula (a); Landsat 7 before applying the BILKO formula (b); LAndsat-8 before applying the BILKO formula (c)); Landsat 5 after applying the BILKO formula (d); Landsat-7 after the application of the BILKO formula (e); Landsat-8 after the application of the BILKO formula (f)

The results of the BILKO algorithm method show that the boundary between land and water body is clear in areas without cloud and fog. In the processed image, the land is black and the sea is white. Coastline digitation is done over the image satellite by applying the BILKO formula. The results of beach digitization carried out on Landsat imagery in 1988, 2003 and 2022 can be seen in Figure 5.

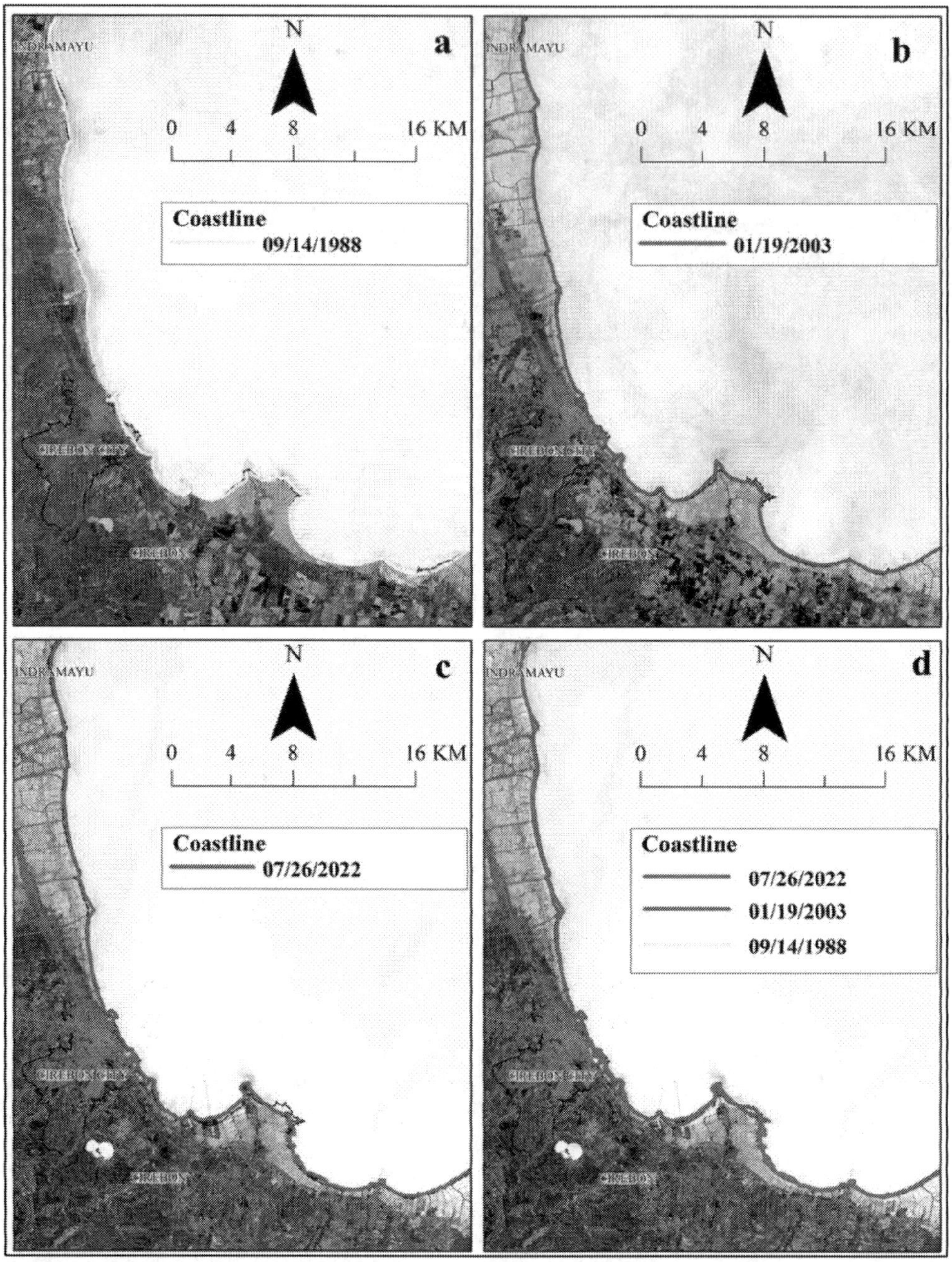

Figure 5. The results of digitizing the coastline on Landsat imagery after applying the BILKO formula: digitizing the coastline on Landsat-5 with the acquisition date of September 14, 1988 (a); digitizing the coastline on Landsat-7 with acquisition date January 19, 2003 (b); digitizing the coastline on Landsat-8 with acquisition time 26 July 2022 (c) and combine digitization of the entire coastline in 1988, 2003 and 2022 (d)

4. Analyzing digital land boundaries

The Digital Shoreline Analysis System (DSAS) v5 software is an add-in to Esri ArcGIS desktop (10.4-10.7+) that enables a user to calculate rate-of-change statistics from multiple historical shoreline positions [11]. Digital Shoreline Analysis System (DSAS) is a remote sensing technology that can be used to automatically detect and calculate coastline changes in an area [12]. DSAS generates transects that are cast perpendicular to the reference baseline at a user-specified spacing alongshore. DSAS v5 supports a baseline located anywhere like offshore, onshore, in the middle of the shoreline data (midshore), or a combination of baseline placements. There are no restrictions on where the reference baseline is drawn, it may be positioned completely to one side of the shoreline data or be placed between the historical shoreline positions. The result of DSAS analysis we can see Figure 6 and all the rate has been classified into 5 classes based on Table 2.

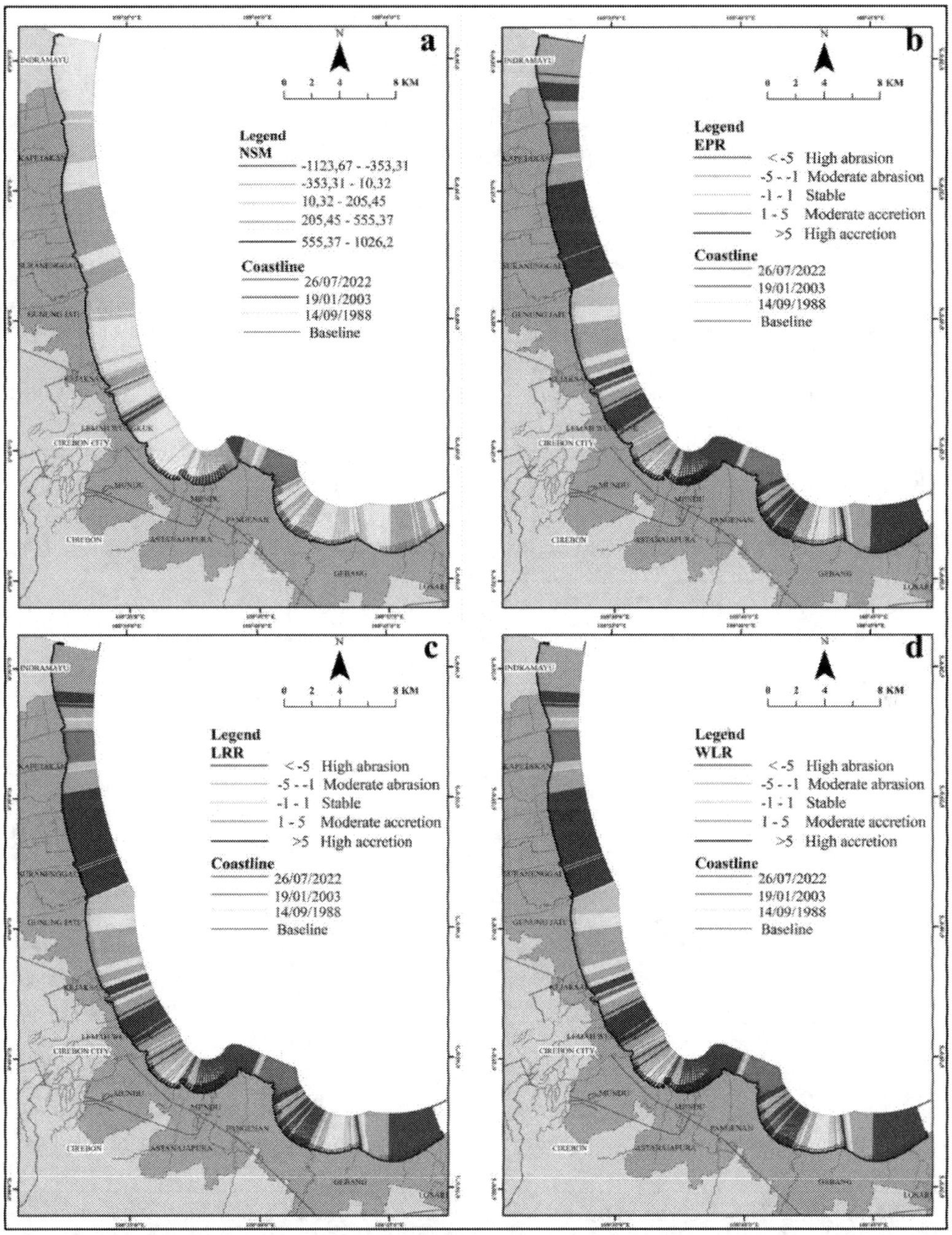

Figure 6. DSAS analysis of the coastlines in 1988, 2003 and 2022 showing NSM maps with a range of values 2649- -1159 with a positive NSM indicating accretion and a negative NSM value indicating abrasion (a); EPR map with value ranges divided into 5 classes (high abrasion <-5 - high accretion >5) (b); LRR map with value range divided into 5 classes (high abrasion <-5 - high accretion >5) (c) and WLR map with value range divided by 5 classes (high abrasion <-5 - high accretion >5) (d)

IOP Conf. Series: Earth and Environmental Science 1245 (2023) 012039 doi:10.1088/1755-1315/1245/1/012039

Table 2. Criteria for coastline change classification[13]

No	Values (m/y)	Class
1	<-5	High abrasion
2	-5 - -1	Moderate abrasion
3	-1 - 1	Stable
4	1 - 5	Moderate accretion
5	>5	High accretion

DSAS measures the distance between the baseline and each shoreline intersection along a transect, and combines date information, and positional uncertainty for each shoreline, to generate the following change metrics: Net Shoreline Movement (NSM), End Point Rate (EPR), Linear Regression Rate (LRR) and Weighted Linear Regression Rate (WLR) [11]. The net shoreline movement (NSM) is the distance between the oldest and the youngest coastlines for each transect. The coastline change envelope (SCE) reports a distance (in meters), not a rate. The SCE value represents the greatest among all the coastlines that intersect. The end point rate (EPR) is calculated by dividing the distance of coastline movement by the time elapsed between the oldest and the most recent coastline. A linear regression rate-of-change statistic can be determined by fitting a least-squares regression line to all coastline points for a transect. In a weighted linear regression, the more reliable data are given greater emphasis or weight towards determining a best-fit line [14].

The highest negative NSM value was found to be -1.123,67 m and the highest positive NSM was found to be 1.026,19 m (Figure 7). The highest abrasion EPR on the same transect with NSM is 33,18 m/y and the highest accretion EPR is 30,3 m/y (Figure 8). The highest abrasion LRR on the same transect with NSM and EPR is 31,16 m/y and the highest accretion EPR is 29,16 m/y (Figure 9). The highest abrasion WLR on the same transect with NSM, EPR and WLR is 31,16 m/y and the highest accretion EPR is 29,16 m/y (Figure 10). A visual representation of the rate of change (1988-2022) of all methods is shown in Figure 11.

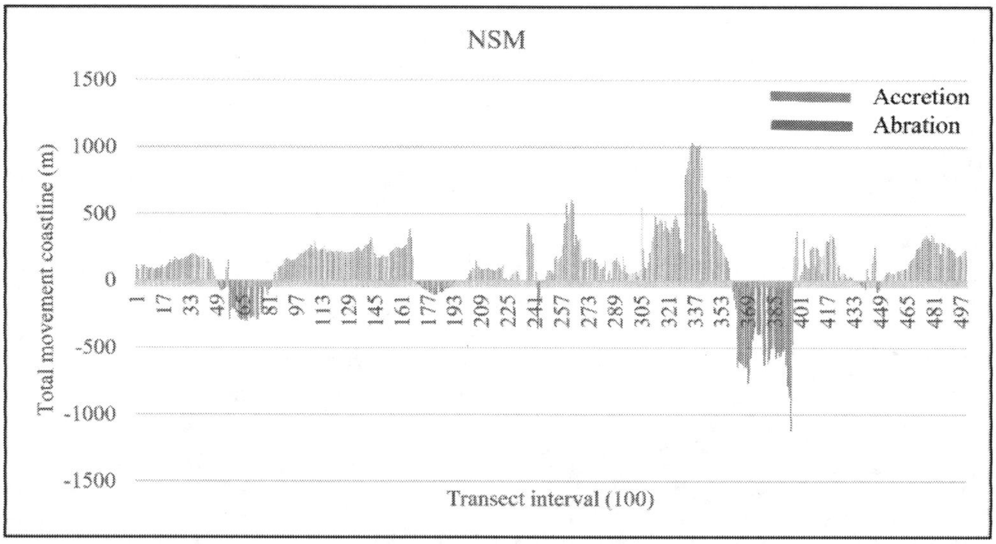

Figure 7. Bar diagram showing the value of NSM on the coastline in 1988-2022 with interval transect 100 and we can see the highest value of NSM is 2.649 m (accretion) and the lowest is 1.159 m (abrasion)

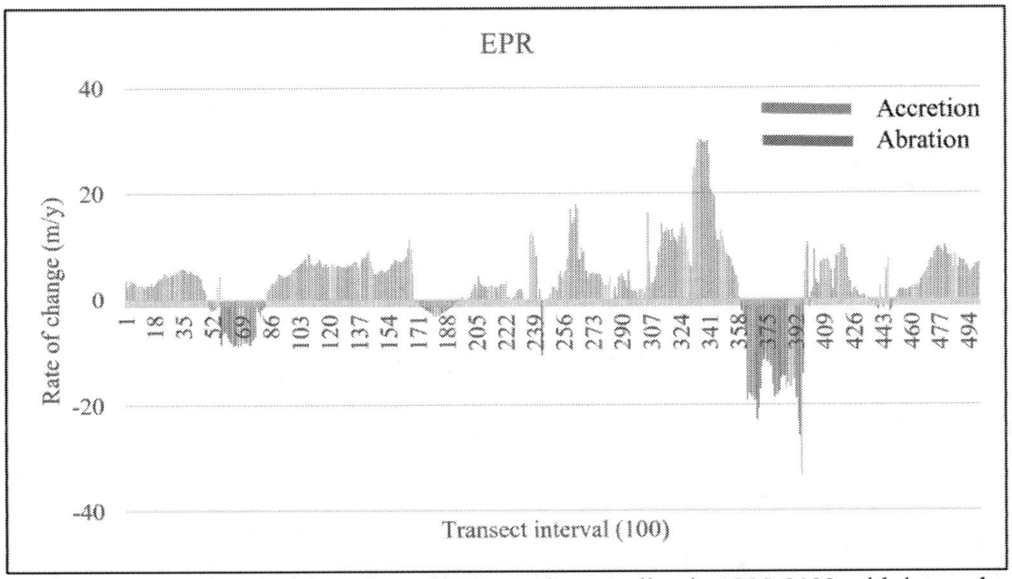

Figure 8. Bar diagram showing the value of EPR on the coastline in 1988-2022 with interval transect 100 and we can see the highest value rate of changes is 30,3 m/y (accretion) and the lowest is 33,18 m/y (abrasion)

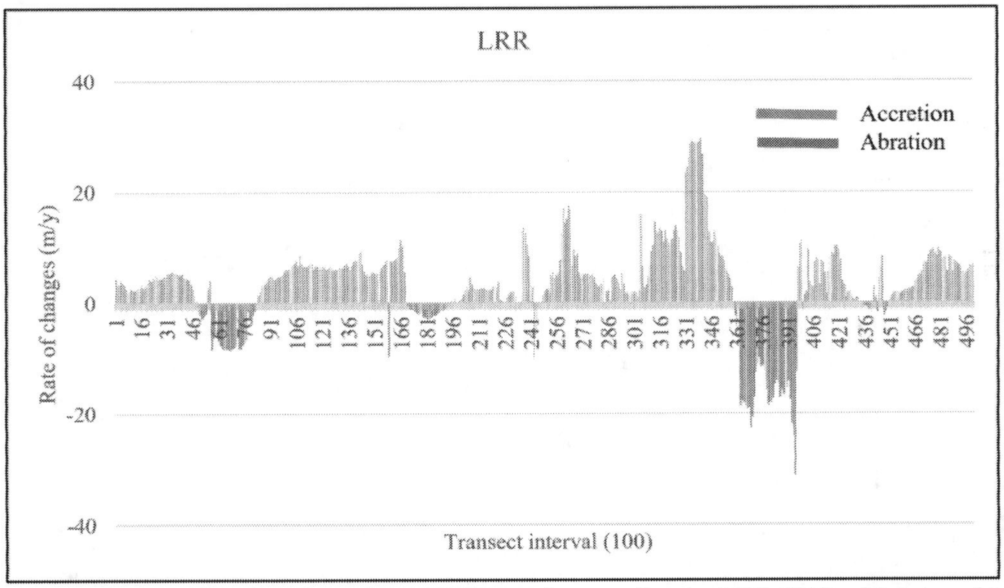

Figure 9. Bar diagram showing the value of LRR on the coastline in 1988-2022 with interval transect 100 and we can see the highest value rate of changes is 29,16 m/y (accretion) and the lowest is 31,16 m/y (abrasion)

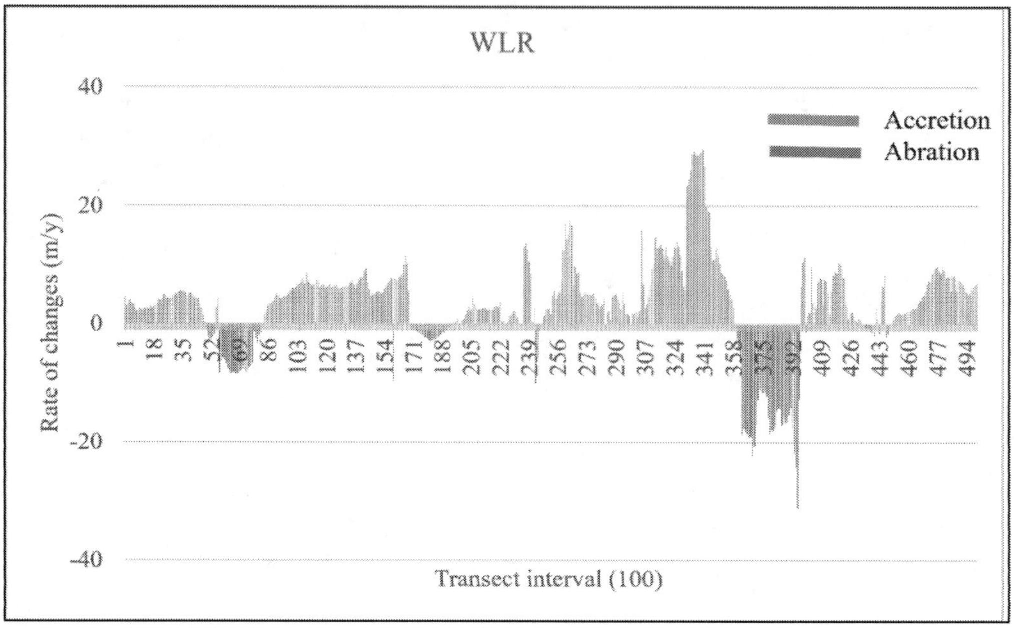

Figure 10. Bar diagram showing the value of WLR on the coastline in 1988-2022 with interval transect 100 and we can see the highest value rate of changes is 29,16 m/y (accretion) and the lowest is 31,16 m/y (abrasion)

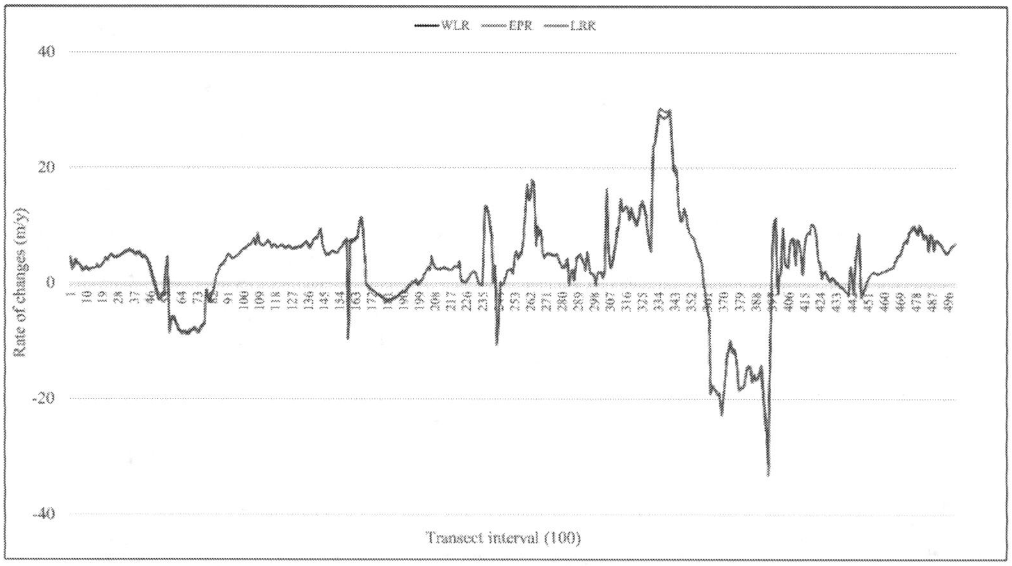

Figure 11. Line graph shows the difference of rate in different method namely EPR, LRR and WLR, which result LRR with WLR are very similar and result of EPR little varies from LRR and WLR

4.1. Coastline change analysis

The results of the DSAS in the study area show that there are 4 processes that occur from the observations of coastline changes namely abrasion, accretion, abrasion to accretion and accretion to abrasion. The dominant process that occurs in the study area is accretion. This is due to the large supply of sediment from the rivers around the Cirebon Beach area such as the Bondet River, Ciberes River, Pemali River, Sukalila River so that it is possible that the dominant process is accretion rather than abrasion. The area of each process that occurs at the study site can be seen in Figure 12 which shows that the process at the study site is dominated by accretion with an area of 710.88 ha while abrasion is with an area of 213.51 ha. The process that begins with abrasion then accretion with an area of 170.28 ha while the process that begins with accretion then abrasion with an area of 71.9 ha.

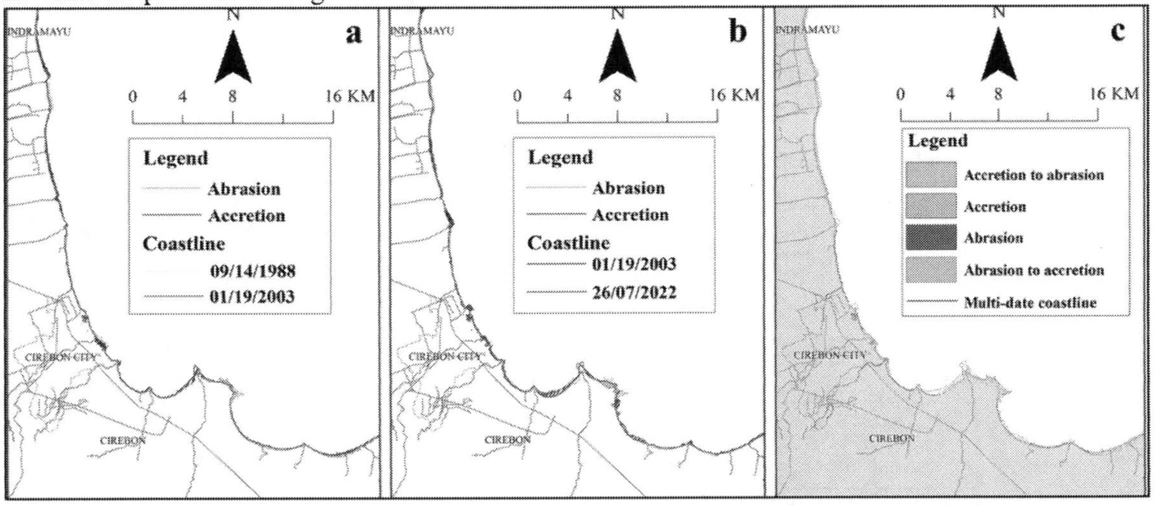

Figure 12. The division of processes that occur is based on the results of DSAS analysis in 1988-2003 with dominant accretion than abrasion (a); 2003-2022 with dominant accretion than abrasion (b) and 1988-2023 with the result dominant accretion than abrasion (c)

4.2. Hydrological indices derived by Landsat-8 images

Hydrological indices are an index that describes the condition of the water content in an area. The hydrological indices are used in this study to accommodate the effect of water content in a vegetation on the recorded image. In the following, several remote sensing hydrological indices bands that are often used are NDWI (Normalized Difference Water Index), MNDWI (Modified Normalized Difference Water Index), WRI (Water Ratio Index) and NDMI (Normalized Difference Moisture Index).

The Normalized Difference Water Index (NDWI) is used to highlight open water features in a satellite image, allowing a water body to “stand out” against the soil and vegetation [15]. The NDWI equation looks like this:

$$NDWI = \left(\frac{Green - NIR}{Green + NIR}\right) \tag{2}$$

The modified normalized difference water index (MNDWI) uses green and SWIR bands for the enhancement of open water features. It also diminishes built-up area features that are often correlated with open water in other index [16]. The NDWI equation looks like this:

$$MNDWI = \frac{Green - SWIR}{Green + SWIR} \tag{3}$$

Table 3. NDWI and MNDWI value ranges and interpretations [17]

Value	Interpretation
0,2 – 1	Water surface
0.0 – 0,2	Flooding, humidity,
-0,3-0	Moderate drought, non-aqueous surfaces
-1-0,3	Drought, non-aqueous surfaces

IOP Conf. Series: Earth and Environmental Science 1245 (2023) 012039 doi:10.1088/1755-1315/1245/1/012039

A total of four spectral reflectance bands are used in this technique. It is the ratio between the total spectral reflectance of two visible bands (i.e., green and red) to the total spectral reflectance of near-infrared and Middle infrared band (MIR) [18]. The WRI equation looks like this:

$$WRI = \frac{(Green + Red)}{(NIR + SWIR)} \quad (4)$$

The Normalized Difference Moisture Index (NDMI) detects moisture levels in vegetation using a combination of near-infrared (NIR) and short-wave infrared (SWIR) spectral bands. It is a reliable indicator of water stress in crops [19]. The NDMI equation looks like this:

$$NDMI = \frac{(NIR - SWIR1)}{(NIR + SWIR1)} \quad (5)$$

Table 4. NDMI value ranges and interpretations [17]

Value	Interpretation
-1 – -0,8	Bare soil
-0,8 – -0,6	Almost absent canopy cover
-0,6 – -0,4	Very low canopy cover
-0,4 – -0,2	Low canopy cover, dry or very low canopy cover, wet
-0,2 – 0	Mid-low canopy cover, high water stress or low canopy cover, low water stress
0 – 0,2	Average canopy cover, high water stress or mid-low canopy cover, low water stress
0,2 – 0,4	Mid-high canopy cover, high water stress or average canopy cover, low water stress
0,4 – 0,6	High canopy cover, no water stress
0,6 – 0,8	Very high canopy cover, no water stress
0,8 – 1	Total canopy cover, no water stress/waterlogging

The results of hydrological indices parameters can be seen in Figure 13. The results of the hydrological index parameters along the Coastal Cirebon show that humidity levels vary from drought (low humidity) to high humidity, but in the eastern part it includes the water surface. This depends on the type of material that makes up the Cirebon coast which is divided into clay and sandy clay. Clay or mud material tends to have a high humidity level because it is impermeable while the humidity level of sandy clay is lower than clay (mud) because it has a greater permeability than clay.

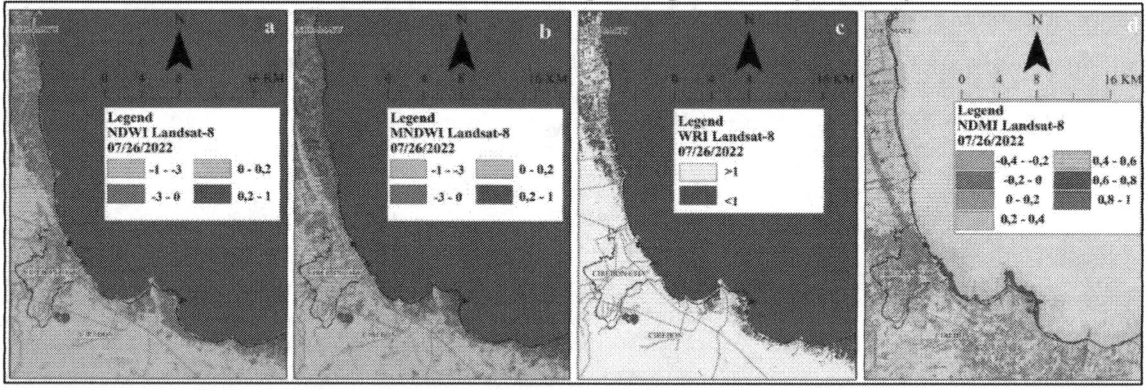

Figure 13. Hydrological indices parameters (NDWI, MNDWI, WRI, NDMI) on Landsat-8 imagery in the study area: NDWI shows a value range of -1-1 with a dominant water surface (a); MNDWI shows a value range of -1-1 dominated by water surface (b); WRI shows a value range of <1 (water body) (c); NDMI with value range -0.4-1 (d)

4.3. Vegetation indices derived by Landsat-8 images

The vegetation indices is the amount of greenness of the vegetation obtained from digital signal processing of the brightness value data of several channels of satellite sensor data. In the following, several remote sensing vegetation indices that are often used are NDVI (Normalized Difference Vegetation Index), SAVI (Soil Adjusted Vegetation Index) and MSAVI (Modified Landsat Soil Adjusted Vegetation Index).

The Normalized Difference Vegetation Index (NDVI) measures the greenness and the density of the vegetation captured in a satellite image. Healthy vegetation has a very characteristic spectral reflectance curve which we can benefit from by calculating the difference between two bands – visible red and near-infrared. The NDVI is that difference expressed as a number – ranging from -1 to 1 [20]. The NDVI equation looks like this:

$$NDVI = \frac{(NIR - Red)}{(NIR + Red)} \tag{6}$$

Table 5. NDVI value ranges and interpretations [17]

Value	Interpretation
<0,1	Bare soil
0,1 – 0,2	Almost absent canopy cover
0,2 – 0,3	Very low canopy cover
0,3 – 0,4	Low canopy cover, low vigour or very low canopy cover, high vigour
0,4 – 0,5	Mid-low canopy cover, low vigour or low canopy cover, high vigour
0,5 – 0,6	Average canopy cover, low vigour or mid-low canopy cover, high vigour
0,6 – 0,7	Mid-high canopy cover, low vigour or average canopy cover, high vigour
0,7 – 0,8	High canopy cover, high vigour
0,8 – 0,9	Very high canopy cover, very high vigour
0,9 – 1	Total canopy cover, very high vigour

Landsat Soil Adjusted Vegetation Index (SAVI) is used to correct Normalized Difference Vegetation Index (NDVI) for the influence of soil brightness in areas where vegetative cover is low. Landsat Surface Reflectance-derived SAVI is calculated as a ratio between the R and NIR values with a soil brightness correction factor (L) defined as 0.5 to accommodate most land cover types [21]. The SAVI equation looks like this:

$$\frac{(NIR - Red)}{(NIR + Red + L)} \times (1 + L) \tag{7}$$

Modified Landsat Soil Adjusted Vegetation Index (MSAVI) minimizes the effect of bare soil on the Soil Adjusted Vegetation Index (SAVI). MSAVI is calculated as a ratio between the R and NIR values with an inductive L function applied to maximize reduction of soil effects on the vegetation signal [22]. The MSAVI equation looks like this:

$$\frac{\left(2 \times NIR + 1 - \sqrt{((2 \times NIR + 1)2 - 8 \times (NIR - Red))}\right)}{2} \tag{8}$$

Table 6. SAVI and MSAVI value ranges and interpretations [17]

Value	Interpretation
-1- 0,2	Bare soil
0,2 - 0,4	The seed germination stage
0,4 - 0,6	The leaf development stage

The results of vegetation indices parameters can be seen in Figure 14. The results of the vegetation index parameter along the Cirebon coast showed that the vegetation level was dense, moderate to sparsely grown with vegetation, canopy cover varying from low to medium. This depends on the type of vegetation that can grow on the Cirebon coastal material. Mangroves usually grow on clay material (mud) so that the vegetation is dense. In addition, land use will also affect the level of canopy cover and vegetation density.

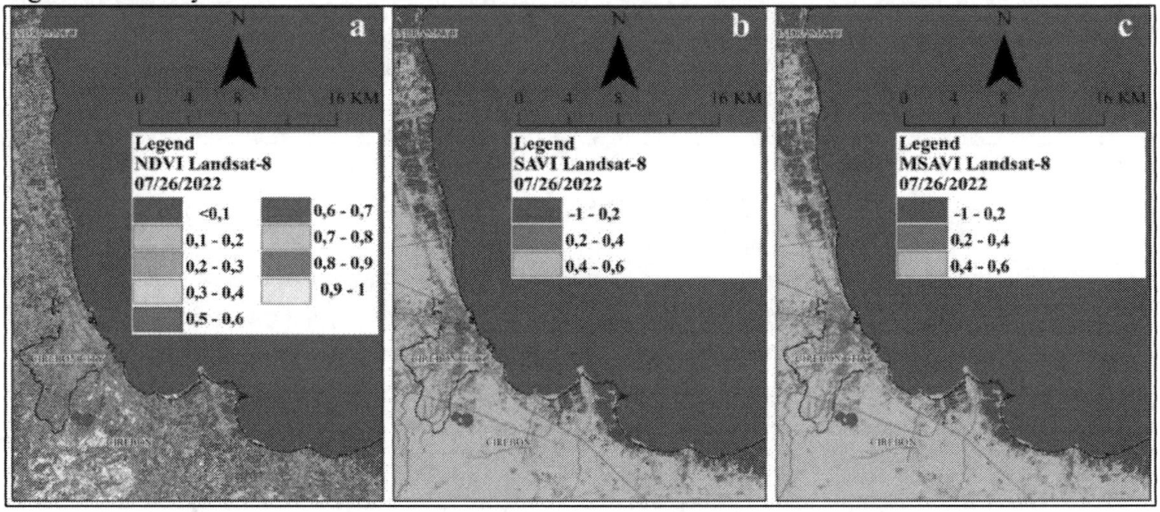

Figure 14. Vegetation indices parameters (NDVI, SAVI, MSAVI) on Landsat-8 imagery: NDVI range value 0,1-0,5 with very low-low canopy cover and low virgour (a); SAVI range value <0,1 bare soil is not covered by other live ground covers (b); SAVI range value <0,1 bare soil is not covered by other live ground covers (c)

5. Discussion

Determination of beach type characteristics based on shoreline change analysis, hydrological indices and vegetation indices is assisted with the Geological Map of the Cirebon Sheet [4] and the Geological Map of the Muara Quarter Sheet [5].

The study area, we know the morphological condition along the Cirebon Coastal, its formation is affected by fluvio-marine and marine. The original fluvio-marine is characterized by the formation of a delta because the material brought from the river is in greater supply than the sea waves and currents that carry the material. The original marine is characterized by the morphology of the coastal plains due to the very sloping topography which forms swamps, mangrove forests and beach sand. Based on the Geological Map of Cirebon scale 1:100.000 [3], the dominant lithology in the study area is alluvium deposits consisting of gravel, sand, clay and coastal deposits consisting of silt from swamp deposits, silt and clay. In addition to alluvium deposits, there are also coastal deposits consisting of mud deposited in swampy areas, silt and gray clay containing mollusks shells, deposited along and around shore.

From the results of the DSAS analysis that has been carried out on the coastline, it can be grouped into 4 categories of processes that occur, namely abrasion, accretion, abrasion to accretion and accretion to abrasion. Based on previous study[23], the calculations of wave flux energy produce a coastal dynamic map show the dominant process that occurs in the Coastal Cirebon Area is accretion, especially in the eastern part. The process of strong accretion, especially at estuary, resulted in this area run into quite rapid shallowing and forming new coastlines. This research is relevant with previous study which shows that the study area is dominated by the accretion process. This is due to the supply of sediment material originating along the coast from the rivers around the Cirebon Beach area such as the Bondet River, Ciberes River, Pemali River, Sukalila River.

The results of the hydrological index parameters along the Coastal Cirebon show that humidity levels vary from drought (low humidity) to high humidity, but in the eastern part it includes the water surface. This depends on the type of material that makes up the Cirebon coast which is divided into clay and sandy clay. Clay or mud material tends to have a high humidity level because it is impermeable while the humidity level of sandy clay is lower than clay (mud) because it has a greater permeability than clay.

The results of the vegetation index parameter along the Cirebon Coast show that the vegetation level was dense, moderate to sparsely grown with vegetation, canopy cover varying from low to medium. This depends on the type of vegetation that can grow on the Cirebon coastal material. Mangroves usually grow on clay material (mud) so that the vegetation is dense. In addition, land use will also affect the level of canopy cover and vegetation density.

Clay material (mud) has impermeable properties so that the processes that occur are dominated by accretion due to the large supply of sediment from rivers that enter the sea. Sandy clay material has a greater permeability than clay and the process is dominated by abrasion because there is no large sediment supply from rivers and the influence of currents and waves from the sea is higher.

Based on these data, it can be concluded that the characteristics of the beach at the study area are divided muddy beaches which are dominated by fluvial processes and coastal structures such as harbor (Figure 15).

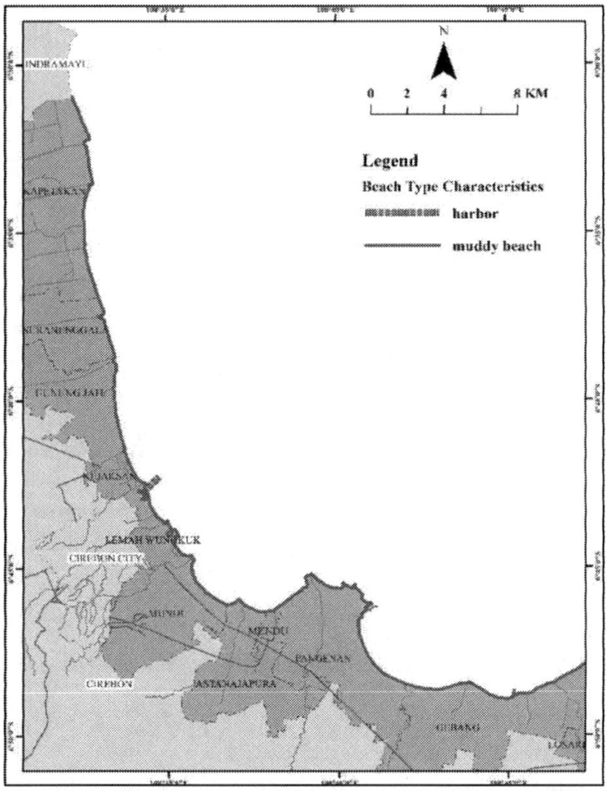

Figure 15. Characteristics of the beach in the study area divided into muddy beaches which are dominated by fluvial processes and coastal structures such as harbor

6. Conclusion

The research was conducted along the Cirebon coast using Landsat imagery data (Landsat-5, Landsat-7 and Landsat-8). The results of image optimization can be proven by the spectral profile which shows a lower blue band value than the green band. The BILKO method is used to describe coastlines that utilize NIR and SWIR bands to separate land and sea. The DSAS method is used to analyze the coastline and the rates of abrasion and accretion, knowing the dominant processes that occur along the coastline in the study area. The results of the DSAS in the study area show that there are 4 processes that occur from the observations of coastline changes namely abrasion, accretion, abrasion to accretion and accretion to abrasion. The dominant process that occurs in the study area is accretion. This is due to the large supply of sediment from the rivers around the Cirebon Beach area such as the Bondet River, Ciberes River, Pemali River, Sukalila River so that it is possible that the dominant process is accretion rather than abrasion.

The dominant lithology in the study area is clay (mud) and sandy clay. Clay material (mud) has impermeable properties so that the processes that occur are dominated by accretion due to the large supply of sediment from rivers that enter the sea. Sandy clay material has greater permeability than clay and the process is dominated by abrasion because there is no large sediment supply from rivers and the influence of currents and waves from the sea is higher.

Based on these data, it can be concluded that the characteristics of the beach at the study site are divided into muddy beaches which are dominated by fluvial processes and coastal structures such as harbor.

Acknowledgement

This research is part of projects titled "Korea-Indonesia Marine Technology Cooperation Research Center (20220512)" and "Ocean and Coastal Basic Survey and Capacity Enhancement in Cirebon, Indonesia (PG53340)" which are funded by the Ministry of Oceans and Fisheries, Korea.

Reference

[1] Poerbondono and Djunarsjah E 2005 *Survei Hidrografi* (PT.Refika Aditama)

[2] Direktorat Jenderal Pengelolaan Ruang Laut 2020 Get to know Sediments

[3] Abdurrahman U, Diastomo H, Avrionesti, Surya M Y, Badriana M R, Suprijo T and Park H 2021 Comprehensive Analysis of Coastal Dynamics in Cirebon Coastal Area (CCA), Indonesia *Journal of Coastal Research* **114**

[4] Silitonga P H, Masria M and Suwarna N 1996 Peta Geologi Bersistem, Indonesia Lembar Cirebon

[5] Sumanang H, Mulyana H, Hidayat S and Basri C 1997 Peta Geologi Kuarter Lembar Muara – Cirebon, Jawa Barat

[6] United States Geological Survey 2022 Earth Explorer

[7] Landsat Missions 2023 Landsat Calibration & Validation *Sattelite and Sensor Calibration*

[8] Cooley T, Anderson G P, Felde G W, Hoke M L, Ratkowski A J, Chetwynd J H, Gardner J A, Adler-Golden S M, Matthew M W, Berk A, Bernstein L S, Acharya P K, Miller D and Lewis P 2002 FLAASH, a MODTRAN4-based atmospheric correction algorithm, its application and validation *IEEE International Geoscience and Remote Sensing Symposium* IEEE International Geoscience and Remote Sensing Symposium. IGARSS 2002 vol 3 (Toronto, Ont., Canada: IEEE) pp 1414–8

[9] Hanifa N R, Djunarsjah E and Wikantika K 2004 Reconstruction of Maritime Boundary between Indonesia and Singapore Using Landsat-ETM Satellite Image

[10] Arum Y P 2021 Analysis of shoreline changes using the bilko method on landsat imagery in Karawang regency (1999-2019) *J. Phys.: Conf. Ser.* **1943** 012009

[11] Woods Hole Coastal and Marine Science Center 2022 Digital Shoreline Analysis System (DSAS) *Computer Software for Calculating Shoreline Change (or positional change of a boundary over time)*

[12] Mutaqin B W 2017 Shoreline changes analysis in Kuwaru coastal area, Yogyakarta, Indonesia: An application of the Digital Shoreline Analysis System (DSAS) *Int. J. SDP* **12** 1203–14

[13] Dey M, S P and Jena B K 2021 A Shoreline Change Detection (2012-2021) and forecasting Using Digital Shoreline Analysis System (DSAS) Tool: A Case Study of Dahej Coast, Gulf of Khambat, Gujarat, India *Indonesia Journal of Geography* **53(2)** 295–309

[14] Himmelstoss E A, Henderson R E, Kratzmann M G and Farris A S 2018 Digital Shoreline Analysis System (DSAS) Version 5.0 User Guide *(USGS, Reston, Virginia)* 110 p

[15] Gao B 1996 NDWI—A normalized difference water index for remote sensing of vegetation liquid water from space *Remote Sensing of Environment* **58** 257–66

[16] Xu H 2006 Modification of normalised difference water index (NDWI) to enhance open water features in remotely sensed imagery *International Journal of Remote Sensing* **27** 3025–33

[17] EOS Data Analytic 2022 Satellite Imagery and Spectral Band Combinations

[18] Shen L and Li C 2010 Water Body Extraction from Landsat ETM+ Imagery Using Adaboost Algorithm *2010 18th International Conference on Geoinformatics* 2010 18th International Conference on Geoinformatics (Beijing, China: IEEE) pp 1–4

[19] Wilson E H and Sader S A 2002 Detection of forest harvest type using multiple dates of Landsat TM imagery *Remote Sensing of Environment* **80** 385–96

[20] Crippen R 1990 Calculating the vegetation index faster *Remote Sensing of Environment* **34** 71–3

[21] Huete A R 1988 A soil-adjusted vegetation index (SAVI) *Remote Sensing of Environment* **25** 295–309

[22] Jiang Z 2007 Interpretation of the modified soil-adjusted vegetation index isolines in red-NIR reflectance space *J. Appl. Remote Sens* **1** 013503

[23] Setyadi D and Aryanto N C 2008 Silting Processes on the Coast and Offshore of Cirebon Due to High Sedimentation Rates from the Land *JSDG* **18 (5)** 299–307

IOP Conf. Series: Earth and Environmental Science 1245 (2023) 012040 doi:10.1088/1755-1315/1245/1/012040

STUDY OF DRIVING FORCE AND SURFACE CURRENT PATTERNS IN LABUAN BAJO FROM HIGH FREQUENCY RADAR DATA FROM OCTOBER TO DECEMBER 2018

Y Baharsyah[1,a], A R Hakim[2], A Ramdhani[3], R Rachmayani[4]

[1] Study program of Oceanography, Institut Teknologi Bandung, Bandung, West Java, Indonesia

[2] Botram Ocea Technology Reseach and Management

[3] Head of Division for Marine Mateorology Information, Meteorological, Climatological, and Geophysical Agency, Jakarta Pusat, DKI Jakarta, Indonesia

[4] Research group, Institut Teknologi Bandung. Bandung, West Java, Indonesia

[a] correspondent author: Yustikabaharsyah@student.itb.ac.id

Abstract. Surface ocean current data in Labuan Bajo from October to December 2018 for determine the characteristics of the surface current and identify the influence of wind and tides in generating surface current. The current data set is derived from measurements of high frequency (HF) radar, which operates on the electromagnetic wave spectrum at high frequency or short wave. Method of data analysis by processing surface current velocity and direction of the zonal and meridional into spatial and temporal graphs. The result showed the average current speed in October was 0.13 m/s, the average current speed in November was 0.12 m/s, and the average current speed in December was 0.09 m/s. The empirical orthogonal function (EOF) Mode 1 has a variance of 56.2% because of the tides, Mode 2 has a variance of 26.6% because of bi-weekly variability, and Mode 3 has a variance of 17.2% because of the intra-seasonal period. This shows that tides have a greater influence on the study region than any other driving force. These results are intended to be useful to the public, particularly in terms of sustainable coastal marine management in Labuan Bajo. Especially in terms of preventing boat accidents caused by ocean currents.

1. Introduction

Generally, ocean current can be classified based on their forming force, namely currents related to density distribution, currents generated by wind, currents generated by ocean waves, and tidal currents. Marine surface current information needed in coastal management, especially related to sedimentation and utilized by marine users such as fishermen, tourism, coastal communities. Ocean current monitoring, in particular, can be used to prevent boat accidents and rescues, which are common in Labuan Bajo. Monitoring sea surface currents can be done using data from high-frequency (HF) radar. HF radar is a type of radar that uses the electromagnetic wave spectrum by employing high frequencies or short waves [6].

The installation of HF radar on Karanga Beach, Labuan Bajo can be utilized in analyzing surface ocean current patterns in Labuan Bajo. HF radar is able to monitor the direction and speed of currents around Labuan Bajo waters [1] Current mapping using hf radar is done in real time with readings taken every hour, making it easy to analyze current conditions. Because it has a wide enough spatial resolution, HF radar can also help map dangerous areas in real time. Analysis of the results of the current processing can be used to improve shipping safety and reduce the number of ship accidents in Labuan Bajo waters. Furthermore, the HF radar can read waves and wind, allowing it to be used as an early warning system for tsunamis or other severe weather conditions [10].

The purpose of this study was to map surface current patterns spatially and temporally and identify the influence of wind and tides in generating surface currents in the waters around Labuan Bajo from October to December 2018 using data from HF radar.

2. Data and Methods

2.1. Description of the Study Site

The research location is in Labuan Bajo Waters, which is located at coordinates 119°32'35.37" E - 119°53'47.92" E and 8°4'15.24" S - 8°35'3" S. The study area can be seen in Figure 1. There are six review points with coordinates, which can be seen in Table 1. Points 1 and 2 are located in shallow waters at a depth of 50 meters; Points 3 and 4 are located offshore around the island of Komodo at a depth of 300 meters; and Points 5 and 6 are located offshore around Labuan Bajo at a depth of 750 meters. The random location of the points is caused by a lack of data at other locations, especially offshore.

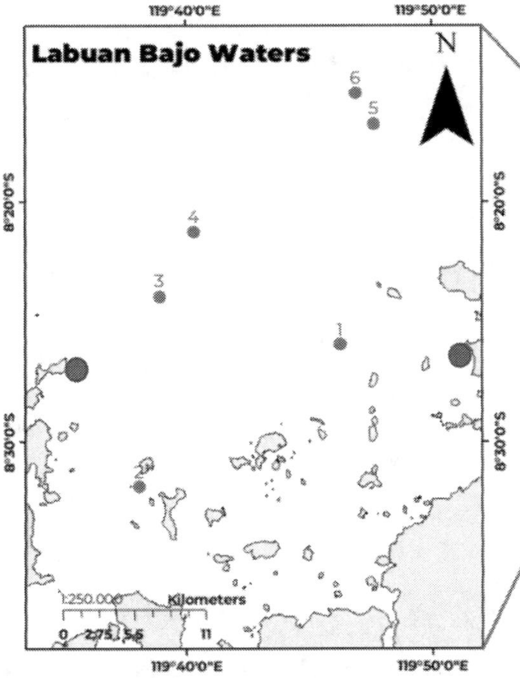

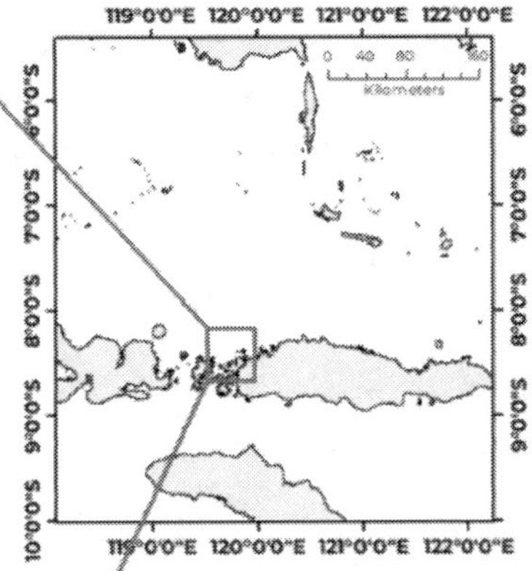

Location	Long (°)	Lat (°)
Point 1	119.76 E	8.45 S
Point 2	119.62 E	8.53 S
Point 3	119.64 E	8.41 S
Point 4	119.66 E	8.35 S
Point 5	119.80 E	8.27 S
Point 6	119.79 E	8.25 S

Figure 1. Research Area in Labuan Bajo, Indonesia and Coordinates point review. Blue circle is the HF-radar location

2.2. Data

The current data used is measurement data from HF radar obtained from stations on the Labuan Bajo coast starting from 3 October to 19 December 2018. The data obtained consists of vector data for hourly current velocity each day, which consists of zonal currents (u) and meridional currents (v) in centimeters per second. In addition, there is also data on the total current velocity (V) in centimeters per second and the current direction in degrees. Flow data has a resolution of 0.0091° x 0.0090°.

The wind data used is sourced from Copernicus Europe Eyes on Earth with an area coverage of 119°E - 120° E and 8° S - 9° S. The data consists of wind speed from October to December 2018 with a 6-hour data interval in units of m/s consisting of a zonal wind component (u) and a meridional wind component (v). Wind data uses a spatial resolution of 0.025° x 0.025°. The water level data used are 2-

minute intervals sourced from the Indonesia Geospatial Information Agency (BIG) from October to December 2018 at Sape Station with coordinates of 119°01'08.41" E and 08°34'04.82" S.

2.3. *Method Data Processing*

2.3.1 *Spectrum Analysis*

The characteristics and driving force of currents in the water, as well as the dominant period in a time series, are determined using spectrum analysis. The spectrum analysis method can be done by using filtering. Filtering is one of the methods used to remove unwanted frequencies from data and save the necessary data [5]. In processing sea surface currents, a high pass filter will be used to separate the tidal currents from the non-tidal currents to obtain the non-tidal currents. Tide is a superposition of infinite harmonic waves with a large period and small frequency, so it needs to be eliminated. This aim to eliminate the influence of the tides on the direction and speed of the currents, assist in determining the driving factors of surface currents, and see the seasonal variations of surface currents.

2.3.2 *Cross Correlation*

Cross correlation is a method used to find a relationship between two time series data x_t and y_t, with y_t data having a relationship with the previous x_t data. In this study, the x_t data is wind and tide data, and the y_t data is ocean current data from HF-radar. The Equation 1 used to calculate the cross correlation for the normalized version.

$$\boldsymbol{r_{xy}} = \frac{\sigma_{xy}}{\sigma_x \sigma_y} \qquad (1)$$

Where $\boldsymbol{r_{xy}}$ is cross correlation, σ_x is cross covariance data x, σ_y is cross covariance data y, and σ_{xy} is cross covariance between the two pieces of data.

According to Evans [2], the criteria for the value of the correlation coefficient are:

- < 0.20 : very weak
- 0.20 - 0.40 : weak
- 0.40 – 0.60 : intermediate
- 0.60 – 0.80 : strong
- 0.80 – 1.00 : very strong

2.3.3 *Empirical Orthogonal Functions*

The empirical orthogonal function (EOF) aims to determine the dominant pattern both spatially and temporally. The EOF can search for a few independent variables that provide maximum information without being excessive. Calculating the EOF value can be done with the following Equation 2:

$$\boldsymbol{X = U\ \sigma V^T(k)} \qquad (2)$$

Where V^T is the result of the transpose matrix,. $\sigma . V$ that shows the eigenvalues of EOF, Column U is the mode of the X matrix, k is component of matrix orthogonal, and column V is the PC (Principal Component) value of each mode [1].

2.3.4 *Fast Fourier Transform*

Fast Fourier Transform (FFT) is a mathematical method that decomposes a continuous function into simple waves [7]. The FFT divides a signal as a function of time into frequency functions that have different amplitudes and frequencies quickly and efficiently. The FFT can be defined by Equation 3.

$$\boldsymbol{S(f) = \int_{-\infty}^{\infty} s(t) e^{-j2\pi f t} dt} \qquad (3)$$

Where $S(f)$ is a signal in the frequency domain, $s(t)$ is a signal in the time domain. And the function $s(t)e^{-j2\pi ft}$ is a signal's constant value.

3. Result and Discussion

3.1 Tidal Data Processing Results

Tidal data processing is carried out to determine the occurrence of spring tides and neap tides in the study area and the types of tides. This processing is also used as supporting data in determining differences in speed and surface currents during spring tide and neap tide conditions. The tidal elevation in Labuan Bajo waters from October to December 2018 is shown in Figure 2, with an average elevation value for 3 months of 1.52 meters. Based on the results of tidal calculations using the least squares method, a Formzhal number value of 0.87 is obtained, meaning that the tides in the study area are generally semidiurnal mixed type tides.

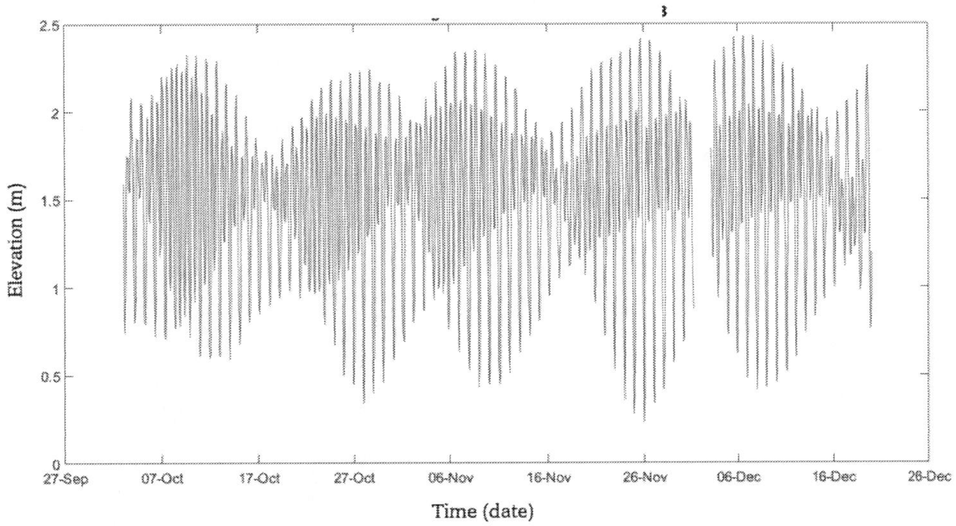

Figure 2. Tidal elevation from October to December 2018.

3.2 The results of wind data processing.

Wind speed and direction can affect the pattern of surface currents in the study area. Wind data is processed to generate time series graphs of wind speed and windrose. The windrose chart showing variations in wind direction and speed from October to December in the study area can be seen in Figure 3.

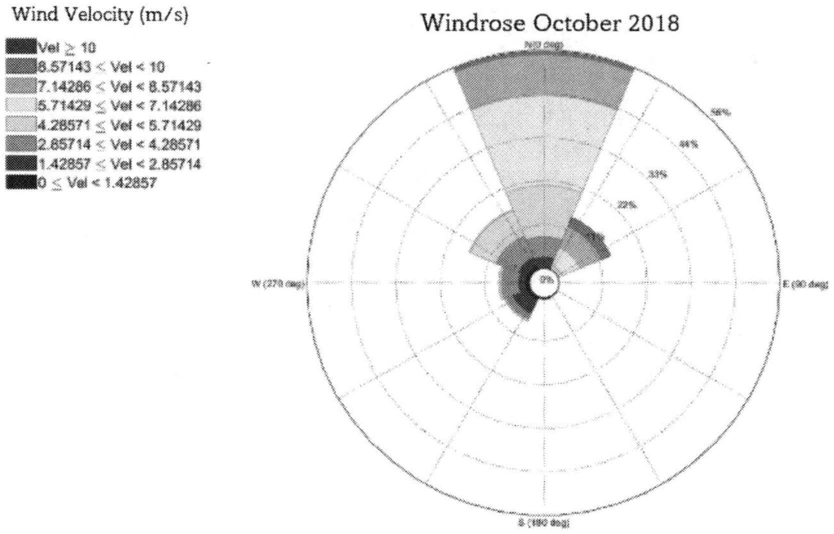
Wind Velocity (m/s)
Vel ≥ 10
8.57143 ≤ Vel < 10
7.14286 ≤ Vel < 8.57143
5.71429 ≤ Vel < 7.14286
4.28571 ≤ Vel < 5.71429
2.85714 ≤ Vel < 4.28571
1.42857 ≤ Vel < 2.85714
0 ≤ Vel < 1.42857
Windrose October 2018
N (0 deg)
E (90 deg)
S (180 deg)
W (270 deg)

A)

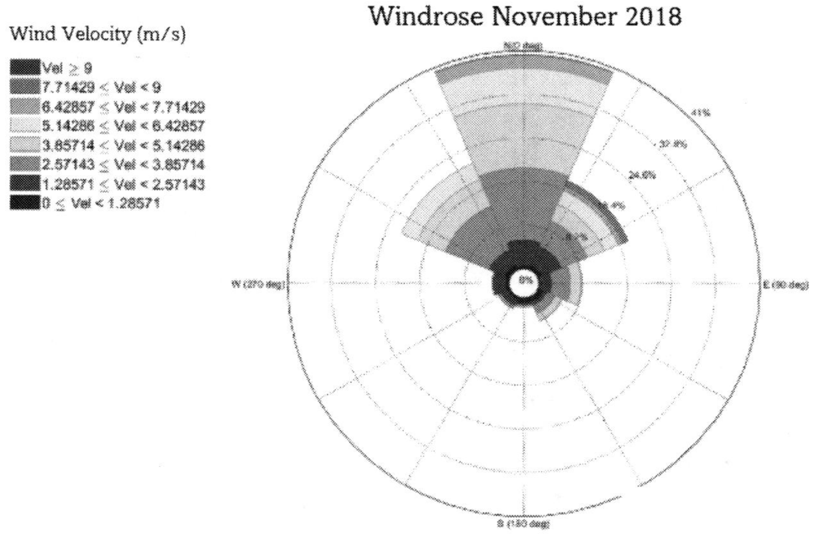
Wind Velocity (m/s)
Vel ≥ 9
7.71429 ≤ Vel < 9
6.42857 ≤ Vel < 7.71429
5.14286 ≤ Vel < 6.42857
3.85714 ≤ Vel < 5.14286
2.57143 ≤ Vel < 3.85714
1.28571 ≤ Vel < 2.57143
0 ≤ Vel < 1.28571
Windrose November 2018
N (0 deg)
41%
24.6%
E (90 deg)
S (180 deg)
W (270 deg)

B)

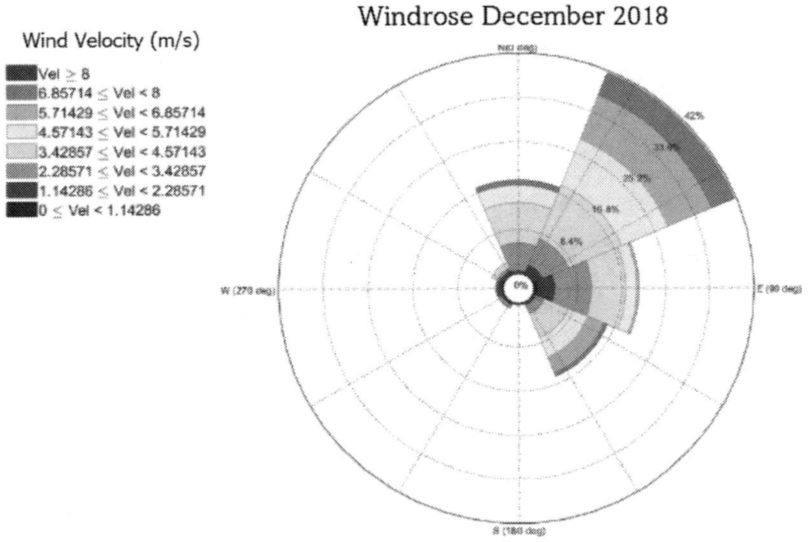

C)

Figure 3. Windrose for A) October, B) November, and C) December 2018.

The transition from the east to west monsoon influences wind direction in October and November [12]. The wind movement in October is predominantly from the north with a percentage of 55%, and the most dominant wind speed is at intervals of 5.7-7.1 m/s. Windrose shows wind movement dominated by moving north with a percentage of 40.13% and dominant wind speeds occurring at intervals of 2.5-3.8 m/s in November. In October and November, the wind pattern is dominated by the west monsoon, which moves from the Asian continent to the Pacific Ocean.

The wind direction in December is influenced by the west monsoon [12]. In December, windrose shows that the wind is predominantly moving from the northeast with a percentage of 41.52%, and the dominant wind speed is at intervals of 3.4-4.5 m/s. The maximum wind speed this month has the largest percentage compared to other months, caused by the influence of the west monsoon, which has been very strong and stable, and the influence of the east monsoon, has weakened.

3.3 Spectral Analysis

A spectral analysis was carried out using the high-pass filter method on current velocity and direction data in the study area for three months at six review points. Figure 4 shows high pass filter processing results from October to December 2018 at six points. The results show three time series graphs consisting of total current (top), tidal current (middle),and residual current (bottom).

IOP Conf. Series: Earth and Environmental Science 1245 (2023) 012040 doi:10.1088/1755-1315/1245/1/012040

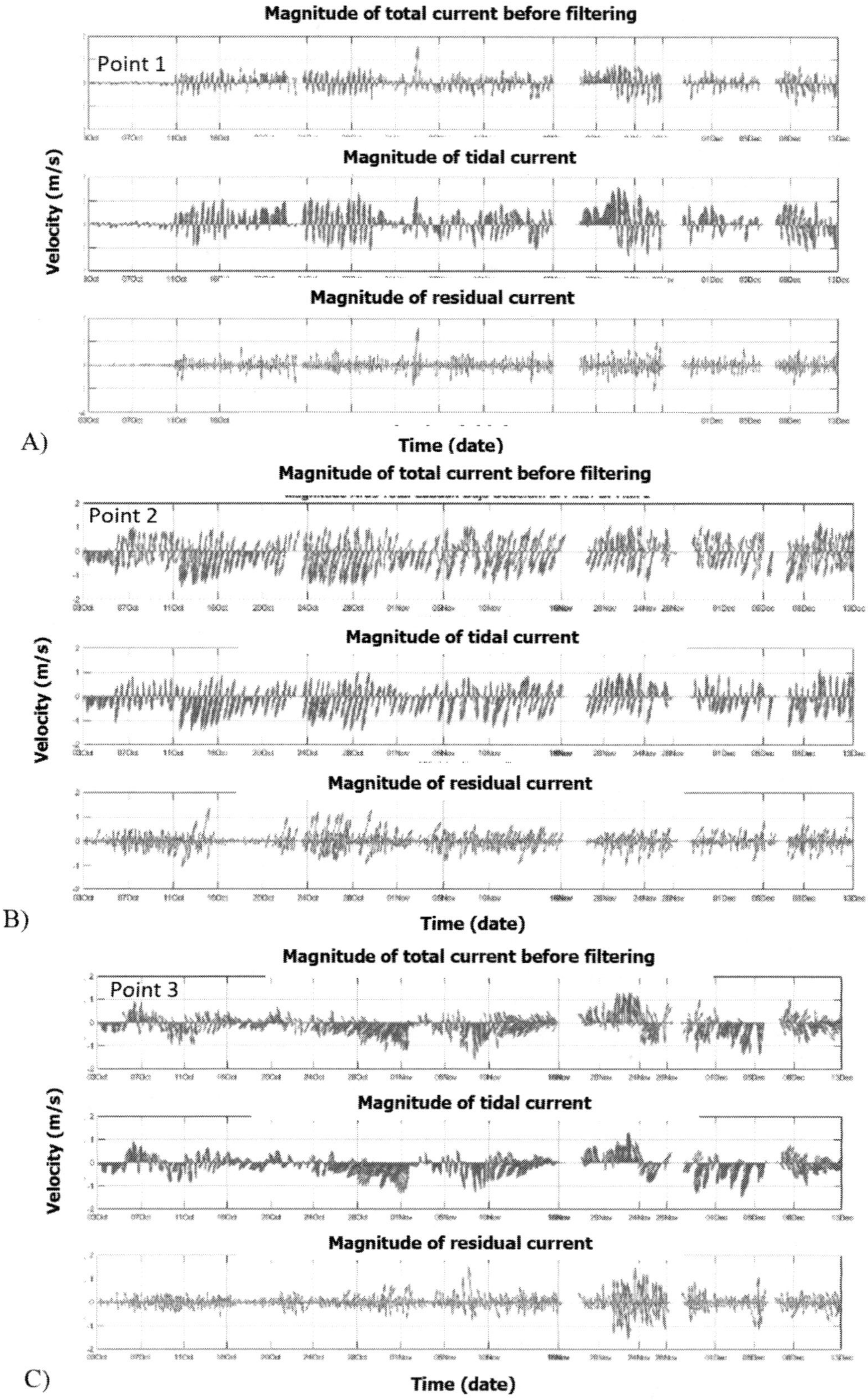

IOP Conf. Series: Earth and Environmental Science 1245 (2023) 012040 doi:10.1088/1755-1315/1245/1/012040

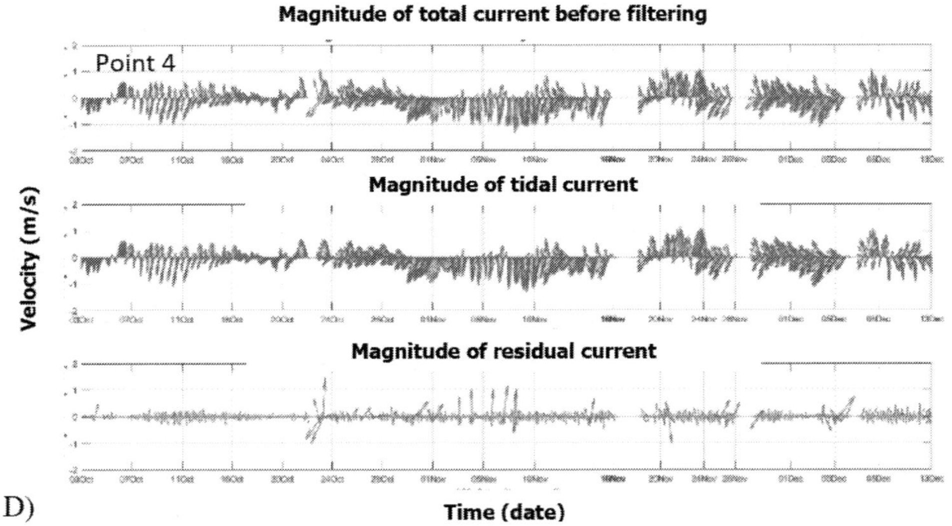

D)

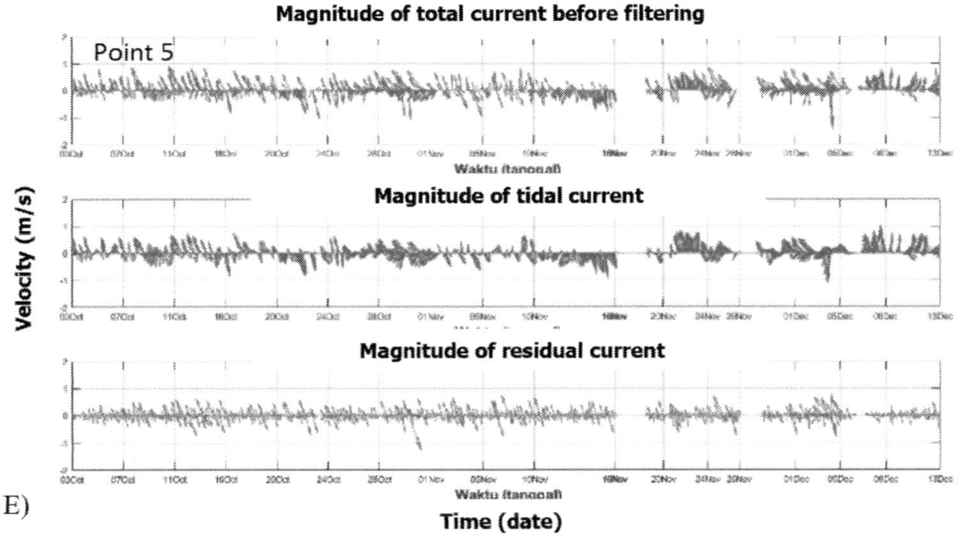

E)

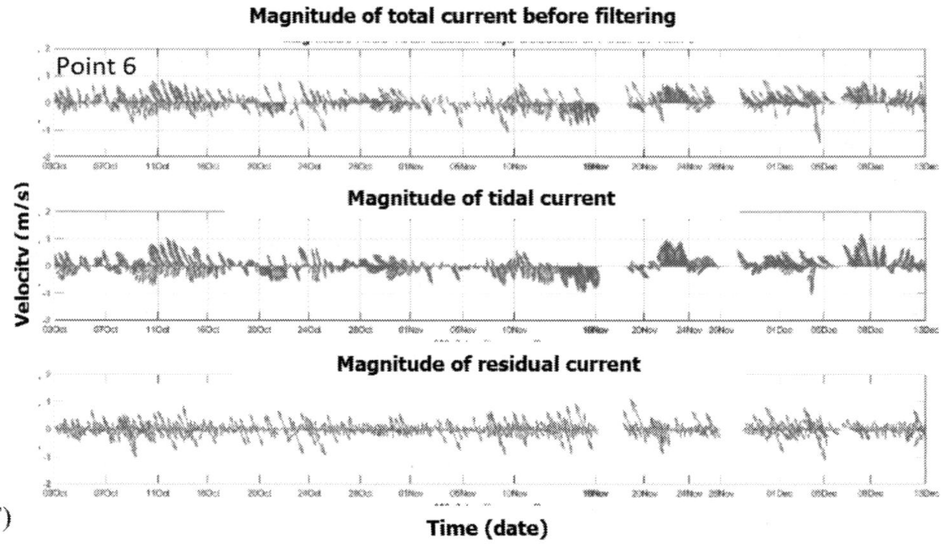

F)

Figure 4. High pass filter processing results for October-December 2018 at A) Point 1, B) Point 2, C) Point 3, D) Point 4, E) Point 5, and F) Point 6.

Each review point has a tidal current that is greater than the residual current that is affected by the wind. However, the residual current at Points 5 and 6 has almost the same speed as the tidal current. This shows that offshore, tidal currents have a decreasing effect with, the influence of residual currents starting to increase. So the effect of residual currents is greater in offshore waters than in shallow waters.

3.4 Cross correlation

To determine the magnitude of the relationship between tides, wind, surface currents, and the lag phase, the cross-correlation method is used. Cross correlation is a technique that compares two time series data sets and objectively determines their compatibility with each other. Lag determines the variable that occurs first and affects other variables. A negative lag value indicates that x occurs before y begins, and a positive lag value indicates that x occurs after y occurs. Residual current data is used, which is correlated with wind speed, to show the correlation between wind and residual current. The blue color shows the relationship between winds and surface currents, while the red color shows the relationship between tides and surface currents. Figure 5 shows correlation of wind and tides to surface currents in 2018 at six points.

IOP Conf. Series: Earth and Environmental Science 1245 (2023) 012040 doi:10.1088/1755-1315/1245/1/012040

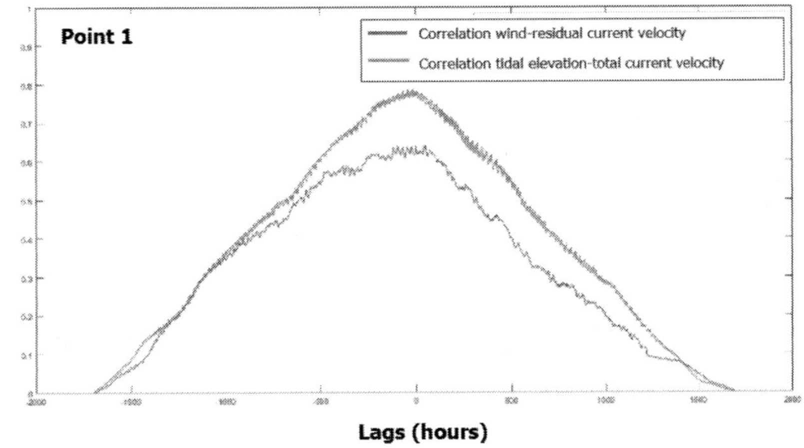

A)

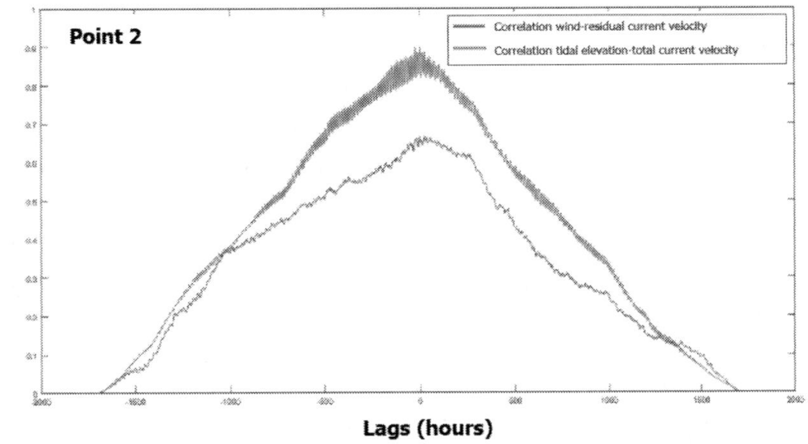

B)

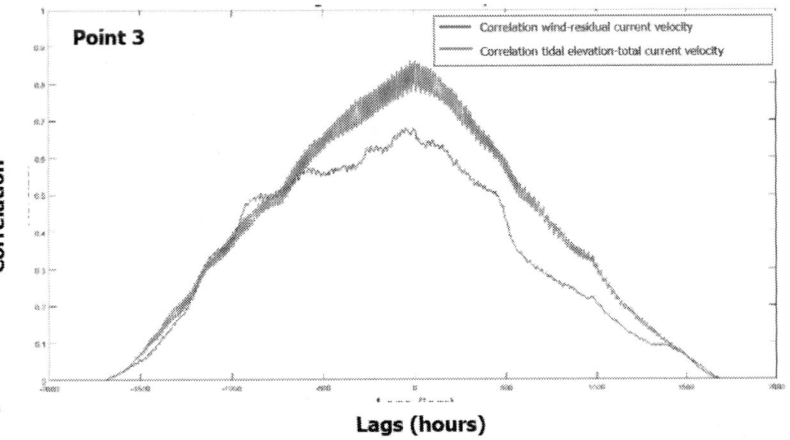

C)

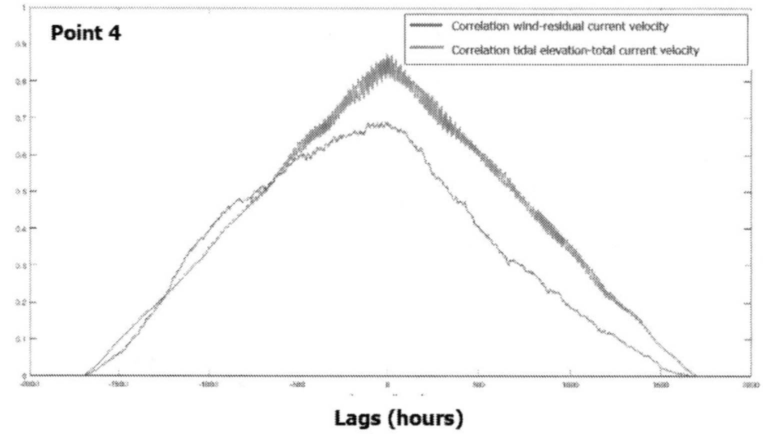

D)

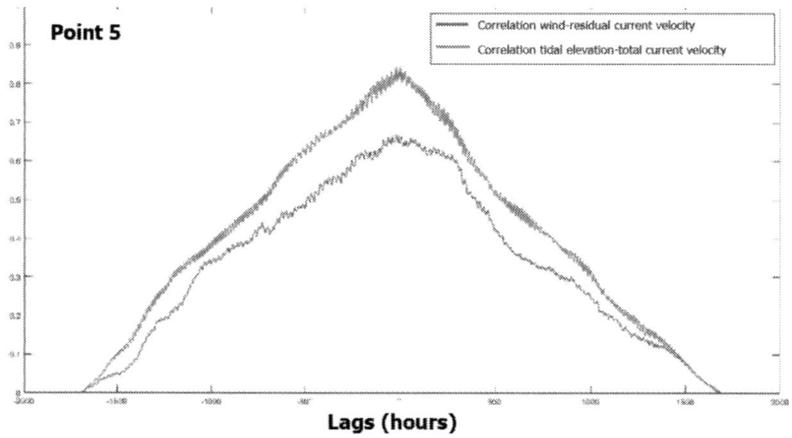

E)

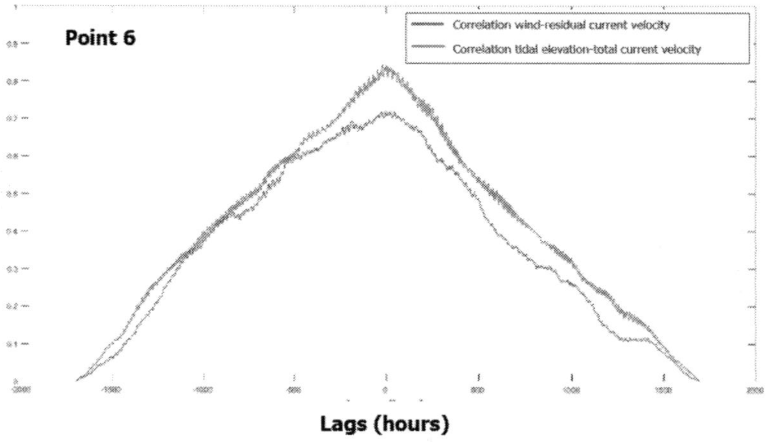

F)

Figure 5. Correlation of wind and tides to surface currents in 2018 at A) Point 1, B) Point 2, C) Point 3, D) Point 4, E) Point 5, and F) Point 6.

The greatest correlation between wind and surface currents is at Point 6 with a value of 0.72, and the largest correlation between tides and surface currents is at Point 2 with a value of 0.90. At each review point (shallow waters and offshore), it shows that tides have a greater correlation value than wind. However, the correlation of wind to surface currents has a greater value at points located offshore than at points located in shallow waters. This shows that the effect of wind is greater offshore than in shallow waters. The correlation of tides and winds to surface currents has a positive value and a large value, indicating that it moves in the direction of the surface currents. Wind always has a faster effect than tides on the formation of surface currents because wind affects surface currents in a short period of time and all at once, as opposed to tides, which have a period of 12 hours [3].

3.5 Current Velocity Time Series Plot

The temporal pattern shows the current velocity in each review to compare current velocity in shallow and open water areas. The data used is the total flow data that has not been filtered. The direction of surface currents in spring tide and neap tide conditions helps to show the effect of wind and tides on the current direction in shallow waters and open waters. Current speed and direction from October to December 2018 during spring tide and neap tide conditions at six point shown in Figure 6.

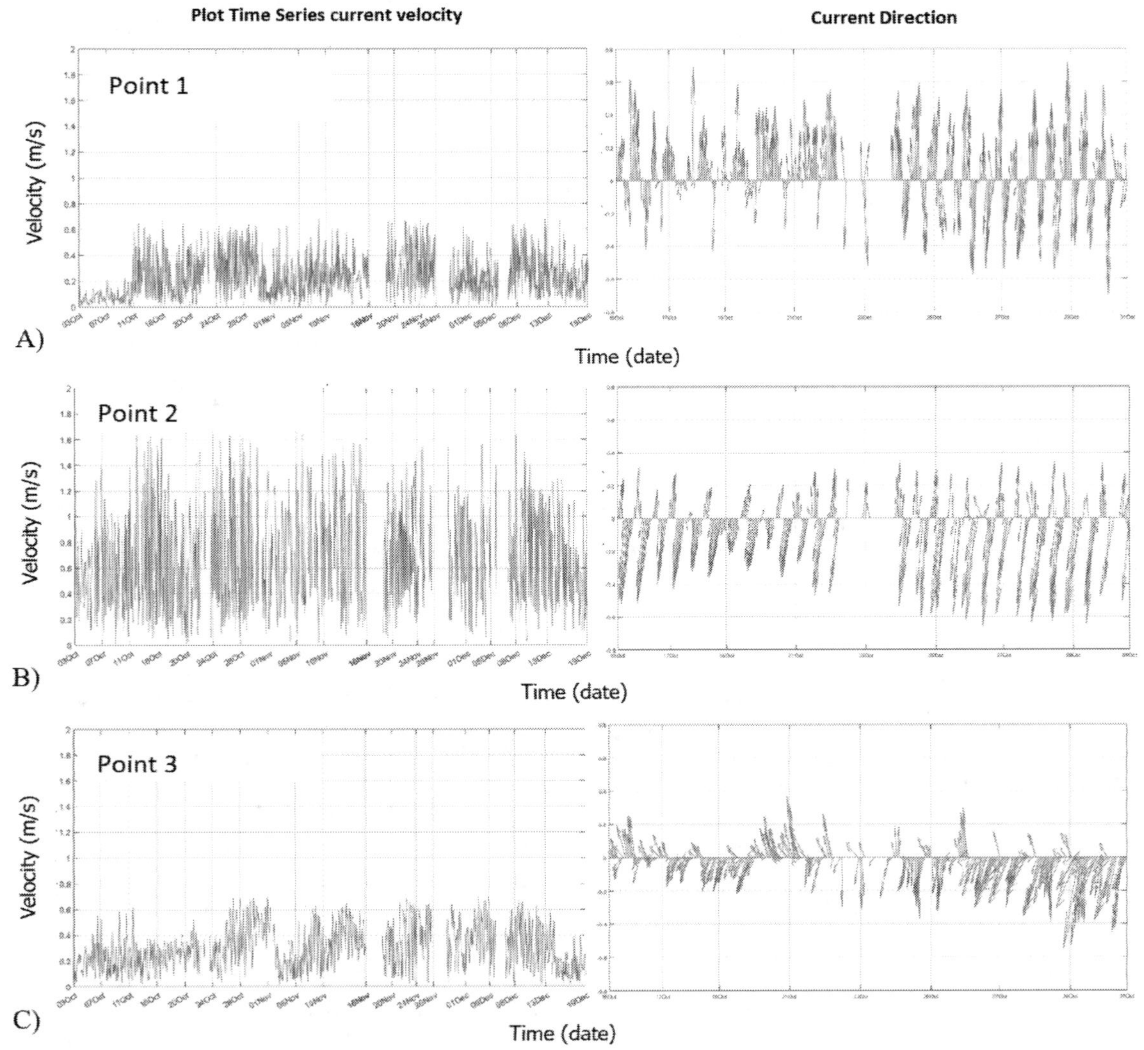

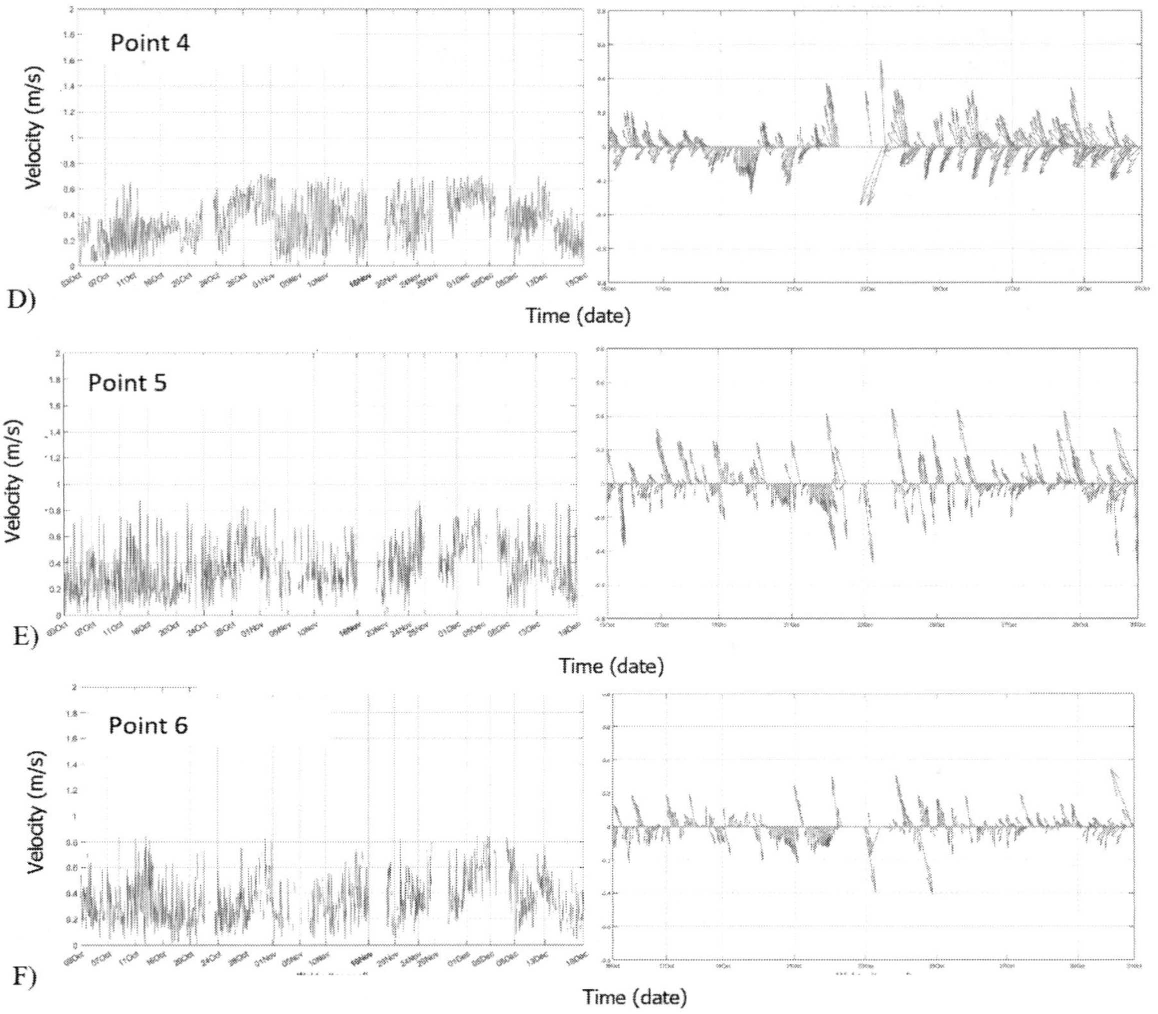

Figure 6. Current speed and direction in October-December 2018 during spring tide and neap tide conditions at A) Point 1, B) Point 2, C) Point 3, D) Point 4, E) Point 5, and F) Point 6

The highest average current velocity value for three months is at Point 2, and the lowest average current velocity value for three months is at Point 1. This can be caused by the disturbance of the dominant current pattern in shallow waters caused by the presence of small islands. Points 3, 4, 5, and 6, which are located offshore, have a small average current velocity difference, indicating that the average current velocity offshore is almost uniform due to deep bathymetry. It also shows that the speed and direction at a point near Komodo Island and a point near Labuan Bajo have the same value. It can be seen that the closer the observation point is to Komodo Island, the higher the current speed every month, and the farther the observation point is to Komodo Island, the smaller the current speed. This is because the effect of tides tends to be greater around the coast or in shallow waters than on the high seas, which causes the current velocity around Komodo Island to have a greater speed than other review points [3]. So that it can be concluded, the influence of the tides on the speed of the current at the review point is greater than that of the wind.

The current direction at points in shallow water (Points 1 and 2) follows the tidal pattern, namely moving north during high tide and south during low tide. Whereas at points located offshore (Points 3,

4, 5, and 6), follow the direction of the wind, namely the current direction, which is predominantly moving to the west and southwest.

3.6 Spatial Pattern of Current Movement

Hourly unfiltered total current velocity data for three months was averaged to produce a spatial pattern of surface current velocity and direction in the study area shown in Figure 7. The highest current speed in October is in the east of Komodo Island, at 0.27 m/s, and the lowest current speed in October is in the west of Labuan Bajo, close to 0 m/s. The average current speed in October is 0.13 m/s. In November, the highest average speed of the current in one month reached 0.30 m/s, and the lowest average speed was 0.02 m/s. The average current speed in November is 0.12 m/s.

In shallow waters, especially between Komodo Island and the islands of East Nusa Tenggara, the currents move irregularly and are faster due to the influence of the topography, which is surrounded by small islands. These small islands act as a barrier that disrupts the dominant current pattern, especially in the eastern part of Komodo Island [11]. Offshore, deep bathymetry causes currents to move more smoothly than in shallow waters [10]. The difference in movement between the current directions in offshore and shallow waters is caused by the influence of dominance on the causal factors [10]. On offshore, current movement is in accordance with the direction of wind movement, and there is no significant difference during high tide, low tide, either towards high tide or towards low tide. However, in shallow waters, the current moves with the tides, that is, away from Komodo Island during high tide and approaching Komodo Island during low tide.

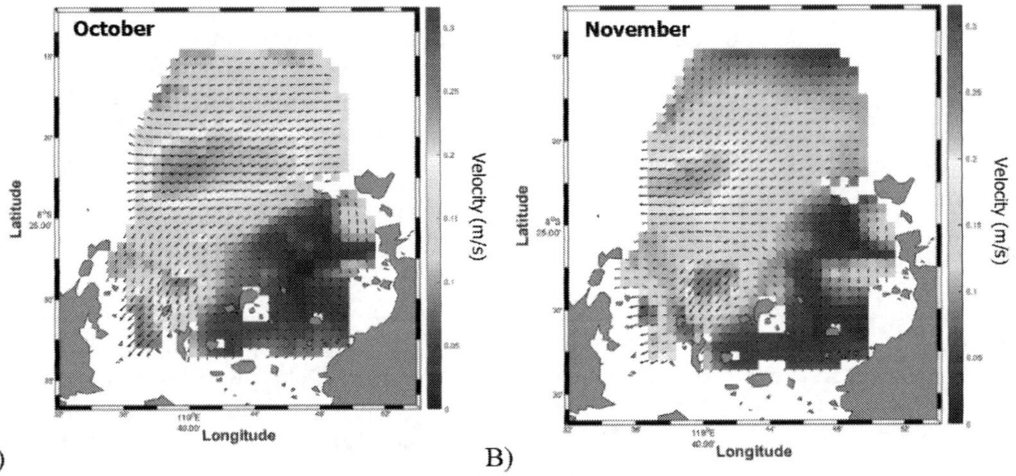

A) B)

IOP Conf. Series: Earth and Environmental Science 1245 (2023) 012040 doi:10.1088/1755-1315/1245/1/012040

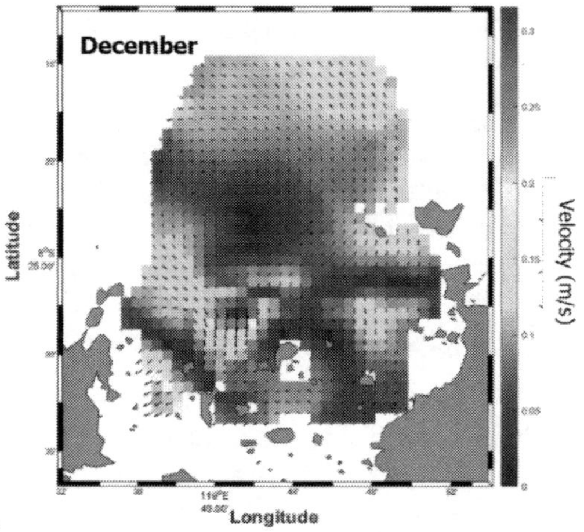

C)

Figure 7 Current direction and surface current velocity (m/s) are shown by spatial color shading in A) October, B) November, and C) December

In December, the highest average current speed reached 0.31 m/s and the lowest average current speed reached 0.01 m/s. The average current speed in December is 0.09 m/s. The current movement pattern in December shows a pattern that is quite different from the current movement in October and November. Offshore, the current rotates counterclockwise with a smaller speed toward the center of rotation and a speed range of 0-0.15 m/s. This rotational motion is caused by the west monsoon winds meeting the southeast trade winds, which move in opposite directions at around 10°S, so that the west monsoon winds are deflected to the north. As a result, surface wind turbulence is formed in the waters around Labuan Bajo, which pushes surface water masses to follow a rotational pattern [4]. Whereas in shallow waters, surface currents show a southward movement from Komodo Island with a speed range of 0-0.31 m/s.

The pattern of surface currents during high and low tide conditions during the spring tide and neap tide is shown in Figure 8. In shallow waters, the current velocity is smaller during high and low tides compared to high tide and low tide, which indicates that the highest sea level elevation will have a current velocity of zero or a small current velocity. The difference in speed values is clearly visible, especially in the waters around Komodo Island and Sebayur Island. In shallow waters, surface currents move away from Komodo Island during high tide and move closer to Komodo Island during low tide, which indicates that current movement is dominated by tides. However, on the offshore, the difference in the speed of the tides and ebb currents is quite small compared to those heading up and down with the same direction of movement, namely towards the west of the Flores Sea. The direction of the current, which always moves to the northwest of Flores both at high and low tide, indicates that surface currents offshore follow the direction of wind movement, but the current speed is still influenced by the tides.

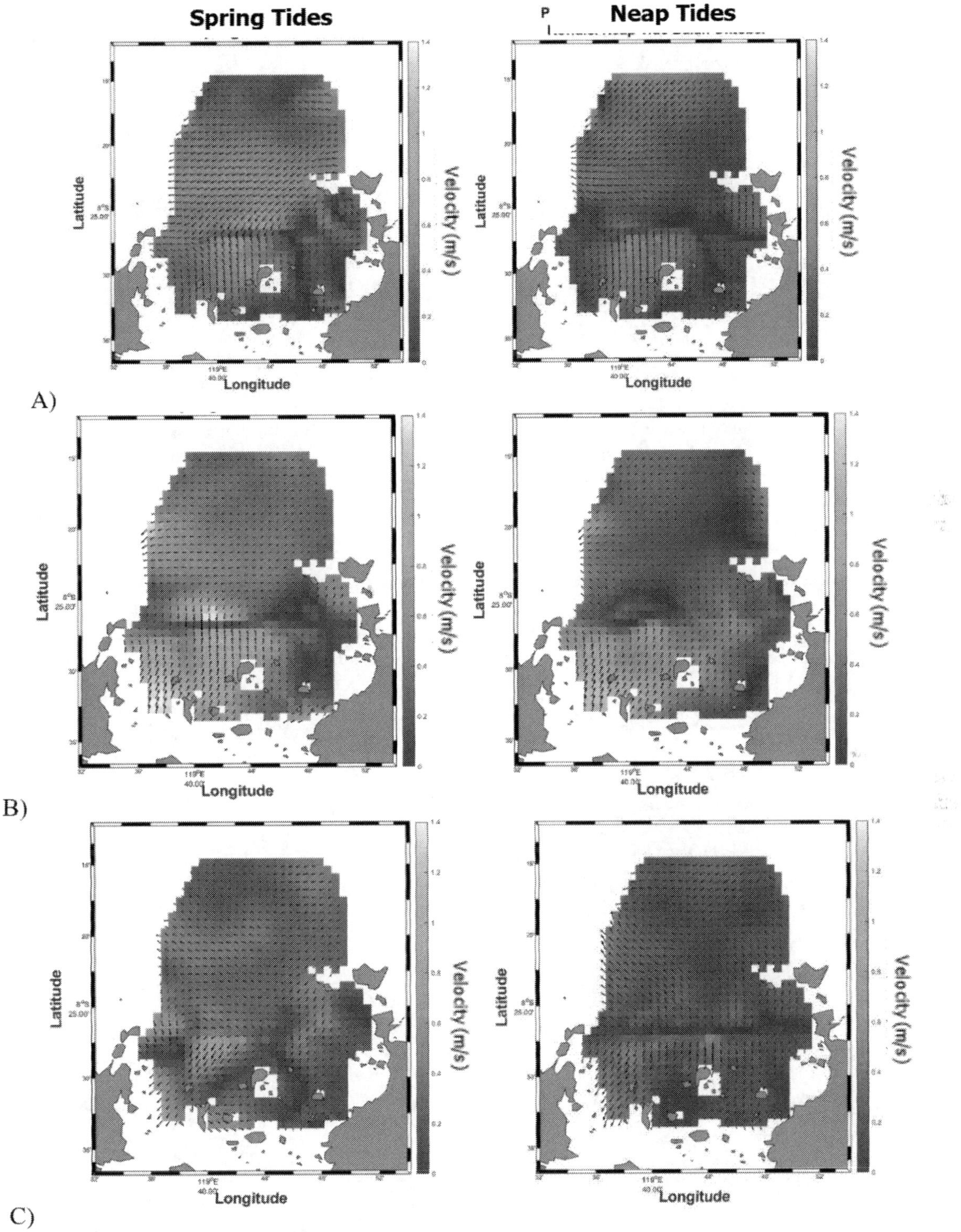
Spring Tides
Neap Tides
Latitude
Longitude
Velocity (m/s)
A)
B)
C)

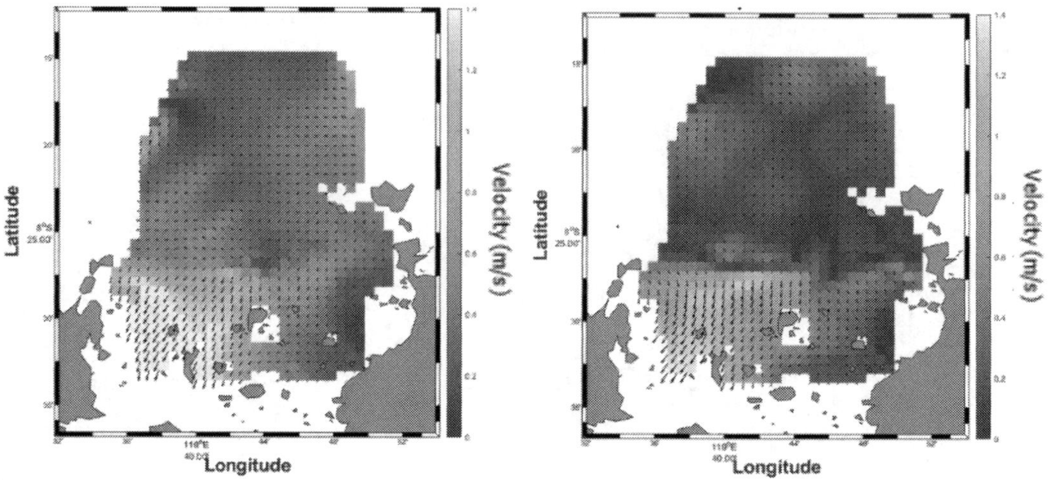

D)

Figure 8 Surface current pattern for spring tide and neap tide conditions in October. A) at high tide, B) towards the high tide, C) at low tide, and D) towards the low tide.

3.7 *Empirical Orthogonal Function (EOF)*

To determine the driving force of surface currents both in shallow waters and offshore, the EOF method is used. The EOF method displays spatially and temporally the dominant variables of the total unfiltered current in three main modes. The EOF spatial structure of surface currents with Mode 1 for three months is shown in EOF Mode 1 with a variance of 56.2%, with the dominant period of Mode 1 located at 12 hours and 24 hours, which indicate periods of highs and lows. The EOF spatial structure of surface currents with Mode 1 for three months is shown in Figure 9.

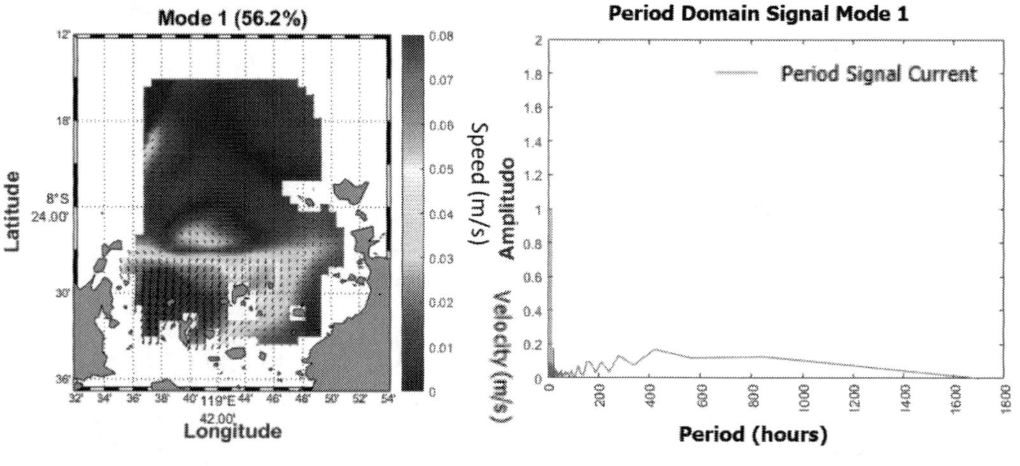

Figure 9. Spatial structure of EOF Mode 1 of meridional currents and zonal currents October-December 2018.

The EOF spatial structure of surface currents with Mode 2 for three months is shown in Figure 10 EOF Mode 2, with a variance of 26.6% has the largest amplitude of 0.06 and is located in the high seas. The dominant period in Mode 2 is 238 hours, or about 10 days, which is classified as a biweekly period. The Mf (lunar fortnightly) tidal component is included in a 2-week period. Based on Stewart [9], the period of Mf is 327.850 hours, or about 13 days. So it can be concluded that the dominant influence on Mode 2 is the component of the fortnightly tide. The EOF spatial structure of surface currents with Mode 3 for three months is shown in Figure 11. EOF Mode 3 with a variance of 17.2% has a dominant period of 566 hours, or about 24 days, which is a monthly period. However, Mode three cannot be further identified due to the short data range.

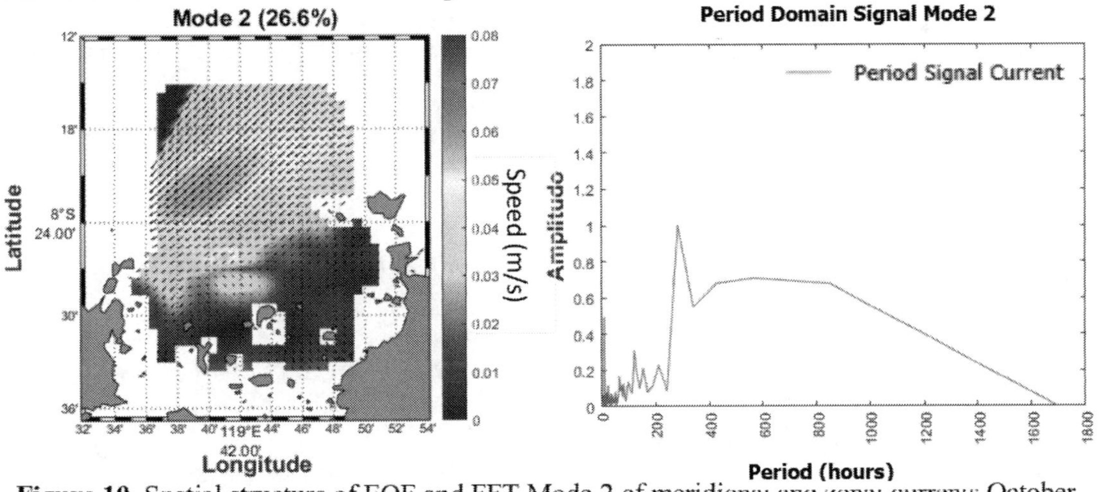

Figure 10. Spatial structure of EOF and FFT Mode 2 of meridional and zonal currents October-December 2018.

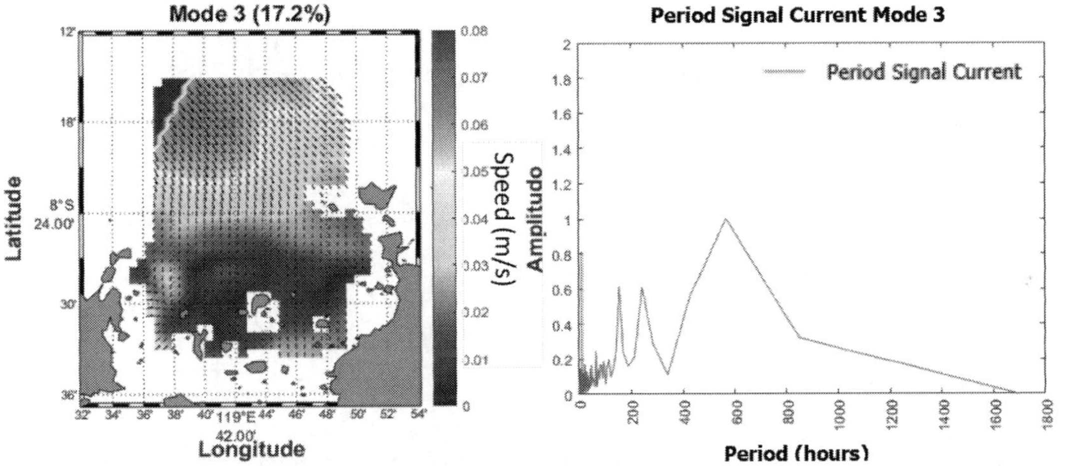

Figure 11. Spatial structure of EOF Mode 3 of meridional currents and zonal currents October-December 2018.

4. Conclusion

The average current speed in October was 0.13 m/s, the average current speed in November was 0.12 m/s, and the average current speed in December was 0.09 m/s. The highest average speed for three months is at Point 2 of 0.70 m/s, and the lowest average speed for three months is at Point 1 of 0.23 m/s. The direction of surface currents in shallow waters indicates the current moves northward during high tide and moves southward during low tide, while the direction of surface currents in offshore waters moves westward and southwest, which is influenced by the wind. This is consistent with Rizki's research [8], which found that the ocean current in Labuan Bajo flow into north at high tide and south at low tide.

The largest correlation between wind and surface currents is offshore (Point 4), at 0.83, and the largest correlation between tides and surface currents is in shallow waters (Point 2), at 0.89. Each review point has a tidal current velocity that is greater than the residual current. Points located offshore (Points 5 and 6) show a greater residual current than other points. EOF Mode 1 has a variance of 56.2%, with a dominant period of 12 hours, which is a semidiurnal tide period, Mode 2 has a variance of 26.6% with a dominant period of 10 days, which is included in the biweekly period, and Mode 3 has a variance of 17.2% with a predominant period of 24 days, which is an intra-seasonal period.

In this study, a three-month time was utilized due to a lack of surface current data from the HF radar, which was recently installed in 2018 at the same time as Covid-19, so the researchers could not come to review directly. It is intended that in the future, this research will be extended over a longer time to examine seasonal factors as well as the effects of El Nino and La Nina. So that the research findings might be expanded, particularly in terms of assisting coastal management in Labuan Bajo and preventing ship mishaps caused by ocean currents.

References

[1] Bjornsson H and Vanegas S A 1997 *A Manual for EOF and SVD Analyses of Climatic Data* **97**-1
[2] Evans J D 1996 *Straigforward Statistics For The Behavioral Science* Pacific
[3] Kusmanto E, Hasanudin M, Setyawan W B 2016 *Oseanologi dan Limnologi di Indonesia* **1(3)** 69-80
[4] Martono. 2012 *Buku Ilmiah PSTA* **2** 36-46.
[5] Mesquita and Michel d S 2005. Data Analysis: Filtering, Croscorrelation, Coherence and Application to Geophysical Data using Matlab.
[6] Paduan J D and Graber H C 1997. Introduction to High-Frequency Radar: Reality and Myth *Oceanography*, **10(2)** 36-39.
[7] Press W H, Vetterling W T, Teukolsky S A, and Flannery B P 1989 Numerical Recipes in Fortran 77 Second Edition *The Art of Scientific Computing*. 490-494.
[8] Rizki T Y 2018 *Applied Technology ad Computing Science Journal* **1(2)** 114-123
[9] Stewart R.H 2003. Introduction to Physical Oceanography. Department of Oceanography, Texas A & M University. 344.
[10] Supriyadi E, Hidayat R, Santikayasa I P, Ramdhani A 2021 *International Journal on Advanced Science Engineering Information Technology* **11** 1521-1534
[11] Webb D J 1993 A simple model of the effect of the Kerguelen Plateau on the strength of the Antarctic Circumpolar Current, Gophys, Astrophys, Fluid Dyn, **70(1-4)** 57-84.
[12] Wyrtki K 1961 *Naga Report 2* 1959-1961.

Acknowledgments

We want to express our gratitude to Meteorology, Climatology, and Geophysical Agency (BMKG) for HF radar data availability in this research.

Effectiveness of vertical error budget model for portable multi-beam echo-sounder in shallow water bathymetric survey

AA Abubakar[1], **and Poerbandono**[2]

[1]Geodesy and Geomatics Engineering, Faculty of Earth Sciences and Technology, Institut Teknologi Bandung, Indonesia
[2]Hydrography Research Group, Faculty of Earth Sciences and Technology, Institut Teknologi Bandung, Indonesia

E-mail: nuraugie@gmail.com

Abstract. With the increasing availability of portable survey equipment and platforms, hydrographic data acquisition now offers a diverse range of device options and installation methods. Tailoring the mounting and installation of hydrographic survey devices to fit the boat or platform dimensions has become common practice. The use of portable equipment is essential, given the wide selection of combined options available in the market. Consequently, sensor offsets and device performance differ for each survey conducted, necessitating the constant verification of accuracy performance.
This paper examines the capabilities of portable multibeam echosounder (MBES) deployments on various types of survey boats. The investigation focuses on error budget modeling and field tests. A model is developed to estimate the propagated uncertainty resulting from different sources of measurement uncertainties. The model primarily considers vertical sources of error in MBES surveys, as they have the most significant impact on hydrographic data quality. To validate the model, field tests are conducted at two different locations using different boat and device configurations. The results demonstrate that the modeled uncertainties align with the measured standard deviations, particularly in the Pramuka area, where the standard deviation of data acquired at the widest beam angle of ±60° is 0.25m, close to the predicted value of 0.28m. However, in the Patimban area, the model overestimates the uncertainty, predicting 0.27m compared to the measured standard deviation of 0.17m.

1. Introduction

Accurate and reliable bathymetric data plays a crucial role in various applications, such as offshore construction, dredging operations, navigation safety, and the study of marine geological and biological systems [1-3]. The acquisition of this data relies on in-situ surveys using Multi-Beam Echo-Sounder (MBES) technology, which is widely employed for nautical chart production, offshore engineering, and mineral resources exploration [4]. MBES systems collect depth data by emitting acoustic pulses and measuring the two-way travel time for a set of predefined beam angles [5]. To ensure high-quality depth data, additional sensor data is incorporated to compensate for errors arising from factors such as sound speed, boat attitude, offsets, and latency [6].

While fixed survey systems permanently installed on vessels offer persistent performance, their use is limited in regions with shallow water depths and difficult access to inland seas [7-8]. Portable systems, mounted on small craft such as fishing boats, have emerged as an alternative for shallow water surveys due to their adaptability and availability [9]. However, portable systems are more susceptible to errors and uncertainties compared to fixed systems, including those associated with sound speed estimation, platform motion, and installation offsets [10]. Thus, it is essential for surveyors to realistically estimate the uncertainty, or error budget, of portable MBES systems.

The widely used approach for predicting vertical uncertainty in bathymetry surveys, developed by Hare [11-12] has been modified to account for additional error sources introduced by Doppler frequency shift and baseline decorrelation when using frequency modulated (FM) pulse [13-14]. Several studies have examined the performance of MBES bathymetric uncertainty prediction models, focusing on depth measurement uncertainty [13][15]. However, only a few studies have considered the total vertical uncertainty (TVU) for reduced depth measurements acquired by autonomous underwater vehicles [16].

To address the need for realistic estimates of reduced depth uncertainty in bathymetry acquired using portable MBES systems, this study remodels the existing prediction model by incorporating a Jacobian matrix. The Jacobian matrix enables the propagation of depth measurement errors, considering the latest insights into error contributors and facilitating the handling of complex equations and correlations [17]. The objective of this study is to evaluate the performance of the model in estimating the TVU of a portable MBES system by comparing modeled reduced depth error budgets with standard deviations from measured datasets obtained in two shallow water areas.

The estimated values obtained from this study can serve as a reference for portable MBES surveys, but they may vary depending on the sensors used and the operating depth. The remainder of this paper is organized as follows: Section 2 provides a description of the model, the approach used for remodeling, and details about the datasets employed. Section 3 presents the results and discusses relevant issues. Finally, the concluding remarks are provided in Section 4.

2. Methodology

2.1. *Depth error budget model description*

Propagation of uncertainty deals with the quantification of the effect of the input variables randomness on the uncertainty of function based on them. Most of the equations in this paper are taken from [12] and [18]. Assumes that a measurement system is modeled as a function of multiple real input quantities and a single real output quantity. This is represented as:

$$y = f\ (x_{i}, x_{i+1\ldots}, x_{n}) \tag{1}$$

In the equation, y denotes the output quantity (measured depth), y represents the real input quantities (individual components in the depth derivation), i is an index for a specific component, and n represents the total count of input quantities. Random errors are treated as independent, with the total determined by squaring the individual errors and computing the square root of their sum:

$$\sigma_{total}^{2} = (\sigma_{i})^{2} + (\sigma_{i+1})^{2} + \ \ldots\ + (\sigma_{n})^{2} \tag{2}$$

MBES systems determined the surface depth by measuring the travel time of a transmitted acoustic signal. The water depth, d, measured from below the transducer to the seafloor can be determined as:

$$d = r \cos P \cos(\theta_s + R + \theta_{mount}) = r \cos P \cos \theta \quad (3)$$

With θ_{mount}, R and P being the across track angle under which the MBES is mounted on the vessel, roll and pitch angle, respectively. $\theta = \theta s + R + \theta mount$ represent rotations around the depth-axis. The factors contributing to the error in the measured depth, as represented by equation (3), include:

- Range error for sounder σ_{d1}
- Beam angle and roll error σ_{d2}
- Pitch error σ_{d3}
- Depth measurement limitation error σ_{d4}
- Steering angle shift due to Doppler Effect error $\sigma_{d5}$and
- Baseline Decorrelation error σ_{d6}.

In the existing model, all the error contributors identified above are mapped into an error in the measured depth by applying propagation of errors to equation (3) then using the root-sum square (RSS) to get the total measured depth error σ_d due to the sounder system given by:

$$\sigma_d = \sqrt{\sigma^2{}_{d_1} + \sigma^2{}_{d_2} + \sigma^2{}_{d_3} + \sigma^2{}_{d_4} + \sigma^2{}_{d_5} + \sigma^2{}_{d_6}} \quad (4)$$

To get the Reduced depth to reference datum uncertainty, the following reduction errors must be added to the measured depth uncertainty quadratically.

- Heave error σ_H
- Dynamic Draught error σ_T and
- Water level error σ_{WL}.

Reduced depth is given by the following equation:

$$D = d - H + T - WL \quad (5)$$

Where D is the reduced depth to reference datum, d is the measured bathymetry beneath the transducer, H is the heave, T is the water level distance from the transducer, WL is the measured water level above the reference datum, which is affected by tide and wave. These components are all independent of each other. The resulting reduced depth uncertainty is obtained as:

$$\sigma_D = \sqrt{\sigma^2{}_d + \sigma^2{}_H + \sigma^2{}_T + \sigma^2{}_{WL}} \quad (6)$$

The general flowchart of this paper is illustrated in figure 1. followed by a detailed description of the errors contributing to reduced depth error will be given in the next section.

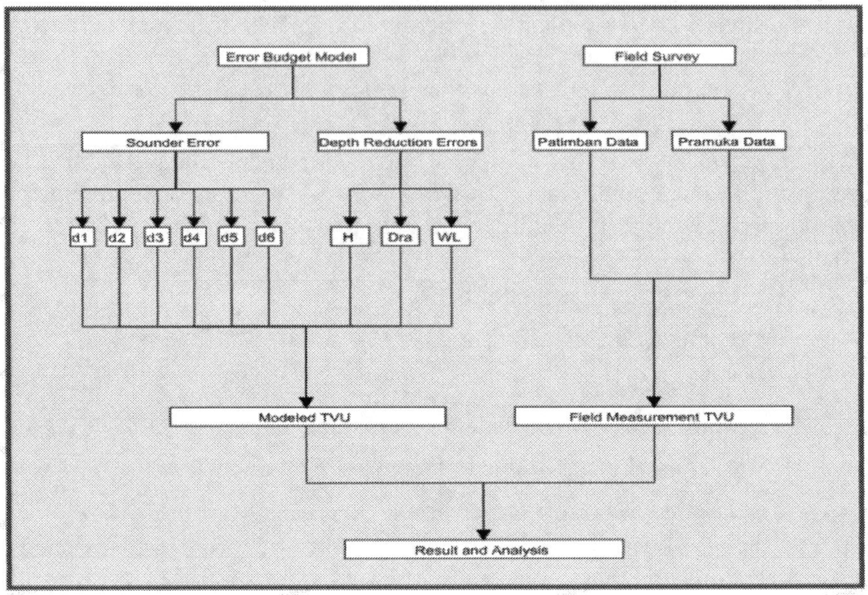

Figure 1. Flow diagram illustrating the comprehensive methodology design for the analysis and evaluation of the system performance.

2.2. *Error budget model of the reduced depth*

The error budget modeling is based on the variances from the measured depth, heave, dynamic draught, and water level measurement. Equations describing the individual uncertainties are provided herein.

2.2.1. *The Measured Depth Uncertainty.* Herein, we apply the Jacobian matrix to map all error contributors in equation (3) to measured depth errors and calculate the total measured depth uncertainty σ_d. The general form of the matrix is given by:

$$\sigma_y = \begin{bmatrix} \frac{\partial f}{\partial x_1} & \frac{\partial f}{\partial x_2} & \cdots & \frac{\partial f}{\partial x_N} \end{bmatrix} \underbrace{\begin{bmatrix} \sigma_{x1} & \sigma_{x1x2} & \dots & \sigma_{x1xN} \\ \sigma_{x2x1} & \sigma_{x_1} & \dots & \sigma_{x2xN} \\ \vdots & \vdots & \ddots & \vdots \\ \sigma_{xNx1} & \sigma_{xNx2} & \cdots & \sigma_{xN} \end{bmatrix}}_{\text{variance-covariance}} \begin{bmatrix} \frac{\partial f}{\partial x_1} \\ \frac{\partial f}{\partial x_2} \\ \vdots \\ \frac{\partial f}{\partial x_N} \end{bmatrix} \quad (7)$$

The matrix of the partial derivatives in equation (7) is known as the Jacobian matrix and is denoted here as J_f. equation (7) can be re-written as:

$$v_y = Jf\ v_x\ J_f{}^T \quad (8)$$

Where

v_y = variance of the computed result, y.

J_f = Jacobian matrix of the measurement function, f, and

v_x = the variance-covariance matrix of the input variables, x.

2.2.2. Heave Uncertainty. To get the total variance of depth due to heave errors, the heave measurement errors, and induced errors are quadratically added, which results in:

$$\sigma_H = \sqrt{\sigma^2{}_{H_{meas}} + \sigma^2{}_{H_{ind}}} \quad (9)$$

Where σ_H is the total heave variance, $\sigma_{Hm}$is the heave measurement error, $\sigma_{Hi}$is the heave induced error.

2.2.3. Dynamic Draught Uncertainty. The dynamic draught variance is calculated by the quadratic sum of the variances of static draught, squat and loading errors, which results in:

$$\sigma_T = \sqrt{\sigma^2{}_{draft} + \sigma^2{}_{squat} + \sigma^2{}_{load}} \quad (10)$$

Where σ_T is the dynamic draught variance, $\sigma_{draught}$ is the static draught error, σ_{squat} is the squat error and σ_{load} is the load error.

2.2.4. Uncertainty for Water Level Measurement. Water level measurement variance σ_{WL} is divided into two components, measurement uncertainty at water level gauge σ_{WLmeas} and spatial and temporal prediction (zoning) uncertainty σWLzone.

$$\sigma_{WL} = \sqrt{\sigma^2{}_{WL_{meas}} + \sigma^2{}_{WL_{pred}}} \quad (11)$$

2.3. *Field Measurement Survey Design*

To assess the agreement between the modelled and measured uncertainties, surveys were conducted at Patimban port development area and Pramuka island in February 2019 and June 2022, respectively. Patimban is located on the northeast coast of the Subang District in the West Java Province, while Pramuka island is situated approximately 50km north of Jakarta, the capital of Indonesia, in the Java Sea as shown in figure 2.

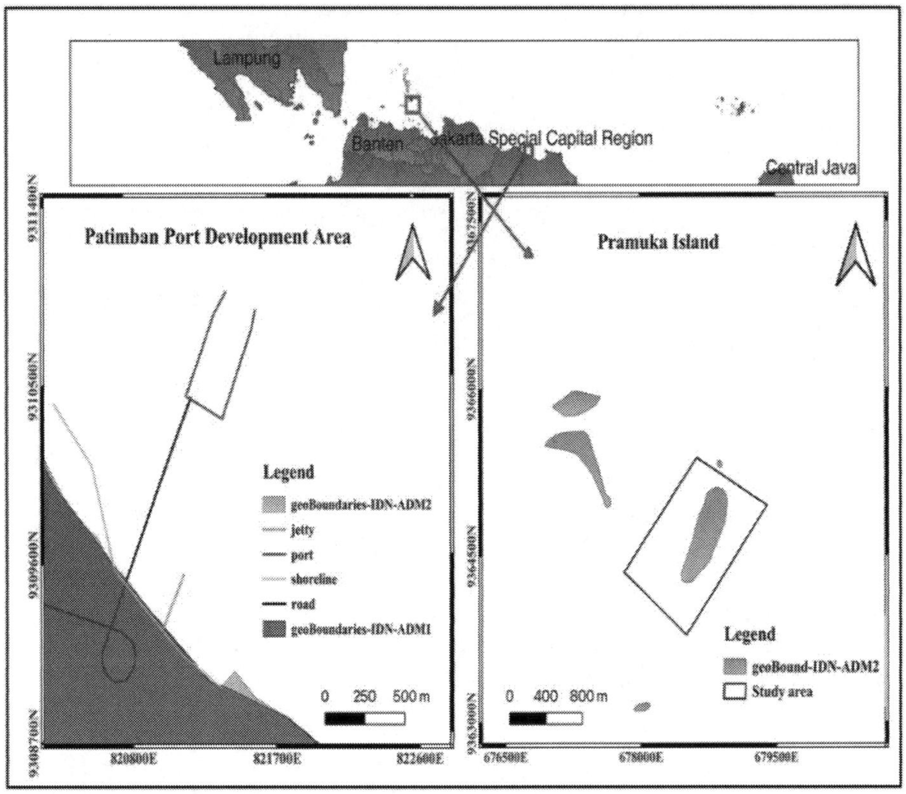

Figure 2. Field measurement locations - Patimban (left) and Pramuka (right)

The data used in this study were collected using the R2 Sonic 2020 and Kongsberg M3 Sonar-500 multibeam echosounders, which were deployed on local fishing boats in Patimban and Pramuka, respectively, operating in Continuous Wave (CW) mode. Additional sensors used during the survey and their specifications can be found in Table 1. and the static offsets of the sensors, as installed at the survey locations, are depicted in Figure 3.

Table 1. Boat dimensions and device specifications for field measurement

	Patimban	**Pramuka**
Vessel Dimension	9.34 x 2.12 meters	12.93 x 2.68 meters
Multibeam Echosounder Manufacturer	R2Sonic R2 SONIC	M3 Sonar Kongsberg
Heave sensor	IMU-108-30	
Manufacturer	SMC	
Heave – fixed component (m)	0.05	
Heave precision–variable component (% wave amplitude)	5	
Roll/Pitch sensor	IMU-108-30	
Manufacturer	SMC	
Roll/Pitch measurement precision (°)	0.3	
Heading sensor	Vector V103	
Manufacturer	Hemisphere GPS	
Heading measurement precision (°)	0.3	
Positioning system	Trimble NetR9	
Manufacturer	Aimar	
Positioning system precision (m)	3	
Sound speed sensor	Base X	
Manufacturer	AML	
Sensor precision (m/s)	0.05	

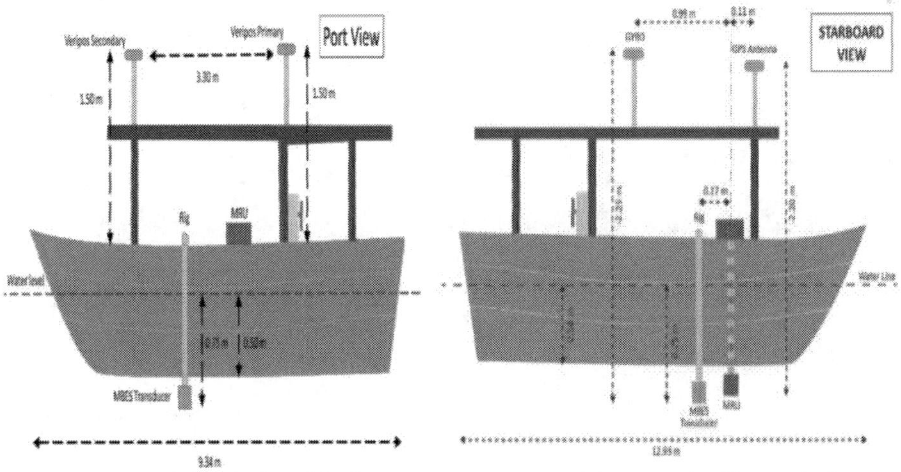

Figure 3. Spatial relationship between sensor positions and offset relative to the center of gravity (cog) - Patimban (left) and Pramuka (right).

2.4. *Field Measurement Standard Deviations*

The acquired measurement data undergoes processing and filtering using Qimera software to remove spikes and produce a cleaned bathymetry. The resulting bathymetric data is then used to calculate the standard deviations, which are compared with the predictions from the model. It should be noted that the model assumes a flat seafloor, and therefore, only the flat portion of the seafloor, depicted by the red area in Figure 4, is considered for determining the standard deviation.

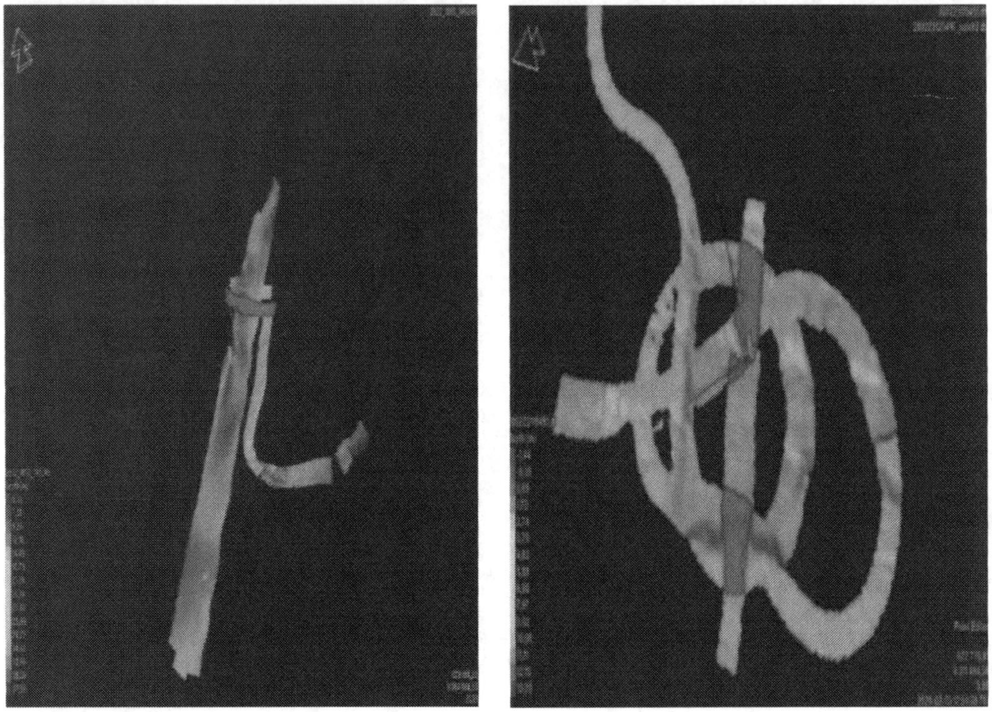

Figure 4. Red portion representing flat area use for measured data standard deviation calculation - Patimban (left) and Pramuka (right)

3. Results and Discussion

3.1. *Computed Reduced Depth Error Budget*

Table 2 shows the parameters used to compute the depth error budget. Most of the parameters are from the sensor specifications [19-20] deployed at the survey areas. TVU was computed as a function of beam angle from $\theta = -60° \ to \ 60°$ based on the average water depth in the survey areas, 6 meters at Patimban and 26 meters at Pramuka.

Table 2. Parameters used for modeled depth error budget computation.

Variable	Symbol	Value	Unit	Rad
Beam angle	θ	-60 - 60	deg	0.0349
Pitch for echo sounder	P	2	deg	0.0349
Roll for echo sounder	R	2	deg	0.0052
Pitch measurement uncertainty	σP_{meas}	0.3	deg	0.052
Roll measurement uncertainty	σR_{meas}	0.3	deg	0.00087
Pitch misalignment uncertainty	$\sigma \Delta Palign$	0.05	deg	0.00087
Pitch misalignment uncertainty	$\sigma \Delta Ralign$	3.05	deg	0.0523
Beamwidth	ψx , ψy	3.0	M	
Range sampling resolution	Δrs	0.01	M	
Range sampling resolution for phase	Δrp	0.01	M	
Pulse length	t	0.02	Ms	
Sound speed profile	v_p	1500	m/s	
Sound speed profile uncertainty	σvp	0.5	m/s	
Surface sound speed	v_s	1530	m/s	
Surface sound speed uncertainty	σvs	0.02	m/s	
Sound speed at seafloor	v	1500	m/s	
Heave measurement uncertainty	σH_{meas}	0.05	M	
Induced-heave uncertainty	$\sigma Hind$	0	M	
Uncertainty for water level measurement	σWL_{meas}	0.01	M	
Uncertainty for water level zoning	σWL_{zone}	0.05	m	
Transducer-VRU offset uncertainty	$\sigma \Delta x$, $\sigma \Delta y$, $\sigma \Delta z$	0.014	M	
Static draught uncertainty	$\sigma static\text{-}draught$	0.03	M	
Squat uncertainty	$\sigma squat$	0.005	M	
Loading change uncertainty	$\sigma load$	0.03	M	

Table 3 and figure 5 display the results of TVU computation, presented as values corresponding to a 95% confidence level. The findings indicate that the smallest TVU values are observed within a beam angle range of 0 $to \pm 15$ degrees, where the bottom detection method shifts from amplitude detection to phase detection. As the beam angle increases, the TVU also increases due to the greater distance between the echo sounder and the seafloor.

IOP Conf. Series: Earth and Environmental Science 1245 (2023) 012041 doi:10.1088/1755-1315/1245/1/012041

Table 3. Modeled depth error budget result.

Patimban: Average depth of 6 meters									
	-60	-45	-30	-15	0	15	30	45	60
v_d	0.21	0.81	0.15	0.14	0.12	0.14	0.16	0.19	0.21
v_D	0.26	0.24	0.22	0.21	0.20	0.21	0.23	0.25	0.27
Pramuka: Average depth of 26 meters									
	-60	-45	-30	-15	0	15	30	45	60
v_d	0.22	0.21	0.19	0.18	0.16	0.18	0.19	0.21	0.22
v_D	0.28	0.26	0.25	0.24	0.23	0.24	0.25	0.26	0.28

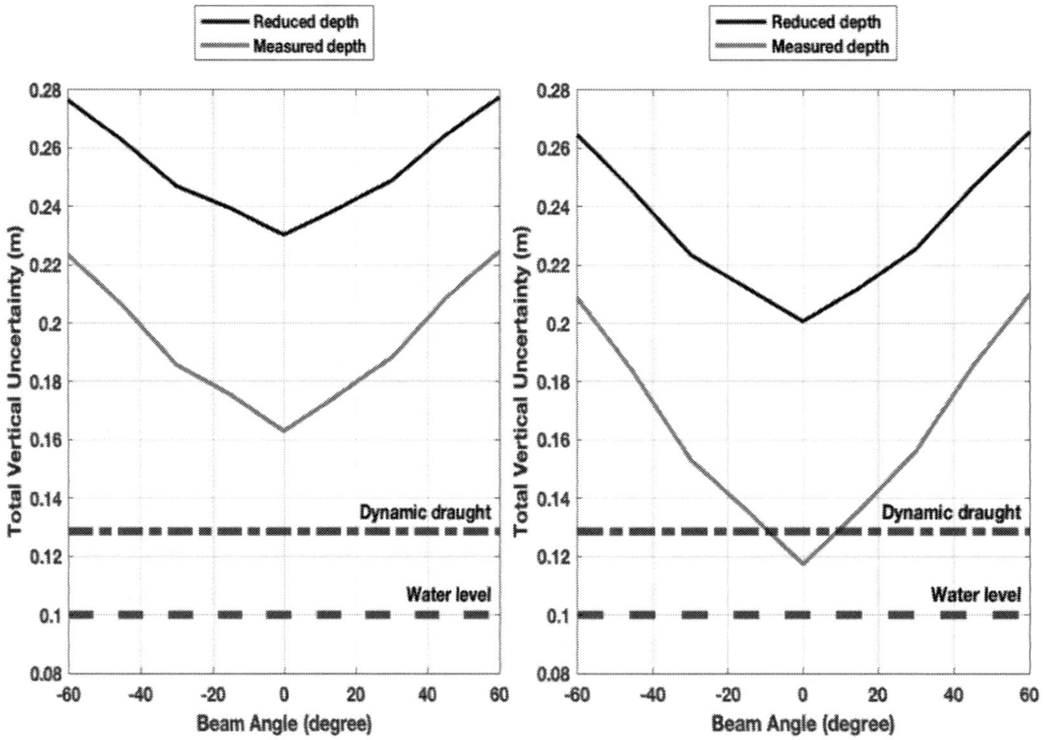

Figure 5. Modeled depth error budget for Patimban (left) and Pramuka (right).

The average depth in Pramuka (26 m) is higher than that of Patimban (6 m). Because of the deeper depth in Pramuka, the range gets longer, and the uncertainty becomes larger than that in Patimban. the uncertainty associated with the water level and draught were constant throughout the swath. The reduced depth uncertainty gets larger in both areas than the measured depth uncertainty because of the effect of water level and draught uncertainties added to the measured depth uncertainty.

3.2. *Comparison of measured and reduced depth error budgets*

The bathymetric uncertainties modeled in this study are based on parameters obtained from the sensor specifications deployed during the survey, the settings used during data acquisition, and environmental factors. A CW type was used during data acquisition, the modeled uncertainties do not account for the contributions of Doppler effect and baseline decorrelation. Table 4-4 compares the model estimate and the measurement standard deviations in the two test areas, showing that there is little difference between the uncertainties, particularly in the Pramuka area where the difference is only 0.03m at the widest beam angle of ±60°. However, a larger discrepancy was observed in the Patimban area, where the model overestimates the uncertainty by 0.09m. Furthermore, the uncertainty values revealed a 1 percent increase in uncertainty for every 20m depth interval. Overall, the model presented in this study performed well in estimating the reduced depth uncertainty of portable hydrographic survey systems.

Table 4. Difference between the modeled depth error budget and the calculated standard deviation

	Model uncertainty at the widest angle (60°) (m)	Meas. std dev (m)	Difference (m)
Patimban	0.26	0.17	0.09
Pramuka	0.28	0.25	0.03

The bar chart in figure 6 offers a useful way to compare and contrast the data presented in the table by providing a visual representation of the predicted and calculated standard deviations in two distinct areas: Patimban (16m) and Pramuka (26m). The chart clearly displays the magnitude of the difference between the predicted and calculated values in each location, with the use of an appropriate color code to enhance readability.

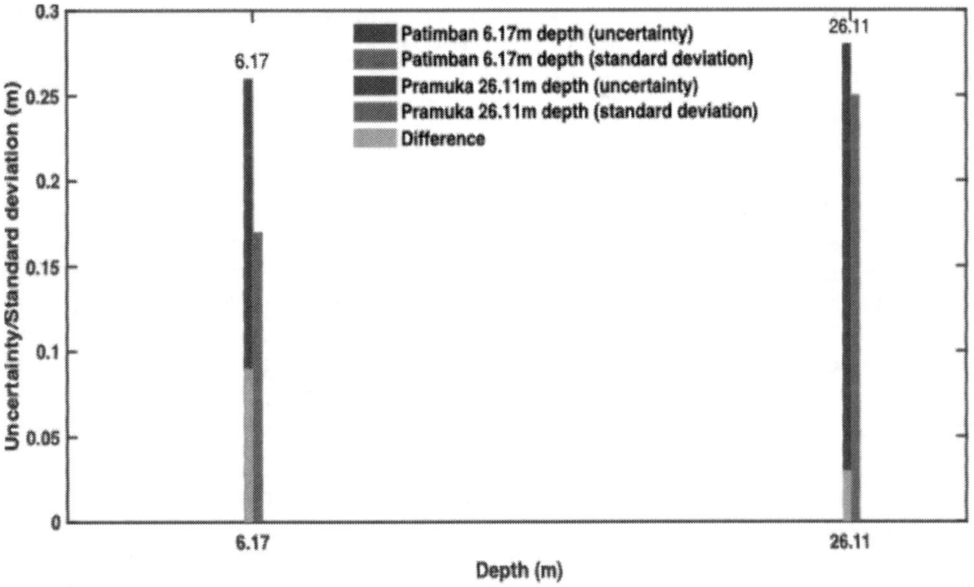

Figure 6. Comparison of modeled and calculated depth error budgets with difference magnitudes.

4. Conclusion

Based on the study's results, the newly proposed error budget provides a dependable estimation of the uncertainty of a dataset obtained through portable systems deployment in the field. The model estimate and the measured depth's uncertainty exhibit a negligible difference in uncertainties in the Pramuka area, indicating that the model performed well in this area.A considerable discrepancy was observed in the Patimban area, where the model overestimated the uncertainty. This discrepancy may be attributed to the presence of various environmental factors in the area. Overall, the model performed well in estimating the reduced depth uncertainty of portable hydrographic survey systems, suggesting that the model can be trusted by surveyors to predict the error budget of the hydrographic survey system they plan to use in the field.

The present study has several limitations that should be acknowledged. firstly, it is important to note that the model used in this study assumes a flat seafloor during the computation of the error budget. However, it is known that there can be significant variations and local slopes in the seafloor topography in the actual measurement area. Another limitation is that the study did not consider other sources of error, such as environmental factors or operator error, which could impact the accuracy of the survey measurements. Additionally, the study only tested two areas, which may not be representative of all survey environments. Future research could expand the study to include a wider range of survey environments and sensors to validate the results and improve the generalizability of the findings.

Importantly, it must be emphasized that the same parameters were employed to calculate the depth error budget in both study areas, solely based on the similarity in technical specifications of the sensors used. However, this approach overlooks potential variations in sensor characteristics and performance, which could significantly introduce biases or inaccuracies in the error budget calculations.

The findings of this study provide valuable insights into the estimation of uncertainty in portable hydrographic survey systems. Surveyors can leverage the proposed error budget model to enhance their decision-making process and improve the accuracy of survey measurements. By considering the limitations and potential areas of improvement identified in this study, further advancements can be made in the field of hydrographic surveying.

Acknowledgement

A big thank you is extended to the students of the Category A Hydrographic Surveyor Profession Education Program of Institut Teknologi Bandung (ITB CAT-A Hydrographic Education) from the batches 2019/2020 and 2020/2021-2021/2022 who conducted the MBES surveys at Patimban and Pramuka Island, respectively.

References

[1] Clarke, J. E. H., Mayer, L. A., & Wells, D. E. (1996). Shallow-Water Imaging Multibeam Sonars: A New Tool for Investigating Seafloor Processes in the Coastal Zone and on the Continental Shelf. In *Marine Geophysical Research* (Vol. 18, Issue 9). Kluwer Academic. Publishers.
[2] Nairn, R. (2010). The Challenge of Hydrographic Surveying and Charting the Antarctic.
[3] Mohammadloo, T. H., Snellen, M., & Simons, D. G. (2020). Assessing the performance of the multi-beam echo-sounder bathymetric uncertainty prediction model. Applied Sciences (Switzerland), 10(13). https://doi.org/10.3390/app10134671

[4] Lamarche, G., & Lurton, X. (2018). Recommendations for improved and coherent acquisition and processing of backscatter data from seafloor-mapping sonars. *Marine Geophysical Research*, *39*(1–2), 5–22. https://doi.org/10.1007/s11001-017-9315-6

[5] Lurton, X., & Augustin, J. M. (2010). A measurement quality factor for swath bathymetry sounders. IEEE Journal of Oceanic Engineering, *35*(4), 852–862. https://doi.org/10.1109/JOE.2010.2064391

[6] Grządziel, A. (2021). The importance of under-keel sound velocity sensor in measuring water depth with multibeam echosounder. *Energies*, *14*(17). https://doi.org/10.3390/en14175267

[7] Suhari, K. T., Karim, H., Gunawan, P. H., & Purwanto, H. (2013). Small ROV marine boat for bathymetry surveys of shallow waters -potential implementation in Malaysia. International seminarArchives of Photogrammetry, Remote Sensing and Spatial Information Sciences - ISPRS Archives, *42*(4W5), 201–208. https://doi.org/10.5194/isprs-archives-XLII-4-W5-2012017

[8] Syetiawan, A., & Gularso, H. (2021). seashore large-scale mapping acceleration using lidar bathymetry Towards a Coastline Base Map in Indonesia. https://www.giminternational.com/content/article/towards-a-coastline-base-map-in-indonesia

[9] Dugan, J. P., Morris, W. D., Vierra, K. C., Piotrowski, C. C., & Farruggia, G. J. (2001). JetskiBased Nearshore Bathymetric and Current Survey System. In *Journal of Coastal Research* (Vol. 17). http://about.jstor.org/terms

[10] Wati, G. N. (2016). error budget analysis for surface and underwater survey system.

[11] Hare, B. R., Eakins, B., & Amante, C. (2011). *modelling bathymetric uncertainty.*

[12]Hare, R., Godin, A., & Mayer, L. (1995). Accuracy estimation of Canadian Swath (multibeam) and Sweep (multi-transducer) sounding systems.

[13] Mohammadloo, H., Snellen, M., & Simons, D. G. (2018). Multi-beam echo-sounder bathymetric measurements: Implications of using frequency modulated pulses. *The Journal of the Acoustical Society of America*, *144*(2), 842–860. https://doi.org/10.1121/1.5050816

[14] Mohammadloo, T. H., Snellen, M., Amiri-Simkooei, A., & Simons, D. G. (2019). assessment of reliability of multi-beam echo-sounder bathymetric uncertainty prediction models.

[15] Afrizal Tengku Ali, T., Shahmy Bin Mohd Said, M., Azwan Abbas, M., Mohd Hashim, N., Asrul Mustafar, M., & Aman Sulaiman, S. (2022). *Investigating the vertical uncertainty in MBES measurements.*

[16] Minami, H. (2013). Estimation of total vertical uncertainty for the bathymetry acquired by autonomous underwater vehicle in deep water. In REPORT OF HYDROGRAPHIC AND OCEANOGRAPHIC RESEARCHES (Issue 50).

[17] Burkholder, E. F. (2017, July 25). Error Propagation (Summary of Mathematical Concepts) (The 3-D Global Spatial Data Model). Boca Raton. http://what-when-how.com/the-3-dglobal-spatial-data-model/error-propagation-summary-of-mathematical-concepts-the-3d-global-spatial-data-model/

[18] Dobbert, M., & Schrijver, B. (2014). *Uncertainty Propagation for Measurements with Multiple Output Quantities.*

[19] R2Sonic. (2020). Multibeam Echosounder Specifications.

[20] Kongsberg. (2016). *M3 SONAR ®-500M* ethernet with vdsl telemetry and synchronization. www.kongsbergmesotech.com

IOP Conf. Series: Earth and Environmental Science 1245 (2023) 012042 doi:10.1088/1755-1315/1245/1/012042

Indonesian "tsunami-generation" in a nutshell: systematic literature review

Afifah Hanum Amahoru[1], Ainur Ridho[1], Radikal Lukafiardi[1], Mizan Bustanul Fuady Bisri[1,2,*]

[1] CARI! (Cerdas Antisipasi Risiko Indonesia), Bandung, Indonesia
[2] GSICS, Kobe University, Kobe, Japan

*mbisri@people.kobe-u.ac.jp

Abstract. Publications related to tsunami-generation phase hold 42.8% of all the publications over all tsunami hazard literatures in Indonesia. Corroborating the fact, tsunami-generation studies are also vital in determining the type of propagation and inundation that may appear in the surrounding coastal regions, which in practice, can also help determining disaster management specifics and risk reduction activities. A literature review regarding tsunami-generation has been done for those purpose by using our CARI! Knowledge Engine, which includes collections of both International and Indonesian journals within the scope of Indonesian locus context. This study unpacks publication trend, the most influential articles, top contributing journals, top main authors, top institutions, and dominant countries as well as the science mapping of tsunami-generation subjects. To gain deeper understanding, we also develop tsunami-generation research framework within the basis of tsunami generation force, time occurrence, monitoring, and measurement approach, as well as hazard analysis approach. From the review, an interesting peak of tsunami-generation publications has been shown in 2020 (39 publications), following the occurrence of several Indonesian major tsunamis in the previous decade and the globally impacting Aceh Tsunami (or the Indian Ocean Earthquake and Tsunami) in 2004. Among all the publications of tsunami-generation force, tsunamigenic earthquake has become the most studied topic (67.7% of all publications), while the most discussed atypical sources mainly come from volcanic eruption (46.67% of all atypical literatures) and submarine landslide (33.33% of all atypical literatures). In terms of tsunami-generation source, it is found that the Sunda Megathrust has been the most-discussed location, which is mentioned within 39 publications, followed by the Palu-Koro Fault with 18 publications and the Krakatau Volcano by 16 publications. Apart from the notice of significant difference between research of historical tsunami and paleotsunami (120:5 publications), some gaps also can be found when discussing the meteotsunami-generation mechanism in Indonesia and the Global Navigation Satellite System's Total Electron Content (GNSS-TEC) method, which, although proven beneficial for the development of tsunami early warning system, accuracy is still the main issue in capturing tsunami signals and positioning actual source location of tsunami. These were some highlighted limitations that arise in our process to understand tsunami-generation mechanisms and the exploitations of the knowledge. Therefore, future studies specifically within these areas are further encouraged.

1. Introduction

Tsunami is a natural disaster that can cause significant damage to coastal communities on a wider area range across oceans. Scientifically, tsunami's life cycle can be divided into three main phases: generation, propagation, and inundation [1]. During the generation phase, a tsunami is created by a sudden displacement of water, which is caused by many factors from typical (i.e., earthquakes) to atypical (i.e., volcanic eruption, landslide, atmospheric disturbance, etc) origins. According to historical records, more than 80% of world tsunamis were generated by earthquakes, and 70% of them, including large tsunamis, were observed around the Pacific Ring of Fire [2].

Situated in the Pacific Ring of Fire, Indonesia is prone to earthquakes and volcanic eruptions. As a consequence, most of the tsunamis in Indonesia are generated by earthquakes, which can occur along the subduction zones, followed by volcanic activity and landslide [3]. Based on history, there were at least 246 tsunami events that occurred in Indonesia in the period between 416-2018 AD [4]. Common locations where tsunamis are generated in Indonesia are those located in the Sunda Trench, Java Trench, and eastern part of Indonesia as shown in figure 1. The Sunda Trench or Sunda Megathrust, which runs along the west coast of Sumatra and Java, is the source of several recent devastating tsunamis, including the 2004 Aceh tsunami and the 2010 Mentawai tsunami. The eastern part of Indonesia also has as many historical records of tsunami events as the western part. However, the main source of the tsunamis in the eastern part of Indonesia is rarely known and there are no detailed sources of its information [3]. Other sources of tsunamis in Indonesia include volcanic eruptions such as the recent 2018 Sunda Strait tsunami caused by the Anak Krakatau Eruption [5].

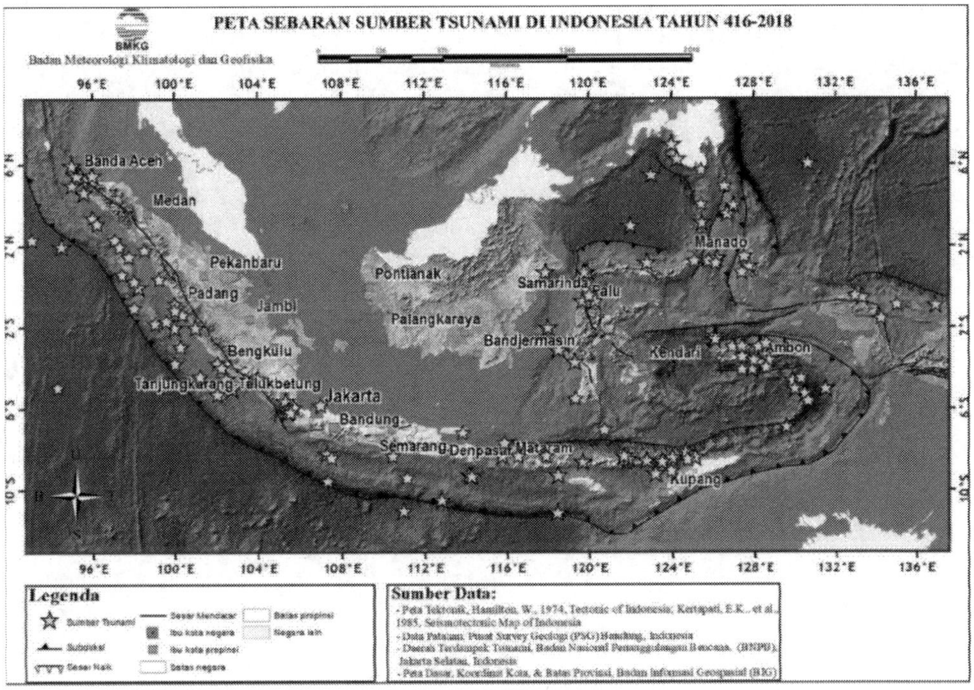

Figure 1. Map of Distribution of Tsunami Sources in Indonesia 416-2018 [4].

Indonesia is particularly vulnerable to tsunamis due to the interconnected nature of its waters, which means that the effects of a tsunami in one area can impact surrounding regions, increasing the likelihood of damage. Coupled with the high population density in coastal areas, the risk of a tsunami disaster in Indonesia considered as high [6]. The generation of tsunami has been one of the important subjects within tsunami research and understanding the concept of tsunami-generation can give insights about the propagation and inundation of a tsunami. For that purpose, this study has been utilizing peer literature reviews regarding tsunami hazard analysis compiled from CARI! Knowledge Engine system including publications that have been published in both International and Indonesian journals within the scope of Indonesian context. In fact, based on the CARI! Knowledge Engine, publications related to the tsunami-generation phase hold 42.8% of publications on the tsunami hazard assessment subject in Indonesia [7]. Therefore, the development of tsunami hazard literacy and understanding of tsunami-generation research landscape are crucial for effective disaster risk management in Indonesia.

To gain a deeper understanding of the trends and themes related to tsunami-generation, this research paper will explore several research questions. These questions include an analysis of publication trends in Indonesia, the most influential articles and top contributing journals, and the top contributing authors, institutions, and countries. Additionally, the research will examine the major themes and topics related to tsunami-generation, suitable frameworks for studying the subject, and future research directions. By answering these research questions, this paper aims to provide a comprehensive overview of the current state of research on tsunami-generation in Indonesia, which will inform future disaster risk management strategies and ultimately contribute to the protection of coastal communities from the devastating effects of tsunamis.

2. Methodology

This research conducted systematic literature reviews (SLRs) by collecting bibliography data on the theme of "tsunami-generation" research to answer the research questions mentioned in the previous section. The SLRs use a specific method or protocol to organize and analyze data, which promotes transparency and replication. The process involves establishing criteria to gather relevant scientific articles and utilizes an inductive reasoning approach. SLRs can give a comprehensive understanding of current knowledge and identify areas for future research in a specific field, theory, or method [8][9].

To do the SLRs, we employed Preferred Reporting Items for Systematic reviews and Meta-Analyses (PRISMA) 2020 statement protocol. PRISMA 2020 offers updated guidance for reporting systematic reviews, considering recent developments in methods for identifying, selecting, evaluating, and combining studies. The detailed review process used PRISMA 2020 for our research is illustrated in figure 2. The procedure used in this review accorporating from identification of literature records, screening, eligibility assessment, and the result of literature data.

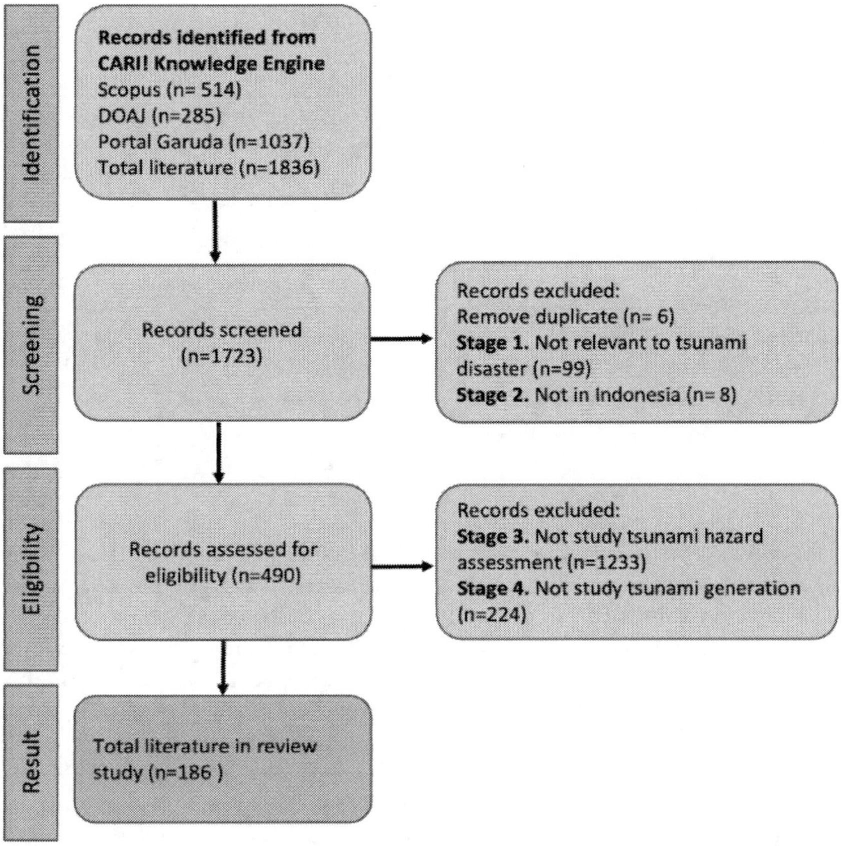

Figure 2. PRISMA 2020 flow diagram of systematic review *used* in this study.

This study utilized CARI! Knowledge Engine database for its collection of literature pertaining to disasters. This database sources literature from Scopus, Directory of Open Access Journal (DOAJ), and Portal Garuda and has the added feature of automatically tagging literature by study location and hazard type which simplifies the process of collection during SLRs. The research keywords used were designed to be broad enough to capture relevant literature while also excluding irrelevant results. The keywords used for each stage during the selection process of tsunami-generation topics are listed in Table 1. The literature was obtained by multistage filtering by keywords from stage 1 to stage 4 with a separate intention rationale of each stage. The search was conducted between November 2022 and January 2023.

Table 1. Research keywords on "Tsunami-Generation".

Stage	Keywords	Rationale	Process
1	"tsunami", "tidal wave", "gelombang pasang"	To identify literature related to tsunami disaster	Automatically by CARI! Knowledge Engine
2	Geographical names in Indonesia e.g., "Sumatra", "Jakarta", ...	To identify literature related to Indonesia	Automatically by CARI! Knowledge Engine
3	"hazard assessment", "hazard model", "simulation", "modelling", ...	To identify literature related to tsunami hazard assessment	Manually by researcher
4	"generation", "source", "earthquake", "volcano", "landslide", "rupture", ...	To identify literature related to tsunami-generation	Manually by researcher

The aim of this review method is to summarize studies that specifically investigate the assessment of "tsunami-generation" literatures in more detail. Literature that does not pertain to the preferred rationales in table 1 were excluded from the study. The study also uses content analysis by developing framework on tsunami-generation research, thus we deliberately tagging each literature based on our proposed conceptual analysis on tsunami analysis approach (e.g., numerical, empirical, or other approaches), historical tsunami occurrences (e.g., historical or paleo tsunami), tsunami-generation sources (e.g., earthquake or volcano locations), and tsunami-generation types (e.g., tectonic, volcanogenic, or others).

To examine the productivity and impact of a specific research area, a performance analysis is used. This approach is similar to profiling participants in research studies but uses more advanced bibliometric metrics for greater accuracy [8]. This study aims to discover published publication trends in number, impactful articles, impactful journals, productivity from main authors, leading institutions, dominant countries, and the methods and context used in tsunami-generation research.

Furthermore, this study leverages the science mapping and network analysis method to gain the understanding of co-analysis covering in literature corpus. Network analysis uses keywords extracted from abstract literature, then network analysis helps classify the studies clusters to show information diversity in the field by identifying differences among keywords. This analysis was assisted by VOS viewer software to create co-occurrence networks of terms extracted from English-language textual data such as titles and abstracts of publications. In this study, abstract of literature written in Bahasa Indonesia were manually translated to English in order to complied with software analysis. Moreover, VOS viewer enables researchers to omit noun phrases that have a low relevance score, allowing them to get rid of general noun phrases. The remaining terms typically represent topics related to the domain of interest, which can be visualized in a co-occurrence network.

Ultimately, this study developed an organizing framework of tsunami-generation research by adopting the previous work of tsunami hazard research. The framework is used to quantitatively and qualitatively analyze tsunami-generation contents based on their interconnection between topics identified from science mapping. Quoting from the November 2022's Indonesian Disaster Knowledge Update (IDKU) of http://caribencana.id, tsunami frameworks can be divided on several concepts, such as tsunami phase, time occurrence, monitoring and measurement approach, and its hazard analysis approach (**Figure 3**). As the generation of tsunami is the first part of tsunami phase, we prefer to exclude tsunami propagation and inundation phase from the discussion and instead, employ the generation force of tsunami as one of the proposed frameworks. The other three concepts are then analyzed within the scope of tsunami-generation which especially within the region of Indonesia.

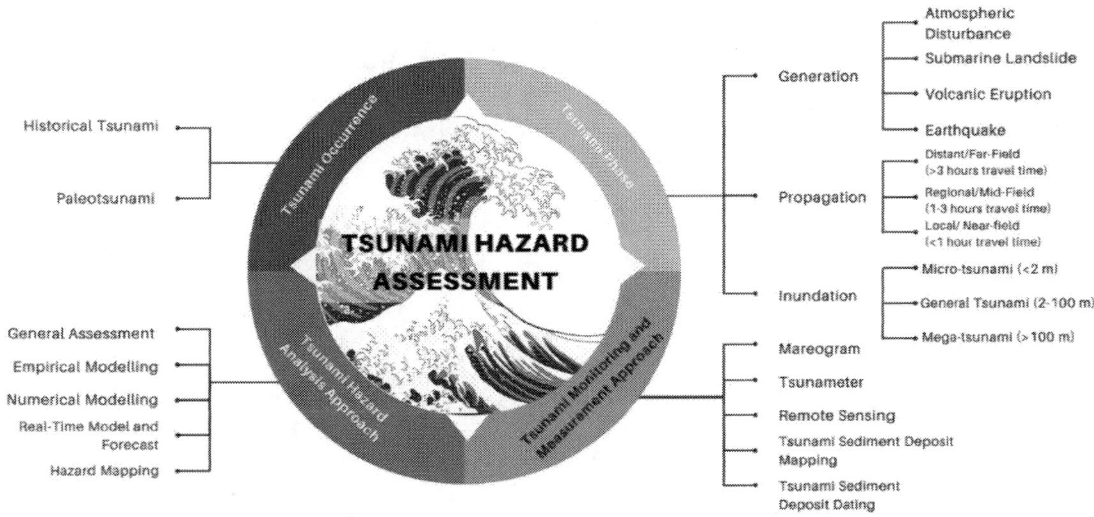

Figure 3. Tsunami-generation framework analysis.

Eventually, wordings from this process are constantly checked with the Tsunami Glossary to ensure universal understanding of tsunami-related terms. It is important to understand that terms generated from the science mapping process might appear to be different than those in the glossary. For the frameworks purpose, similar terms presented in the glossary will be used.

Within the whole processes, Sankey diagram may be used to analyze relationship between subtopics of tsunami-generation research theme (portrayed by boxes). This may be visualized proportionally based on the number of literatures. The larger size of the box and the wider lines connecting two boxes indicate a greater number of publications accounted for them. In this paper, word cloud is also used to understand topics related with tsunami-generation.

3. Result findings and discussion

After conducting the PRISMA protocol, this study has analysed a review of 186 literatures comprises of 166 journal articles and 20 proceedings which are published between 1995 and 2022. The dataset literature consists of 123 pieces of literatures sourced from Scopus, 36 pieces sourced from Portal Garuda, and 27 pieces from DOAJ directories. According to the written language used in the literatures, it is found that 157 literatures has been written in English, and 29 literatures written in Bahasa Indonesia. These Indonesian publications are then translated to English to ensure the inclusion of the publications within the corpus. The review analysis in this study has started from bibliometric analysis of performance and science mapping, followed by a deep dive into content analysis using the developed tsunami-generation framework.

3.1. Performance analysis

The performance analysis study has successfully discovered the quantitative metrics of the publication trends, the most impactful articles, top contributed journals, top contributing main authors, top contributing institutions, contributed countries, and the methods and context used in tsunami-generation research.

3.1.1. Publication trend. The publication trend of research on tsunami-generation in Indonesia over the years is shown in Figure 4. The data shows that the earliest article on this topic in Indonesia was published in 1995 [10], and that the number of publications has steadily increased in the last 30 years.

There were only a few publications in the early years, but the field has grown significantly, especially after 2009s, with the highest record of 39 tsunami-generation publications published in 2020. Notably, from the figure we can see that the research activity in this field has increased significantly following the Indian Ocean Tsunami (IOT) of 2004, as well as the following Mentawai tsunami of 2010, and even more so after the Palu tsunami and Sunda Strait tsunami of 2018.

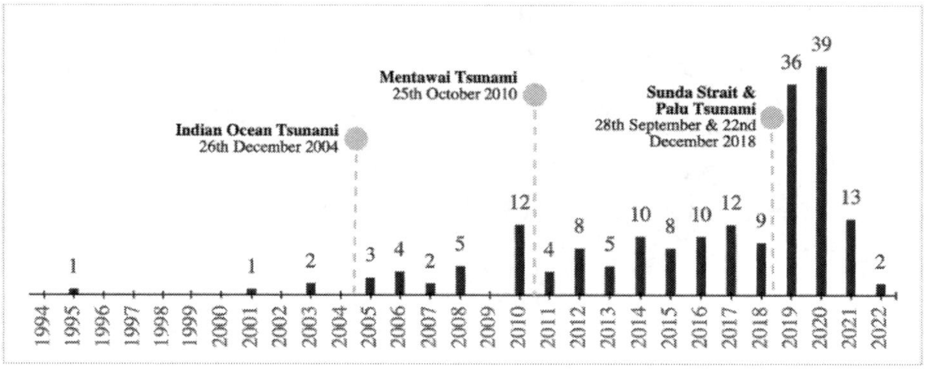

Figure 4. Number of published literatures per year since 1994 to 2022.

Badan Meteorologi Klimatologi dan Geofisika (BMKG) has recorded 189 tsunami events from year 416 until 2018 that took place in the Indonesian region [4]. Following the major tsunami caused by Krakatoa Eruption in August 1883, there were numerous tsunamis eventuate after this event but have less impacts until the Indian Ocean Tsunami (IOT) of 26th December 2004 took place. The IOT became the deadliest tsunami event that occur in modern era, due to its casualties of 227.898 persons in countries adjacent to Indian Ocean [11][12]. Approximately, there are 169.000 casualties counted that are coming from Indonesia. This event triggered many researchers around the world to study this tsunami event from many disciplines and topics including tsunami-generation even after a decade.

3.1.2. Top influential articles and top contributing journals. The top ten influential articles indicated by articles citation number about tsunami-generation in Indonesia has been presented in below. The table shows M. Chlieh's article [13] as the most cited article in the field with the total of 344 citation since published in 2007 or average of 22,9 citations per year. Interestingly, the top 5 cited articles [13][14][15][16][17] are all investigated about tsunami-generation caused by Great Sumatran megathrust in Indian Ocean, also all of authors managing a research collaboration among the expert from different affiliation. The next most cited articles studied probabilistic tsunami hazard assessment authored by Horspool [18], numerical simulation of flores tsunami by Imamura [10] and Sulawesi tsunami by Heidarzadeh [19], and source mechanism of tsunami triggered from volcano eruption by Paris [20].

Table 2. Top ten most influential articles about tsunami hazard assessment in Indonesia.

Authors	Article title	Journal	Year	TC[a]	C/Y[b]
M. Chlieh; J.P. Avouac; V. Hjorleifsdottir; T.R.A. Song; C. Ji; K. Sieh; A. Sladen; H. Hebert; L. Prawirodirdjo; Y. Bock; J. Galetzka	Coseismic Slip and Afterslip of the Great Mw 9.15 Sumatra–Andaman Earthquake of 2004	Bulletin of the Seismological Society of America	2007	344	22.9
C. Subarya, M. Chlieh, L. Prawirodirdjo, J.P. Avouac, Y. Bock, K. Sieh, A.J. Meltzner, D.H. Natawidjaja & R. McCaffrey	Plate-boundary deformation associated with the great Sumatra–Andaman earthquake	Nature	2006	298	18.6
D.H. Natawidjaja, K. Sieh, M. Chlieh, J. Galetzka, B.W. Suwargadi, H. Cheng, R.L. Edwards, J.P. Avouac, S.N. Ward	Source parameters of the great Sumatran megathrust earthquakes of 1797 and 1833 inferred from coral microatolls	Journal of Geophysical Research: Solid Earth	2006	159	9.9
E.M. Hill, J.C. Borrero, Z. Huang, Q. Qiu, P. Banerjee, D.H. Natawidjaja, P. Elosegui, H.M. Fritz, B.W. Suwargadi, I.R. Pranantyo, L. Li, K.A. Macpherson, V. Skanavis, C.E. Synolakis, K. Sie	The 2010 Mw 7.8 Mentawai earthquake: Very shallow source of a rare tsunami earthquake determined from tsunami field survey and near-field GPS data	Journal of Geophysical Research: Solid Earth	2008	118	11.8
E.A. Okal, C.E. Synolakis	Far-field tsunami hazard from mega-thrust earthquakes in the Indian Ocean	Geophysical Journal International	2008	116	8.3
N. Horspool, I. Pranantyo, J. Griffin, H. Latief, D. H. Natawidjaja, W. Kongko, A. Cipta, B. Bustaman, S. D. Anugrah, and H. K. Thio	A probabilistic tsunami hazard assessment for Indonesia	Natural Hazards Earth System Sciences	2014	72	10.3
P. Shearer, R. Bürgmann	Lessons Learned from the 2004 Sumatra-Andaman Megathrust Rupture	Annual Review of Earth and Planetary Sciences	2010	57	4.74
F. Imamura, E. Gica, T. Takahashi & N. Shuto	Numerical simulation of the 1992 Flores tsunami: Interpretation of tsunami phenomena in northeastern Flores Island and damage at Babi Island	Pure and applied geophysics	1995	55	2.0
M. Heidarzadeh, A. Muhari & A.B. Wijanarto	Insights on the Source of the 28 September 2018 Sulawesi Tsunami, Indonesia Based on Spectral Analyses and Numerical Simulations	Pure and Applied Geophysics	2019	50	16.6
R. Paris, A.D. Switzer, M. Belousova, A. Belousov, B. Ontowirjo, P.L. Whelley & M. Ulvrova	Volcanic tsunami: a review of source mechanisms, past events and hazards in Southeast Asia (Indonesia, Philippines, Papua New Guinea)	Natural Hazards	2014	50	7.1

[a]TC = Total citaition
[b]C/Y = Citation per year

The literature corpus of 186 articles on tsunami-generation were published across 98 journals, Figure 5 below only shows the journal with a minimum of three articles published within (contains 16,3% or 92 articles). The size of the text indicating the number of articles published within and their colour indicating the total citation number of each journal. The most contributing journal is *Pure and Applied Geophysics (PAG)* with 18 articles, followed by Natural *Hazards and Earth System Sciences (NHESS), Natural Hazards,* and *Science of Tsunami Hazards* with each 8 articles published. However, in terms of influence, *Natural Hazards and Earth System Sciences (NHESS)* leads the position with 365 citations each, followed by *Journal of Geophysical Research: Solid Earth (JGR)* with 338 citations, while *Pure and Applied Geophysics (PAG)* have 174 citations. Even though *NHESS* and *JGR* have much less published articles related to tsunami-generation in Indonesia compared to *PAG*, the articles become most impactful, this is consistent from their h-index which considerably higher than the other journals.

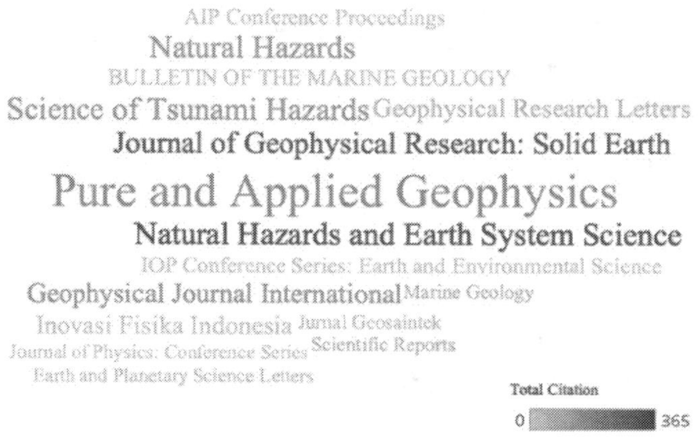

Figure 5. Top contributing published journal

3.1.3. Top main authors, top institutions, and dominant countries. The top 15 contributively institution for tsunami-generation research in Indonesia is presented on ***Figure*** *6* below, where the size of text indicating the published articles and the colour gradation indicating the total citation number. Based on published articles number, Universitas Negeri Surabaya become the most productive institution with a total of 14 articles. According to the influential manner, GFZ German Research Centre for Geosciences followed by Lembaga Ilmu Pengetahuan Indonesia lead the position with total citation number of 248 citation from 6 articles and 179 citations from 5 articles, respectively. Taken collectively the top 15 institutions contributing of 81 (44,5%) articles published that have amassed 948 citations in the field.

Figure 6. Main author's institution affiliation

Based on main author's country of institution affiliation, Indonesian researcher overtop the ranking with 83 articles published as shown in **Figure** 7. Other than Indonesia, the most contributing country is Japan, followed by United States which authored 17 and 15 articles, respectively. Indonesian researcher also leads the citation number ranking with total of 762 citations made by 34 different institutions within the country, followed by United States with total of 702 citations made by 10 different institutions within the said country. The region of institution's affiliation having tsunami-generation study in Indonesia is dominated by Europe (7 countries), Asia (9 countries), and Pacific (2 countries). Further distribution on the researcher's affiliation locations can be seen on the inset within **Figure** 7.

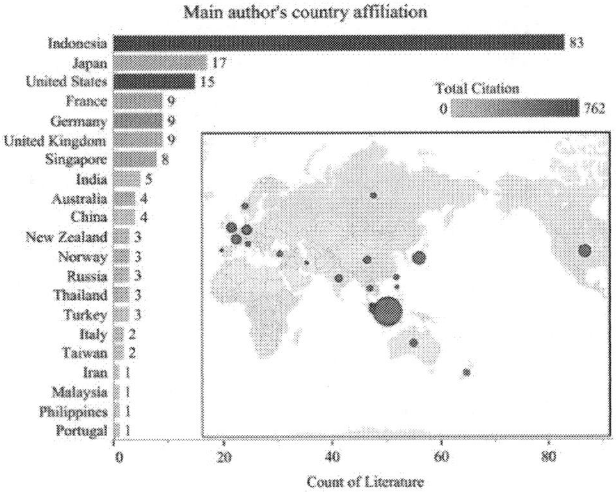

Figure 7. Main author's country affiliation of selected literatures on "Tsunami-Generation".

The top ten contributing authors for tsunami-generation in Indonesia is presented with Table 3. The table shows that Chlieh M. from California Institute of Technology, United States and Subarya C. from National Coordinating Agency for Surveys and Mapping, Indonesia are the most influential is evidenced by the number of citations (TC) in their publications. While the most productive authors with many citations are Natawidjaja D.H. from Lembaga Ilmu Pengetahuan Indonesia, Indonesia, Løvholt F. from Norwegian Geotechnical Insitute, Norway, and Mustafa B. from Universitas Andalas, Indonesia with 3 total publications (TP) with overall 174 total citations, 2 total publications and 83 total citations, 2 total publications and 72 total citations respectively.

In general, most of Indonesian researchers tend to have studied tsunami-generation across regions when compared to researchers from abroad who have studied more in the Sumatra and Sulawesi regions (Figure 8). As shown in **Figure 8**, the proportion of research region in the Sumatra, Sulawesi, and nation-wide/conceptual is studied more widely when compared to the regions of Java, Maluku & Papua, and Bali & Nusa Tenggara. The regions of Sumatra and Sulawesi have become the focus of research due to the devastating tsunamis that occurred in Aceh in 2004, Mentawai in 2010, and Palu in 2018, highlighting the inseparable relationship between the region and the need for further study. The research locus in the Java region was driven by the 2006 Pangandaran and 2018 Krakatau tsunamis. There is only a small proportion of the total studies that examine tsunami-generation with a research locus in central and eastern Indonesia, especially the Bali & Nusa Tenggara and Maluku & Papua regions.

Most of Indonesian researcher affiliations based in Java region, followed by the national institution affiliations such as central government agencies (BMKG, LIPI, Badan Geologi, etc). There are very few research affiliations come from Sumatra and Sulawesi, despite these two areas are prone to tsunamis and devastating tsunamis have occurred in their respective regions. No affiliations of Indonesian researchers were found from the Bali & Nusa Tenggara regions and the Maluku & Papua regions. This condition is due to the difference in the number of universities and research institutions, especially those concerned with disaster management issues, which are more common in the Java region than in the Sumatra and Sulawesi regions. Similar to the previous explanations, the Sumatra region is still the main focus for researchers from local univeristies in Indonesia, followed by national-wide and conceptual study category and then research about tsunami hazard in Java, Sulawesi, Bali & Nusa Tenggara, and Maluku & Papua.

IOP Conf. Series: Earth and Environmental Science 1245 (2023) 012042 doi:10.1088/1755-1315/1245/1/012042

Table 3. Top ten most influential main author about tsunami hazard assessment in Indonesia.

Author	Affiliation and country	TP[a]	TCP[b]	TC[c]	TC/TP[d]	TC/TCP[e]
M. Chlieh	California Institut of Technology, United States	1	1	344	344	344
C. Surbaya	National Coordinating Agency for Surveys and Mapping, Indonesia	1	1	298	298	298
D.H. Natawidjaja	Lembaga Ilmu Pengetahuan Indonesia, Indonesia	3	2	174	58	87
W. Hanka	GFZ German Research Center for Geosicences, Germany	1	1	137	137	137
Hill E.M.	Nanyang Technological University Singapore	1	1	118	118	118
EA. Okal	Northwestern University, United States	1	1	116	116	116
J. Lander	National Oceanic and Atmospheric Administration, United States	1	1	91	91	91
F. Løvholt	Norwegian Geotechnical Institute, Norway	2	2	83	41.5	41.5
B. Mustafa	Universitas Andalas, Indonesia	2	2	72	36	36
N. Horspool	Geoscience Australia, Australia	1	1	72	72	72

[a]TP = Total publication
[b]TCP = Total cited publication
[c]TC = Total citations
[d]TC/TP = Average citations per publication
[e]TC/TCP = Average citations per cited publication

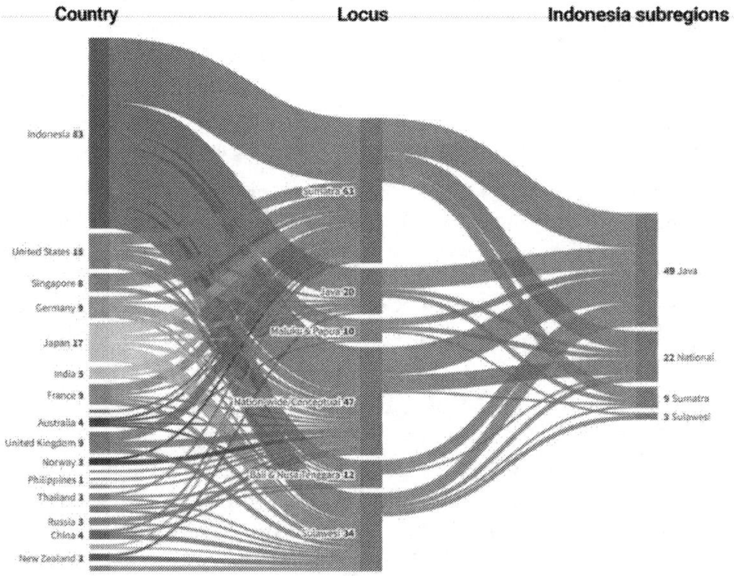

Figure 8. Sankey diagram showing the origin of researcher affiliation by country (left part), locus study (middle part), and origin of Indonesian researcher affiliation by Indonesia subregions (right part)

IOP Conf. Series: Earth and Environmental Science 1245 (2023) 012042 doi:10.1088/1755-1315/1245/1/012042

3.2. Science mapping

The science mapping process has successfully generated network maps of literature in tsunami-generation, in addition we also split the analysis within the network and temporal basis to better gained understanding of the evolution of produced knowledge in the topic within the years.

3.2.1. Network analysis. The overview of the network produced by VOS Viewer has generated the nodes of keywords and their related links within tsunami-generation field, which can be seen in Figure 9. The circle nodes' sizes symbolize the frequency of keywords occurrences, while the nodes that are closer together indicate a higher occurrence of the keywords appearing together, and the thickness of the link connecting them shows the significance of the co-occurrence. Four clusters of networks are found which consisted of central node which is tsunami (blue; 389 occurrences), earthquake (green; 424 occurrences), model (red; 116 occurrences), and parameter (yellow; 59 occurrences). Network analysis reveals the co-occurrence of keywords in a literature and effectively maps the strengths and traces the relationship structure in the theme. In general view, study about tsunami is very linked to earthquake hazard.

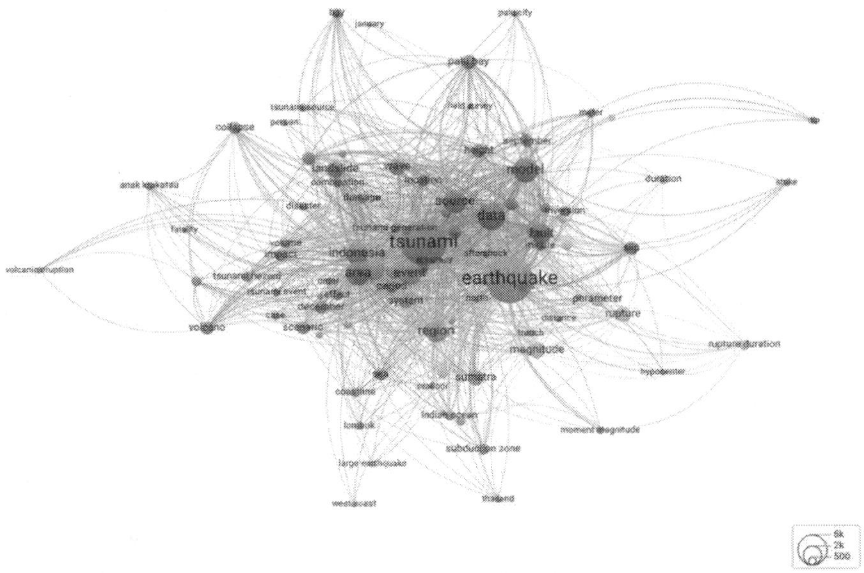

Figure 9. Keyword network colour representing the generated cluster from selected literature.

In pursuance of more detailed analysis, selecting the nodes that have greater linkage is used in successive analysis. Looking at Figure 10a of blue cluster, tsunami generated by volcano force is become the primary focus, as shown by the terms of volcano eruption, anak krakatau, flank collapse, and volcano itself. The picture in Figure 10b found that the yellow clusters have primary focus on parametric and characteristic of tsunami caused by earthquake, as shown by the terms of parameter, magnitude, duration, rupture, and direction. While the picture on Figure 10c which highlighted the green cluster composed of tsunami generated by earthquake in subduction zone, as shown by terms of subduction zone, Sumatra, and Indian ocean. The red cluster on Figure 10d emphasized the tsunami generated by submarine landslides, as shown by terms of submarine landslides, landslides, and palu bay. Roughly, the network analysis has succeeded to acquire the various source mechanisms of tsunami-generation study among the literatures. Although it fails to explain the various research methods used in the literature.

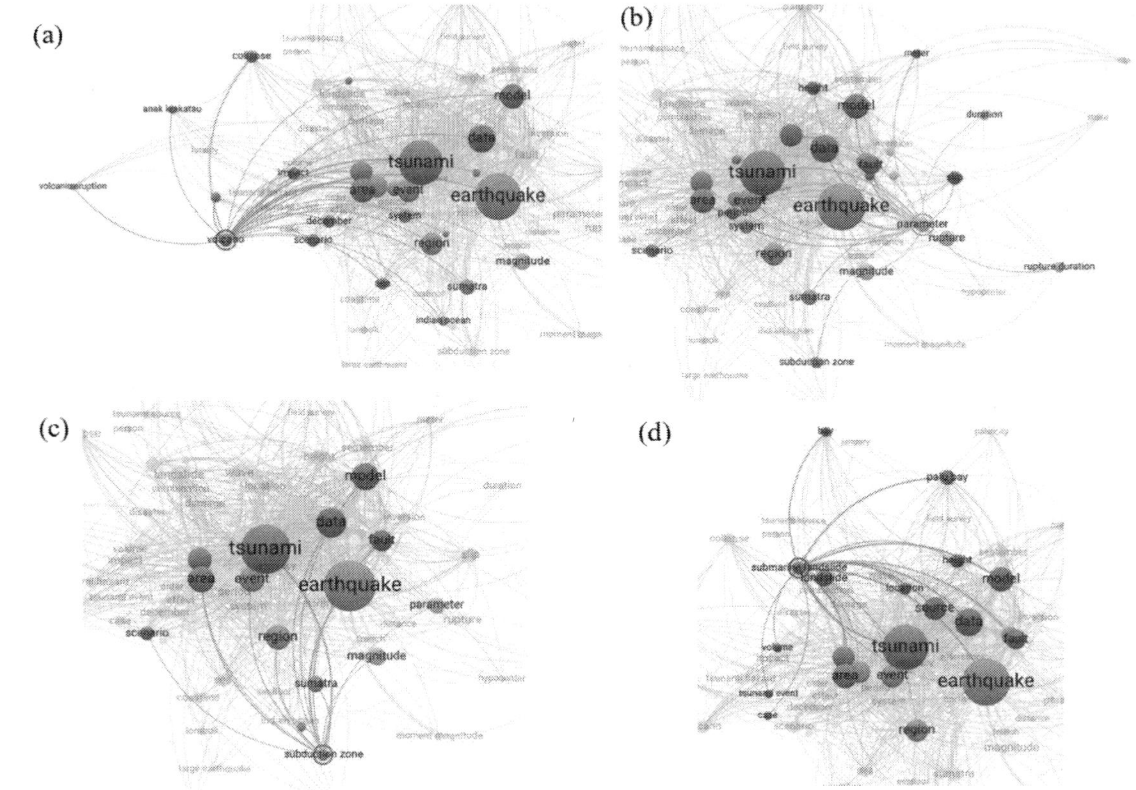

Figure 10. (a)-(d) Focused network keyword for different selected nodes.

3.2.2. Temporal analysis. The average number of literatures published per year and its earliest published literature shown in Figure 11 tends to focus on tsunami that generated by earthquake and located in Sumatra and Indian Ocean. Moving to the mid-2016s, research on tsunami-generation tend to focus on the parameter, model, and developing model. On the latest published literature, research tend to focus on the recent big tsunami event such as The Sunda Strait Tsunami and Palu Bay Tsunami in 2018. Other important topics that can be captured within this time are around the submarine landslides, volcanic-generated tsunami, and events around Anak Krakatau. In this section, topic dynamics within several time periods will be discussed to see the changes that occurs overtime within the tsunami-generation field.

Assessing popular topics within time periods requires several time markers to highlight the most used topic words between each period. For this, the majority of research on tsunami-generation has been examined on the past tsunami event that happen across Indonesian water. The three recent big tsunamis become the most studied topic with just below 30 literatures, which are: tsunami event of Palu 2018, Aceh 2004, and Krakatau (Sunda strait) 2018, respectively. Some literatures included more than one tsunami event as their studies, in order to conduct comparison of tsunami-generation mechanism or different tsunami event that occur on the same location.

For the temporal analysis, the most relevant markers are set for the top-3 major tsunami events in Indonesia shown in Figure 12. Thus, the first marker is set on the time of Indian Ocean Tsunami, which happened on the 26th of December 2004; the second is the Mentawai Earthquake and Tsunami on the 225th of October 2010; and the last were the Sunda Strait Tsunami and Palu Tsunami in 2018, which were happened on the 22nd of December and the 28th of September, respectively. Because all those events were occurred during the second half of their respective year, the year 2004, 2010, and 2018 are marked as the end of a period and their next years are counted as the start of the next period.

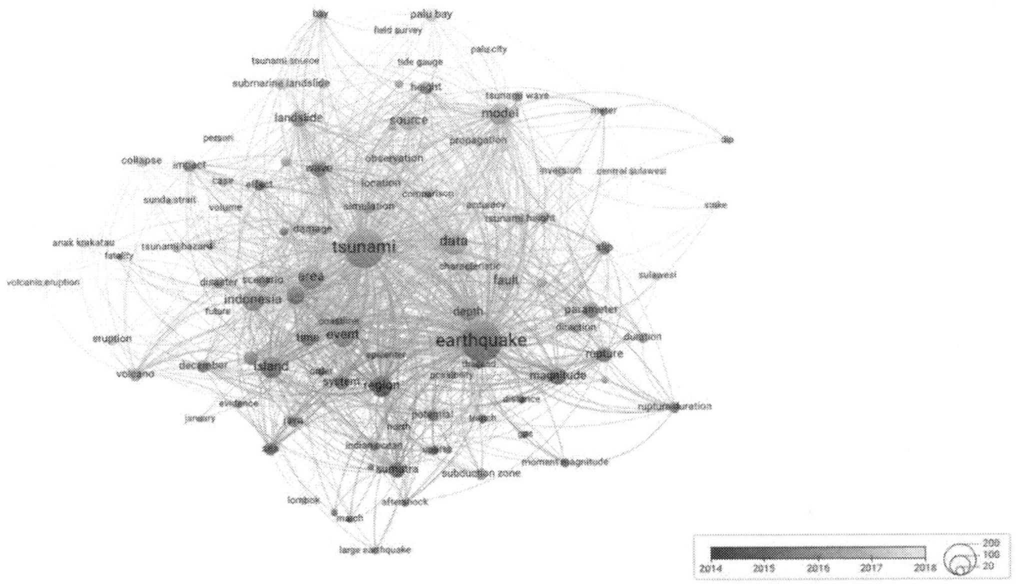

Figure 11. Keyword network color representing the average number of literatures per year.

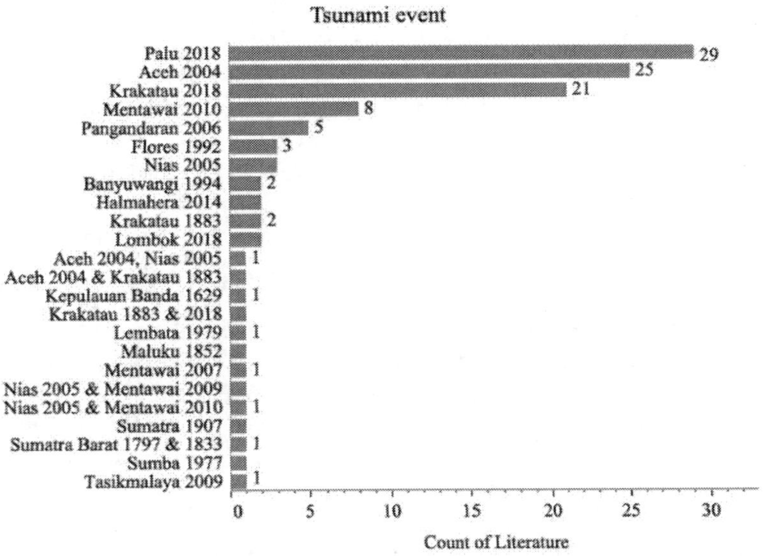

Figure 12. Number of literatures according to tsunami events (n=115).

Among 186 literatures, only 4 papers counted before 2004, which are only present in 1995, 2001, and 2003. Between 2005 and 2010, there are 26 papers counted, with 12 of them came from 2010 and and none from 2009. A total of 66 papers were counted within a period of 2011 until 2018 where the number of literatures varies annually, between 4 to 12 papers per year. From 2019 onwards, the number of papers were skyrocketed to a maximum of 39 in 2020 (Figure 4), with a total of 90 literatures counted to date.

For each of the period, word clouds are used to determine major topics discussed within each time. The series of figures below (Figure 13, Figure 14, Figure 15, Figure 16) shows the variety of word clouds over the course of four periods of time. Each of them has undergone several word filters, which ranging from the stop words until the common words in tsunami research. However, these common words filter are not generalized. For example, the word 'Tsunami' in the first period has been removed due to the very high frequency of the word compared to other words (Figure 13a). Thus, to help maximizing the visual of the other words, 'Tsunami' has been removed (Figure 13b). Similar case happened with the fourth period where 'Tsunami' in Figure 16a was removed, resulting in the word cloud in Figure 16b.

Apart from 'Tsunami', common words such as 'year', 'zone', 'part', 'area', 'new', 'main', 'period', 'data', 'due', 'meter', 'kilometer', 'condition', 'consist', 'high', 'low', 'result', 'occur', 'associate', 'value', 'include', 'including', 'region', 'along', cause', 'causing', 'go', 'event', 'different', 'carried', 'use', 'one', 'height', 'elevation', 'effort', 'involve', 'compute', 'water', 'sea', 'composite', all were eliminated from each word cloud due to its irrelevant meaning to the tsunami-generation topics. The name of months like 'September', or 'December' is also eliminated due to the same reason, along with numbers and directions such as 'two', and 'west'. In this case, we also make sure to eliminate any kinds of stop words [21] that might occur.

Before 2004 (Figure 13), tsunami-generation topic was not focused on any specific source mechanism. The most talked topic was about the main generations of tsunami in Indonesia [10][22][23][24], which are circling around earthquake and landslides. Publications of tsunami-generation during this time were inclined towards the kind of propagated waveforms generated from the source, and historical tsunami catalog. The inundation risk was also being discussed along with the generation topic, also the risk components in "Riangkroko" and other Sulawesi coast areas were analyzed. Until 2004, the highest frequency of historical tsunami events in Indonesia came from the Makassar Region [22], hence the large "Makassar" and "Strait" words in the picture.

Unlike the years before the Indian Ocean Tsunami of 2004, tsunami caused by "earthquake" was the most popular topics in the following years until 2010 (Figure 14). This topic was on par with the "tsunami" topic itself, since the 2004 tsunami was generated from fault ruptures over the Andaman Sea, near Sumatra Island [25][17][11]. From the figure, the tsunamigenic discussion was also extended to the Thai coastal area, covering topics such as the megathrust study, earthquake magnitude, the seafloor uplift, tsunami elevation, and effects to corals due to sedimentation. Discussions circling around tsunami risk from its propagation and inundation like the period before 2004 were less present, and so the comparative study to tsunamis around the world like those in Pacific or the Indian tsunamis itself.

After the 2010 Mentawai Earthquake and Tsunami, it is evident that both the two key terms in the previous period were still the focus in this period (Figure 15). Similar with the times before the Indian Ocean Tsunami takes place, the propagation waves and their parameters are one of the topics that were popular, while the discussions about earthquake's geological parameters were more becoming after the last period. During this period, there was also an introduction of the Indonesian Tsunami Early Warning System (TEWS) that was established in 2008 [26], and this time, the introduction of tsunami observation analysis using satellite imageries appeared and were becoming one of the popular topic discussions in Indonesia.

After being hit by two major tsunamis in Indonesia [27][28], earthquake-generated tsunamis are still the main topics to be discussed till date (Figure 16). The introduction of tsunami in Sulawesi [28] shows the major increase in the discussions about tsunami in Palu, one of the province capitals in Sulawesi. The liquefied gravity-flow induced tsunami gives "liquefaction" the same amount of attention as tectonic-generated tsunami. Due to the occurrences of both tsunami that are located in straits (Sunda Strait and Makassar Strait), the coastal areas are almost equally highlighted the "wave" discussion. The high number of literatures published might also be one of the tsunami awareness campaigns coming from of the establishment of Sendai Framework for Disaster Risk Reduction on the 14-18 March 2015 [29] and the UN GA 70/203 World Tsunami Awareness Day on the 5th of November 2015 [30]. Nevertheless, the graph in Figure 4 shows a spiking interest from people to learn more about "tsunami-generation" as an important part of tsunami hazard assessment in Indonesia.

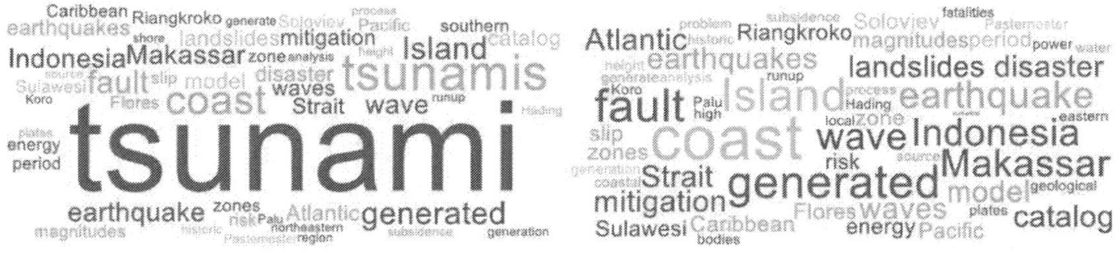

Figure 13. Word clouds that show the time period before 2004 with 'Tsunami' topic included (a) and without 'Tsunami' topic (b).

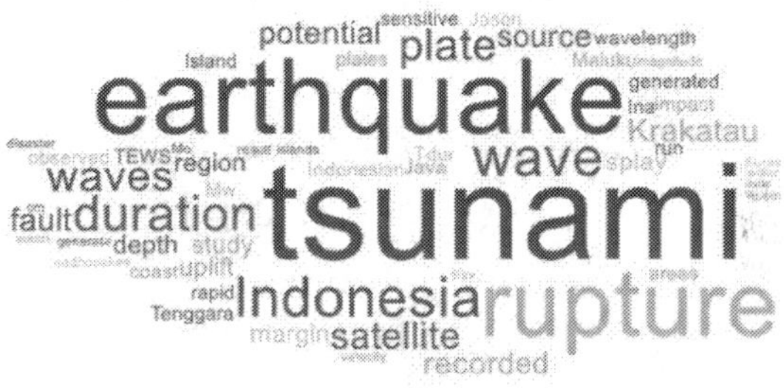

Figure 14. Word clouds that show the time period between 2005 and 2010.

Figure 15. Word clouds that show the time period between 2011 and 2018.

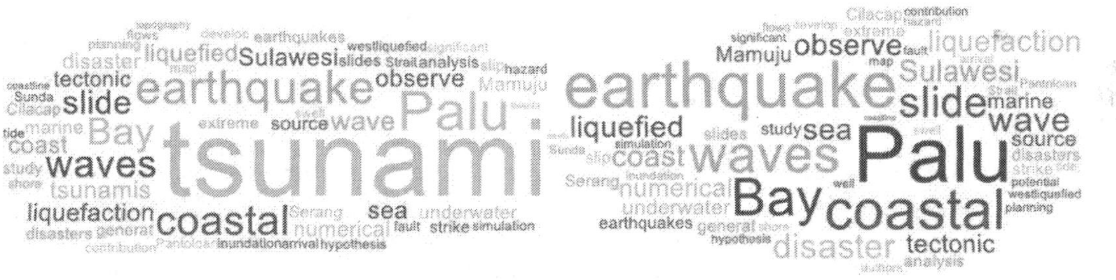

Figure 16. Word clouds that show the time period within 2019 and beyond with 'Tsunami' topic included (a) and without 'Tsunami' topic (b).

3.3. Tsunami-generation research framework

Since the tsunami phase context has been excluded from the discussion, the remaining contexts of time occurrence, monitoring and measurement approach, and hazard analysis approach are analysed along with the tsunami-generation force. These frameworks below are applicable for Indonesia, and the terms identification for each category will be discussed thoroughly.

3.3.1. Tsunami-generation force. As portrayed in the network analysis section, several tsunami-generation forces have been identified, which are: earthquakes, volcano eruptions, submarine landslides,

IOP Conf. Series: Earth and Environmental Science 1245 (2023) 012042 doi:10.1088/1755-1315/1245/1/012042

and flank collapse or the aerial landslides. In some literatures [2][31], there are two major classifications of tsunami-generation, which are typical and atypical tsunami.

The typical source of tsunami is the type of tsunami that is produced by tectonic earthquakes. In some literatures, tsunamis this type is mentioned as the tsunamigenic earthquakes [32][33][34]. Looking at Figure 17a, it is evident that this type of tsunami has been leading by 126 literatures, or 67.7% of all tsunami-generation literatures in this study. The atypical sources of tsunami, on the other hand, is classifies as landslides (submarine or aerial), atmospheric disturbance, and volcanic eruptions. Tsunamis originated from landslides are called tsunamigenic landslides [35], while tsunamis generated by atmospheric disturbance are called meteorological tsunami or meteotsunami [2][36], and those coming from volcanic eruptions are called the volcanogenic tsunami [37][38]. In total, these tsunamis constitute 60 literatures (32.26% of all literatures in this study), where volcano eruptions been studied within 28 literatures (46.67% of atypical literatures) or the most popular atypical tsunami studies within the field (Figure 17a). This is followed by submarine landslides by 20 literatures (33.33% of atypical literatures). The 'combination' findings in Figure 17a denotes the study of tsunami-generation that investigates one until three combinations of force, while the 'all' findings shows the literatures that study all the tsunami-generation forces.

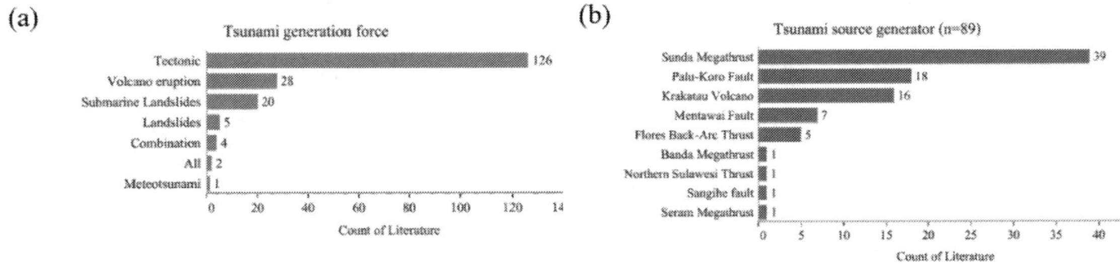

Figure 17. (a) number of literatures according to tsunami-generation force (n=186) and (b) according to tsunami source generator (n=89).

A thorough study for specific tsunami source-generator identifies 89 literatures with a specific tsunami-generating source (Figure 17b), while the other 97 literatures are mostly discussing either several tsunamis at once, focus on lessons learned for Indonesians from tsunamis outside of Indonesia [39][17], or the study of tsunami-generation in general. Eventually, the figure shows some source generators that are circling around the tectonic fracture zones (Figure 17b). Among those, Sunda Megathrust has been the most frequent source that are discussed among scholars by 39 literatures, which is followed by Palu-koro fault by 18 literatures. The third popular source studied among scholars is the Krakatau Volcano by 16 literatures. Apparently, Krakatau is the only volcano in Indonesia that is found to be studied for its tsunami-generating mechanism. Although meteotsunami is not frequent in Indonesia, some studies abroad may mentioned the meteotsunami as the result of typhoons [40] or tropical cyclones, which may possibly be happen over the northern and southern Indonesia due to the frequent cyclones.

It is also noteworthy to mention that the sources in Figure 17b may or may not be the factor that generates tsunami. Some of them, for example, the Palu-Koro fault in Sulawesi, has generated tsunamigenic landslides that creates wave-related hazard along the Palu Bay, Indonesia [41][28][19]. Note that the Palu-Koro fault is the location and eventually, source generator for the following tsunamigenic landslides, not the one that causes tsunami. Despite that, the findings in Figure 17b corroborates the dominance of tectonic fracture zones in generating tsunamigenic earthquakes in Indonesia (Figure 17a), followed by volcanic eruptions.

3.3.2. Time occurrence. Studying tsunami based on their time occurrence is important for scholars to understand which methodology to use to analyse the specific tsunami. The data for recent tsunami can easily be found from several data sharing platform (e.g., NGDC/WDS Global Historical Tsunami Database [42]), while for ancient tsunamis where the documentation of any sort is unable to be found, many methods such as radiocarbon dating are being used [34][43][44].

In general, there are two major classes of tsunami type based on the time occurrence, which are historical tsunami and paleotsunami. The historical tsunami denotes the tsunami that has been documented through eyewitness or instrumental observation within the historical record. In practice, historical tsunami ranges from several approaches: the manual documentation [45], which dated back from the first tsunami's written documentation from observation; and instrumental observation [46], which dated back from the 19th century onwards, from the time when the first automatic gauge recording was available. The paleotsunamis, on the other hand, are tsunamis occurred prior to the historical record or for which there are no written observations. To successfully identify this type of tsunami, scientists need to use specific measurement methods, one of which is the tsunami deposit dating [47][48][49] to obtain the paleotsunami characteristics.

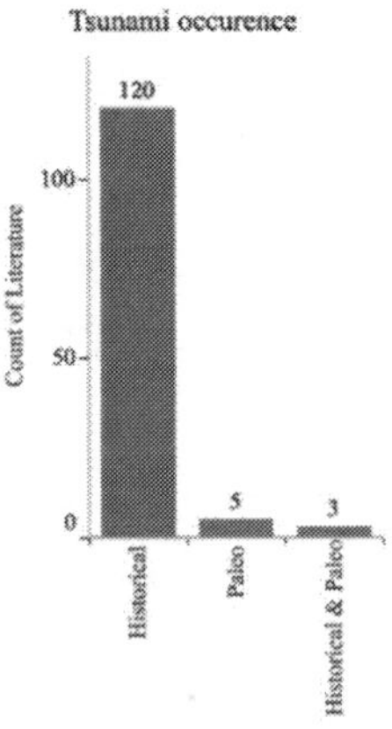

Figure 18. number of literatures according to tsunami occurrence (n=128)

Our study highlights the significant differences between research discussing historical tsunamis and paleotsunamis (Figure 18). As shown in the figure, there are 120 research publications that discuss historical tsunamis, with 5 research publications related to paleotsunamis, and 3 publications discussing both types. This finding indicates that there is a large knowledge gap between historical tsunamis and paleotsunamis. One reason might come from the fact that more recent tsunami events have advantages in the use of manual or instrumental observation [46], thus are easier to study due to the availability of sequential dataset. Meanwhile, the data needed to study paleotsunamis are more difficult to obtain due to the required field observation [34]. In addition, the recent major tsunami events that have occurred in Indonesia over the past two decades might have encouraged researchers to study historical tsunamis.

3.3.3. Monitoring and measurement approach. Identifying tsunami can be done using several methods or physical tools that function manually or automatically. These tools are used as monitoring purpose by measuring data and store them in the cloud, which could also be done in both manual and automatic modes. These tools are classified as mareograph, remote sensing, tsunami deposit mapping and tsunami deposit dating. From all the literatures, there are 68 that are specifically focus on monitoring and measurement approach (***Figure** 19*). Among these, remote sensing has been used by 41,18% of monitoring literatures, followed by tsunami deposit mapping by 23.53%, mareogram by 22.06%, and tsunami deposit dating by 11.75% of monitoring literatures.

As a tool to measure and monitor time series of sea level, mareogram has been around since 1830 [45] in the form of mechanical float and stilling gauges until this day with the advancement of seafloor bottom pressures. In Indonesia, studies using mareogram data has occurred since 2011, where all of them implementing gauges data in their study. An advanced example of a mareograph that is built specifically for rapid tsunamigenic earthquake detection is the Deep-ocean Assessment and Reporting of Tsunamis (DART) that records tsunami signal from bottom pressure recorder. This device*, which is widely known as the tsunameter,* is located in the rim close to subduction zones, and data transfer from the bottom to ground controller is done through the connected surface buoy and satellite receiver [50][51]. This real-time tsunami monitoring system has been positioned in strategic locations throughout the ocean, which includes the surrounding of Indonesian waters [52].

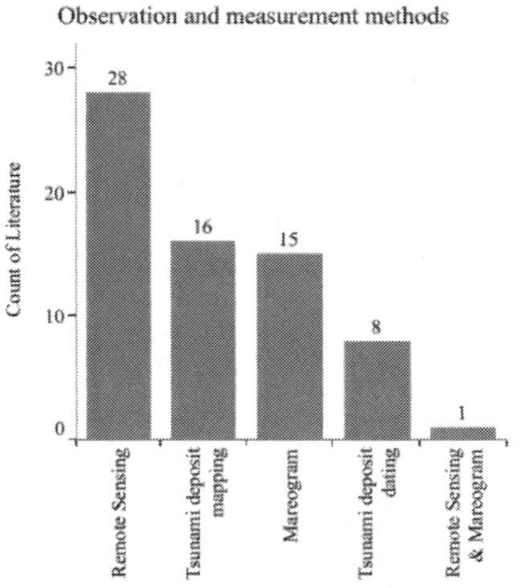

Figure 19. Number of literatures according to observation and measurement methods (n=68).

As early as the founding of mareograph, a remote observation in the form of aerial photography was first demonstrated in 1850 [46]. Ever since, remote sensing area have been developed in the forms of coastal high frequency radar for wave monitoring and ocean monitoring system using satellites. In Indonesia, remote sensing studies related to tsunami-generation has been around since 2007. The recent fruitful implementation of tsunami detection using remote sensing methodology is the use of Ionospheric Total Electron Content (TEC) disturbances data derived from Global Navigation Satellite System's (GNSS) ground-based stations. Using signal analysis, tsunami wave can be extracted to find the time and location of tsunami occurrence [53]. This GNSS-TEC method has also been tried out in Indonesia [54][52] to develop the Ionospheric Tsunami Power Index (ITPI) using data from nine tsunami events.

Unlike both mareogram and remote sensing techniques that can only be useful for modern tsunamis, the tsunami sediment deposit identification and mapping can be done to anticipate future tsunami-generator forces that may occur around a certain area. This method could work for both historical and paleotsunami data and usually done to anticipate earthquakes. Tsunami deposit mapping has been used several times since 2005 in Indonesia, as a general assessment of tsunamigenic earthquake using field surveys of surface ruptures, bathymetry mapping, and seismic-reflection survey [55]. In Indonesia, it has been demonstrated as a post-disaster survey of Palu Tsunami 2018.

As previously mentioned, paleotsunami assessment is usually conducted using tsunami deposit dating. This includes radiocarbon dating of deep-sea core sampling of sediment deposit, coral sample, and seashell sample analysis [43][44][56][47][48][49]. In Indonesian tsunami-generation literatures, a total of 8 literatures has been identified since 2013 regarding tsunami deposit dating in Indonesia, where the radiocarbon dating has been implemented to determine source mechanism and the time occurrence of ancient tsunamis.

3.3.4. Hazard analysis approach. In assessing tsunami-generation research, some approaches to analyse tsunami is used to understand source mechanism underlying the propagation of the tsunami itself. The analysis ranges from the general assessment and hazard mapping to a series of modelling approaches.

The general assessment counts any information related to tsunami-generation assessment in a general setting, which includes tsunami-generation characteristic analysis [48][24] using a set of known parameters from previous assessments, such as paleo tsunami study or a case-area study. The papers with a focus on the making of tsunami hazard map is considered as the hazard mapping category [57]. Between the modelling methods, empirical model comprises of specific estimation parameters using a set of statistical analysis. This includes the inverse estimation of source's parameters and estimation of parameters' change, such as earthquake's rupture length, speed reduction and travel time delay [58][59].

The numerical model, on the other hand, is using advanced mathematical model to simulate tsunami's parameter on a wider setting (two or three-dimensional, wider basin, detailed resolution, etc.). On the implementation, numerical model can be as simple as the Deterministic Tsunami Hazard Assessment (DTHA) [60] that only model one scenario using a fixed set of input parameters or can be as complicated and advanced as the Probabilistic Tsunami Hazard Assessment (PTHA) [18][10] that can model several scenarios at once. A travel time maps, coastal amplitude maps, velocity, and energy directivity map is one of the results to be expected from using PTHA.

A real-time model and forecast relate to the rapid but reliable model to provide real-time sea-level changes using data gathered from deep sea platform located around the subduction zones on the rim of the ocean basin [50]. Further, this connects to the discussion about tsunami early warning and early detection analysis. Another form of real-time forecast related to tsunamigenic landslide is the Dinsar application that can detect flank collapse before the occurrence of tsunami [61].

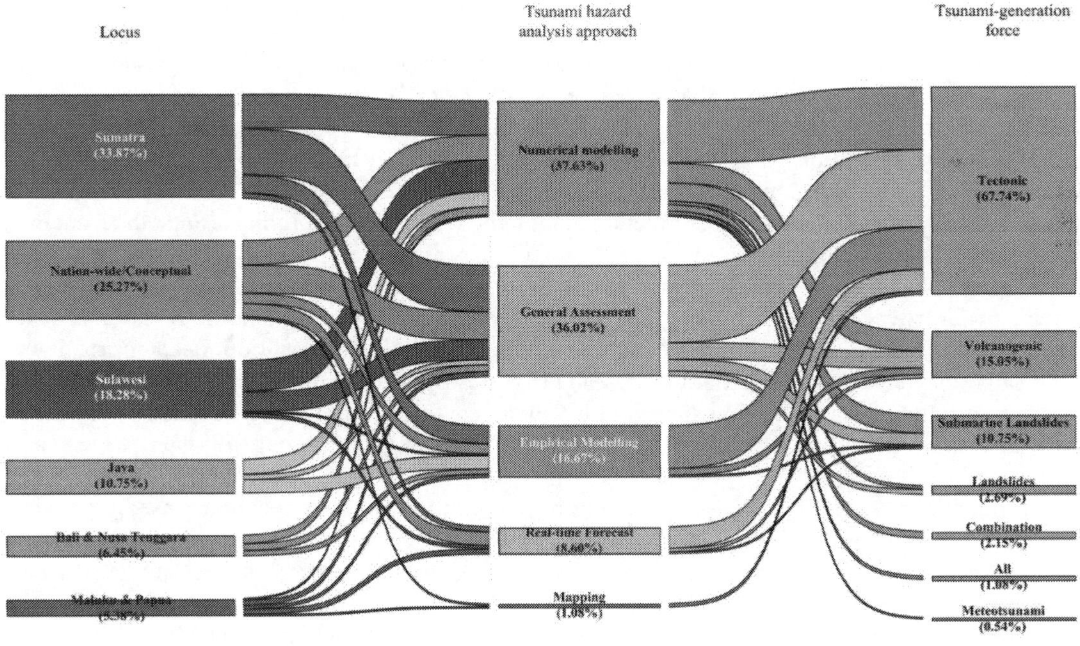

Figure 20. Sankey diagram of study location, tsunami modeling, and generation force (n=186).

On **Figure** 20, sankey diagram illustrates the distribution and relationship of research literature between studied location, tsunami modelling, and tsunami-generation force. Research that examined in Sumatra region tends to use general assessment of tsunami-generation studies, accompanied by numerical modelling approach. While research on national or conceptual level generally tends to use both approaches equally. In contrast, research on Sulawesi region tends to use more numerical modelling approach to study tsunami-generation compared to general assessment approach. In Java region, studies are more drawn to the approach of numerical modelling accompanied by empirical modelling, while research in Bali & Nusa Tenggara, Maluku & Papua is comparatively equal among tsunami modelling approach. Emphasizing on tsunami hazard analysis, all approaches are used to highly investigate tsunami-generation force by tectonic mechanism.

4. Conclusion

The generation of tsunami has been one of the important subjects within tsunami research as it holds the key to map and analyze the propagation and inundation area of a tsunami, which can later be the base of disaster management system and disaster risk reduction activities. From times, there has been some improvements in literatures of tsunami-generation, where 2020 marked the peak of publications in the field, by 39 publications. As a typical source of tsunami, tsunamigenic earthquakes have been the most studied articles where 67.7% of the literatures covering the subject, while the highly discussed atypical sources, volcano eruptions, has a lot of mentioning that covers 46.67% of all the atypical tsunami literatures in Indonesia. The next popular atypical studies are within the topic of submarine landslides which covers 33.33% of all atypical studies. Eventually, the focus study of the typical tsunami is mainly circling around the Sunda Megathrust (39 publications), while the three most popular generation sources for atypical tsunami were around Sumatra and Sulawesi regions (18 in Palu Koro Fault, 16 in Krakatau, and 7 in Mentawai Fault).

While these locations are known mostly for their historical tsunamis (shown by 120 publications within historical tsunamis against only 5 publications within paleotsunamis), more paleotsunamis in the regions also store some important tsunami-generation characteristics that can be traced around the megathrust area. The research review regarding paleotsunami was evident to be low in Indonesia and thus this gap can be reduced by increasing the interest and opportunities related to paleo research.

Over the eastern Indonesia, historical tsunami database from year 416 until 2018 shows as frequent tsunami appearances as those in the western areas, but our study shows that the tsunami-generation literatures linked to the locus were very low compared to those on the western side. As the information about tsunami generations and their cycles can raise some ideas regarding disaster management, conducting future studies over tsunami-generation in the eastern side of Indonesia is strongly encouraged to promote disaster risk reduction among the fellow citizens living around the tsunami-prone areas.

In Indonesia, the development of monitoring tools is always ever-progressing, especially in terms of the use of remote sensing and mareogram. One thing that is mostly escaped the researcher's sight are the monitoring and analysis of meteotsunami, which is due to their generation mechanisms that are yet to be fully understood. Considering the vast areas of atmospheric disturbance surrounding the northern and southern Indonesia that is constantly hit by cyclones, this field of research will be very interesting to be studied in the future as it can lessen the probability of misdiagnose the disaster, which can better the mitigation activities.

Further improvements have also been seen regarding the analyzing technique used by researchers. One of which to be highlighted is the use of GNSS TEC where this method has been shown to be very promising to support real-time forecasting of tsunami-generation in the future. However, due to the limitations of the number of GNSS ground-based stations, a less accurate tsunami source location and the probable tsunami propagation to the surroundings can only be achieved. This could be a challenge to Indonesian scholars to build more ground stations for tsunami early warning system purpose.

Over the decades, monitoring and analyzing tsunamis has been inseparable. It has been shown from our study that the development of both processes can still be developed further. The examples from the literature reviews have shown the valuable efforts scholars did to understand the tsunami generation and from those, increase the effectiveness of tsunami early warning system. Limitations may arise, but those are the more reason why these monitoring and analyzing technique needs to be further studied and developed.

References

[1] F. Dias and D. Dutykh, "DYNAMICS OF TSUNAMI WAVES BT - Extreme Man-Made and Natural Hazards in Dynamics of Structures," 2007, pp. 201–224.
[2] United Nations Educational Scientific and Cultural Organization, "Tsunami glossary. Fourth edition." UNESCO, 2019. [Online]. Available: http://hdl.handle.net/1834/42083
[3] I. Pranantyo, "Tsunami Hazard in Eastern Indonesia: Source Identification and Reconstruction for Historical Case Studies," 2020. doi: 10.25911/5e709973ede85.
[4] B. M. K. dan G. Pusat Gempabumi dan Tsunami, Kedeputian Bidang Geofisika, *Katalog Tsunami Indonesia per-wilayah Tahun 416-2018*. Badan Meteorologi Klimatologi dan Geofisika, 2019. [Online]. Available: https://cdn.bmkg.go.id/Web/Katalog-Tsunami-Indonesia-416-2018-per-Wilayah.pdf
[5] B. W. Mutaqin, F. Lavigne, D. S. Hadmoko, and M. N. Ngalawani, "Volcanic Eruption-Induced Tsunami in Indonesia: A Review," *IOP Conf. Ser. Earth Environ. Sci.*, vol. 256, no. 1, p. 12023, 2019, doi: 10.1088/1755-1315/256/1/012023.
[6] F. Løvholt *et al.*, "Tsunami risk reduction – are we better prepared today than in 2004?," *Int. J. Disaster Risk Reduct.*, vol. 10, pp. 127–142, 2014, doi: https://doi.org/10.1016/j.ijdrr.2014.07.008.
[7] A. N. Aini, A. H. Amahoru, A. Ridho, D. P. Adikarma, and R. Lukafiardi, "Indonesia Disaster Knowledge Update - November 2022: Research Publications about Tsunami Hazard Assessment in Indonesia," *Cerdas Antisipasi Risiko Bencana (CARI!)*, 2022. https://knowledge.caribencana.id/special_brief/brief-november-2022 (accessed Feb. 26, 2023).
[8] S. Kraus *et al.*, "Literature reviews as independent studies: guidelines for academic practice," *Rev. Manag. Sci.*, vol. 16, no. 8, pp. 2577–2595, 2022, doi: 10.1007/s11846-022-00588-8.
[9] M. J. Page *et al.*, "The PRISMA 2020 statement: An updated guideline for reporting systematic reviews," *Int. J. Surg.*, vol. 88, p. 105906, 2021, doi: https://doi.org/10.1016/j.ijsu.2021.105906.
[10] F. Imamura, E. Gica, T. Takahashi, and N. Shuto, "Numerical simulation of the 1992 Flores tsunami: Interpretation of tsunami phenomena in northeastern Flores Island and damage at Babi Island," *pure Appl. Geophys.*, vol. 144, no. 3, pp. 555–568, 1995, doi: 10.1007/BF00874383.
[11] P. Shearer and R. Burgmann, "Lessons Learned from the 2004 Sumatra-Andaman Megathrust Rupture," *Annu. Rev. Earth Planet. Sci.*, vol. 38, pp. 103–131, May 2010, doi: 10.1146/annurev-earth-040809-152537.
[12] J. Telford, J. Cosgrave, and R. Houghton, *Joint evaluation of the international response to the Indian Ocean tsunami: Synthesis Report*. London: Tsunami Evaluation Coalition (TEC), 2006.
[13] M. Chlieh *et al.*, "Coseismic Slip and Afterslip of the Great M_w 9.15 Sumatra–Andaman Earthquake of 2004," *Bull. Seismol. Soc. Am.*, vol. 97, Jan. 2007, doi: 10.1785/0120050631.
[14] C. Subarya *et al.*, "Plate-boundary deformation associated with the great Sumatra-Andaman earthquake," *Nature*, vol. 440, pp. 46–51, Apr. 2006, doi: 10.1038/nature04522.
[15] D. Natawidjaja *et al.*, "Source parameters of the great Sumatran megathrust earthquakes of 1797 and 1833 inferred from coral microatolls," *J. Geophys. Res.*, vol. 111, Jun. 2006, doi: 10.1029/2005JB004025.
[16] E. Hill *et al.*, "The 2010 Mw 7.8 Mentawai earthquake: Very shallow source of a rare tsunami earthquake determined from tsunami field survey and near-field GPS data," *J. Geophys. Res.*, vol. 117, Jun. 2012, doi: 10.1029/2012JB009159.
[17] E. A. Okal and C. E. Synolakis, "Far-field tsunami hazard from mega-thrust earthquakes in the Indian Ocean," *Geophys. J. Int.*, vol. 172, no. 3, pp. 995–1015, Mar. 2008, doi: https://doi.org/10.1111/j.1365-246X.2007.03674.x.
[18] N. Horspool *et al.*, "A probabilistic tsunami hazard assessment for Indonesia," *Nat. Hazards Earth Syst. Sci., 14, 3105–3122, 2014*, vol. 2, Nov. 2014, doi: 10.5194/nhess-14-3105-2014.

[19] M. Heidarzadeh, A. Muhari, and A. B. Wijanarto, "Insights on the Source of the 28 September 2018 Sulawesi Tsunami, Indonesia Based on Spectral Analyses and Numerical Simulations," *Pure Appl. Geophys.*, vol. 176, no. 1, pp. 25–43, 2019, doi: 10.1007/s00024-018-2065-9.
[20] R. Paris *et al.*, "Volcanic tsunami: a review of source mechanisms, past events and hazards in Southeast Asia (Indonesia, Philippines, Papua New Guinea)," *Nat. Hazards*, vol. 70, no. 1, pp. 447–470, 2014, doi: 10.1007/s11069-013-0822-8.
[21] P. Upadhyay, "Removing stop words with NLTK in Python." https://www.geeksforgeeks.org/removing-stop-words-nltk-python/
[22] G. Prasetya, W. de Lange, and T. Healy, "The Makassar Strait Tsunamigenic Region, Indonesia," *Nat. Hazards*, vol. 24, pp. 295–307, Nov. 2001, doi: 10.1023/A:1012297413280.
[23] W. Wisyanto, "Analisis Potensi Bahaya Tsunami di Pantai Selatan Jawa Timur melalui Identifikasi Kemiripan Bentuk Tapchan," *Alami J. Teknol. Reduksi Risiko Bencana*, vol. 8, no. 2, 2003.
[24] J. Lander, L. Whiteside, and P. Lockridge, "Two decades of global tsunamis 1982-2002," *Sci. Tsunami Hazards*, vol. 21, Jan. 2003.
[25] S. Vongvisessomjai and P. Suppataratarn, "2004 Sumatra Tsunami," *Songklanakarin J. Sci. Technol.*, 2005.
[26] Kementrian Sekretariat Negara Republik Indonesia, "Presiden Yudhoyono Resmikan Peluncuran InaTEWS," 2008. https://www.setneg.go.id/baca/index/presiden_yudhoyono_resmikan_peluncuran_inatews (accessed Feb. 26, 2023).
[27] S. Grilli *et al.*, "Modelling of the tsunami from the December 22, 2018 lateral collapse of Anak Krakatau volcano in the Sunda Straits, Indonesia," *Sci. Rep.*, vol. 9, Aug. 2019, doi: 10.1038/s41598-019-48327-6.
[28] H. Takagi, M. Pratama, S. Kurobe, M. Esteban, R. Aranguiz, and K. Bowei, "Analysis of generation and arrival time of landslide tsunami to Palu City due to the 2018 Sulawesi Earthquake," *Landslides*, vol. 16, Mar. 2019, doi: 10.1007/s10346-019-01166-y.
[29] U. Nations Office for Disaster Risk Reduction, "Sendai Framework for Disaster Risk Reduction 2015 - 2030," 2015. [Online]. Available: https://www.undrr.org/publication/sendai-framework-disaster-risk-reduction-2015-2030#:~:text=The Sendai Framework for Disaster,Investing in disaster reduction for
[30] United Nations, "World Tsunami Awareness Day 5 November," 2015. https://www.un.org/en/observances/tsunami-awareness-day (accessed Feb. 26, 2023).
[31] T. Srinivasa Kumar *et al.*, "Tsunami Early Warning Services BT - Social and Economic Impact of Earth Sciences," V. K. Gahalaut and M. Rajeevan, Eds. Singapore: Springer Nature Singapore, 2023, pp. 351–375. doi: 10.1007/978-981-19-6929-4_18.
[32] Y. Fauzi, Hartono, W. Kongko, and K. S. Brotopuspito, "A study on the potential of tsunamigenic earthquakes in Java Subduction Zones," *IOP Conf. Ser. Earth Environ. Sci.*, vol. 485, no. 1, p. 12051, 2020, doi: 10.1088/1755-1315/485/1/012051.
[33] S. Pribadi, A. Pagacancang, N. Puspito, and G. Ibrahim, "Characteristics of Earthquake-Generated Tsunamis in Indonesia Based on Source Parameter Analysis," *J. Math. Fundam. Sci.*, vol. 45, pp. 189–207, Jul. 2013, doi: 10.5614/j.math.fund.sci.2013.45.2.8.
[34] Aswan *et al.*, "Age determination of paleotsunami sediments around Lombok Island, Indonesia, and identification of their possible tsunamigenic earthquakes," *Earthq. Sci.*, vol. 30, no. 2, pp. 107–113, 2017, doi: 10.1007/s11589-017-0179-2.
[35] S. Brune, A. Y. Babeyko, C. Gaedicke, and S. Ladage, "Hazard assessment of underwater landslide-generated tsunamis: a case study in the Padang region, Indonesia," *Nat. Hazards*, vol. 53, no. 2, pp. 205–218, 2010, doi: 10.1007/s11069-009-9424-x.
[36] N. A. Akbar, E. Utami, W. S. Putro, Z. L. Dupe, A. Cahyadi, and H. Achiari, "A Preliminary Study of Meteotsunami Using Fuzzy Logic Algorithm over Sunda Strait, Indonesia," in *2020 3rd International Conference on Information and Communications Technology (ICOIACT)*,

2020, pp. 499–503. doi: 10.1109/ICOIACT50329.2020.9332144.
[37] R. Williams, P. Rowley, and M. Garthwaite, "Reconstructing the Anak Krakatau flank collapse that caused the December 2018 Indonesian tsunami," *Geology*, vol. 47, Aug. 2019, doi: 10.1130/G46517.1.
[38] G. Manimaran and A. Ravindran, "A Caution Forthcoming Volcanogenic Tsunami in Indian Ocean," *Res. J. Chem. Environ.*, vol. 11, pp. 29–35, Jul. 2007.
[39] F. Løvholt, D. Kühn, H. Bungum, C. Harbitz, and S. Glimsdal, "Historical tsunamis and present tsunami hazard in eastern Indonesia and the southern Philippines," *J. Geophys. Res.*, vol. 117, p. B09310, Sep. 2012, doi: 10.1029/2012JB009425.
[40] L. Shi, M. Olabarrieta, D. S. Nolan, and J. C. Warner, "Tropical cyclone rainbands can trigger meteotsunamis," *Nat. Commun.*, vol. 11, no. 1, p. 678, 2020, doi: 10.1038/s41467-020-14423-9.
[41] K. Pakoksung, A. Suppasri, F. Imamura, A. Cipta, A. Omang, and A. Muhari, "Simulation of the Submarine Landslide Tsunami on 28 September 2018 in Palu Bay, Sulawesi Island, Indonesia, Using a Two-Layer Model," *Pure Appl. Geophys.*, vol. 176, Aug. 2019, doi: 10.1007/s00024-019-02235-y.
[42] National Oceanic and Atmospheric Administration, "NGDC/WDS Global Historical Tsunami Database." https://www.ngdc.noaa.gov/hazard/tsu_db.shtml
[43] R. C. P., K. Rajendran, V. Andrade, and S. Srinivasalu, "Ages and relative sizes of pre-2004 tsunamis in the Bay of Bengal inferred from geologic evidence in the Andaman and Nicobar Islands: AGES AND SIZES OF PRE-2004 TSUNAMIS," *J. Geophys. Res. (Solid Earth)*, vol. 118, pp. 1345–1362, Apr. 2013, doi: 10.1002/jgrb.50122.
[44] D. Hilman, "SIKLUS MEGA-TSUNAMI DI WILAYAH ACEH-ANDAMAN DALAM KONTEKS SEJARAH," *J. Ris. Geol. dan Pertamb.*, vol. 25, p. 49, Jun. 2015, doi: 10.14203/risetgeotam2015.v25.107.
[45] National Oceanography Centre, "Tide gauge instrumentation - Tide gauges." https://web.archive.org/web/20150824020453/http://noc.ac.uk/science-technology/climate-sea-level/sea-level/tides/tide-gauges/tide-gauge-instrumentation
[46] Humboldt State Geospatial Online, "History of Remote Sensing." http://gsp.humboldt.edu/olm/Courses/GSP_216/online/lesson1/history.html (accessed Feb. 26, 2023).
[47] C. Rubin *et al.*, "Highly variable recurrence of tsunamis in the 7,400 years before the 2004 Indian Ocean tsunami," *Nat. Commun.*, vol. 8, p. 16019, Jul. 2017, doi: 10.1038/ncomms16019.
[48] J. Malik *et al.*, "Tsunami records of the last 8000 years in the Andaman Island, India, from mega and large earthquakes: Insights on recurrence interval," *Sci. Rep.*, vol. 9, p. 18463, Dec. 2019, doi: 10.1038/s41598-019-54750-6.
[49] B. Mutaqin *et al.*, "Evidence of Unknown Paleo-Tsunami Events along the Alas Strait, West Sumbawa, Indonesia," *Geosciences*, vol. 11, p. 46, Jan. 2021, doi: 10.3390/geosciences11020046.
[50] G. Mungov, M. Eblé, and R. Bouchard, "DART® Tsunameter Retrospective and Real-Time Data: A Reflection on 10 Years of Processing in Support of Tsunami Research and Operations," *Pure Appl. Geophys.*, vol. 170, no. 9, pp. 1369–1384, 2013, doi: 10.1007/s00024-012-0477-5.
[51] C. Meinig, S. Stalin, A. Nakamura, and H. Milburn, "Real-Time Deep-Ocean Tsunami Measuring, Monitoring, and Reporting System: The NOAA DART II Description and Disclosure," Jan. 2005.
[52] National Oceanic and Atmospheric Administration, "DART® (Deep-ocean Assessment and Reporting of Tsunamis)." https://nctr.pmel.noaa.gov/Dart/
[53] J. Liu *et al.*, "The Source Detection of 28 September 2018 Sulawesi Tsunami by Using Ionospheric GNSS Total Electron Content Disturbance," *Geosci. Lett.*, vol. 7, p. 11, Aug. 2020,

doi: 10.1186/s40562-020-00160-w.
[54] N. Widjajanti, B. Muslim, C. J. Kumalasari, and T. Ulma, "Ionospheric Tsunami Power Index (ITPI) Testing for Tsunami Detection," *Sci. Tsunami Hazards*, vol. 40, no. 4, pp. 345–354, 2021.
[55] D. Natawidjaja *et al.*, "The 2018 Mw7.5 Palu 'supershear' earthquake ruptures geological fault's multi-segment separated by large bends: Results from integrating field measurements, LiDAR, swath bathymetry, and seismic-reflection data," *Geophys. J. Int.*, vol. 224, Oct. 2020, doi: 10.1093/gji/ggaa498.
[56] J. Patton *et al.*, "A 6600 year earthquake history in the region of the 2004 Sumatra-Andaman subduction zone earthquake," *Geosphere*, vol. 11, Nov. 2015, doi: 10.1130/GES01066.1.
[57] R. Riyadi, "ANALISIS PETA BENTUK RUPA BUMI DALAM MENENTUKAN LOKASI DARI PENGARUH TSUNAMI DI KABUPATEN CILACAP," *IJTIMAIYA J. Soc. Sci. Teach.*, no. Vol 3, No 2 (2019): IJTIMAIYA : Journal of Social Science Teaching, 2019, [Online]. Available: https://journal.iainkudus.ac.id/index.php/Ijtimaia/article/view/6293/pdf
[58] N. Muhammad, W. Lestari, and F. Syaifuddin, "Analisa Struktur Regional Penyebab Gempa dan Tsunami Berdasarkan Anomali Gravitasi dan Dinamika Lempeng," *J. Geosaintek*, vol. 3, p. 75, May 2017, doi: 10.12962/j25023659.v3i2.2960.
[59] M. Madlazim, "Estimasi Durasi, Arah dan Panjang Rupture serta Lokasi-Lokasi Gempa Susulan Menggunakan Perhitungan Cepat," *J. Penelit. Fis. dan Apl.*, vol. 1, Dec. 2011, doi: 10.26740/jpfa.v1n2.p8-18.
[60] I. Pranantyo, M. Heidarzadeh, and P. Cummins, "Complex tsunami hazards in eastern Indonesia from seismic and non-seismic sources: Deterministic modelling based on historical and modern data," *Geosci. Lett.*, vol. 8, p. 20, May 2021, doi: 10.1186/s40562-021-00190-y.
[61] K. K. I Gede Boy Darmawan, Zelica Krismalia Manurung, Muhammad Nurul, Winona Putri Prihadita, "APLIKASI DINSAR UNTUK IDENTIFIKASI DEFORMASI PERMUKAAN GUNUNG ANAK KRAKATAU PADA PERISTIWA LONGSOR SEBELUM TSUNAMI SELAT SUNDA," *J. Geosaintek*, vol. 7, 2021.

Long-term trend and variability of ocean heat content in the Indonesian Maritime Continent

Subekti Mujiasih[*(1)(2)], **Mochamad Furqon Azis Ismail**[(1)(3)], **Abdul Basit**[(1)], **Herlina Ika Ratnawati**[(1)], **Rahaden Bagas Hatmaja**[(1)], **Johanis Dominggus Lekalette**[(1)]

[1]Research Center for Climate and Atmosphere, National Research and Innovation Agency (BRIN), Jl. Cisitu Sangkuriang, Bandung 40135, West Java, Indonesia

[2]GeoHydrodynamics and Environment Research (GHER), Department of Astrophysics, Geophysics and Oceanography (AGO), University of Liège, Sart-Tilman B5a, 4000 Liège, Belgium

[3]Humboldt Fellow, GEOMAR Helmholtz Centre for Ocean Research Kiel, Kiel 24148, Germany

*sube002@brin.go.id

Abstract. The evolution of 64 years of the ocean heat content for the upper 300 meters (OHC300) in the Indonesian Maritime Continent, viz the Banda Sea, the Sulu Sea, the Java Sea, and the Karimata Strait, was investigated using the combination of ocean and atmosphere reanalysis data. The study aims to understand its long-term trend variability and association with El Niño-Southern Oscillation (ENSO) and Indian Ocean Dipole (IOD). The result showed a positive trend in all basins except the Banda Sea. The most significant rise was the Sulu Sea. The IOD fluctuation was more pronounced in the Western part of Indonesia. In contrast, the ENSO variation was more apparent in the Eastern part of Indonesia. The coincided La Nina events and the negative IOD significantly increased OHC—conversely, the coinciding of the El Niño events and the positive IOD caused a decrease in OHC magnitude. The significant factor increasing OHC in the maritime continent is the ENSO fluctuation, followed by the IOD fluctuation.

Keyword: ocean heat content, Banda Sea, Sulu Sea, Java Sea, Karimata Strait

1. Introduction

The Earth's climate system has an energy imbalance of nearly 1 W m^{-2} (Trenberth et al., 2014). The Earth's energy imbalance has been mainly attributed to increased greenhouse gases (GHGs) and aerosols driven forward by anthropogenic activity during the industrial period. The GHGs and aerosols trap more radiation and hence, contribute to global warming (IPCC, 2014). The world's oceans absorb more than 90% of this extra heat, which causes an increase in the ocean heat content (OHC) and sea level rise (Trenberth & Fasullo, 2012). Because of the ocean's prominent role in global heat storage changes, the OHC dynamics significantly impact Earth's energy imbalance on interannual and longer periods.

One of the climate indicators that correctly indicates that the world is warming up is the OHC in the upper 300 m, hereafter OHC300 (Xue et al., 2012). The OHC300 averaged over the regional seas exhibits substantial decadal variability rather than consistent warming. In addition, trends in OHC300 over the past 50 years are spatially quite varied, showing significant effects of oceanic heat redistribution caused by changes in ocean circulation (Doney et al., 2007). The uneven spatial variations in the OHC300 in the tropical oceans will likely impact natural modes of variability and regional climate conditions, particularly in the Indonesian Maritime Continent (IMC). The IMC is a large continent in Southeast Asia that consists of the Indonesian Archipelago, straits, and marginal seas as Figure 1. Observations reveal that seasonal warming variability in the IMC controlled by the heat exchange at the air-sea interface (Ismail et al., 2023, Ismail & Karstensen., 2023). The previous study focused on the OHC derived from Argo data in west Sumatera compared to Indian Ocean Dipole (IOD) in the period 1999-2011 (Radjawane et al., 2015), OHC derived from temperature and salinity of HYCOM model in Lombok area and compared to El Niño-Southern Oscillation (ENSO) and IOD for the period 2011-2015 (Chairuasni et al., 2020).

Considering the role of the maritime continent in the global ocean circulation and limitation of study of OHC, in this study, we investigate the spatial and temporal characteristics of the decadal variability of the OHC300 and estimate the possible relationship and mutual correspondence of the oceanic and atmospheric components such as ENSO and IOD based on the OCEAN5 reanalysis outputs. This work attempts to explore the decadal variability of OHC300 in the IMC using a more extended temporal period from 1958 – 2021.

2. Data and Method

2.1. Data

2.1.1. OHC300. We used OHC300 from Copernicus data, the upper Global Ocean Heat Content (0-300m) from Reanalysis & Multi-Observations Reprocessing with the most prolonged period, January 1958 - December 2021 (64 years). The source was https://data.marine.copernicus.eu/product/GLOBAL_OMI_OHC_area_averaged_anomalies_0_300/description. The data was used to see the trend, the variability for 64 years and association with the IOD and the ENSO index. We chose the Sulu Seas, the Banda Seas, Karimata Strait, and the Java Sea in order to represent the Indonesian maritime continent area.

2.1.2. ENSO. The ENSO data originated from the National Oceanic and Atmospheric Administration (NOAA) (https://origin.cpc.ncep.noaa.gov/products/analysis_monitoring/ensostuff/ONI_v5.php). It shows a threshold of 3 months running mean of ERSST.v5 SST anomalies in the Niño 3.4 region (5°N-5°S, 120°-170°W) of +/- 0.5 degrees centered 30-year base periods updated every five years. If the threshold is more than 0.5 degrees, it means warm on the west coast of South America. While the threshold is below 0.5 degrees, it means cold on the west coast of South America (Huang et al., 2017)

To understand more detail, we included one example of El Niño and La Nina event in the year 1998 in Figure 1 (https://origin.cpc.ncep.noaa.gov/products/analysis_monitoring/ensocycle/ensocycle.shtml). During El Niño, the west coast of South America experienced an increased sea surface temperature (about 1-2 degrees above average), which is indicated by red (figure 1). On the contrary, during La Nina, the west coast of South America experienced a decrease in sea surface temperature (about 1-2 degrees below average), indicated by the blue color.

How about the impact on the Indonesia area? During El Niño, Indonesia experiences a decreased temperature (cold) and is dry. On the contrary, during La Nina, Indonesia experiences an increase in

SST (warm) and tends to be wet. Regarding this impact, we signed red for warm SST (during La Nina event) and green for cold SST (during the El Niño event) in Figure 3e.

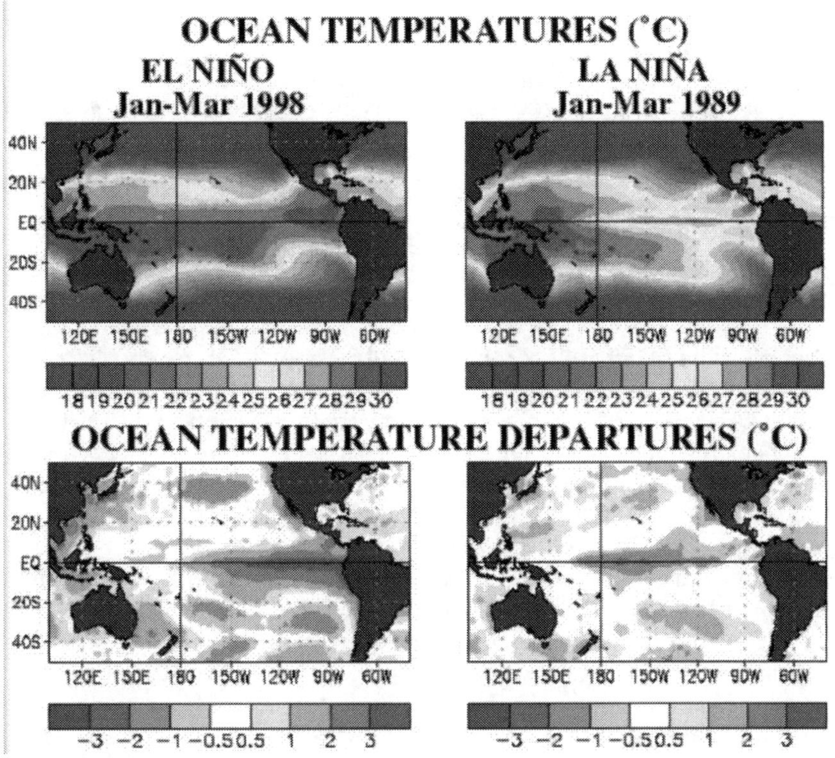

Figure 1. SST condition and the ENSO index during El Niño and La Nina event January – March 1998. (Source : https://origin.cpc.ncep.noaa.gov/products/analysis_monitoring/ensocycle/ensocycle.shtml)

2.1.3. IOD. Intensity of the IOD is represented by anomalous SST gradient between the western equatorial Indian Ocean (50E-70E and 10S-10N) and the south eastern equatorial Indian Ocean (90E-110E and 10S-0N). This gradient is named as Dipole Mode Index (DMI). When the DMI is positive then, the phenomenon is refereed as the positive IOD and when it is negative, it is refereed as negative IOD (https://psl.noaa.gov/gcos_wgsp/Timeseries/DMI/). To understand the effect of IOD on OHC300, we used IOD in the same period, January 1958 - December 2021, which originated from https://www.cpc.ncep.noaa.gov/products/international/ocean_monitoring/IODMI/DMI_month.html/. The data was calculated using ERSST.V5 (Huang et al., 2017). It was estimated from the subtraction between the western Indian Ocean (WTIO) and the south-eastern Indian Ocean (SETIO). WTIO is SSTA averaged in 50°E-70°E, 10°S-10°N. SETIO is SSTA averaged in 90°E-110°E, 10°S-0°N. Then, DMI is WTIO subtracted by SETIO. Positive and negative periods are based on a threshold of +/- 0.4°C for the DMI and centered 30-year base periods updated every ten years. Currently, the climatology used was from 1991-2020. A positive IOD means warm in the WTIO areas or cold in the SETIO. In other words, the western Indonesia area experiences a decrease in SST and tends to be dry. A negative IOD represents cold in the WTIO areas or warm in the SETIO areas. Western Indonesia area experiences an increase in SST and tends to be wet. Regarding the impact on the Indonesia areas, we signed cold (warm) in the SETIO by green (red) in Figure 3f.

2.2. Method

In this study, the OHC was computed using the following equation:

$$OHC = \rho C \int_0^z T(z)dz \quad (1)$$

Where ρ is the density of seawater (1026 kg m^{-3}), C is the specific heat of sea water (3990 J kg^{-1} K^{-1}), z is the depth limit of the calculation (300 meters), and T is the temperature at each depth (°K).

2.2.1. Trend

We analyzed a trend using a function trend of the Climate data toolbox (Greene et al., 2019) which calculates the linear trend of a data series by least squares. The sampling rate is 12 months. Because we want to know the trend per decade, we multiply the sampling rate by 10.

In this study, we investigate not only in time series but also in spatial. So, a spatial trend analysis OHC was calculated for all spatial grid points inside the area of interest (Fig. 2) over the entire data period from January 1958 to December 2021. While a long-term trend, we average the OHC of all grids every year for each area, namely Banda Sea, Sulu Sea, Java Sea and Karimata Strait (Fig 3). Note that a seasonal cycle in the data has been removed. The unit is Joule m^{-2} $decade^{-1}$ (Jm^{-2} $decade^{-1}$).

2.2.2. Time series analysis and association with the IOD and the ENSO using correlation.

We calculate the OHC anomalies each year by subtracting the long-term mean of OHC from OHC's mean per year. We applied it to each area, such as Banda, Sulu, Java, and Karimata Strait (Figure 3). The mean of OHC per year is the average of OHC from all grids of each area per year. At the same time, the long-term mean is the average of OHC for all grids over time (1958-2021).

Furthermore, we analyze the OHC anomalies associated with the IOD and the ENSO. Regarding IOD and the ENSO, we calculated the association using the Pearson correlation coefficient (https://nl.mathworks.com/help/matlab/ref/corrcoef.html). The correlation coefficient of two random variables measures their linear dependence. The value is between -1 to 1. A value of -1 means that the two variables have a perfect negative linear correlation, with 0 being no correlation and + 1 indicating a perfect positive correlation (Williams et al., 2020). If each variable has *N* scalar observations, then the Pearson correlation coefficient is defined as

$$\rho(A,B) = \frac{1}{N-1}\sum_{i=1}^{N}\left(\frac{A_i-\mu_A}{\sigma_A}\right)\left(\frac{B_i-\mu_B}{\sigma_B}\right) \quad (2)$$

Where μ_A and σ_A are the mean and standard deviation of A, respectively, and μ_B and σ_B are the mean and standard deviation of B. In our case, A can be the OHC anomalies, and B can be the IOD or the ENSO index.

Furthermore, we analyze the OHC anomalies associated with the coincided IOD and ENSO (El Niño and La Niña event).

2.2.3. Hovmoller diagram

To understand the evolution of OHC spatially, we used the Hovmoller diagram, such as time-longitude and time-latitude. Time-longitude enables us to see the development of the OHC pattern from one longitude to another over time. In comparison, time-latitude allows us to see the evolution of the OHC pattern from one latitude to another by time.

3. Result and Discussion

3.1. *Trend analysis*

3.1.1. *Analysis of spatial trend*

Figure 2 shows a spatial trend of OHC from 1958-2021 with the removed seasonal cycle. Four boxes represented the Banda Sea (red), the Sulu Sea (yellow), the Java Sea (green), and the Karimata Strait (white), respectively. The multiplier of the contour values is 10^7 Jm^{-2} decade^{-1}. The white contour means a spatial trend from 1958-2021 below -1. The dark blue contour contains a value between -1 and 1. The light blue contour represents values above 1 of the trends. The trend in the Banda Sea demonstrates a decreasing value. The Sulu Sea tends up to 1. The Karimata Strait tends to increase, as well as the Java Sea. The Karimata Strait is dominated by trend values ranging from -1 to 1. The Java Sea varies between -1 and more than 1. The Sulu Sea is huge variability about below -1 until above 1. The Banda is dominated by less than -1. The magnitude range of each area is quite similar to the long-term trend in the time series (Figure 3).

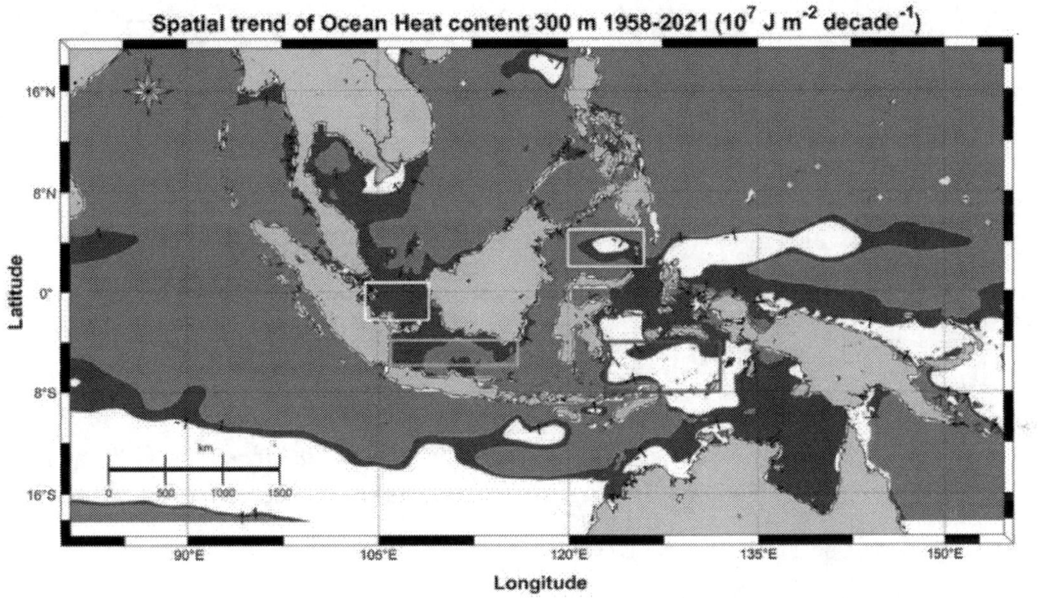

Figure 2. A rectangular colour indicates each concern area of the spatial trend of OHC300. Karimata Strait (white), Java Sea (green), Sulu Sea (yellow), and Banda Sea (red). White shape contour (< -1), dark blue (between -1 and 1), light blue (> 1).

3.1.2. *Analysis of long-term trend from time series*

Besides, Figure 3 also shows a trend value for each area displayed in the legend (black text). A positive trend of OHC300 covers all basins except the Banda Sea. The positive trend of the Sulu Sea, the Java Sea, and the Karimata Strait is 1.4693 x 10^7 Jm^{-2} decade^{-1}, 1.2877 x 10^7 Jm^{-2} decade^{-1}, and 0.21193 x 10^7 Jm^{-2} decade^{-1}, respectively. At the same time, the Banda Sea has a negative value (-1.6336 x 10^7 Jm^{-2} decade^{-1}), meaning that it experiences a decrease. The most significant rise is the Sulu Sea. Generally, the pattern of OHC300 of the Banda Sea and Sulu Sea are similar—likewise, Karimata Strait and Java Sea. Due to the adjacent locations, the Java Sea and the Karimata Strait are identical from 1997 to 2021.

3.2. Analysis of time series
Figure 3 shows the time series of the OHC300 from 1958-2021 for four areas: Banda Sea (a) and Sulu Sea (b), Java Sea (c), and Karimata Strait (d). Every time series has varied negative values (green) and positive values (red) with unit Jm^{-2}. However, a multiplier value is different depending on the calculation result and to plot the graph and analyses easier. A negative OHC and positive cold OHC mean cold and warm OHC, respectively. Figure 3a and Figure 3b show a similarity between warm and cold periods, namely warm years (1962-1980 and 2008-2014) and cold years (1980-2006). Whereas cold years in the Java Sea and Karimata strait (Figure 3c and 3d) took longer (1974-1998). The warm years are 1958-1974 and 1998-2021.

3.2.1. Correlation analysis
We calculated a correlation coefficient from the time series of OHC anomalies in Figure 3 and displayed it in **Table 1**. The table shows a correlation between the OHC300 anomalies and the ENSO index, the OHC300 anomalies and the IOD. The OHC and the ENSO index in east Indonesia, such as the Banda Sea and Sulu Sea, are more correlated than in west Indonesia, such as the Java Sea and Karimata Strait. On the opposite, the correlation between the OHC and IOD in East Indonesia is less correlated than the correlation in West Indonesia. That was reasonable because the area position, the more distance, the lesser effect. The most significant correlation with the ENSO index is the Sulu Sea, followed by the Banda Sea, Karimata Strait, and Java Sea. In comparison, the most considerable correlation with IOD is the Java Sea, followed by Karimata Strait, Sulu Sea, and Banda Sea, respectively.

Table 1. Correlation between OHC300 per area against the ENSO and the IOD

Area	Correlation	
	ENSO	IOD
Banda Sea	-0.6546	-0.1185
Sulu Sea	-0.6937	-0.1375
Java Sea	0.1	-0.4339
Karimata Strait	0.1962	-0.232

3.2.2. Analysis OHC versus ENSO
The Banda is theoretically influenced by ENSO fluctuation because the area is in eastern Indonesia, which is adjacent directly to the Pacific Ocean (Gordon & Susanto, 2001). The study shows that SST in the Banda Sea correlates well with the El Niño and La Niña events. The evidence of interannual SST variation has been observed in some areas of the Indonesian Seas such as in the Banda Sea (Gordon and Susanto, 2001). It is found that lower/higher SST is identified during the El-Nino/La-Nina periods. Inter-annual SST variability of the Indonesian waters are not only influenced by interannual variability of wind forcing but also interannual variability of thermocline temperatures (Gordon and Susanto, 2001). They suggested that the shallowing/deepening thermocline advected from the Pacific Ocean into the Indonesia Seas is the main factor in inducing lower/higher SST during the El-Nino/La-Nino. The shallowing/deepening thermocline is related to the upwelling/downwelling wave propagation from the Pacific Ocean scattering into the Indonesian Archipelago (Iskandar, 2010). Because the OHC is derived from sea temperature, the association between the OHC and the ENSO index exists. It was also valid for the Sulu Sea.
Fluctuation of OHC in the Banda and Sulu Seas tend to be interannual cycle. The OHCs in the Banda and Sulu Seas are colder than normal during the El Niño event. Meanwhile, during the La Niña event,

the Banda and Sulu Seas are warmer than normal. Figure 3a and Figure 3b show that during El Niño, the OHCs in eastern Indonesia is colder than normal such as 1958-1959, 1965-1967, 1969-1970, 1982-1983, 1987-1988, 1990-1995, 1997-1998, 2002-2005, 2014-2016, and 2019-2020. Generally, the El Niño event lasts 1-3 years, with a cycle every 2-5 years. A decreasing OHC in these areas followed the El Niño event cycle.
Conversely, these areas experienced an increasing OHC during the La Niña events. During the circumstances, eastern Indonesia is warmer than normal, namely 1962, 1964-1965, 1967-1968, 1971,1973-1976, 1978, 1981, 1984-1986, 1988-1989, 1996-1997, 1999-2001, 2007-2008, 2011-2013, 2017-2018, and 2021. The event lasts 1-4 years and every 1-5 years. An increasing OHC in these areas followed La Niña events. Furthermore, El Niño and La Niña generally influenced Indonesia in an interannual cycle (a more extended period).

Fluctuation of OHC in the Java Seas and Karimata Strait tends to be intra-annual until interannual. The OHCs in the Java Sea and Karimata Strait are warmer than normal during the El Niño event. Meanwhile, during the La Niña event, the Java Seas and Karimata Strait are colder than normal. Figure 3c and Figure 3d show that during El Niño, the OHCs in western Indonesia is warmer than normal such as 1958-1959, 1965-1967, 1969-1970, 1982-1983, 1987-1988, 1992, 1997-1998, 2002-2003, 2009-2010, 2015-2016, and 2018-2020. Generally, the El Niño event lasts 1-3 years, with a cycle every 2-5 years. An increasing OHC in these areas followed the El Niño event cycle.
Conversely, these areas experienced a decreasing OHC during the La Niña events. During the circumstances, western Indonesia is colder than normal, namely 1962, 1964-1965, 1967-1968, 1971,1973-1976, 1978, 1981, 1984-1986, 1988-1989, 1996-1997, 1999-2002, 2006-2008, 2010-2012, 2017-2018. The event lasts 1-4 years and every 1-5 years. A decreasing OHC in these areas followed La Niña events. Furthermore, El Niño and La Niña generally influenced Indonesia in an interannual cycle (a more extended period).

3.2.3. Analysis OHC versus IOD

As known, SETIO, including the Indian Ocean, is connected to the Java Sea and Karimata Strait through the Sunda Strait. The connection contributes to the effect of IOD fluctuation in these areas. If positive IOD, the Java Sea and Karimata Strait are colder than the climatology. If negative IOD, the Java Sea and Karimata Strait are warmer than the climatology. Figure 3f shows that, generally, the duration of the IOD event (positive and negative) was 4-6 months (intra-seasonal), repeating every three months. During negative IOD, the Java Sea and Karimata Strait were warmer than the average, such as 1958-1960, 1964, 1983, 1989, 1991, 1998, 2016, and 2021. In contrast, during the positive IOD, the Java Sea and Karimata Strait were colder than the average, such as 1961, 1967, 1973, 1982,1991, 1994, 1998, 2003, 2007, 2015, and 2019.

However, the effect of IOD negative/positive on the Banda and Sulu Seas seems slightly correlated to the Java Sea and Karimata Strait. Only IOD, about 1 or -1, will have the opposite effect on these areas.

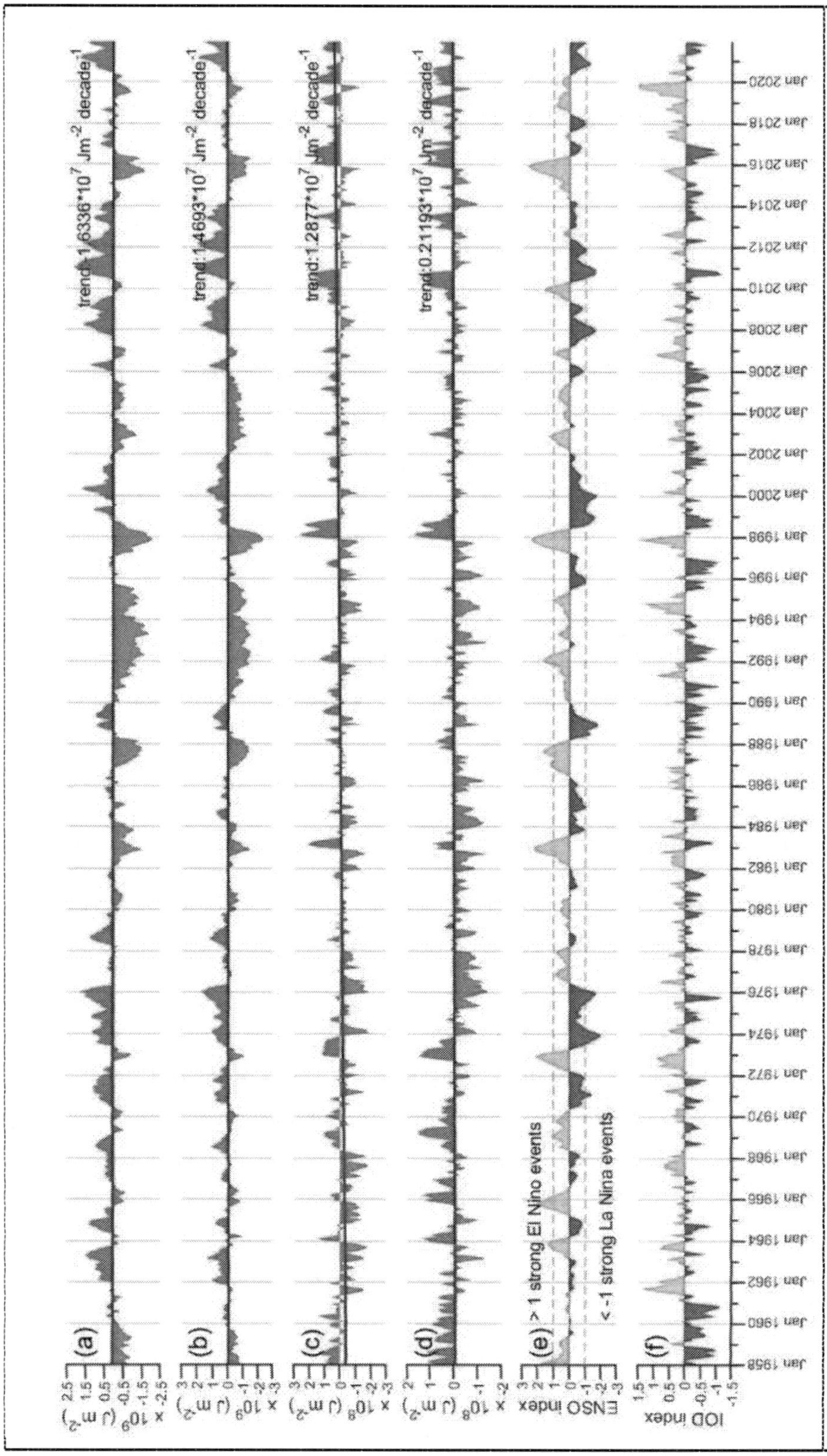

Figure 3. Comparison of time series of OHC300 for each area against the ENSO and IOD and the trend.

3.3. *Hovmoller diagram*

3.3.1. The association between the OHC against the coincided IOD and ENSO (El Niño and La Niña event).

We applied a time-latitude analysis of the OHC anomaly to investigate the impact of a coincidence ENSO with the IOD fluctuation and OHC. Figure 4 shows interannual variability along latitude for the period 2002-2021. We only display the last 20 years to analyse quickly, meaning that spatial variability in a short period can be seen clearly. Previously, we hypothesized that the possible warmest OHC happened when the negative ENSO (La Niña) coincided with the negative IOD event. From Figure 3, the coincided event between the La Niña and the Negative IOD is 2011, 2013, 2017, and 2021. Compared to Figure 4, the warmest is the year 2011, during the La Niña event coincided with the Negative IOD, although it did not cover the whole months of 2011. It occurred in all areas, with eastern Indonesia being more influenced than western Indonesia. Although the magnitude of negative IOD is quite large, if the magnitude of the ENSO negative is small, the impact only happened in the Java Sea and Karimata Strait. In addition, during the negative IOD (2011, 2013, 2017, and 2021), as in Figure 4, an increasing OHC lies on the north Karimata Strait. We suggested it is the impact of La Niña.

In contrast, we argued that this coldest OHC occurred due to the influence of the period of positive ENSO (El Niño) and positive IOD events. Figure 3 shows the coincided event in 2004, 2007, 2015, and 2019. The coldest was in 2015 during the El Niño coincided with the Positive IOD. Same as above, the Karimata Strait does not experience much because the magnitude of the Positive IOD is relatively tiny. The other coincided event is 2019; the magnitude of positive IOD is quite significant, and the positive ENSO value is relatively small, but the OHC in Karimata Strait is still warm. The situation's impact is only in Banda, Sulu, and Java. The positive IOD value is relatively high. However, it did not have a strong influence. We conclude that IOD significantly affects the Java Sea and Karimata Strait. The ENSO influenced the Banda Sea and Karimata Strait. However, when they coincide, the ENSO event is more controlling than the IOD.

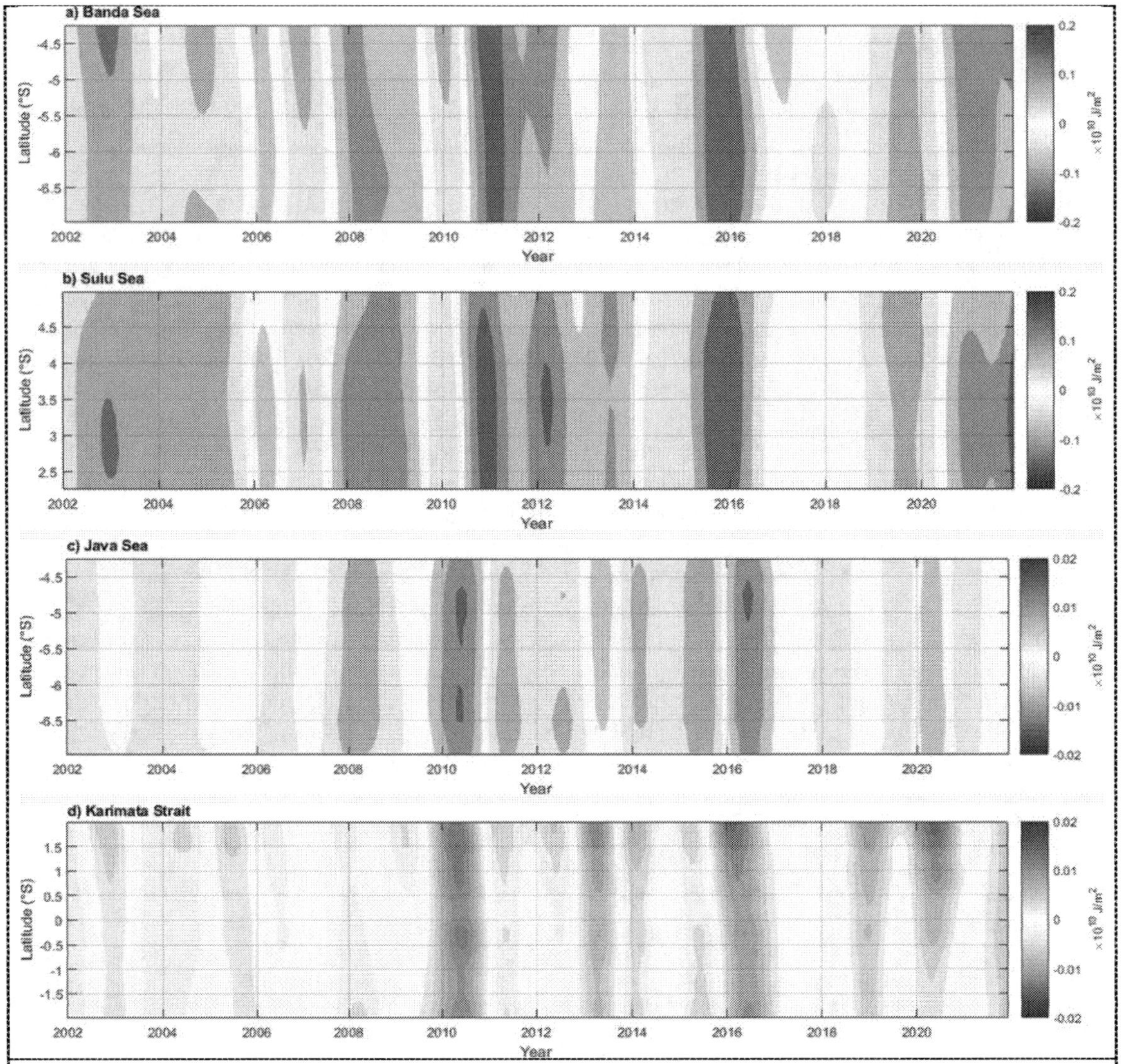

Figure 4. Anomaly of OHC by a longitude averaged. OHC anomaly based on period from 1958 to 2021 (64 years).

4. Conclusion

Based on analysis above, we conclude that the trend of both spatial and time series shows a similar pattern which is a positive trend in all basins except the Banda Sea. The most significant rise was the Sulu Sea. The IOD fluctuation more influenced west Indonesia. In contrast, the ENSO variation more influenced east Indonesia. The coincided La Niña event and Negative IOD significantly increased OHC—conversely, the coinciding of the El Niño event and the Positive IOD caused a decrease in OHC magnitude. The significant factor increasing OHC in the maritime continent is the ENSO fluctuation, followed by the IOD fluctuation.

Acknowledgment

The OHC300 data was downloaded from the Copernicus Climate Change Service (C3S) Climate Data Store (https://doi.org/10.24381/cds.67e8eeb7). Subekti Mujiasih was supported by the University of Liege. Mochamad Furqon Azis Ismail was supported by the Alexander von Humboldt-Stiftung. Subekti Mujiasih, Mochamad Furqon Azis Ismail and Rahaden Bagas Hatmaja contributed as the main contributor. Abdul Basit, Herlina Ika Ratnawati, and Johanis Dominggus Lekalette contributed as associate contributor to this paper.

References

[1] Trenberth K E, Fasullo J T and Balmaseda M A 2014 Earth's Energy Imbalance *Journal of Climate.* **27 (9)**, 3129–3144. https://doi.org/10.1175/JCLI-D-13-00294.1

[2] IPCC 2014 *The Physical Science Basis*: Working Group I Contribution to the Fifth Assessment Report of the Intergovernmental Panel on Climate Change.

[3] Trenberth K E and Fasullo J T 2012 Tracking Earth's Energy: From El Niño to Global Warming. *Surveys in Geophysics*, **33(3)**, 413–426. https://doi.org/10.1007/s10712-011-9150-2

[4] Xue Y, Balmaseda M A, Boyer T, Ferry N, Good S, Ishikawa I, Kumar A, Rienecker M, Rosati A J and Yin Y 2012 A comparative analysis of upper-ocean heat content variability from an ensemble of operational ocean reanalyses *Journal of Climate.* **25(20)**, 6905–6929. https://doi.org/10.1175/JCLI-D-11-00542.1

[5] Doney S C, Yeager S, Danabasoglu G, Large W G and McWilliams J C 2007. Mechanisms Governing Interannual Variability of Upper-Ocean Temperature in a Global Ocean Hindcast Simulation. *Journal of Physical Oceanography,* **37(7)**, 1918–1938. https://doi.org/10.1175/JPO3089.1

[6] Ismail M F A and Karstensen J: *Seasonal variation in the barrier layer of the Banda Sea*, EGU General Assembly 2023, Vienna, Austria, 24–28 Apr 2023, EGU23-7758, https://doi.org/10.5194/egusphere-egu23-7758, 2023.

[7] Ismail M F A, Karstensen J, Ribbe J, Arifin T, Chandra H, Akhwady R, Yulihastin E, Basit A, and Budiman A S 2023 Seasonal mixed layer temperature and salt balances in the Banda Sea observed by an Argo float. *Geoscience Letters,* **10(1),** 10. https://doi.org/10.1186/s40562-023-00266-x

[8] Levitus S, Antonov J, and Boyer T 2005 Warming of the world ocean, 1955–2003. *Geophysical Research Letters*, **32(2).** https://doi.org/https://doi.org/10.1029/2004GL021592

[9] Levitus S, Antonov J I, Boyer T P, Baranova O K, Garcia H E, Locarnini R A, Mishonov A V, Reagan J R, Seidov D and Yarosh E S 2012 World ocean heat content and thermosteric sea level change (0–2000 m), 1955–2010. *Geophysical Research Letters*, **39(10).**

[10] Radjawane I M, Bernawis L I, Priyono B, Fadli M and Putuhena H S 2015 Interannual variation of ocean heat content in outer Indonesian waters in warming ocean (Case study: West Sumatra waters). *AIP Conference Proceedings*, 1677(August). https://doi.org/10.1063/1.4930695

[11] Chairuasni L N, Ramadhan F L, Bernawis L I, Rachmayani R, and Putri M R 2020 The Ocean Heat Content of Lombok Strait Water Masses in 2011 - 2015 *IOP Conference Series: Earth and Environmental Science* 618 (1) https://doi.org/10.1088/1755-1315/618/1/012014

[12] ERSST.V5 Huang B, Peter W Thorne et al 2017: Extended Reconstructed Sea Surface Temperature version 5 (ERSSTv5), Upgrades, validations, and intercomparisons. *J. Climate.*

[13] Greene C A, Thirumalai K, Kearney K A, Delgado J M, Schwanghart, W, Wlfenbarger N S, Blankenship D D 2019. The Climate Data Toolbox for MATLAB. *Geochemistry, Geophysics, Geosystems*, *20*(7), 3774–3781. https://doi.org/10.1029/2019GC008392

[14] Williams, B., Halloin, C., Löbel, W., Finklea, F., Lipke, E., Zweigerdt, R., & Cremaschi, S.

(2020). Data-Driven Model Development for Cardiomyocyte Production Experimental Failure Prediction. In *Computer Aided Chemical Engineering* (Vol. 48). https://doi.org/10.1016/B978-0-12-823377-1.50274-3

[15] Gordon, A. L., & Susanto, R. D. (2001). Banda Sea surface-layer divergence. *Ocean Dynamics*, *52*(1), 2–10. https://doi.org/10.1007/s10236-001-8172-6

[16] Iskandar I 2010 Seasonal and interannual patterns of sea surface temperature in Banda Sea as revealed by self-organizing map *Continental Shelf Research*, *30*(9), 1136–1148. https://doi.org/10.1016/j.csr.2010.03.003

Mapping of Tsunami Evacuation Routes and Shelters in Batukaras Village, Pangandaran Regency, Indonesia

Farras Rizqy Hafizh[1], Berto Noven Marsekal[2], Candida Aulia De Silva Nusantara[3], and Wiwin Windupranata[3]

[1]Master Program of Geodesy and Geomatics Engineering, Institut Teknologi Bandung, Jalan Ganesha 10, Bandung 40132, Indonesia
[2]Study Program of Geodesy and Geomatics Engineering, Institut Teknologi Bandung, Jalan Ganesha 10, Bandung 40132, Indonesia
[3]Hydrography Research Group, Institut Teknologi Bandung, Jalan Ganesha 10, Bandung 40132, Indonesia

Corresponding email: farrasrizqyhafizh@outlook.com

Abstract. Batukaras is a village located on the Southern Coast of Java Island. The village is part of Pangandaran Regency, one of the top tourist destinations in West Java. Every year, the village attracts national and international tourists alike to Batukaras beach, famous as a surfing destination. However, the village has seen records of tsunami as recently as 2006, where a tsunami from a 7.7 magnitude earthquake struck the village and caused the death of hundreds and economic losses valued at millions of rupiahs. A series of activities aimed to increase disaster preparedness in the village has been done ever since. Recently, a team from ITB Research and Community Service Institute (LPPM) has made an effort to identify IOC-UNESCO Tsunami Ready indicators in Batukaras Village. Part of the effort is to identify tsunami evacuation routes in the village. Field observation was done to locate evacuation shelters and signages around the village, and the data is presented in the form of a tsunami evacuation map. Our study resulted in a successful mapping of evacuation routes and shelters in Batukaras Village. Furthermore, the data can be used towards the fulfillment of the IOC-UNESCO Tsunami Ready indicators in Batukaras Village.

Keywords: Batukaras, Tsunami, Disaster preparedness, Hazard

1. Introduction

Batukaras Village is an administrative unit located in Cijulang District, Pangandaran Regency, West Java Indonesia. According to data from Statistics Indonesia, Pangandaran Regency is the most visited tourist destination in West Java. In 2021, the regency attracted 3,604,128 national and international tourists [1]. The village is located in an enclosed bay facing the Indian ocean. Tourism is one of the main pillars of the local economy, alongside fishery and agriculture. Due to its unique geographical location, Batukaras beach is popular among surfers compared to other beaches in Pangandaran Regency. Coastal ecotourism such as mangrove tourism is part of the wider efforts conducted by the government of Pangandaran Regency to empower the local economy [2]. The agriculture and fishery being prominent industries in Batukaras' economy contributing of up to 26,73% in Gross Domestic Product of Pangandaran [3].

On July 17th, 2006, a tsunami earthquake of 7.7 magnitude occurred on the south of Java, resulting in a large tsunami striking the southern coast of Java. More than 600 deaths were observed and more than 75,000 people were displaced in the days following the event [4]. Batukaras Village was one of the villages affected in the disaster. Relatively few people felt the shake of the earthquake in the coastal areas. At the time there were no tsunami warning systems or evacuation signages in place, when the tsunami occurred there was a lack of public awareness [5]. In the end, this event becomes an important reminder of the need for disaster mitigation and preparedness in Batukaras Village community. Since the event, the local government in Pangandaran has enacted the Community-based Settlement Rehabilitation Project (CSRRP), aimed not only for community recovery but also resilience [6].

Tsunami evacuation shelters have emerged as one of the mitigation efforts in the event of a tsunami disaster in several cities of Indonesia [7]. Such shelters must consider several factors such as the topography, tsunami evacuation route, and maximum inundation height [8]. Recent efforts carried out by local authorities has resulted in the installation of evacuation route signs leading to evacuation shelters placed strategically around the village [9]. There are five temporary evacuation shelters and one permanent shelter located in the vicinity of Batukaras beach. The temporary shelters serve as the earliest point of evacuation in the event of a tsunami, increasing survival rate during disaster [10]. After the initial inundation of tsunami, subsequent recovery efforts are directed towards the final evacuation shelter. These shelters are located in elevated topography around the beach.

However, there has been no significant previous effort to identify the locations of existing evacuation route signs and evacuation shelters in Batukaras Village. Therefore, as part of an effort to identify IOC-UNESCO Tsunami Ready indicators in Batukaras Village done by ITB Research and Community Service Institute (LPPM) team, mapping of tsunami evacuation routes and public shelters was conducted in July 2022. In this study, identification of routes, signage, and shelters was obtained from field observation and presented in a tsunami evacuation map for Batukaras Village. The work presented provides a new approach in the planning of evacuation routes by using network analysis method, providing a more efficient approach in support of tsunami mitigation efforts compared to one without spatial analysis.

2. Methodology

In order to conduct mapping of tsunami evacuation routes and shelters in Batukaras, first data collection in the field must be done. A three-day visit to the village was materialized in July 2022 to gather data of existing evacuation shelters in Batukaras Village. The location and condition of each evacuation shelter was recorded using smartphone geotagging applications. Tracking of evacuation signage located in Batukaras village was also conducted during the visit.

A tsunami evacuation simulation by foot was done in order to determine the travel time from designated evacuation starting points to the shelters. The starting points were determined according to one of two criteria: points farthest from the final evacuation point that is within the inundation zone, and/or points that are of high tourist activity in the village. The visit was part of an ongoing effort by ITB Research and Community Service Institute (LPPM) to support Batukaras' attempt to become IOC-UNESCO recognized Tsunami Ready Community.

After the field data acquisition has been conducted, for each evacuation starting point the optimum destination shelter can be determined based on the shortest travel time by means of spatial analysis. An Origin-Destination (OD) matrix is first developed to determine the possible routes from the evacuation starting points to the evacuation shelter. OD matrix is commonly used in the study of transport process [11], in this case the movement of humans in the event of tsunami evacuation. The OD matrix provides all the possibilities of the number of trips between two spatial points [12]. The Dijkstra algorithm is used in determining the optimal evacuation route to the evacuation shelter. The algorithm determines the shortest part between two end points based on the number of arcs and its weights [13]. Based on the determination of shortest paths from these algorithms, a map showing the evacuation routes is produced and aimed to be implemented as part of disaster mitigation efforts in Batukaras village.

3. Results and Discussion

3.1. Evacuation Shelters and Signage

After the field visit, five Temporary Evacuation Shelters (TES) and one Final Evacuation Shelter (FES) have been identified in Batukaras village. These temporary shelters are designed as first destination for evacuees as they flee the coastline. After the initial inundation has receded, evacuees are directed to the FES located in Tanjakan Heras. Here, a more permanent evacuation center may be installed that includes lodging, public kitchen, and medical facilities. Figure 1 shows the map of TES and FES around Batukaras Village, with Table 1 details of which can be found in Table 1. All evacuation shelters were found to be in good condition and able to serve as a temporary destination. Furthermore, mapping of evacuation signages was also conducted, shown in Figure 1 as orange signage icons. 16 out of 19 evacuation signs were found to be in good condition, the rest was found to be missing.

Table 1. List of Evacuation Shelters

Name of Evacuation Shelter	*Type*
Tanjakan Heras	FES
Gunung Tumpeng	TES
Kabuyutan	TES
Bukit	TES
Bukit Belakang	TES
Batukaras Surf Resort	TES

3.2. Evacuation Routes

During the field visit to Batukaras Village, determination of evacuation routes was made based on existing evacuation signages pointing to TES and FES in the village. Five spots around Batukaras Village are identified as evacuation starting points based on one of two criteria: farthest distance from FES or locations of high tourist activity in the village. A list of starting points is shown in Table 2. Route of evacuation can be found in Figure 1 shown in orange lines on the map overlaid on local road networks. Additionally, evacuation simulation by foot was conducted along the evacuation routes, duration of each route traversed in the simulation can be found in Table 3. Here, only one shelter was shown as the optimum destination for each starting point based on the shortest duration of evacuation simulation and the rough capacity of each tsunami evacuation point.

Table 2. Route and duration of evacuation

Start	*End*	*Duration*
B	FES Tanjakan Heras	18min 57secs
C	FES Tanjakan Heras	24min 44secs
D	Bukit	7min 23secs
E	Bukit Belakang	0min 27secs

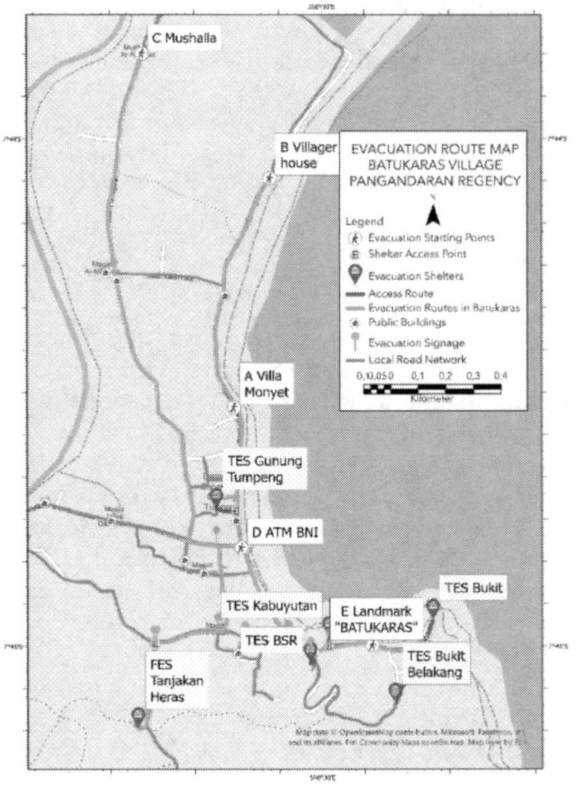

Figure 1. Evacuation Route Map in Batukaras

Table 3. List of evacuation start points

Code	*Name and Description*	*Criteria*
A	Villa Monyet – tourist lodging	High tourist activity
B	Villager House – farthest settlement up north from FES	Farthest distance from FES
C	Mushalla – farthest settlement up north from FES	Farthest distance from FES
D	BNI ATM – center for tourist activity	High tourist activity
E	Batukaras Lettering Landmark – adjacent to Batukaras beach	High tourist activity

3.3. Network Analysis

Network analysis was carried out using several stages, including digitizing aerial photographs of Batukaras Village, making OD matrixes for each evacuation starting point, analyzing the shortest distance from each evacuation starting point to the nearest temporary evacuation shelter (TES), and implementing the closest route to the nearest TES from each building on the mapped area.

The digitization process was carried out using Quantum GIS 3.26 software. Digitization is carried out on several objects, including buildings, roads, evacuation starting points, evacuation signs, and temporary evacuation site points. Aerial photography is sourced from a local survey conducted in the same year. Digitization results can be seen in Figure 2.

IOP Conf. Series: Earth and Environmental Science 1245 (2023) 012044 doi:10.1088/1755-1315/1245/1/012044

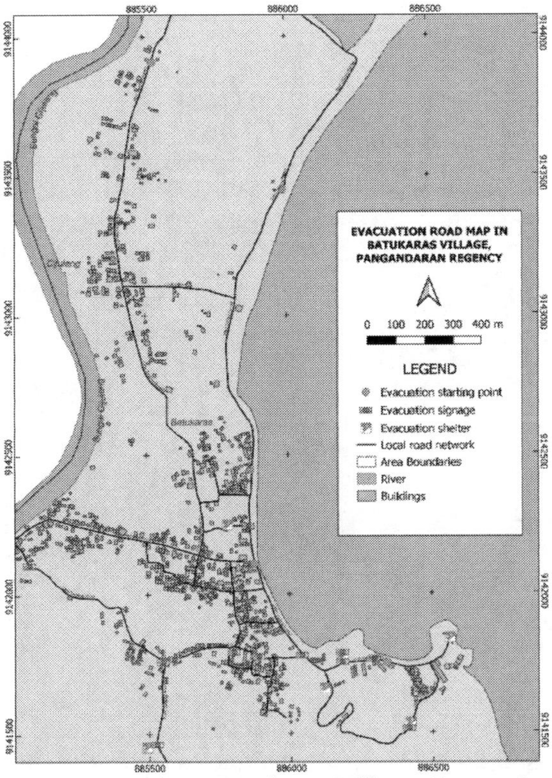

Figure 2. Digitizing result on Batukaras Village, Pangandaran Regency

The Origin-Destination Matrix aims to determine which routes can be taken to get to all evacuation places from all evacuation starting points. The OD Matrix created will be selected based on the closest distance to the evacuation site using a SQL query. From all of the routes determined, the nearest OD matrix will be determined. The OD matrix can be seen in Figure 3a and the nearest OD matrix can be seen in Figure 3b. The results show that for most of the evacuation starting points, the matrix points to the Gunung Tumpeng evacuation shelter as the optimum destination for evacuation, albeit for evacuation starting point E where the Kabuyutan site is selected as the optimum evacuation shelter.

The application of Djikstra algorithm is used to obtain the shortest route that can be taken to get to the evacuation site. The Dijkstra algorithm estimates the shortest route by considering the distance between vertices as weights [14]. The shortest route taken from the initial point of evacuation to the evacuation site can be seen in Figure 4a.

In this study, an additional approach is conducted by considering buildings as evacuation starting points in addition to the previously predetermined start points. The same framework is then implemented on each building in the mapped area to obtain the OD matrix closest to the evacuation site and the shortest route to the nearest evacuation site. The OD matrix closest to the evacuation site can be seen in Figure 4b. The shortest evacuation route to the nearest evacuation site can be seen in Figure 5.

The building-based approach to optimum shelter determination shows similar results to the evacuation starting point-based approach presented earlier. Here one can observe that Gunung Tumpeng TES is selected as the optimum destination based on the OD matrix for much of the area in Batukaras village, in particular areas north and central to the village. In the southern part of the village, the Gunung Tumpeng TES is selected as the optimum evacuation site for most of the buildings. One could observe that this scenario would cause overcapacity if implemented in the TES. The algorithm does not take into account the capacity of the shelter, hence the load. However, the routes shown in Figure 5 show a clearer plan compared to Figure 4a.

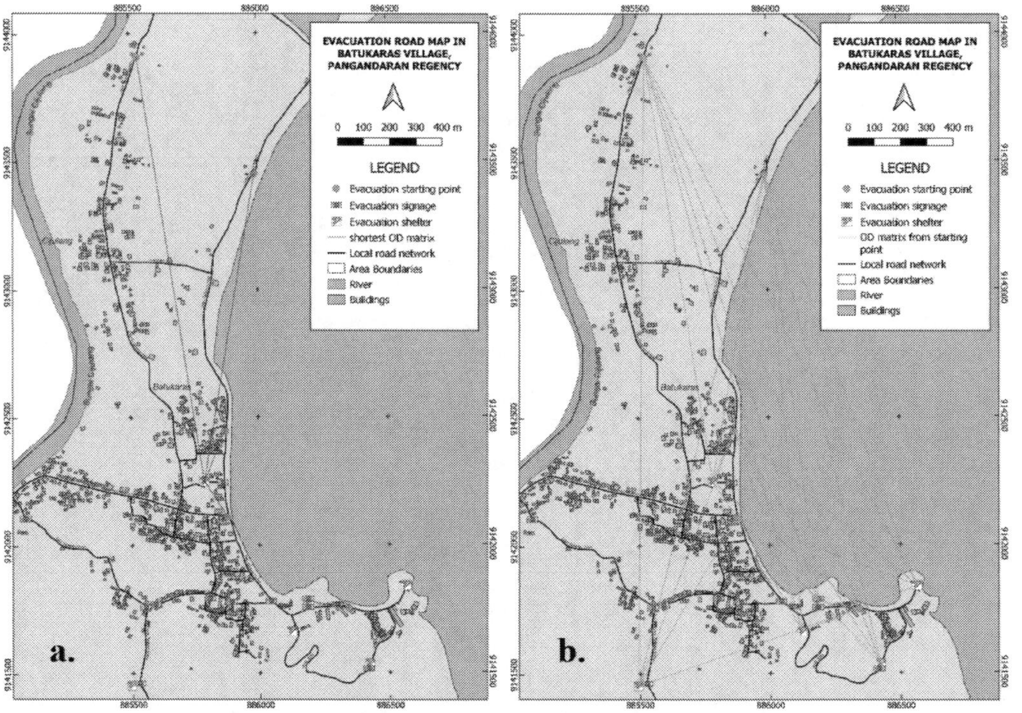

Figure 3a. OD matrix result from start points in Batukaras Village
Figure 3b. Shortest OD matrix from start to nearest evacuation shelter

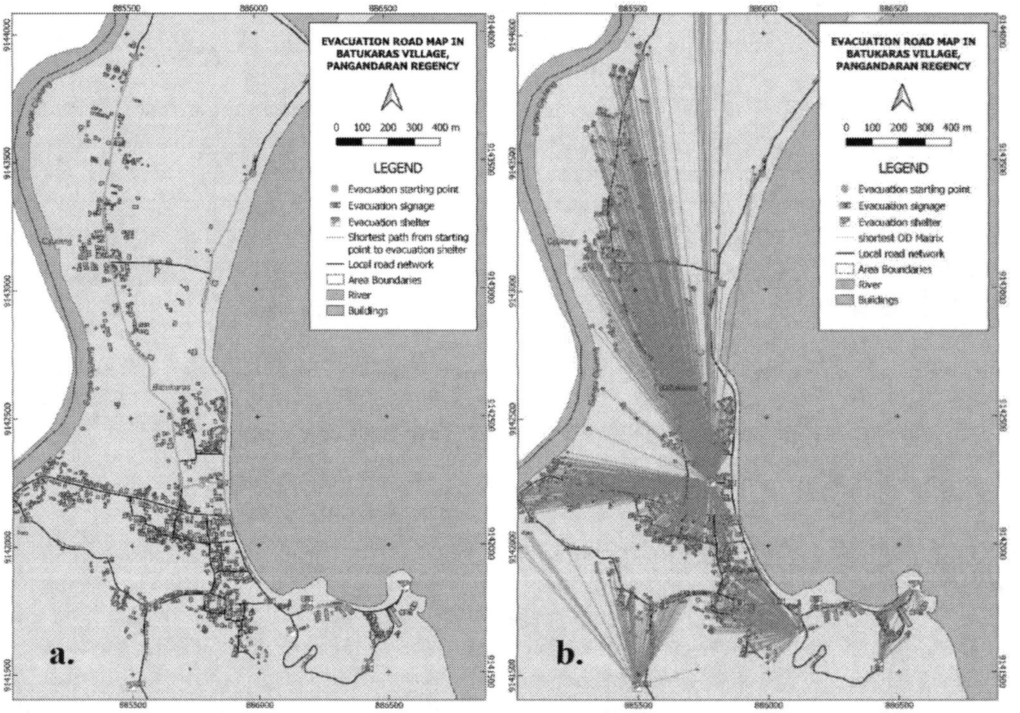

Figure 4a. Shortest route to nearest evacuation shelter
Figure 4b. Shortest OD Matrix from buildings to nearest evacuation shelter

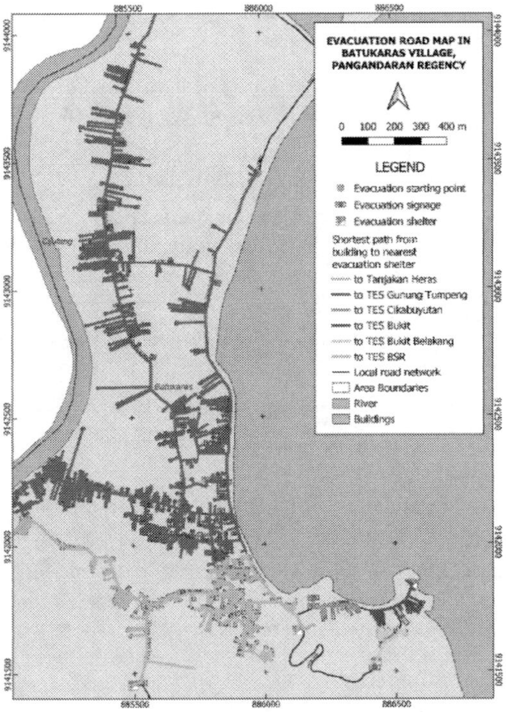

Figure 5. Shortest route from buildings to the nearest evacuation shelter

4. Conclusion

Based on the data from field observation done during a visit to Batukaras Village, mapping of tsunami evacuation routes and shelters was successfully conducted. Five temporary evacuation shelters and one final evacuation shelter have been located in the village, as well as routes leading to the shelters from five evacuation start points around the village as well as from buildings. However, this study still has several limitations, primarily in the parameters used as the consideration for the network analysis. The capacity of each evacuation shelter has not yet been taken into consideration. Distance from the shore as well as elevation of the topography should also be included in the analysis for future studies. Regardless, it is believed that the resulting evacuation map would be sufficient used for preliminary disaster preparedness efforts in Batukaras, including fulfillment of the IOC-UNESCO Tsunami Ready indicators in Batukaras Village.

References

[1] Statistics Indonesia (Badan Pusat Statistik) Provinsi Jawa Barat, "Jumlah Kunjungan Wisatawan Ke Objek Wisata 2018-2021," [Online]. Available: https://jabar.bps.go.id/indicator/16/220/1/jumlah-kunjungan-wisatawan-ke-objek-wisata.html. [Accessed 01 November 2022].

[2] A. E. Putri, U. L. S. Khadijah and E. Novianti, "Community empowerment in the development of mangrove tourism in Batu Karas of Pangandaran, West Java," *Geo Journal of Tourism and Geosites* , vol. 31, no. 3, pp. 972-978, 2020.

[3] Statistics Indonesia - Badan Pusat Statistik (BPS) Kabupaten Ciamis, "Produk Domestik Regional Bruto Kabupaten Pangandaran Menurut Lapangan Usaha 2017-2021," 2022.

[4] J. Mori, W. D. Mooney, S. K. Afnimar, A. I. A. and S. Widiyantoro, "The 17 July 2006 Tsunami Earthquake in West Java, Indonesia," *Seismological Research Letters,* vol. 78, pp. 201-207, 2007.

[5] D. Mardiatno, M. N. Malawani and R. M. Nisaa', "The future tsunami risk potential as a consequence of building development in Pangandaran Region, West Java, Indonesia," *International Journal of Disaster Risk Reduction,* vol. 46, 2020.

[6] M. B. F. Bisri, "Pangandaran Village Resiliency Level due to Earthquake and Tsunami Risk," in *Ninth*

IEEE International Symposium on Parallel and Distributed Processing with Applications Workshops, Busan, 2011.

[7] F. Ashar, D. Amaratunga and R. Haigh, "The Analysis of Tsunami Vertical Shelter in Padang City," in *4th International Conference on Building Resilience*, Salford Quays, 2014.

[8] A. Pimanmas, P. Joyklad and P. Warnitchai, "Structural Design Guideline for Tsunami Evacuation Shelter," *Journal of Earthquake and Tsunami,* vol. 4, pp. 269-284, 2010.

[9] D. Prameswari, "Tak Perlu Khawatir Liburan di Pangandaran, BPBD Siapkan Rambu dan Jalur Evakuasi - TIMES Indonesia," 26 December 2020. [Online]. Available: https://www.timesindonesia.co.id/peristiwa-daerah/318548/tak-perlu-khawatir-liburan-di-pangandaran-bpbd-siapkan-rambu-dan-jalur-evakuasi. [Accessed 01 November 2022].

[10] S. Park, J. W. v. d. Lindt, R. Gupta and D. Cox, "Method to determine the locations of tsunami vertical evacuation shelters," *Natural Hazards,* vol. 63, p. 891–908, 2012.

[11] B. Huang, Comprehensive Geographic Information Systems, Elsevier, 2018.

[12] V. Dragu and E. A. Roman, "The Origin–Destination Matrix Development," in *MATEC Web of Conferences 290*, 2019.

[13] H. Wang, Y. Yu and Q. Yuan, "Application of Dijkstra algorithm in robot path-planning," in *2011 Second International Conference on Mechanic Automation and Control Engineering*, Hohhot, 2011.

[14] M. Noto and H. Sato, "A method for the shortest path search by extended Dijkstra algorithm," in *Smc 2000 conference proceedings. 2000 ieee international conference on systems, man and cybernetics. 'cybernetics evolving to systems, humans, organizations, and their complex interactions*, Nashville, 2000

Acknowledgements
The authors would like to thank ITB Research and Community Service Institute (LPPM) for their generous support in funding the ongoing efforts in Batukaras. We would also like to recognize National Research and Innovation Agency (BRIN), Meteorology, Climatology and Geophysics Agency (BMKG), Regional Disaster Management Agency (BPBD) Pangandaran Regency, U-Inspire, Batukaras Village, and the Community Early Awareness Forum (FKDM) Batukaras Village for their support in the study.

Evaluation of GNSS reflectometry method for sea level estimation in Indonesia

Lisa A. Cahyaningtyas (a*), Dudy D. Wijaya (a), Bambang Setyadji (a), Aditya R. Kartadikaria (c), Hansan Park (d), Sidik T. Wibowo (d), Rega Himawan (e)

(a) Geodesy, Institut Teknologi Bandung, Bandung, 40132, Indonesia; (b) Oceanography, Institut Teknologi Bandung, Bandung, 40132, Indonesia; (c) Marine Technology Cooperation Research Center – Cirebon Center, 45611, Indonesia; (d) Division for Horizontal Control Network and Geodynamics, 16911, Geospatial Information Agency (BIG), Jl. Raya Jakarta-Bogor Km. 46, Cibinong, Indonesia; (e) PT Wisesa Berkah Bumi, Jakarta, 133220, Indonesia

Email: lisaayuc@gmail.com

Abstract. Over the last decade, GNSS-R (Global Navigation Satellite System reflectometry) has emerged as a technique for observing sea level height using data from GNSS observations. GNSS-R estimates sea level height from the phase center antenna using the reflection of the extracted GNSS signal at sea level. With a large number of tidal stations equipped with GNSS antennas, GNSS-R has the potential to be implemented in Indonesia. GNSS-R observations can also cover sea surface areas tens to hundreds of kilometres away from where the antenna is located over the coast. Furthermore, the installation of a GNSS antenna at a safe height on land allows for the observation of sea levels under extreme conditions such as during storms and cyclones. To employ GNSS-R effectively, several factors must be considered, including signal-to-noise ratio (SNR) data analysis, data processing, filtering control variables, and increased frequency extraction. This study will focus on trying to use GNSS-R method by identifying the best control variables for each study area and evaluating the results. This study used three months of GPS and GLONASS satellite SNR data from two stations, Barus and Morotai. The separated multipath data was then analyzed using the Lomb-Scargle Periodogram (LSP) based on its frequency. The results of GNSS-R sea surface height observations were validated using tide gauge data at both stations. Based on the test results, the RMSE values were 8.7 cm and 8.4 cm at Barus and Morotai stations, respectively. GNSS-R results strongly correlate with tide gauge data, with correlation coefficients of 95% and 98% at Barus and Morotai stations, respectively. Based on these findings, the GNSS-R method can be used to complement tidal sensor data by applying proper quality control.

1. Introduction

Multipath or reflected signals are one of the major error sources in high-accuracy GNSS (Global Navigation Satellite System) positioning, and there are numerous studies to mitigate the effects [1], [2], [3]. In contrast, [4] introduced a method for environmental sensing using GNSS reflected signals. [5] utilized signal-to-noise ratio (SNR) to investigate interference patterns of reflected GNSS signals and measure the distance between the antenna phase center and seawater as a reflected surface. Since then, research related to the use of reflected signals, also known as GNSS reflectometry (GNSS-R), has been carried out to measure sea level from the processing of the SNR that was acquired by a geodetic GNSS receiver [6], [7], [8].

The advantages of using GNSS-R include allowing sea level observations refer to a geocentric frame [9]; reaching a large area (with a radius of up to hundreds of kilometers) rather than a single point [10], [11]; and being installed at a safe height on land to avoid extreme weather in coastal areas [12]. In Indonesia, there are 139 tidal stations managed by Geospatial Information Agency (BIG) until 2018 and some of them completed with GNSS installation [13]. The GNSS installation above tidal station has potential and more advantage to complete the measurement from tide gauges. The GNSS-R measurements showed high correlations with tide gauges (exceeding 0.9) and delivered centimeter-level accuracy [8], [14], [15]. The comparison between tidal constituents from GNSS-R results and tide gauges data also has a good agreement [16], [17] to get astronomical tide models. [12] used GNSS-R for detecting storm surges in Hong Kong by comparing GNSS-R results and the astronomical tides model.

To improve temporal resolution, however, several factors must be considered: (1) analysis of Signal-to-Noise Ratio (SNR) data [18]; (2) processing control variables and data filtering [19]; (3) increased frequency extraction [20]; and (4) data with a longer observation time [21]. This study will focus on the application of the GNSS-R method by investigating the control variables for optimal multipath signal separation based on GNSS data from Barus and Morotai stations in Indonesia.

2. Data and method

2.1. Data

This research uses data from two stations, Barus (CBRS) and Morotai (CMOR). CBRS is located in the southern part of North Sumatra Province at 2° 0' 18.053" LU and 98° 23' 52.445" E. CMOR, meanwhile, is located in the southern portion of Morotai Island, North Maluku, at the coordinates 2° 0' 59.624" N and 128° 16' 49.386" E. Observational data from CMOR were used from 1 February to 30 April 2022. (doy 061 – 121), whereas CBRS data were used from 1 January to 31 March 2022 (doy 001 – 090). Table 1 shows the data used for this study.

Table 1. Summarizes the data that used in this research.

Data	Interval	Data format	Source
Observation satellite	30 second	*.yyo (RINEX 2.11)	Geospatial Information Agency (BIG)
Water level (tide gauges)	1 hour	*.txt	Geospatial Information Agency (BIG)
Levelling	-	*.xls	Geospatial Information Agency (BIG)
Precise ephemeris	-	*.sp3	CDDIS**
Broadcast ephemeris	-	*.yyn	CDDIS***

** https://cddis.nasa.gov/Data_and_Derived_Products/GNSS/orbit_products.html
***https://cddis.nasa.gov/Data_and_Derived_Products/GNSS/broadcast_ephemeris_data.html

2.2. Method

The GNSS-R method estimates sea level height using reflected signals (multipath) originating from the sea surface [22]. Due to the distance between the receiver and the satellite as well as the antenna gain pattern, SNR data is dominated by long-period wave variations. Low-order polynomials eliminate long-period variations at low elevations (30°), taking multipath effects [9].

Multipath can be simply modelled for planar and horizontal reflectors. As explained by [1] ,the path delay δ is dependent on two variables: the distance between the antenna and the plane of reflection; the satellite's elevation angle. The additional distance traveled by the multipath signal relative to the direct signal is written as (see Equation 1):

$$\delta = 2h \sin e \tag{1}$$

where h is the distance between the antenna and the reflecting surface and e is the satellite's elevation angle to the horizontal plane (see Figure 1).

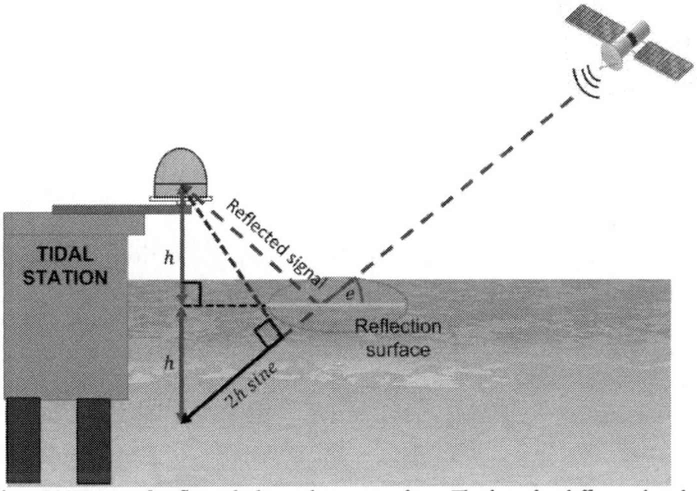

Figure 1. The scheme when GNSS signal reflected above the sea surface. The h is the different height between the sea surface and the antenna phase center. The e is the satellite elevation angle.

The composite SNR resulting from the direct signal and one multipath signal can be determined using the law of cosines. SNR is a function of the amplitude of the direct signal A_d, the amplitude of the multipath signal A_m, and the relative phase of the multipath θ, which can be written as [9], [14], [18], [22]:

$$SNR^2 = A_d^2 + A_m^2 + 2A_dA_m \cos\theta \tag{2}$$

In this study, the polynomial used for the detrended process is a 4th order polynomial. To produce a dSNR (detrended SNR) arc, a lower order polynomial is fitted with each satellite arc and then removed. dSNR is a cosine function that represents the effect of the reflected signal on the SNR observation data [19], [23]:

$$dSNR_{com} = A\cos\left(\frac{4\pi H_R \sin e}{\lambda} + \phi\right) \tag{3}$$

where A is the amplitude, H_R is the reflector height, the satellite elevation angle is e, and ϕ is the phase.

The LSP, also known as least square spectral analysis, can be used to determine the reflector height over time. The dominant frequency of the multipath can be determined based on the spectral analysis of the dSNR as a function of the $\sin(e)$ of elevation angle. Assuming a constant reflector height in one satellite arc and a horizontal reflector, it is possible to convert the dominant frequency to a high reflector [24].

The reflector height determined from the LSP analysis is the distance between the reflecting plane, which is the sea surface, and the phase center of the antenna. To derive sea level height, the reflector height must be adjusted to correspond with the zero-tide palm on the tide sensor. This height difference is determined by the Geospatial Information Agency using a precise levelling measurement technique. Table 2 displays the height differences (offsets) for each station used in this study.

Table 2. The height difference between zero-palm and GNSS antenna.

Station	Location	Station code	Offset (m)
Morotai	Maluku Utara	0158MORO03	7.235
Barus	Sumatera Utara	0169BRUS01	8.038

3. Results and discussions

The observation area is chosen based on the area of signal reflection, which is determined by the satellite's azimuth and elevation angle. At both stations, the First Fresnel Zone (FFZ) describes the signal reflection. The FFZ results are dependent on the satellite's elevation above the reflecting plane. The satellite's elevation decreases the size of the FFZ. When the satellite's elevation increases, the multipath effect decreases.

CBRS uses various reflected signals from the satellite elevation angle of 5°-20° at azimuths of 30°-160° and 208°-350°, as depicted in Figure 2. At CMOR, the signal used to analyze multipath signal data arises from an azimuth of 180° to 360°. In this azimuth, as depicted in Figure 2, the FFZ of the satellite, with an elevation angle of 5°-25°, is located above the sea surface reflection plane.

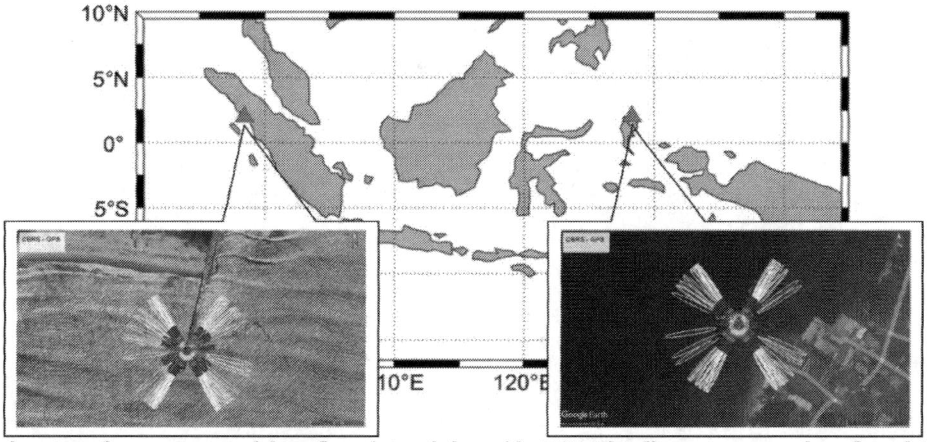

Figure 2. The location of two stations and the reflected signal showed by FFZ. The ellipses represent the reflected GPS L1 signals for elevation angles 5° (yellow), 10° (blue), 15° (red), 20° (green), and 25° (cyan).

As described in the methods subsection (see subsection 2.2), multipath separation is obtained by removing the direct signal. In this research, the trend of direct signals can be obtained by detrending for processing using a 4th-order fitting polynomial. The results from spectral analysis of the satellite arc showed no differences in reflector height for polynomials of degree 0-8 [15]. So, using 4th-order polynomial with not affect the SNR-analysis results. The multipath separation results are then processed with an LSP to determine the reflector height. LSP is utilized to estimate the normalized spectral strength or periodogram from a collection of successive SNR data, for each predetermined time interval.

Several obstacles, such as double peaks and excessive noise, are found during LSP spectral analysis. It is necessary to implement a minimum frequency amplitude strength value and a peak-to-noise ratio value in order to prevent visible noise in the presence of low frequency peaks that cause the reflector height to be zero. The average amplitude of all multipath SNR data from the LSP analysis can be used to calculate this value. Additionally, the range of approximated SNR data frequency values can be applied. This frequency range can be determined from the SNR analysis results by evaluating the noise frequency oscillations that exist before and after the dominant frequency in one spectrum of the SNR data series.

Figures 3 and 4 depict the sea level height results at CBRS and CMOR following an LSP analysis and quality control. The quality control is applied to the data from both stations during the LSP process by adjusting the peak-to-noise ratio (>2.1 for both stations), minimum amplitude (>4 for both stations), and reflector height (4-10 for both stations) to get valid reflector height. The red dots represent the sea level from GNSS-R using GPS and GLONASS observation data. The resulting sea level heights at both stations show a medium level of temporal variation over a given time interval.

Figure 3 shows that the temporal variation of GNSS-R sea level results at CBRS is not equally distributed. The number of sea level data using GNSS-R derived from three month's observational data was 933. The majority of these data were collected during periods of low tide, for example on doy 009 and 017 and doy 065 and 080.

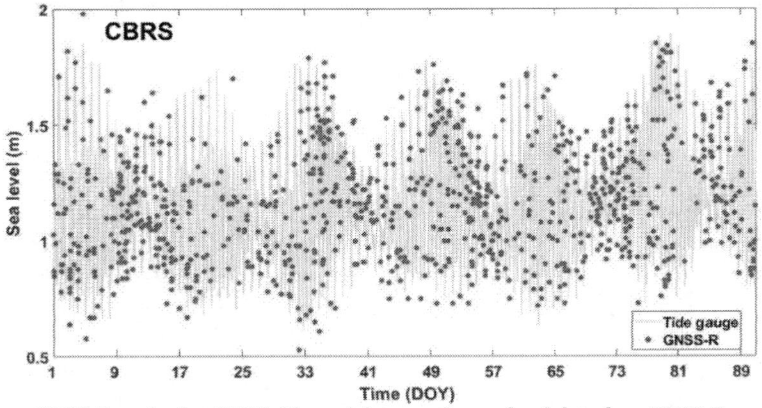

Figure 3. The result from GNSS-R method at CBRS. The red dots are the sea level data from GNSS-R.

The result of GNSS-R measurements of sea level height at CMOR is depicted in Figure 4. The distribution of the data obtained from CBRS and CMOR are comparable. The quantity of data obtained from CMOR were 1008, broadly dispersed between doy 072 and 080.

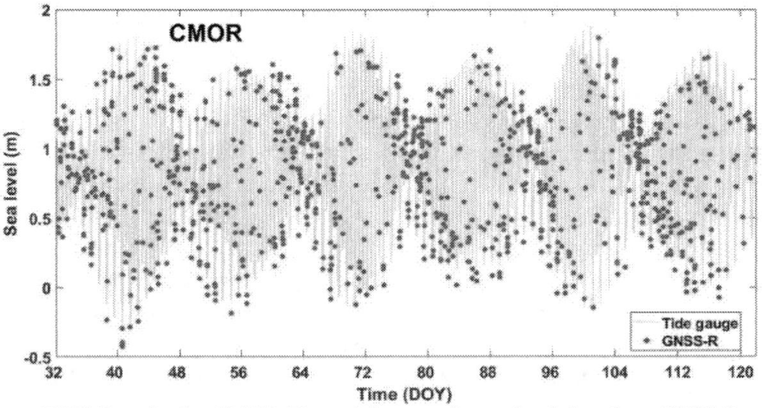

Figure 4. The result from GNSS-R method at CMOR. The red dots are the sea level data from GNSS-R.

The quality of the resulting data will be evaluated by comparing the GNSS-R sea level height measurements from the two stations with the tide gauge measurements. It is possible to determine the high quality of the GNSS-R sea level results by examining the accuracy and correlation between the GNSS-R result data and the tidal sensors. The accuracy is represented by the Root Mean Square error (RMSe) value, whereas the correlation between the two data is indicated by the R-square (R^2). Table 3 summarizes the RMSe calculation results and correlation at each station.

Table 3. RMSE, correlation with tide gauge, and the number of observations from at CBRS and CMOR.

Station	RMSe (cm)	Correlation (%)	Number of observations
CBRS	8.7	94	933
CMOR	8.4	98.2	1008

The RMSe values for CBRS and CMOR were 8.7 cm and 8.4 cm, respectively. While the correlation between GNSS-R and tide gauge data at both stations is 95.5% at CBRS and 98.8% at CMOR. At both stations, high correlation values indicate an agreement between the tidal sensor data and GNSS-R data. According to the number of data produced, CMOR data have greater temporal variations.

The difference between the correlation values at the two observation stations indicates the influence of local sea conditions and characteristics. The tides in the observation location impact the temporal resolution of the GNSS-R-derived sea level height data. It can be concluded from the distribution of data at the two stations that a large tidal range reduces the quantity of data collected. Consequently, this temporal resolution can influence the RMSe values and the obtained correlations.

The activities in nearby locations also influence the GNSS-R sea level results. The large number of ships that rely on the port and are close to the antenna receiver introduces more noise to the data. This noise is caused by multipath reflected not only from the sea surface but also from nearby buildings and objects capable of reflecting satellite signals.

The application of quality control affects the quantity of observational data collected as well. The quantity of parameters applied during LSP processing causes a significant amount of wasted data. The implementation of quality control parameters, on the other hand, can generate data with high reliability and accuracy.

4. Conclusion

The aim of this study was to evaluate the results of the GNSS-R at two sites, CBRS and CMOR. By using three months of L1 GPS and GLONASS data from CBRS and CMOR, it is possible to determine sea level height data with a strong correlation to data from the tide gauges. Using a low-order (order 4) polynomial fit, GNSS-R sea level observation data can be derived from multipath separation results with direct signals. At both stations, the quality of GNSS-R sea level height measurements relative to tidal sensors from the calculations of RMSe is 8 cm. The correlation coefficients for CBRS and CMOR are 95% and 98%, respectively. These results indicate that the GNSS-R method can be used to measure sea level height over Indonesian coasts. To achieve optimal results, however, optimal quality control is required by analysing the value for every parameter at study area. It will help to make strategy when analysing the result. In addition, the use of multiple satellites and multiple frequencies will result in enhanced temporal variations.

Acknowledgement

This research is part of projects titled “Korea-Indonesia Marine Technology Cooperation Research Center (20220512)” and “Ocean and Coastal Basic Survey and Capacity Enhancement in Cirebon, Indonesia (PG53340)” which are funded by the Ministry of Oceans and Fisheries, Korea.

References

[1] Y. Georgiadou and A. Kleusberg, “On carrier signal multipath effects in relative gps positioning.pdf,” *manuscripta Geod.*, vol. 13, pp. 172–179, 1987.

[2] K. D. Park *et al.*, “Development of an antenna and multipath calibration system for Global Positioning System sites,” *Radio Sci.*, vol. 39, no. 5, pp. 1–13, 2004, doi: 10.1029/2003RS002999.

[3] A. Bilich, K. M. Larson, and P. Axelrad, “Modeling GPS phase multipath with SNR: Case study from the Salar de Uyuni, Boliva,” *J. Geophys. Res. Solid Earth*, vol. 113, no. 4, pp. 1–12, 2008, doi: 10.1029/2007JB005194.

[4] M. Martin-Neira, “A passive reflectometry and interferometry system (PARIS): application to ocean altimetry,” *ESA J.*, vol. 17, no. 4, pp. 331–355, 1993.

[5] K. D. Anderson, “Determination of Water Level and Tides Using Interferometric Observations of GPS Signals,” 1999.

[6] J. S. Löfgren, R. Haas, and J. M. Johansson, “Monitoring coastal sea level using reflected GNSS signals,” *Adv. Sp. Res.*, vol. 47, no. 2, pp. 213–220, Jan. 2011, doi: 10.1016/j.asr.2010.08.015.

[7] K. M. Larson, J. S. Löfgren, and R. Haas, “Coastal sea level measurements using a single geodetic GPS receiver,” *Adv. Sp. Res.*, vol. 51, no. 8, pp. 1301–1310, 2013, doi: 10.1016/j.asr.2012.04.017.

[8] N. Roussel *et al.*, “Sea level monitoring and sea state estimate using a single geodetic receiver,” *Remote Sens. Environ.*, vol. 171, pp. 261–277, Dec. 2015, doi: 10.1016/j.rse.2015.10.011.

[9] K. M. Larson, R. D. Ray, F. G. Nievinski, and J. T. Freymueller, “The accidental tide gauge: A GPS reflection case study from kachemak bay, Alaska,” *IEEE Geosci. Remote Sens. Lett.*, vol. 10, no. 5, pp. 1200–1204, 2013, doi: 10.1109/LGRS.2012.2236075.

[10] N. Roussel *et al.*, “Simulations of direct and reflected wave trajectories for ground-based

GNSS-R experiments," *Geosci. Model Dev.*, vol. 7, no. 5, pp. 2261–2279, 2014, doi: 10.5194/gmd-7-2261-2014.

[11] X. Wang, Q. Zhang, and S. Zhang, "Azimuth selection for sea level measurements using geodetic GPS receivers," *Adv. Sp. Res.*, vol. 61, no. 6, pp. 1546–1557, 2018, doi: 10.1016/j.asr.2018.01.002.

[12] D. Peng, E. M. Hill, L. Li, A. D. Switzer, and K. M. Larson, "Application of GNSS interferometric reflectometry for detecting storm surges," *GPS Solut.*, vol. 23, no. 2, pp. 1–11, 2019, doi: 10.1007/s10291-019-0838-y.

[13] Badan Informasi Geospasial, "INA-TIDES," Bogor, 2018.

[14] X. Wang, Q. Zhang, and S. Zhang, "Sea level estimation from SNR data of geodetic receivers using wavelet analysis," *GPS Solut.*, vol. 23, no. 1, Jan. 2019, doi: 10.1007/s10291-018-0798-7.

[15] F. Geremia-Nievinski *et al.*, "SNR-based GNSS reflectometry for coastal sea-level altimetry: results from the first IAG inter-comparison campaign," *J. Geod.*, vol. 94, no. 8, Aug. 2020, doi: 10.1007/s00190-020-01387-3.

[16] J. S. Löfgren, R. Haas, and H. G. Scherneck, "Sea level time series and ocean tide analysis from multipath signals at five GPS sites in different parts of the world," *J. Geodyn.*, vol. 80, pp. 66–80, Oct. 2014, doi: 10.1016/j.jog.2014.02.012.

[17] K. M. Larson, R. D. Ray, and S. D. P. Williams, "A 10-year comparison of water levels measured with a geodetic GPS receiver versus a conventional tide gauge," *J. Atmos. Ocean. Technol.*, vol. 34, no. 2, pp. 295–307, 2017, doi: 10.1175/JTECH-D-16-0101.1.

[18] J. S. Löfgren, *Local sea level observations using reflected GNSS signals*. 2014.

[19] W. Zhou, L. Liu, L. Huang, Y. Yao, J. Chen, and S. Li, "A New GPS SNR-based Combination Approach for Land Surface Snow Depth Monitoring," *Sci. Rep.*, vol. 9, no. 1, Dec. 2019, doi: 10.1038/s41598-019-40456-2.

[20] X. Wang, Q. Zhang, and S. Zhang, "Water levels measured with SNR using wavelet decomposition and Lomb–Scargle periodogram," *GPS Solut.*, vol. 22, no. 22, Jan. 2018, doi: 10.1007/s10291-017-0684-8.

[21] D. Peng, L. Feng, K. M. Larson, and E. M. Hill, "Measuring coastal absolute sea-level changes using GNSS interferometric reflectometry," *Remote Sens.*, vol. 13, no. 21, 2021, doi: 10.3390/rs13214319.

[22] A. Bilich and K. M. Larson, "Mapping the GPS multipath environment using the signal-to-noise ratio (SNR)," *Radio Sci.*, vol. 42, no. 6, 2007, doi: 10.1029/2007RS003652.

[23] A. Syetiawan, D. D. Wijaya, and I. Meilano, "Quality control in GNSS reflectometry method for tide observations," *Telkomnika (Telecommunication Comput. Electron. Control.*, vol. 19, no. 6, pp. 1935–1947, 2021, doi: 10.12928/TELKOMNIKA.v19i6.21670.

[24] J. S. Löfgren and R. Haas, "Sea level measurements using multi-frequency GPS and GLONASS observations," *EURASIP J. Adv. Signal Process.*, vol. 2014, no. 1, 2014, doi: 10.1186/1687-6180-2014-50.

Determination of Indonesia's Marine Deflection of Vertical from Multi-Mission Altimetry Satellite Data

Ahmad Fauzan[1*], Dudy Darmawan Wijaya[1], Fathin Nurzaman[1], Faruq Khadami[2] and Hansan Park[3,4]

[1] Department of Geodesy and Geomatics Engineering, Faculty of Earth Science and Technology, Bandung Institute of Technology
[2] Department of Earth Science, Faculty of Earth Science and Technology, Bandung Institute of Technology
[3] Korea-Indonesia Marine Technology Cooperation Research Center, Cirebon, Indonesia
[4] Korea Institute of Ocean Science & Technology, Korea

*Corresponding author: ahmdfauznn@gmail.com

Abstract. The deflection of vertical (DOV) is important for studying the changes in the Earth's mass, improving the accuracy of geoid models, and the accuracy of the Earth's gravity field. Compared to traditional methods such as astrogeodesy and astrogravimetry measurements, altimetry satellites have the advantage of observing DOV due to uniform global distribution, denser measurement points, consistent precision, less time required, high temporal resolution, and can be used in all weather conditions. In this study, multi-mission altimetry satellites including TOPEX/Poseidon, Jason-1, Jason-2, Jason-3, ENVISAT, Geosat Follow On (GFO), Sentinel-3A, Sentinel-3B, Sentinel-6A, Cryosat-2, and SARAL were used to determine DOV in Indonesia's marine with a spatial resolution of 1' x 1'. The Hwang et al. (1998) method is used to calculate DOV from satellite altimetry data by least-squares adjustment with weights using the second-order inverse distance weighting (IDW) method for each grid. The results show that the variance of the north-south and west-east components resulting from the gridding results are 0.05 arcseconds and 1.2 arcseconds. The evaluation results with the global model from EGM2008 show that the standard deviation values for the north-south and west-east components are 0.63 arcseconds and 1.96 arcseconds, respectively. From the evaluation, it can be concluded that the gridded DOV model is close to EGM2008, but there are still challenges for improvement, especially for the west-east DOV component.

1. Introduction

The deflection of vertical (DOV) is important for studying the changes in Earth's mass, improving the accuracy of geoid models, and the accuracy of the Earth's gravity field [1,2]. DOV represents the angular deviation formed between the direction of gravity at any point on the Earth's surface and the normal direction of the reference ellipsoid [2]. Based on the data collection approach, DOV is divided into land DOV and marine DOV. This research will focus on marine DOV. The marine DOV can be obtained by traditional methods such as astrogeodesy, astrogravimetry, and gravimetry. However, the interference of ocean waves, currents, tides, and the need for a lot of energy and a lot of time make it difficult to be conducted [3,4]. Currently, satellite altimetry is the most important technology for obtaining marine

DOV due to its uniform global distribution, denser measurement points, consistent precision, less time required, high temporal resolution, and all-weather use [5-10].

With its radar altimeter, an altimetry satellite can measure sea surface height (SSH) [11]. DOV can be obtained from SSH in three ways proposed by Sandwell [12], Olgiati et al. [13], and Hwang et al. [14]. By comprehensively comparing the three methods, the Hwang et al. method [14] was chosen because it is theoretically more precise [15]. Subtraction between two adjacent SSH points then divided by the spherical distance between the two points can effectively attenuate longwave errors such as satellite orbital errors, signal propagation errors, and topographic errors [16,17]. DOV will later be used in the determination of ocean gravity. There are several methods to derive the ocean gravity from DOV, such as the inverse Vening Meinesz (IVM) formula [17], Laplace equation [18], and least-squeres collocation (LSC) method [19]. IVM is a commonly used method today to obtain ocean gravity from altimetry satellite data [6].

In this study, Indonesia and its surroundings (90°W-150°W, -20°S-20°N) are selected as the study area and multi-mission altimetry satellite data are used to calculate DOV with a resolution of 1' x 1'. There are 11 altimetry satellites used, including TOPEX/Poseidon, Jason-1, Jason-2, Jason-3, ENVISAT, Geosat Follow On (GFO), Sentinel-3A, Sentinel-3B, Sentinel-6A, Cryosat-2, and SARAL. The global DOV model from EGM2008 is used to evaluate the results of the DOV calculations that have been done.

2. Data

2.1 Altimetry Satellite

SSH data of altimetry satellites were obtained from the Radar Alimeter Database System (RADS). RADS is an altimetry database first developed by Delft University of Technology and later followed by the National Oceanic and Atmospheric Administration (NOAA) and the European Organization for the Exploitation of Meteorological Satellites (EUMETSAT) [20]. The database provides mathematical operation services so that users can make corrections to SSH freely according to their needs. The list of SSH corrections used in this study can be seen in table 1.

Table 1. List of corrections used in SSH. For some corrections, more than one model is written and sorted in order of priority, if the topmost order is not available, the next order is used.

Correction	Model
Dry Toposphere	• ECMWF operational analysis • ERA-Interim
Wet Troposphere	• On-board microwave radiometer • ECMWF operational analysis • ERA-Interim
Ionosphere	• Dual-frequency altimeter range measurement
Dynamic Atmospheric Correction	• MOG2D DAC from ERA-Interim • MOG2D DAC • MOK2D DAC
Solid Earth Tide	• Cartwright-Taylor-Edden solid earth tide model
Ocean Tide	• FES2014b ocean tide
Loading Tide	• FES2014a load tide • Long-period equilibrium tide
Pole Tide	• IERS(EOP)05C04 EOP and predictions
Sea Satate Bias	• Non-parametric sea state bias models for Ku-band and C-band by CLS • NOAA hybrid sea state bias
Ref_frame_offset	• RADS default

There are 11 altimetry satellite missions that are used as the main data in this study. For certain satellite missions, more than 1 phase will be used. There are at least 4 types of phases on altimetry satellites, such as the Excat Repeat Mission (ERM), Geodetic Mission (GM), Interleaved Mission (IM), and Drifting Phase (DP). In the ERM phase, altimetry satellites are designed to go through the specified paths repeatedly according to the repetition period. While the GM phase is the opposite where it will not return to the path it has been on, the distance between paths will be as close as possible. IM is considered as ERM due to repetition while DP of SARAL is considered as GM. The satellites can be seen in table 2.

Table 2. List of altimetry satellite missions used

Satellite Mission	Phase (Cycle)	Revisit Time (Days)	Ground Track Separation at Equator (km)	Periode
TOPEX/Poseideon (TP)	ERM (002 – 363)	10	315	Oct 2, 1992 – Jul 21, 2005
	IM (369 – 480)	10	315	Sep 19, 2002 – Sep 23, 2005
Jason-1 (J1)	ERM (003 – 259)	10	315	Feb 03, 2002 – Jan 15, 2009
	IM (263 – 372)	10	315	Feb 19, 2009 – Feb 04, 2012
	GM			May 06, 2012 – Jun 19, 2013
Jason-2 (J2)	ERM (001 – 302)	10	315	Jul 11, 2008 – Sep 11, 2016
	IM (306 – 326)	10	315	Oct 16, 2016 – May 02, 2017
	GM			Jul 10, 2017 – Jul 17, 2018
Jason-3 (J3)	ERM (001 – 226)	10	315	Feb 16, 2016 – Mar 27, 2022
	IM (301 – 319)	10	315	Apr 28, 2022 – Oct 23, 2022
ENVISAT (N1)	ERM (008-093)	30 – 35	80	Jul 22, 2002 – Sep 13, 2010
Geosat Follow On (GFO)	ERM (037 – 215)	17	164	Jan 07, 2000 – Apr 29, 2008
Sentinel-3A (3A)	ERM (002 – 090)	27	104	Mar 12, 2016 – Sep 13, 2022
Sentinel-3B (3B)	ERM (020 – 071)	27	104	Des 16, 2018 – Sep 23, 2022
Sentinel-6A (6A)	ERM (005 – 027)	10	315	Des 26, 2020 – Oct 21, 2022
Cryosat-2 (C2)	GM	369	7.7	Jul 15, 2010 – Nov 4, 2022
SARAL (SA)	ERM (001 – 034)	35	75	Mar 13, 2013 – May 11, 2016
	DP			Aug12, 2016 – Oct 27, 2022

2.2 EGM2008

EGM2008 is a spherical harmonic model of the Earth's gravitational potential developed by Pavlis et al. [21]. The data used in making the model is a combination of terrestrial, altimetry, and airborne gravity data. One of the functional models is the DOV model. The EGM2008 DOV model in the research area is used as a comparison to the gridding DOV model.

3. Metode

3.1 Mean Sea Surface Height (MSSH)

In [22], Bao et al. explained that there are 4 types of procedures in deriving sea surface gradients that will be used in calculating DOV, including: (1) applying various types of corrections to the altimetry satellite-derived SSH, calculating the gradient along-track by numerical subtraction, and transforming it into the meridional component and the prime vertical component of DOVs at the crossing point; (2) adjusting the SSH and interpolating the slope on the intersecting track along the track between the two crossing points; (3) grids along the sea surface gradient path are transformed into north-south and west-east DOV components by least-squares where all available sea surface gradients along-track are used as observations and different weights are assigned to each gradient; (4) sea surface gradients can also be derived from Mean Sea Surface Height (MSSH) and independent sea surface topography models. In this study, the fourth procedure is used by modeling the MSSH using 11 multi-mission satellite altimetry data (see table 1) so that the data density is very high and the time variation of the ocean signal and data noise can be reduced [17,23].

In the MSS calculation process, collinear analysis was first performed on SSH data so that sea level variability can be eliminated and can be combined with other missions to produce longer time series such as ENVISAT and SARAL missions. For EGM and IM data, collinear analysis can be done by creating time series on the same path in each cycle. As for GM and DP where there is no repetition of paths, Sea Level Anomaly (SLA) generated by ERM data is used. Furthermore, ERM and IM data were homogenized by analyzing their crossing points with reference missions (T/P, J1, J2, J3, and 6A). Then the subtraction analysis is carried out to all satellites including the satellite itself. For GM and DP data, homogenization is performed by performing subtraction analysis at the crossover point with all previously homogenized ERM data. Once all the data has been collinearly analyzed and homogenized, it is ready for the along-track DOV calculation.

3.2 Along-Track DOV

By definition, the DOV along the track is the geoid gradient with the opposite sign [17]. Since MSSH data is used in this study, the DOV can be obtained by calculating the gradient between two adjacent MSSH points along the path as follows.

$$\varepsilon = -\frac{MSSH_j - MSSH_i}{d_{ij}} \tag{1}$$

Where ε is along-track DOV, $MSSH_i$ and $MSSH_j$ are the MSSH at point i and MSSH at point j respectively that are located consecutively and on the same track, and d_{ij} is the distance between the two points.

3.3 Gridded DOVs' Derivation

The DOV consists of two directional components, the meridian or north-south component and the prime vertical or west-east component. The relationship can be seen in the following equation.

$$\varepsilon = \xi cos\alpha + \eta sin\alpha \tag{2}$$

Where ε is DOV, α the DOV azimuth that has a direction along the trajectory of the altimetry data used, ξ the north-south component of the DOV, and η the west-east component of the DOV. By calculating the entire DOV of all the multi-mission altimetry satellite data used and its azimuth, the north-south and east-west components can be calculated in the following way.

$$\varepsilon_i + v_i = \xi cos\alpha_i + \eta sin\alpha_i, i = 1,2, \ldots, n \quad (3)$$

With n being the number of observation points around the grid point to be calculated, v_i, α_i, and ε_i, are the residual, azimuth, and DOV of the i-th observation, respectively. From equation (3) the following matrix can be formed.

$$V + L = AX \quad (4)$$

Where $V = \begin{pmatrix} v_1 \\ v_2 \\ \vdots \\ v_n \end{pmatrix}$, $L = \begin{pmatrix} \varepsilon_1 \\ \varepsilon_2 \\ \vdots \\ \varepsilon_n \end{pmatrix}$, $A = \begin{bmatrix} cos\alpha_1 & sin\alpha_1 \\ cos\alpha_2 & sin\alpha_2 \\ \vdots & \vdots \\ cos\alpha_n & sin\alpha_n \end{bmatrix}$, and $X = \begin{pmatrix} \xi \\ \eta \end{pmatrix}$.

With least-squares adjustment, equation (4) can be solved as follows.

$$X = (A^T WA)^{-1} A^T WL \quad (5)$$

$$W = \begin{bmatrix} w_1 & 0 & \ldots & 0 \\ 0 & w_2 & & 0 \\ \vdots & & \ddots & \ldots \\ 0 & 0 & \ldots & w_n \end{bmatrix} \qquad w_i = \frac{1}{d_i^2} \quad (6)$$

Where W is weight matrix and d_i is the distance of the i-th point to the calculated grid point. The weighting method used is second-order inverse distance weighting (IDW). The weight given by IDW is inversely proportional to the distance between the prediction location and the sample location. The weight will decrease as the distance increases. Using least-squares adjustment, the variances of the north-south and east-west components can be known.

$$\sigma^2 = \frac{V^T WV}{p-q} \quad (7)$$

$$Q_{xx} = (A^T WA)^{-1} \quad (8)$$

$$\sum_{xx} = \sigma^2 Q_{xx} = \begin{bmatrix} \sigma_\xi^2 & \sigma_{\xi\eta}^2 \\ \sigma_{\eta\xi}^2 & \sigma_\eta^2 \end{bmatrix} \quad (9)$$

Where σ^2 is the variance of the estimated true error, p is number of measurements, q is number of parameters, and $\sum_{xx}$ *is* variance matrix whose main diagonals $\sigma_\xi^2$and σ_η^2 are the variances of the north-south and west-east components of the DOV, respectively.

In this study, least-squares adjustment with weights using the second-order IDW method was used for each grid. With this technique, the calculation of the DOV component of each grid does not only include points within the grid, but a sub-region is created with the same origin point as the grid point with a size larger than the grid in order to provide the benefit of providing more points and so that the resulting surface is continuous because there are overlapping calculation areas between adjacent grids. The size of the sub-region used in this study is 5 arc-minutes or about 9 km at the equator. After the least-squares is performed with the combination, interpolation is performed to fill in the empty grid due to the sparse number of points such as those found in areas close to the coast. Interpolation is done with the SURFACE function in GMT.

4. Results and Analysis

Using the least squares adjustment with weights using the second-order IDW method and a sub-region size of 5 arc-minutes, the DOV results for the north-south and west-east components are obtained as shown in figure 1.

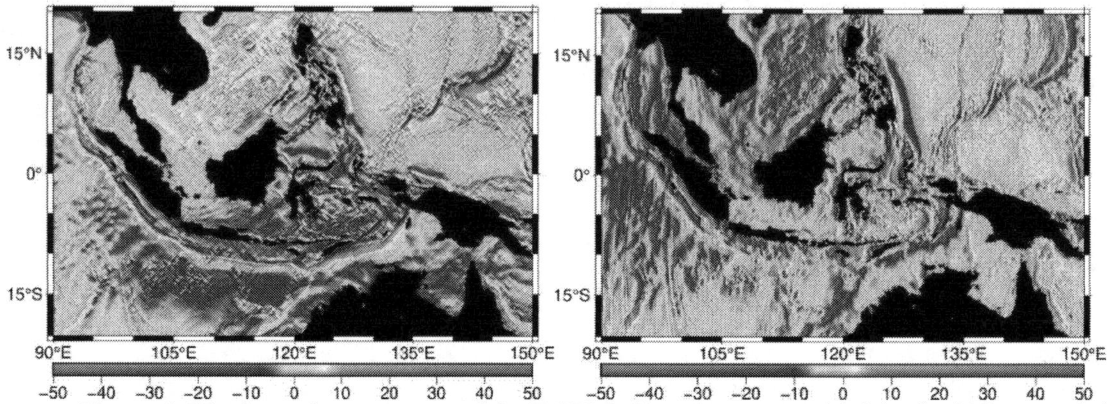

Figure 1. North-south component of the DOV (left) and west-east (right) of gridding result in arcsecond

Table 3 Variance (in arcsecond) for the north-south and west-east component of the DOV

Deflection of Vertical	σ^2
North-South	0.05
West-East	1.12

It can be seen that the range of DOV values is -50 to 50 arcseconds. The variance value for the north-south component is 0.05 arcseconds and the variance for the west-east component is 1.2 arcseconds (see table 3). This shows that the gridding using least-squares adjustment with weights using the second order IDW and choosing a sub-region size of 5 arc-minutes process performed has run well so that the results produced are considered good. For a better understanding of the gridding results, a comparison was made with the EGM2008 global model. The maximum, minimum, average, and standard deviation values were subtracted from each grid between the two models. The results can be seen in table 4.

Table 4. Comparing DOV gridding results with EGM2008 (unit: arcsecond)

Deflection of Vertical	Min	Max	Mean	Standard Deviation
North-South	-30.35	24.25	0.006	0.93
West-East	-84.72	84.72	0.008	1.96

From table 4 we can see that for the north-south component of the DOV model, the standard deviation is 0.93 arcseconds. This shows that the north-south component of the DOV model is very close to the north-south component of the EGM2008 DOV model. Then for the west-east component, the standard deviation is 1.96 arcseconds which shows it is quite close but no closer to the EGM2008 model than the north-south component. Visualization of the residual, which is the difference between the gridded DOV model and the EGM2008 model, can be seen in figure 2 and visualization its distribution of values can be seen in figure 3.

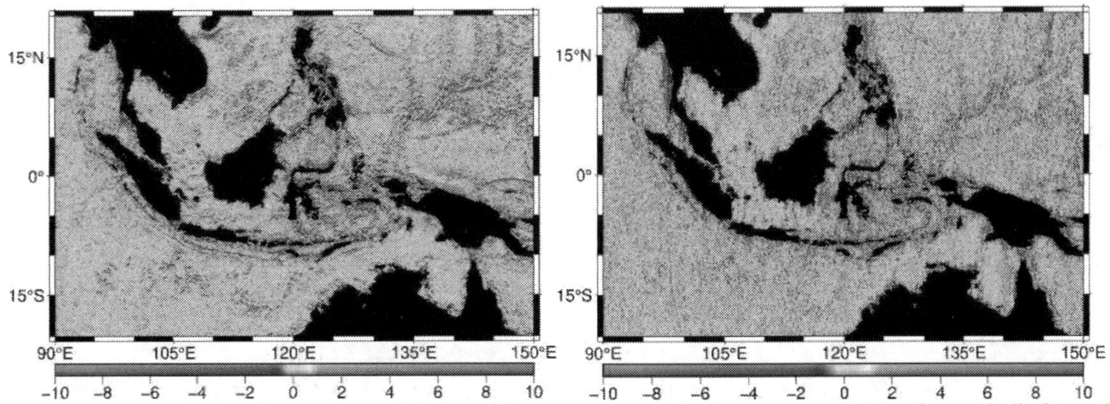

Figure 2 Residuals between gridding and EGM2008 results in arcsecond. North-south (left) and west-east (right) DOV components

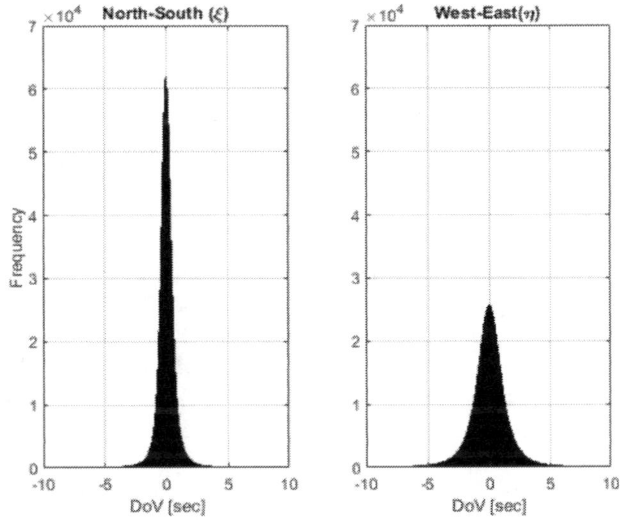

Figure 3 Distribution of residuals between gridding and EGM2008 results. North-south (left) and west-east (right) components of the DOV

From figure 3, the residual values are distributed to zero, for the north-south component of the DOV is about 3 arcseconds and west-east is about 6 arcseconds, two times larger than the north-south component. This is because the inclination of the altimetry, which is usually north, causes the north-south component of the DOV derived from altimetry to be more accurate than the west-east component [24]. These results show that overall the derived DOV model is close to the EGM2008 global.

5. Conclusion

With 11 multi-mission altimetry satellite data of various phases including TOPEX/Poseidon, Jason-1, Jason-2, Jason-3, ENVISAT, Geosat Follow On (GFO), Sentinel-3A, Sentinel-3B, Sentinel-6A, Cryosat-2, and SARAL, the Hwang et al. method [14], and least-squares adjustment with weights using the second-order IDW method, a 1' x 1' resolution model of marine DOV in Indonesia and its surroundings can be derived. The results show that the variance of the north-south and west-east components resulting from the gridding results are 0.05 arcseconds and 1.2 arcseconds. The evaluation results with the global model from EGM2008 show that the standard deviation values for the north-south and west-east components are 0.63 arcseconds and 1.96 arcseconds, respectively. From the evaluation, it can be concluded that the gridded DOV model is close to EGM2008, but there are still challenges for improvement, especially for the west-east DOV component due to the inclination of the altimetry.

Acknowledgment
This research is part of projects titled "Korea-Indonesia Marine Technology Cooperation Research Center (20220512)" and "Ocean and Coastal Basic Survey and Capacity Enhancement in Cirebon, Indonesia (PG53340)" which are funded by the Ministry of Oceans and Fisheries, Korea. The EGM2008 data is obtained from Muhammad Syahrullah Fathulhuda.

References
[1] S. Zhang, D. T. Sandwell, T. Jin, and D. Li 2017 Inversion of marine gravity anomalies over southeastern China seas from multi-satellite altimeter vertical deflections *J. Appl. Geophys.*, vol. **137** pp. 128–137
[2] J. Y. Guo 2016 Temporal-spatial distribution of oceanic vertical deflections determined by TOPEX/poseidon and Jason-1/2 missions *Earth Sci. Res. J.*, vol. **20**, no. 2
[3] H. Wang, Y. Wang, and Y. Lu 2008 High precision vertical deflection over China marginal sea and global sea derived from multi-satellite altimeter *Geo-Spatial Inf. Sci.* vol. **11** no. 4 pp. 289–293
[4] H. Ji 2021 On Deflections of Vertical Determined from HY-2A/GM Altimetry Data in the Bay of Bengal *IEEE J. Sel. Top. Appl. Earth Obs. Remote Sens.* vol. **14** pp. 12048–12060
[5] Z. Ling, L. Zhao, T. Zhang, G. Zhai, and F. Yang 2022 Comparison of marine gravity measurements from shipborne and satellite altimetry in the arctic ocean *Remote Sens.*, vol. **14** no. 1 pp. 1–11
[6] S. Liu, Y. Li, Q. Sun, J. Wan, Y. Jiao, and J. Jiang 2021 Evaluation of Marine Gravity Anomaly Calculation Accuracy by Multi-Source Satellite Altimetry Data *Front. Earth Sci.* vol. **9** pp. 1–11
[7] V. S. Nguyen, V. T. Pham, L. Van Nguyen, O. B. Andersen, R. Forsberg, and D. Tien Bui 2020 Marine gravity anomaly mapping for the Gulf of Tonkin area (Vietnam) using Cryosat-2 and Saral/AltiKa satellite altimetry data *Adv. Sp. Res.* vol. **66** no. 3 pp. 505–519
[8] D. Yu, C. Hwang, O. B. Andersen, E. T. Y. Chang, and L. Gaultier 2021 Gravity recovery from SWOT altimetry using geoid height and geoid gradient *Remote Sens. Environ.* vol. **265** p. 112650
[9] S. Zhang 2020 Inversion of global marine gravity anomalies with vertical deflection method deduced from Laplace equation *Cehui Xuebao/Acta Geod. Cartogr. Sin.* vol. **49** no. 4 pp. 452–460
[10] C. Zhu, J. Guo, C. Hwang, J. Gao, J. Yuan, and X. Liu 2019 How HY-2A/GM altimeter performs in marine gravity derivation: Assessment in the South China Sea *Geophys. J. Int.* vol. **219** no. 2 pp. 1056–1064
[11] E. Garcia 2014 Retracking cryosat-2, envisat and jason-1 radar altimetry waveforms for improved gravity field recovery *Geophys. J. Int.* vol. **196** no. 3, pp. 1402–1422
[12] D. T. Sandwell 1992 Antarctic marine gravity field from high-density satellite altimetry *Geophys. J. Int.* vol. **109** no. 2 pp. 437–448
[13] A. Olgiati 1995 Gravity anomalies from satellite altimetry: comparison between computation via geoid heights and via deflections of the vertical *Bull. Géodésique* vol. **69** no. 4 pp. 252–260
[14] C. Hwang, E.-C. Kao, and B. Parsons 1998 Global derivation of marine gravity anomalies from Seasat, Geosat, ERS-1 and TOPEX/Poseidon altimeter data *Geophys. J. Int.* vol. **134** no. 2 pp. 449–459
[15] X. Liu, G. Hui, J. Guo, T. Zhang, and M. Song 2023 Inversion of Deflection of the Vertical in the South China Sea Using ICESat-2 Sea Surface Height Data *Remote Sens.* vol. **15** no. 1
[16] C. Zhu *et al.* 2020 Marine gravity determined from multi-satellite GM/ERM altimeter data over the South China Sea: SCSGA V1.0 *J. Geod.* vol. **94**, no. 5

[17] C. Hwang, H. Y. Hsu, and R. J. Jang 2002 Global mean sea surface and marine gravity anomaly from multi-satellite altimetry: Applications of deflection-geoid and inverse Vening Meinesz formulae *J. Geod.* vol. **76** no. 8 pp. 407–418

[18] D. T. Sandwell and W. H. F. Smith 2009 Global marine gravity from retracked Geosat and ERS-1 altimetry: Ridge segmentation versus spreading rate *J. Geophys. Res. Solid Earth* vol. **114** no. 1

[19] C. Hwang and B. Parsons 1995 Gravity anomalies derived from Seasat, Geosat, ERS-1 and TOPEX/POSEIDON altimetry and ship gravity: a case study over the Reykjanes Ridge *Geophys. J. Int.*, vol. **122**, no. 2, pp. 551–568

[20] R. Scharroo, E. Leuliette, M. Naeije, C. Martin-Puig, and N. Pires 2016 RADS version 4: An efficient way to analyse the multi-mission altimeter database *Eur. Sp. Agency, (Special Publ. ESA SP* vol. **SP-740** pp. 2–5

[21] N. K. Pavlis, S. A. Holmes, S. C. Kenyon, and J. K. Factor 2012 The development and evaluation of the Earth Gravitational Model 2008 (EGM2008) *J. Geophys. Res. Solid Earth* vol. **117** no. 4 pp. 1–38

[22] L. Bao, H. Xu, and Z. Li 2013 Towards a 1 mGal accuracy and 1 min resolution altimetry gravity field *J. Geod.* vol. **87** no. 10 pp. 961–969

[23] O. B. Andersen and P. Knudsen 2009 DNSC08 mean sea surface and mean dynamic topography models *J. Geophys. Res. Ocean.* vol. **114** no. 11 pp. 1–12

[24] R. F. Annan and X. Wan 2021 Recovering Marine Gravity Over the Gulf of Guinea From Multi-Satellite Sea Surface Heights, *Front. Earth Sci.*, vol. **9**, pp. 1–11

Coastline Change Analysis using Landsat-8 on Rebana Metropolitan Area, West Java

Tri Kies Welly [1], Wiwin Windupranata[2]

[1]Magister Program of Geodesy and Geomatics Engineering, Faculty of Earth Sciences and Technology, Institut Teknologi Bandung, Indonesia;

[2]Hydrography Research Group, Faculty of Earth Science and Technology, Institut Teknologi Bandung, Indonesia.

Corresponding author: 25121022@mahasiswa.itb.ac.id

Abstract. The Rebana metropolitan area is an industrial area development covering Cirebon-Patimban-Kertajati to support accelerated economic growth, especially in the West Java region. In order to minimize failures in supporting the development targets of the Rebana metropolitan area, it is necessary to carry out spatial planning with consideration of the hazard aspects in the development area. One of the aspects that need to be studied further is the dynamics of the coastal area such as coastline changes that occur due to erosion or sedimentation. Analysis of erosion and sedimentation is essential for further studies to be carried out regarding the factors that cause erosion and sedimentation and how to protect and prevent them in the future. This research utilizes remote sensing techniques and integrated geographic information systems (GIS) to identify coastline changes from 2017-2022 in the coastal areas of Subang, Indramayu, and Cirebon districts which are included in the Rebana metropolitan area. In this study, Landsat-8 satellite imagery is used to identify and quantify Spatio-temporal changes that occur in the coastal area of the Rebana metropolitan area in 6 years. The Normalized Difference Wetness Index (NDWI) algorithm is used to separate wet and dry objects which are interpreted as bodies of water and land. Thresholding the results of the NDWI algorithm along with the on-screen digitization approach was carried out for coastline extraction. The calculation of coastline changes is carried out using the Digital Shoreline Analysis System (DSAS) so that the rate of change of coastline in the study area is obtained. During the study period, the results showed that the average erosion rate in the Rebana metropolitan area was -3.66 m/year and the average sedimentation rate was +6.28 m/year, where the greatest erosion rate occurred in Kapetakan sub-district, Cirebon Regency with an average rate -7.63 m/year, and the greatest sedimentation rate occurs in Pusakanagara sub-district, Subang Regency with an average rate +24.64 m/year. Karangampel Subdistrict in Indramayu Regency has 79.89%, the widest erosion area of its total coastline length, and Balongan Subdistrict, Indramayu Regency has the widest sedimentation, 91.37% of the total coastline length. The land use that is most affected by erosion and accretion in the coastal area of the Rebana metropolitan area is fishpond.

Keywords: Coastline Change, DSAS, NDWI, Rebana Metropolitan Area

1. Introduction

Every year there is an increase in population in Indonesia [1]. Demand will increase as the population rise, makes growth in the industrial sector must be accompanied by population growth to meet supply [2]. The industrial sector also plays a vital role as the prime mover and supporter of the national economy [3].

However, industrial planning requires careful planning to carry out industrial development strategically. Several parameters must be considered in industrial development, including socio-economic, physical-environmental, infrastructure and urban development, and potential hazards [4–6]. Some hazard parameters that need to be reviewed from industrial development are erosion and accretion, especially for industrial planning in coastal area [7].

One area of Indonesia that has the potential for a large increase in industrial in the coastal area is Rebana Metropolitan. Rebana Metropolitan is the north/northeast region of West Java Province which consists of seven cities in which there are Patimban Port, Cirebon Port and Kertajati West Java International Airport (BIJB), which functions as a connectivity and logistics center. The strategic area of the Rebana Metropolite can support the increase in the industrial economy [8].

However, the Rebana metropolitan area, which is located in the lowlands in northern Java, has considerable potential for coastline dynamics. Areas dominated by sedimentary mud are very easy to change position and cause changes in the shape of the beach [9], [10]. Changes in the coastal area can be seen from the description of the differences in the coastline. This change is indicated to have occurred due to erosion and accretion along the coast which can be known through remote sensing imagery in the form of satellite imagery, namely in 2017 and 2022 [11], [12] . The results of remote sensing data analysis are then processed with a geographic information system to analyze coastal environmental conditions.

Aim of this research is to determine changes in the coastline in the coastal Rebana Metropolitan area which is estimated to become an industrial center. Coastal land management is analyzed to see the impact and causes of changes in the coastline that occur. It is hoped that this study will identify the factors that cause erosion and sedimentation as well as ways to protect and prevent them so that economic development in Metropolitan Rebana can be sustainable.

2. Methodology

2.1. Study Area

The area study in this research focuses on the coastal area of the Rebana Metropolitan area, which is in the province of West Java. The case study area comprises 20 sub-districts in three districts adjacent to the coastline along the Rebana Metropolitan Area [8]. The study area can be seen in Figure 1.

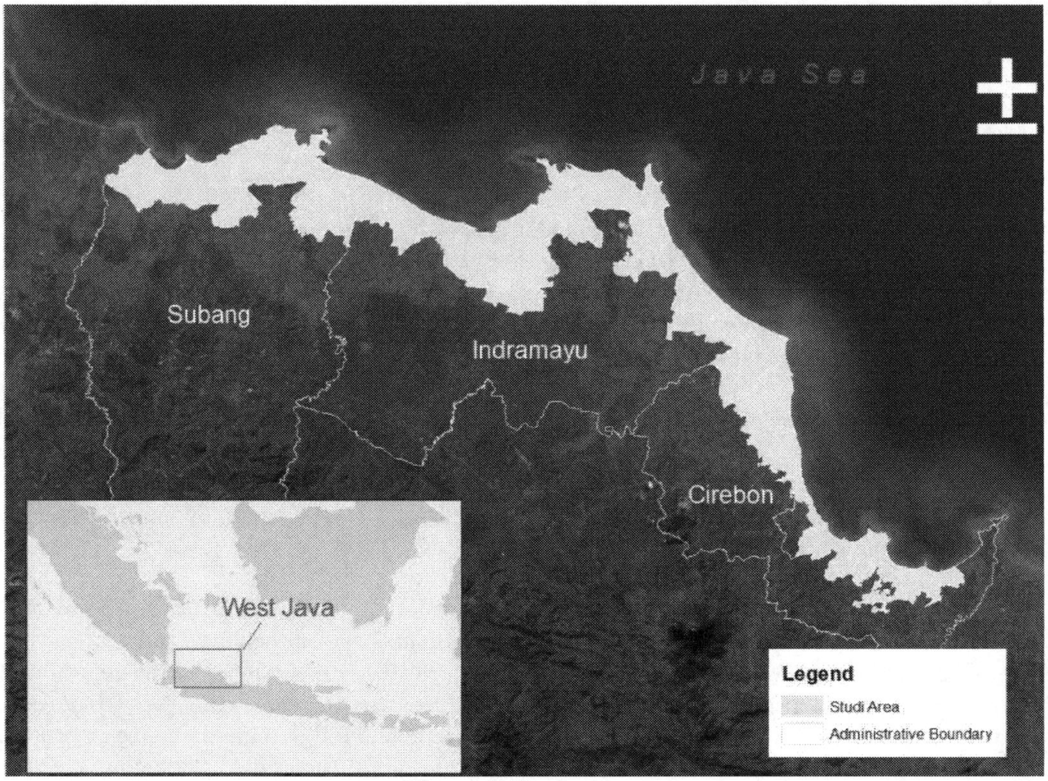

Figure 1. Study Area

2.2. Data

Data used in this study can be seen in Table 1. In general, this study uses five data consisting of administrative data, Ocean tide data, Landsat-8, Digital Elevation Model (DEM), and Landuse.

Table 1. Data used in this study

No	Data	Product	Spatial Resolution	Referensce
1	Administrative Data	Topographic Map of Indonesia (RBI)	1:25.000	[11]
2	Ocean tide data	TPXO 9 Tide Models	1/6 degree	[12]
3	Landsat-8	USGS Landsat 8 Multispectral	30 m	[13]
4	Digital Elevation Model (DEM)	National Data Digital Elevation Model (DEMNAS)	0.27 arcsecond	[11]
5	Landuse	Topographic Map of Indonesia (RBI)	1:25.000	[11]
6	Evidence of Coastline Changes in the Field	Documentation	-	Field observations

The satellite data used in this study consisted of 6 (five) years of Landsat 8 image recording data. Onboard Operational Land Imager (OLI) and Thermal Infrared Sensor (TIRS) with a resolution of 30 m. the data used is image acquisition in 2017-2022. The choice of data acquisition date in Table 2 is based on the availability of data each year and the quality of satellite data as seen from the percentage of cloud cover in coastal areas.

Table 2. Data and quality of satellite imagery selected in the study

Data	Data Acquisition Date	Land Cloud Cover (%)	Geometric RMSE Model (m)	Geometric RMSE Model X (m)	Geometric RMSE Model Y (m)
1	9/30/2017	11.20%	6.845	3.832	5.672
2	9/17/2018	00.42%	6.018	4.101	4.405
3	11/7/2019	02.93%	6.215	4.140	4.635
4	10/8/2020	03.49%	6.308	4.160	4.742
5	9/16/2021	06.74%	6.318	4.545	4.388
6	4/13/2022	18.23%	7.391	5.355	5.093

2.3. Methods

The general methods in this study can be seen in Figure 2. The method can be divided into four sections that will be explained below.

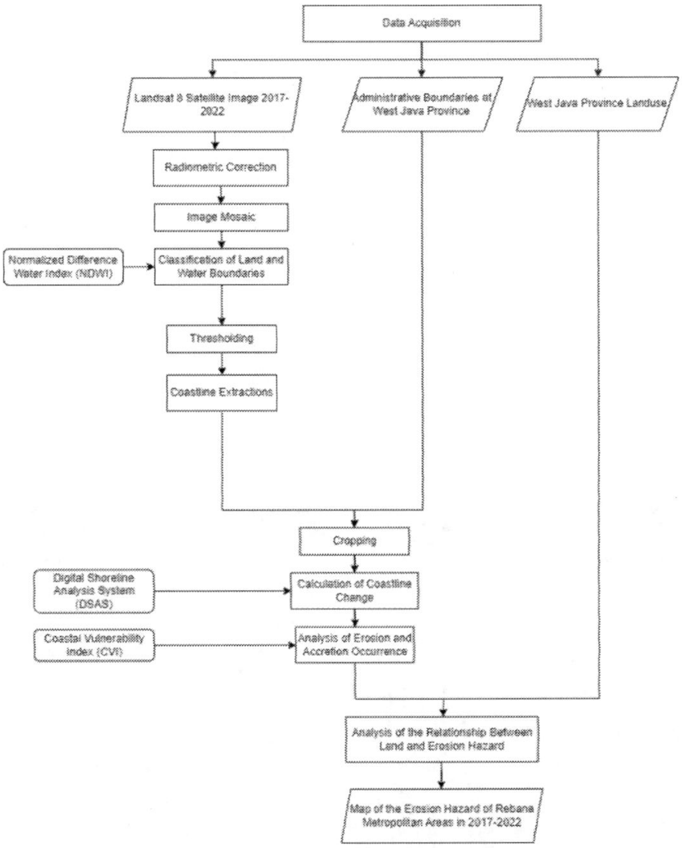

Figure 2. Methodology Flowchart

2.3.1. Radiometric Correction

Radiometric correction is intended to eliminate or minimize radiometric errors due to external aspects in the form of atmospheric disturbances during the recording process. [15]. This process also converts

DN (Digital Number) to ToA (Top of Atmospheric) or radian value. The radiometric correction process can be carried out using various applications, one of which was carried out in this study using QGIS software. The radiometric correction equation used for Landsat is used in this study [16]. the equation for radiometric correction is in equation 1 below.

$$\rho\lambda = \frac{(M\rho + Qcal + A\rho)}{\sin \theta se} \quad (1)$$

$\rho\lambda$= Reflectance value (which has been corrected for the angle of the sun).
$M\rho$= Rescaling constant.
$Qcal$= Pixel Value (Digital Number).
Aρ= Addition Constant.
θse= The sun's elevation angle during image recording.

2.3.2. Coastline Extractions

a) Normalized Difference Water Index (NDWI)

The Normalized Difference Water Index (NDWI) method is a method which is used to compare humidity degrees on satellite imagery. The NDWI method uses Band 3 (green) to assess the strength of plants and water bodies, and Band 5 (NIR) emphasizes the amount of biomass The NDWI index results can range from -1 to +1. The equation for calculating NDWI is in equation 2 below. The selection of this wavelength is done for:

1) maximizing the reflection of water features by using green wavelengths;
2) minimizing low NIR reflectance by water features; And
3) take advantage of the high NIR reflectance by terrestrial vegetation and soil features.

$$\text{NDWI} = \frac{\text{Green}-\text{NIR}}{\text{Green}+\text{NIR}} \quad (2)$$

When equation (2) is used to process multispectral satellite imagery, the water feature has a positive value. While ground and terrestrial vegetation features have a zero or negative value, because the NIR reflectance is usually higher than the green channel [17]. The equation used for using the NDWI algorithm can be seen in equation 2 [18]. the results of the NDWI algorithm can be seen in Figure 3b.

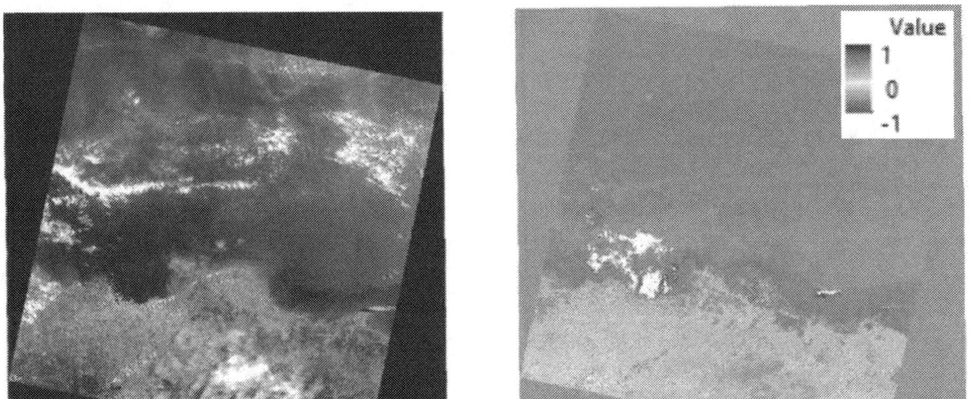

(a) (b)

Figure 3. Sentinel 2 Image RGB Band Composite (a), NDWI Algorithm Results (b).

b) Thresholding

The results of processing using the NDWI algorithm produce spectral values between -1 to 1, which indicate the degree of wetting in the image processing results. Wetness classification using NDWI can be seen in Tabel 3 below [19].

Table 3. Wetness level based on NDWI value

Class	NDWI	Wetness Level
1	-1 < NDWI < 0	Non-Water Body
2	0 < NDWI < 0.33	Moderate Wetness
3	0.33 < NDWI < 1	High Wetness

Thresholding aims to make boundaries based on the classification in the table above so that boundaries between dry land and wetlands which in this study are called land and sea will be obtained. land classification is at a value of -1 to 0, while waters is a value of 0 to 1 [19]. The thresholding results for NDWI processing based on the classification in table 2 can be seen in Figure 4 below.

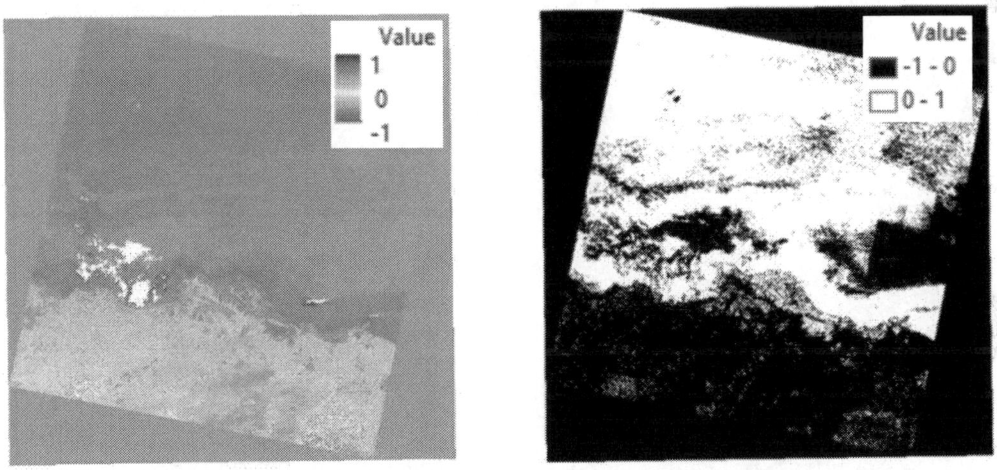

(a) (b)

Figure 4. NDWI before Thresholding (a) and NDWI after Thresholding (b).

c) Delineation

The results of land and sea detection are then digitized, either manually or automatically. However, in this study, digitization was carried out in a combination of automatic and manual to ensure that the coastline is formed better and avoids noise around the coastline. The results of land and sea delineations are symbolized by coastlines in 2017-2022 as shown in figure 5 below.

Figure 5. Delineation result in the coastal area 2017-2022.

2.3.3. *Coastline Change Analysis using the Digital Shoreline Analysis System (DSAS)*

The stages of analysis using DSAS are establishment of a coastline, making baselines, transect analysis, and calculate the coastline change. The coastline change rate was analyzed using the End Point Rate (EPR) statistical approach using the Digital Shoreline Analysis System (DSAS) tool. The EPR method calculates the rate of change of coastline by dividing the distance between the oldest and the most recent coastline by the time. EPR calculations are performed on each transect that is formed with a distance of 15 m between transects. the distance between transects is chosen based on half of the pixel size of the Landsat 8 image.Figure 6 is an illustration of the results of processing using DSAS.

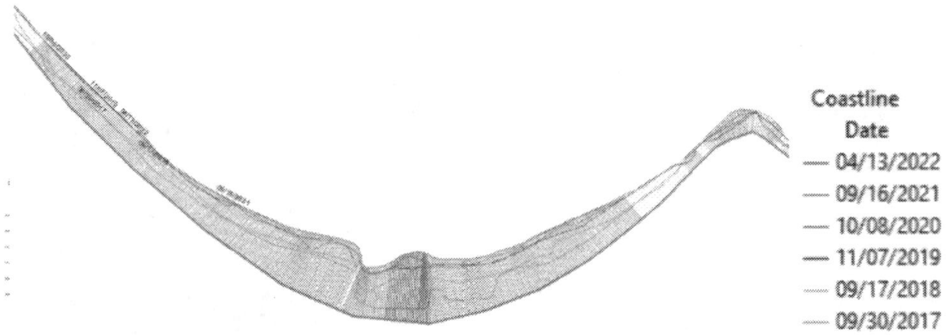

Figure 6. Example of coastline dynamics in Subang Regency

2.3.4. *Erosion Hazard Classification*

The classification of coastal erosion hazard is determined based on the Coastal Vulnerability Index (CVI), where erosion and accretion hazards are divided into five classes that can be seen in Table 4. The danger of coastal erosion will be greater if the coastline changes towards the land and lower if there is a change in the coastline towards the sea (accretion).

Table 4. CVI Indicators [21]

Ranking of Coastal Vulnerability Index					
Variable	Very Low	Low	Moderate	High	Very High
	1	2	3	4	5
Geomorphology	Rocky, eliffed coasts, Fiands	Medium cliffs, Indented coasts	Low cliffs, Glacial drift	Cobble beaches, Estuary	Barrier beaches, Sand beaches, Mud flats, Deltas, Mangrove, Coral reefs.

Coastal slope (%)	>0.115	0.115-0.055	0.055-0.035	0.025-0.022	<0.022
Relative sea-level change (mm/yr)	<1.8	1.8-2.5	2.5-3.0	3.0-3.4	>3.4
Shoreline erosion/ accretion (m/yr)	>2.0	1.0-2.0	-1.0-+1.0	-1.1—(-2.0)	<-2.0
Mean tide range (m)	>6.0	4.1-6.0	2.0-4.0	1.0-1.9	<1.0
Mean wave height (m)	<0.55	0.55-0.85	0.85-1.05	1.05-1.25	>1.25

Erosion hazard classes are divided into 5 classes, namely very low with a rate of above 2 m/year, low class with a rate of 1.0 to 2.0 m/year, moderate class with a rate of 1.0 to -1.0, high class with a rate of -1.1 to 2.0 m/year and very high class with a rate of more than -2.0 m/year. A positive value indicates a change in the coastline towards the sea (increasing land area) while a negative value indicates a change in the coastline towards land (decreasing land area).

3. Result and Discussion

The results of this study are divided into two, an analysis of coastline changes in the Rebana metropolitan area and the relationship between land cover in the Rebana metropolitan area and phenomena that occur along the coast. Clearly, the analysis of the results of this study is explained as follows.

3.1. coastline dynamics

The dynamics of the coastline in the Rebana metropolitan area, divided into 20 sub-districts, is obtained using DSAS. The baseline is used as a reference in calculating the length of the coastline, which is taken based on the coastline on the results of the 2017 coastline extraction as the first data in this study. The distance between transects along the coastline is 15 m, with the number of transects attached in the Table 5.

Table 5. The results of making transects along the coastline

Sub-District	**Transect Total**		**Baseline Length**
	Erosion	**Accretion**	
Pusakanagara	225	894	1119.00
Sukra	295	219	514.00
Patrol	312	185	497.00
Kadanghaur	382	566	948.00
Losarang	298	333	631.00
Cantigi	362	915	1277.00
Pasekan	931	566	1497.00
Indramayu	191	197	388.00
Balongan	32	339	371.00
Juntinyuat	201	524	725.00
Karangampel	143	36	179.00
Krangkeng	226	373	599.00
Kapetakan	71	51	122.00
Suranenggala	96	197	293.00
Gunung Jati	359	397	756.00

Sub-District	Transect Total		Baseline Length
	Erosion	Accretion	
Kejaksan	37	130	167.00
Lemah Wungkuk	150	126	276.00
Mundu	196	165	361.00
Astanajapura	43	234	277.00
Pangenan	526	275	801.00

From the results of coastline extraction, the sub-district with the longest coastline was obtained, namely the Pasekan sub-district, Indramayu Regency, which is 12.69% of the total coastline length in the Rebana metropolitan area. Pasekan sub-district obtained longest coastline affected by erosion, which is 931 m long. Meanwhile, Cantigi sub-district obtained coastline most affected by accretion. From the results of processing using DSAS, the average coastline change rate is obtained from the calculation of the EPR, which can be seen in the Figure 7.

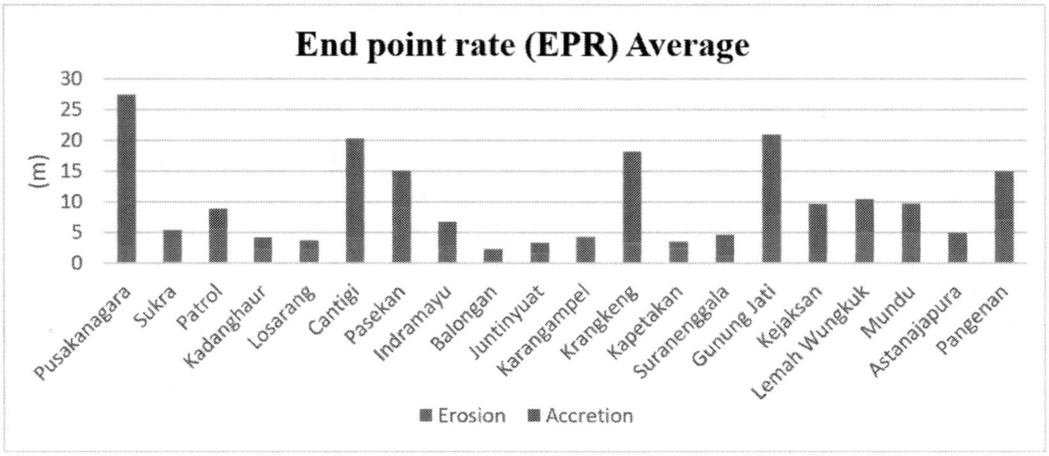

Figure 7. End Point Rate (ERP) Average

The results of the average EPR calculation show that the district with the largest average erosion rate is experienced by Pasekan District, Indramayu Regency, which is equal to 8.97 m/year. Meanwhile, the district with the highest average accretion rate was Pusakanagara District, Subang Regency with a maximum value of 24.64 m3/year. Pasekan District experienced the greatest erosion phenomenon with the majority of the coastal area being used as fish ponds. The lack of mangroves and water barriers, coupled with the sloping topography with sandy mud, causes continous erosion. Meanwhile, the Pusakanagara area experienced the largest average accretion because it has many upstream rivers in its area that carry sediment particles from the river to its coastal areas. The existence of mangroves in the coastal area also plays a role in binding sediments, causing an increase in land area in the Pusakanagara sub-district.

3.2. relationship between landuse and erosion hazardcoastline dynamics

The results of the calculation of coastline changes using DSAS are then carried out to classify the erosion hazard based on the Coastal Vulnerability Index (CVI), which is divided into five classes. The results of the erosion hazard classification in the coastal area of the Rebana metropolitan area can be seen based on the Figure 9 below.

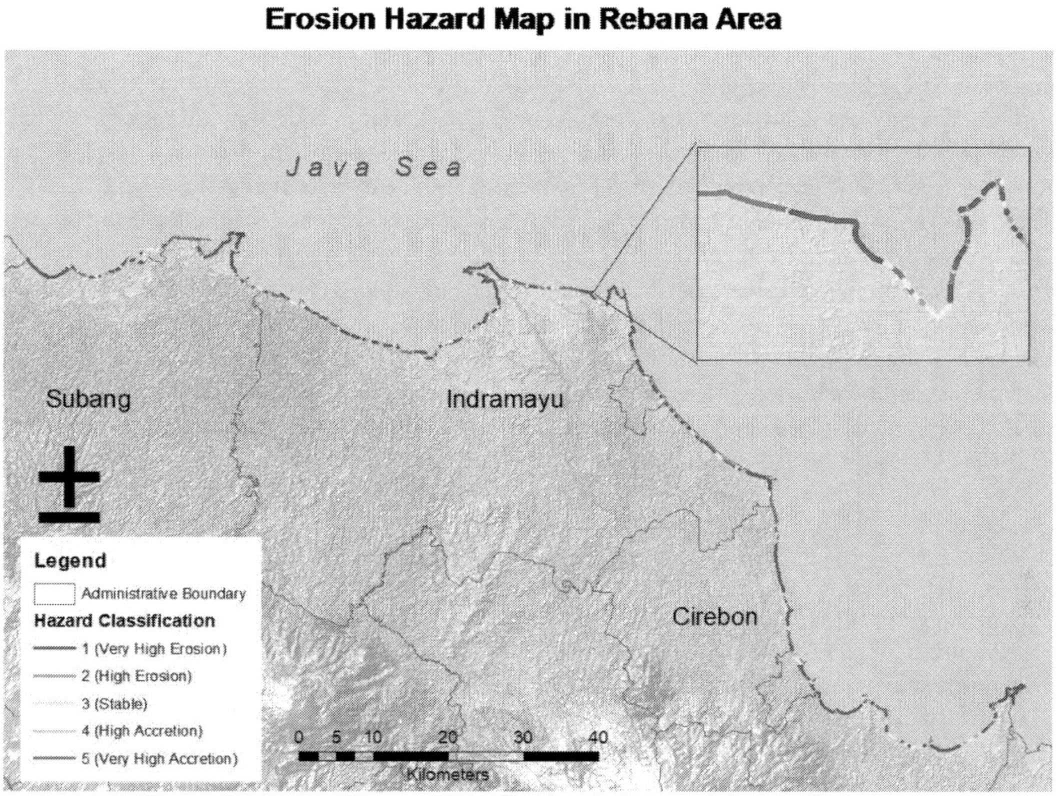

Figure 8. Erosion Hazard Classification in Rebana Metropolitan Area

Based on the results of the classification of erosion hazard in the coastal area of the Rebana metropolitan area, the total length of land use affected by by erosion is obtained. In detail, the types of land use affected by erosion and accretion can be explained in the Table 6 below.

Table 6. Hazard classification in percent

Landuse	Hazard Classification				
	1 (%)	2 (%)	3 (%)	4 (%)	5 (%)
Built-Up Area	26.51	14.94	47.11	4.47	6.97
Water Bodies	32.64	5.13	35.12	12.27	14.84
Rice Fields	15.96	14.26	38.29	6.54	24.96
Fish Ponds	35.40	7.08	21.89	7.27	28.37

From Table 6, the results of land cover with a high erosion hazard impact (class 5) have an impact on the fish pond area of 28.37% or 1164252.77 m along the coastline. While land use with high accretion hazard impacts (class 1) also has an impact on the fish pond area of 35.40% or 4104386.87 m along the coastline. This is probably caused by the slope of the land in which the land use is relatively flat. Not only is there a natural accretion phenomenon in the downstream area of the river and fish ponds, human activities that take advantage of the accretion phenomenon in the fish pond area, such as adding new fish ponds from the results of sedimentation in the surrounding area are also one of the causes of rapid coastline changes in fish pond area.

4. Conclusion

The conclusions obtained from the results of this study are as follows.

1. During the study period, the results showed that the average erosion rate in the Rebana metropolitan area was -3.66 m/year and the average sedimentation rate was +6.28 m/year, where the greatest erosion rate occurred in Pasekan District, Indramayu Regency with an average rate 8.97 m/year m/year, and the greatest sedimentation rate occurs in Pusakanagara sub-district, Subang Regency with an average rate +24.64 m/year. Karangampel Subdistrict in Indramayu Regency has 79.89%, the widest erosion area of its total subdistrict coastline length, and Balongan Subdistrict, Indramayu Regency has the widest sedimentation, 91.37% of the total sub district coastline length.
2. The land use most affected by erosion and accretion in the coastal area of the Rebana metropolitan area is ponds, which occur due to natural phenomena and the result of human activities.
3. The phenomenon of erosion and accretion that occurs in this study cannot be explained with certainty whether caused by natural factors or human factors, so additional parameters are needed to ascertain the causes of erosion and accretion such as hydrological morphodynamic models.
4. Detected coastline changes of less than 15 m are still ambiguous whether they occur due to errors in digitizing or are indeed shoreline changes, so for further research it is necessary to study shoreline changes using high-resolution imagery.

References

[1] Statistics Indonesia, 2018, Indonesia Population Projection 2015-2045 Result of SUPAS 2015 (Revised Edition) Jakarta: BPS RI. .

[2] Karabegović I, 2018 The Role of Industrial And Service Robots In The 4th Industrial Revolution–" Industry 4.0" *Acta Technica Corviniensis-Bulletin of Engineering* **11**, 2 p. 11–16.

[3] Coordinating Ministry for Economic Affairs, 2022, Membangun Sektor Industri yang Mandiri dan Berdaulat untuk Menjadi Kekuatan Ekonomi Dalam Negeri. [Online]. Available: https://www.ekon.go.id/publikasi/detail/4139/membangun-sektor-industri-yang-mandiri-dan-berdaulat-untuk-menjadi-kekuatan-ekonomi-dalam-negeri#:~:text=Sektor%20industri%20memainkan%20peran%20penting,tantangan%20akibat%20pandemi%20Covid%2D19. [Accessed: 28-Feb-2023].

[4] Johar A Jain S S and Garg P K, 2013 Land suitability analysis for industrial development using GIS *Journal of Geomatics* **7** p. 101–106.

[5] Ruiz M C Romero E Pérez M A and Fernández I, 2012 Development and application of a multi-criteria spatial decision support system for planning sustainable industrial areas in Northern Spain *Autom Constr* **22** p. 320–333.

[6] Arabsheibani R Kanani Sadat Y and Abedini A, 2016 Land suitability assessment for locating industrial parks: a hybrid multi criteria decision-making approach using Geographical Information System *Geographical Research* **54**, 4 p. 446–460.

[7] Steiner F McSherry L and Cohen J, 2000 Land suitability analysis for the upper Gila River watershed *Landsc Urban Plan* **50**, 4 p. 199–214.

[8] Bappeda Jabar, 2020, Mengenal Rebana Metropolitan, Masa Depan Ekonomi Jawa Barat. [Online]. Available: Mengenal Rebana Metropolitan, Masa Depan Ekonomi Jawa Barat. [Accessed: 28-Feb-2023].

[9] HERIATI A and HUSRIN S, 2017 Perubahan Garis Pantai di Pesisir Cirebon Berdasarkan Analisis Spasial AIDA HERIATI, SEMEIDI HUSRIN *Reka Geomatika* **2017**, 2 p. 52–60.

[10] Prawiradisastra S, 2003 Permasalahan Abrasi di Wilayah Pesisir Kabupaten Indramayu *Alami* **8**, 2.

[11] Karabegović I, Apr. 2018 The Role Of Industrial And Service Robots In The 4th Industrial Revolution – "Industry 4.0" *A Cta Technica Corviniensis – Bulletin of Engineering* **9**, 2.

[12] Johar A Jain S S and Garg P K, 2013 Land suitability analysis for industrial development using GIS *Article in Journal of Geomatics* **101**, 2.

[13] Badan Informasi Geospasial (BIG), 2021, Basemap Rupa Bumi Indonesia (RBI). [Online]. Available: http://basemap.big.go.id/rbi/. [Accessed: 02-Nov-2021].

[14] BNPB, 2018, Benar, Tsunami Menerjang Pantai di Selat Sunda. [Online]. Available: https://bnpb.go.id/berita/benar-tsunami-menerjang-pantai-di-selat-sunda#:~:text=Dampak%20tsunami%20menyebabkan%20korban%20jiwa,semua%20daerah%20terdampak%20di%20data. [Accessed: 20-Jan-2023].
[15] USGS, 2018, Landsat 8.
[16] S H S, 2009 *Penginderaan Jauh dan Pengenalan Sistem Informasi Geografi Untuk Bidang Ilmu Kebumian* Bandung: ITB.
[17] Agitya S H, 2016, Analisis akurasi citra modis dan citra landsat 8 menggunakan algoritma normalized burn ratio untuk pemetaan area terbakar (Studi Kasus : Provinsi Riau), Institut Teknologi Sepuluh Maret, Surabaya.
[18] Hernoza F Susilo B and Erlansari A, 2020 Pemetaan Daerah Rawan Banjir Menggunakan Penginderaan Jauh Dengan Metode Normalized Difference Vegetation Index, Normalized Difference Water Index Dan Simple Additive Weighting (Studi Kasus: Kota Bengkulu) *Rekursif Journal* **8**, 2.
[19] McFeeters S K, May 1996 The use of the Normalized Difference Water Index (NDWI) in the delineation of open water features *Int J Remote Sens* **17**, 7 p. 1425–1432.
[20] Xu H, Jul. 2006 Modification of normalised difference water index (NDWI) to enhance open water features in remotely sensed imagery *Int J Remote Sens* **27**, 14 p. 3025–3033.
[21] USGS, 1999, Coastal Vulnerability Index (CVI). .

AUTHOR INDEX